T0181710

Lecture Notes in Computer Science 12745

More information about this subseries at http://www.springer.com/series/7407

Maciej Paszynski · Dieter Kranzlmüller ·
Valeria V. Krzhizhanovskaya ·
Jack J. Dongarra · Peter M. A. Sloot (Eds.)

Computational Science – ICCS 2021

21st International Conference
Krakow, Poland, June 16–18, 2021
Proceedings, Part IV

 Springer

Editors
Maciej Paszynski (iD)
AGH University of Science and Technology
Krakow, Poland

Valeria V. Krzhizhanovskaya (iD)
University of Amsterdam
Amsterdam, The Netherlands

Peter M. A. Sloot (iD)
University of Amsterdam
Amsterdam, The Netherlands

ITMO University
St. Petersburg, Russia

Nanyang Technological University
Singapore, Singapore

Dieter Kranzlmüller (iD)
Ludwig-Maximilians-Universität München
Munich, Germany

Leibniz Supercomputing Center (LRZ)
Garching bei München, Germany

Jack J. Dongarra (iD)
University of Tennessee at Knoxville
Knoxville, TN, USA

ISSN 0302-9743 ISSN 1611-3349 (electronic)
Lecture Notes in Computer Science
ISBN 978-3-030-77969-6 ISBN 978-3-030-77970-2 (eBook)
https://doi.org/10.1007/978-3-030-77970-2

LNCS Sublibrary: SL1 – Theoretical Computer Science and General Issues

Preface

Welcome to the proceedings of the 21st annual International Conference on Computational Science (ICCS 2021 - https://www.iccs-meeting.org/iccs2021/).

In preparing this edition, we had high hopes that the ongoing COVID-19 pandemic would fade away and allow us to meet this June in the beautiful city of Kraków, Poland. Unfortunately, this is not yet the case, as the world struggles to adapt to the many profound changes brought about by this crisis. ICCS 2021 has had to adapt too and is thus being held entirely online, for the first time in its history.

These challenges notwithstanding, we have tried our best to keep the ICCS community as dynamic and productive as always. We are proud to present the proceedings you are reading as a result of that.

ICCS 2021 was jointly organized by the AGH University of Science and Technology, the University of Amsterdam, NTU Singapore, and the University of Tennessee.

The International Conference on Computational Science is an annual conference that brings together researchers and scientists from mathematics and computer science as basic computing disciplines, as well as researchers from various application areas who are pioneering computational methods in sciences such as physics, chemistry, life sciences, engineering, arts, and humanitarian fields, to discuss problems and solutions in the area, identify new issues, and shape future directions for research.

Since its inception in 2001, ICCS has attracted an increasing number of attendees and higher quality papers, and this year is not an exception, with over 350 registered participants. The proceedings have become a primary intellectual resource for computational science researchers, defining and advancing the state of the art in this field.

The theme for 2021, **"Computational Science for a Better Future,"** highlights the role of computational science in tackling the current challenges of our fast-changing world. This conference was a unique event focusing on recent developments in scalable scientific algorithms, advanced software tools, computational grids, advanced numerical methods, and novel application areas. These innovative models, algorithms, and tools drive new science through efficient application in physical systems, computational and systems biology, environmental systems, finance, and other areas.

ICCS is well known for its excellent lineup of keynote speakers. The keynotes for 2021 were given by

- **Maciej Besta**, ETH Zürich, Switzerland
- **Marian Bubak**, AGH University of Science and Technology, Poland | Sano Centre for Computational Medicine, Poland
- **Anne Gelb**, Dartmouth College, USA
- **Georgiy Stenchikov**, King Abdullah University of Science and Technology, Saudi Arabia
- **Marco Viceconti**, University of Bologna, Italy

- **Krzysztof Walczak**, Poznan University of Economics and Business, Poland
- **Jessica Zhang**, Carnegie Mellon University, USA

This year we had 635 submissions (156 submissions to the main track and 479 to the thematic tracks). In the main track, 48 full papers were accepted (31%); in the thematic tracks, 212 full papers were accepted (44%). A high acceptance rate in the thematic tracks is explained by the nature of these tracks, where organisers personally invite many experts in a particular field to participate in their sessions.

ICCS relies strongly on our thematic track organizers' vital contributions to attract high-quality papers in many subject areas. We would like to thank all committee members from the main and thematic tracks for their contribution to ensure a high standard for the accepted papers. We would also like to thank *Springer, Elsevier,* and *Intellegibilis* for their support. Finally, we appreciate all the local organizing committee members for their hard work to prepare for this conference.

We are proud to note that ICCS is an A-rank conference in the CORE classification.

We wish you good health in these troubled times and look forward to meeting you at the conference.

June 2021

Maciej Paszynski
Dieter Kranzlmüller
Valeria V. Krzhizhanovskaya
Jack J. Dongarra
Peter M. A. Sloot

Organization

Local Organizing Committee at AGH University of Science and Technology

Chairs

Maciej Paszynski
Aleksander Byrski

Members

Marcin Łos
Maciej Woźniak
Leszek Siwik
Magdalena Suchoń

Thematic Tracks and Organizers

Advances in High-Performance Computational Earth Sciences: Applications and Frameworks – IHPCES

Takashi Shimokawabe
Kohei Fujita
Dominik Bartuschat

Applications of Computational Methods in Artificial Intelligence and Machine Learning – ACMAIML

Kourosh Modarresi
Paul Hofmann
Raja Velu
Peter Woehrmann

Artificial Intelligence and High-Performance Computing for Advanced Simulations – AIHPC4AS

Maciej Paszynski
Robert Schaefer
David Pardo
Victor Calo

Biomedical and Bioinformatics Challenges for Computer Science – BBC

Mario Cannataro
Giuseppe Agapito

Mauro Castelli
Riccardo Dondi
Italo Zoppis

Classifier Learning from Difficult Data – CLD^2

Michał Woźniak
Bartosz Krawczyk

Computational Analysis of Complex Social Systems – CSOC

Debraj Roy

Computational Collective Intelligence – CCI

Marcin Maleszka
Ngoc Thanh Nguyen
Marcin Hernes
Sinh Van Nguyen

Computational Health – CompHealth

Sergey Kovalchuk
Georgiy Bobashev
Stefan Thurner

Computational Methods for Emerging Problems in (dis-)Information Analysis – DisA

Michal Choras
Robert Burduk
Konstantinos Demestichas

Computational Methods in Smart Agriculture – CMSA

Andrew Lewis

Computational Optimization, Modelling, and Simulation – COMS

Xin-She Yang
Leifur Leifsson
Slawomir Koziel

Computational Science in IoT and Smart Systems – IoTSS

Vaidy Sunderam
Dariusz Mrozek

Computer Graphics, Image Processing and Artificial Intelligence – CGIPAI

Andres Iglesias
Lihua You
Alexander Malyshev
Hassan Ugail

Data-Driven Computational Sciences – DDCS

Craig Douglas

Machine Learning and Data Assimilation for Dynamical Systems – MLDADS

Rossella Arcucci

MeshFree Methods and Radial Basis Functions in Computational Sciences – MESHFREE

Vaclav Skala
Marco-Evangelos Biancolini
Samsul Ariffin Abdul Karim
Rongjiang Pan
Fernando-César Meira-Menandro

Multiscale Modelling and Simulation – MMS

Derek Groen
Diana Suleimenova
Stefano Casarin
Bartosz Bosak
Wouter Edeling

Quantum Computing Workshop – QCW

Katarzyna Rycerz
Marian Bubak

Simulations of Flow and Transport: Modeling, Algorithms and Computation – SOFTMAC

Shuyu Sun
Jingfa Li
James Liu

Smart Systems: Bringing Together Computer Vision, Sensor Networks and Machine Learning – SmartSys

Pedro Cardoso
Roberto Lam

João Rodrigues
Jânio Monteiro

Software Engineering for Computational Science – SE4Science

Jeffrey Carver
Neil Chue Hong
Anna-Lena Lamprecht

Solving Problems with Uncertainty – SPU

Vassil Alexandrov
Aneta Karaivanova

Teaching Computational Science – WTCS

Angela Shiflet
Nia Alexandrov
Alfredo Tirado-Ramos

Uncertainty Quantification for Computational Models – UNEQUIvOCAL

Wouter Edeling
Anna Nikishova

Reviewers

Ahmad Abdelfattah
Samsul Ariffin Abdul
    Karim
Tesfamariam Mulugeta
    Abuhay
Giuseppe Agapito
Elisabete Alberdi
Luis Alexandre
Vassil Alexandrov
Nia Alexandrov
Julen Alvarez-Aramberri
Sergey Alyaev
Tomasz Andrysiak
Samuel Aning
Michael Antolovich
Hideo Aochi
Hamid Arabnejad
Rossella Arcucci
Costin Badica
Marina Balakhontceva

Bartosz Balis
Krzysztof Banas
Dariusz Barbucha
Valeria Bartsch
Dominik Bartuschat
Pouria Behnodfaur
Joern Behrens
Adrian Bekasiewicz
Gebrail Bekdas
Mehmet Belen
Stefano Beretta
Benjamin Berkels
Daniel Berrar
Sanjukta Bhowmick
Georgiy Bobashev
Bartosz Bosak
Isabel Sofia Brito
Marc Brittain
Jérémy Buisson
Robert Burduk

Michael Burkhart
Allah Bux
Krisztian Buza
Aleksander Byrski
Cristiano Cabrita
Xing Cai
Barbara Calabrese
Jose Camata
Almudena Campuzano
Mario Cannataro
Alberto Cano
Pedro Cardoso
Alberto Carrassi
Alfonso Carriazo
Jeffrey Carver
Manuel Castañón-Puga
Mauro Castelli
Eduardo Cesar
Nicholas Chancellor
Patrikakis Charalampos

Henri-Pierre Charles
Ehtzaz Chaudhry
Long Chen
Sibo Cheng
Siew Ann Cheong
Lock-Yue Chew
Marta Chinnici
Sung-Bae Cho
Michal Choras
Neil Chue Hong
Svetlana Chuprina
Paola Cinnella
Noélia Correia
Adriano Cortes
Ana Cortes
Enrique
  Costa-Montenegro
David Coster
Carlos Cotta
Helene Coullon
Daan Crommelin
Attila Csikasz-Nagy
Loïc Cudennec
Javier Cuenca
António Cunha
Boguslaw Cyganek
Ireneusz Czarnowski
Pawel Czarnul
Lisandro Dalcin
Bhaskar Dasgupta
Konstantinos Demestichas
Quanling Deng
Tiziana Di Matteo
Eric Dignum
Jamie Diner
Riccardo Dondi
Craig Douglas
Li Douglas
Rafal Drezewski
Vitor Duarte
Thomas Dufaud
Wouter Edeling
Nasir Eisty
Kareem El-Safty
Amgad Elsayed
Nahid Emad

Christian Engelmann
Roberto R. Expósito
Fangxin Fang
Antonino Fiannaca
Christos
  Filelis-Papadopoulos
Martin Frank
Alberto Freitas
Ruy Freitas Reis
Karl Frinkle
Kohei Fujita
Hiroshi Fujiwara
Takeshi Fukaya
Wlodzimierz Funika
Takashi Furumura
Ernst Fusch
David Gal
Teresa Galvão
Akemi Galvez-Tomida
Ford Lumban Gaol
Luis Emilio
  Garcia-Castillo
Frédéric Gava
Piotr Gawron
Alex Gerbessiotis
Agata Gielczyk
Adam Glos
Sergiy Gogolenko
Jorge
  González-Domínguez
Yuriy Gorbachev
Pawel Gorecki
Michael Gowanlock
Ewa Grabska
Manuel Graña
Derek Groen
Joanna Grzyb
Pedro Guerreiro
Tobias Guggemos
Federica Gugole
Bogdan Gulowaty
Shihui Guo
Xiaohu Guo
Manish Gupta
Piotr Gurgul
Filip Guzy

Pietro Hiram Guzzi
Zulfiqar Habib
Panagiotis Hadjidoukas
Susanne Halstead
Feilin Han
Masatoshi Hanai
Habibollah Haron
Ali Hashemian
Carina Haupt
Claire Heaney
Alexander Heinecke
Marcin Hernes
Bogumila Hnatkowska
Maximilian Höb
Jori Hoencamp
Paul Hofmann
Claudio Iacopino
Andres Iglesias
Takeshi Iwashita
Alireza Jahani
Momin Jamil
Peter Janku
Jiri Jaros
Caroline Jay
Fabienne Jezequel
Shalu Jhanwar
Tao Jiang
Chao Jin
Zhong Jin
David Johnson
Guido Juckeland
George Kampis
Aneta Karaivanova
Takahiro Katagiri
Timo Kehrer
Christoph Kessler
Jakub Klikowski
Alexandra Klimova
Harald Koestler
Ivana Kolingerova
Georgy Kopanitsa
Sotiris Kotsiantis
Sergey Kovalchuk
Michal Koziarski
Slawomir Koziel
Rafal Kozik

Bartosz Krawczyk
Dariusz Krol
Valeria Krzhizhanovskaya
Adam Krzyzak
Pawel Ksieniewicz
Marek Kubalcík
Sebastian Kuckuk
Eileen Kuehn
Michael Kuhn
Michal Kulczewski
Julian Martin Kunkel
Krzysztof Kurowski
Marcin Kuta
Bogdan Kwolek
Panagiotis Kyziropoulos
Massimo La Rosa
Roberto Lam
Anna-Lena Lamprecht
Rubin Landau
Johannes Langguth
Shin-Jye Lee
Mike Lees
Leifur Leifsson
Kenneth Leiter
Florin Leon
Vasiliy Leonenko
Roy Lettieri
Jake Lever
Andrew Lewis
Jingfa Li
Hui Liang
James Liu
Yen-Chen Liu
Zhao Liu
Hui Liu
Pengcheng Liu
Hong Liu
Marcelo Lobosco
Robert Lodder
Chu Kiong Loo
Marcin Los
Stephane Louise
Frederic Loulergue
Hatem Ltaief
Paul Lu
Stefan Luding

Laura Lyman
Scott MacLachlan
Lukasz Madej
Lech Madeyski
Luca Magri
Imran Mahmood
Peyman Mahouti
Marcin Maleszka
Alexander Malyshev
Livia Marcellino
Tomas Margalef
Tiziana Margaria
Osni Marques
M. Carmen Márquez
   García
Paula Martins
Jaime Afonso Martins
Pawel Matuszyk
Valerie Maxville
Pedro Medeiros
Fernando-César
   Meira-Menandro
Roderick Melnik
Valentin Melnikov
Ivan Merelli
Marianna Milano
Leandro Minku
Jaroslaw Miszczak
Kourosh Modarresi
Jânio Monteiro
Fernando Monteiro
James Montgomery
Dariusz Mrozek
Peter Mueller
Ignacio Muga
Judit Munoz-Matute
Philip Nadler
Hiromichi Nagao
Jethro Nagawkar
Kengo Nakajima
Grzegorz J. Nalepa
I. Michael Navon
Philipp Neumann
Du Nguyen
Ngoc Thanh Nguyen
Quang-Vu Nguyen

Sinh Van Nguyen
Nancy Nichols
Anna Nikishova
Hitoshi Nishizawa
Algirdas Noreika
Manuel Núñez
Krzysztof Okarma
Pablo Oliveira
Javier Omella
Kenji Ono
Eneko Osaba
Aziz Ouaarab
Raymond Padmos
Marek Palicki
Junjun Pan
Rongjiang Pan
Nikela Papadopoulou
Marcin Paprzycki
David Pardo
Anna Paszynska
Maciej Paszynski
Abani Patra
Dana Petcu
Serge Petiton
Bernhard Pfahringer
Toby Phillips
Frank Phillipson
Juan C. Pichel
Anna
   Pietrenko-Dabrowska
Laércio L. Pilla
Yuri Pirola
Nadia Pisanti
Sabri Pllana
Mihail Popov
Simon Portegies Zwart
Roland Potthast
Malgorzata
   Przybyla-Kasperek
Ela Pustulka-Hunt
Alexander Pyayt
Kun Qian
Yipeng Qin
Rick Quax
Cesar Quilodran Casas
Enrique S. Quintana-Orti

Ewaryst Rafajlowicz
Ajaykumar Rajasekharan
Raul Ramirez
Célia Ramos
Marcus Randall
Lukasz Rauch
Vishal Raul
Robin Richardson
Sophie Robert
João Rodrigues
Daniel Rodriguez
Albert Romkes
Debraj Roy
Jerzy Rozenblit
Konstantin Ryabinin
Katarzyna Rycerz
Khalid Saeed
Ozlem Salehi
Alberto Sanchez
Aysin Sanci
Gabriele Santin
Rodrigo Santos
Robert Schaefer
Karin Schiller
Ulf D. Schiller
Bertil Schmidt
Martin Schreiber
Gabriela Schütz
Christoph Schweimer
Marinella Sciortino
Diego Sevilla
Mostafa Shahriari
Abolfazi
   Shahzadeh-Fazeli
Vivek Sheraton
Angela Shiflet
Takashi Shimokawabe
Alexander Shukhman
Marcin Sieniek
Nazareen
   Sikkandar Basha
Anna Sikora
Diana Sima
Robert Sinkovits
Haozhen Situ
Leszek Siwik

Vaclav Skala
Ewa
   Skubalska-Rafajlowicz
Peter Sloot
Renata Slota
Oskar Slowik
Grazyna Slusarczyk
Sucha Smanchat
Maciej Smolka
Thiago Sobral
Robert Speck
Katarzyna Stapor
Robert Staszewski
Steve Stevenson
Tomasz Stopa
Achim Streit
Barbara Strug
Patricia Suarez Valero
Vishwas Hebbur Venkata
Subba Rao
Bongwon Suh
Diana Suleimenova
Shuyu Sun
Ray Sun
Vaidy Sunderam
Martin Swain
Jerzy Swiatek
Piotr Szczepaniak
Tadeusz Szuba
Ryszard Tadeusiewicz
Daisuke Takahashi
Zaid Tashman
Osamu Tatebe
Carlos Tavares Calafate
Andrei Tchernykh
Kasim Tersic
Jannis Teunissen
Nestor Tiglao
Alfredo Tirado-Ramos
Zainab Titus
Pawel Topa
Mariusz Topolski
Pawel Trajdos
Bogdan Trawinski
Jan Treur
Leonardo Trujillo

Paolo Trunfio
Ka-Wai Tsang
Hassan Ugail
Eirik Valseth
Ben van Werkhoven
Vítor Vasconcelos
Alexandra Vatyan
Raja Velu
Colin Venters
Milana Vuckovic
Jianwu Wang
Meili Wang
Peng Wang
Jaroslaw Watróbski
Holger Wendland
Lars Wienbrandt
Izabela Wierzbowska
Peter Woehrmann
Szymon Wojciechowski
Michal Wozniak
Maciej Wozniak
Dunhui Xiao
Huilin Xing
Wei Xue
Abuzer Yakaryilmaz
Yoshifumi Yamamoto
Xin-She Yang
Dongwei Ye
Hujun Yin
Lihua You
Han Yu
Drago Žagar
Michal Zak
Gabor Závodszky
Yao Zhang
Wenshu Zhang
Wenbin Zhang
Jian-Jun Zhang
Jinghui Zhong
Sotirios Ziavras
Zoltan Zimboras
Italo Zoppis
Chiara Zucco
Pavel Zun
Pawel Zyblewski
Karol Zyczkowski

Contents – Part IV

Computational Science in IoT and Smart Systems

Computational Methods for Emerging Problems in (dis-)Information Analysis

The Methods and Approaches
of Explainable Artificial Intelligence

Mateusz Szczepański[1,2]($\boxtimes$), Michał Choraś[1,2], Marek Pawlicki[1,2],
and Aleksandra Pawlicka[1]

[1] ITTI Sp. z o.o., Poznań, Poland
`mateusz.szczepanski@itti.com.pl`
[2] UTP University of Science and Technology, Bydgoszcz, Poland

Abstract. Artificial Intelligence has found innumerable applications, becoming ubiquitous in the contemporary society. From making unnoticeable, minor choices to determining people's fates (the case of predictive policing). This fact raises serious concerns about the lack of explainability of those systems. Finding ways to enable humans to comprehend the results provided by AI is a blooming area of research right now. This paper explores the current findings in the field of Explainable Artificial Intelligence (xAI), along with xAI methods and solutions that realise them. The paper provides an umbrella perspective on available xAI options, sorting them into a range of levels of abstraction, starting from community-developed code snippets implementing facets of xAI research all the way up to comprehensive solutions utilising state-of-the-art achievements in the domain.

Keywords: xAI · AI · Intelligent systems · Explainability

1 Introduction

Since **Artificial Intelligence (AI) models** have become sophisticated enough to outclass many competing approaches in their respective fields, their popularity has been on the rise [1]. With initiatives such as **autonomous vehicles**, **various recommendation systems** (e.g., used by Netflix or Google Sybil), **personal assistants** and many more, intelligent systems are being instilled in everyone's lives.

This increasing ubiquity, along with the black-box nature of the best performing solutions, has led to some serious concerns [1–3], such as the questions of finding whether the model is unbiased [3], guaranteeing the security of the AI models [4], ensuring the model's decisions are right [5], or deciding whether to trust a system, the decisions of which cannot be understood [1].

The need to answer those questions has initiated the concept of **Explainable Artificial Intelligence (xAI)** [1]. Its main concern is to deliver the tools and methods that allow human operators to understand the driving forces behind

© Springer Nature Switzerland AG 2021
M. Paszynski et al. (Eds.): ICCS 2021, LNCS 12745, pp. 3–17, 2021.
https://doi.org/10.1007/978-3-030-77970-2_1

the decisions made by AI [6]. The field also relies on the achievements of other disciplines, such as psychology or sociology [2].

Following the expansion of deep learning solutions, the search for rational explanations to the decisions taken by Artificial Intelligence has gained wider recognition [6]. This very year, a number of papers in the field have been published. Some of them present a general overview of the concept [7,8], while others focus on specific, particular features of the Explainable AI [9,10]. Finally, scientific papers which recommend using xAI in a particular field, or prove how beneficial this kind of application would be, have been published, e.g., [11–13], etc.

At present, the discipline is expanding in a dynamic way, enjoying its renaissance [3] and attracting the attention of the biggest corporations, such as Google [14] and IBM [15].

In other words, the accuracy obtained by AI is not the only factor that must be considered at this moment. The ability to understand the decision processes driving AI seems to be of crucial importance, too [5]. This subject has recently started to attract a wider audience [2]. Therefore, the following paper aims to become a starting point for exploring Explainable Artificial Intelligence, the main approaches and available solutions. It is structured as follows: firstly, the notion of Explainable Artificial Intelligence is introduced, with the criteria for explanations and some practical issues. Then, an overview of xAI taxonomies solutions is performed, and lastly, an umbrella perspective of the solutions that utilise xAI is given. The above approach is summarised in the conclusion section that follows.

2   Explainable Artificial Intelligence

The following subsection goes into the details of explainable artificial intelligence, and its advantages over the classical, black box approach to AI are illustrated.

2.1   The Issue About the Black-Box Artificial Intelligence

In psychology, there is the term of the *"Clever Hans effect"* [16]. The name comes from a horse which was famous for its ability to answer questions and solve arithmetic equations, communicating the results by tapping its hoof. However, it later turned out that instead of being a genius, the animal could simply read the cues from the body language of the person asking questions, and stopped tapping accordingly [1]. Today, the "Clever Hans effect" refers to a situation when, in the course of a flawed experiment, the questioner cues the desired behaviour in an unintentional way.

As scientists have learned, this effect is not limited to animals and humans, but also applies to artificial intelligence models as well. There have been observed the cases of models that were successful in performing their tasks only when very specific conditions were met (e.g., a model recognised boats provided that there was water in the picture, too) [1]. This issue may carry adverse implications.

One of the main concerns of today is related to the application of AI in predictive policing. For example, it has been brought to the public's attention that some discriminatory practices generated "dirty data". The data, having been directly ingested by the predictive policing system, posed the risk of reinforcing and amplifying deeply ingrained biases [17]. This in turn might easily have led to disrespecting individual rights, human dignity and undermining justice [18]. In fact, it has indeed been observed that the intelligent criminal justice system had been deciding whether a person deserved parole or not based on their ethnicity [19]. This particular incident has since become a valid argument illustrating the need for artificial intelligence to be transparent, especially in high stake decision processes. An unexplainable system is unverifiable, and therefore untrustworthy. Probably no end user would wish to trust such a system with their lives. Actually, the matter caused so much controversy that a few jurisdictions in the US have ceased their use of predictive policing, whilst in Europe it is being argued that it would be better to pause the use of it until the systems become explainable and transparent enough [17].

2.2 Exploring Explainable Artificial Intelligence

As stated before, the AI-based systems need to be transparent. So much so, in some cases the transparency has been required by law [3]. Therefore, new solutions needed to be found. Thus, the essence of Explainable Artificial intelligence has become that, **given an audience, an explainable Artificial Intelligence is one that produces details or reasons to make its functioning clear or easy to understand** [20]. In order to start discussing explainability, one should then first define the term. In the literature, there exist several terms which, in the context of AI are often used interchangeably to describe a very similar concept, i.e. "explainability", "interpretability", "understandability", "comprehensibility", "intelligibility" and "transparency". However, there are slight differences between them, or rather, the terms have somewhat different undertones, and there is still an ongoing discussion concerning what they actually mean and what they differ in [1–3,20].

In order to clarify this issue, Table 1 presents the meaning of the synonyms in detail. For the sake of this paper, the term "explainability" was selected, due to its broadest scope, active nature and its already established position in the subject literature.

2.3 The Criteria for Explanations

Generally, all of those considerations lead to the objective of determining **what constitutes a good explanation**. To begin with, as Carvalho et al. highlight, an important distinction must be made between the aim of achieving a *correct* explanation and the *best* explanation. Generally speaking, there are **non-pragmatic and pragmatic theories of explanation**. The former group is concentrated on achieving **correct** explanations, while the latter searches for **good** explanations [3].

The non-pragmatic theories usually assume that there is only one, true reason behind the actions of an intelligent system. Their aim is to unveil this reason,

Table 1. The terms used when discussing explainability in the context of AI

Term	Definition
Explainability	Refers to the extent to which human users are able to comprehend and literally explain the mechanisms that drive the learning of an AI/ML system [21] It is an active feature of a model; the term refers to the actions taken by the model to clarify its inner working [22]
Interpretability	Related to the aspects concerning observing the outputs of an AI system. The more predictable the changes of the system outputs when having switched algorithmic parameters, the higher the system's interpretability. Otherwise stated, it concerns the extent to which humans are able to forecast the results produced by an AI system, relying on various inputs [21]. It is a passive feature of the model [22]
Understandability	Used to describe the situation where the user is able to comprehend and generate explanations of how the model works (its way of functioning), without being offered any description of the processes within the learning model [22,23]
Intelligibility	In the context of AI, it is understood in a very similar way to understandability [22,23]
Comprehensibility	Is used to describe the capability of the learning model to outline the knowledge it has learnt in a manner that the user can understand [22]
Transparency	A transparent model is one which does not need any other interface or process to be understood, i.e. it is understandable by itself [22]

but whether or not it is understandable for an audience, is beyond their concern. On the other hand, the pragmatic theories include the listener as an important part of the whole process. Explanation must be formulated in the manner that the audience can understand and use.

The pragmatic theory adds a powerful tool to the theoretical arsenal of xAI researchers and designers: the **Rashomon effect** [3]. It states that an event can have multiple explanations; i.e., more than one explanation can actually be found, and a person can select the one that fits their goals best, while still keeping some level of *"truthfulness"*. However, though certainly useful, it still leaves the matter of selecting the *"best"* explanation from all of the *"good"* ones.

There have been a number of attempts to solve this issue [1]. General guidelines, as well as more objective measures of quality have been suggested. For example, Hansen and Rieger present the *"xAI Desiderata"*, proposed by Swartout and Moore in 1993:

1. **Fidelity**: the explanation must be a reasonable representation of what the system actually does.

2. **Understandability**: Involves multiple usability factors including terminology, user competencies, levels of abstraction and interactivity.
3. **Sufficiency**: Should be able to explain function and terminology, and be detailed enough to justify decision.
4. **Low Construction Overhead**: The explanation should not dominate the cost of designing AI.
5. **Efficiency**: The explanation system should not slow down the AI significantly[1].

Though developed in the context of expert systems, it still remains true for modern AI systems. It also presents a challenge for the community, because designing a solution that adheres to all the principles is not an easy task. As regards the **Quantitative Interpretability Indicators** [24,25], that is the indicators that can be measured and compared, there have been the attempts to formulate those, preferably in a universal manner. The **Axiomatic Explanation Consistency Framework** [25,26] is one of such endeavours. It measures to what degree an explanation method achieves the objective of attaining explanation consistency, and is based upon three axioms [3]:

- **Identity** - Identical objects must have identical explanations.
- **Separability** - Nonidentical objects cannot have identical explanations.
- **Stability** - Similar objects must have similar explanations.

2.4 A Range of Practical Issues

Besides the above-mentioned theoretical aspects, there exist a number of other practical issues. At present, most top performing models are Artificial Neural Networks (ANN). These work by utilising layers of connected computation units called neurons [27]. Though each one on its own is only able to solve simple mathematical problems, together they form complex equations capable of diagnosing cancer, for instance [28]. This ability to generate more abstract concepts based on the simpler ones [29] is what gives Neural Networks their power, but is also the main reason for why achieving their explainability is a non-trivial task. There can be thousands or millions of neurons that interact with one another. Somehow, they are able to form some sort of representations that allow performing advanced tasks. How can those concepts be grasped, though? And even if one is able to frame the concept, the question remains of how to present it to people in an understandable way. Finally, there are also the issues of accuracy loss and a drastic increase of additional overhead.

3 An Overview of xAI Taxonomies

In the recent years, many approaches to explainability have been developed. Many attempts at taxonomising the domain have also been undertaken. One of those attempts can be found in [30]. A comprehensive and in-depth survey on xAI can be found in [31], where authors place considerable effort to handle the formalisms and multidisciplinarity of the field. A brief attempt at a user-centered taxonomy was placed in [32]. A preliminary taxonomy of human subject evaluation can be found in [33]. There is also a comprehensive taxonomy of xAI presented in [20], which includes the methods for both shallow and **Deep Learning (DL)**.

To begin with, the main division present within xAI should be pointed out, i.e., the distinction between the models that inherently have some level of explainability and the ones that need to utilise external means to achieve it. Arrieta et al. present further decomposition of the first category based on the domain, within which the model is transparent [20]. They highlight three main classes:

1. **Simulatable models** - the models that can be fully comprehended and simulated by humans,
2. **Decomposable models** - the models that every part of which, i.e., input, parameter and calculation, can be explained,
3. **Algorithmically transparent models** - the process that generates the output can be understood by a man [20].

Generally, linear models, decision trees rule-base systems etc. are inherently transparent, with the degree varying across the mentioned domains. Nevertheless, with the increasing complexity, these explainable properties can be lost. For example, in case of decision trees, when they get too deep and wide, it becomes quite difficult to follow the paths that a system uses to generate predictions [34].

Unfortunately, most models do not possess this natural transparency; therefore, external methods are needed. Those techniques fall into the wide category of the **post-hoc explanations**. They **"aim at communicating understandable information about how an already developed model produces its predictions for any given input"** [20]. In other words, they make opaque system explainable to some degree.

The post-hoc methods are further split into the **model agnostic** and **model specific** ones. The former means that a method can be used by different Machine Learning models, while the latter marks those designed to explain specific algorithms. Of course, those can be divided even further. The authors of [20] propose to organise the agnostic methods as follows:

- **Feature relevance explanation** - the techniques based on measuring the importance that each feature has for the model's prediction,
- **Explanation by simplification** - the methods where a new, simpler model is built. It resembles the original and keeps a similar performance score, but the level of its complexity has been lowered,
- **Visual Explanation** - as the name suggests, the algorithms belonging to this category employ some form of graphical representation to explain an opaque model.

A good example of an agnostic method is the **Local Interpretable Model-Agnostic Explanation (LIME)** [35], which trains an interpretable linear model around the prediction. It falls into the category of *"explanation by simplification"* and has achieved a significant popularity [1]. Another popular agnostic method is **Shapley Additive exPlanations (SHAP)** [36]. It is a game-theory based framework that calculates an additive feature importance score for each prediction using the Shapley values [20].

As already mentioned, the model specific approaches are designed for particular algorithms. Although they lose the flexibility offered by the agnostic approaches, they may allow for a higher level of fidelity and accuracy. All in all, they were made to leverage the traits of the model they explain. Though the tools are being searched for which can explain shallow models, such as **Support Vector Machines (SVM)**, the main focus is on something else. Since the top performing artificial intelligence systems are usually based on deep learning, it should be no surprise that the methods designed to explain them attract the most attention [20]. There is a variety of approaches dedicated to them. It should be mentioned though that many agnostic methods prove useful for explaining various aspects of deep networks, e.g., the SHAP [36] or LIME [20]. Nonetheless, there are the methods that make sense only with ANN. **Layer-wise relevance propagation (LRP)** [37] is an example of such a method [1]. Founded theoretically on Deep Taylor decomposition, it propagates the output backwards through the network in order to calculate the impact of the input. Like in the case of image recognition, it is expected that the pixels representing the object one wants to detect have a higher score than the others. Of course, it is not the only one. In addition, there are the attribution methods, such as Grad-CAM, hybrid approaches, the systems which combine other deep learning algorithms to automatically generate textual explanation, and many other ways to achieve explainability of DL systems [20]. The final section of this paper will present several of them.

4   An Overview of xAI Solutions

4.1   xAI Methods

Developers and scientists have been looking for practical solutions that will fulfil the pressing need for xAI in modern intelligent systems [1]. This search has ultimately led to the creation of many new algorithms, together with the ways to use them in practice.

To begin with, there are standalone methods developed that are available to the community. Those usually take the form of a source code which the developer can download from the portals like GitHub. In some cases, standard copy-paste procedures are enough to use them as part of the program. This is a rather "*low-level*" approach. When there is the need for more of them, it can quickly become cumbersome and unpractical; even more so if each one of them has its own set of dependencies.

4.2   xAI Libraries and Frameworks

One level of abstraction above the code fragments there are modules, libraries and frameworks. Those provide the practitioners with whole collections of methods in a single package. iNNvestigate [38] is a good example. This library can be simply imported using Python's package manager pip. It allows a developer to quickly use algorithms such as PatternNet, PatternAttribution [39], and different variants of LRP [40]. Another representative for this category is Skater [41]. It provides completely different methods from iNNvestigate, like bfPartial Dependence or LIME.

The last example for this category is the **AI Explainability 360 Open Source Toolkit** from IBM. It presents itself as one of the best frameworks currently available for the practitioners. It offers a diverse selection of algorithms like **ProtoDash** [42] or **Contrastive Explanation Method (CEM)** [43], even improving some of them [15]. Additionally, in contrast to the libraries mentioned earlier, it also provides some metrics to evaluate the quality of the explanation, though it is still quite limited. All of this is backed up by an extensive amount of materials, tutorials and guidelines, which makes it easier to start working with xAI.

4.3 xAI as Part of the System

A popular alternative for frameworks is designing and implementing solutions integrated into a specific system. The main benefit of this approach comes with full customisation, allowing to cater for the specific needs of stakeholders and their product. Explainability is therefore a natural part of the whole and should seamlessly integrate with the rest of the solution. On the other hand, the main disadvantage is the need for additional resources necessary to develop an xAI module from scratch. Additionally, this solution requires the personnel to have expert knowledge about the subject. Therefore, it is suggested to follow this path only if there is a viable reason to do so.

An example from the financial technology market is Flowcast [44]. The solution offers machine learning products for money lending companies. Smartcredit is one of them and is supposed to help in making decisions about financing thin-file **small and medium-size enterprises (SMEs)**, i.e., companies with small amount of traditional financial data used by banks in classic loan application process. This often leads to the rejection of such applicants, although some of them are potential good clients. The creators of the solution claim that this market offers 540 billion dollars' worth of financeable opportunities. Therefore, their system was designed to collect information from non-traditional sources like transaction data, to help the lender get a better picture of an SME company and assess the risks more accurately. Their platform supports a selection of ML algorithms, one of them being a variant of the boosted trees algorithm [45]. As explainability is crucial in the finance sector and the mentioned algorithm is naturally opaque, they had to find a way to clearly explain system assessments. Thus, they use SHAP along with **Natural Language Processing (NLP)** to generate plain-text sentences explaining the output in layman's terms. This is supposed to provide the description of why the system made such a decision, what must be done to change it and the level of confidence in it. They highlight it that the risk professionals employing their platform can access up to top ten reasons why each decision was made. The quality of those explanations is tested by focus groups comprised of risk management professionals and consumers.

The concluding examples of system with an integrated xAI module come from the area of cybersecurity. There is work in the domain of xAI geared towards explaining the decisions of Artificial Neural Networks used as an intrusion detection system. The solution leverages aggregations of decision trees to find the closest explanations for a classified sample [46] To protect network environments from unwanted, malevolent activity, **intrusion detection systems (IDS)** are deployed. As mentioned earlier, the systems that employ some form of ML have become very popular. The ones with best performance usually utilise some form of deep learning. As it was explained in [3], this opaqueness raises concerns and fosters lack of trust. This is a serious issue in the field of cybersecurity, where a wrong decision can lead to dramatic consequences. An expert needs clear understanding, in order to be able to make the right decisions. The authors of [6] present a way to help with that. On a sample dataset, they have trained two deep neural networks to act as IDS. Then, they attached an explainability module that uses the earlier-mentioned SHAP algorithm. The explanation is provided using simple charts that clearly show the features and their contributions. Additionally, the paper introduces a new way to show global relations between feature values and classes. It still needs extensive testing to prove both feasibility and resistance to sophisticated types of attacks. As a final note, it should be clarified that the authors of [6] present their solution as a framework. In this paper, the framework is treated as a collection of ready-made algorithms and tools that support some way of developing a piece of software. Therefore, because this solution would still have to be implemented and integrated into an IDS by a developer, it was placed in this subsection.

As all of the examples above illustrate, *"xAI as part of the system"* is, even with its shortcomings, a valid and fairly popular approach. However, it is not a proper solution if one does not have the knowledge and resources necessary to use xAI this way. Similarly, this is not the best solution for those who only want to validate a model or gain some additional insights into the data without a 'deep dive' into the domain. The last subsection proposes solutions to this issue.

4.4   xAI as a Service

In this section, a promising way of delivering xAI to the companies, developers and scientists is discussed. It is called **"xAI as a Service" (xAI-S)**. As mentioned earlier, implementing the explainable part manually has its unique benefits. However, in most cases it would need excessive resources and would not prove to be as worthy in the long run. Following, there are several examples of *xAI lending* services.

One company offering such service is called **DarwinAI**. On their website [47], they present **The Gensynth Platform**. It is designed to help developers build deep learning models faster, by automatic generation of high performance neural networks that can be deployed in many environments. The fact that it also offers explainability is even more important from the point of view of this work. Their materials show that this is achieved by **Generative Synthesis**. The crux of it is to use another AI model, which will learn how the observed ANN works and generate a compact version of it. Thanks to this, a mathematical model explaining the decision process can be constructed. So far, it has been applied by companies such as Intel, Nvidia and Audi.

Fiddler Labs also offer their own system that helps to achieve explainability [48]. However, while the DarwinAI tool seems to focus more on supporting quick development of deep learning models, Fiddler is all about xAI. While it offers a way to understand AI predictions using methods such as SHAP or **Integrated Gradients** [49], it does not limit itself to them. The official materials highlight other capabilities of the platform, like continuous monitoring of the deployed models. It can be utilised to detect abnormalities in deployed models or catch data anomalies by rising settable alerts. The system also investigates feature relationships, for example by comparing distributions across dataset splits or explaining performance within a specific subset. Last but not least, it allows to test *"What-if scenarios"* i.e. check how different values of input features impact model's decisions [48]. All of that is complemented by the inclusion of human feedback in the workflow, and modern user interface.

The final example illustrates that even the biggest corporations are developing an interest in xAI and the possibilities it offers. *Google Explainable AI* is a part of the Google Cloud platform and has been released in beta version. It is a collection of ready-made tools and frameworks, rather than a streamlined solution, providing a supplement for other products offered by the Google AI Platform. Nevertheless, some of the solutions, like the **What-If Tool**, can be used within a range of environments. Owing to its diverse nature, it is hard to unambiguously classify this whole collection into one category. Nonetheless, these tools are developed to support an existing development service, so the platform roughly falls into the same category as Fiddler and Darwin AI. The mentioned frameworks and tools offer a range of advantages. The mentioned What-If Tool, for example, allows checking feature attribution, test different scenarios to see their impact on the model, examine it for fairness, compare it with others and more; all of that delivered in the form of an interactive dashboard. The official website presents the full list of features and tools available, along with in-depth descriptions, guidelines and tutorials [14]. The platform integrates xAI implementations of Axiomatic Attribution for Deep Networks [49], Sampled Shapley [50], eXplanation with Ranked Area Integrals (XRAI) [51] and others.

There are of course many more startups, products and frameworks that either offer or utilise explainability, which are not included in this section; Rulex [52], Kyndi [53], H20.ai [54], to name just a few [55].

4.5 Current Initiatives and Research Projects

Apart from business and development solutions mentioned in previous sections, there are also several research initiatives and projects looking into the future of AI and xAI.

Obviously, most research projects worked on explainability for image recognition and image retrieval tasks. However, there are projects that touch upon many other domains, like physics etc.

Explainability is one of the challenges recognized by the SPARTA, a Horizon 2020 cybersecurity pilot project, funded by the European Commission. In particular, SAFAIR Programme (Secure and Fair AI Systems for Citizens) of the H2020 SPARTA project focuses on security, explainability, and fairness of AI/ML systems, especially in the cybersecurity domain [56]. Explainability is also one of the factors closely interlinked with ELSA (ethical, legal, societal) activities of SPARTA. Both the SPARTA project and the SAFAIR Programme have started in 2019, and the results are expected by 2022.

Explainability in cybersecurity domain is very challenging (not as visually comprehensible as heatmaps of images), but such aspects are also in the agenda of SIMARGL (Secure Intelligent Methods for Advanced Recognition of Malware and Stegomalware) project working on malware and stegomalware detection mechanisms.

Another H2020 project dealing with the explainability of AI solutions is the Transparent, Reliable and Unbiased Smart Tool for AI (TRUST). It aims at creating an AI platform which is going to be trustworthy and collaborative, and employ explainable by design models and learning models. All the while, the learning process that is going to be adopted is said to be "human-centric" and integrate cognition [57].

In her plenary talk, [58] explored the research field of xAI, used to "overcome the shortcomings of pure statistical learning" and provide the results in the form that could be comprehended by human users [58].

5   Conclusions

This paper discusses the concept of xAI and describes some of its noteworthy solutions.

Explainability is worth being brought to AI models for a variety of reasons:

- **Explainability helps to root out the** *"Clever Hans"* **models** [1]. An opaque model is by its nature difficult to debug or verify. In that case, only the results are visible, not the process. It forces a developer to follow tedious and inefficient approaches in order to find possible inconsistencies. This slows down the whole development and makes it unstable, which in turns increases the risk of obtaining faulty models. However, if the decision process is clear to the designer, many potential problems immediately become apparent.
- **Explainability is a cornerstone of reliability**. This statement results directly from the previous one. Deployed models face challenges such as **Concept Drift** (changes in the hidden context that can induce more or less radical changes in the concept of interest [59]) and **Data Decay** [60]. These are caused by the change of data relevance and its dynamics over time. To alleviate those, both the model and data have to be regularly verified to stay relevant, and, consequently, reliable.
- **Explainability brings trust in the system's decisions** [1]. People will not use the tools they do not trust. It is especially true when the stakes are high. A physician deciding about the treatment needs to know the reasons for

the system reaching such a diagnosis in order to verify it and decide whether they should agree with it or not. Either way, a clear picture is necessary to make a decision based on AI system's output.

– **Explainability reduces bias and supports fairness** [3]. By understanding the principles behind system's decision, it is possible to identify unwanted biases that are present in a dataset. This helps to build models that support our modern ethics, instead of deepening unfair treatment based on race, gender or orientation.
– **Explainability allows to gain additional insights into the domain** [1,31]. An explainable system can detect and unveil unknown relations present in the data. This may lead to new discoveries and studies, making the transparent AI models valuable for the scientific community.

Raising awareness about AI explainability and implementing it across various sectors is still an ongoing process. Even though the questions of explaining AI models to people without losing their accuracy are not easy, the scientific community keeps searching for answers. Since xAI enjoys its renaissance, many new approaches were developed in the recent years [20]. Though a perfect one does not exist, a lot of them show promise and have already proved to be useful.

We hope, this work may serve as a reference point to understand the various tools and xAI solutions/problems better.

Acknowledgment. This work is funded under the SPARTA project, which has received funding from the European Union's Horizon 2020 research and innovation programme under grant agreement No 830892.

References

1. Samek, W., Montavon, G., Vedaldi, A., Hansen, L.K., Müller, K.-R. (eds.): Explainable AI: Interpreting, Explaining and Visualizing Deep Learning. LNCS (LNAI), vol. 11700. Springer, Cham (2019). https://doi.org/10.1007/978-3-030-28954-6
2. Miller, T.: Machine learning interpretability: a survey on methods and metrics. Electronics **8**, 832 (2019)
3. Carvalho, D.V., Pereira, E.M., Cardoso, J.S.: Explanation in artificial intelligence: insights from the social sciences. Artif. Intell. **267**, 1–38 (2019)
4. Pawlicki, M., Choraś, M., Kozik, R.: Defending network intrusion detection systems against adversarial evasion attacks. FGCS **110**, 148–154 (2020)
5. Choraś, M., Pawlicki, M., Puchalski, D., Kozik, R.: Machine learning – the results are not the only thing that matters! What about security, explainability and fairness? In: Krzhizhanovskaya, V.V., et al. (eds.) ICCS 2020. LNCS, vol. 12140, pp. 615–628. Springer, Cham (2020). https://doi.org/10.1007/978-3-030-50423-6_46
6. Wang, M., Zheng, K., Yang, Y., Wang, X.: An explainable machine learning framework for intrusion detection systems. IEEE Access **8**, 73127–73141 (2020)
7. Vilone, G., Longo, L.: Explainable Artificial Intelligence: a Systematic Review (2020)
8. Xie, N., Ras, G., van Gerven, M., Doran, D.: Explainable Deep Learning: A Field Guide for the Uninitiated (2020)

9. Stoyanovich, J., Van Bavel, J.J., West, T.V.: The imperative of interpretable machines. Nat. Mach. Intell. **2**(4), 197–199 (2020)
10. Roscher, R., Bohn, B., Duarte, M.F., Garcke, J.: Explainable Machine Learning for Scientific Insights and Discoveries, CoRR (2019)
11. Tjoa, E., Guan, E.: A survey on explainable artificial intelligence: toward medical XAI. IEEE Trans. Neural Netw. Learn. Syst. (2020)
12. Ghosh, A., Kandasamy, D.: Interpretable artificial intelligence: why and when. Am. J. Roentgenol. **214**(5), 1137–1138 (2020)
13. Reyes, M., et al.: On the interpretability of artificial intelligence in radiology. Radiol. Artif. Intell. **2**(3), e190043 (2020)
14. https://cloud.google.com/explainable-ai
15. Arya, V., et al.: One explanation does not fit all: a toolkit and taxonomy of AI explainability techniques (2019)
16. Samhita, L., Gross, H.: The "Clever Hans phenomenon" revisited. Commun. Integr. Biol. **6**(6), 27122 (2013)
17. Greene, T.: AI now: predictive policing systems are flawed because they replicate and amplify racism. TNW (2020)
18. Asaro, P.M.: AI ethics in predictive policing: from models of threat to an ethics of care. IEEE TSM **38**(2), 40–53 (2019)
19. Wexler, R.: When a computer program keeps you in jail: how computers are harming criminal justice. New York Times (2017)
20. Arrieta, A.B., et al.: Explainable artificial intelligence (XAI): concepts, taxonomies, opportunities and challenges toward responsible AI. Inf. Fusion **58**, 82–115 (2020)
21. Choraś, M., Pawlicki, M., Puchalski, D., Kozik, R.: Machine learning – the results are not the only thing that matters! what about security, explainability and fairness? In: Krzhizhanovskaya, V.V., et al. (eds.) ICCS 2020. LNCS, vol. 12140, pp. 615–628. Springer, Cham (2020). https://doi.org/10.1007/978-3-030-50423-6_46
22. Gandhi, M.: What exactly is meant by explainability and interpretability of AI? Analytics Vidhya (2020)
23. Taylor, M.E.: Intelligibility is a key component to trust in machine learning. Borealis AI (2019)
24. Doshi-Velez, F., Kim, B.: Considerations for evaluation and generalization in interpretable machine learning. In: Escalante, H.J., et al. (eds.) Explainable and Interpretable Models in Computer Vision and Machine Learning. TSSCML, pp. 3–17. Springer, Cham (2018). https://doi.org/10.1007/978-3-319-98131-4_1
25. Doshi-Velez, F., Been, K.: Towards a rigorous science of interpretable machine learning. arXiv preprint arXiv:1702.08608 (2017)
26. Honegger, M.: Shedding Light on Black Box Machine Learning Algorithms: Development of an Axiomatic Framework to Assess the Quality of Methods that Explain Individual Predictions. arXiv preprint arXiv:1808.05054 (2018)
27. Russel, S., Norvig, P.: Artificial Intelligence: A Modern Approach (2010)
28. Liu, S., Zheng, H., Feng, Y., Li, W.: Prostate cancer diagnosis using deep learning with 3D multiparametric MRI. In: Medical Imaging2017: Computer-Aided Diagnosis (2017)
29. Goodfellow, I., Bengio, Y., Courville, A.: Deep Learning (2016)
30. Lipton, Z.C.: The mythos of model interpretability. In: International Conference "In Machine Learning: Workshop on Human Interpretability in Machine Learning" (2016)
31. Adadi, A., Berrada, M.: Peeking inside the black-box: a survey on explainable artificial intelligence (XAI). In: ICCS, vol. 6 (2018)

32. Weina, J., Carpendale, S., Hamarneh, G., Gromala, D.: Bridging AI developers and end users: an end-user-centred explainable AI taxonomy and visual vocabularies. In: IEEE Vis (2019)
33. Chromik, M., Schuessler, M.: A taxonomy for human subject evaluation of black-box explanations in XAI. In: ExSS-ATEC@ IUI (2020)
34. Blanco-Justicia, A., Domingo-Ferrer, J.: Machine learning explainability through comprehensible decision trees. In: Holzinger, A., Kieseberg, P., Tjoa, A.M., Weippl, E. (eds.) CD-MAKE 2019. LNCS, vol. 11713, pp. 15–26. Springer, Cham (2019). https://doi.org/10.1007/978-3-030-29726-8_2
35. Ribeiro, M., Singh, S., Guestrin, C.: Why should i trust you?: explaining the predictions of any classifier. In: Conference of the North American Chapter of the Association for Computational Linguistics: Demonstrations, San Diego, CA (2016)
36. Lundberg, S., Lee, S.I.: A unified approach to interpreting model predictions. In: Advances in Neural Information Processing Systems (2017)
37. Bach, S., Binder, A., Montavon, G., Klauschen, F., Müller, K.R., Samek, W.: On pixel-wise explanations for non-linear classifier decisions by layer-wise relevance propagation. PLoS ONE **10**(7), 0130140 (2015)
38. Alber, M., et al.: iNNvestigate neural networks!, arXiv (2018)
39. Kindermans, P.-J., et al.: Learning how to explain neural networks: patternnet and patternattribution (2017)
40. Montavon, G., Binder, A., Lapuschkin, S., Samek, W., Müller, K.-R.: Layer-wise relevance propagation: an overview. In: Samek, W., Montavon, G., Vedaldi, A., Hansen, L.K., Müller, K.-R. (eds.) Explainable AI: Interpreting, Explaining and Visualizing Deep Learning. LNCS (LNAI), vol. 11700, pp. 193–209. Springer, Cham (2019). https://doi.org/10.1007/978-3-030-28954-6_10
41. https://github.com/oracle/Skater. Accessed 30 Dec 2020
42. Gurumoorthy, K.S., Dhurandhar, A., Cecchi, G., Aggarwal, C.: Efficient data representation by selecting prototypes with importance weights. In: ICD. IEEE (2019)
43. Dhurandhar, A., et al.: Explanations based on the missing: towards contrastive explanations with pertinent negatives. In: Advances in Neural Information Processing Systems (2018)
44. https://flowcast.ai. Accessed 30 Dec 2020
45. https://resources.flowcast.ai/resources/big-data-smart-credit-white-paper/. Accessed 18 Mar 2021
46. Szczepański, M., Choraś, M., Pawlicki, M., Kozik, R.: Achieving explainability of intrusion detection system by hybrid oracle-explainer approach. In: IJCNN (2020)
47. https://darwinai.com. Accessed 30 Dec 2020
48. https://www.fiddler.ai. Accessed 30 Dec 2020
49. Sundararajan, M., Taly, A., Yan, Q.: Axiomatic attribution for deep networks, arXiv preprint arXiv:1703.01365 (2017)
50. Maleki, S., Tran-Thanh, L., Hines, G., Rahwan, T., Rogers, A.: Bounding the estimation error of sampling-based Shapley value approximation, arXiv:1306.4265 (2013)
51. Kapishnikov, A., Bolukbasi, T., Viégas, F., Terry, M.: Xrai: better attributions through regions. In: IEEE International Conference on Computer Vision (2019)
52. https://www.rulex.ai. Accessed 30 Dec 2020
53. https://kyndi.com. Accessed 30 Dec 2020
54. https://www.h2o.ai. Accessed 30 Dec 2020
55. https://www.ventureradar.com. Accessed 30 Dec 2020
56. https://www.sparta.eu/programs/safair/. Accessed 18 Mar 2021

57. https://cordis.europa.eu/project/id/952060. Accessed 30 Dec 2020
58. Zanni-Merk, C.: On the Need of an Explainable Artificial Intelligence (2020)
59. Widmer, G., Kubat, M.: Learning in the presence of concept drift and hidden contexts. Mach. Learn. **23**(1), 69–101 (1996)
60. https://peterasaro.org/writing/Asaro_PredictivePolicing.pdf. Accessed 30 Dec 2020

Fake or Real? The Novel Approach to Detecting Online Disinformation Based on Multi ML Classifiers

Martyna Tarczewska, Anna Marciniak⬤, and Agata Giełczyk(✉)⬤

University of Science and Technology, Bydgoszcz, Poland
agata.gielczyk@utp.edu.pl

Abstract. Background: the machine learning (ML) techniques have been implemented in numerous applications and domains, including health-care, security, entertainment, and sports. This paper presents how ML can be used for detecting fake news. The problem of online disinformation has recently become one of the most challenging issues of computer science. Methods: in this research, a fake news detection method based on multi classifiers (CNN, XGBoost, Random Forest, Naive Bayes, SVM) has been developed. In the proposed method, two classifiers cooperate; consequently, they obtain better results. Realistic, publicly available data was used in order to train and test the classifiers, Results: in the article, several experiments were presented; they differ in the implemented classifiers, and some improved parameters. Promising results (accuracy = 0.95, precision = 0.99, recall = 0.91, and F1-score = 0.95) were reported. Conclusion: the presented research proves that machine learning is a promising approach to fake news detection.

Keywords: Fake news · Online disinformation · Machine learning

1 Introduction

According to the Collins dictionary, fake news can be defined as 'false, often sensational, information disseminated under the guise of news reporting'[1]. Despite the fact that fake news existed for many years, its impact has recently increased. This trend can be easily observed, e.g., by means of the Google Trends tool[2]. It shows that the phrase 'fake news' has rapidly become more popular since November 2016. Traditionally, fake news was known as rumors or propaganda, mostly used in order to make political or economic gains. The main goal of creating fake news has remained unchanged. However, currently it can spread more easily thanks to the popularity of social networks. The current pandemic reality has led to a serious outbreak of misinformation. It can be very dangerous in social, health-care and political aspects, like in the case of the fake news concerning the COVID-19 pandemic and its connection with the 5G transmission [2], etc.

---

[1] https://www.collinsdictionary.com/dictionary/english/fake-news.
[2] https://trends.google.com/trends/explore?date=today%205-y&q=fake%20news.

© Springer Nature Switzerland AG 2021
M. Paszynski et al. (Eds.): ICCS 2021, LNCS 12745, pp. 18–27, 2021.
https://doi.org/10.1007/978-3-030-77970-2_2

Several subtypes of fake news can be listed [12]:

- rumor - an item of circulating information the veracity status of which is yet to be verified at the time of posting;
- hoax - a deliberately fabricated falsehood made to masquerade as truth;
- click-bait - a piece of low-quality journalism which is intended to attract traffic and monetize via advertising revenue;
- disinformation - fake or inaccurate information which is intentionally false and spread deliberately;
- misinformation - fake or inaccurate information which is spread unintentionally;
- fake news - a news article that is intentionally and verifiable false.

Moreover, it is possible to find some pieces of information which can be classified as satire. Unlike subtler forms of deception, satire may feature more obvious cues that reveal its disassociation from the truth. In fact, satire is meant to be recognized as a joke, at least by some readers.

Due to the variety of fake news types, the methods used in order to classify it are also diverse. The typology of fake news detection approaches is presented in Fig. 1. One of the most popular approaches to detecting fake news is NLP, consisting in analyzing the text of the news/tweet/post [5]. In such approaches, pattern recognition systems are trained in order to discover lexical [10], linguistic [8], psycholinguistic [23], syntactic [4] and semantic [9] features.

The general concept behind the authors' reputation system is to evaluate the source of the information - it can be a publisher, a www address or an IP address. In such an approach, some websites or information providers (e.g., CNN or BBC) can be assumed to be reliable. A sample system focusing on the author's credibility was presented in [20].

Another approach is to implement network analysis, which refers to the network and graph theory. In this approach, the relations between the news' author and the user who reposts or shares it are discovered, as presented in [19].

Since images have become a dominant and powerful communication channel, the last but not least group of methods used in order to detect the fake news is based on image analysis. The ML-based approach to fake news detection by recognizing image forgery is presented in [11] and in [14].

The remainder of the paper is structured as follows: Sect. 2 presents the current state of the art. In Sect. 3, the proposed solution is described in detail. Section 4 contains and discusses the obtained results. Section 5 provides threats to validity, and conclusions.

2 Related Work

Amongst the approaches to detecting fake news, convolutional neural networks (CNN), support vector machine (SVM), random forest (RF) and the XGBoost classifier are currently the most commonly used ones.

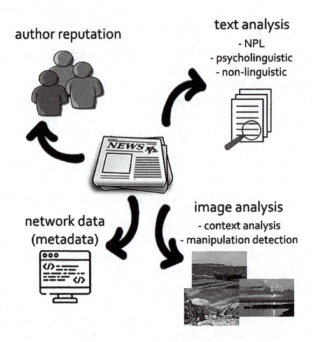

Fig. 1. The typology of fake news detection approaches [5]

The authors in [18] developed a system for fake news detection with the use of supervised learning methods. The authors compared several algorithms, including k-Nearest Neighbors (kNN), Naive Bayes (NB), Random Forests (RF), Support Vector Machine with RBF kernel (SVM), and XGBoost (XGB). In this case, the best results were obtained with RF and XGB - AUC 0.85 and 0.86, respectively; F1-score 0.81 for both models.

The authors in [1] tested various machine learning methods on four different datasets. Altogether, 13 methods were tested, including CNN, LSTM and XGB. Among those three methods, XGB achieved the best accuracy (over 89% each time). Among the other tested methods, the best results were obtained by RF and linear SVM classifiers, with accuracy over 91% (in 3 out of 4 datasets) and over 90% (in 3 out of 4 datasets), respectively.

The authors in [22] developed a novel, hybrid CNN to integrate metadata with text. Authors compared two approaches: text-only models (including SVM, logistic regression, Bi-LSTM and CNN) and test and meta-data hybrid models (hybrid CNN). The better results were obtained with the hybrid CNN approach, both in the test and in the validation dataset.

In [7], the authors proposed a deep neural network approach, where CNN and LSTM models were used. Both single models and their combinations have been tested and compared with previously developed models, including SVM and the model described above. However, the authors' approach did not result in better accuracy (97.84%) than the previously published models [24].

The authors in [13] studied fake news detection with different degrees of fakeness by integrating several sources. The authors proposed a Multi-source Multi-class Fake news Detection framework (MMFD), with automated feature extraction (performed with the use of CNN and LSTM), incorporated multi-source fusion and fakeness discrimination. Moreover, the authors compared MMFD to SVM, RF, kNN and Wang method (described above [22]). In each case, the MMFD achieved better accuracy.

The authors in [17] developed a system for automatic fake news detection; they focused on the preparation of new data for further analysis, which consists of the evaluation of the linguistic features, creating a machine learning model and making a comparison to human performance. Linear SVM classifier and five-fold cross-validation were used to create the fake news detector and R, caret and e1071 packages were used to conduct machine learning classification. The results achieved by the models were comparable to those achieved by humans.

The authors in [21] focused on fake news detection using deep learning architecture. In this model, the authors included both the CNN and LSTM neural networks and combined them with principle component analysis (PCA) and Chi-Square. By using this approach, the authors achieved the fake news detection accuracy of 97.8%.

In [15], the approach based on kNN was developed. The authors used the dataset which has been collected from Buzz Feed News organization and which is commonly used in scientific methods. It contains Facebook posts. Using the proposed approach, it was possible to obtain accuracy reaching 79%.

The article [3] presented a hybrid architecture is which is based on Bidirectional LSTM and Convolutional Neural Network. Using both types of classifiers enabled to incorporate news content and information concerning the user profile as well. The proposed hybrid architecture performed better than individual architecture and it gave overall accuracy of 42.2%. The experiments were conducted using the Liar dataset. Authors pointed that the similarity of classes in the dataset (pants-fire, false, barely-true were claimed to be almost same) was the biggest challenge in this research.

Authors in [16] also proposed a novel hybrid method. Their model combined the Convolutional Neural Network and the Recurrent Neural Network for fake news classification. It was successfully validated on two fake news datasets (ISO and FA-KES), achieving detection results that was significantly better than other non-hybrid baseline methods. One of the key points of the proposed method was the pre-processing, namely Word2Vec provided by Google and GloVe, pre-trained word embedding.

The interesting solution was introduced in [25], where the explainable fake news detection tool was presented. In this approach the XGBoost classifier was implemented in order to detect the online disinformation. The usability of the proposed solution was demonstrated on a real-world dataset crawled from PolitiFact, where thousands of verified political news have been collected.

3   Presented Approach

3.1   Dataset and Pre-processing

In this research, a publicly available dataset was used, which can be downloaded from the Kaggle website (https://www.kaggle.com/c/fake-news/data). A single row in the dataset contains the following elements: id - unique id for a news article, title - the title of a news article, author - the author of the news article, text - the text of the article and label - that marks the article as potentially unreliable. The label equal to 1 shows that the article is fake news, whereas 0 means that it is reliable news. The initial pipeline of the proposed method is presented in Fig. 2. First of all, the pre-processing needs to be performed. Thus, the article body is converted to the lower case, and the stop words which are commonly used words (such as 'the', 'a', 'an', 'in'), and do not significantly impact of the whole text's sense, are deleted.

The dataset contains over 20k labeled rows. The dataset is well balanced - half of the articles were marked as fake, and the other half as real. During the experiments, the dataset was divided into the training set (80%) and the testing set (20%). Thus, the training data was obtained, which also was balanced - 2k fake news and 2k reliable news. The experiments were performed using the 5-fold cross validation.

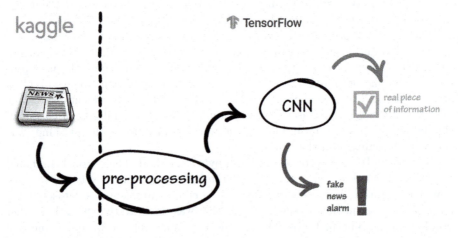

Fig. 2. The pipeline of the CNN-based method - the starting point for further experiments

3.2   Machine Learning

All the experiments were performed using the Keras API that works with Tensorflow. These tools enabled using the machine learning methods. As mentioned in the state-of-the-art review, there are several ML methods that are widely implemented in fake news detection. In order to perform this research, the following

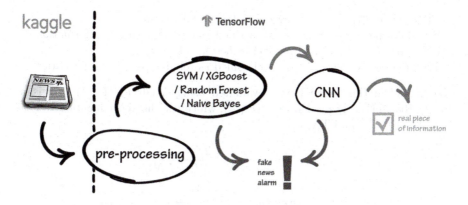

Fig. 3. The pipeline of the proposed method

ones were selected: Convolutional Neural Network (CNN), eXtreme Gradient Boosting (XGBoost), Support Vector Machine (SVM), Naive Bayes (NB) and Random Forest (RF).

First of all, the CNN training and testing were run with the default parameters of the network. This kind of approach would be the starting point for the next algorithms; its general pipeline is presented in Fig. 2. The next step was to improve the parameters of the CNN so that it could give higher accuracy. Thus, 128 convolutional layers with activation type=relu were added, the dropout as modified to 0.2 and 10 dense layers with activation type=relu were added. The proposed improvements were performed according to the state-of-the-art review and the authors' experience. The CNN with improved parameters is further called the boosted CNN.

The next step of the proposed method was to implement a number of methods: XGBoost, SVM, NB, and RF as a single classifier in place of CNN.

The last part of the research was to add the additional classifier which would initially scan the articles. This approach is presented in Fig. 3. The first classifier verifies the article. If the article receives the label 'fake', it is finally marked as the false piece of information (red arrow in Fig. 3). Otherwise, the next classifier analyzes the article (green arrow in Fig. 3). The decision of the second classifier is final and the article gets the label fake or reliable. The additional classifiers were again: XGBoost, SVM, NB and RF, whereas the final decision was made by the boosted CNN.

4   Results and Discussion

Since the fake news detection problem can be understood as a binary classification, confusion matrices were used in order to evaluate and compare the ML-based methods. Four measures were defined as follows:

- TP - true positives - fake news classified as fake news;
- FP - false positives - fake news classified as reliable pieces of information;
- FN - false negatives - real news classified as fake news;
- TN - true negatives - real news classified as reliable pieces of information.

Each model in this research was evaluated using Accuracy (Eq. 1), Precision (Eq. 2), Recall (Eq. 3) and F1-score (Eq. 4), which use the above mentioned measures TP, FP, FN and TN.

$$Acc = \frac{TP + TN}{TP + TN + FP + FN} \tag{1}$$

$$precision = \frac{TP}{TP + FP} \tag{2}$$

$$recall = \frac{TP}{TP + FN} \tag{3}$$

$$F1 - score = 2 \cdot \frac{precision \cdot recall}{precision + recall} \tag{4}$$

Table 1. Obtained results

Classifier	Accuracy	Precision	Recall	F1-score
CNN (default)	0.8889	0.8696	0.9006	0.8846
CNN (boosted)	0.9213	0.9150	0.9248	0.9194
XGBoost	0.8992	0.8778	0.9135	0.8953
SVM	0.6311	0.6096	0.6276	0.6184
Naive Bayes	0.5810	0.4801	0.5893	0.5291
Random Forest	0.7853	0.7112	0.8269	0.7647
CNN + XGBoost	**0.9487**	**0.9941**	**0.9117**	**0.9511**
CNN + SVM	0.8328	0.9736	0.7603	0.8538
CNN + Bayes	0.7921	0.9565	0.7205	0.8219
CNN + Random Forest	0.9458	0.9941	0.9068	0.9485

The obtained results are presented in Table 1. As seen in it, by modifying selected parameter of CNN it was possible to improve the results. When it comes to the comparison of the single classifiers (CNN excluded), the most promising results were obtained by XGBoost (Acc = 90%, Prec = 88%, Rec = 91% and F1 = 90%). Each classifier used with CNN gave the improved results. It is also remarkable that the combination CNN + Random Forest is very promising, even though Random Forest used as a single classifier was not impressive (Acc = 79%, Prec = 71%, Rec = 83% and F1 = 76%). Nevertheless, the highest values of Accuracy, Precision, Recall and F1-score were achieved by connecting CNN with

XGBoost, namely Acc = 0.95, Prec = 0.99, Rec = 0.91 and F1 = 0.95. This result is the most encouraging and thus, marked in bold in the Table 1. The obtained results are also presented in a visual way as the confusion matrices in Fig. 4. The selected experiments results have been presented there: default CNN, boosted CNN, XGBoost, CNN+XGBoost, RF and CNN+RF. The results' improvement is especially visible between RF and CNN+RF (the third row), where the number of FP was decreased significantly.

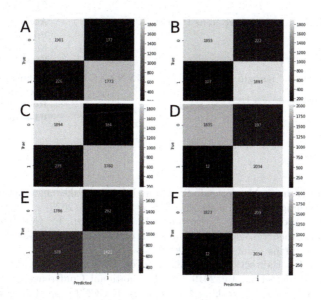

Fig. 4. Confusion matrices for the selected experiments - A: CNN default, B: CNN boosted, C: XGBoost, D: CNN + XGBoost, E: RF and F: CNN + RF

5   Conclusions

In this paper, an efficient and accurate ML-based approach to the fake news detection has been presented. The obtained results were promising, as seen in Table 2. Consequently, the approach enables obtaining results that are similar to the results of the current state-of-the-art approaches. However, it is essential to mention that the fake news detecting methods are hardly comparable due to the variety of the fake news datasets.

The proposed solution may be extended in the future, e.g., by implementing another type of classifier or introducing some more pre-processing methods. Other possible extension that could be done in the nearest future are rebuilding the pipeline of the proposed solution and adding the block of explanation. This kind of approach could give both the fake/real assessment and the explanation why the algorithm decided in such a way.

It is also remarkable, as proposed in [6], that automatic fake news detection tools should be designed to augment human judgement, not to replace it. The human aspect would be especially helpful in recognizing satire and jokes.

Table 2. Obtained results in related works and in the proposed method

Reference	Dataset	Method	Result
De Sarkar et al. [7]	LIAR	CNN	Accuracy = 98% Precision = 93% Recall = 84% F1-score = 88%
Reis et al. [18]	BuzzFace	XGBoost	AUC = 86% F1 = 81%
Ahmad et al. [1]	ISOT Fake News Dataset	XGBoost	Accuracy = 98% Precision = 99% Recall = 99% F1-score = 99%
Tarczewska et al.	**Kaggle**	**CNN+XGBoost**	**Accuracy = 95%** **Precision = 99%** **Recall = 91%** **F1-score = 95%**

References

1. Ahmad, I., Yousaf, M., Yousaf, S., Ahmad, M.O.: Fake news detection using machine learning ensemble methods. Complexity **2020** (2020)
2. Ahmed, W., Vidal-Alaball, J., Downing, J., López Seguí, F.: COVID-19 and the 5g conspiracy theory: social network analysis of Twitter data. J. Med Internet Res. **22**(5) (2020). https://doi.org/10.2196/19458
3. Balwant, M.K.: Bidirectional LSTM based on POS tags and CNN architecture for fake news detection. In: 2019 10th International Conference on Computing, Communication and Networking Technologies (ICCCNT), pp. 1–6. IEEE (2019)
4. Capistrano, J.L.C., Suarez, J.J.P., Naval Jr, P.C.: SALSA: detection of cybertrolls using sentiment, aggression, lexical and syntactic analysis of tweets. In: Proceedings of the 9th International Conference on Web Intelligence, Mining and Semantics, pp. 1–6 (2019)
5. Choraś, M., et al.: Advanced machine learning techniques for fake news (online disinformation) detection: a systematic mapping study. Appl. Soft Comput. 107050 (2020)
6. Conroy, N.K., Rubin, V.L., Chen, Y.: Automatic deception detection: methods for finding fake news. Proc. Assoc. Inf. Sci. Technol. **52**(1), 1–4 (2015)
7. De Sarkar, S., Yang, F., Mukherjee, A.: Attending sentences to detect satirical fake news. In: Proceedings of the 27th International Conference on Computational Linguistics, pp. 3371–3380 (2018)
8. Dey, A., Rafi, R.Z., Parash, S.H., Arko, S.K., Chakrabarty, A.: Fake news pattern recognition using linguistic analysis. In: 2018 Joint 7th International Conference on Informatics, Electronics & Vision (ICIEV) and 2018 2nd International Conference on Imaging, Vision & Pattern Recognition (icIVPR), pp. 305–309. IEEE (2018)
9. Gaglani, J., Gandhi, Y., Gogate, S., Halbe, A.: Unsupervised whatsapp fake news detection using semantic search. In: 2020 4th International Conference on Intelligent Computing and Control Systems (ICICCS), pp. 285–289. IEEE (2020)
10. Giglou, H.B., Razmara, J., Rahgouy, M., Sanaei, M.: LSACoNet: a combination of lexical and conceptual features for analysis of fake news spreaders on Twitter. In: CLEF (2020)

11. Gragnaniello, D., Marra, F., Poggi, G., Verdoliva, L.: Analysis of adversarial attacks against cnn-based image forgery detectors. In: 2018 26th European Signal Processing Conference (EUSIPCO), pp. 967–971. IEEE (2018)

12. Guo, B., Ding, Y., Sun, Y., Ma, S., Li, K., Yu, Z.: The mass, fake news, and cognition security. Front. Comput. Sci. **15**(3), 1–13 (2020). https://doi.org/10.1007/s11704-020-9256-0

13. Karimi, H., Roy, P., Saba-Sadiya, S., Tang, J.: Multi-source multi-class fake news detection. In: Proceedings of the 27th International Conference on Computational Linguistics, pp. 1546–1557 (2018)

14. Kasban, H., Nassar, S.: An efficient approach for forgery detection in digital images using Hilbert-Huang transform. Appl. Soft Comput. **97**, 106728 (2020)

15. Kesarwani, A., Chauhan, S.S., Nair, A.R.: Fake news detection on social media using k-nearest neighbor classifier. In: 2020 International Conference on Advances in Computing and Communication Engineering (ICACCE), pp. 1–4. IEEE (2020)

16. Nasir, J.A., Khan, O.S., Varlamis, I.: Fake news detection: a hybrid CNN-RNN based deep learning approach. Int. J. Inf. Manag. Data Insights **1**(1), 100007 (2021)

17. Pérez-Rosas, V., Kleinberg, B., Lefevre, A., Mihalcea, R.: Automatic detection of fake news. arXiv preprint arXiv:1708.07104 (2017)

18. Reis, J.C., Correia, A., Murai, F., Veloso, A., Benevenuto, F.: Supervised learning for fake news detection. IEEE Intell. Syst. **34**(2), 76–81 (2019)

19. Shu, K., Bernard, H.R., Liu, H.: Studying fake news via network analysis: detection and mitigation. In: Agarwal, N., Dokoohaki, N., Tokdemir, S. (eds.) Emerging Research Challenges and Opportunities in Computational Social Network Analysis and Mining. LNSN, pp. 43–65. Springer, Cham (2019). https://doi.org/10.1007/978-3-319-94105-9_3

20. Sitaula, N., Mohan, C.K., Grygiel, J., Zhou, X., Zafarani, R.: Credibility-based fake news detection. In: Shu, K., Wang, S., Lee, D., Liu, H. (eds.) Disinformation, Misinformation, and Fake News in Social Media. LNSN, pp. 163–182. Springer, Cham (2020). https://doi.org/10.1007/978-3-030-42699-6_9

21. Umer, M., Imtiaz, Z., Ullah, S., Mehmood, A., Choi, G.S., On, B.W.: Fake news stance detection using deep learning architecture (CNN-LSTM). IEEE Access **8**, 156695–156706 (2020)

22. Wang, W.Y.: "Liar, liar pants on fire": a new benchmark dataset for fake news detection. arXiv preprint arXiv:1705.00648 (2017)

23. Wawer, A., Wojdyga, G., Sarzyńska-Wawer, J.: Fact checking or psycholinguistics: how to distinguish fake and true claims? In: Proceedings of the Second Workshop on Fact Extraction and VERification (FEVER), pp. 7–12 (2019)

24. Yang, F., Mukherjee, A., Dragut, E.: Satirical news detection and analysis using attention mechanism and linguistic features. arXiv preprint arXiv:1709.01189 (2017)

25. Yang, F., et al.: XFake: explainable fake news detector with visualizations. In: The World Wide Web Conference, pp. 3600–3604 (2019)

Transformer Based Models in Fake News Detection

Sebastian Kula[1,2(✉)], Rafał Kozik[1], Michał Choraś[1], and Michał Woźniak[1,3]

[1] UTP University of Science and Technology, Bydgoszcz, Poland
[2] Kazimierz Wielki University, Bydgoszcz, Poland
skula@ukw.edu.pl
[3] Wrocław University of Science and Technology, Wrocław, Poland

Abstract. The article presents models for detecting fake news and the results of the analyzes of the application of these models. The precision, f1-score, recall metrics were proposed as a measure of the model quality assessment. Neural network architectures, based on the state-of-the-art solutions of the Transformer type were applied to create the models. The computing capabilities of the Google Colaboratory remote platform, as well as the Flair library, made it feasible to obtain reliable, robust models for fake news detection. The problem of disinformation and fake news is an important issue for modern societies, which commonly use state-of-the-art telecommunications technologies. Artificial intelligence and deep learning techniques are considered to be effective tools in protection against these undesirable phenomena.

Keywords: Fake news detection · Transformers · Natural language processing · Deep learning · SocialTruth

1 Introduction

The dynamic development of social media, instant messaging, internet information portals, and means of electronic communication resulted in a significant decrease in the influence and importance of the traditional mass media, such as radio, television and paper (printed) press. Modern societies use the technical innovations offered by technology in the ICT area with great enthusiasm.

These dynamic changes are accompanied by new phenomena that may potentially have a destructive impact on society, and may undermine the credibility of institutions, governments and companies. Fake news is one such undesirable phenomenon that is present in the ICT revolution. Fake news is defined as deliberate disinformation, as an action aimed at causing disorder and information chaos through false or partially true messages. This phenomenon, initially unnoticed, is growing. The scale of the impact of this type of practices on society is evidenced by the fact that it is commonly accepted that fake news influenced the results of political elections or referenda in politically significant and important countries. In these countries, despite them having high levels of democratic standards and well-established electoral mechanisms, large social groups

M. Paszynski et al. (Eds.): ICCS 2021, LNCS 12745, pp. 28–38, 2021.
https://doi.org/10.1007/978-3-030-77970-2_3

were successfully manipulated through fake news. Fake news can also be used to manipulate public health, economic marketing, product sales, and public safety. Considering the above, many institutions, governments, authorities responsible for security, as well as companies for which ethical standards and the ethos of the institution are important, significantly increased their interest in the phenomenon of fake news and effective methods of preventing it.

In the fight against fake news, machine learning methods are very successful, allowing the creation of models that are not only accurate and effective, but also practical, ensuring the detection of an undesirable phenomenon almost in real time. Automatic real-time disinformation detection is a must. The enormous amount of information that reaches the average recipient every day means that the models that do not allow for swift detection of disinformation are not applicable.

Among the machine learning methods, the methods based on deep learning show considerable promise. The DL (Deep Learning) methods have been created based on the learning processes taking place in the human brain. It is expected that through DL, it will be possible to create models that will not only detect the already identified types of disinformation, but will be able to act in advance and identify new, currently unknown techniques of disinformation. The big advantage of the DL methods is that they can be trained without feature engineering or with relatively little application of the feature engineering. This allows for the presumption that a neural network based on DL will automatically detect the features, which are characteristic for fake news, and that it will do it better than a human.

Combating disinformation requires the use of NLP (Natural Language Processing) techniques, which replace a journalist, linguist or media expert in the process of evaluating the credibility of information. The NLP procedures imitate human processes; one such process is intuitive reasoning. The intuitive reasoning is considered to be a key element in specialists with extensive professional experience. It is a largely subconscious process resulting from a long process of learning and gaining experience. The phenomenon of intuitive reasoning in humans is reproduced in computer deep learning algorithms, where a relatively large amount of data allows the detection of patterns typical of these data, without the need to engineer the features before training the neural network.

As regards fake news, the DL-based models allow detection of the features that are associated with written texts, such as stylistics, phraseology, syntax, semantics, pragmatics, morphology, i.e., the features that are characteristic of the author of fake news. Thus, if the author is a human, it should be assumed that it will be possible to detect the literary features that are constant for a given author.

The article presents methods based on deep learning models and Transformers architectures. The main goal of this work was to create reliable models allowing the verification of short texts, such as article titles, available on the web. To meet this challenge, the focus was on the analysis of the architectures that are considered to be revolutionary, state-of-the-art methods for many NLP tasks.

The article describes related works in the Sect. 2, the description of Transformers architectures is presented in the Sect. 3, the Sect. 4 is a description of the proposed methods using the Flair library, the conducted experiments are included in the Sect. 5. Results and application of the models for verification of scraped web pages titles are presented in Sects. 6 and 7, respectively.

2   Related Works

Along with the development of AI (Artificial Intelligence) methods, applied in various research and engineering domains, the methods related to NLP are also developing dynamically. Text classifiers, which use better and more precise ML methods, play a special and essential role in defending against fake news. Until recently, the most commonly applied methods used in text classification were algorithms derived from the Naive Bayes, SVM (Support Vector Machines), CNN (Convolution Neural Networks) and RNN (Recurrent Neural Networks) algorithms, which are a type of the DNN (Deep Neural Network). Currently, the methods based on Transformers architectures are achieving outstanding results.

In [21], the authors use the SVM method and the feature selection method to reduce the data size. Their method has been validated using already classified datasets. The authors of the article [5] used the Naive Bayes and SVM methods to classify 327884 pieces of information from Twitter as believable and unbelievable, reporting very high accuracy, up to 99.9%. In another work, an existing classified dataset was used to create an overview of five ML models detecting fake news [3]. The authors analyzed classifiers based on Naive Bayes, Logistic Regression, Linear SVM, Stochastic Gradient Classifier and Random Forest Classifiers. The obtained results showed that SVM and Logistic Regression have the best performance on the applied dataset in the model [3]. The authors of [17] focused on deep learning by applying LSTM (Long Short-Term Memory), CNN and pre-trained GloVe embeddings in their model. A similar hybrid architecture based on the Flair library [4] was presented in [10], where pre-trained GloVe at the word embeddings level and RNN at the document embeddings level were applied. A big change in the classifiers of texts was brought about by the development of Transformer techniques and architectures using attention models. Models based on these techniques are ahead of other, previously used models in terms of the obtained metrics results; simultaneously, they are strongly parallel, which results in the possibility of training on parallel platforms, including the GPUs. This is of essential importance when training the neural network and optimizing the computing cost, which is high in the case of big data. Transformer based methods, specifically BERT, were applied in [9], where BERT base architectures were applied for fake news detection. In [13], BERT was applied to build models relating to the credibility of the texts, based on the database, which contains 20015 news articles, labeled as fake or true [19]; the accuracy of 98% was achieved.

The recent and broad systematic mapping study of the fake news detection techniques is presented in [12].

3 Transformers Architectures

The emergence of the self-attention mechanism and Transformer architectures meant that in NLP algorithms, the context in the sentence is much more important than the words themselves. This mechanism caused it that the binary representation of the word (token) is not constant and changes depending on the surroundings of the words (tokens) in the sentence. The use of Transformers for NLP-related tasks was proposed in [16], where they are presented as departing from recursion in favor of the attention mechanism.

Transformer-based methods are present in many architectures; this article uses the following architectures: BERT [7], RoBERTa [11], DistilBERT [14], xlNet [20], DistilGPT2 [18]. They differ mainly in the size and number of the layers of the neural network applied. The base version of BERT architecture contains 109 million parameters in the case of the cased corpus applied for training, and 110 million parameters for the situation, when it is trained on the uncased corpus; BERT large, in turn, contains 335 million parameters [18]. DistilBERT architectures, trained from the uncased corpus contain 66 million parameters, DistilGPT2 82 million, RoBERTa large 355 million parameters, xlNet trained from the corpus cased 340 million parameters [18]. Table 1 lists the details of the Transformer-based architectures.

Table 1. Parameters values of transformer architectures [18]

Transformer-based architecture	Number of parameters in millions	Number of layers	Hidden states size	Number of self-attention heads
BERT base cased	109	12	768	12
BERT base uncased	110	12	768	12
BERT large cased	335	24	1024	16
BERT large uncased	336	24	1024	16
DistilBERT uncased	66	6	768	12
DistilGPT2	82	6	768	12
RoBERTa large	355	24	1024	16
xlNet large cased	340	24	1024	16

4 Transformer Based Classifiers

In this work, the Flair [4] library was applied to create text classifier models that detect fake news. This library allows, in addition to choosing the pre-trained architecture, to define the embeddings technique, whether it be at the word, sentence or document level. Selection in the Flair is made with the use of the TransformerDocumentEmbeddings command, which causes the sentence-level embedding to be extracted, and with the use of the DocumentRNNEmbeddings command that document-level embeddings is extracted [4]. Both embeddings techniques were used in the work to create models.

The choice of the embeddings technique modifies the architecture of the neural network. The application of the document level embeddings, i.e., producing vector representations of the entire document, results in adding an additional layer in the architecture [4]. In the presented work, this additional layer is the GRU (Gated Recurrent Units) layer. The second key element modifying the architecture, applied to create the classifier models is the selected pre-trained, Transformer-based embeddings architecture. In the work, selected Transformer-based architectures, listed in Table 1 were applied with maintaining the values of the parameters of these architectures presented in the table. The architecture applied to create the classifier models is therefore dynamic and not the same for all models. In the article, the following designations for the architectures used to create models, based on sentence level embeddings and Transformer-based architectures are introduced: distilbert-base-uncased, distilgpt2, roberta-large, xlnet-large-cased, bert-large-cased_TDE. Architecture based on BERT and document level embeddings is marked as bert-large-cased_DRE. It does not differ from the bert-large-cased_TDE regarding the implemented Transformer-based architecture, which is BERT; the difference between architectures is the additional GRU layer.

The model creation process was as follows: the first step consisted in selecting ready-made databases containing article titles that were already classified as fake and true, next the pre-processing of all data in datasets, setting the dynamic architecture by the selecting the pre-trained Transformer-based architecture and choosing embedding technique, the network training based on the selected version of the dynamic architecture, and the last step was the creation of the model for classification. The described process is shown in Fig. 1. The process is repeatable and depicts all the steps that were performed for all the models created.

5 Experiment Setup

This section details the experiments conducted, including the description of the data applied and the hyperparameters values set in the training routine.

5.1 Collections

Two collections were created for the experiments, based on the data repository FakeNewsNet, KaiDMML [15]. Both collections contain article titles, grouped as fake and true. Collection 1 contains the titles of news content that were collected using the fact-checking website Politifact [8], and collection 2 contains data collected using GossipCop [1]. Collection 1 contains a total of 1056 items, of which 432 are labeled as fake and 624 are labeled as real (true). The titles of the articles in this collection are relatively short, because the maximum length of the word sequence is 340 characters. The primary sources of origin of the articles vary,

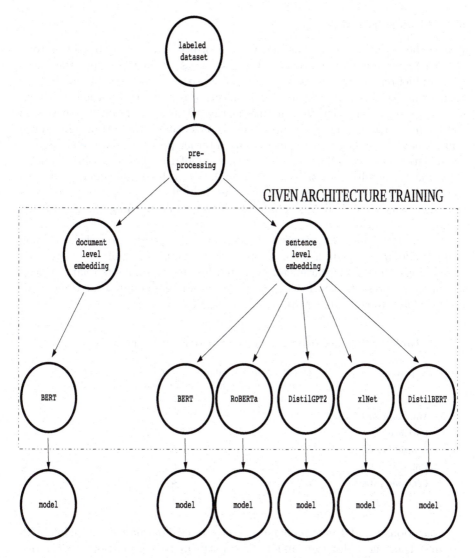

Fig. 1. The proposed processing pipeline

but 7.5% of the items labeled as true came from the www.youtube website, and 3.5% of the items classified as fake came from the www.yournewswire website.

Collection 2 is larger, with 16,817 items classified as true and 5323 items classified as fake. The number of article titles in the fake category is over three times smaller than the number of titles in the true category. Like in the collection 1, the web sources are differentiated with the leading portals; in the case of the titles of articles in the true category, 9.3% are derived from people.com portal, and in the case of fake articles, 8.6% are derived from the hollywoodlife.com portal.

5.2   Models Training

Before the training, pre-processing was performed and hyperparameters were set. There are different approaches to pre-processing; in this paper the maximum reduction was chosen, assuming that the titles of the articles mainly have an informative function and the linguistic correctness or punctuation were considered as secondary elements in this context. The assumption was based on the premise that the linguistic correctness is of secondary importance in spoken and colloquial language. Social media and internet portals often use simplified, abbreviated or even colloquial speech. As part of the pre-processing, punctuation marks, possible emoticons, website addresses, http and e-mail addresses were removed from the collections. As a result of the reduction, pure text was obtained, which required less computing power while training the neural network.

The adopted values of hyperparameters vary, depending on the collection and the adopted embeddings technique (either sentence-level or document-level embeddings). The hyperparameters adopted for the experiments related to collection 1 are presented in Table 2, and the parameters adopted for the experiments related to collection 2 are presented in Table 3.

Table 2. Hyperparameters values of experiments for the collection 1

Hyperparameter	Hyperparameter values for document-level embeddings	Hyperparameter values for sentence-level embeddings
Max number of epochs	5	5
Patience	5	3
Anneal factor	0.5	0.5
Batch size	32	32
Learning rate	0.1	3e-05

Collections 1 and 2 were prepared for the cross validation procedure by dividing into training, validation and testing parts in the proportion 0.8/0.1/0.1. However, the collection 2 has been reduced to 8743 items for the training, 1094 items for the validation and 1094 items for the testing, for the purpose of balancing. The adopted cross validation procedure is the default procedure built into the Flair library [4]. This procedure requires dividing the corpus into training, testing and validation sets. The testing set is used only to conduct tests and to calculate metrics on the trained and selected as the best one model. The set of validation is used to indicate the best model from all the models obtained after each training epoch. The training set is used to train, i.e. modify the parameters of the neural network. During each training epoch, 10 iterations were performed using the training set. Five models were made for each collection; for the collection 1 models based on the following architectures were prepared: roberta-large, bert-large-cased_TDE, distilgpt2, xlnet-large-cased

Table 3. Hyperparameters values of experiments for the collection 2

Hyperparameter	Hyperparameter values for document-level embeddings	Hyperparameter values for sentence-level embeddings
Max number of epochs	15	15
Patience	5	3
Anneal factor	0.5	0.5
Batch size	32	32
Learning rate	0.1	3e-05

using the sentence-level embeddings technique and bert-large-cased_DRE with the document-level embeddings technique. Models based on the following architectures were created for the collection 2: bert-large-cased_TDE, distilbert-base-uncased, distilgpt2, roberta-large with the sentence-level embeddings technique and bert-large-cased_DRE with the document-level embeddings technique.

The work related to pre-processing and training of the neural network was performed on the Google Colaboratory remote platform by applying the GPU TeslaT4, CUDA version 10.1, RAM 12.72 GB, 68.4 GB HDD, pandas version 1.1.5 and the Flair version 0.6.1.

The computational times required to train the neural network for models for the collection 1 are from a minimum of 91 s to a maximum of 437 s, and for the collection 2 from 1245 s to 9909 s. Detailed data on neural network training times are shown in Table 4 and 5. On their basis, it was noticed that more computation time is required for the architectures with more parameters, and simultaneously, a significant reduction in the computational time was obtained for the hybrid method, with the use of the document level embeddings technique.

6 Results

The obtained models were verified by analyzing the following metrics: precision, recall, f1-score, obtained in the model testing procedure. The conducted analysis proved, that in the vast majority of cases, the obtained results significantly outperformed the results presented in the original work for the subject of the KaiDMML dataset [15]. Table 6 depicts the results for the collection 1, and Tab. 7 for the collection 2; in both tables the results which outperform the results in [15] are marked in bold.

In order to confirm the effectiveness of the created models in detecting false information and the usefulness of the models in practical applications, additional practical tests were carried out. They consisted in downloading the titles of articles by applying the web scraper technique from selected websites, and their verification in terms of the content of disinformation or fake news. Sixty-seven titles were downloaded from the newyorker.com, borowitz-report webpage [6], and 141 article titles from the Deutsche Welle webpage [2]. The created model,

Table 4. Computation time needed for models training, based on collection 1; the comparison between various architectures applied for the training

Architecture	Training time [s]
xlnet-large-cased	437
roberta-large	336
distilgpt2	93
bert-large-cased_TDE	353
bert-large-cased_DRE	91

Table 5. Computation time needed for models training, based on collection 2; the comparison between various architectures applied for the training

Architecture	Training time [s]
roberta-large	8337
distilgpt2	2309
distilbert-base-uncased	3384
bert-large-cased_TDE	9909
bert-large-cased_DRE	1245

based on the pre-trained roberta-large architecture and sentence-level embedding technique classified 83.58% of the titles on the website of borowitz-report as untrue, and 4.96% of the titles on the website of Deutsche Welle also as untrue. Such a large difference in the results confirms the practical effectiveness of the created model, which clearly indicated the satirical website as a source of information classified as fake news. The obtained value of the precision metric of the test was 0.8889.

Table 6. Resulted metrics for testing of models based on the collection 1 for the label fake (the comparison between architectures, xlnet-large-cased, roberta-large, distilgpt2, bert-large-cased_TDE, bert-large-cased_DRE); the results which outperform the values in [15] are marked in bold

Architecture	Precision	Recall	f1-score
xlnet-large-cased	**0.8444**	**0.8837**	**0.8636**
roberta-large	**0.8889**	**0.9302**	**0.9091**
distilgpt2	**0.8788**	0.6744	**0.7632**
bert-large-cased_TDE	**0.9412**	0.7442	**0.8312**
bert-large-cased_DRE	**0.8780**	**0.8372**	**0.8571**

Table 7. Resulted metrics for testing of models based on the collection 2 for the label fake (the comparison between architectures, roberta-large, distilgpt2, distilbert-base-uncased, bert-large-cased_TDE, bert-large-cased_DRE); the results which outperform the values in [15] are marked in bold

Architecture	Precision	Recall	f1-score
roberta-large	**0.7823**	0.8308	**0.8058**
distilgpt2	**0.7968**	0.7519	**0.7737**
distilbert-base-uncased	**0.7914**	0.7914	**0.7914**
bert-large-cased_TDE	**0.7939**	0.7594	**0.7762**
bert-large-cased_DRE	**0.7637**	0.7350	**0.7490**

7   Conclusion

The paper presents effective methods of detecting fake news and disinformation, based on Transformer architectures. The major contribution is the creation of disinformation detection models based on the Flair library, which enables to design various architectures to train neural networks. The architectures were created through the implementation of pre-trained Transformers based architectures and the application of the sentence level or the document level embeddings. The presented experiments have proved that applying remote platforms and state-of-the-art NLP approaches can successfully detect disinformation. The results of the experiments showed that the Transformer based models outperform the models reported so far.

For the future work, experiments are planned on much larger databases, which will be created from scratch by obtaining contents from publicly available websites using the web scraper techniques. Data classification will be made on the basis of the prevailing opinions about sources, i.e., the addresses of websites.

Acknowledgement. This work is supported by SocialTruth project (http://socialtruth.eu), which has received funding from the European Union's Horizon 2020 research and innovation programme under grant agreement No. 825477.

References

1. Gossip Cop. https://www.gossipcop.com/. Accessed 03 Jan 2021
2. TOP STORIES. https://www.dw.com/en/. Accessed 03 Jan 2021
3. Agarwal, V., Sultana, H.P., Malhotra, S., Sarkar, A.: Analysis of classifiers for fake news detection. Procedia Comput. Sci. **165**, 377–383 (2019). 2nd International Conference on Recent Trends in Advanced Computing ICRTAC -DISRUP - TIV INNOVATION, 11–12 November 2019
4. Akbik, A., Bergmann, T., Blythe, D., Rasul, K., Schweter, S., Vollgraf, R.: FLAIR: an easy-to-use framework for state-of-the-art NLP. In: Ammar, W., Louis, A., Mostafazadeh, N. (eds.) NAACL-HLT (Demonstrations), pp. 54–59. Association for Computational Linguistics (2019)

5. Aphiwongsophon, S., Chongstitvatana, P.: Detecting fake news with machine learning method. In: 2018 15th International Conference on Electrical Engineering/Electronics, Computer, Telecommunications and Information Technology (ECTI-CON), pp. 528–531 (2018)
6. Borowitz, A.: Satire from the Borowitz Report. https://www.newyorker.com/humor/borowitz-report/. Accessed 03 Jan 2021
7. Devlin, J., Chang, M.-W., Lee, K., Toutanova, K.: BERT: pre-training of deep bidirectional transformers for language understanding. In: Proceedings of the 2019 Conference of the North American Chapter of the Association for Computational Linguistics: Human Language Technologies, vol. 1 (Long and Short Papers), Minneapolis, Minnesota, pp. 4171–4186. Association for Computational Linguistics, June 2019
8. Poynter Institute. Politifact. https://www.politifact.com/. Accessed 03 Jan 2021
9. Kula, S., Choraś, M., Kozik, R.: Application of the BERT-based architecture in fake news detection. In: Herrero, Á., Cambra, C., Urda, D., Sedano, J., Quintián, H., Corchado, E. (eds.) CISIS 2019. AISC, vol. 1267, pp. 239–249. Springer, Cham (2021). https://doi.org/10.1007/978-3-030-57805-3_23
10. Kula, S., Choraś, M., Kozik, R., Ksieniewicz, P., Woźniak, M.: Sentiment analysis for fake news detection by means of neural networks. In: Krzhizhanovskaya, V.V., et al. (eds.) ICCS 2020. LNCS, vol. 12140, pp. 653–666. Springer, Cham (2020). https://doi.org/10.1007/978-3-030-50423-6_49
11. Liu, Y., et al.: RoBERTa: a robustly optimized BERT pretraining approach. CoRR, abs/1907.11692 (2019)
12. Choraś, M., et al.: Advanced machine learning techniques for fake news (online disinformation) detection: a systematic mapping study. Appl. Soft Comput. **101**, 107050 (2021)
13. Rodríguez, Á.I., Iglesias, L.L.: Fake news detection using deep learning. CoRR, abs/1910.03496 (2019)
14. Sanh, V., Debut, L., Chaumond, J., Wolf, T.: DistilBERT, a distilled version of BERT: smaller, faster, cheaper and lighter. CoRR, abs/1910.01108 (2019)
15. Shu, K., Mahudeswaran, D., Wang, S., Lee, D., Liu, H.: Fakenewsnet: a data repository with news content, social context and dynamic information for studying fake news on social media. arXiv preprint arXiv:1809.01286 (2018)
16. Vaswani, A., et al.: Attention is all you need. arxiv:1706.03762Comment, 15 pages, 5 figures (2017)
17. Volkova, S., Shaffer, K., Jang, J.Y., Hodas, N.O.: Separating facts from fiction: linguistic models to classify suspicious and trusted news posts on twitter. In: Barzilay, R., Kan, M.-Y. (eds.) ACL (2), pp. 647–653. Association for Computational Linguistics (2017)
18. Wolf, T., et al.: Transformers: state-of-the-art natural language processing. In: Liu, Q., Schlangen, D. (eds.) Proceedings of the 2020 Conference on Empirical Methods in Natural Language Processing: System Demonstrations, EMNLP 2020 - Demos, Online, 16–20 November 2020, pp. 38–45. Association for Computational Linguistics (2020)
19. Yang, Y., Zheng, L., Zhang, J., Cui, Q., Li, Z., Yu, P.S.: TI-CNN: convolutional neural networks for fake news detection. CoRR, abs/1806.00749 (2018)
20. Yang, Z., Dai, Z., Yang, Y., Carbonell, J., Salakhutdinov, R., Le, Q.V.: Xlnet: generalized autoregressive pretraining for language understanding. CoRR, abs/1906.08237 (2019)
21. Yazdi, K.M., Yazdi, A.M., Khodayi, S., Hou, J., Zhou, W., Saedy, S.: Improving fake news detection using k-means and support vector machine approaches. Int. J. Electrical Electronic Commun. Sci. 13.0(2) (2020)

Towards Model-Agnostic Ensemble Explanations

Szymon Bobek[1,2]([envelope]) [iD], Paweł Bałaga[1], and Grzegorz J. Nalepa[1,2] [iD]

[1] Jagiellonian Human-Centered Artificial Intelligence Laboratory (JAHCAI)
and Institute of Applied Computer Science, Jagiellonian University,
31-007 Kraków, Poland
{szymon.bobek,grzegorz.j.nalepa}@uj.edu.pl
[2] AGH University of Science and Technology, Kraków, Poland

Abstract. Explainable Artificial Intelligence (XAI) methods form a
large portfolio of different frameworks and algorithms. Although the
main goal of all of explanation methods is to provide an insight into
the decision process of AI system, their underlying mechanisms may differ. This may result in very different explanations for the same tasks. In
this work, we present an approach that aims at combining several XAI
algorithms into one ensemble explanation mechanism via quantitative,
automated evaluation framework. We focus on model-agnostic explainers
to provide most robustness and we demonstrate our approach on image
classification task.

Keywords: Explainable artificial intelligence · Machine learning ·
Image processing

1 Introduction

Explainable Artificial Intelligence (XAI) has become an inherent component of
data mining (DM) and machine learning (ML) pipelines in the areas where the
insight into decision process of an automated system is important. Although
the explainability (or intelligibility) is not a new concept in AI [16], it has been
most extensively developed over the last decade. This is possibly due to the huge
successes in black-box ML models such as deep neural networks in sensitive application contexts like medicine, industry 4.0 etc., but also a legal need of providing
accountability and transparency to the reasoning process of AI systems [4]. A
variety of algorithms for generating justifications for AI decisions and lack of
explanations format standards, make it hard to integrate XAI methods into the
standard ML/DM pipeline. Moreover, assessing quality of generated explanations is also non trivial task, as there is lack of unified metrics for evaluating
XAI methods in an automated, quantitative manner.

The integration and evaluation of different ML methods into one pipeline is
done via unified interfaces and metrics such as accuracy, F1 score, area under
the ROC curve and many others. Different metrics may be relevant for different

© Springer Nature Switzerland AG 2021
M. Paszynski et al. (Eds.): ICCS 2021, LNCS 12745, pp. 39–51, 2021.
https://doi.org/10.1007/978-3-030-77970-2_4

ML/DM tasks (recall over precision in medical diagnosis, F1 over accuracy in imbalanced datasets, etc.). The same issue arises with explanability. Metrics such as stability, or consistency or comprehensibility may be relevant depending on who is the addressee of the explanation and what is a domain of explanations, or even what is the stage of the ML system development. The variety of explanation mechanisms makes their validation and inclusion into DM/ML process a non-trivial process.

Considering all of the above, the main goal of the work presented in this paper is to deliver a framework for calculating evaluation metrics for various XAI algorithms, and exploit these metrics in order to build an ensemble explanation mechanism that will combine explanations generated by different algorithms into one, comprehensive solution that can be easily included into the standard DM/ML pipeline. This approach can also be used to select the best explanation framework with respect to arbitrary selected criteria (metrics). We demonstrate our solution on the artificially generated, reproducible dataset and real-life scenario involving image classification task.

The rest of the paper is organised as follows. The overview of the achievements in the field of XAI with respect to assessing their quality was given in Sect. 2. The overview of our solution is given in Sect. 3. In Sect. 4 we present the results of our approach when applied to image classification taks. Finally, in Sect. 5 we discuss the limitation of the approach and describe future works.

2   Explainable Artificial Intelligence

The XAI methods are one of the most rapidly developed mechanisms in the last decade that main goal is to add transparency and accountability to machine learning (ML) and data mining (DM) models [1].

However, the analysis of explanations generated by the algorithms such as LIME [13], SHAP [8] or Anchor [14] is most often reduced to feature selection. This is caused by the fact that such an analysis is a tedious task that involves generating multiple explanations with possibly multiple algorithms and then confronting them and assessing their quality by an expert judgement. Tools and methods for comparable analysis of results of explanations, and selecting or combining explanations are not fully investigated. Aforementioned frameworks provide basic methods for investigating explanations that are based mostly on visual presentation of results in a form of box plots, violin plots and other classical approaches. There are attempts at visualizing explanations which enhance the intelligibility of the explanations itself. However, these are mostly model-specific methods such as saliency maps for DNN [11] or task specific visualization [10].

What is more, although there are metrics used for assessing quality of explanations [9], the assessment is not automated by any framework, nor combined into a unified framework that will allow for reliable comparison of different explanation mechanisms.

There were several attempts at providing methodological approaches for evaluation and verification of given explanation results [9,18]. Among many qualitative approaches there are also ones that allow for quantitative evaluation. In [15] measures such as fidelity, consistency and stability were coined, that can be used for a numerical comparison of methods. In [22] the aforementioned measures were used to improve overall explanations. In [2] a measure that allows to capture stability or robustness of explanations was introduced. Another explanation framework that implements evaluation metrics is given in [21]. Authors present a local explainer with evaluation metrics: stability and correctness. However in neither of the above cases the evaluation is used in further context, limiting their usefulness only to quantify the explanations given by the particular framework. In [23] authors exploit the context of features within a training instance to improve explanations generated with LIME. In [6] a context of an instance that is being explained is generated for the purpose of up-sampling and generating explanations. A more advanced approach was discussed in [17], where an interactive explanation architecture was presented that allows for interactive verification and ad-hoc personalization of the explanations.

Further works on exploiting explanation mechanism as a part of ML/DM workflow include several papers. In [3] authors introduce ExplainExplore system which is an interactive explanation system to explore explanations that fit the subjective preference of data scientists. It leverages the domain knowledge of the data scientist to find optimal parameter settings and instance perturbations, and enable the discussion of the model and its explanation with domain experts. However it does not operationalize it into fully automated system, still relying in core aspect on human-in-the-loop component. Another example of auditing framework was presented in [12]. The framework is intended to contribute to closing the accountability gap in the development and deployment of large-scale artificial intelligence systems by including explainability into the process of auditing AI systems. Yet, it is more of a methodological approach rather than an automated system. Approach described in [7] shows a method for combining many local high quality explanations into one, which makes it similar to the work presented in our work, however the authors method usage is limited to tree-based models only. More generic approach was presented in [19], where authors present a Python toolbox that provides functionality for inspecting fairness, accountability and transparency of all aspects of the machine learning process: data (and their features), models and predictions. However, this framework is more focused on inspecting datasets rather than explanations generated for models trained with the dataset. Furthermore, similarly to other frameworks, it does not support combining explanations of arbitrary XAI algorithm into one explanation, nor compare them as long as they do not provide the same explanation format.

Taking into consideration the full landscape of the aforementioned methods and frameworks and their limitations, the motivation for our work arised. We aimed at filling in the gap between XAI methods and ML/DM pipeline by providing a framework that will allow not only to quantify explanations, but also use this measures to combine explanations into better ones, or to allow for automatic selection of the best model-explainer pair. More specifically, our goal was to introduce cross-platform solution that is independent of the XAI algorithm, under the assumption that it provides measure of importance of features in the decision process. The following sections provides more detailed description of our framework.

3   Ensemble Explanation Framework

In this work we focus on three metrics delivered by the *InXAI* framework[1] developed by us. It can be used along with explanation frameworks either to choose best explanation mechanism that fits project requirements (high stability, high consistency, etc.), or to generate unified explanations according to specified objective metric. Although the description of the framework is out of the scope of this paper, it is worth mentioning the it follow the scikit-learn[2] interface, which allows the XAI methods to be included in the ML/DM pipeline not only in theoretical, but also practical way.

3.1   Metrics of Explainability

In this paper we focus for simplicity only on three metrics of explainability implemented in the *InXAI* framework: consistency, stability and area under the loss curve.

For the sake of further discussion we assume following notation. The importance of feature i and instance j delivered by explanation model e for machine learning model m will be denoted as $\Phi_{i,j}^{e \rightarrow m}$. If we skip subscripts, we assume marginal value over missed subscripts. Therefore, a complete explanation matrix for every feature and every instance generated by explanation e for model m will be denoted as $\Phi^{e \rightarrow m}$.

Consistency. Consistency measures how explanations generated for predictions of different ML models are similar to each other. Therefore, it is more related to stability of ML models with respect to decision making rather than to explanation mechanisms directly. Assuming that $M(X)$ is a set of ML models with high accuracy, Eq. (1) depicts the consistency measure:

$$C(\Phi^{e \rightarrow m_1}, \Phi^{e \rightarrow m_2}, \ldots, \Phi^{e \rightarrow m_n}) = \frac{1}{\max\limits_{a,b \in m_1, m_2, \ldots, m_n} ||\Phi_j^{e \rightarrow m_a} - \Phi_j^{e \rightarrow m_b}||_2 + 1} \quad (1)$$

Stability. Stability (or robustness) assures generation of similar explanations for similar input. To obtain a numerical value to this property, modified notion of Lipschitz continuity has been proposed in [2]:

$$\hat{L}(\Phi^{e \rightarrow m}, X) = \max\limits_{x_j \in N_\epsilon(x_i)} \frac{||x_i - x_j||_2}{||\Phi_i^{e \rightarrow m} - \Phi_j^{e \rightarrow m}||_2 + 1} \quad (2)$$

---

[1] See: https://github.com/sbobek/inxai.
[2] See: https://scikit-learn.org.

where $N_\epsilon(x_i)$ is a set such as:

$$N_\epsilon(x_i) = \{x_j \in X \,|\, \|x_i - x_j\| < \epsilon\} \tag{3}$$

This optimization problem finds parameter describing most differing explanations $f(x)$ for points in a vicinity of x_i, dictated by the set $N_\epsilon(x_i)$ proportional to the distance between the neighbours.

Area Under the Loss Curve. Area under the loss curve (AUCx) depicts the loss in accuracy (or other selected metric) when features are perturbed gradually according to their inverse importance returned by explanation algorithm.

Therefore, if the AUCx is high, it may imply that the importance of the features was set incorrectly, as perturbation caused large loss in accuracy. The loss in accuracy is defined as a difference in baseline accuracy obtained by a non-perturbed dataset and the accuracy obtained from perturbed dataset. In our work we used the trapezoidal rule to calculate it over set of accuracy losses for different perturbation rates.

It is worth noting that current version of the framework does only cover XAI algorithms that provide explanations in a form of feature importance assigned to particular features. This makes it more difficult to apply the framework to rule-based systems that does not provide such information out of the box. Currently we assume binary feature importance for rule-based explainer, meaning that features that were used for explanation have importance 1, while others features importance are 0. However, this feature is not yest provided out of the box, and needs to be programmed by the user.

3.2   Ensemble Score

Calculating ensemble score is not limited to the metrics defined above and can be easily extended and modified as we will show in Sect. 4. The main goal of *ES* score is to capture the weighted importance of different metrics into one value. The definition of *ES* for a set of metrics M and weights w, was given in Eq. (4).

$$ES(M, w) = \sum w_i \cdot M_i \tag{4}$$

Having the ensemble score, we calculate a new, combined vector of explanation Φ^{ens} as a weighted sum of ensemble scores and associated with them original explanations $\Phi^{e_1}, \Phi^{e_2}, \ldots \Phi^{e_n}$. The weights are assigned arbitrary depending on the desired influence of a particular metric to the ensemble explanation.

Therefore, the final ensemble explanation is given by the Eq. (5).

$$\Phi^{ens} = \frac{ES(M, w) \cdot [\gamma_1 \Phi^{e_1}, \gamma_2 \Phi^{e_2}, \ldots, \gamma_n \Phi^{e_n}]}{\sum_{i=1}^{n} ES_i(M, w)} \tag{5}$$

Where $\gamma_1, \gamma_2, \ldots, \gamma_n$ are scaling factors that make it possible to compare and combine explanations obtained from different XAI frameworks. Note that classic per-column or per-sample scaling will corrupt the internal dependencies between importance and features. Therefore the scaling is performed over the whole matrix of explanation generated by the same model. In our approach we used min-max normalization that is given in Eq. (6).

$$\Phi' = \frac{\Phi - \min(\Phi)}{\max(\Phi) - \min(\Phi)} \tag{6}$$

The following plots and results were generated for the dataset presented in Fig. 1 along with two ML models and their decision boundaries used for calculating consistency measures.

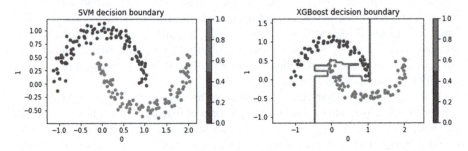

Fig. 1. Dataset and two classifiers with their decision boundaries used with InXAI framework for ensemble explanation generation.

The *ES* can also serve as a confidence measure of the explanation for single instance, or explanation framework with respect to selected metrics. Based on this confidence the ensemble explanations are created. Figure 2 presents *ES* for LIME and SHAP. The transparency of a data point depicts the uncertainty of the explanation.

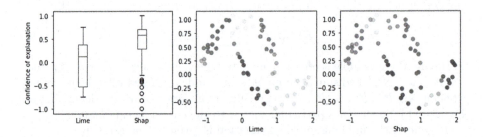

Fig. 2. Confidence of explanations where the maximum weight was put on the consistency metric.

The metrics for an ensemble explainer built with Eq. (5) and its two components (LIME and SHAP) were given in Fig. 3. It is worth noting that the high
value of weight for consistency, improves the consistency measure (upper row)
in the ensemble. The same can be observed with stability (middle row). However, comparing to SHAP, only some of explanations were improved in terms of
stability, while the overall (global) measure remained intact, or even worsen (see
Sect. 5 for details of this phenomenon). With the maximum weight set on AUCx
metric, we observe the overall reduction in the curve area. However, due to the
fact that both explainers are correct with respect to this measure, the difference
is not that apparent.

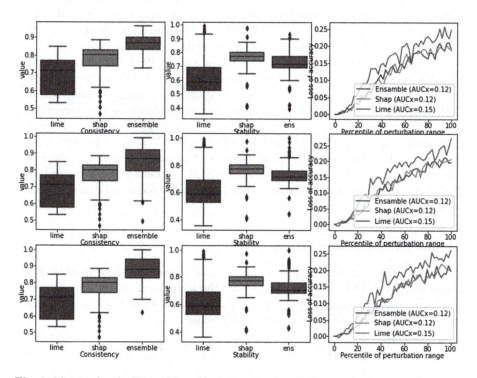

Fig. 3. Metrics for the Ensemble explanations generated for sample dataset. The upper
row presents results for weights $[0.8, 0.1, 0.1]$ assigned to consistency, stability and
AUCx metrics. Middle row gives results for weights $[0.1, 0.8, 0.1]$. Bottom row presents
results for weights $[0.1, 0.1, 0.8]$. Single point represents metric value of single datapoint
(local), while the global measure can be considered a spread of local values.

In the next section we demonstrate the solution on the example of a system
for emotion classification based on face expressions photographs.

4   Image Classification Use Case Scenario

A strong advantage of the Ensemble Score defined in Eq. 4 is that, since it is problem-agnostic, it can accommodate any metric. Similarly, the formula given by Eq. 5 does not depend on model-specific assumptions. It provides a flexible foundation for building ensemble explanations involving a variety of methods. This section shows how this generic framework can be utilized to generate ensemble explanations for the problem of image classification.

Classification of affective images relies on recognizing facial features in photographs of human subjects. Model-agnostic methods such as SHAP, LIME or ANCHOR are all capable of explaining predictions related to such images, however, due to differences in how they operate, their outcomes are not easily comparable. We demonstrate that it is possible to find analogies between results produced by these methods and to integrate them into an ensemble.

A common characteristic of SHAP, LIME or ANCHOR is that all these methods allow identifying subareas of the image that contribute to a given prediction. To reduce dimensionality of the feature space to a computationally feasible level, images must be preprocessed by a segmentation algorithm of choice. Effectively, feature importance $\Phi_{i,j}^{e \to m}$ determines how individual superpixels in the segmented image contribute to model prediction. $\Phi_{j}^{e \to m}$ denotes feature importance vector for explanation e, which consists of as many elements as there are superpixels for instance j. In this work the SLIC segmentation algorithm was used as it performed well on the task of isolating facial features in images in our dataset.

4.1   Metric Definition for Images

Since Eq. 4 allows arbitrary definition of metric set M we introduce an alternative formulation of stability and consistency for the purpose of this example.

Firstly, we define an auxiliary metric called similarity in Eq. 7, which evaluates how close two explanations are to each other:

$$\text{Sim}(\Phi_i^{e \to m}, \Phi_j^{e \to m}) = 1 - \frac{||\Phi_i^{e \to m} - \Phi_j^{e \to m}||_1}{S} \tag{7}$$

where S specifies the number of features (image segments). It is assumed that explanations are mapped to range $[0, 1]$ so as to guarantee similarity to be a normalized value between 0 and 1. Value of 1 identifies two explanations as identical, whereas 0 corresponds to no similarity.

Alternative definition of consistency is given by Eq. 8. In this formulation consistency is to be interpreted as average similarity between assessed explanation $\Phi^{e \to m_0}$ and a set of reference explanations $\Phi^{e \to m_1}$ through $\Phi^{e \to m_n}$ obtained for independently trained ML models.

$$C(\Phi^{e \to m_0}, \Phi^{e \to m_1}, \Phi^{e \to m_2}, \dots, \Phi^{e \to m_n}) = \frac{\sum_{k=1}^{n} \text{Sim}(\Phi_j^{e \to m_0}, \Phi_j^{e \to m_k})}{n} \tag{8}$$

Stability measure can also be defined in terms of similarity between the assessed explanation $\Phi_i^{e\to m}$ and explanations obtained for perturbed instances in the vicinity of x_i, as given by Eq. 9.

$$\hat{L}(\Phi^{e\to m}, X) = \sum_{x_j \in N_\epsilon(x_i)} \frac{\text{Sim}(\Phi_i^{e\to m}, \Phi_j^{e\to m})}{|N_\epsilon(x_i)|} \tag{9}$$

Consistency and stability metrics proposed in this section are fully compatible with Eq. 4. In this example the Ensemble Score was calculated with equal weights for both consistency and stability.

4.2   Model Assumptions

The image classifier was built on top of a neural embedding network. The core of the network was based on Inception Resnet V1 model that was adjusted and fine-tuned for facial expression classification task. Training and computation of explanations was performed on RGB images of size 160×160 pixels. Images were sourced from a dataset [20] where no explicit label information was provided for individual instances, because the dataset was designed for triplet learning. Therefore, the classifier was built on top of labels generated artificially according to the following procedure instead:

1. Compute embeddings $e_1, \ldots, e_n$ for all n instances in the training set.
2. Perform k-means clustering with the number of clusters selected by optimizing silhouette score.
3. Build a K-NN classifier that maps any given embedding to index of the cluster that it fits best. For sake of this research $K = 200$ was used.

Summarizing, cluster indices were used directly as labels for the purpose of classifying unknown instances.

Note that from the perspective of the ensemble framework demonstrated further in this work, implementation details of the image classifier are not critical. However, defining a classifier was necessary to produce explanations with the underlying methods: SHAP, LIME and ANCHOR. The ensemble approach by itself can merge any set of explanations, if only they quantify importance of each feature of every explained instance.

In order to enable calculating consistency multiple models were trained, each with a different embedding space size and hyperparameter configuration. Due to these configuration differences, independent classifiers had to be built separately for each model. To ensure a fair comparison, cluster number optimization (according to silhouette score) was conducted only once for a specific reference model, and assumed equal for all other models. As a result, count of different class labels the same across all explanation models. To compute stability we sampled such instances from neighborhood $N_\epsilon(x_i)$, for which it was known that the explanation should remain unaltered.

In the next step a selection of instances was chosen from the validation dataset [20] and fed to SHAP, LIME or ANCHOR frameworks independently.

Each framework-specific explanation was evaluated in terms of stability, consistency and, most importantly, Ensemble Score.

4.3   Explanation Scaling and Aggregation

Here we demonstrate how independent explanations obtained from SHAP, LIME and ANCHOR can be combined into one ensemble explanation.

SHAP and LIME feature importance attributed to each superpixel is a real number. On the other hand, ANCHOR determines which image segments have crucial contribution towards a specific model prediction; that is, were they not part of the image, the prediction would have been different. ANCHOR explanations can be seen as a vector of binary values, where 1 is assigned to the crucial features and 0 corresponds to features that have no impact on prediction.

To enable aggregation of results originating from different XAI frameworks, explanations need to be scaled. Recommended choice of scaling factor for explanations generated by k-th model is given by Eq. 10.

$$\gamma_k = \frac{1}{||\Phi^{e \to k}||_{max}} \tag{10}$$

As a consequence, it is guaranteed that – independently of the XAI framework – scaled feature importance values fit in a normalized range $[-1, +1]$ and that feature with the strongest contribution is assigned importance of ± 1. Sign depends on whether the maximum contribution is positive or negative.

Inserting the original explanations, their Ensemble Scores and scaling coefficients γ into Eq. 5 yields the final ensemble explanation.

4.4   Results

We present example results obtained according to the framework described in this work. Figure 4 provides ensemble explanations for an array of facial images picked randomly from the validation dataset [20]. Visualization is based on a color mapping, where color intensity corresponds to importance of specific image features. Green-colored areas have a positive contribution towards particular explained label, and red overlay signifies negative contribution.

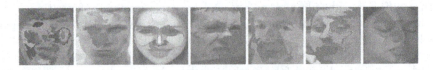

Fig. 4. Example ensemble explanations for facial expression images

Resulting ensemble explanations are visually appealing combinations of several underlying XAI approaches. Note that positive contributions are generally dominating, which is partly due to the fact that ANCHOR never attributes negative contribution to features.

Ability to utilize strengths of multiple explanation models simultaneously is the primary advantage of the presented approach. We found that, in multiple cases, facial features captured by the ensemble explanations remained in stronger agreement with human intuition than in underlying explanation methods assessed individually. Probably this is because feature importance in ensemble explanations was derived as a weighted average of respective feature importance values in the underlying methods according to Eq. 5. Therefore potential errors in importance values might be canceled out by aggregating multiple partial solutions. The risk that the ensemble overestimates or underestimates contribution of a feature is lower than in case of any individual underlying explanation model.

On the other hand, values of stability, consistency and Ensemble Score obtained in our study were characterized with relatively low variance across example images considered in this research. A presumable root cause is that image areas where feature importance was close to zero (neutral impact on prediction) were usually much larger than areas where feature contribution was strongest. It is a desired characteristic because it makes explanations more specific and understandable, i.e. focused on critical features. However, the downside was that the similarity measure between two explanations was influenced predominantly by neutral areas, resulting in low variance of the similarity measure, on which stability, consistency and ensemble score are built. To alleviate this issue, alternative, more sensitive formulations of the similarity measure are also possible and can easily be used as a drop-in replacement in Eq. 8 and Eq. 9.

Another point is that quantifying stability, consistency and Ensemble Score allowed objective validation of the explanation model used in our research. Since each metric takes into account a different set of factors, using a combination of such metrics made the validation process more robust. For example, note that a faulty model might also yield high-stability explanations, although it is less likely to produce high-consistency results.

In conclusion, it was shown that the approach introduced in this work successfully unifies various XAI frameworks that were not initially designed with compatibility on mind. It is also inherently possible to extend this approach on methods other than SHAP, LIME or ANCHOR. The proposed framework has high potential to be used in automated assessment and comparison of different explanation models according to a set of well-defined objective measures. However, to fully utilize the automation potential, there is a need for comprehensive tests to confirm that ensemble explanation visualizations are consistent with human perception.

5   Summary and Future Works

The methods of Explainable Artificial Intelligence methods form a large portfolio of versatile frameworks and algorithms providing insights into the decision process of an AI system. However, their underlying mechanisms may be very different. Thus it may result in very different explanations for the same tasks. In this paper, we presented an original approach that aims at combining several XAI algorithms into one ensemble explanation mechanism via quantitative, automated evaluation framework. We focused on model-agnostic explainers to provide most robustness. We provided an illustrative demonstration of our approach on image classification task.

Weights such as stability works in most of the cases only locally and can be used to weight single instance explanation. This means that combining several explanations with high stability does not assure the resulting ensemble will also have the high stability, as the neighbourhood of explanations was altered. We plan to use SMAC [5] or similar Bayesian optimizer to optimize explanations with respect to the selected metric in a way that they will be optimized globally.

We also plan to conduct observational studies with domain experts and real-life use-cases to validate the feasibility of our solution. Finally we will be evaluating this approach on different datasets, including industrial ones.

Acknowledgements. The paper is funded from the XPM project funded by the National Science Centre, Poland under CHIST-ERA programme (NCN UMO-2020/02/Y/ST6/00070).

References

1. Adadi, A., Berrada, M.: Peeking inside the black-box: a survey on explainable artificial intelligence (XAI). IEEE Access **6**, 52138–52160 (2018)
2. Alvarez-Melis, D., Jaakkola, T.S.: On the robustness of interpretability methods (2018)
3. Collaris, D., van Wijk, J.J.: Explainexplore: visual exploration of machine learning explanations. In: 2020 IEEE Pacific Visualization Symposium (PacificVis), pp. 26–35 (2020)
4. Goodman, B., Flaxman, S.: European union regulations on algorithmic decision-making and a "right to explanation". arXiv preprint arXiv:1606.08813 (2016)
5. Hutter, F., Hoos, H.H., Leyton-Brown, K.: Sequential model-based optimization for general algorithm configuration (extended version).Technical report TR-2010-10, University of British Columbia, Department of Computer Science (2010). http://www.cs.ubc.ca/~hutter/papers/10-TR-SMAC.pdf
6. Liu, N., Shin, D., Hu, X.: Contextual outlier interpretation. arXiv preprint arXiv:1711.10589 (2017)
7. Lundberg, S.M., et al.: Explainable AI for trees: from local explanations to global understanding. CoRR, abs/1905.04610 (2019)
8. Lundberg, S.M., Lee, S.-I.: A unified approach to interpreting model predictions. In: Proceedings of the 31st International Conference on Neural Information Processing Systems, NIPS 2017, pp. 4768–4777. Curran Associates Inc. (2017)
9. Mohseni, S., Zarei, N., Ragan, E.D.: A multidisciplinary survey and framework for design and evaluation of explainable AI systems (2020)
10. Mujkanovic, F., Doskoč, V., Schirneck, M., Schäfer, P., Friedrich, T.: timeXplain - a framework for explaining the predictions of time series classifiers (2020)

11. Pope, P.E., Kolouri, S., Rostami, M., Martin, C.E., Hoffmann, H.: Explainability methods for graph convolutional neural networks. In: 2019 IEEE/CVF Conference on Computer Vision and Pattern Recognition (CVPR), pp. 10764–10773 (2019)

12. Raji, I.D., et al.: Closing the AI accountability gap: defining an end-to-end framework for internal algorithmic auditing. CoRR, abs/2001.00973 (2020)

13. Ribeiro, M.T., Singh, S., Guestrin, C.: "Why should i trust you?": explaining the predictions of any classifier. In: Proceedings of the 22nd ACM SIGKDD International Conference on Knowledge Discovery and Data Mining, KDD 2016, pp. 1135–1144. Association for Computing Machinery, New York (2016)

14. Ribeiro, M.T., Singh, S., Guestrin, C.: Anchors: high-precision model-agnostic explanations. In: Thirty-Second AAAI Conference on Artificial Intelligence. AAAI Publications (2018)

15. Robnik-Šikonja, M., Bohanec, M.: Perturbation-based explanations of prediction models. In: Zhou, J., Chen, F. (eds.) Human and Machine Learning. HIS, pp. 159–175. Springer, Cham (2018). https://doi.org/10.1007/978-3-319-90403-0_9

16. Schank, R.C.: Explanation: a first pass. In: Kolodner, J.L., Riesbeck, C.K. (eds.) Experience, Memory, and Reasoning, pp. 139–165. Lawrence Erlbaum Associates, Hillsdale (1986)

17. Sokol, K., Flach, P.: One explanation does not fit all. KI - Künstliche Intelligenz **34**(2), 235–250 (2020)

18. Sokol, K., Flach, P.A.: Explainability fact sheets: a framework for systematic assessment of explainable approaches. CoRR, abs/1912.05100 (2019)

19. Sokol, K., Santos-Rodríguez, R., Flach, P.A.: FAT forensics: a python toolbox for algorithmic fairness, accountability and transparency. CoRR, abs/1909.05167 (2019)

20. Vemulapalli, R., Agarwala, A.: A compact embedding for facial expression similarity. CoRR, abs/1811.11283 (2018)

21. Verma, M., Ganguly, D.: LIRME: locally interpretable ranking model explanation. In: Proceedings of the 42nd International ACM SIGIR Conference on Research and Development in Information Retrieval, SIGIR 2019, pp. 1281–1284. Association for Computing Machinery, New York (2019)

22. Yeh, C.-K., Hsieh, C.-Y., Suggala, A.S., Inouye, D.I., Ravikumar, P.: On the (in)fidelity and sensitivity for explanations (2019)

23. Zhang, Z., Yang, F., Wang, H., Hu, X.: Contextual local explanation for black box classifiers. CoRR, abs/1910.00768 (2019)

Computational Methods in Smart Agriculture

Bluetooth Low Energy Livestock Positioning for Smart Farming Applications

Maciej Nikodem[✉][iD]

Department of Computer Engineering, Wrocław University of Science
and Technology, Wybrzeże Wyspiańskiego 27, 50-370 Wrocław, Poland
`maciej.nikodem@pwr.edu.pl`

Abstract. Device localization provides additional information and context to IoT systems, including Agriculture 4.0 and Smart Farming. However, enabling localization incurs additional requirements and trade-offs that often do not fit into application constraints – use of specific radio technologies, increased communication, computational, and energy costs. This paper presents a localization method that was designed for Smart Farming and applies to a wide range of radio technologies and IoT systems. The method was verified in a real-life IoT system dedicated to monitor cow health and behavior. In a large multi-path environment, with a large number of obstacles, using only 10 anchors, the system achieves an average localization error equal to 6.3 m. This allows to use the proposed approach for animal tracking and activity monitoring which is beneficial for well-being assessment.

Keywords: Smart Farming · Indoor localization · Signal strength · Bluetooth Low Energy

1 Introduction

The number of IoT systems and devices rapidly increases, and IoT applications become more and more popular. For a large number of such applications, the ability to localize the devices is of additional value as it gives context to the application and increases usability and functionality. Since the deployment of global navigation satellite systems (GNSSs) the task of determining the device location becomes straightforward. However, its use requires to include GNSS receivers that increase device energy consumption, demand larger batteries, and incur additional costs. Additionally, the GNSS does not work well in indoor environments and is practically unavailable underground and in multi-story buildings. Consequently, the IoT devices and communication technologies use other localization approaches including methods based on the angle of arrival, time of flight, or time difference of arrival. These methods may not require additional hardware however, incur increased costs in terms of energy, communication bandwidth, or

© Springer Nature Switzerland AG 2021
M. Paszynski et al. (Eds.): ICCS 2021, LNCS 12745, pp. 55–67, 2021.
https://doi.org/10.1007/978-3-030-77970-2_5

processing time. As a result, enabling localization in an IoT system is a trade-off between constraints, desired parameters, and functionalities of the system. This includes the aforementioned costs and the resulting localization accuracy.

Animal localization is a challenging topic for Smart Farming and precision agriculture. First of all, it allows to localize individual animal in the grazing fields or sheds, thus simplifying the everyday work of farmers and veterinary doctors. Additionally, location information over time provides information about animal activity, which is an indicator of health status. As presented in [1, 8, 9], changes in the activity may signal diseases or estrus and can be often detected before emergence of clinical symptoms. Contemporary radio technologies used in IoT systems enable localization through either radio signal strength (e.g. [4, 11]) or time of flight measurements (e.g. [5, 15]). While the former group gives less accurate localization, the latter is more energy demanding and requires more complex deployments. Additionally, the use of signal strength localization ensures applicability to a wide range of various radio technologies and can be adjusted to use other radio signal quality indicators (e.g. [6]).

This paper presents an animal localization algorithm based on signal strength measurements and dedicated for indoor environments. The algorithm provides accurate localization using small number of reference points (anchors) and requiring limited measurements during setup. The accuracy of the proposed algorithm was verified in a dairy cow farm located in Zagrodno, Poland, where a dedicated cow health monitoring system is deployed [13]. Using 10 anchors, in a $1\,400\,m^2$ shed the average localization accuracy equals 6.3 m.

2   Related Work

Signal strength localization has gained much attention in the last years [3, 7, 10]. This was the result of the fact that recently the number of low power IoT devices increased significantly, and this method does not incur additional complexity and cost.

Indoor localization using signal strength is often focused on office buildings where the number of anchors (reference devices with known localization) is large and they are deployed densely. The article by Wang et al. [14] is an example of such applications. The authors analyze localization in the office space of 44 by 22 m. The results achieved are quite accurate with an average error equal to 1.8 m and maximum error slightly exceeding 10 m. However, this is achieved with up to 34 anchors deployed with an average distance between the neighboring anchors of 5 m. Additionally, the localization method proposed in [14] is based on fingerprinting and requires mapping the whole area before the localization can be used.

Signal strength-based localization is also interesting for industrial and agriculture applications, including Smart Farming A good example of such system is a localization in an industrial workshop [3] or the dairy farm [2, 4, 11]. In the latter example, the goal of localization is to monitor the activity of the animals and shorten the time required to locate an individual cow on a large farm or field. For these applications accuracy of a few meters is acceptable. Article by Trogh et al. [12] presents an indoor localization solution for dairy cows. The proposed approach uses Bluetooth Low Energy (BLE) and is based on path loss model. The model is derived from the floor plan of the localization area and does not require any measurements before the localization system can be used. During the evaluation, 11 anchors were deployed in 30 by 13 m area, allowing to estimate cow location with an average error of 4.2 m (median 3.3 m). The results are good but the area is relatively small and is located in the center of the barn, far away from large obstacles (e.g. walls) thus minimizing the adverse effects of propagation phenomena.

Bloch et al. [2] investigate a localization system for dairy cows. The system estimates location based on RSSI measurements and two different approaches: log-distance path loss model, and fingerprinting of the localization area. The results of experiments conducted in $420\,m^2$ barn show slightly better results for the method based on fingerprinting. The resulting average localization error was equal to 3.3 m while the maximal error was slightly above 10 m. Although the accuracy is considered good, it was achieved with dense deployment of the anchors – the total amount of 10 anchors were deployed in the area, with an average distance between neighboring anchors of 10 m.

Takahiro et al. [11] analyzed localization of animals in an open field of approximately $11\,400\,m^2$. Each animal wore four BLE transmitters and a GNSS receiver attached to its collar. Transmitted BLE signals were received by 20 receivers deployed in the area perimeter and used for estimating the animal location, using a neural network. The average localization error achieved was slightly above 6 m. This is considered a good result taking into account the size of the area and a relatively small number of anchors. However, this was achieved with redundancy and in preferable radio propagation conditions, as the experiments were conducted in an open field and as many as four BLE devices per animal were used.

Signal strength-based localization was also analyzed by Cardoso et al. [4] who attempted to determine a position in an open square area of $440\,m^2$. Using only four anchors the average localization error of the proposed algorithm was equal to 2.48 m, and the maximum error to 6.5 m. However, the proposed algorithm was evaluated only in a few locations along the diagonal of the area with some of the locations close to the anchors. Because test locations were not spread across the localization area, the results are not considered reliable.

Hindermann et al. [5] investigated the use of ultra-wideband (UWB) radios and time-of-flight measurements for localization in a 12 by 25 m barn. They have used only four anchors and achieved a localization error that was 0.5 m in most scenarios and test locations. The results are impressively good and allow not only to track activity but also to analyze the interactions between the animals. However, the approach uses radio technology that has relatively high energy demands and requires additional hardware, cabling, and costs. Time-of-flight based localization is also available in a commercial solution from Smartbow [15]. In the experiments conducted in approx. $2\,100\,m^2$ area using 17 anchors the resulting accuracy was below 2.93 m during 95% of time. Similar to the previous approach, this technology also requires additional hardware, cabling, and costs.

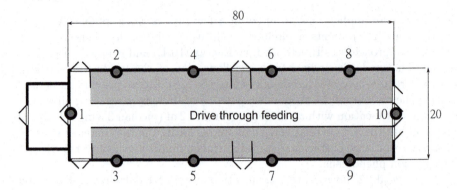

Fig. 1. The map of the area used for the evaluation of the proposed localization algorithm. Gray areas mark sections where animals reside and can walk freely. Green circles mark the locations of the anchors numbered 1 to 10. (Color figure online)

The aforementioned localization approaches have limited applicability to large-scale smart farming applications. This is either due to the use of specific radio technologies, the requirement to derive radio propagation models that are inaccurate in indoor applications, or the use of complex and time-consuming procedures to set up the localization system (build a radio map of the area for fingerprint-based methods). This article proposes a different approach that uses a limited measurement campaign to derive a geometrical model which translates RSSI values to rings – a range of distances from the anchors. This is in contrast to the propagation models where RSSI value determines a single distance which is "good on average". The contribution of the paper is as follows:

- proposal of a geometrical approach to signal strength based localization, that can be used with various radio technologies,
- evaluation of the proposed localization method in real-life system (animal monitoring) deployed over a large indoor area.

3   System Architecture and Localization Algorithm

The proposed localization algorithm is designed to be integrated with typical IoT architecture. The end devices (tags) report measured data to neighboring anchors. Anchors receive the transmitted data and take radio signal measurements including RSSI and possibly other parameters (e.g. link quality indicator - LQI). Anchors aggregate the data and the measurements over time and transmit this information (together with the tag identifier and timestamp) to central server located in the cloud. When measurements from anchors are collected, the server runs a localization algorithm to determine tag location.

The experimental evaluation presented in this paper was run in the system of the aforementioned structure deployed to monitor the behavior, activity and well-being of dairy cows [13]. The test barn of 20 by 80 m (Fig. 1) hosts approximately 80 mature cows all wearing the monitoring tags. The tags are BLE

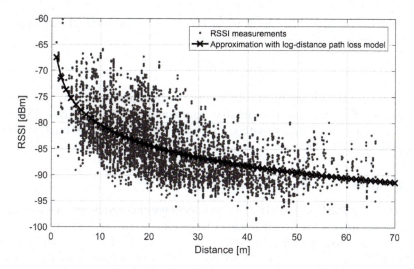

Fig. 2. RSSI values measured in the experiment for different locations and distances from the anchors. Black line marks log-distance path loss model approximating the measured values.

devices mounted on collars, take periodic measurements of animal activity, calculate aggregates, and transmit them inside advertisement packets broadcasted periodically every 250 ms. Ten anchors deployed in the barn receive the aggregated measurements and record radio signal parameters – RSSI and strength of the radio signal when receiving the radio message. The signal parameters are then aggregated separately for each tag in non-overlapping, 30 s wide time windows. The aggregates include the average value and standard deviation of RSSI measurements, the number of messages received from a tag, and total signal strength (TSS) which is an accumulated strength of all the radio messages received from a tag. Aggregates are transmitted and stored on the server which runs the localization procedure.

3.1   Localization Based on Radio Propagation Model

A typical localization approach based on RSSI attempts to determine distance vs. RSSI dependency using a selected propagation model. Unfortunately, the simple propagation models do not adhere to indoor environments where signal propagation is complex, affected by obstacles, and various propagation phenomena. For example, consider the RSSI measurements presented in Fig. 2. It can be noticed that the same value of RSSI was measured for significantly different distances. Due to high variance of the distances (for a given RSSI value) an approximation with a propagation model would result in a single relation that is good on average but fails to accurately estimate the real distance. For the measurements presented in Fig. 2, using the least square method, the approximated log-distance path loss model equals:

$$\hat{d} = 10^{\frac{\text{RSSI}+67.9961}{-12.5215}},$$ (1)

where RSSI is the measurement and $\hat{d}$ is the estimated distance to the anchor. The position of the tag can be estimated from multilateration [5] given distances to at least 3 anchors. Unfortunately, inaccuracies in distance estimation result in large localization errors.

3.2 Proposed Localization Procedure

To avoid the limitations and achieve good accuracy this article proposes to use a more general approach where a range of RSSI measurements determines a range of possible distances. Considering the previous example (Fig. 2) it can be noted that for RSSI values exceeding -75 dBm the distance is almost always smaller than 20 m. Similar for RSSI between -80 dBm and -75 dBm the distance varies from approx. 4 to 25–30 m. Similar dependencies can be observed for smaller RSSI values. Simultaneously, a RSSI vs distance relationship can be divided into three sections depending on the vicinity to the anchor: immediate, near, and far zone. For BLE transmission the measured RSSI values in the immediate zone are above approximately -80 dBm. In this zone, the signal strength drops quickly with the distance. In the near zone the RSSI varies between approximately -80 dBm and -90 dBm, and the RSSI vs distance relationship flattens. In the far zone the RSSI drops below -90 dBm and there is almost no correlation between RSSI and the distance.

Based on the aforementioned observations the proposed RSSI vs. distance model defines ranges of distances (rings) that depend on RSSI value. The possible RSSI values are divided into disjoint ranges and for each range, the corresponding distances are assigned. (Fig. 3). The distances are then approximated with the probability distribution that is used to calculate the minimum and maximum value of the distance for this RSSI range. Because for each RSSI range the distribution of the distances was close to normal distribution the minimum and maximum distances were calculated based on the mean and the standard deviation of the distances:

$$\begin{aligned} D_{i,\min} &= \text{mean}(\mathbf{D}_i) - 1.5 \cdot \text{std}(\mathbf{D}_i), \\ D_{i,\max} &= \text{mean}(\mathbf{D}_i) + 1.5 \cdot \text{std}(\mathbf{D}_i), \end{aligned}$$ (2)

where $\mathbf{D}_i$ denotes the distances for the i-th RSSI range, mean and std are an average and a standard deviation respectively. It is worth to notice that the resulting distance ranges (rings) overlap. This is because of the large variance of RSSI measurements for each distance. The actual model for the evaluation scenario was derived in a small measurement campaign and is presented in Table 1. The model differs for various anchors because their location and impact of the environment on the RSSI measurements is different. Additionally, for some of the anchors (e.g. 1, 5, 10) the minimal distance does not increase with lowering

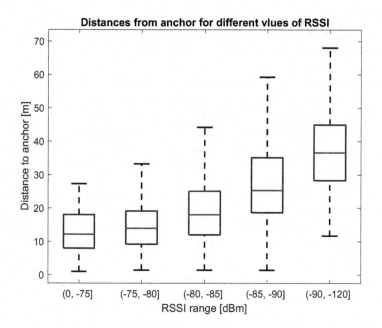

Fig. 3. The values of the distance for disjoint RSSI ranges. The RSSI measurements were collected at different locations by 10 anchors.

RSSI. This is likely because the measured RSSI is sometimes attenuated by the animal wearing the tag (as the tag is mounted on a side of a neck).

The localization procedure takes the RSSI_i and TSS_i measurements for every anchor $i = 1, 2, \ldots, n$. Based on the model the algorithm uses the RSSI_i to determine the $\text{D}_{i,\min}$ and $\text{D}_{i,\max}$ distances to the i-th anchor. Afterwards, for each (x, y) coordinate of the localization area, the algorithm defines a discrete function $f(x, y)$. The function is a weighted sum of TSS_i values if (x, y) coordinates belong to the i-th ring (if the tag distance to the i-th anchor falls between $\text{D}_{i,\min}$ and $\text{D}_{i,\min}$):

$$f(x, y) = \sum_{i=1}^{n} w(x, y) \cdot \text{TSS}_i, \tag{3}$$

where

$$w(x, y) = \begin{cases} 1 \text{ iff } \text{D}_{i,\min} \leq d_i(x, y) \leq \text{D}_{i,\max}, \\ 0 \text{ otherwise}, \end{cases} \tag{4}$$

and $d_i(x, y)$ denotes the Euclidean distance from the i-th anchor to (x, y) point.

The resulting function $f(x, y)$ defines the likelihood of the tag location – in particular, it is unlikely that the tag is located in the region where the function value is small. Therefore, in the next step, the algorithm finds a threshold T that is used to distinguish (x, y) coordinates with the highest likelihood. Based on experimental evaluation the threshold is equal to the 85-th percentile of $f(x, y)$ values for all (x, y). Using the threshold T algorithm calculates:

Table 1. The RSSI-ring model for the anchors. The measured value of RSSI determines a range of possible distances to the anchor ($D_{i,min}$–$D_{i,max}$). The ranges define rings around anchors pointing estimated location of the tag.

Anchor	Distance ranges [m] for various RSSI measurements [dBm]				
i	... ≥ -75	$-75 > ... \geq -80$	$-80 > ... \geq -85$	$-85 > ... \geq -90$	$-90 > ...$
1	9–15	2–26	6–37	10–47	25–63
2	5–16	4–27	4–36	10–48	24–53
3	5–9	4–17	4–30	11–40	27–54
4	6–21	5–25	7–26	10–35	19–40
5	4–25	2–21	7–23	12–38	20–40
6	3–24	8–30	13–35	16–36	28–32
7	6–20	4–29	5–28	7–34	17–33
8	10–17	7–26	7–25	13–45	29–52
9	5–11	5–18	5–31	8–48	21–51
10	4–22	4–37	10–57	21–61	42–63

$$\hat{f}(x,y) = \begin{cases} f(x,y) & \text{iff } f(x,y) > T, \\ 0 & \text{otherwise.} \end{cases} \tag{5}$$

The final location of the tag is determined as a center of mass of $\hat{f}(x,y)$, i.e.:

$$\hat{x} = \frac{\sum_{(x,y)} x \cdot \hat{f}(x,y)}{\sum_{(x,y)} \hat{f}(x,y)}, \quad \hat{y} = \frac{\sum_{(x,y)} y \cdot \hat{f}(x,y)}{\sum_{(x,y)} \hat{f}(x,y)}. \tag{6}$$

4   Evaluation and Analysis

The evaluation was conducted for 33 tags mounted on animals, during normal operation of the farm, without affecting the natural behavior of the cows. During the experiment 67 animal locations were measured using a laser range finder, with an estimated accuracy below 1m. In each test location, all anchors have reported to the server several aggregated measurements containing the average value of RSSI, total signal strength (TSS), and a number of raw RSSI measurements. Overall the server received almost 5000 data points from all cows and all locations, that were used for evaluation of the proposed localization method.

Figure 4 shows one of the test points. The blue and red marks in the figure denote real and estimated location of the animal, respectively. The error in location estimation equals 5.3 m. It can be seen that the highest value of the $f(x,y)$ function (white areas in the figure) span along the X axis (width of the barn) causing the resulting $(\hat{x}, \hat{y})$ to be biased towards the center of the barn. A similar bias is observed for all the test points and is a consequence of the proposed localization approach – first, the rings are wide and overlap largely;

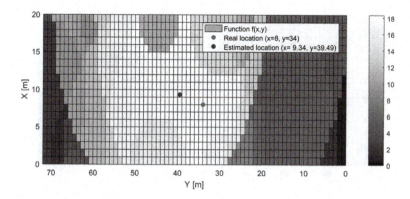

Fig. 4. Example of localization using single RSSI-ring model for each anchor. The colors of the surface represent values of the $f(x, y)$ function, red and blue marks denote real and estimated location of the tag, respectively. (Color figure online)

Table 2. The localization error of the proposed approach when using different number and sets of anchors. The reported results were achieved for a single RSSI-ring model for each anchor (cf. Table 1)

Anchors		Euclidean error [m]			
Number	List	Min	Average	Median	Max
10	1–10	0.6	8.3	7.8	24.7
8	2–9	0.6	8.5	7.5	37.7
6	1, 4–7,10	0.9	9.2	8.6	23.5
6	1–3, 8–10	0.7	9.4	9	23.1
4	2, 3, 8, 9	1.8	9.8	9.4	25.2

second, because the anchors are located on the edges of the area, the center of the ring's mass tends to be located in the center of the barn.

Table 2 presents statistics on the accuracy of the proposed localization method in terms of Euclidean error. The results are presented for a different number and set of anchors used in the localization procedure. This simulates localization accuracy for scenarios where fewer anchors are available in the system. The best results are achieved when all 10 anchors are used. In this case, the mean error is slightly above 8 m, and for over 50% of the test points (median) the error does not exceed 7.8 m. The accuracy of the localization drops as the number of anchors is reduced. While the mean error when using 8 anchors (except anchors 1 and 10) increases by approx. 2.5%, the error for 6 and 4 anchors is higher by 16% and 37%, respectively. Also the median and the maximal error increases as the number of anchors drops.

The localization accuracy can be further improved if we take into account that dairy cows on a farm are divided into groups. Due to the different feeding, the groups reside in different sections of the barn and never mix. Consequently,

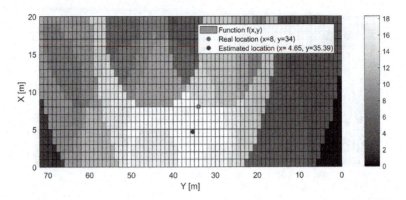

Fig. 5. Example of localization with improved approach – using two RSSI-ring models for each anchor

Table 3. Comparison of the localization error in the improved approach (two RSSI-ring models per anchor) and the log-distance path loss model based localization

Method	Euclidean error [m]			
	Min	Average	Median	Max
Path loss based	0.43	10.3	9.9	33.18
Our (first approach)	0.6	8.3	7.8	24.7
Our (improved approach)	0.8	6.3	5.4	20.8

for each cow in the barn we know if she resides in the top or bottom section of the barn (cf. Fig. 1). This allows building separate RSSI-ring models for cows on the same and on the other side of the barn with respect to anchor location. As a result, each anchor will have two RSSI-ring models: one for animals on the same side of the barn and the other for the animals on the other side. Figure 5 presents localization result for the same test location as in Fig. 4, but using the improved approach. The maximal value of $f(x, y)$ function is now limited to a significantly smaller area and is almost entirely included with the bottom section of the barn. For the test location presented in Fig. 5 the resulting localization accuracy improves as the error drops from 5.3 to 3.6 m (32%). The improvement is also observed for all the test locations lowering the average, median, and maximal errors to 6.3, 5.4, and 20.8 m, respectively (Table 3).

The reported errors are also significantly better compared to the traditional approach based on the log-distance path loss model (Table 3). For example, if one uses the log-distance path loss model (1) and multilateration procedure from four closest anchors (with the smallest values of RSSI), then the average and median errors for the experimental scenario equal 10.3 and 9.9 m, respectively. This means that the improved algorithm reduces the average and mean localization error by approximately 39% and 45%, respectively.

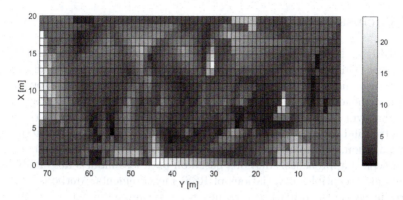

Fig. 6. Localization error in different locations of the area for the improved approach (two RSSI-ring models for each anchor). The colors denote the Euclidean error. (Color figure online)

Table 4. Comparison of the localization error for different methods based on signal strength measurements

Approach	Location	Area size [m²]	# Anchors	Area per anchor [m²]	Average error [m]
[4]	Outdoor	440	4	110	2.48
[11]	Outdoor	11 400	20	570	6
[12]	Indoor	390	11	35	4.2
[2]	Indoor	420	10	42	3.3
Path loss model	Indoor	1 440	10	144	10.3
Our improved	Indoor	1 440	10	144	6.3

The localization error is smaller for tags located closer to the center of the barn and increases as they approach the edges. This is presented in Fig. 6 that shows a localization error for the whole area. The surface is an extrapolation of the errors in all the measurement points. Locations are the least accurate when tags are located close to anchor no 1 (left of the localization area). This is possibly due to the fact that anchor number 1 was located 8 m away from the edge of the localization area. Consequently, the ranges estimated for this anchor were always large and could have affected the localization result.

Table 4 shows a comparison of the localization error of the proposed approach and solutions presented in the literature, which are based on signal strength measurements. Because the density of anchors significantly affects the accuracy therefore apart from comparing average error we also compare the average area covered by each anchor, which is calculated as a ratio of the area and the number of anchors deployed. Compared to other RSSI-based systems for indoor localization of farm animals [2,12] the proposed approach yields slightly larger errors, but operates in a significantly larger area, using a small number of anchors, and does not require time-consuming fingerprinting. Lower average error is achieved in outdoor deployments [4,11] where radio propagation conditions are preferable and all RSSI-based methods perform better. Additionally, the reported results were achieved using complex system [11] and through evaluation in specific test locations [4].

5   Conclusions

Using BLE and RSSI for indoor localization is one of the most challenging approaches. This is because the averaged RSSI measurements are affected by the radio channel used, attenuation due to obstacles, multi-path propagation, and other phenomena. This makes it impossible to derive a radio propagation model that will accurately relate the value of RSSI to the distance between the transmitter and receiver. Large area, a large number of animals obstructing signal propagation, and unfavorable location of the tags (on the side of animal neck) harden typical approaches. Methods based on path loss models are highly inaccurate and susceptible to variations of RSSI measurements. Methods based on fingerprinting, on the other hand, require time-consuming measuring campaigns that are troublesome and impractical for real-life, large scale applications. The proposed approach overcomes the limitations of previous methods. The geometric localization is less susceptible to variations and inaccuracies in RSSI and it requires a relatively small measurement campaign compared to fingerprinting. Additionally, the proposed method uses small number of anchors and achieves accuracy that is acceptable for a large range of applications, including animal tracking and monitoring their activities.

Acknowledgment. The author would like to thank Michał Zdunek (vet) and Agro-Tak Zagrodno dairy cow farm for access to the evaluation area, data collected by health monitoring system, and support during localization experiments.

References

1. Antanaitis, R., Žilaitis, V., Kucinskas, A., Juozaitienė, V., Leonauskaite, K.: Changes in cow activity, milk yield, and milk conductivity before clinical diagnosis of ketosis, and acidosis. Veterinarija ir Zootechnika **70**, 3–9 (2015)
2. Bloch, V., Pastell, M.: Monitoring of cow location in a barn by an open-source, low-cost, low-energy bluetooth tag system. Sensors **20**, 3841 (2020)
3. Cannizzaro, D., et al.: A comparison analysis of BLE-based algorithms for localization in industrial environments. Electronics **9**, 44 (2019). https://doi.org/10.3390/electronics9010044
4. Cardoso, A., Pereira, J., Nóbrega, L., Gonçalves, P., Pedreiras, P., Silva, V.: SheepIT: Activity and Location Monitoring (2018)
5. Hindermann, P., Nüesch, S., Frueh, D., Rüst, A., Gygax, L.: High precision real-time location estimates in a real-life barn environment using a commercial ultra wideband chip. Comput. Electron. Agric. **170**, 105250 (2020). https://doi.org/10.1016/j.compag.2020.105250
6. Huircán, J.I., et al.: ZigBee-based wireless sensor network localization for cattle monitoring in grazing fields. Comput. Electron. Agric. **74**(2), 258–264 (2010). https://doi.org/10.1016/j.compag.2010.08.014. http://www.sciencedirect.com/science/article/pii/S0168169910001584
7. Kunhoth, J., Karkar, A.G., Al-Maadeed, S., Al-Ali, A.: Indoor positioning and wayfinding systems: a survey. HCIS **10**(1), 1–41 (2020). https://doi.org/10.1186/s13673-020-00222-0

8. Luo, J., et al.: Sensor network for monitoring livestock behaviour. In: 2020 IEEE Sensors, pp. 1–4 (2020). https://doi.org/10.1109/SENSORS47125.2020.9278693

9. Macmillan, K., Mohanathas, G., Plastow, G., Colazo, M.: Performance and optimization of an ear tag automated activity monitor for estrus prediction in dairy heifers. Theriogenology **155**, 197–204 (2020). https://doi.org/10.1016/j.theriogenology.2020.06.018

10. Simões, W.C.S.S., Machado, G.S., Sales, A.M.A., de Lucena, M.M., Jazdi, N., de Lucena, V.F.: A review of technologies and techniques for indoor navigation systems for the visually impaired. Sensors **20**(14), 3935 (2020). https://doi.org/10.3390/s20143935

11. Takahiro, Y., et al.: A study on outdoor localization method by recurrent deep learning based on time series of received signal strength from low power wireless tag. IEICE Commun. Express **8**, 572–577 (2019). https://doi.org/10.1587/comex.2019GCL0065

12. Trogh, J., Plets, D., Martens, L., Joseph, W.: Bluetooth low energy based location tracking for livestock monitoring (2017)

13. Unold, O., et al.: IoT-based cow health monitoring system. In: Krzhizhanovskaya, V.V., et al. (eds.) ICCS 2020. LNCS, vol. 12141, pp. 344–356. Springer, Cham (2020). https://doi.org/10.1007/978-3-030-50426-7_26

14. Wang, Y., Yang, Q., Zhang, G., Zhang, P.: Indoor positioning system using euclidean distance correction algorithm with bluetooth low energy beacon. In: 2016 International Conference on Internet of Things and Applications (IOTA), pp. 243–247 (2016). https://doi.org/10.1109/IOTA.2016.7562730

15. Wolfger, B., Jones, B., Orsel, K., Bewley, J.: Technical note: evaluation of an ear-attached real-time location-monitoring system. J. Dairy Sci. **100**, 2219–2224 (2016). https://doi.org/10.3168/jds.2016-11527

Monitoring the Uniformity of Fish Feeding Based on Image Feature Analysis

Piotr Lech[1] , Krzysztof Okarma[1]([✉]) , Agata Korzelecka-Orkisz[2] ,
Adam Tański[2] , and Krzysztof Formicki[2]

[1] Department of Signal Processing and Multimedia Engineering,
Faculty of Electrical Engineering, West Pomeranian University of Technology
in Szczecin, Sikorskiego 37, 70-313 Szczecin, Poland
{piotr.lech,krzysztof.okarma}@zut.edu.pl
[2] Department of Hydrobiology, Ichthyology and Biotechnology of Reproduction,
Faculty of Food Sciences and Fisheries, West Pomeranian University of Technology
in Szczecin, Kazimierza Królewicza 4, 71-550 Szczecin, Poland
{agata.korzelecka-orkisz,adam.tanski,krzysztof.formicki}@zut.edu.pl

Abstract. The main purpose of the conducted research is the development and experimental verification of the methods for detection of fish feeding as well as checking its uniformity in the recirculating aquaculture systems (RAS) using machine vision. A particular emphasis has been set on the methods useful for rainbow trout farming.

Obtained results, based on the analysis of individual video frames, convince that the estimation of feeding uniformity in individual RAS-based farming ponds is possible using the selected local image features without the necessity of camera calibration. The experimental results have been achieved for the images acquired in the RAS-based rainbow trout farming ponds and verified using some publicly available video sequences from tilapia and catfish feeding.

Keywords: Image analysis · Image features · Fish feeding · Aquaculture · Recirculating aquaculture systems

1 Motivation

The motivation of the paper is the necessity of optimization of rainbow trout feeding which should be controlled automatically, preferably using a machine vision approach utilizing image data captured by cameras mounted over the farming ponds.

The feeding frequency is an important factor influencing the growth rate. Feeding can be done manually or mechanically through automatic feeders. The main advantage of using automatic machines is reducing the time and labor consumption, whereas the disadvantage is the reduction of fish condition controls.

M. Paszynski et al. (Eds.): ICCS 2021, LNCS 12745, pp. 68–74, 2021.
https://doi.org/10.1007/978-3-030-77970-2_6

Providing each of the fish with the correct portion of food is a big challenge. Uneven growth of fish is a negative phenomenon at every stage of the production, even in the case of the last tranche of the fish to be sold. Hence, it is extremely important to develop methods of controlling the feeding process in terms of even food supply. One of the technical possibilities, considered in the paper, is the use of machine vision methods.

2 Observation

Regardless of the construction of farming ponds for fattening (natural, artificial, or rectangular cylindrical), visually observable phenomena related to feeding are identical when fish aggressively taking food are considered.

The course of feeding over time can be illustrated in a simplified manner in three phases (Fig. 1):

- phase 1 – interest in food (stimulation): more and more fish appear in the observed area in a short period, the strongest individuals dominate,
- phase 2 – optimal feeding time: different and declarative depending on the type of food, a season of the year; it is characteristic that other fish join and this state is practically as long as the food is provided, however, the feeding time is strictly defined and depends on adopted strategy,
- phase 3 – rest: the feeding was interrupted, the activity of the fish ceased.

It has been assumed that the control of the uniformity of feeding may be achieved by the analysis of some local image features calculated for N individual regions to detect the presence of each of three states independently.

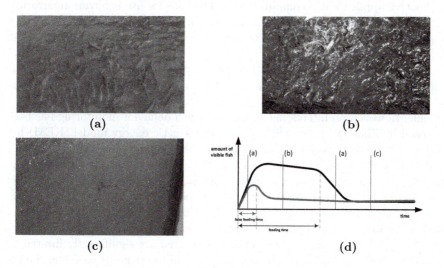

Fig. 1. Sample images illustrating three main types of rainbow trout behavior in an open rectangular pond: stimulation (a), feeding (b) and rest phase (c), and a simplified illustration of fish behavior over time (d) in these phases.

3   Related Machine Vision Methods

A growing availability of cameras and wide possibilities of image data acquisition, and real-time analysis make it possible to apply some machine vision solutions not only for detection and tracking fish but also for the analysis of their behavior. One of the examples is the method proposed by Spampinato et al. [6] where the detection, tracking and counting fish using an underwater camera has been made using the video texture analysis, fish detection and CamShift tracking modules.

Another recently proposed idea, employed for underwater videos, utilizes deep learning methods for fish detection and classification [2], mainly for underwater autonomous operation in weakly illuminated deep-sea environment.

Since an essential element necessary for determining the optimal farming strategy is the observation of changes in fish behavior, a few computer vision methods have been proposed for this purpose [1]. In the paper written by Spampinato et al. [7] the fish motion trajectories are determined and clustered to identify an abnormal behavior using an underwater camera.

On the other hand, fish behavior analysis may be conducted statistically without the use of more or less sophisticated detection, classification and tracking methods. In the case of RAS-based fish breeding, individual species may be troublesome for detection and tracing, especially during feeding, differently than e.g. in deep-sea environment.

A statistical analysis conducted by Papadakis et al. [4] makes it possible to assess the behavioral changes in a natural underwater environment, using a combination of 70 textural features, based on Gabor filters and the histogram and Gray Level Co-Occurrence Matrix (GLCM), with 50 shape features extracted using Fourier descriptors and affine Curvature Scale Space transform.

Another application of computer vision methods for fish behavior monitoring, regarding the water quality and toxin concentration, influencing the tail-beat frequency and wall hitting rate in the aquarium, has been presented by Xiao et al. [8].

As stated in one of the recent papers [3], "the robustness and reliability of multi-target tracking methods for groups remains a challenge in the field of computer vision" (p. 2), making their application for fish behavior monitoring troublesome. However, an interesting approach to fish behavior monitoring has been proposed by Zhao et al. [9], where a modified kinetic energy model (KEM) has been used to detect special behaviors in the RAS instead of tracking.

An overview of some alternative methods, e.g. based on near infrared imaging [10], passive and active acoustics, including sonars, as well as some other sensors, may be found in the recent survey paper written by Li et al. [3].

Regardless of the above mentioned limitations of tracking of individual fish species, in many cases the applications of computer vision methods are also related to the use of underwater cameras and often are significantly limited by the water clarity. Therefore, a method proposed in the paper is based on the use of a camera located above the water and the analysis of the water surface and a fish shoal as a whole, making it possible to detect the feeding phase also for a cloudy water.

4   Proposed Approach

Considering the difficulties in installing underwater cameras, especially in concrete tanks, as well as cleaning and servicing the tank, some significant limitations occur, causing a necessity to rely on the analysis of image sequences captured only by cameras installed over the pond, observing the water surface. Some other important reasons prompting us to use the top surface observations are possible cable breakage during use of an automatic target system and, even more importantly, injuring fish in contact with underwater camera housings.

An additional factor is the limited and varying water transparency since all the experiments have been made for an open-air rectangular concrete tank and additionally verified using some other publicly available video sequences downloaded from an Internet resources.

The main goal of the conducted experiments has been related to the determination of possible use of statistical features for the classification of rainbow trout behavior regarding the feeding phase (sample images are shown in Fig. 1) without the use of underwater cameras and sophisticated deep-learning methods. These assumptions have been made due to the limited availability of training data as well as considering the expected full explainability of the developed solution. Considering also the necessity of monitoring the feeding uniformity, the use of handcrafted local features is of special interest. Therefore, the division of each video frame into regions has been assumed.

The initial experiments have been made using the following features calculated after color-to-grayscale conversion: image entropy, variance, and four statistical Haralick features used for texture analysis calculated using the symmetrical normalized Gray-Level Co-occurrence Matrix (GLCM), namely contrast, correlation, homogeneity, and energy.

The best results of initial experiments using global features have been obtained for correlation but a proper classification is still impossible in some cases, however the use of local correlation leads to satisfactory results.

The proposed approach to the determination of the rainbow trout behavior in the feeding phase should be further combined with an additional analysis of local features making it possible to determine the duration of fish activity characteristic for feeding in all regions. Such a duration may be expressed as the number of video frames for which a specified region is classified as representing the feeding phase. Hence, the uniformity of the feeding may be estimated by the variance of the number of such frames for each region. The number and the size of the analyzed regions may be adjusted depending on the configuration and parameters of the camera, shape of the tank as well as the specificity and method of providing food. The idea of the use of the local features is demonstrated in Fig. 2 for the division into 4×4 grid. As it may be noticed, significantly higher local correlation results are obtained for the fragments where food has been provided. Hence, such an approach has been further verified for video sequences captured during various phases of fish feeding, also for a higher number of regions.

To improve the stability of the obtained results, particularly for varying lighting conditions, an additional preprocessing has been proposed based on the

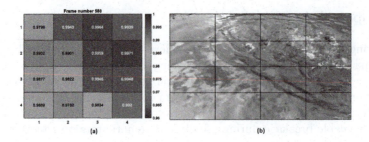

Fig. 2. Visualization of the local correlation results (a) for a sample video frame (b) illustrating the non-uniformity of feeding.

application of the Contrast Limited Adaptive Histogram Equalization (CLAHE) [5]. This method may be applied for the whole image or each region independently, however, obtained experimental results in both cases lead to similar conclusions.

To demonstrate the validity of the proposed approach, some experiments have been conducted for video sequences captured by a camera mounted over an open-air rectangular concrete pond for various stages of fish feeding.

For visualization purposes a few representative video frames have been chosen illustrating the feeding phase as well as rest and stimulation phases (Fig. 3). To demonstrate the differences among the obtained local correlation values, some results obtained for raw video frames as well as those with CLAHE preprocessing have been presented, assuming the application of histogram equalization for each region independently. Analyzing the heatmaps presented in Fig. 3, the dependence between the observed local fish activity and the obtained local correlation values may be easily noticed. It is also worth noting that the colormaps used in various images are different. As may be observed in the right column of Fig. 3, the local stimulation caused by a thrown pebble causes an increase of the local fish activity and higher local correlation values in the top left corner.

Comparing the local values presented in Fig. 3, a distinction between the local stimulation of feeding phase can be efficiently made, particularly for the application of the CLAHE algorithm. Analyzing the results for raw images, local correlation values for the stimulation in some cases exceed 0.95 whereas for the feeding phase they might be below 0.96 as marked on the heatmaps. Therefore, in such case a proper distinction should be done using the average values for several consecutive video frames or using the average local values.

Applying additionally the CLAHE-based image preprocessing for the whole image, the correlation values for stimulation decrease and the determination between the stimulation and feeding is easier. Nonetheless, the best results regarding this issue have been achieved using the region-independent use of the CLAHE method, as illustrated in the bottom rows of Fig. 3.

As it may be observed, the highest value for the local stimulation in the top left corner of the right image in Fig. 3 is 0.928 whereas the lowest one for feeding (bottom row for the middle image) is 0.9431. However, the determination of the

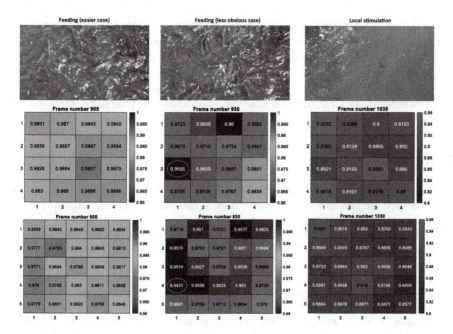

Fig. 3. Sample feeding and stimulation phase images (top row) and the corresponding heatmaps of the local correlation values – for raw images (middle row), and for CLAHE applied independently for each region using 5×5 grid (bottom).

feeding uniformity seems to be easier for raw images, assuming the preceding classification of the fish behavior as discussed above.

The proposed solution has also been verified for some other video sequences downloaded from Internet resources. The obtained results have confirmed its usefulness and universality also with the use of 6×6 regions grid.

5   Conclusions and Future Work

The paper is focused on the development and verification of vision methods, making it possible to detect and control the fish feeding process, involving the detection of feeding and estimation of its uniformity. The analysis concerns the consecutive video frames focusing on actual values of image features. The next stage of research is related to the development of behavioral patterns of feeding, taking into account the dynamics of the process over time, extending the presented methods based mainly on single frame features.

The proposed solution is different than most approaches developed by other researchers, usually utilizing underwater cameras or tracking of individual fish, as it is based on the analysis of water surface, considering a shoal of fish as a whole. Hence, a comparison with some alternative solutions is troublesome. On the other hand, the presented approach may be considered as a universal solution of relatively low computational complexity, possible to apply in open-air ponds, also for lower water clarity. Regardless of its robustness to varying light, our further research will concentrate on its further extension towards a more comprehensive fish behavior analysis for the controlled lighting conditions.

Acknowledgments. The research was conducted within the project no 00002-6521.1-OR1600001/17/20 financed by the "Fisheries and the Sea" program.

References

1. An, D., Hao, J., Wei, Y., Wang, Y., Yu, X.: Application of computer vision in fish intelligent feeding system—a review. Aquac. Res. **52**(2), 423–437 (2020). https://doi.org/10.1111/are.14907
2. Han, F., Yao, J., Zhu, H., Wang, C.: Underwater image processing and object detection based on deep CNN method. J. Sens. **2020**, 1–20 (2020). https://doi.org/10.1155/2020/6707328
3. Li, D., Wang, Z., Wu, S., Miao, Z., Du, L., Duan, Y.: Automatic recognition methods of fish feeding behavior in aquaculture: a review. Aquaculture **528** (2020). https://doi.org/10.1016/j.aquaculture.2020.735508
4. Papadakis, V.M., Papadakis, I.E., Lamprianidou, F., Glaropoulos, A., Kentouri, M.: A computer-vision system and methodology for the analysis of fish behavior. Aquacult. Eng. **46**, 53–59 (2012). https://doi.org/10.1016/j.aquaeng.2011.11.002
5. Pizer, S.M., et al.: Adaptive histogram equalization and its variations. Comput. Vis. Graph. Image Process. **39**(3), 355–368 (1987). https://doi.org/10.1016/s0734-189x(87)80186-x
6. Spampinato, C., Chen-Burger, Y.H., Nadarajan, G., Fisher, R.B.: Detecting, tracking and counting fish in low quality unconstrained underwater videos. In: Proceedings of the Third International Conference on Computer Vision Theory and Applications. SciTePress (2008). https://doi.org/10.5220/0001077705140519
7. Spampinato, C., Giordano, D., Salvo, R.D., Chen-Burger, Y.H.J., Fisher, R.B., Nadarajan, G.: Automatic fish classification for underwater species behavior understanding. In: Proceedings of the First ACM International Workshop on Analysis and Retrieval of Tracked Events and Motion in Imagery Streams - ARTEMIS 2010. ACM Press (2010). https://doi.org/10.1145/1877868.1877881
8. Xiao, G., Feng, M., Cheng, Z., Zhao, M., Mao, J., Mirowski, L.: Water quality monitoring using abnormal tail-beat frequency of crucian carp. Ecotoxicol. Environ. Saf. **111**, 185–191 (2015). https://doi.org/10.1016/j.ecoenv.2014.09.028
9. Zhao, J., et al.: Spatial behavioral characteristics and statistics-based kinetic energy modeling in special behaviors detection of a shoal of fish in a recirculating aquaculture system. Comput. Electron. Agric. **127**, 271–280 (2016). https://doi.org/10.1016/j.compag.2016.06.025
10. Zhou, C., et al.: Near infrared computer vision and neuro-fuzzy model-based feeding decision system for fish in aquaculture. Comput. Electron. Agric. **146**, 114–124 (2018). https://doi.org/10.1016/j.compag.2018.02.006

A New Multi-objective Approach to Optimize Irrigation Using a Crop Simulation Model and Weather History

Mikhail Gasanov[1]($\boxtimes$) , Daniil Merkulov[1] , Artyom Nikitin[1] ,
Sergey Matveev[2,3] , Nikita Stasenko[1] , Anna Petrovskaia[1] ,
Mariia Pukalchik[1] , and Ivan Oseledets[1,2]

[1] Skolkovo Institute of Science and Technology, Bolshoy Boulevard 30,
bld. 1, 121205 Moscow, Russia
`Mikhail.Gasanov@skoltech.ru`
[2] Marchuk Institute of Numerical Mathematics, RAS, Gubkin st 8,
119333 Moscow, Russia
[3] Faculty of Computational Mathematics and Cybernetics, Lomonosov Moscow State
University, Leninskiye Gory 1, 119991 Moscow, Russia
`http://www.skoltech.ru/`, `http://www.inm.ras.ru/`, `https://cs.msu.ru/`

Abstract. Optimization of water consumption in agriculture is necessary to preserve freshwater reserves and reduce the environment's burden. Finding optimal irrigation and water resources for crops is necessary
to increase the efficiency of water usage. Many optimization approaches
maximize crop yield or profit but do not consider the impact on the environment. We propose a machine learning approach based on the crop simulation model WOFOST to assess the crop yield and water use efficiency.
In our research, we use weather history to evaluate various weather scenarios. The application of multi-criteria optimization based on the non-
dominated sorting genetic algorithm-II (NSGA-II) allows users to find
the dates and volume of water for irrigation, maximizing the yield and
reducing the total water consumption. In the study case, we compared
the effectiveness of NSGA-II with Monte Carlo search and a real farmer's
strategy. We showed a decrease in water consumption simultaneously
with increased sugar-beet yield using the NSGA-II algorithm. Our approach yielded a higher potato crop than a farmer with a similar level of
water consumption. The NSGA-II algorithm received an increase in yield
for potato crops, but water use efficiency remained at the farmer's level.
NSGA-II used water resources more efficiently than the Monte Carlo
search and reduced water losses to the lower soil horizons.

Keywords: Water use efficiency · Machine learning · Multi-objective
optimization · Sustainable agriculture

M. Gasanov and D. Merkulov—These authors contributed equally to the work.

M. Paszynski et al. (Eds.): ICCS 2021, LNCS 12745, pp. 75–88, 2021.
https://doi.org/10.1007/978-3-030-77970-2_7

1 Introduction

Global population growth leads to urbanization and intensification of food production. This intensification of food production causes an rise in water consumption, which yields a negative impact on the environment and may result in
a reduction of freshwater quality [19]. The lack of availability of water resources
is one of the main limiting factors in regions with low yields [16]. Efficient water
resources for agricultural purposes is necessary to ensure food security and to
reduce this environmental impact.

One can describe water resource efficiency as the amount of water spent to
produce a certain amount of crop yield [18]. There are several factors that affect
the efficiency of water resources usage [17,22]. A part of the water is involved
in plant growth, development, and transpiration, so this part is considered to
be used efficiently. Another part of the water is not accessible to plant roots
due to evaporation, migration with surface runoff, and deep percolation. So we
can define water loss (part of irrigation water that cannot be transformed into
economic gain) and efficiently used water (water that is transformed into yield).
For irrigation agriculture, water use efficiency may vary from 13 to 18% of the
water supplied [27]. Gleick estimates that approximately 63% of all water for
irrigation is lost due to deep percolation and runoff [9]. Thus, it is necessary to
reduce water loss for sustainable agriculture and for the conservation of water
resources [11]. It is worth noting that the high level of water migration from the
root zone can cause mineral fertilizers to percolate into the groundwater, which
causes eutrophication and additional stress on the nearest water systems and
their inhabitants [31]. It can also affect the migration of pesticides to groundwater [14], increasing environmental risks. Therefore, there is a need to minimize
the amount of inaccessible water for the plant and deep percolation to reduce
the impact of inefficient crop irrigation on the environment.

Conducting field experiments to find the best agricultural management practices is time-consuming, as it requires evaluating all possible combinations of
agricultural practices. Crop simulation models are widely used to plan agricultural practices, such as planting and harvesting crops, fertilizing, and watering.
Crop simulation models allow users to evaluate various agricultural activities and
predict crop yields [7]. The most widespread used crop simulation models are
the following: DNDC [8], APSIM [10], AGROTOOL [2], DSSAT [12], MONICA
[20], AquaCrop [25], WOFOST [26], and others. These models have many differences in their ideology, utilized equations, choice of programming languages
for the software implementations, the minimum set of input parameters, and
spatial/temporal resolution.

Rapid computations allows users to supply optimization algorithms with a
simulation model as an objective function and to improve agricultural practices
automatically. A previous study's multi-objective differential evolution algorithm
(MDEA) was applied to minimize water use and to maximize South African
regions' income [1]. Yousefi et al. suggested that Multi-Objective Particle Swarm
Optimization reduces the negative impacts of using treated water and it maximizes crop benefits [32]. In paper[21], the authors use multi-criteria optimization

to maximize crop profit and reduce irrigation volume based on the CROPWAT model which showed the possibility of reducing irrigation volume by a quarter. García-Vila and Fereres used weather history and the AquaCrop model to optimize irrigation strategy on the farm-scale level in conditions of water scarcity [6]. Recently, research tested using DSSAT crop simulation system and the U-NSGA-III optimization algorithm to maximize crop yield and to minimize nitrogen leaching by selecting optimal irrigation water amounts and nitrogen fertilizers [15].

However, most papers consider a single crop and optimize agricultural practice for a single year based on weather data, which may not be very useful given the lack of opportunity to predict the weather for the whole vegetation season in the future [23]. In most papers, the researchers also consider the optimization of continuous irrigation and fertilization parameters, such as water volume or fertilizer amount, but they do not consider the dates of the agricultural practices. We analyze how to combine crop simulation model WOFOST and multi-objective optimization to maximize crop yield and minimize water loss to address this gap.

We use the NASA POWER weather history and compute mean crop yield and mean water loss for the last 20 years to evaluate different weather scenarios [24]. We compare the performance of our approach based on the NSGA-II optimization algorithm [4] against the Monte Carlo search with the farmer's agriculture practice to assess the proposed solution.

2   Materials and Methods

In this section, we describe the materials and methods that we use in our research.

2.1   Crop Simulation Model WOFOST

Crop simulation models (CSM) describe the dynamics of the atmosphere-soil-plants system's main processes that affect crop productivity. They can evaluate crop system productivity depending on weather, irrigation, and fertilizer application. Such models allow a user to avoid conducting long-term field experiments experiments and to select optimal agricultural practice. The rapid calculations of such models, about 0.5–2 s on a personal computer, allows a researcher to use such models in optimization problems as a function of the agricultural field productivity.

We utilize the WOrld FOod Studies (WOFOST) crop simulation model developed in Wageningen University to identify crops' productivity and irrigation water loss [30]. The WOFOST crop model describes dynamic growth processes, photosynthesis, transpiration, respiration, and biomass partitioning. We chose the WOFOST model because it is adapted and calibrated for European crops and environmental conditions. Figure 1 shows the experiment's scheme and the WOFSOT model application to optimize irrigation.

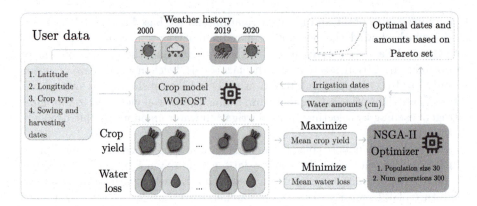

Fig. 1. Based on user data, we receive the weather history from NASA POWER. The optimizer generates dates and water amount for irrigation. Then we simulate the crop growth selected by a user for each of the weather history and get the yield and water loss for each year. The average yield values are passed to the optimizer and affect choosing the following combinations in the population. Finally, we select specific dates and water amount for irrigation from the Pareto front with the weighted sum method (see Sect. 2.3)

The WOFOST crop model requires weather data for each day of the growing season to perform simulations. We received data from the NASA POWER database for the crop simulations [24]. The NASA Energy System provides weather data from 1983 to the present with a delay of three months. NASA POWER allows users to collect weather history data with daily time resolution and grid resolution of half a degree of arc of longitude by half a degree of arc of latitude. The WOFOST model requires several meteorological observations for each day, such as incoming global radiation (W/m^2), daily minimum temperature $(°C)$, daily maximum temperature $(°C)$, daily average vapor pressure (hPa), daily total precipitation (cm/day), daily average wind speed at an altitude of 2 m (m/sec). NASA's POWER data contains weather omissions for some dates that average 1–2% of all dates. We use the pandas' package *fillforward* method in python to fill data gaps.

The WOFOST model accepts input data in a YAML file format containing necessary information about crop parameters, cultivar, soil conditions, and weather data in a CSV format. To assess the crop's productivity, we used the variable - total weight of storage organs (TWSO, t/ha). To estimate the volume of deep percolation water, we used the total amount of water lost to deeper soil (LOSST, cm). The WOFOST model has been used for more than 25 years and has various implementations in Fortran, python, and R. We used a python implementation of PCSE/WOFSOT model[1].

---

[1] https://github.com/ajwdewit/pcse.

2.2   Multi-objective Optimization

Notations and Terms. In optimization problems, variables are changed to maximize or minimize the objective function. In agriculture, many tasks require finding the optimal solution, such as choosing irrigation dates and the amount of water to be watered, the amount of fertilizer application, and the date of application. Typically, the farmer tries to minimize their losses by avoiding inefficient use of fertilizers and water resources, fuel consumption, and to maximize their yield and crop quality.

However, in the real-world, minimizing one cost could immediately lead to maximization of another cost, which depends on the same set of variables. For example, one can use a massive amount of water or fertilizer to increase yield, but this can sometimes be barely profitable because of their water or fertilizer expenses. One way to deal with multiple objectives that could conflict with each other is through multi-objective optimization.

Suppose, we have T loss functions $\mathcal{L}^i(\boldsymbol{\theta}), \forall i = \overline{1,T}$:

$$\min_{\theta} \mathcal{L}(\boldsymbol{\theta}) = \min_{\theta} \left(\mathcal{L}^1(\boldsymbol{\theta}), \ldots, \mathcal{L}^T(\boldsymbol{\theta}) \right)^{\top} \tag{1}$$

In such a setting, we need to specify a way to compare a vector of objectives. The typical way to do this is by introducing the concept of Pareto dominance and Pareto optimality.

Definition 1 (Pareto optimality).

1. *A point $\boldsymbol{\theta}_1$ dominates $\boldsymbol{\theta}_2$ for a multi-objective optimization problem (1), if $\mathcal{L}^i(\boldsymbol{\theta}_1) \leq \mathcal{L}^i(\boldsymbol{\theta}_2) \, \forall i = \overline{1,T}$ and at least one inequality is strict.*
2. *A point $\boldsymbol{\theta}_*$ is called Pareto-optimal solution, if there is no other point $\boldsymbol{\theta}$ that dominates it.*

This article exploits multi-objective optimization to determine optimal irrigation dates and optimal water volume for irrigation to maximize crop yield and minimize water loss (water inaccessible to plants). In this setting we have $T = 2$ functions to minimize concurrently: $\mathcal{L}^1(\boldsymbol{\theta})$ - mean water loss for different weather scenarios, and $\mathcal{L}^2(\boldsymbol{\theta})$ - mean crop yield for different weather scenarios (taken with negative sign). The vector of parameters $\boldsymbol{\theta}$ contains irrigation dates and amounts of water, which is needed to be spent in the corresponding day. In our experiments we use 7 irrigation dates, therefore $\boldsymbol{\theta}$ is 14-dimensional vector.

Multi-objective Optimization Algorithms. A lot of effective multi-objective optimization algorithms were developed [5,13,33]. However, due to the specific structure of the loss functions, which are the outputs of a WOFOST model, a function's value is the only information we have. Since no gradients or any other higher-order details are available, we are restricted to use zero-order algorithms. In this work, we compare the following approaches:

- **Monte Carlo optimizer.** We randomly generate $\boldsymbol{\theta}$ vectors and choose the best (in terms of Pareto optimization) point for our problems.

– **NSGA-II optimizer.** We use non-dominated sorting genetic algorithm (also known as NSGA-II) [4] method from PyMoo [3] package for Multiobjective optimization in python.

The experimental setup and optimization parameters are described in the Sect. 2.6.

2.3   Choice of Point from Pareto Front

As a solution of problem (1) we typically have a set of Pareto optimal points, which is called *Pareto front*. The points cannot be compared directly between each other, since they all are optimal in some sense. At this stage, the best $\boldsymbol{\theta}$ needs to be chosen according to some prior information. There are several approaches that can be made for this choice [28, 29].

Figures 3b and 4b show the Pareto front with all of the points being Pareto-optimal. One could consider the full spectrum of proposed solutions, but it may be convenient to select only a particular one. A farmer might want to know how to save money with a proposed approach or how to deal with strictly limited water resources.

We propose an inclusive way to address this problem. After the optimization procedure we have a set of m possible solutions $\{\boldsymbol{\theta}_j\}_{j=1}^m$ with corresponding loss functions values $\{\mathcal{L}^1(\boldsymbol{\theta}_j)\}_{j=1}^m$ and $\{\mathcal{L}^2(\boldsymbol{\theta}_j)\}_{j=1}^m$. We normalize the values of each objective function to zero mean and unit variance, denoting it as $\left\{\hat{\mathcal{L}}^1(\boldsymbol{\theta}_j)\right\}_{j=1}^m$ and $\left\{\hat{\mathcal{L}}^2(\boldsymbol{\theta}_j)\right\}_{j=1}^m$. This step allows users to deal with the objectives of different scales. After standardization, we select a point with minimal sum of the normalized loss value functions.

$$\min_{j\in(1,\dots,m)} \hat{\mathcal{L}}^1(\boldsymbol{\theta}_j) + \hat{\mathcal{L}}^2(\boldsymbol{\theta}_j) \tag{2}$$

In this approach, we treat each objective equally. The results from this choice are presented on the Figs. 3b and 4b. However, the approach could be easily transformed to a weighted choice when you multiply objectives in the problem (2) to some coefficients, which could be interpreted as important. Note that it would be better to use some a-priori information to determine a specific choice of the irrigation dates and amounts of water among the Pareto set in each particular case.

For clarity we mean this specific method of choosing point from Pareto front, while we compare methods between each other. In all figures by "Ours" we mean a single point from Pareto front, produced by NSGA-II method, selected using the routine described above, just as well as for the Monte Carlo method. Since, the way of comparison is not unique, we also present the entire set of intermediate points for each method.

2.4   Weather Averaging

The irrigation schedule is affected by uncertain weather factors. For example, precipitation affects the choice of irrigation dates. Since we cannot predict the weather for an entire growing season ahead, we cannot know the irrigation schedule's optimal dates in advance. However, we can consider weather history over the past few decades as various possible weather scenarios. We can assume that the weather next year may be similar to the past decades' weather scenarios. We can also assume that we can find the irrigation dates and irrigation volume for a particular location that will increase crop yield on average for various weather scenarios and minimize average water loss. For example, we can take the last 20–30 years and consider the weather as different climate scenarios for a given geographical region.

The general plan of the experiment is presented in Fig. 1. The user specifies geographic coordinates of crop and planning of agricultural management practices, such as planting and harvesting crop dates, dates and amounts of irrigation, dates of fertilizer application, and fertilizer amounts. We use NASA's POWER data to receive weather data for recent years based on based on their geographical coordinates. At the next step, we initialize the NSGA-II optimizer to search for optimal irrigation dates and water volumes. For calculations, we use irrigation dates as discrete integer values ranging from the planting date (day 0) to the harvesting date, usually in the range of 120–150 days. We use water volume values between 0 and 150 mm of water per hectare. The optimizer offers solutions in the form of a combination of irrigation dates and water volumes, which we add to the input data for the WOFOST model. After using these inputs, we run simulations for each year for the last 20 years of available weather. For each year, we compute the crop yield value and the volume of lost water. Then the obtained values for 20 years are averaged and returned to the optimizer. Based on these two target values, the optimizer offers new combinations of irrigation dates and water volume. As the number of iterations increases, the values progress towards to optimal solutions. As a result, we receive a Pareto-set of optimal combinations of irrigation dates and water volumes. The specific choice of the solution is described in Sect. 2.3.

2.5   Case Study

To assess the method's performance, we have chosen agricultural fields in the Moscow region, Russia (Fig. 2). The fields are located on the Oka River banks in the floodplain. The soil was characterized as Sandy loam with the following characteristics: bulk density - 1.4 g/cm3̂, clay content - 13.9%, silt content - 13.1%, sand content - 73%, surface hardness - 0,87 MPa and subsurface hardness - 3.81 MPa. In these fields, farmers grow vegetable crops such as beets, potatoes, onions, and carrots. For the experiment, we chose sugar-beet (*Beta vulgaris*) and potato (*Solánum tuberósum*). We received information from farmers about sowing and harvesting operations and the proposed irrigation operations for sugar-beet and potatoes fields for the 2019 year. Irrigation was done seven times

Fig. 2. Map of the investigated region. We marked the experimental fields with potato and sugar-beet near the Oka river with color.

per season (June 10 and 20, July 1, 10, 20 and 29, August 15) for both crops with a water amount of 2 cm/ha. Sometimes farmers have to shift the dates by 1–2 days due to weather and other conditions. We took this into account and conducted ten simulations randomly changing the dates of watering, planting and harvesting for part of the experiment with farmer data. During the season, farmers, on average, contribute 190 kg/ha of nitrogen fertilizers for both crops.

2.6 Numerical Experiments Setup

We performed all of our numerical simulations on the Google Cloud platform (4 vCPUs, 4 GB memory). The average time of a WOFOST model run takes 10 s for 20 different weather scenarios. For the NSGA-II algorithm, we used 300 generations and a population size of 30 for each generation, so the algorithm performs 9000 estimations of computing crop yield and water loss. To maintain equality, we ran the Monte Carlo search with the number of iterations of 9000. The whole optimization procedure took around 25 h for the single run. We conducted 10 runs with different random initializations and calculated mean and standard deviations for the reported metrics.

3 Results and Discussion

In this section, we describe the results and compare the performance of our approach against Monte Carlo search. We ran the NSGA-II and the Monte Carlo search algorithms ten times with random initialization to evaluate their performance. As it was discussed above in Sect. 2.3, we compare the performance of the methods with respect to the specific choice of the point from Pareto front. Thus, we received ten values of yield and water loss for potato and sugar beet and calculated the mean and standard deviation of loss functions. We ran the calculation ten times with a random date deviation of 1–2 days for the farmer's irrigation scheme to calculate the mean and standard deviation. The Tables 1, 2 below contains mean values as well as standard deviations for the selected parameters.

Table 1. Potato

	Yield (t/ha)	Water loss (cm)
Farmer	12.74 ± 0.03	**23.99 ± 0.21**
Monte Carlo	13.95 ± 0.18	35.54 ± 2.26
Ours	**14.11 ± 0.09**	26.84 ± 1.40

Table 2. Sugar-beet

	Yield (t/ha)	Water loss (cm)
Farmer	11.97 ± 0.1	31.73 ± 0.17
Monte Carlo	**12.17 ± 0.03**	42.38 ± 2.97
Ours	12.16 ± 0.03	**28.22 ± 0.98**

The results presented in Table 1 with potatoes experiments show that the NSGA-II algorithm consistently achieves higher yields than Monte Carlo. Results of experiments with sugar beet presented in the Table 2 show that the Monte Carlo and NSGA-II produced approximately the same crop yield level. However, the NSGA-II algorithm chose strategies with significantly lower water loss.

3.1 Potato Crop

Figure 3 compare our approach based on NSGA-II to the Monte Carlo search for the potato crop. Scatter-plot Fig. 3a demonstrates the objective values received by NSGA-II, Monte Carlo search, and values achieved by a farmer's strategy for the single run. Scatter-plot Fig. 3b shows Pareto front achieved by NSGA-II and optimal solution selected by a weighted-sum method. We considered the importance of crop yield and water loss equally to select the optimal solution. Using the same approach, we selected the optimal solution from the objective values generated by the Monte Carlo search. Additionally, we plotted objective values based on farmer strategy.

These numerical experiments on optimizing potato irrigation based on weather history are shown in Fig. 3. Points from both NSGA-II and the Monte Carlo search algorithms are approaching a limit of the yield of approximately 14.5 t/ha. We can assume that this is the maximum yield under the given conditions of soil, weather, and agricultural practices. In the seasonal water loss minimization problem, the NSGA-II algorithm is superior to the Monte Carlo search. Most of the water loss values are lower than the values obtained by the Monte Carlo search on the graph. It is interesting to mention that there are points from the NSGA-II Pareto front, which dominate a farmer's choice, while this property is not valid for the Monte Carlo Pareto front.

In the experiment to optimize potato irrigation, our approach achieved a mean potato yield of 14.11 t/ha, which is 9.7% higher than the farmer's solution with 12.74 t/ha. The Monte Carlo search achieved a yield of 13.95 t/ha, which is 8.6% percent higher than the farmer's.

On the other hand, farmer irrigated the field more efficiently (23.99 cm) than our approach (26.84 cm) and the Monte Carlo search (35.54 cm). However, our method only increased the mean water loss by 10%, whereas the Monte Carlo search by as much as 33%.

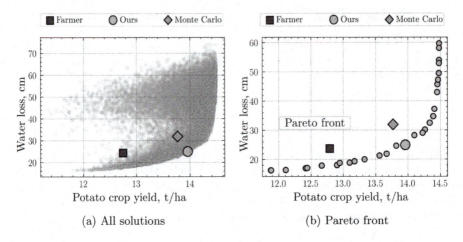

(a) All solutions   (b) Pareto front

Fig. 3. NSGA-II performs better at searching for strategies with low water loss. Objective values selected by weighted-sum methods denoted by the larger icons show our method's advantage over the Monte Carlo search and the farmer's strategy for crop yield objective. The inflection on the line of objectives values achieved by NSGA-II shows that the increase in irrigation associated with water loss does not increase productivity.

The plot also shows the Pareto front, where we can identify the inflection when the yield values of the order of 14.5 t/ha and water loss of 30 cm are reached. Therefore, we can conclude that a further increase in irrigation does not increase potato yield.

3.2 Sugar-Beet Crop

The values of the algorithm solutions for optimizing sugar-beet irrigation are shown in Fig. 4. Figure 4a represents all the objective function values obtained by our algorithm, Monte Carlo's search and the farmer strategy for the single run. Figure 4b represents the Pareto front of our solution and the values selected based on our method's weighted sum method, Monte Carlo, and the farmer's value. The maximal yield obtained for both algorithms was about 12.15 t/ha. For the sugar-beet, the inflection on the Pareto front is more pronounced.

NSGA-II and the Monte Carlo achieved approximately the same yield values for sugar-beet. The mean crop yield value was 12.17 t/ha for both algorithms that 1.6% higher than the result of the farmer's strategy with 12.00 t/ha. However, our method (28.22 cm) reduced water loss by 11%, while the Monte Carlo search (42.38) strategy increased water loss by 33.5%.

One of the multi-criteria optimization tasks is to reduce seasonal water losses. We compared the distribution of mean values of total irrigated water and the distribution of deep percolation water losses over 20 years of weather scenarios for all solutions produced by NSGA-II and Monte Carlo search for the single run. Results were obtained based on our approach and on the Monte Carlo search

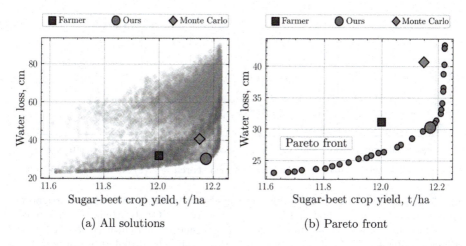

(a) All solutions

(b) Pareto front

Fig. 4. NSGA-II performs better at searching for strategies with low water loss. Objective values selected by weighted-sum methods denoted with the larger icons show our method's superiority over the Monte Carlo search and the farmer's strategy for crop yield and water loss. The inflection on the line of objectives values achieved by NSGA-II shows that the increase in irrigation, associated with water loss, does not increase productivity.

for sugar-beet and potato. The scatter-plots in Fig. 5 illustrate the mean water loss dependence over 20 years of weather scenarios based on the total seasonal irrigation. Figure 5b demonstrates a scatter-plot for the Monte Carlo search. The scatter-plots in Fig. 5 show the mean water loss dependence over 20 years of weather scenarios on the mean of total seasonal irrigation over 20 years of weather scenarios. Because of the random selection of irrigation water values, total seasonal irrigation and water loss have a normal distribution.

Figures 5a and 5c illustrate the result of the optimizer's performance, which shifts the seasonal water loss values' distribution to smaller values as water loss minimization undergoes. For both crops, seasonal water irrigation distribution and decreases the seasonal water loss values are similar. Because of the optimizer, these distributions are shifted towards smaller values. However, we can note differences in the distributions for potato and sugar-beet. The distribution of sugar-beet irrigation water has not moved as much to the lower values as for potatoes. However, the distribution of sugar-beet water losses has shifted significantly to the lower values. Such differences may be related to different plant physiology and root system features defined in the crop model.

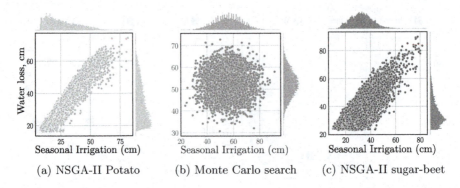

(a) NSGA-II Potato   (b) Monte Carlo search   (c) NSGA-II sugar-beet

Fig. 5. Scatter-plots and distributions of values for water loss (cm) and total irrigation amount (cm) for our method and Monte Carlo search. NSGA-II attempts to minimize water loss and water volume for irrigation and generates agricultural practices that decrease water loss in front of the Monte Carlo search. The distributions of objective values for NSGA-II are shifted to low values, which is positive for agricultural purposes.

4   Conclusions

Multi-objective irrigation optimization based on crop model WOFOST and evolutionary algorithm NSGA-II has demonstrated its efficiency in finding the optimal irrigation strategy. We have shown the effectiveness of using the approach on the case study example with sugar-beet and potato crops. The results with potatoes experiments show that the NSGA-II algorithm consistently achieves higher yields. In experiments with sugar-beet, the Monte Carlo and NSGA-II produced approximately the same level of yield. However, the NSGA-II algorithm chose strategies with significantly lower water loss. Based on our numerical experiments, we see the advantage of using evolutionary multi-objective optimization over the Monte Carlo search. The use of the algorithm reduces seasonal water loss and increases crop yield based on long-term weather data. The source code and the results are available in our GitHub repository[2].

Acknowledgement. This work is supported by the Russian Science Foundation (project No. 20-74-10102). Vectors used in plots and graphical abstract from vecteezy project.

References

1. Adeyemo, J., Otieno, F.: Differential evolution algorithm for solving multi-objective crop planning model. Agric. Water Manag. **97**(6), 848–856 (2010)
2. Badenko, V., Terleev, V., Topaj, A.: Agrotool software as an intellectual core of decision support systems in computer aided agriculture. In: Applied Mechanics and Materials, vol. 635, pp. 1688–1691. Trans Tech Publications (2014)

---

[2] https://github.com/EDSEL-skoltech/multi_objective_irrigation

3. Blank, J., Deb, K.: Pymoo: multi-objective optimization in python. IEEE Access **8**, 89497–89509 (2020)
4. Deb, K., Pratap, A., Agarwal, S., Meyarivan, T.: A fast and elitist multiobjective genetic algorithm: NSGA-II. IEEE Trans. Evol. Comput. **6**(2), 182–197 (2002)
5. Désidéri, J.A.: Multiple-gradient descent algorithm (MGDA) for multiobjective optimization. C.R. Math. **350**(5–6), 313–318 (2012)
6. García-Vila, M., Fereres, E.: Combining the simulation crop model AquaCrop with an economic model for the optimization of irrigation management at farm level. Eur. J. Agron. **36**(1), 21–31 (2012)
7. Gasanov, M., et al.: Sensitivity analysis of soil parameters in crop model supported with high-throughput computing. In: Krzhizhanovskaya, V.V., et al. (eds.) ICCS 2020. LNCS, vol. 12143, pp. 731–741. Springer, Cham (2020). https://doi.org/10.1007/978-3-030-50436-6_54
8. Giltrap, D.L., Li, C., Saggar, S.: DNDC: a process-based model of greenhouse gas fluxes from agricultural soils. Agric. Ecosyst. Environ. **136**(3–4), 292–300 (2010)
9. Gleick, P.H.: Water in crisis. Pacific Institute for Studies in Development, Environment & Security. Stockholm Environment Institute, Oxford University Press, 473p **9**, 1051-0761 (1993)
10. Holzworth, D.P., et al.: APSIM-evolution towards a new generation of agricultural systems simulation. Environ. Model. Softw. **62**, 327–350 (2014)
11. Hsiao, T.C., Steduto, P., Fereres, E.: A systematic and quantitative approach to improve water use efficiency in agriculture. Irrig. Sci. **25**(3), 209–231 (2007)
12. Jones, J.W., et al.: The DSSAT cropping system model. Eur. J. Agron. **18**(3–4), 235–265 (2003)
13. Katrutsa, A., Merkulov, D., Tursynbek, N., Oseledets, I.: Follow the bisector: a simple method for multi-objective optimization. arXiv preprint arXiv:2007.06937 (2020)
14. Kellogg, R.L., Nehring, R.F., Grube, A., Goss, D.W., Plotkin, S.: Environmental indicators of pesticide leaching and runoff from farm fields. In: Ball, V.E., Norton, G.W. (eds.) Agricultural Productivity, pp. 213–256. Springer, Boston (2002). https://doi.org/10.1007/978-1-4615-0851-9_9
15. Kropp, I., et al.: A multi-objective approach to water and nutrient efficiency for sustainable agricultural intensification. Agric. Syst. **173**, 289–302 (2019)
16. Lal, R.: Food security in a changing climate. Ecohydrol. Hydrobiol. **13**(1), 8–21 (2013)
17. Mbava, N., Mutema, M., Zengeni, R., Shimelis, H., Chaplot, V.: Factors affecting crop water use efficiency: a worldwide meta-analysis. Agric. Water Manag. **228**, 105878 (2020)
18. Morison, J., Baker, N., Mullineaux, P., Davies, W.: Improving water use in crop production. Philos. Trans. R. Soc. B Biol. Sci. **363**(1491), 639–658 (2008)
19. Mueller, N.D., Gerber, J.S., Johnston, M., Ray, D.K., Ramankutty, N., Foley, J.A.: Closing yield gaps through nutrient and water management. Nature **490**(7419), 254–257 (2012)
20. Nendel, C., et al.: The MONICA model: testing predictability for crop growth, soil moisture and nitrogen dynamics. Ecol. Model. **222**(9), 1614–1625 (2011)
21. Pandey, A., Ostrowski, M., Pandey, R.: Simulation and optimization for irrigation and crop planning. Irrig. Drain. **61**(2), 178–188 (2012)
22. Pretty, J.: Agricultural sustainability: concepts, principles and evidence. Philos. Trans. R. Soc. B: Biol. Sci. **363**(1491), 447–465 (2008)
23. Richardson, L.F.: Weather Prediction by Numerical Process. Cambridge University Press, Cambridge (2007)

24. Sparks, A.H.: nasapower: a NASA POWER global meteorology, surface solar energy and climatology data client for R. J. Open Source Softw. **3**(30), 1035 (2018)
25. Steduto, P., Hsiao, T.C., Raes, D., Fereres, E.: AquaCrop-the FAO crop model to simulate yield response to water: I. concepts and underlying principles. Agron. J. **101**(3), 426–437 (2009)
26. Van Diepen, C.V., Wolf, J., Van Keulen, H., Rappoldt, C.: WOFOST: a simulation model of crop production. Soil Use Manag. **5**(1), 16–24 (1989)
27. Wallace, J.S., Gregory, P.J.: Water resources and their use in food production systems. Aquat. Sci. **64**(4), 363–375 (2002)
28. Wierzbicki, A.P.: The use of reference objectives in multiobjective optimization. In: Fandel, G., Gal, T. (eds.) Multiple Criteria Decision Making Theory and Application, pp. 468–486. Springer, Heidelberg (1980). https://doi.org/10.1007/978-3-642-48782-8_32
29. Wierzbicki, A.P.: A mathematical basis for satisficing decision making. Math. Model. **3**(5), 391–405 (1982)
30. de Wit, A., et al.: 25 years of the WOFOST cropping systems model. Agric. Syst. **168**, 154–167 (2019)
31. Withers, P.J., Neal, C., Jarvie, H.P., Doody, D.G.: Agriculture and eutrophication: where do we go from here? Sustainability **6**(9), 5853–5875 (2014)
32. Yousefi, M., Banihabib, M.E., Soltani, J., Roozbahani, A.: Multi-objective particle swarm optimization model for conjunctive use of treated wastewater and groundwater. Agric. Water Manag. **208**, 224–231 (2018)
33. Yu, T., Kumar, S., Gupta, A., Levine, S., Hausman, K., Finn, C.: Gradient surgery for multi-task learning. arXiv preprint arXiv:2001.06782 (2020)

Computational Optimization, Modelling and Simulation

Expedited Trust-Region-Based Design Closure of Antennas by Variable-Resolution EM Simulations

Slawomir Koziel[1,2](✉) (iD), Anna Pietrenko-Dabrowska[2] (iD), and Leifur Leifsson[3] (iD)

[1] Engineering Optimization and Modeling Center, School of Science and Engineering, Reykjavík University, Menntavegur 1, 101, Reykjavík, Iceland
koziel@ru.is

[2] Faculty of Electronics Telecommunications and Informatics, Gdansk University of Technology, Narutowicza 11/12, 80-233 Gdansk, Poland
anna.dabrowska@pg.edu.pl

[3] Department of Aerospace Engineering, Iowa State University, Ames, IA 50011, USA
leifur@iastate.edu

Abstract. The observed growth in the complexity of modern antenna topologies fostered a widespread employment of numerical optimization methods as the primary tools for final adjustment of the system parameters. This is mainly caused by insufficiency of traditional design closure approaches, largely based on parameter sweeping. Reliable evaluation of complex antenna structures requires full-wave electromagnetic (EM) analysis. Yet, EM-driven parametric optimization is, more often than not, extremely costly, especially when global search is involved, e.g., performed with population-based metaheuristic algorithms. Over the years, numerous methods of lowering these expenditures have been proposed. Among these, the methods exploiting variable-fidelity simulations started gaining certain popularity. Still, such frameworks are predominantly restricted to two levels of fidelity, referred to as coarse and fine models. This paper introduces a reduced-cost trust-region gradient-based algorithm involving variable-resolution simulations, in which the fidelity of EM analysis is selected from a continuous spectrum of admissible levels. The algorithm is launched with the coarsest discretization level of the antenna under design. As the optimization process converges, for reliability reasons, the model fidelity is increased to reach the highest level at the final stage. The proposed algorithm allows for a significant reduction of the computational cost (up to sixty percent with respect to the reference trust-region algorithm) without compromising the design quality, which is corroborated by thorough numerical experiments involving four broadband antenna structures.

Keywords: Antenna design · EM-driven optimization · Gradient search · variable-resolution simulations · Model management

1 Introduction

The use of full-wave electromagnetic (EM) simulations have become a commonplace in the design of modern antenna structures. This is primarily because EM analysis is the

© Springer Nature Switzerland AG 2021
M. Paszynski et al. (Eds.): ICCS 2021, LNCS 12745, pp. 91–104, 2021.
https://doi.org/10.1007/978-3-030-77970-2_8

only tool capable of rendering reliable evaluation of increasingly complex designs, as well as accounting for mutual coupling effects, or the presence of environmental components (e.g., connectors) or nearby devices. In recent years, the emergence and rapid development of new application areas, e.g., the internet of things (IoT) [1], 5G wireless communications [2], wearable [3] or implantable devices [4], increased the complexity of antenna topologies even further. Such intricate designs, described by large numbers of optimizable variables, can no longer be tuned by means of supervised parameter sweeping. This enforces the employment of numerical optimization algorithms for antenna parameter refinement. Still, EM-driven optimization is unavoidably associated with high computational expenditures, which may turn unacceptable. Even the cost of the local gradient search [5, 6] may be sizeable. As far as global optimization is concerned, the number of required EM simulations required by the optimizer to converge may exceed several thousand [7], which significantly hinders applicability of popular procedures such as differential evolution [8, 9] evolutionary algorithms [10, 11], or particle swarm optimizers [12, 13].

A considerable research effort has been directed towards mitigation of the forenamed issues. One distinguishable group of techniques are algorithmic enhancements, in which gradient-based procedures are sped up by dedicated mechanisms. These include sparse Jacobian updates based on design relocation monitoring [14], sensitivity variation tracking [15], the employment of updating formulas [16], or a combination thereof [17]. Another option is the incorporation of adjoint sensitivities [18]. An entirely different approach is offered by surrogate-assisted frameworks, where a fast representation of the system at hand (referred to as a surrogate or a metamodel) replaces expensive EM simulations when making predictions about possibly improved parameter sets. The surrogates may be data-driven (e.g., kriging [19], radial basis functions [20], Gausssian process regression [21], or neural networks [22], to name but a few), or physics-based [23]. Nevertheless, in antenna design, the practical usage of data-driven models is hampered to a large extent by the curse of dimensionality, as well as significant nonlinearity of antenna characteristics. In consequence, the verification case studies reported in available literature typically feature two to just six adjustable parameters [24, 25].

Physics-based models may not be as popular as data-driven ones, still, they seem to be an attractive alternative in situations, where setting up the latter is hardly possible. Physics-based models incorporate the problem-specific knowledge in the form of low-fidelity representations of the system at hand, e.g., equivalent circuits [26], or coarse-mesh EM simulations [27]. To construct the surrogate, the low-fidelity model undergoes a correction (or enhancement) using a relatively small number of high-fidelity data samples. Popular optimization methods incorporating physics-based surrogates include space mapping [28], adaptive response scaling [29], and feature-based technology [30].

Computational efficiency of the optimization procedures may be enhanced by the employment of variable-fidelity simulations. In practice, however, two levels of model discretization are typically used. This paper introduces a novel optimization framework involving variable-resolution EM simulations embedded into the trust-region gradient search algorithm. The discretization levels are selected from the predefined spectrum, ranging from the lowest fidelity (still of a practical utility) to that ensuring sufficiently accurate representation of the antenna characteristics. The specific fidelity level is decided upon based on the optimization process convergence status, as well as the improvement of the objective function value in the consecutive iterations. At the beginning of the entire process, the coarsest discretization level is adopted, which enables a relatively cheap exploitation of the knowledge about the antenna under study, and expedited exploration of the parameter space. The reliability of the design process is ensured by gradually increasing the model fidelity, eventually reaching the finest assumed resolution. Our methodology is demonstrated using a benchmark set of four broadband antennas, all optimized to improve their impedance matching. The efficacy of the approach is compared to the standard trust-region procedure, and several state-of-the art accelerated techniques. The computational savings are around sixty percent with respect to the reference, without significantly deteriorating the design quality.

2   Antenna Design by Variable-Resolution EM Simulations

This section delineates the proposed optimization framework exploiting variable-resolution simulations, in which the model fidelity is adjusted based on the convergence status of the optimization process. In the initial stage of our procedure, the antenna discretization density is set to the lowest value from the predefined spectrum of admissible levels. Next, it is gradually increased as the optimization process converges. The section is organised as follows. The antenna optimization task is formulated in Sect. 2.1. Section 2.2 provides a brief description of the conventional trust-region algorithm with numerical derivatives. The description of variable-resolution simulation models (Sect. 2.3), as well as the overall optimization framework (Sect. 2.4) concludes the section.

2.1   Simulation-Driven Antenna Design

Design closure refers to the stage of the antenna development process in which its topology has been already established, and the geometry parameters ensuring the best achievable performance are to be identified. This tuning requires a definition of a suitable metric quantifying the design quality. Toward this end, we employ a merit function $U(x)$, where x refers to the vector of antenna designable parameters. The optimization task is formulated as

$$x^* = \arg \min_{x} U(x) \qquad (1)$$

subject to the inequality $g_k(x) \leq 0$, $k = 1, ..., n_g$, and equality constraints $h_k(x) = 0$, $k = 1, ..., n_h$. The definition of the objective function U reflects the design goals so that its lower values correspond to better designs. Here, we are interested in minimizing the antenna in-band reflection. Hence, the adopted merit function takes the following form $U(x) = S(x) = \max\{f \in F : |S_{11}(x,f)|\}$, where f denotes a frequency from the intended antenna operating range, and S_{11} denotes the (complex) reflection coefficient. As evaluation of the constraints corresponding to antenna electrical and/or field properties involves full-wave simulation, handling them explicitly is not straightforward. A convenient way of dealing with them is offered by a penalty function approach [31], where the design closure task is reformulated as follows:

$$x^* = \arg\min_x U_P(x), \quad U_P(x) = U(x) + \sum_{k=1}^{n_g+n_h} \beta_k c_k(x) \tag{2}$$

In (2), U_P is a sum of the original merit function U and the penalty terms. The violation of each constraint is quantified by the factor $c_k(x)$, with β_k being the penalty coefficients.

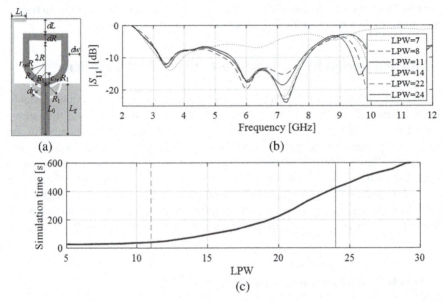

(a) (b) (c)

Fig. 1. Variable-resolution simulations: (a) a broadband monopole antenna: (b) the family of the antenna reflection responses simulated at various discretization levels (parametrized using the LPW parameter); (c) the dependence of the simulation time on LPW (averaged over several designs). The high- (—) and the minimum usable low-fidelity model (- - -) mesh densities are marked with vertical lines.

2.2 Trust-Region Gradient-Based Search

The core of our framework is a conventional trust-region (TR) algorithm [32], which is briefly recollected below. The optimization problem (1) (or (2), if the penalty function approach is applied) is solved in a local sense. During this process a series of approximations $x^{(i)}$, $i = 0, 1, \ldots$, to the optimum solution x^* is yielded. Each consecutive approximation $x^{(i)}$ of x^* is obtained by optimizing the linear expansion model $U_L^{(i)}$ of the relevant antenna characteristics at the current iteration point $x^{(i)}$. Here, we consider the antenna reflection response S_{11}, therefore we have

$$S_L^{(i)}(x,f) = S_{11}(x^{(i)},f) + \mathbf{G}_S(x^{(i)},f) \cdot (x - x^{(i)}) \tag{3}$$

In (3), the gradient of S_{11} at $x^{(i)}$ is denoted as $G_S(x^{(i)},f)$. $U_L^{(i)}$ is defined similarly as U_P (except for $S_L^{(i)}(x)$ replacing S_{11}). Other types of responses (e.g., gain), are handled in a similar manner. The approximations to the optimum solution are rendered as:

$$x^{(i+1)} = \arg \min_{x; \; -d^{(i)} \leq x - x^{(i)} \leq d^{(i)}} U_L^{(i)}(x) \tag{4}$$

The gradient G_S is typically estimated through finite differentiation (FD) at the cost of n additional EM analyses, where n stands for the antenna parameter number. The sub-problem (6) is solved within the interval $[x^{(i)} - d^{(i)}, x^{(i)} + d^{(i)}]$, referred to as the trust region. Here, the initial size vector $d^{(0)}$ is made proportional to the bounds on the design variables so as to avoid variable scaling, and to allow for similar treatment of variables of significantly different ranges. This is because the antenna geometry parameters may range from less than a millimeter in the case of gaps up to as much as tens of millimeters, when the dimensions of the ground plane are considered. The improvement in the objective function value for the candidate design found by (6) leads to its acceptance. Otherwise, the iteration is repeated with a reduced TR vector size [32].

The basic version of the TR algorithm requires performing full FD update of the antenna sensitivities in each algorithm iteration. Recently, several expedited variations of TR algorithm have been reported that employ sparse sensitivity updating schemes [14–16]. Each of these methods exploits a different mechanism allowing for omitting FD in certain cases, including a relative design relocation control [14], gradient changes monitoring [15], or selective Broyden updates [16]. The proposed procedure has been benchmarked against two of these methods, i.e., [14] and [16], along with the original TR algorithm with full-FD sensitivity update (see Sect. 3).

2.3 Variable-Resolution EM Simulations

In this work, the optimization process is accelerated by employing variable-resolution EM simulations. Variable-fidelity methods have been widely used for expediting the design of antenna structures [25, 29, 33]. Yet, they are typically limited to two surrogate levels: low- and high-fidelity models (or, in other words, coarse and fine ones). In antenna design, coarse models are most often based on coarse-mesh EM simulations [34], whereas in microwave engineering, equivalent circuits are frequently employed [27]. The coarse model, upon applying a suitable correction, is capable of rendering reliable predictions of the system output and can be used to find the approximate optimal solution of the fine model. Among the methods of this class, space mapping [25] and response correction techniques [29] may be listed as representative examples.

Reliability and computational efficiency of the variable-fidelity optimization framework strongly depend on the appropriate selection of underlying low-fidelity model [35]. Let us consider an example. The family of reflection characteristics simulated at several levels of antenna discretization for an ultra-wideband antenna is shown in Fig. 1. Here, the discretization level is parametrized with the use of the LPW (lines per wavelength), which is employed for mesh density control in CST Microwave Studio, the commercial software package used for antenna evaluation. Selecting too low LPW causes excessive discrepancies between the corresponding model response and that of the high-fidelity model, thereby, making the model unusable. Coarser discretization is advantageous from the point of view of computational savings, yet, it may lead to deterioration of the design quality. These factors have to be taken into consideration while adjusting suitable model discretization level.

Analysis of the antenna response family, such as that presented in Fig. 1(b), allows us to establish an admissible LPW range for the device under study: from L_{min} (corresponding to the coarsest, yet still practically useful discretization) up to L_{max} (ensuring accurate representation of the antenna output). We aim at expediting the optimization process by exploiting the simulation models from the said range $L_{min} \leq L \leq L_{max}$, thereby improving computational efficiency of the entire optimization process. The adopted model management scheme allowing for suitable adjustment of the discretization level has been described in Sect. 2.4.

2.4 Model Management Scheme

This section outlines the adopted model management scheme that governs the fidelity of the EM model throughout the optimization run based on its convergence status. The assumption is that the model resolution is set with the use of the sole parameter $L \in (L_{min}, L_{max})$, where L_{min} and L_{max} refer to the lowest acceptable and the fine discretization level, respectively. In our approach, the decision making scheme has been developed to satisfy the following requirements. For computational efficiency reasons, the optimization process should be initialized with the lowest admissible discretization level, thereby allowing for inexpensive exploitation of the problem-specific knowledge when searching for a better design. Whereas for reliability reasons, the last stages of the optimization procedure should be carried out at the highest discretization level. The intermediate discretization levels should be adjusted based on the algorithm convergence status: (i) $\|x^{(i+1)} - x^{(i)}\|$ (convergence in argument), and (ii) $U_P(x^{(i+1)}) - U_P(x^{(i)})$ (objective function improvement). To improve the stability of the optimization process, the transition between consecutive discretization levels should be as smooth as possible regarding the assumptions mentioned above.

In addition, the following algorithm termination conditions are applied (the algorithm terminates if either of them is satisfied): (i) $\|x^{(i+1)} - x^{(i)}\| < \varepsilon_x$; (ii) $\|d^{(i)}\| < \varepsilon_x$; and (iii) $|U_P(x^{(i+1)}) - U_P(x^{(i)})| < \varepsilon_U$. In the numerical experiments of Sect. 3, the aforementioned thresholds are set to $\varepsilon_x = \varepsilon_x = 10^{-3}$. We also define an auxiliary variable

$$Q^{(i)}(\varepsilon_x, \varepsilon_U) = \max \left\{ \frac{\varepsilon_x}{\|x^{(i+1)} - x^{(i)}\|}, \frac{\varepsilon_U}{|U_P(x^{(i+1)}) - U_P(x^{(i)})|} \right\} \tag{5}$$

The proposed convergence-based model management scheme works as follows: in the ith iteration, the discretization parameter $L^{(i)}$ is adjusted according to the following rule (which ensures discretization parameter monotonicity)

$$L^{(i+1)} = \begin{cases} L_{min} & \text{if } Q^{(i)}(\varepsilon_x, \varepsilon_U) \leq M \\ \max \left\{ L^{(i)}, L_{min} + (L_{max} - L_{min})[Q^{(i)}(\varepsilon_x, \varepsilon_U) - M]^{\frac{1}{\alpha}} \right\} \end{cases} \tag{6}$$

In the numerical experiments of Sect. 3, we adopt $M = 10^{-2}$ and $\alpha = 3$. Therefore, the initial increase of parameter L is rather quick, and it is launched two decades before the algorithm convergence (in terms of the norm-wise distance between the consecutive iteration points). As the simulation time strongly depends on the parameter L, the former seems to be reasonable: it brings substantial computational savings at the beginning of the entire process without significant detriment to its accuracy.

In order to ensure that at the end of the optimization run the EM model is evaluated at the highest discretization level, a safeguard mechanism is implemented enforcing that $L^{(i+1)}$ ultimately reaches L_{max}. This is because the sole use of the formula (6) does not ensure eventual switching to L_{max}, e.g., in the case of unsuccessful iterations causing the TR size vector to get smaller than the termination threshold and premature termination of the algorithm. Upon algorithm termination the following condition is applied

$$\text{IF } L^{(i)} < L_{max} \text{ THEN } L^{(i+1)} = L_{max} \quad \text{AND} \quad d^{(i+1)} = M_d d^{(i)} \frac{\varepsilon_x}{||d^{(i)}||} \quad (7)$$

In our experiments, we adopt multiplication factor $M_d = 10$. Thus, the termination condition is bypassed by (9), and subsequent iterations are carried out with $L^{(i+1)} = L_{max}$. Clearly, if the value L_{max} has been already reached, the above safeguard mechanism is not triggered.

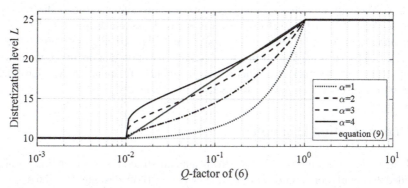

Fig. 2. Discretization level profiles for several values of the control parameter α (cf. (8)) along with a piece-wise profile (cf. (9)). The plots have been created for $L_{min} = 10$, $L_{max} = 25$, and $M = 10^{-2}$.

Let us now discuss the selection of the control parameter α. Several profiles of the discretization levels for various values of α are shown in Fig. 2, and the additional piece-wise linear profile from L_{min} to L_{max} described by the parameter-less equation

$$L^{(i+1)} = \begin{cases} L_{min} \text{ if } Q^{(i)}(\varepsilon_x, \varepsilon_U) \leq M \\ \max\left\{L^{(i)}, L_{min} + (L_{max} - L_{min})\left[1 - \frac{\log(Q^{(i)}(\varepsilon_x,\varepsilon_U))}{\log M}\right]\right\} \end{cases} \quad (8)$$

Observe that (10) is approximated most accurately by $\alpha = 3$ (which also more flexible).

In our approach, in order to speed-up the optimization process, the evaluation of the antenna response sensitivities is carried out at a lower fidelity level assessed as $L_{FD} = \max\{L_{min}, \lambda L^{(i)}\}$, where $0 \leq \lambda \leq 1$ is an algorithm control parameter (set to $\lambda = 2/3$ in numerical experiments of Sect. 3). This additional acceleration mechanism capitalizes on the fact that the models of different fidelities are typically well correlated even though they might be misaligned, and this correlation improves with an increase in $L^{(i)}$). This allows for rendering antenna response gradients in a reliable manner.

2.5   Optimization Framework with Variable-Resolution EM Simulations

The proposed optimization algorithm utilizes, as the search engine, the trust-region routine recollected in Sect. 2.2 in conjunction with the convergence-based model management scheme outlined in Sect. 2.4. The algorithm control parameters include: (i) ε_x, ε_U – termination thresholds; (ii) M – the threshold for launching discretization level increase; (iii) α – control parameter governing discretization level profile; (iv) λ – control parameter for L_{FD} adjustment (estimation of the antenna gradient); (v) M_d –multiplication factor serving to increase the TR size in (8), when closer to convergence.

The following general rules for setting the values of the aforementioned parameters apply. The termination thresholds ε_x and ε_U are to be set by the user so as to reflect the assumed resolution level of the optimization procedure. The remaining parameters are set to their default values $M = 10^{-2}$, $\alpha = 3$, $\lambda = 2/3$, and $M_d = 10$ (as elaborated on in Sect. 2.4). The extreme values L_{min} and L_{max} of the admissible spectrum of discretization levels (see Sect. 2.3) are to be selected by the user based on visual inspection of the family of antenna responses. L_{min} should be the lowest yet still practically useful discretization level, i.e., such that is capable of properly yielding all important details of the antenna response (e.g., its resonances). Whereas L_{max} should correspond to the high-fidelity model whose discretization level ensures sufficient accuracy.

3   Demonstration Examples

The benchmark set comprises four broadband antenna structures shown in Fig. 3. Table 1 provided the details concerning their parameter vectors (both designable and fixed ones), as well as the description of the substrate each structure is implemented on [36–39]. All antennas are evaluated using the time-domain solver of CST Microwave Studio; all simulation models incorporate the SMA connectors. The intended operating frequency range is 3.1 GHz to 10.6 GHz. The design goal has been defined as minimization of the maximum in-band reflection within UWB band. The objective function is defined as $U(x) = \max\{3.1\,\text{GHz} \leq f \leq 10.6\,\text{GHz}: |S_{11}(x,f)|\}$. Table 1 also gathers the ranges of the admissible values of lines-per-wavelength (LPW) parameter (see Sect. 2.3) for all antennas, along with the corresponding simulation times. The ranges are defined by L_{min} (the lowest practically usable discretrization), and L_{max} (the value for high-fidelity model).

The proposed optimization algorithm is benchmarked against the following three procedures: (i) the conventional TR algorithm [32] (Algorithm 1), (ii) the accelerated TR version [14] (Algorithm 2), in which sparse sensitivity updating scheme is based on relative design relocation, as well as (iii) the expedited version reported in [16] (Algorithm 3), employing the Broyden formula instead of FD for the selected variables. All benchmark procedures utilize solely high-fidelity EM simulations, i.e., the discretization level in each iteration equals $L^{(i)} = L_{max}$ (for the respective antennas).

In Algorithm 2, proposed in [14], some of FD-based sensitivity updates are omitted for the variables that exhibit small relative change with respect to the current TR region size. In addition, the optimization history is monitored so that to ensure that the relevant portion of the sensitivity matrix is updated through FD once in few iterations. The algorithm control parameter N defines the maximum allowable number of update-free iterations. Here, the numerical results have been obtained for $N = 3$. For a more detailed account of Algorithm 2, see [14]. The acceleration mechanism of the Algorithm 3 consists in replacing FD a rank-one Broyden formula (BF) for the selected variables whose directions are aligned well enough with the most recent design relocation. The alignment threshold is the algorithm control parameter: as it increases, BF is applied less frequently, thereby, enforcing more frequent FD updates and possibly leads to design quality enhancement. More details on Algorithm 3 can be found in [16].

All the considered algorithms are the local procedures and, in general, the presented optimization tasks are multimodal. Therefore, carrying out the search from different initial designs typically yields distinct local optima. Multimodality is mainly caused by a parameter redundancy occurring for the considered antenna structures, which, in turn, is a result of the modifications introduced to their geometries aiming at size reduction.

The average performance of all the algorithms is assessed with the use of the following factors: (i) computational efficiency expressed in terms of the number of equivalent EM evaluations, (ii) design quality estimated as the average value of the objective function across the performed optimization runs, and (iii) result repeatability quantified by standard deviation of the objective function values across the entire set. As a consequence of the problem multimodality, the standard deviation is non-zero even for the reference TR algorithm (Algorithm 1), being presumably the most reliable procedure of the entire benchmark set. Therefore, the observed deterioration of the design repeatability should be compared with that of Algorithm 1.

Table 1. Verification structures

An-tenna	Geometry parameters [mm]	Substrate	Lowest-fidel. model L_{min}	Simulation time [s]	High-fidelity model L_{min}	Simulation time [s]
I [38]	$x = [l_0 \ g \ a \ l_1 \ l_2 \ w_1 \ o]^T$ $w_0 = 2o + a, \ w_f = 1.7$	RF-35: $\varepsilon_r = 3.5$ $h = 0.76$ mm	10	42	21	150
II [39]	$x = [L_0 \ dR \ R \ r_{rel} \ dL \ dw \ Lg \ L_1 \ R_1 \ dr \ c_{rel}]^T$ $w_0 = 1.7$	RF-35: $\varepsilon_r = 3.5$ $h = 0.76$ mm	11	41	24	424
III [40]	$x = [L_g \ L_0 \ L_s \ W_s \ d \ dL \ d_s \ dW_s \ dW \ a \ b]^T$ $W_0 = 3.0$	FR4: $\varepsilon_r = 4.3$ $h = 1.55$ mm	10	46	20	265
IV [41]	$x = [L_0 \ L_1 \ L_2 \ L \ dL \ L_g \ w_1 \ w_2 \ w \ dw \ L_s \ w_s \ c]^T$ $w_0 = 1.7$	RO4350: $\varepsilon_r = 3.48$ $h = 0.76$ mm	10	37	25	97

The values of the control parameters of the proposed procedure have been set to (see also Sect. 2.4): $M = 10^{-2}$ (discretization level increase threshold), $\alpha = 3$ (constrol parameter for adjusting discretization level profile), $\lambda = 2/3$ (FD discretization level

Table 2. Antenna I and II: optimization results and benchmarking

Algorithm	Antenna I					Antenna II				
	Cost[1]	Cost savings[2]	max $\lvert S_{11}\rvert$[3]	Δ max $\lvert S_{11}\rvert$[4]	std max $\lvert S_{11}\rvert$[5]	Cost[1]	Cost savings[2]	max $\lvert S_{11}\rvert$[3]	Δ max $\lvert S_{11}\rvert$[4]	std max $\lvert S_{11}\rvert$[5]
1	97.6	–	−11.9	–	0.4	111.2	–	−14.9	–	0.6
2 [14]	45.1	53.8 %	−11.1	0.8	1.0	58.3	48%	−13.7	1.2	1.3
3 [16]	53.0	46 %	−10.7	1.2	2.7	75.9	32%	−14.3	0.6	1.0
This work	48.2	51 %	−11.2	0.7	0.7	25.8	77%	−13.8	1.1	1.0

[1] Number of equivalent high-fidelity EM simulations (averaged over 10 algorithm runs).
[2] Relative computational savings in percent w.r.t. the reference TR algorithm.
[3] Objective function value (maximum in-band reflection in dB), averaged over 10 algorithm runs.
[4] Degradation of max$\lvert S_{11}\rvert$ w.r.t. the reference algorithm in dB, averaged over 10 algorithm runs.
[5] Standard deviation of max$\lvert S_{11}\rvert$ in dB across the set of 10 algorithm runs.

Table 3. Antenna III and IV: optimization results and benchmarking

Algorithm	Antenna III					Antenna IV				
	Cost[1]	Cost savings[2]	max $\lvert S_{11}\rvert$[3]	Δ max $\lvert S_{11}\rvert$[4]	std max $\lvert S_{11}\rvert$[5]	Cost[1]	Cost savings[2]	max $\lvert S_{11}\rvert$[3]	Δ max $\lvert S_{11}\rvert$[4]	std max $\lvert S_{11}\rvert$[5]
1	111.0	–	−13.9	–	1.0	139.7	–	−17.6	–	1.2
2 [14]	73.1	34 %	−12.8	1.1	1.3	91.2	34 %	−16.3	1.3	2.5
3 [16]	80.0	28 %	−11.9	1.9	2.0	89.2	36 %	−15.1	2.5	2.6
This work	42.3	62 %	−11.3	2.6	1.0	97.2	31 %	−17.0	0.6	2.1

[1] Number of equivalent high-fidelity EM simulations (averaged over 10 algorithm runs).
[2] Relative computational savings in percent w.r.t. the reference TR algorithm.
[3] Objective function value (maximum in-band reflection in dB), averaged over 10 algorithm runs.
[4] Degradation of max$\lvert S_{11}\rvert$ w.r.t. the reference algorithm in dB, averaged over 10 algorithm runs.
[5] Standard deviation of max$\lvert S_{11}\rvert$ in dB across the set of 10 algorithm runs.

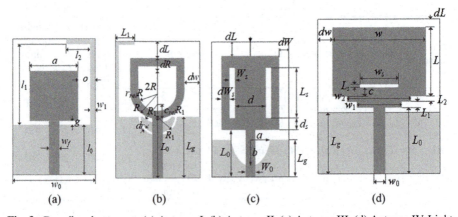

Fig. 3. Broadband antennas: (a) Antenna I, (b) Antenna II, (c) Antenna III, (d) Antenna IV. Light gray shade indicates ground plane.

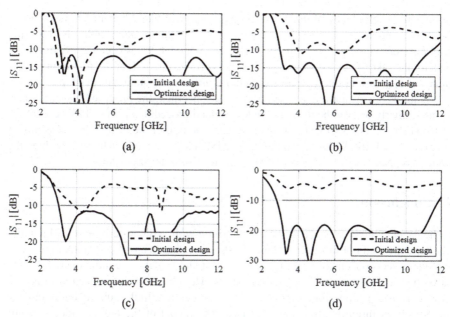

Fig. 4. Initial (- - -) and optimized (—) reflection responses of the benchmark antennas at the selected designs obtained using the proposed multi-fidelity framework: (a) Antenna I, (b) Antenna II, (c) Antenna III, (d) Antenna IV. Horizontal lines indicate design specifications.

L_{FD} control parameter), $M_d = 10$ (multiplication factor for increasing the TR size when closer to convergence). For all the algorithms, the following termination thresholds $\varepsilon_x = \varepsilon_U = 10^{-3}$ have been adopted.

The numerical results for Antennas I through IV have been obtained for ten independent algorithm runs starting from random initial designs, and gathered in Tables 2 and 3, respectively. The Tables include the optimization cost calculated as the equivalent number of high-fidelity antenna evaluations, the cost savings w.r.t. the reference TR algorithm, as well as the objective function value, its deterioration w.r.t. Algorithm 1 and the standard deviation. The comparison of the latter should take into account the value obtained for the conventional algorithm. The cost of the proposed algorithm employing variable-resolution EM simulations is computed based on the time evaluation ratios between the low- and the high-fidelity models. Figure 4 presents the selected reflection characteristics for the initial and optimized designs.

The results indicate that the proposed optimization framework outperforms the benchmark routines in terms of the computational efficiency for all the considered antenna structures. The average speedup exceeds 55 percent (from around 30 percent for Antenna IV up to almost 80 percent for Antenna II). The cost savings are comparable to that of the expedited versions [14] and [16] using solely high-fidelity simulations (in the case of Antennas I and IV), or are even considerably higher (in the case of Antennas II and III). This is a consequence of the evaluation ratios between the high- and lowest-fidelity models, which are larger for Antennas I and IV (10.3 and 5.8, respectively), and smaller for Antennas II and III (3.6 and 2.6, respectively). This implies that the computational efficiency of the proposed procedure will likely benefit from expanding the discretization level range. Actually, here, the maximum admissible discretization level L_{max} does not coincide with the high-fidelity model. It has been merely treated as such for the sake of consistency with the results provided in [14] and [16]. Setting larger values of L_{max} would likely result in reaching further computational speedup.

As far as the design quality degradation is concerned, the proposed algorithm performs similarly or even better than the accelerated procedures [14] or [16] (except for Antenna III). In our approach, the average deterioration of the design quality is minor, not exceeding 1 dB. Additionally, the obtained standard deviation values (describing the solution repeatability) is similar to that of the reference TR procedure.

4   Conclusion

In the paper, a novel trust-region-based algorithm with variable-resolution EM simulation for expedited optimization of antenna structures has been proposed. Our approach exploits a decision making routine, in which model fidelity is continuously adjusted based on the optimization convergence status. The algorithm is initiated with lowest (least expensive) admissible level of model discretization. In the subsequent iterations, the model resolution is gradually increased. The model of the highest fidelity is utilized only as the algorithm gets closer to the optimum, in order to ensure reliability of the entire process. This allows for achieving a significant computational speedup of around eighty percent (in comparison to the conventional trust-region routine) owing to a low-cost exploitation of the problem-specific knowledge embedded in models of lower discretization levels at early stages of the optimization run. Our methodology has been comprehensively validated using the benchmark set comprising four broadband antennas. The proposed framework also outperforms the recently reported accelerated trust-region-based routines exploiting sparse sensitivity updates in terms of the design quality. At the same time, the computational efficiency is comparable or even better. The future work will include introducing acceleration mechanisms similar to those employed in the benchmark techniques, which will possibly result in additional computational speedup.

Acknowledgement. The authors would like to thank Dassault Systemes, France, for making CST Microwave Studio available. This work was supported in part by the Icelandic Centre for Research (RANNIS) Grant 217771 and by National Science Centre of Poland Grant 2020/37/B/ST7/01448.

References

1. Jha, K.R., Bukhari, B., Singh, C., Mishra, G., Sharma, S.K.: Compact planar multistandard MIMO antenna for IoT applications. IEEE Trans. Antennas Propag. **66**(7), 3327–3336 (2018)
2. Nie, Z., Zhai, H., Liu, L., Li, J., Hu, D., Shi, J.: A dual-polarized frequency-reconfigurable low-profile antenna with harmonic suppression for 5G application. IEEE Antennas Wirel. Propag. Lett. **18**(6), 1228–1232 (2019)
3. Yan, S., Soh, P.J., Vandenbosch, G.A.E.: Wearable dual-band magneto-electric dipole antenna for WBAN/WLAN applications. IEEE Trans. Antennas Propag. **63**(9), 4165–4169 (2015)

4. Wang, J., Leach, M., Lim, E.G., Wang, Z., Pei, R., Huang, Y.: An implantable and conformal antenna for wireless capsule endoscopy. IEEE Antennas Wirel. Propag. Lett. **17**(7), 1153–1157 (2018)
5. Ohira, M., Miura, A., Taromaru, M., Ueba, M.: Efficient gain optimization techniques for azimuth beam/null steering of inverted-F multiport parasitic array radiator (MuPAR) antenna. IEEE Trans. Antennas Propag. **60**(3), 1352–1361 (2012)
6. Wang, J., Yang, X.S., Wang, B.Z.: Efficient gradient-based optimization of pixel antenna with large-scale connections. IET Microwaves Antennas Propag. **12**(3), 385–389 (2018)
7. Lalbakhsh, A., Afzal, M.U., Esselle, K.: Multiobjective particle swarm optimization to design a time-delay equalizer metasurface for an electromagnetic band-gap resonator antenna. IEEE Antennas Wirel. Propag. Lett. **16**, 912–915 (2017)
8. Goudos, S.K., Siakavara, K., Vafiadis, E., Sahalos, J.N.: Pareto optimal Yagi-Uda antenna design using multi-objective differential evolution. Prog. Electromagn. Res. **105**, 231–251 (2010)
9. Zaharis, Z.D., Gravas, I.P., Lazaridis, P.I., Glover, I.A., Antonopoulos, C.S., Xenos, T.D.: Optimal LTE-protected LPDA design for DVB-T reception using particle swarm optimization with velocity mutation. IEEE Trans. Antennas Propag. **66**(8), 3926–3935 (2018)
10. Oliveira, P.S., D'Assunção, A.G., Souza, E.A.M., Peixeiro, C.: A fast and accurate technique for FSS and antenna designs based on the social spider optimization algorithm. Microw. Opt. Technol. Lett. **58**, 1912–1917 (2016)
11. Haupt, R.L.: Antenna design with a mixed integer genetic algorithm. IEEE Trans. Antennas Propag. **55**(3), 577–582 (2007)
12. Lalbakhsh, A., Afzal, M.U., Esselle, K.P., Smith, S.L.: Wideband near-field correction of a Fabry-Perot resonator antenna. IEEE Trans. Antennas Propag. **67**(3), 1975–1980 (2019)
13. Zaharis, Z.D., et al.: Exponential log-periodic antenna design using improved particle swarm optimization with velocity mutation. IEEE Trans. Magn. **53**(6), 1–4 (2017)
14. Koziel, S., Pietrenko-Dabrowska, A.: Reduced-cost EM-driven optimization of antenna structures by means of trust-region gradient-search with sparse Jacobian updates. IET Microwaves Antennas Propag. **13**(10), 1646–1652 (2019)
15. Koziel, S., Pietrenko-Dabrowska, A.: Expedited feature-based quasi-global optimization of multi-band antennas with Jacobian variability tracking. IEEE Access. **8**, 83907–83915 (2020)
16. Koziel, S., Pietrenko-Dabrowska, A.: Expedited optimization of antenna input characteristics with adaptive Broyden updates. Eng. Comput. **37**(3), 851–862 (2019)
17. Koziel, S., Pietrenko-Dabrowska, A.: Reduced-cost design closure of antennas by means of gradient search with restricted sensitivity updates. Metrol. Meas. Syst. **26**(4), 595–605 (2019)
18. Koziel, S., Ogurtsov, S., Cheng, Q.S., Bandler, J.W.: Rapid EM-based microwave design optimization exploiting shape-preserving response prediction and adjoint sensitivities. IET Microwaves Antennas Propag. **8**(10), 775–781 (2014)
19. Hassan, A.K.S.O., Etman, A.S., Soliman, E.A.: Optimization of a novel nano antenna with two radiation modes using kriging surrogate models. IEEE Photonic J. **10**(4), 4800807 (2018)
20. Barmuta, P., Ferranti, F., Gibiino, G.P., Lewandowski, A., Schreurs, D.M.M.P.: Compact behavioral models of nonlinear active devices using response surface methodology. IEEE Trans. Microwave Theory Techn. **63**(1), 56–64 (2015)
21. Jacobs, J.P.: Characterization by Gaussian processes of finite substrate size effects on gain patterns of microstrip antennas. IET Microwaves Antennas Propag. **10**(11), 1189–1195 (2016)
22. Rawat, A., Yadav, R.N., Shrivastava, S.C.: Neural network applications in smart antenna arrays: a review. AEU Int. J. Electron. Commun. **66**(11), 903–912 (2012)
23. Toktas, A., Ustun, D., Tekbas, M.: Multi-objective design of multi-layer radar absorber using surrogate-based optimization. IEEE Trans. Microw. Theory Techn. **67**(8), 3318–3329 (2019)

24. Lv, Z., Wang, L., Han, Z., Zhao, J., Wang, W.: Surrogate-assisted particle swarm optimization algorithm with Pareto active learning for expensive multi-objective optimization. IEEE/CAA J. Automatica Sinica. **6**(3), 838–849 (2019)
25. Cervantes-González, J.C., Rayas-Sánchez, J.R., López, C.A., Camacho-Pérez, J.R., Brito-Brito, Z., Chávez-Hurtado, J.L.: Space mapping optimization of handset antennas considering EM effects of mobile phone components and human body. Int. J. RF Microwave CAE **26**(2), 121–128 (2016)
26. Bandler, J.W., et al.: Space mapping: the state of the art. IEEE Trans. Microwave Theory Techn. **52**(1), 337–361 (2004)
27. Koziel, S., Ogurtsov, S.: Antenna Design by Simulation-Driven Optimization. Surrogate-Based Approach. Springer, New York (2014)
28. Feng, F., Zhang, J., Zhang, W., Zhao, Z., Jin, J., Zhang, Q.J.: Coarse- and fine-mesh space mapping for EM optimization incorporating mesh deformation. IEEE Microwave Wireless Comput. Lett. **29**(8), 510–512 (2019)
29. Koziel, S., Unnsteinsson, S.D.: Expedited design closure of antennas by means of trust-region-based adaptive response scaling. IEEE Antennas Wirel. Propag. Lett. **17**(6), 1099–1103 (2018)
30. Koziel, S.: Fast simulation-driven antenna design using response-feature surrogates. Int. J. RF & Microwave CAE **25**(5), 394–402 (2015)
31. Ullah, U., Koziel, S., Mabrouk, I.B.: Rapid re-design and bandwidth/size trade-offs for compact wideband circular polarization antennas using inverse surrogates and fast EM-based parameter tuning. IEEE Trans. Antennas Propag. **68**(1), 81–89 (2019)
32. Conn, A.R., Gould, N.I.M., Toint, P.L.: Trust Region Methods, MPS-SIAM Series on Optimization (2000)
33. Su, Y., Li, J., Fan, Z., Chen, R.: Shaping optimization of double reflector antenna based on manifold mapping. In: International Applied Computational Electromagnetics Society Symposium International Applied Computational Electromagnetics Society Symposium. (ACES), Suzhou, China, pp. 1–2 (2017)
34. Bandler, J.W., et al.: Space mapping: the state of the art. IEEE Trans. Microwave Theory Tech. **52**(1), 337–361 (2004)
35. Koziel, S., Ogurtsov, S.: Model management for cost-efficient surrogate-based optimization of antennas using variable-fidelity electromagnetic simulations. IET Microwaves Antennas Propag. **6**(15), 1643–1650 (2012)
36. Koziel, S., Bekasiewicz, A.: Low-cost multi-objective optimization of antennas using Pareto front exploration and response features. In: International Symposium on Antennas and Propagation, Fajardo, Puerto Rico (2016)
37. Alsath, M.G.N., Kanagasabai, M.: Compact UWB monopole antenna for automotive communications. IEEE Trans. Antennas Propag. **63**(9), 4204–4208 (2015)
38. Haq, M.A., Koziel, S.: Simulation-based optimization for rigorous assessment of ground plane modifications in compact UWB antenna design. Int. J. RF Microwave CAE **28**(4), e21204 (2018)
39. Suryawanshi, D.R., Singh, B.A.: A compact UWB rectangular slotted monopole antenna. In: IEEE International Conference on Control, Instrumentation, Communication and Computational Technologies (ICCICCT), pp. 1130–1136 (2014)

Optimum Design of Tuned Mass Dampers for Adjacent Structures via Flower Pollination Algorithm

Sinan Melih Nigdeli[1(✉)], Gebrail Bekdaş[1], and Xin-She Yang[2]

[1] Department of Civil Engineering, Istanbul University-Cerrahpaşa, Avcılar, Istanbul, Turkey
{melihnig,bekdas}@istanbul.edu.tr
[2] Department of Design Engineering and Maths, Middlesex University, Middlesex, London, UK
x.yang@mdx.ac.uk

Abstract. It is a very known issue that tuned mass dampers (TMDs) on an effective system for structures subjected to earthquake excitations. TMDs can be also used as a protective system for adjacent structures that may pound to each other. With a suitable optimization methodology, it is possible to find an optimally tuned TMD that is effective in reducing the responses of structure with an additional protective feature that reduces the amount of required seismic gap between adjacent structures by using an objective function. This function considers the displacement of structures with respect to each other. As the optimization methodology, the flower pollination algorithm (FPA) is used in finding the optimum parameters of TMDs of both structures. The method was evaluated on two 10-story adjacent structures and the optimum results were compared with harmony search (HS) based methodology.

Keywords: Adjacent buildings · Optimization · Control · Tuned mass dampers · Flower pollination algorithm

1 Introduction

During the major strong earthquakes, one of the reasons for damage of structures is the pounding of building blocks. These damages may lead to termination of the use buildings by the occurred high damages that can be retrofitted or not. The worst-case that are observed in the historical earthquakes is the collapse of the building with fatalities. To avoid this danger, the regular way to protect the adjacent structures is to provide a seismic gap. Sometimes, this seismic gap cannot be provided in the required amount, or the effect that occurred in structures during earthquakes may be bigger than the expected amount.

According to Jeng and Tzeng [1], five major types of pounding existence as follows:

- Mid-column pounding
- Heavier adjacent building pounding
- Taller adjacent building pounding

M. Paszynski et al. (Eds.): ICCS 2021, LNCS 12745, pp. 105–115, 2021.
https://doi.org/10.1007/978-3-030-77970-2_9

– Eccentric pounding
– End building pounding

Mid-column pounding is the most seen case in the collapse of the structures after earthquakes. It is the pounding of the heavy story level of a building to the mid-point of the columns of the other building. The damage of slender columns is dangerous for the total collapse of the structures.

Heavier adjacent pounding is dangerous due to a heavy structure collide with a lighter adjacent one. It is a majorly seen type since the behavior of heavy and light structures can differ during earthquakes because of very different value of critical periods. The same behavior difference can be the same for one high-rise and low-rise adjacent building. This situation is called taller adjacent building pounding.

Due to torsional irregularity of structure, eccentric pounding may occur in a side of the structure due to the increasing effect of displacement of the corner points.

In the end, building pounding, series of structure blocks act as a series of colliding pendulums.

To prevent pounding several control methods have been proposed. These structural control types are passive, active, and semi-active or hybrid systems. As a passive system, nonlinear hysteric damper interconnecting adjacent structure [2], bumper-type collision shear walls [3], viscoelastic dampers connecting the adjacent structures [4], rubber shock absorbers [5], viscous damper with different retrofit schemes [6], passive damper [7] mass dampers [9] were presented. As semi-active systems, Magnetorheological (MR) damper [10, 11] and variable damping semi-active (VSDA) systems [12] were used. Kim and Kang developed a hybrid system by controlling the damping force of MR dampers [13]. The proposed active control system for the adjacent structures includes hydraulics actuators using linear quadratic Gaussian (LQG) controllers by Xu and Zhang [14] active control system using preference-based optimum design approach [15].

As a recent development, Guenidi et al. [16] proposed shared TMDs that are using passive or MR dampers as elements for connection to adjacent structures. Wu et al. [17] investigated adjacent inelastic reinforced concrete frame structures connected with viscous fluid dampers. Baili et al. [18] connected adjacent single degree of freedom (SDOF) systems with spring-dashpot-inerter control systems. Azimi and Yeznabad [19] investigated semi-active MR dampers for adjacent structures by proposing a swarm-baed parallel control algorithm. Also, Lin et al. [20] proposed a modified crow search algorithm-based fuzzy control for adjacent structures using MR dampers. Nigdeli and Bekdaş employed a hybrid harmony search algorithm to optimize adjacent structures using a vibration absorber system [21]. Wang et al. [22] proposed to link adjacent structures via tuned liquid column damper-inerter (TLCDI).

The optimization of all control systems is needed for the effective performance of the system. Especially, the systems for control of adjacent structure involves the consideration of the behavior of all structure or a complex model of linked structures. In that case, it will be a complex problem that can be solved via metaheuristic algorithms. For that reason, the seismic gap between these structures can be reduced by the use of optimum TMDs for both structures.

In this study, the flower pollination algorithm developed by Yang [23] is proposed for this control problem in the optimization of TMDs. In the optimization methodology, time-domain solutions are evaluated to consider an objective function that is the relative displacement of structure with respect to the other one. The methodology is applied to 10-story adjacent structures that have different behavior. The results of the method are compared with the harmony search-based methodology that is also applied for the same models by using the same numerical analysis criteria.

2 The Optimization Methodology

For structures under seismic excitations, it is needed to develop numerical iterations for the time-history analysis. Due to that, it is not possible to develop an equation of the response of the structure. The only and detailed way is to solve sets of coupled differential equations related to the motion of the structure that is also subjected to a ground acceleration due to an earthquake. The content of this excitation also includes different frequencies. Another factor is also related to the damping of the structure, and it is a restriction in finding a simple equation. With the increase of the degree of structural system, it is harder to solve the coupled equations. Since it is not possible to derive a simple equation, the case of finding the minimum of this equation for a set of design variables is not possible.

In the case of the present paper, two adjacent buildings will be investigated to add TMDs to the structure. By this passive control system, the objective is to reduce the maximum displacement of the structure with respect to the other. Via this reduction, the pounding of the structure blocks will be prevented, since the amount of the required seismic gap reduces. The objective function ($f(X)$) that is depending on the analysis results by considering the properties of TMDs as the design variables is shown in Eq. 1. From x_1 to x_N (x_1 to x_N for structure 2), the displacements of the stories of structure 1 are shown.

$$f(X) = \max\left(|[x_1 x_2 \ldots x_N]^T - [x_1 x_2 \ldots x_N]^T|\right) \tag{1}$$

The set of design variables are shown as Eq. 2 for i^{th} individual of the population (p) used in the optimization.

$$X_i = \{m_d \ T_d \ \xi_d \ m_d \ T_d \ \xi_d\} \quad i = 1 \ to \ p \tag{2}$$

The case study of adjacent structures with TMDs is shown in Fig. 1. In Table 1, the design constraints and variables related to the problem are listed. It must be noted that the italic symbols represent the response of the second structure.

Instead of stiffness (k_d, k_d) and damping coefficient (c_d, c_d), the periods (T_d, T_d) and damping ratios (ξ_d, ξ_d) of TMDs are respectively defined as Eq. 3 and Eq. 4 are considered as the design variables of the problem.

$$T_d = 2\pi \sqrt{\frac{m_d}{k_d}} \tag{3}$$

$$\xi_d = \frac{c_d}{2m_d \sqrt{\frac{k_d}{m_d}}} \tag{4}$$

In the optimization process, after the definition of the design constants and range of design variables, the initial solution matrix including sets of design variables is randomly generated. Afterward, the analysis of adjacent structures is done using Matlab with Simulink [24] to find the value of f(X) for all sets of variables for future comparison with the updated design variables by using algorithm rules.

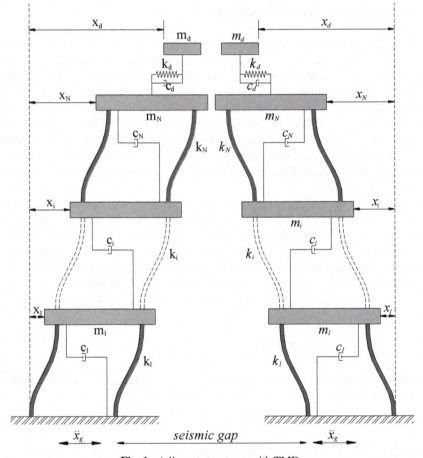

Fig. 1. Adjacent structures with TMDs

For the analysis, the equations of motion of N-story adjacent structures with TMDs on the top are written as Eq. 5.

$$M\ddot{x} + C\dot{x} + Kx = -M\{1\}\ddot{x}_g \tag{5}$$
$$M\ddot{x} + C\dot{x} + Kx = -M\{1\}\ddot{x}_g(t)$$

M, C and K (M, C and K for structure 2) are respectively mass (Eq. 6), damping (Eq. 7) and stiffness (Eq. 8) matrices. The vector of structural displacements (Eq. 9) is

Table 1. The design constants and design variables ranges

Symbol	Type	Definition
mi	Constant	Mass of the ith story for structure 1 (i = 1 to N)
c_i	Constant	Stiffness of the ith story for structure 1 (i = 1 to N)
k_i	Constant	Damping coefficient of the ith story for structure 1 (i = 1 to N)
m_i	Constant	Mass of the ith story for structure 2 (i = 1 to N)
c_i	Constant	Stiffness of the ith story for structure 2 (i = 1 to N)
k_i	Constant	Damping coefficient of the ith story for structure 2 (i = 1 to N)
$\ddot{x}_g$	Constant	Ground acceleration (defined as data)
m_d	Variable	Mass of TMD on the structure 1
T_d	Variable	Periods of TMD on the structure 1
ξ_d	Variable	Damping ratio of TMD on the structure 1
m_d	Variable	Mass of TMD on the structure 2
T_d	Variable	Periods of TMD on the structure 2
ξ_d	Variable	Damping ratio of TMD on the structure 2

shown with x and x for structures 1 and 2, respectively. {1} represents a vector of ones.

$$M = \text{diag}[m_1 m_2 \ldots m_N m_d] \quad M = \text{diag}[m_1 m_2 \ldots m_N m_d] \tag{6}$$

$$C = \begin{bmatrix} (c_1 + c_2) & -c_2 & & & & \\ -c_2 & (c_2 + c_3) & -c_3 & & & \\ & & \cdot & \cdot & & \\ & & \cdot & \cdot & \cdot & \\ & & & \cdot & \cdot & \cdot \\ & & & & -c_N & (c_N + c_d) & -c_d \\ & & & & & -c_d & c_d \end{bmatrix}$$

$$C = \begin{bmatrix} (c_1 + c_2) & -c_2 & & & & \\ -c_2 & (c_2 + c_3) & -c_3 & & & \\ & & \cdot & \cdot & & \\ & & \cdot & \cdot & \cdot & \\ & & & \cdot & \cdot & \cdot \\ & & & & -c_N & (c_N + c_d) & -c_d \\ & & & & & -c_d & c_d \end{bmatrix} \tag{7}$$

$$K = \begin{bmatrix} (k_1 + k_2) & -k_2 \\ -k_2 & (k_2 + k_3) & -k_3 \\ & & \cdot & \cdot \\ & & \cdot & \cdot & \cdot \\ & & & \cdot & \cdot & \cdot \\ & & & & -k_N & (k_N + k_d) & -k_d \\ & & & & & -k_d & k_d \end{bmatrix}$$

$$K = \begin{bmatrix} (k_1 + k_2) & -k_2 \\ -k_2 & (k_2 + k_3) & -k_3 \\ & & \cdot & \cdot \\ & & \cdot & \cdot & \cdot \\ & & & \cdot & \cdot & \cdot \\ & & & & -k_N & (k_N + k_d) & -k_d \\ & & & & & -k_d & k_d \end{bmatrix} \tag{8}$$

$$\mathrm{x} = [\mathrm{x_1 x_2 \ldots x_N x_d}]^T \quad x = [x_1 x_2 \ldots x_N x_d]^T \tag{9}$$

After the generation of an initial solution matrix, the iterative optimization process starts. FPA is a nature-inspired metaheuristic algorithm that imitates the process of pollen transfer for flowering plants. FPA includes two optimization types that are global and local optimization. These optimization phases are chosen according to a switch probability (sp).

The main idea of the pollination process of flowers to be used as an optimization algorithm is associated with flower constancy. It is the tendency of a specific pollinator to a specific flower type. Each solution of design variables is associated as a flower.

As a type of optimization phase, global optimization or namely, global pollination involves two types of pollination. These are biotic pollination and cross-pollination. In biotic pollination, the carry process of pollens is done via pollinators that are living organisms like bees, insects, etc. Cross-pollination is the pollination of different plants. Since these two types involve different plants and the pollen transfer is done in long distances, it is named global pollination. It is formulated as Eq. 10.

$$X_i^{t+1} = X_i^t + L(X_i^t - g^*) \tag{10}$$

In Eq. 10, a Levy distribution (L) shows the effect of the random flight of pollinators. Also, the best existing solution (g^*) is used in the generation of a new solution (X_i^{t+1} for i^{th} individual and $t + 1^{th}$ iteration) using the existing one (X_i^t).

In local optimization or namely local pollination, self-pollination and abiotic pollination are imitated. In self-pollination, the reproduction is done for the same plant as self-fertilization. In abiotic pollination, the carry of pollens is done by natural events like winds, diffusion in water, etc. It is formulated as Eq. 11.

$$X_i^{t+1} = X_i^t + \varepsilon(X_j^t - X_k^t) \tag{11}$$

Since it is local pollination, two existing solutions (X_j^t and X_k^t) are used with a linear distribution (ε) taken as a random number between 0 and 1.

The newly generated solution is checked for the objective function value, and the new ones are selected instead of the existing ones if the value of f(X) is smaller than the existing ones for the modified new solutions. This process continues for several iterations.

3 The Numerical Example

As the numerical validation of the method, two adjacent structure with ten stories were investigated. The properties of these structures are listed in Table 2. The first structure has 1 s critical period, and it has the same properties for all stories. The second structure has 2 s critical period, and it has different properties for all stories. The first structure has heavier masses than the second one. Also, the first structure is more rigid comparing to the second structure.

During the optimization, six different earthquake records shown in Table 3 are used. The excitation with the maximum objective function value is considered. These records were downloaded by Pacific Earthquake Engineering Research Center (PEER) [25] database.

Table 2. Properties of adjacent structures

Story	Structure 1 [26]			Structure 2 [27]		
	m_i (t)	k_i (kN/m)	c_i (kNs/m)	m_i (t)	k_i (kN/m)	c_i(kNs/m)
10	360	650000	6200	98	34310	442.599
9				107	37430	482.847
8				116	40550	523.095
7				125	43670	563.343
6				134	46790	603.591
5				143	49910	643.839
4				152	53020	683.958
3				161	56140	724.206
2				170	52260	674.154
1				179	62470	805.863

The optimum results are reported in Table 4 for HS and FPA, respectively. During the optimization, sp (harmony memory considering rate in HS) is taken as 0.5, and the optimization is done for a population number of 20. The ranges of T_d are selected between 0.8 and 1.2 times of the critical period of the uncontrolled structure. Also, the range for ξ_d is 0.01–0.20. The range of m_d is equal to 1%–5% of the total mass of the structures.

The maximum objective function value for uncontrolled structures is 1.4361 m and it occurs under the Rinaldi record of the 1994 Northridge earthquake. This value reduces

Table 3. The ground motions

Earthquake	Date	Station	Component	PGA (g)	PGV (cm/s)	PGD (cm)
Cape Mendocino	1992	Petrolia	PET090	0.662	89.7	29.55
Kobe	1995	0 KJMA	KJM000	0.821	81.3	17.68
Erzincan	1992	95 Erzincan	ERZ-NS	0.515	83.9	27.35
Northridge	1994	Rinaldi	RRS228	0.838	166.1	28.78
Northridge	1994	24514 Sylmar	SYL360	0.843	129.6	32.68
Loma Prieta	1989	16 LGPC	LGP000	0.563	94.8	41.18

Table 4. The optimum results

	FPA	HS
m_d (t)	180	180
m_d (t)	69.25	69.07
T_d (s)	0.90	0.92
T_d (s)	1.6175	1.87
ξ_d	0.01	0.02
ξ_d	0.01	0.04
f(X) (m)	1.1219	1.1366

to 1.2119 m and 1.1366 m for FPA and HS optimized TMDs, respectively. The optimal design obtained by the FPA is similar to that by HS, but FPA is more effective in reducing the displacement.

4 Conclusions

The optimum result for the reduction of the objective functions shows great differences for both algorithms. As known, the mass of TMD is maximum for the best design of TMD, but it is limited for the axial force capacity of the structure. In that situation, FPA is effective to find TMD with maximum allowed mass, while HS also finds the maximum for the first structure and a near maximum one for the second structure.

In Fig. 2, the objective function values for different excitations are given for all stories of the structure. Since the optimization is done for the maximum value, the value of the 10th story is the considered one for the optimization. The best effect of the optimum TMD system is seen for the critical excitation with the most effect on relative displacements of the structure. For the Rinaldi record of the Northridge earthquake, the objective function values reduce by 22% and 21% for FPA and HS, respectively.

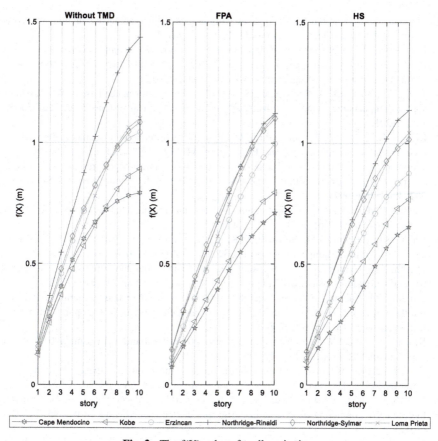

Fig. 2. The f(X) values for all excitations

As seen from the findings, FPA is more effective than HS for finding a precise value. It is also seen that TMD application reduces the value of the seismic gap required between the structures.

The performance of TMD is validated for all excitations that are used in this study, but the best performance is seen for the most critical excitation. In that situation, it can be said that TMDs for adjacent structures are most effective on the responses with the maximum effect.

References

1. Jeng, V., Tzeng, W.L.: Assessment of seismic pounding hazard for Taipei City. Eng. Struct. **22**, 459–471 (2000)
2. Ni, Y.Q., Ko, J.M., Ying, Z.G.: Random seismic response analysis of adjacent buildings coupled with non-linear hysteretic dampers. J. Sound Vib. **246**(3), 403–417 (2001)
3. Anagnostopoulos, S.A., Karamaneas, C.E.: Use of collision shear walls to minimize seismic separation and to protect adjacent buildings from collapse due to earthquake-induced pounding. Earthquake Eng. Struct. Dyn. **37**, 1371–1388 (2008)

4. Matsagar, V.A., Jangid, R.S.: Viscoelastic damper connected to adjacent structures involving seismic isolation. J. Civ. Eng. Manag. **11**(4), 309–322 (2005)
5. Polycarpou, P.C., Komodromos, P., Polycarpou, A.C.: A nonlinear impact model for simulating the use of rubber shock absorbers for mitigating the effects of structural pounding during earthquakes. Earthquake Eng. Struct. Dyn. **42**, 81–100 (2013)
6. Tubaldi, E., Barbato, M., Ghazizadeh, S.: A probabilistic performance-based risk assessment approach for seismic pounding with efficient application to linear systems. Struct. Saf. **36–37**, 14–22 (2012)
7. Bigdeli, K., Hare, W.: Tesfamariam S. Configuration optimization of dampers for adjacent buildings under seismic excitations. Eng. Optim. **44**, 1491–1509 (2012)
8. Trombetti, T., Silvestri, S.: Novel schemes for inserting seismic dampers in shear-type systems based upon the mass proportional component of the Rayleigh damping matrix. J. Sound Vib. **302**, 486–526 (2007)
9. Nigdeli, S.M., Bekdas, G.: Optimum tuned mass damper approaches for adjacent structures. Earthquakes Struct. **7**(6), 1071–1091 (2014)
10. Bharti, S.D., Dumne, S.M., Shrimali, M.K.: Seismic response analysis of adjacent buildings connected with MR dampers. Eng. Struct. **32**, 2122–2133 (2010)
11. Sheikh, M.N., Xiong, J., Li, W.H.: Reduction of seismic pounding effects of base-isolated RC highway bridges using MR damper. Struct. Eng. Mech. **41**(6), 791–803 (2012)
12. Cundumi, O., Suarez, L.E.: Numerical investigation of a variable damping semiactive device for the mitigation of the seismic response of adjacent structures. Comput. Aided Civil Infrastruct. Eng. **23**, 291–308 (2008)
13. Kim, G.C., Kang, J.W.: Seismic response control of adjacent building by using hybrid control algorithm of MR. Procedia Eng. **14**, 1013–1020 (2011)
14. Xu, Y.L., Zhang, W.S.: Closed-form solution for seismic response of adjacent buildings with linear quadratic Gaussian controllers. Earthquake Eng. Struct. Dyn. **31**, 235–259 (2002)
15. Park, K.-S., Ok, S.-Y.: Optimal design of actively controlled adjacent structures for balancing the mutually conflicting objectives in design preference aspects. Eng. Struct. **45**, 213–222 (2012)
16. Guenidi, Z., Abdeddaim, M., Ounis, A., Shrimali, M.K., Datta, T.K.: Control of adjacent buildings using shared tuned mass damper. Procedia Eng. **199**, 1568–1573 (2017)
17. Wu, Q.Y., Zhu, H.P., Chen, X.Y.: Seismic fragility analysis of adjacent inelastic structures connected with viscous fluid dampers. Adv. Struct. Eng. **20**(1), 18–33 (2017)
18. Basili, M., De Angelis, M., Pietrosanti, D.: Defective two adjacent single degree of freedom systems linked by spring-dashpot-inerter for vibration control. Eng. Struct. **188**, 480–492 (2019)
19. Azimi, M., Yeznabad, A.M.: Swarm-based parallel control of adjacent irregular buildings considering soil-structure interaction. J. Sens. Actuator Netw. **9**(2), 18 (2020)
20. Lin, X., Chen, S., Lin, W.: Modified crow search algorithm–based fuzzy control of adjacent buildings connected by magnetorheological dampers considering soil–structure interaction. J. Vibr. Control **27**(1–2), 57–72 (2020). https://doi.org/10.1177/1077546320923438
21. Nigdeli S.M., Bekdaş G.: Hybrid harmony search algorithm for optimum design of vibration absorber system for adjacent buildings. In: Nigdeli S.M., Kim J.H., Bekdaş G., Yadav A. (eds) Proceedings of 6th International Conference on Harmony Search, Soft Computing and Applications. ICHSA 2020. Advances in Intelligent Systems and Computing, vol 1275. Springer, Singapore (2021). https://doi.org/10.1007/978-981-15-8603-3_8
22. Wang, Q., Qiao, H., De Domenico, D., Zhu, Z., Tang, Y.: Seismic response control of adjacent high-rise buildings linked by the Tuned Liquid Column Damper-Inerter (TLCDI). Eng. Struct. **223**, 111169 (2020)

23. Yang, X.-S.: Flower pollination algorithm for global optimization. In: Durand-Lose, J., Jonoska, N. (eds.) UCNC 2012. LNCS, vol. 7445, pp. 240–249. Springer, Heidelberg (2012). https://doi.org/10.1007/978-3-642-32894-7_27
24. Mathworks, : MATLAB R2010a. The MathWorks Inc., Natick, MA, USA (2010)
25. Pacific Earthquake Engineering Research Center (PEER). Strong Ground Motion Databases. https://peer.berkeley.edu/peer-strong-ground-motion-databases
26. Singh, M.P., Matheu, E.E., Suarez, L.E.: Active and semi-active control of structures under seismic excitation. Earthquake Eng. Struct. Dynam. **26**(2), 193–213 (1997)
27. Sadek, F., Mohraz, B., Taylor, A.W., Chung, R.M.: A method of estimating the parameters of tuned mass dampers for seismic applications. Earthquake Eng. Struct. Dynam. **26**(6), 617–635 (1997)

On Fast Multi-objective Optimization of Antenna Structures Using Pareto Front Triangulation and Inverse Surrogates

Anna Pietrenko-Dabrowska[1] ⬡, Slawomir Koziel[2,1(✉)] ⬡, and Leifur Leifsson[3] ⬡

[1] Faculty of Electronics Telecommunications and Informatics,
Gdansk University of Technology, Narutowicza 11/12, 80-233 Gdansk, Poland
anna.dabrowska@pg.edu.pl
[2] Engineering Optimization & Modeling Center, School of Science and Engineering,
Reykjavík University, Menntavegur 1, 101 Reykjavík, Iceland
koziel@ru.is
[3] Department of Aerospace Engineering, Iowa State University, Ames, IA 50011, USA
leifur@iastate.edu

Abstract. Design of contemporary antenna systems is a challenging endeavor, where conceptual developments and initial parametric studies, interleaved with topology evolution, are followed by a meticulous adjustment of the structure dimensions. The latter is necessary to boost the antenna performance as much as possible, and often requires handling several and often conflicting objectives, pertinent to both electrical and field properties of the structure. Unless the designer's priorities are already established, multi-objective optimization (MO) is the preferred way of yielding the most comprehensive information about the best available design trade-offs. Notwithstanding, MO of antennas has to be carried out at the level of full-wave electromagnetic (EM) simulation models which poses serious difficulties due to high computational costs of the process. Popular mitigation methods include surrogate-assisted procedures; however, rendering reliable metamodels is problematic at higher-dimensional parameter spaces. This paper proposes a simple yet efficient methodology for multi-objective design of antenna structures, which is based on sequential identification of the Pareto-optimal points using inverse surrogates, and triangulation of the already acquired Pareto front representation. The two major benefits of the presented procedure are low computational complexity, and uniformity of the produced Pareto set, as demonstrated using two microstrip structures, a wideband monopole and a planar quasi-Yagi. In both cases, ten-element Pareto sets are generated at the cost of only a few hundreds of EM analyses of the respective devices. At the same time, the savings over the state-of-the-art surrogate-based MO algorithm are as high as seventy percent.

Keywords: Antenna systems · Electromagnetic simulation · Design optimization · Multi-objective design · Inverse modeling

© Springer Nature Switzerland AG 2021
M. Paszynski et al. (Eds.): ICCS 2021, LNCS 12745, pp. 116–130, 2021.
https://doi.org/10.1007/978-3-030-77970-2_10

1   Introduction

Design of modern antenna structures is a complex process involving several stages that include, among others, conceptual development, topology evolution (typically supported by parametric studies), as well as design closure, i.e., the final adjustment of antenna parameters. Nowadays, antenna geometries become more and more complex in order to meet the increasing performance requirements related to particular application areas such as wireless communications [1, 2], internet of things (IoT) [2], or wearable [4], and implantable devices [5]. More often than not, design specifications include additional functionalities such as multi-band operation [6], tunability [7], or circular polarization [8]. Proper tuning of antenna dimensions is instrumental in achieving the best possible performance, yet it is challenging. For reliability reasons, it has to be carried out using full-wave electromagnetic (EM) simulation tools, which entails considerable computational expenses.

The problem is exacerbated by the necessity of handling several objectives, which are typically conflicting so that the enhancement of one leads to degradation of others. A representative example are compact antennas where reduction of physical dimensions has detrimental effects on both electrical and field properties of the structure (e.g., [9]). Consequently, practical design requires identification of the trade-off solutions. This can only be achieved using numerical optimization techniques. However, conventional algorithms, both local (gradient-based [10], pattern search [11]), and population-based metaheuristics (e.g., differential evolution [12], particle swarm optimizers [13]) are only capable of processing scalar objectives. To allow the employment of conventional methods, multi-objective problems are often reformulated, using e.g., objective aggregation [14], or objective prioritization [15]. Rendering comprehensive information about available design trade-offs requires genuine multi-objective optimization (MO) [16]. Undoubtedly, the most popular solution approaches are MO versions of population-based metaheuristic algorithms, e.g., evolutionary algorithms [17], differential evolution [18], particle swarm optimization [19], and many others [20–23]. The advantage of population-based methods is that the Pareto set can be generated within a single algorithm run. However, the computational cost of these algorithms is high. As a matter of fact, it is normally prohibitive when executed at the level of EM simulations.

Nature-inspired procedures can be accelerated by incorporating surrogate modelling methods [24]. Unfortunately, due to high nonlinearity of antenna characteristics, the curse of dimensionality becomes the major obstacle so that construction of reliable surrogates within the entire parameter space of interest is only possible for structures described by few parameters [25]. This can be mitigated to a certain extent using machine learning approaches, where the initial surrogate is gradually refined using additional EM data acquired with the use of appropriate infill criteria [26]. Widely used modelling techniques include kriging [27], Gaussian process regression [28], and support vector regression [29]. Another possibility has been offered by constrained modelling procedures that limit the surrogate model domain to a relevant regions of the space (e.g., those containing the Pareto front in the case of MO problems) [30, 31]. All the aforementioned techniques are stochastic. Recently, deterministic surrogate-based MO frameworks have been developed as well, including point-by-point Pareto front exploration [32], generalized bisection [33], as well as sequential domain patching (SDP)

[34]. The most important advantage of deterministic algorithms is in eliminating the need to construct globally accurate surrogates (most operations are executed using local metamodels [35]).

This work discusses a novel deterministic surrogate-assisted framework for MO of antennas. The foundation of the presented approach is sequential generation of the Pareto-optimal designs using inverse surrogates and triangulation of the already available representation of the Pareto front. The framework is capable of handling any number of objectives, and allows for a rendition of uniformly distributed Pareto sets. Furthermore, it is computationally efficient, which is demonstrated using two antenna examples: a broadband monopole, and a planar Yagi antenna. In both cases, ten element sets of trade-off designs are obtained at the cost of only a few hundreds of EM simulations of the respective structures. It is also shown that our approach is competitive to state-of-the-art surrogate-assisted methods in terms of the CPU cost of the MO process, but also the uniformity of the obtained Pareto set.

2 Multi-objective Design of Antenna Structures Using Inverse Surrogates and Pareto Set Triangulation

The purpose of this section is to formulate the MO procedure being the subject of this work. We discuss the basic components of the algorithm with the emphasis on the inverse surrogate modeling and Pareto set triangulation, as a way of generating the initial designs to obtain additional trade-off solutions, further tuned using the customized local refinement procedure.

2.1 Antenna Design for Multiple Performance Figures. Problem Formulation

It is assumed that the antenna structure of interest is to be designed with respect to N_{obj} figures of interest (objectives), F_k, $k = 1, ..., N_{obj}$, all to be minimized. Here, the MO process is understood as identification of Pareto-optimal points [36] representing the best possible trade-offs between the considered objectives. The objective vector will be denoted as $F = [F_1 \ F_2 \ ... \ F_{Nobj}]^T$.

The antenna outputs (typically, frequency responses such as reflection coefficient, gain, etc.), are obtained by means of full-wave electromagnetic (EM) analysis. The aggregated vector of antenna characteristics is denoted as $R(x)$, where x stands for adjustable parameters (typically, antenna dimensions). As mentioned in Sect. 1, direct MO of antenna structures at the level of EM analysis tends to be expensive in computational terms.

2.2 Inverse Surrogate. Triangulation of Pareto-Optimal Solution Set

We will denote by $x^{(k)}$, $k = 1, ..., p$, the elements of the Pareto set identified by iteration k of the MO algorithm; $F^{(k)} = F(x^{(k)}) = [F_1^{(k)} \ ... \ F_{Nobj}^{(k)}]^T$ stand for the corresponding objective vectors. Among them, the first N_{obj} Pareto-optimal points are obtained by solving the single-objective tasks

$$x^{(k)} = \arg \min_{x \in X} F_k(R(x)) \tag{1}$$

These vectors determine the span of the Pareto front and are the basis to find the remaining solutions. The process is iterative, and involves triangulation of the existing set, as well as auxiliary inverse surrogate models.

Consider the inverse surrogate (metamodel) $s(\boldsymbol{F})$: $F \rightarrow X$, where F and X are the objective and parameter spaces of the MO problem, respectively. The training data to render the surrogate is $\{\boldsymbol{F}^{(k)}, \boldsymbol{x}^{(k)}\}_{k=1, ..., p}$. Note that s is referred to as inverse because its set of values is the parameter space of the antenna at hand. In other words, the surrogate makes predictions concerning the Pareto-optimal designs corresponding to the specific objective vectors $\boldsymbol{F}$. In contrast, typically considered forward models are used to predict antenna responses corresponding to specific parameter vectors $\boldsymbol{x}$. Here, the surrogate is set up using kriging interpolation [37].

The second tool utilized in the proposed methodology is triangulation of the Pareto set $\{\boldsymbol{F}^{(k)}\}_{k=1, ..., p}$, the result of which is a set of simplexes $\boldsymbol{S}^{(j)}, j = 1, ..., K_p$. In this work, the simplexes are considered in the objective space, and represented by vertices $\boldsymbol{S}^{(j)} = \{\boldsymbol{F}^{(j.1)}, ..., \boldsymbol{F}^{(j.Nobj)}\}$, where $\boldsymbol{F}^{(j.r)} \in \{\boldsymbol{F}^{(k)}\}_{k=1, ..., p}$, for $r = 1, ..., N_{obj}$. In order to avoid degenerate simplexes, Delaunay triangulation is employed [38].

2.3 Infill Points and Refinement Procedure

The set of simplexes $\boldsymbol{S}^{(j)}$ constructed in Sect. 2.2 can be considered as a partitioning of the current Pareto set. Additional Pareto-optimal points are found using a sequential sampling process as described below. Let $A(\boldsymbol{S}^{(j)})$ stand for the volume of $\boldsymbol{S}^{(j)}$, and

$$j_{\max} = \arg \max_{1 \leq j \leq N_{obj}} \{A(\boldsymbol{S}^{(j)})\} \tag{2}$$

be the index of the largest volume simplex. Using (2), the new objective vector $\boldsymbol{F}_{tmp}$ is established as

$$\boldsymbol{F}_{tmp} = \frac{1}{N_{obj}} \sum_{k=1}^{N_{obj}} \boldsymbol{F}^{(j_{\max}.k)} \tag{3}$$

More specifically, $\boldsymbol{F}_{tmp} = [F_{tmp.1} ... F_{tmp.Nobj}]^T$ is the centre of the simplex featuring the largest volume among the set $\boldsymbol{S}^{(j)}, j = 1, ..., K_p$.

At this point, the inverse surrogate s introduced in Sect. 2.2 is employed to find the representation $\boldsymbol{x}_{tmp}$ of $\boldsymbol{F}_{tmp}$ in the parameter space as

$$\boldsymbol{x}_{tmp} = s(\boldsymbol{F}_{tmp}) \tag{4}$$

An alternative initial design is also produced as the centre of $\boldsymbol{S}^{(j_{max})}$ in the parameter space, i.e.,

$$\boldsymbol{x}_{tmp.alt} = \frac{1}{N_{obj}} \sum_{k=1}^{N_{obj}} \boldsymbol{x}^{(j_{\max}.k)} \tag{5}$$

The vectors $\boldsymbol{x}^{(j_{max}.k)} \in X$, $k = 1, ..., N_{obj}$, correspond to $\boldsymbol{F}^{(j_{max}.)} \in F$. Analytically, $\boldsymbol{x}_{tmp.alt}$ is identified similarly as in (4) but using the linear model established using the (parameter space) vertices of the simplex $\boldsymbol{S}^{(j_{max})}$. The 'ultimate' initial design is then selected as the better of the two, x_{tmp} and $x_{tmp.alt}$, in terms of the smaller value of the objective F_1. The alternative vector (5) is considered because of possibly poor predictive power of the surrogate s at certain stages of the MO process, which is primarily the effect of low cardinality of the Pareto set.

The next stage of the optimization process is design refinement. It is necessary because the initial design (whether x_{tmp} or $x_{tmp.alt}$) is only an approximation of the true Pareto optimal point, and it normally needs to be relocated towards the Pareto front. Here, it is realized by solving a local optimization task formulated as

$$x^{(p+1)} = \arg \min_{\substack{x, F_2(x) \leq F_{tmp.2} \\ M \\ F_{N_{obj}}(x) \leq F_{tmp.N_{obj}}}} F_1(R(x)) \tag{6}$$

According to (6), we aim at minimizing the first objective without degrading the remaining ones (as compared to their values F_{tmp} at the initial design). In this work, (6) is solved by means of trust-region gradient-based procedure. More specifically, a series $x^{(p+1.i)}$, $i = 0, 1, \ldots$, of approximations to $x^{(p+1)}$ is generated as

$$x^{(p+1.i+1)} = \arg \min_{\substack{x, x^{(p+1.i)} - d^{(i)} \leq x \leq x^{(p+1.i)} + d^{(i)} \\ F_2(x) \leq F_{tmp.2} \\ M \\ F_{N_{obj}}(x) \leq F_{tmp.N_{obj}}}} F_1(L^{(i)}(x)) \tag{7}$$

In (7), $L^{(i)} = R(x^{(p+1.i)}) + J_R(x^{(p+1.i)}) \cdot (x - x^{(p+1.i)})$ is the first-order Taylor model of R at $x^{(p+1.i)}$, established using the Jacobian matrix J_R. The latter is estimated by means of finite differentiation (for $i = 0$). Subsequently, J_R is updated using the rank-one Broyden formula [39], which is sufficient as $x^{(p+1.0)} = x_{tmp}$ is normally close to $x^{(p+1)}$. Furthermore, the predictive power of the metamodel s will improve over time due to reduced distances between the Pareto-optimal vectors $x^{(k)}$. A conceptual illustration of the MO process has been provided in Fig. 1.

Upon solving (6), the vector $x^{(p+1)}$ complements the Pareto set, which is then used to enhance the surrogate s with the updated training data set $\{F^{(k)}, x^{(k)}\}_{k=1,\ldots,p+1}$. This concludes the pth iteration of the MO procedure. The algorithm is terminated upon yielding the required number of Pareto-optimal points.

2.4 Optimization Framework

The flow diagram of the overall MO procedure has been shown in Fig. 2. As mentioned before, the first step is to acquire the extreme Pareto-optimal points $x^{(k)}$, $k = 1, \ldots,$ N_{obj}, obtained by solving the single-objective task (1) (cf. Sect. 2.2). This data is used to construct the metamodel s. The surrogate is applied to generate the initial design x_{tmp}. The latter is the image $s(F_{tmp})$ of the objective vector calculated as the centre of the largest simplex produced through triangulation of the Pareto set available so far in the MO process. The final stage of the algorithm iteration is design refinement (cf. (6), (7)), where the vector x_{tmp} is 'pushed' towards the Pareto front through minimization of the first objective, while imposing constraints on the remaining ones. The CPU cost of the refinement step is low because of employing sparse Jacobian updates. The termination condition is based on identification of the required number of parameter vectors.

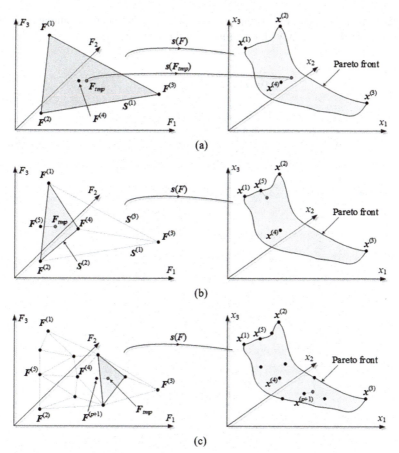

Fig. 1. Multi-objective optimization of antennas using Pareto front triangulation and inverse surrogates. The left and the right panels show the objective and the parameter spaces, respectively: (a) first iteration: the extreme Pareto-optimal designs F_k, $k = 1, 2, 3$, are triangulated in the objective space to produce the initial point $s(F_{tmp})$; this design is refined (cf. (5), (6)) to obtain the new Pareto-optimal point $x^{(4)}$ and its representation in the objective space $F^{(4)}$; (b) second iteration of the algorithm, where the initial (objective space) design F_{tmp} is allocated in the centre of the largest simplex (here, $S^{(2)}$); (c) one of the further iterations of the procedure.

One of the intrinsic advantages of the presented MO procedure is that it is fully deterministic. In particular, no randomized optimization techniques are involved, which also allows us to estimate the cost of the search process beforehand. The fact of utilizing the already existing knowledge about the Pareto set in the form of the inverse surrogate further improves the efficacy of the method.

Another advantage of the proposed approach is that it permits a uniform coverage of the Pareto front. The underlying assumption here is connectivity of the Pareto front (i.e., that it does not contain several disjoint regions or subsets). While such an assumption does not hold in general, it is normally the case for many practical antenna design tasks.

The latter is mainly ensured by a continuous dependence between the antenna geometry parameters and the frequency characteristics of the structure.

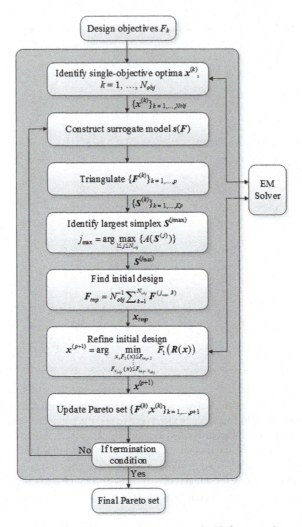

Fig. 2. Flow diagram of the proposed MO framework

3   Demonstration Examples

This section demonstrates the MO procedure introduced in Sect. 2 using two examples of microstrip antennas: a broadband monopole and a planar Yagi. Both structures are optimized with respect to two objectives each: size reduction and matching improvement (monopole), as well as matching improvement and in-band gain enhancement (Yagi).

Benchmarking with respect to state-of-the-art surrogate-assisted MO algorithm is also provided.

3.1 Case I: Broadband Monopole Antenna

Consider an ultra-wideband (UWB) monopole antenna [40]. The structure, shown in Fig. 3, is implemented on FR4 substrate ($\varepsilon_r = 4.3$, $h = 1.55$ mm) and described by eleven independent parameters $x = [L_g\ L_0\ L_s\ W_s\ d\ dL\ d_s\ dW_s\ dW\ a\ b]^T$; $W_0 = 2.0$ mm is fixed to ensure 50 Ω input impedance. All parameters are in millimetres. The computational model is evaluated in CST Microwave Studio (~600,000 mesh cells, simulation time 3 min), and contains the SMA connector.

The monopole antenna is optimized with respect to two objectives: minimization of the maximum in-band reflection (F_1), and minimization of antenna footprint $A(x)$ (F_2). The frequency range of interest is 3.1 GHz to 10.6 GHz, whereas the size is defined as the area of the substrate $A(x) = (a + 2o)(l_0 + l_1 + w_1)$. Furthermore, the only part of the Pareto front that is of interest consists of the designs for which $F_1 \leq -10$ dB, which is the standard acceptance level when considering antenna impedance matching.

The MO process has been conducted as described in Sect. 2. In the first step, the two single-objective optima have been found using the gradient-based algorithm [41]: $x^{(1)} = [9.07\ 13.39\ 9.93\ 0.43\ 2.03\ 9.17\ 0.80\ 2.29\ 3.02\ 0.29\ 0.59]^T$ mm, $x^{(2)} = [9.81\ 13.26\ 7.82\ 0.23\ 4.36\ 0.00\ 0.97\ 1.20\ 0.00\ 0.80\ 0.62]^T$ mm. Subsequently, eight more designs have been identified, leading to the Pareto set shown in Fig. 4 (see also Fig. 5 for antenna reflection responses at the selected designs). The breakdown of the CPU cost of the MO process can be found in Table 1. The overall expenses amount to 575 full-wave antenna simulations, which includes 403 EM analyses to find the designs $x^{(1)}$ and $x^{(2)}$. The average cost is 20 analyses per design (excluding extreme Pareto-optimal point generation).

Benchmarking was carried out using the surrogate-based procedure [40], which employs the kriging surrogate rendered in the interval $[l^*\ u^*]$ with $l^* = \min\{x^{(1)}, x^{(2)}\}$ and $u^* = \max\{x^{(1)}, x^{(2)}\}$; the initial Pareto set is obtained by optimizing the metamodel using a multi-objective evolutionary algorithm (MOEA) [42]. The final designs are produced using output space mapping [43].

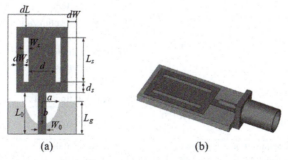

(a)   (b)

Fig. 3. Verification case I: broadband monopole antenna [40]: (a) antenna geometry (ground plane marked using the light-grey shade), (b) perspective view including the SMA connector.

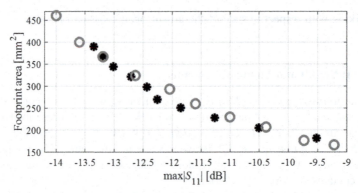

Fig. 4. Verification case I: Pareto set found by means of the proposed MO algorithm (o), the set identified using the surrogate-assisted technique of [24] (*).

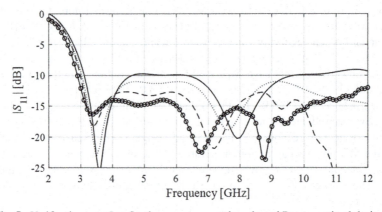

Fig. 5. Verification case I: reflection responses at the selected Pareto-optimal designs.

Table 1. Verification case I: optimization cost

This work (inverse modeling & refinement)		Surrogate-assisted procedure [24] (benchmark)	
Cost item	Cost#	Cost item	Cost#
Single-objective optimization runs (designs $x^{(1)}$ and $x^{(2)}$)	$403 \times R$	Single-objective optimization runs (designs $x^{(1)}$ and $x^{(2)}$)	$403 \times R$
		Data acquisition for kriging surrogate	$1000 \times R$
Design refinement	$162 \times R$	MOEA optimization*	N/A
		Design refinement	$30 \times R$
Total cost#	$575 \times R$ (29 h)	Total cost#	$1433 \times R$ (72 h)

* The cost of MOEA optimization is negligible compared to other stages of the process.
# The cost is expressed in terms of the equivalent number of EM simulations (marked as $\times R$).

According to the methodology of [24], restricting the domain of the surrogate to $[\boldsymbol{l}^*$ $\boldsymbol{u}^*]$ allows for mitigating the problem of dimensionality to some extent. Notwithstanding, the computational cost of the benchmark algorithm is as high as 1433 EM simulations, two thirds of which are related to training data acquisition (1000 samples). This was necessary to ensure sufficient accuracy of the metamodel (7.7% of the average RMS error). Based on this data, it can be observed that the proposed methodology allows for sixty percent computational savings. Additionally, our approach leads to a fairly uniform coverage of the Pareto front, as well as broader span of the Pareto set as compared to the benchmark.

3.2 Case II: Planar Yagi Antenna

Consider the planar Yagi antenna [44] shown in Fig. 6. The structure is implemented on RT6010 substrate ($\varepsilon_r = 10.2$, $h = 0.635$ mm), and described by eight independent parameters $\boldsymbol{x} = [s_1\ s_2\ v_1\ v_2\ u_1\ u_2\ u_3\ u_4]^T$. The fixed parameters are $w_1 = w_3 = w_4 = 0.6$, $w_2 = 1.2$, $u_5 = 1.5$, $s_3 = 3.0$, and $v_3 = 17.5$ (dimensions in mm). Similarly as in the previous example, the computational model is implemented in CST Microwave Studio and evaluated using its time domain solver (~600,000 mesh cells, simulation time 4 min).

The intended operating frequency range of the antenna is 10 GHz to 11 GHz. The structure is optimized with respect to the following two objectives: minimization of the in-band reflection (F_1), and maximization of the average end-fire gain (F_2). The designs $\boldsymbol{x}^{(1)} = [4.38\ 3.56\ 8.90\ 4.16\ 4.08\ 4.74\ 2.15\ 1.50]^T$, and $\boldsymbol{x}^{(2)} = [5.19\ 6.90\ 7.10\ 5.08\ 3.54\ 4.78\ 2.23\ 0.93]^T$ have been found using the gradient-based search procedure [41].

The results have been shown in Figs. 7 and 8, as well as Table 2. The overall cost of the MO process is 290 EM simulations of the antenna at hand, which includes 160 analyses required for rendering the designs $\boldsymbol{x}^{(1)}$ and $\boldsymbol{x}^{(2)}$. The surrogate-assisted procedure [24] has been used for the sake of comparison. The cost of the benchmark procedure is 1190 EM simulations, 1000 of which were used to construct the kriging surrogate (the average RMS error of the metamodel is 3.8 and 3.6% for the antenna reflection and gain, respectively). Consequently, the proposed methodology yields almost seventy percent computational savings. It can be observed that the results are consistent with those obtained in Sect. 3.1: our approach allows for generating a uniform coverage of the Pareto front, and with a larger span than for the benchmark.

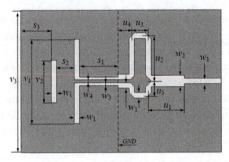

Fig. 6. Verification case II: planar Yagi antenna [44].

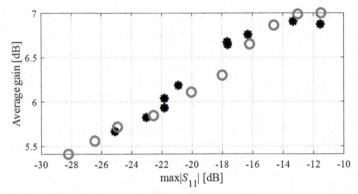

Fig. 7. Verification case II: Pareto set found using the proposed MO approach (o), the set identified using the surrogate-assisted technique of [24] (*).

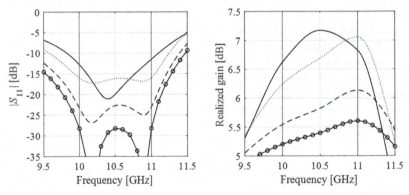

Fig. 8. Verification case II: reflection (left) and end-fire gain (right) responses for the selected Pareto-optimal designs.

Table 2. Verification case II: optimization cost

This work (inverse modeling & refinement)		Surrogate-assisted procedure [24] (benchmark)	
Cost item	Cost#	Cost item	Cost#
Single-objective optimization runs (designs $x^{(1)}$ and $x^{(2)}$)	$160 \times R$	Single-objective optimization runs (designs $x^{(1)}$ and $x^{(2)}$)	$160 \times R$
		Data acquisition for kriging surrogate	$1000 \times R$
Design refinement	$130 \times R$	MOEA optimization*	N/A
		Design refinement	$30 \times R$
Total cost#	$290 \times R$ (19.5 h)	Total cost#	$1190 \times R$ (80 h)

*The cost of MOEA optimization is negligible compared to other stages of the process.
#The cost is expressed in terms of the equivalent number of EM simulations (marked as $\times R$).

4 Conclusions

In the paper, a deterministic framework for multi-objective design of antenna structures has been proposed. The foundation of our technique is a sequential generation of the Pareto set elements using triangulation of already rendered points, as well as the inverse surrogates. A local gradient-based refinement is also involved to improve the design quality. The major benefits of the presented approach include no need to engage stochastic search procedures (in particular, population-based metaheuristics), low computational cost, and uniform coverage of the Pareto front. These features have been corroborated using two examples of microstrip antennas optimized for matching improvement, reduction of the footprint area, and maximization of the in-band gain. In both cases, ten-element Pareto sets have been obtained at the cost of a few hundreds of EM analysis of the respective structures, which yields about seventy percent savings as compared to the state-of-the-art surrogate-based technique. The proposed framework can be considered an alternative to available techniques for efficient and reliable MO of antennas, especially when handling miniaturized structures, where one of the objectives is a reduction of the physical dimensions of the radiator.

Acknowledgement. The authors would like to thank Dassault Systemes, France, for making CST Microwave Studio available. This work is partially supported by the Icelandic Centre for Research (RANNIS) Grant 206606051 and by National Science Centre of Poland Grant 2018/31/B/ST7/02369.

References

1. Ren, Z., Zhao, A., Wu, S.: MIMO antenna with compact decoupled antenna pairs for 5G mobile terminals. IEEE Ant. Wirel. Prop. Lett. **18**(7), 1367–1371 (2019)
2. Zhao, A., Ren, Z.: Size reduction of self-isolated MIMO antenna system for 5G mobile phone applications. IEEE Ant. Wirel. Prop. Lett. **18**(1), 152–156 (2019)

3. Houret, T., Lizzi, L., Ferrero, F., Danchesi, C., Boudaud, S.: DTC-enabled frequency-tunable inverted-F antenna for IoT applications. IEEE Ant. Wirel. Prop. Lett. **19**(2), 307–311 (2020)
4. Gao, G., Yang, C., Hu, B., Zhang, R., Wang, S.: A wide-bandwidth wearable all-textile PIFA with dual resonance modes for 5 GHz WLAN applications. IEEE Trans. Ant. Prop. **67**(6), 4206–4211 (2019)
5. Wang, J., Leach, M., Lim, E.G., Wang, Z., Pei, R., Huang, Y.: An implantable and conformal antenna for wireless capsule endoscopy. IEEE Ant. Wirel. Prop. Lett. **17**(7), 1153–1157 (2018)
6. Yang, G., Zhang, S., Li, J., Zhang, Y., Pedersen, G.F.: A multi-band magneto-electric dipole antenna with wide beam-width. IEEE Access **8**, 68820–68827 (2020)
7. Tan, L., Wu, R., Poo, Y.: Magnetically reconfigurable SIW antenna with tunable frequencies and polarizations. IEEE Trans. Ant. Prop. **63**(6), 2772–2776 (2015)
8. Kaddour, A., Bories, S., Bellion, A., Delaveaud, C.: 3-D-printed compact wideband magne-toelectric dipoles with circular polarization. IEEE Ant. Wirel. Prop. Lett. **17**(11), 2026–2030 (2018)
9. Koziel, S., Cheng, Q.S., Li, S.: Optimization-driven antenna design framework with multiple performance constraints. Int. J. RF Microwave CAE **28**(4), e21208 (2018)
10. Koziel, S., Pietrenko-Dabrowska, A.: Expedited feature-based quasi-global optimization of multi-band antennas with Jacobian variability tracking. IEEE Access. **8**, 83907–83915 (2020)
11. Kolda, T.G., Lewis, R.M., Torczon, V.: Optimization by direct search: new perspectives on some classical and modern methods. SIAM Rev. **45**, 385–482 (2003)
12. Zhao, W.J., Liu, E.X., Wang, B., Gao, S.P., Png, C.E.: Differential evolutionary optimization of an equivalent dipole model for electromagnetic emission analysis. IEEE Trans. Electromagn. Comp. **60**(6), 1635–1639 (2018)
13. Lalbakhsh, A., Afzal, M.U., Esselle, K.P.: Multiobjective particle swarm optimization to design a time-delay equalizer metasurface for an electromagnetic band-gap resonator antenna. IEEE Ant. Wirel. Prop. Lett. **16**, 915 (2017)
14. Marler, R.T., Arora, J.S.: The weighted sum method for multi-objective optimization: new insights. Struct. Multidisc. Opt. **41**, 853–862 (2010)
15. Ullah, U., Koziel, S., Mabrouk, I.B.: Rapid re-design and bandwidth/size trade-offs for compact wideband circular polarization antennas using inverse surrogates and fast EM-based parameter tuning. IEEE Trans. Ant. Prop. **68**(1), 81–89 (2019)
16. Mirjalili, S., Dong, J.S.: Multi-objective Optimization Using Artificial Intelligence Techniques. Springer Briefs in Applied Sciences and Technology, New York (2019)
17. Carvalho, R., Saldanha, R.R., Gomes, B.N., Lisboa, A.C., Martins, A.X.: A multi objective evolutionary algorithm based on decomposition for optimal design of Yagi-Uda antennas. IEEE Trans. Magn. **48**(2), 803–806 (2012)
18. Goudos, S.K., Gotsis, K.A., Siakavara, K., Vafiadis, E.E., Sahalos, J.N.: A multi-objective approach to subarrayed linear antenna design based on memetic differential evolution. IEEE Trans. Ant. Prop. **61**(6), 3042–3052 (2013)
19. Zhang, Y., Liu, X., Bao, F., Chi, J., Zhang, C., Liu., P.: Particle swarm optimization with adaptive learning strategy. Knowl. Based Syst. **196** (2020). Article number 105789
20. Maddio, S., Pelosi, G., Righini, M., Selleri, S.: A multi-objective invasive weed optimization for broad band sequential rotation networks. In: IEEE International Symposium Antenna Proposition, Boston, pp. 955–956 (2018)
21. Zhu, D.Z., Werner, P.L., Werner, D.H.: Multi-objective lazy ant colony optimization for frequency selective surface design. In: IEEE International Symposium Antenna Proposition, Boston, pp. 2035–2036 (2018)
22. Ranjan, P., Mahto, S.K., Choubey, A.: BWDO algorithm and its application in antenna array and pixelated metasurface synthesis. IET Microwaves Ant. Prop. **13**(9), 1263–1270 (2019)

23. Zhang, C., Fu, X., Peng, S., Wang, Y., Chang, J.: New multi-objective optimisation algorithm for uniformly excited aperiodic array synthesis. IET Microwaves Ant. Prop. **13**(2), 171–177 (2019)
24. Koziel, S., Bekasiewicz, A.: Multi-objective Design of Antennas Using Surrogate Models. World Scientific, Singapore (2016)
25. De Villiers, D.I.L., Couckuyt, I., Dhaene, T.: Multi-objective optimization of reflector antennas using kriging and probability of improvement. In: International Symposium Antenna Proposition, San Diego, pp. 985–986 (2017)
26. Xiao, S., Liu, G.Q., Zhang, K.L., Jing, Y.Z., Duan, J.H., Di Barba, P., Sykulski, J.K.: Multi-objective Pareto optimization of electromagnetic devices exploiting kriging with Lipschitzian optimized expected improvement. IEEE Trans. Magn. **54**(3), 1 (2018). Paper ID 7001704
27. Xia, B., Ren, Z., Koh, C. S.: Utilizing kriging surrogate models for multi-objective robust optimization of electromagnetic devices. IEEE Trans. Magn. **50**(2), 693 (2014). Paper 7017104
28. Jacobs, J.P.: Characterization by Gaussian processes of finite substrate size effects on gain patterns of microstrip antennas. IET Microwaves Ant. Prop. **10**(11), 1189–1195 (2016)
29. Lv, Z., Wang, L., Han, Z., Zhao, J., Wang, W.: Surrogate-assisted particle swarm optimization algorithm with Pareto active learning for expensive multi-objective optimization. IEEE J. Automatica Sinica. **6**(3), 838–849 (2019)
30. Koziel, S., Sigurdsson, A.T.: Multi-fidelity EM simulations and constrained surrogate modeling for low-cost multi-objective design optimization of antennas. IET Microwaves Ant. Prop. **12**(13), 2025–2029 (2018)
31. Koziel, S., Pietrenko-Dabrowska, A.: Rapid multi-objective optimization of antennas using nested kriging surrogates and single-fidelity EM simulation models. Eng. Comp. **37**(4), 1491–1512 (2019)
32. Koziel, S., Kurgan, P.: Rapid multi-objective design of integrated on-chip inductors by means of Pareto front exploration and design extrapolation. Int. J. Electromagn. Waves Appl. **33**(11), 1416–1426 (2019)
33. Unnsteinsson, S.D., Koziel, S.: Generalized Pareto ranking bisection for computationally feasible multi-objective antenna optimization. Int. J. RF Microwave CAE **28**(8), e21406 (2018)
34. Amrit, A., Leifsson, L., Koziel, S.: Fast multi-objective aerodynamic optimization using sequential domain patching and multi-fidelity models. J. Aircraft **57**, 388 (2020)
35. Liu, Y., Cheng, Q.S., Koziel, S.: A generalized SDP multi-objective optimization method for EM-based microwave device design. Sensors **19**(14), 3065 (2019)
36. Deb, K.: Multi-objective Optimization Using Evolutionary Algorithms. Wiley, New York (2001)
37. Forrester, A.I.J., Keane, A.J.: Recent advances in surrogate-based optimization. Prog. Aerosp. Sci. **45**, 50–79 (2009)
38. Borouchaki, H., George, P.L., Lo, S.H.: Optimal Delaunay point insertion. Int. J. Numerical Methods Eng. **39**(20), 3407–3437 (1996)
39. Broyden, C.G.: A class of methods for solving nonlinear simultaneous equations. Math. Comp. **19**, 577–593 (1965)
40. Haq, M.A., Koziel, S.: Simulation-based optimization for rigorous assessment of ground plane modifications in compact UWB antenna design. Int. J. RF Microwave CAE **28**(4), e21204 (2018)
41. Conn, A.R., Gould, N.I.M., Toint, P.L.: Trust Region Methods. MPS-SIAM Series on Optimization (2000)
42. Fonseca, C.M.: Multiobjective genetic algorithms with application to control engineering problems. Ph.D. thesis, Department of Automatic Control and Systems Engineering, University of Sheffield, Sheffield (1995)

43. Koziel, S., Cheng, Q.S., Bandler, J.W.: Space mapping. IEEE Microwave Mag. **9**(6), 105–122 (2008)
44. Kaneda, N., Deal, W.R., Qian, Y., Waterhouse, R., Itoh, T.: A broad-band planar quasi Yagi antenna. IEEE Trans. Ant. Propag. **50**, 1158–1160 (2002)

Optimizations of a Generic Holographic Projection Model for GPU's

Mark Voschezang[1,2]([envelope]) [ORCID] and Martin Fransen[2] [ORCID]

[1] Informatics Institute, University of Amsterdam, Amsterdam, The Netherlands
mark.voschezang@icloud.com
[2] Nikhef, Amsterdam, The Netherlands
martinfr@nikhef.nl

Abstract. Holographic projections are volumetric projections that make use of the wave-like nature of light and may find use in applications such as volumetric displays, 3D printing, lithography and LIDAR. Modelling different types of holographic projectors is straightforward but challenging due to the large number of samples that are required. Although computing capabilities have improved, recent simulations still have to make trade-offs between accuracy, performance and level of generalization. Our research focuses on the development of optimizations that make optimal use of modern hardware, allowing larger and higher-quality simulations to be run. Several algorithms are proposed; (1) a brute force algorithm that can reach 20% of the theoretical peak performance and reached a 43× speedup w.r.t. a previous GPU implementation and (2) a Monte Carlo algorithm that is another magnitude faster but has a lower accuracy. These implementations help researchers to develop and test new holographic devices.

Keywords: Computer-generated holograms (CGH) · Digital holography · GPU computing (GPGPU) · CUDA · Monte Carlo integration

1 Introduction

Holography was invented as early as 1947 by Dennis Gabor, but high-resolution holographic projectors are still not commercially viable [9]. Holograms are the product of the interference of light-waves [10]. They may find use in applications such as volumetric displays, 3D printing or LIDAR. Figure 1 shows examples of two holographic projections.

Modern holographic projectors give full control over an array of light sources and can display arbitrary distributions of light. The "pixels" of such a projector emit light of a specific wavelength and can be tuned in intensity and phase. A major limitation of projectors is the limited pixel pitch of projectors. Due to the small wavelength of visible light ($0.4\ \mu m$ to $0.7\ \mu m$), holographic projectors would need a pixel pitch of less than $0.2\ \mu m$ in order to correctly reproduce the correspondingly small structures in a light field. A larger pixel pitch will cause under-sampling artefacts in the projections. Typical under-sampling effects can

© Springer Nature Switzerland AG 2021
M. Paszynski et al. (Eds.): ICCS 2021, LNCS 12745, pp. 131–144, 2021.
https://doi.org/10.1007/978-3-030-77970-2_11

Fig. 1. Two holographic projections; a tilted ring (left) and a question mark symbol (right). The diamond shaped spot is the light source of the projector.

be reduced by placing the pixels at a-periodic or random intervals [13,21]. In order to study these effects, a prototype of a holographic projector is being developed, shown in Fig. 2. The role of simulations is to predict the effect that certain components will have on the projections (e.g. the size and positions of apertures).

The behaviour of light is a common subject to study and there are various models that can be used under different circumstances [10]. This paper relies on a single, generic model, namely a point-source model of the electric field component of light. The projector consists of a number of discrete point light sources of which the resulting light field is calculated at discrete points in the projection volume. The light intensity distribution at various positions is obtained by superimposing the received light components at those positions. Such a component can be described by

$$E(t) = \frac{A}{\delta} \exp\left(i\left(\phi + \frac{2\pi\delta}{\lambda} + \omega t\right)\right),$$

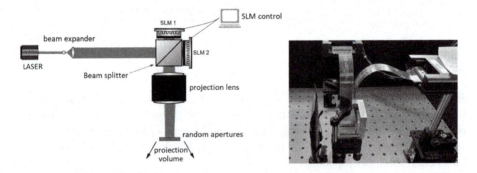

Fig. 2. A schematic (left) and a picture (right) of an experimental setup of a holographic projector. A beam splitter first splits the light emitted from a laser into two beams that are reflected by the Spatial Light Modulators (SLM 1 and 2). The reflected light is then superimposed at the beam splitter and directed through a projection lens and a mask that contains apertures at specific positions.

where δ is the distance from the source, A is the source strength, ϕ the phase offset of the light source, λ is the wavelength and ω is the frequency of the light. Because the time-dependent component is the same everywhere in the projection volume it can be omitted; only the phase offset of the source ϕ and the phase shift over distance $\frac{2\pi\delta}{\lambda}$, are taken into account. The superposition of these components can be formulated as either a summation or an integral (w.r.t space). The result can be expressed as a *phasor* or as a polar coordinate. The computational complexity is quadratic w.r.t the number of sample points and the computations have to be done in double precision to prevent significant rounding errors. The spatial frequency of two beams under an angle α is given by $\frac{2\sin\frac{\alpha}{2}}{\lambda}$. Simulating large volumes using this approach with at least two samples per wavelength requires an enormous amount of computing power.

Recently, multiple optimizations for holographic simulations for have been developed for modern hardware [17,23]. The majority of these models use assumptions that reduce the computational complexity of the problem. An example of this is the Fresnel approximation, which assumes that sample points lie in the near field [10,27]. Compressive holography is a more complex method that uses compressive sensing to approximate the projection distribution with a relatively low number of sample points [6,26,28]. These models are applicable to a limited number of projector configurations. As a result, these models cannot be used to study arbitrary projector designs. Hence our choice of model makes a qualitative difference.

Two algorithms, that first optimize computational efficiency and then complexity, have been developed. The corresponding implementations will aid physicists to run larger and more detailed simulations, which they can use to develop and test new holographic devices.

2   Background

This section gives a brief overview of relevant background material.

2.1   GPU Computing

Over the past decades, the performance and capabilities of computing hardware has increased [3]. Whereas CPU's are generally optimized to have a low latency, for example by maximizing clock frequency, GPU's are designed to do a narrow scope of tasks very efficiently [15]. They achieve this by making use of massive parallelism.

The extend to which applications can make optimal use of GPU hardware differs. A standard metric to compare performance is throughput, defined as the number of FLoating-point OPerations per Second (FLOPS). State-of-the-art linear algebra algorithms can reach 61.4%, 86.8% and over 90% of the theoretical peak performance for various GPU architectures [1,12,25]. In case of smaller, sub-optimal input sizes the efficiency decreases, for example to 30% [2]. Depending on the sparsity of the input, state-of-the-art sparse matrix algorithms may only reach a fraction (<0.08 %) of the theoretical peak performance [19]. These results shows that making optimal use of GPU hardware can be difficult, even for fundamental applications.

This research focuses on NVIDIA GPU's and the CUDA Programing Model. CUDA revolves around *kernels* which are Single Instruction Multiple Thread (SIMT) functions [7]. In other words, a single function is performed in parallel on many different threads or cores, with different input values per thread. Kernels are executed on a *grid* (representing the whole GPU), which contains *blocks* of threads (representing multiprocessors and cores). These abstractions allow arbitrary grid and block sizes to be used for different kernels, independent from the GPU that the program will run on. The number of grids, blocks and threads can be higher than the actual number of corresponding hardware units. Schedulers on the GPU act as multiplexers, making use of interleaved execution of tasks to maximize the utilization of hardware. In addition, the parallel execution of kernels and data transfers in CUDA can be managed using *streams*. Streams can be described as independent virtual containers that allow data transfers and kernel executions to happen in parallel.

2.2   Numerical Integration

Integrals can be approximated using two general techniques that are based on summations [8]. The first technique discretizes space and sums the resulting components, for example using a middle Riemann sum. The main disadvantage of this method is that the granularity of the discretization may cause systematic errors. The second technique is called Monte Carlo sampling and uses randomly distributed sample points [22]. For a D-dimensional unit hypercube it has the form

$$\int f(\mathbf{x})\,d\mathbf{x} \approx \frac{1}{N}\sum_{n=1}^{N} f(\mathbf{u}_n) \qquad \mathbf{u}_n \sim \mathcal{U}(0,1)^D.$$

The summation converges, with high probability, for large enough N. This means that it is possible to make a tradeoff between accuracy and computational cost by using fewer sample points.

Variance Reduction. It can be shown that the mean absolute error of an MC estimate grows with a factor $\frac{\sigma}{\sqrt{N}}$, where σ is the standard deviation of the estimator [16]. This means that the convergence of MC estimates can be improved by constructing an estimator with lower variance. There exist a variety of methods that achieve this, but they usually rely on a priori knowledge about the underlying distribution. This paper focuses on general solutions and no prior knowledge is assumed. Therefore the technique is stratified sampling is chosen [22]. It uses a conditional variable to spread out the samples over the sample space, making each sample set more representative of the original distribution.

An alternative variance reduction method is importance sampling [22]. This approach focuses the sampling on the areas that contribute the most to the estimation. This technique has successfully been applied to holography, albeit for an older generation of hardware [5].

3 Methods

This section introduces a number of algorithms that compute superpositions, gradually increasing in complexity. **Kernels** compute *partial* superpositions for multiple source datapoints at multiple target positions. They are combined with **estimators** that use these kernels to make estimations of *full* superpositions.

A number of CUDA-variables are used. Threads and blocks are indexed with a *threadIdx* and a *blockIdx* and can be divided in multiple dimensions (x, y, z). The number of threads per block is denoted by *blockDim* and the number of blocks per grid is denoted by *gridDim*. For convenience we define *gridSize* as the total number of threads per grid.

3.1 Superposition Kernels

All superposition kernels compute the partial superpositions of a set of source phasors (with three-dimensional positions $\mathbf{u}$) at certain target positions $\mathbf{v}$. They use a SIMT approach where each GPU thread applies the following function to a subset of the input data:

$$f(A, \phi, \mathbf{u}, \mathbf{v}) := \frac{A}{\delta} \exp\left(i \left(\phi \pm \frac{2\pi\delta}{\lambda} \right) \right) \qquad \delta \equiv |\mathbf{u} - \mathbf{v}|. \tag{1}$$

These partial results are written to a matrix $\mathbf{Y}$. Note that the matrix $\mathbf{Y}$ may not fit in GPU memory; this problem is solved in the next section by splitting the dataset up in chunks. After a superpositon kernel has terminated a second kernel is used sum the rows of this matrix, resulting in a vector that contains the full superpositions. This second kernel is a simple matrix-vector product, for which we have used the library function *cublasZgemv* [18].

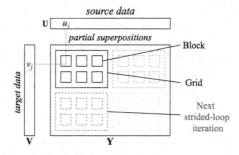

Fig. 3. The data-structures used in the superposition kernels.

Naive Kernel. The first thread dimension (x) is mapped to the source data and the second dimension (y) is mapped to the target data. The Naive kernel iterates over the source and target datapoints using a *strided* loop, allowing the kernel to be used for arbitrary input sizes. This approach is shown graphically in Fig. 3 and in pseudocode in Algorithm 1.

Algorithm 1: Naive Superposition Kernel

Kernel Superposition($\mathbf{a}, \boldsymbol{\phi}, \mathbf{U}, \mathbf{V}, \mathbf{Y}$) :
Result: Written to $y_{m,n} \in \mathbf{Y} \in \mathbb{C}^{M \times N}$
input : source phasors with polar coordinates $\mathbf{a}, \boldsymbol{\phi} \in \mathbb{R}^N$,
           source positions $\mathbf{u}_n \in \mathbf{U}$ for $n = 1, \ldots, N$,
           target positions $\mathbf{v}_m \in \mathbf{V}$ for $m = 1, \ldots, M$
$\langle t_x, t_y \rangle \leftarrow$ GlobalThreadIdx();
for $n \leftarrow t_x$; $\ n \leq N$; $\ n \leftarrow n + gridSize.x$ **do**
     `//` $a_n, \phi_n, \mathbf{u}_n$ `can be cached here`
     for $m \leftarrow t_y$; $\ m \leq M$; $\ m \leftarrow m + gridSize.y$ **do**
         $y_{m,n} \leftarrow f(a_n, \phi_n, \mathbf{u}_n, \mathbf{v}_m)$;   `// Equation 1`

Advanced Kernels. The memory complexity for reading and writing is $\mathcal{O}(NM)$ for the naive kernel (excluding any additional caching). The memory complexity of the subsequent summation kernel is $\mathcal{O}(NM)$ as well because it uses the output of the superposition kernel as input. We will consider two optimizations that reduce the writing frequency and refer to them as the Reduced kernel and the Shared kernel. They are formalized together in Algorithm 2, where the Boolean *shared_memory* is used to distinguish the two approaches.

Algorithm 2: Reduced & Shared Superposition Kernel

Kernel Superposition($\mathbf{a}, \boldsymbol{\phi}, \mathbf{U}, \mathbf{V}, \mathbf{Y}$) :
Result: Written to $y_{m,n} \in \mathbf{Y}$
$\langle t_x, t_y \rangle \leftarrow$ GlobalThreadIdx();
for $m \leftarrow t_y$; $\ m \leq M$; $\ m \leftarrow m + gridSize.y$ **do**
     if $t_x > N$ **then** continue to next iteration $y' \leftarrow 0$;   `// local`
     `temporary memory`
     for $n \leftarrow t_x$; $\ n \leq N$; $\ n \leftarrow n + gridSize.x$ **do**
         $y' \leftarrow y' + f(a_n, \phi_n, \mathbf{u}_n, \mathbf{v}_m)$;
     if *shared_memory* **then**
         $y' \leftarrow$ block-level reduction of y';   `// using shared memory`
         if $t_y = 1$ **then** $y_{m, 1+\text{blockIdx.x}} \leftarrow y'$
     else
         $y_{m, t_x} \leftarrow y'$;

The *Reduced* kernel aggregates partial superposition results locally, at the thread-level. This reduces the writing frequency. The inner and outer loops are switched s.t. the inner loop becomes a direct summation of partial superpositions.

As a result, writing to global memory happens only once per summation and there is no additional local memory required.

The *Shared* kernel reduces the writing frequency further by using shared data to aggregate the partial results of all threads in a block. The amount of block-level reduction is a trade-off between workload per thread, synchronization per block and global memory access. The library CUB provides a number of reductions that can be used inside CUDA kernels [11,14].

3.2   Estimators

The partial superpositions computed by the kernels are combined by an estimator. The estimators split the dataset up and send them to the GPU in batches, using different (concurrent) streams, as shown in Fig. 4. The **deterministic estimator** does this sequentially and is straightforward to implement. The stochastic estimators are more complex and are given below. This paper describes three variants. They all use a discrete source dataset, but they can easily be transformed to the case of a continuous source distribution.

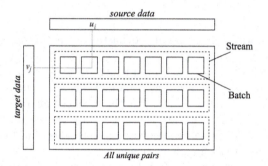

Fig. 4. Data traversal using batches and streams. Each batch corresponds to a single kernel, as shown in Fig. 3. Multiple streams can be active at the same time but the batches within a stream are executed in series (from left to right).

The **stochastic estimators** improve performance by using Monte Carlo (MC) sampling to reduce the number of source datapoints that are used. A trivial solution would be to use the deterministic estimator with a random subset of the source dataset. However, this would increase the risk of overfitting (over-generalization) because all target datapoints would use the same subset of data. The purpose of the stochastic estimators is enforce that target datapoints can sample independently from the source dataset.

Figure 5 shows a simplified representation of a stochastic estimator for a single stream. First the source data is shuffled and distributed over the available streams. Each stream independently computes the partial superpositions of multiple batches using either one of the kernels described in the previous section. This is repeated until a stream has seen enough source datapoints. The partial results are then summed and checked for convergence. The stream is terminated when either the corresponding superpositions have converged or when the maximum number of iterations has been reached.

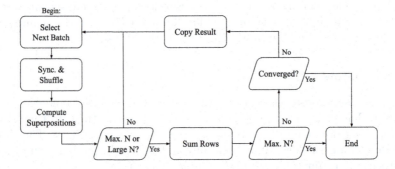

Fig. 5. A flowchart of the Stochastic Estimator, for a single stream. This scheme is repeated until all batches are completed. The step "Sync. & Shuffle" is optional and can be omitted if the target datapoint are chosen during kernel execution.

Sampling Methods. The general stochastic estimator algorithm can make use of two different sampling methods. We will refer to these as True MC and Batched MC. Both of these can be implemented using either conditional and unconditional sampling. The relevant combinations are visualized in Fig. 6.

The *True MC* estimator adheres to the traditional MC method. Instead of iterating over a range of indices, the indices are chosen randomly inside the superposition kernel. The disadvantage of this method is that it results in random access of global memory, which is inefficient.

The *Batched MC* estimator is slightly more complex. Datapoints are grouped together s.t. each kernel accesses a pre-specified subset of data. This avoids the

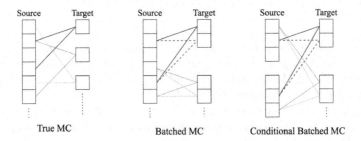

Fig. 6. Monte Carlo estimators. The lines indicate which source datapoints are used by which target datapoints. For Batched MC the *target* datapoints are grouped and for conditional MC the *source* datapoints are grouped as well.

random access pattern inside the superposition kernel, without increasing the memory usage. The available input dataset is distributed between the streams and then shuffled. This independent reshuffling means that the sampling is done with replacement. The main issue with this approach is that the target sample points within batches become correlated due to their shared input sample set.

The *Conditional Batched MC* estimator mitigates this effect by using conditional sampling. This forces the samples to become more representative of the whole dataset.

Convergence. The stochastic estimators combine two relatively simple convergence criteria, which are both based on a current estimate and a previous estimate. The first criteria computes the absolute difference between the (normalized) amplitudes of the two estimations and compares it to a threshold. The second criteria makes use of the fact that a sum of random vectors grows with the square root of the number of elements in that sum [1]. It is assumed that certain, far-away phasors are distributed as random vectors. The second convergence criteria normalizes the amplitudes of the estimations by dividing them by the square root of the sample sizes.

4   Experiments and Results

This section contains a number of experiments that test the performance of the kernels and estimators. The three kernels are abbreviated as #1: Naive kernel, #2: Reduced kernel (with thread-level aggregation) and #3: Shared kernel (with block-level aggregation using shared data). The estimators are the deterministic estimator, which uses the full dataset, and the three stochastic estimators, which use random subsets of data. The implementations are available online [24].

The experiments are run on a machine with an Intel Xeon E5-1650 CPU and a NVIDIA Quadro GV100 GPU (generation Volta), with a peak performance of 7.4 TFLOPS (double precision). The level of compiler-optimization is set to default (-O3).

The following metrics are used. *Runtime* is measured after initialization of all CPU datastructures but before initialization of the GPU datastructures. *Speedup* is defined as the mean runtime of a baseline implementation divided by the runtime of an optimized implementation. *Efficiency* or throughput is measured in double precision FLOPS [2]. The total number of floating-point operations for

---

[1] A sum of N random vectors can be represented by a two-dimensional random walk of N steps. The corresponding distribution is derived by (and named after) Rayleigh and is given by $p(\ell) = \frac{2\ell}{N} \exp\left(-\ell^2/N\right)$, where ℓ is the length or absolute value of the phasor-sum[20]. The expected value of this distribution is proportional to $\sqrt{N}$. This distribution requires the input phasors to have unit length, but more refined solutions that allow for arbitrary amplitude and phase distributions exist as well [4].

[2] The exact number of floating point operations is difficult to determine because compilers can restructure code and certain implementations are hardware-dependent. For example fused multiply-add operations and hardware-units for special arithmetic such as the square root [7].

the superpositions of N source datapoints and M target datapoints is defined as $29NM$.

The validation of the model and implementations is done using a prototype of a holographic projector. These experiments are important but are not discussed here.

4.1 Accuracy

The accuracy of the stochastic estimators depends on the input values and parameters such as convergence threshold and batch size. A single typical input distribution is used to show the qualitative differences that can be caused by the different estimators. The irradiance of the resulting projections is shown in Fig. 7. The projection generated by the deterministic estimator projection is considered to be the ground truth. Overall, the stochastic estimators produce

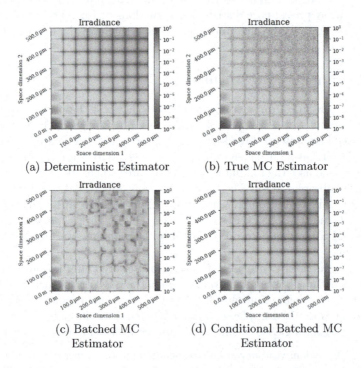

(a) Deterministic Estimator

(b) True MC Estimator

(c) Batched MC
Estimator

(d) Conditional Batched MC
Estimator

Fig. 7. Projections generated by four different estimators, shown with a logarithmic color-scale. The projection plane is parallel to a projector with 512×512 pixels (at a distance of 35 cm) and limited to the top-right quadrant w.r.t the center of the projector screen. The projection is sampled using 1024×1024 sample points and the convergence threshold is $\epsilon = 10^{-4}$.

accurate results in high amplitude regions, but start to diverge in low-amplitude regions. The True MC projection contains uniform-like noise. The unconditional Batched MC projection shows intra-batch correlations that are not present in the ground truth. The Conditional Batched MC projection is the most accurate, but still contains differences with the ground truth.

4.2   Kernel Performance

Figure 8 shows the performance of the three kernels as function of the number of source datapoints. A linear least-squares regression model is fitted and was significant with a p-value of 0.001. Kernels #2 and #3 clearly outperform kernel #1. The difference between kernels #2 and #3 is small, but kernel #2 does have a lower slope.

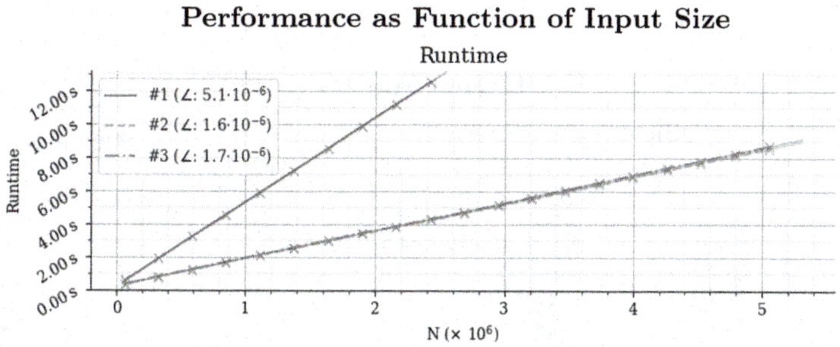

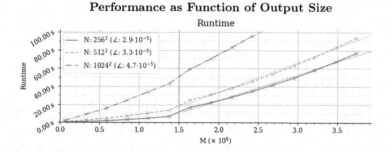

Fig. 8. The performance of the Naive Kernel (#1), the Reduced Kernel (#2) and the Shared Kernel (#3). The slope of linear fit is denoted by $\angle$. The parameters are: $M = 256^2$, gridDim = blockDim = $\langle 8, 8 \rangle$, thread size = $\langle 64, 8 \rangle$.

Fig. 9. Performance as function of output size for the Reduced Kernel (#2). The slope of linear fit using the data after the red vertical line is denoted by $\angle$. The parameters are: gridDim = blockDim = $\langle 8, 8 \rangle$, thread size = $\langle 256, 16 \rangle$. (Color figure online)

Figure 9 shows the performance of kernel #2 as function of the number of target datapoints. In contrast to the previous results, the performance is no longer linear. The runtime increases monotonically, but becomes slightly convex after $1.5 \cdot 10^6$ target datapoints. However, the independence of the target dataset allows the computations to be split up into independent chunks with a size that maximizes the efficiency per chunk. Using this technique would still result in linear computational growth.

An existing implementation, written in Matlab, is used as a baseline. It runs on the same hardware but uses single-precision data, which gives it an performance advantage. Figure 10 shows the efficiency and speedup of the three algorithms in comparison with the Matlab implementation. The Reduced kernel has the best performance, with a speedup of 43.9, followed by the Shared kernel with a speedup of 41.1. The efficiency of the Reduced is 15.3 TFLOPS, which is 20.7% of the theoretical peak performance of this GPU. The rest of the experiments will use the kernel #2.

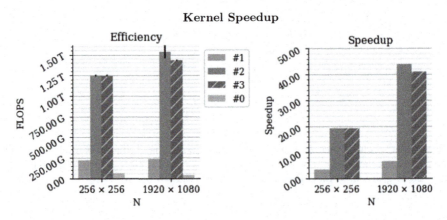

Fig. 10. The performance of the three CUDA kernels in comparison with the Matlab implementation (labelled as #0). The right graph does not contain error bars. The shared parameters are $M = 1024 \times 1024$ and gridDim = $\langle 16, 16 \rangle$. For algorithm #1 the remaining parameters are thread size = $\langle 32, 32 \rangle$, blockDim = $\langle 8, 8 \rangle$ and for algorithm #2 and #3 the parameters are thread size = $\langle 512, 256 \rangle$, blockDim = $\langle 16, 16 \rangle$.

4.3   Stochastic Estimator Performance

Figure 11 shows the performance of the unconditional estimators for different convergence thresholds. Batched MC is roughly an order of magnitude faster than True MC. The lowest threshold (10^{-16}) is used to prevent convergence and thus represent the worst-case scenario. The speedups for a convergence threshold of 0.01 w.r.t a non-converging result is 15.3 for True MC and 10.6 for Batched MC. For larger input sizes the relative speedups should be even higher.

Estimator Performance

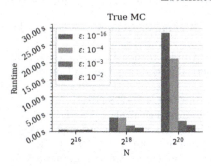
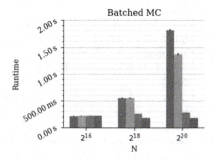

Fig. 11. Comparison of unconditional MC estimators for different input sizes and convergence thresholds. The parameters are: $M = 256^2$, gridDim $= \langle 8, 8 \rangle$, blockDim $= \langle 16, 16 \rangle$.

5   Conclusion

This research proposes two optimizations of simulations of holographic projectors. Because the underlying model does not rely on complex assumptions it can be used to model arbitrary projectors. The proposed (double-precision) implementation reaches a speedup of a factor 43.9 compared to a previous (single-precision) implementation, using the same hardware. Moreover, it reached over 20% of the theoretical peak performance, which compares favorably to state-of-the-art GPU implementations in other domains. This also shows that additional speedups that are larger than a factor five cannot be reached with the same hardware (and the same mathematical model).

A second implementation circumvents this hardware-related upper bound by reducing the number of operations per estimation. This is achieved using Monte Carlo integration. Although this comes with a decrease in accuracy, it can be used to give rough estimates. The results can subsequently be verified using the first implementation.

References

1. Abdelfattah, A., Baboulin, M., et al.: High-performance tensor contractions for GPUs. Procedia Comput. Sci. **80**, 108–118 (2016)
2. Abdelfattah, Ahmad, Haidar, Azzam, Tomov, Stanimire, Dongarra, Jack: Performance, design, and autotuning of batched GEMM for GPUs. In: Kunkel, Julian M.., Balaji, Pavan, Dongarra, Jack (eds.) ISC High Performance 2016. LNCS, vol. 9697, pp. 21–38. Springer, Cham (2016). https://doi.org/10.1007/978-3-319-41321-1_2
3. Asanovic, K., Bodik, R., Catanzaro, B.C., et al.: The landscape of parallel computing research: a view from Berkeley. Technical report UCB/EECS-2006-183, EECS Department, University of California, Berkeley (2006)
4. Beckmann, P.: Statistical distribution of the amplitude and phase of a multiply scattered field. J. Res. Natl. Bur. Stand. **66D**(3), 231–240 (1962)

5. Bokor, N., Papp, Z.: Monte Carlo method in computer holography. Opt. Eng. **36**, 1014–1020 (1997)
6. Brady, D.J., Choi, K., Marks, D.L., Horisaki, R., Lim, S.: Compressive holography. Opt. Express **17**(15), 13040–13049 (2009)
7. Corporation, N.: CUDA C++ Programming Guide, version 11.1.1, October 2020
8. Davis, P.J., Rabinowitz, P.: Methods of numerical integration. Courier Corporation (2007)
9. Gabor, D.: A new microscopic principle. Nature **161**, 777–778 (1948)
10. Hecht, E., Zajac, A.: Optics. Addison Wesley, Reading (1974)
11. Kogge, P.M., Stone, H.S.: A parallel algorithm for the efficient solution of a general class of recurrence equations. IEEE Trans. Comput. **100**(8), 786–793 (1973)
12. Li, X., Liang, Y., Yan, S., et al.: A coordinated tiling and batching framework for efficient GEMM on GPUs. In: Proceedings of the 24th Symposium on Principles and Practice of Parallel Programming, pp. 229–241 (2019)
13. Li, Z.: Principle and characteristics of 3D display based on random source constructive interference. Opt. Express **22**(14), 16863–16875 (2014)
14. Merrill, D.: Cub documentation (2020). Accessed 01 Aug 2020
15. Nickolls, J., Dally, W.J.: The GPU computing era. IEEE Micro **30**(2), 56–69 (2010)
16. Niederreiter, H.: Random number generation and quasi-Monte Carlo methods. SIAM (1992)
17. Nishitsuji, T., Shimobaba, T., et al.: Review of fast calculation techniques for computer-generated holograms with the point-light-source-based model. IEEE Trans. Industr. Inf. **13**(5), 2447–2454 (2017)
18. NVIDIA Corporation: The API Reference guide for cuBLAS (v11.0.3) (2020)
19. Pal, S., Beaumont, J., Park, et al.: Outerspace: an outer product based sparse matrix multiplication accelerator. In: 2018 IEEE International Symposium on High Performance Computer Architecture (HPCA), pp. 724–736. IEEE (2018)
20. Rayleigh, J.W.S.B.: The Theory of Sound, vol. 2. Macmillan (1896)
21. Rivenson, Y., Stern, A., Javidi, B.: Compressive Fresnel holography. J. Display Technol. **6**(10), 506–509 (2010)
22. Ross, S.M.: Simulation. Academic Press, New York (2012)
23. Tsang, P., Poon, T.C., Wu, Y.: Review of fast methods for point-based computer-generated holography. Photon. Res. **6**(9), 837–846 (2018)
24. Voschezang, M.: Holographic Projector Simulations (2020). github.com/voschezang/Holographic-Projector-Simulations/tree/snapshot-stochastic-estimators
25. Younge, A.J., Walters, J.P., Crago, S., Fox, G.C.: Evaluating GPU pass through in XEN for high performance cloud computing. In: 2014 IEEE International Parallel & Distributed Processing Symposium Workshops, pp. 852–859. IEEE (2014)
26. Zhang, H., Cao, L., Zhang, H., Zhang, W., Jin, G., Brady, D.J.: Efficient block-wise algorithm for compressive holography. Opt. Express **25**(21), 24991–25003 (2017)
27. Zhao, T., et al.: Accelerating computation of CGH using symmetric compressed look-up-table in color holographic display. Opt. Express **26**(13), 16063–16073 (2018)
28. Zhao, Y., Cao, L., et al.: Accurate calculation of computer-generated holograms using angular-spectrum layer-oriented method. Opt. Express **23**(20), 25440–25449 (2015)

Similarity and Conformity Graphs in Lighting Optimization and Assessment

Artur Basiura[ID], Adam Sędziwy[✉][ID], and Konrad Komnata[ID]

Department of Applied Computer Science, AGH University of Science and Technology, Al. Mickiewicza 30, 30-059 Kraków, Poland
{abasiura,sedziwy,kkomnata}@agh.edu.pl

Abstract. Lighting affects everyday life in terms of safety, comfort and quality of life. On the other side it consumes significant amounts of energy. Thanks to the effect of scale, even a small unit reduction of a power efficiency yields the significant energy and cost savings.

Unfortunately, planning a highly optimized lighting installation is a task of the high complexity, due to a huge number of variants to be checked. In such circumstances it becomes necessary to use a formal model, applicable for automated bulk processing, which allows finding the best setup or estimating resultant installation power in an acceptable time, i.e., in hours rather than days. This paper introduces such a formal model relying on the *similarity* and *conformity graph* concepts. The examples of their practical application in outdoor lighting planning are also presented. Applying those structures allows reducing substantially a processing time required for planning large scale installations.

Keywords: Graph methods · Similarity graph · Conformity graph · Optimization · Complexity

1 Introduction

The growing civilization needs require making decisions quickly. Additionally, such choices should be based on the real data. In many cases, however, it is not possible to obtain those data on demand. In such situations the lacking information is estimated by human's intuition or expertise. A time pressure and lack of hard data cause our decisions to be influenced by a social environment, short-term needs or opinions of others. As a result, this choice is not optimal in many cases. These problems are referred to as the cognitive traps and they are discussed in multiple works [1]. The matter becomes more complicated if decisions impact the future and when that influence is broad, for example in the scale of a city, region or country. One of such areas is a public lighting which influences not only a quality of life and safety of people but also the energy balances of municipalities. We can observe a growing power consumption, which is caused, among others, by the increasing use of light [5,8,20]. This also means that the percentage of electric energy we use for public lighting has a significant share in the overall volume of greenhouse gases being produced [16].

© Springer Nature Switzerland AG 2021
M. Paszynski et al. (Eds.): ICCS 2021, LNCS 12745, pp. 145–157, 2021.
https://doi.org/10.1007/978-3-030-77970-2_12

The use of efficient LED (*light-emitting diode*) light sources gives a power usage reduction of the order of 40–60% [11], compared to the high-intensity discharge (sodium) lamps. The right choice of optimal installation parameters, however, can result in a much greater, spectacular reduction reaching up to 80% [15,16]. Those savings can play a key role in terms of the further investments. In the case of medium sized cities, the cost of electricity for lighting often exceeds 1 million euro which gives €2,000 of annual savings, at this rate of power reduction. This budget can be used for financing other related works. For that reason it seems reasonable to carry out a citywide investment to maximize the savings which can cover either some further works in other areas or a current investment, when made in the ESCO (*energy service company*) financing model. It is therefore crucial to assess very quickly the cost of an investment itself and the potential rate of return. Unfortunately, preparing all required photometric projects and thus estimating investment costs is often a long-term process burdened with additional costs. Analysis of investments that have taken a place in the city of Cracow, Poland, in recent years, shows that the time of preparation of a single photometric project for 3,741 lighting points takes 3–10 weeks. By extrapolating it to the total number of city lights in Cracow (approx. 70,000) we can estimate the time required for financial analysis of a citywide investment at the level of at least 56 weeks of continuous (24/7) calculations, which is over the year. In the real life cases such times are utterly unacceptable.

It is neither possible to point precisely which city area should be the subject to a planned investment and in what order, nor to select the scope of a modernization (e.g., only fixtures and arms are replaced, while poles remained unchanged). Usually an estimation is made on the basis of available fixture powers, which is not the best approach in many cases. It is crucial to develop some indirect methods allowing to estimate as quick as possible, expected outlays and the return on investment (ROI). This paper introduces concepts of similarity and conformity graphs, which can be used to estimate investment risk and for such a quick estimation.

The paper is organized as follows. In the next section the state of the art is presented. In Sect. 3 the notions of a base graph (Sect. 3.1), similarity graph (Sect. 3.2) and conformity graph (Sect. 3.3) are introduced. The case study demonstrating application of proposed models to a real-life case, is presented and discussed in Sect. 4. The final section contains conclusions and proposed directions of the further research.

2 State of the Art

Creating optimized, energy-efficient lighting installations was considered in numerous scientific works. One can distinguish two approaches: the first is focused on optimization of an installation parameters (e.g., fixture model, pole height) [4,9,10,12–14,17,19], while the second one is based on lighting control tuning, i.e. adapting lighting levels to the varying conditions such as traffic flow intensity or weather conditions [2,3,21–23]. The main criterion is a final installation power. The critical issue of such an optimization, however, is the time required for completing such a project (due to related computational complexity). Also practical methods used for its completion are revealed rarely.

One of the few works which attempt to present these factors is the work [7], in which authors propose a genetic algorithm to determine exact parameters of an installation, i.e., locations of poles and pole heights. It was achieved thanks to the appropriate definition of a chromosome which contained exact pole locations and fixture mounting angles, in addition to a fixture type and a pole height. The chromosome length is dependent on a number of light points in an optimized layout, in this case, and thus it has a considerable impact on a calculation time. For the initial population of 300 chromosomes, six types of fixtures (usually one considers thousands of models, produced by several vendors) and four potential pole heights, the algorithm execution time (50 generations) was about 2.5 h while. Obviously, with increasing number of fixture models and enlarged street area, the computation time raised to 4 h (see Table 1).

Table 1. Times required for finding the optimal design, when using a genetic algorithm (see [7]).

	Situation 1 (Parking)	Situation 2 (Handball court)
Problem space	6 fixture types	14 fixture types
Illuminated surface	800 m^2	1056 m^2
Number of "generations"	50	10
Time taken to find a solution	2.5 h	4 h

The long solution search times, as those seen above, enforce developing more efficient calculation methods. The above example shows that the GA-based approach fails when preparing a city-scale lighting design: the required time is not acceptable. When analyzing large investments (tens of thousands of streetlights), processing time is a factor of the great importance for quick decision making. The optimization methods presented in [15,18] allow to shorten this time. It is not enough, however, for making a quick choice. Therefore, the in-depth research work on this area becomes crucial. Developing algorithms which allow preparing a project with less accuracy but with a known estimated power, in a few hours instead, can bring practical benefits. The structures of similarity and conformity graphs, extending the graph concept introduced in [6,18], are proposed in the next section. Methods of their processing are also discussed.

3   Graph Models

Before defining main graph structures, i.e., *similarity* and *conformity graphs*, it is necessary to introduce the generic structure storing information related to a lighting infrastructure. It is referred to as a *base graph*.

3.1 Base Graph

Definition 1. *Base graph (abbrev. BG) is a graph of the form:*

$$G = (V, E, \Sigma, \Gamma, type, attr),$$

where:

- *V is a finite, non-empty set of graph nodes,*
- *E is a finite set of edges,*
- *Σ is a set of node types,*
- *Γ is a set of edge types, where $\Sigma \cap \Gamma = \emptyset$,*
- *$type : V \cup E \rightarrow \Sigma \cup \Gamma$ is a function that returns the type of a given node/edge: $type(V) = \Sigma$, $type(E) = \Gamma$,*
- *$attr$ is a function that returns a set of attribute types for a given node/edge type.*

In order to represent a lighting installation the following types, shown in Table 2, are ascribed to vertices of an infrastructure BG.

Table 2. Exemplary node types (elements of the Σ set) of an infrastructure graph

Physical entity	Node type	Description
Street/area	U	Type representing the illuminated region
Segment	S	Street subarea (when street geometry varies)
Lighting point	L	Street luminaire
Fixture Type	F	Fixture
Pole	P	Luminaire's pole
Arm	R	Luminaire's arm

Each node can be incident with an edge of a type from Γ which represents a relationship between two nodes. For instance B - "belongs to", "illuminates", "depends on" etc.

Example. The example of a scene consisting of a street and its lighting infrastructure, is shown in Fig. 1. It is compound of four street segments (S1, S2, S3, S4) having a common layout but different lighting installations, in terms of geometry properties. Each segment is assigned with at least one group of lamps (CL1, CL2A, CL2B, CL3, CL4). Those groups are the subject to optimization. The entire scene is represented by the base graph shown in Fig. 2. To improve the readability we neglected edge labels on that.

As shown in Fig. 1 segments S1 and S3 are *very similar* not only in terms of the street geometry but also due to the *similar* lighting installation layouts (e.g., nearly identical lamp spacings). Thus an optimal setup found for the installation

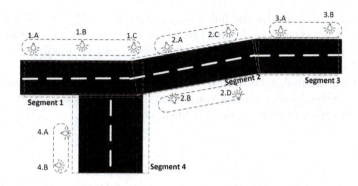

Fig. 1. The sample lighting situation

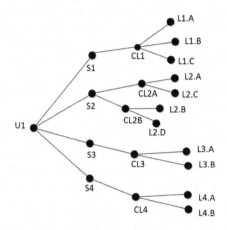

Fig. 2. Base graph representation of the scene shown in Fig. 1

{1.A, 1.B, 1.C} would be applicable to {3.A, 3.B} as well. There arises a question how to assess whether two lighting situations are *similar* to each other and how to quantify this *similarity*. In other words: does there exist any metrics, in the space of base graphs, which would be applicable for lighting situation comparison.

The answer to this question is affirmative. In the next subsections we introduce the notion of a similarity graph.

3.2 Similarity Graph

A similarity graph is a base graph which contains edges of a specific type, $K \in \Gamma$, referred to as *similarity edges*, connecting nodes of the same type (say, two S-type nodes), with an attributing function *attr*, such that $attr(K) = g \in \mathcal{D}^{\Sigma}$, where $\mathcal{D} = \{f | f : V \times V \longrightarrow [0, 1] \wedge f(x, y) = f(y, x)\}$ is a set of functions which quantify similarity of node attribute values.

Example. Let us consider the following example to clarify this idea. Suppose $e = \{u, v\} \in E$, $type(e) = K$ and $type(u) = type(v) = S$ with a road width $W \in attr(S)$. As said, a type of the edge e is K and its attribute value is a function f which for two vertices of the same type, being endpoints of e (here: u and v) returns a number between 0 (no similarity between u and v) and 1 (full similarity between u and v). If a considered attribute is a road width, W, for the segment type (S) then f can be defined as:

$$f(u, v) = e^{-|w_u - w_v|},$$

where w_u, w_v denote segment widths for u and v respectively.

It should be emphasized that the form of an f function strongly depends on a context. Even for a single object type, e.g., S, it can have various forms, dependently on an attribute which is actually considered. In particular, one can apply $f \in \mathcal{D}$ which depends on several attributes of a node, say road width, number of lanes, surface properties and average daily traffic flow.

The formal definition of a similarity graph is given below:

Definition 2. *Similarity graph (abbrev. SG) is a base graph such that there exists $K \in \Gamma$ and*

1. $e = \{u, v\} \in E \subseteq \mathcal{P}_2(V) \wedge type(e) = K \implies type(u) = type(v)$,
2. $attr(K) = g \in \mathcal{D}^\Sigma$, where $\mathcal{D} = \{f | f : V \times V \longrightarrow [0, 1] \wedge f(x, y) = f(y, x)\}$.

An edge e satisfying 1 is referred to as a similarity edge. $\mathcal{P}_2(V)$ denotes all two element subsets of 2^V and $V, E, \Sigma, \Gamma, type, attr$ were defined in Definition 1.

A sample similarity graph for the representation shown in Fig. 2 is presented in Fig. 3. The values of some exemplary function f are marked alongside similarity edges (dashed lines). Only similarity edges connecting nodes of the type S are considered here.

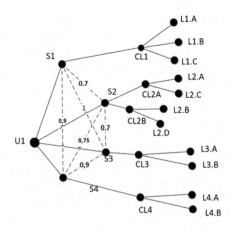

Fig. 3. Similarity graph with similarities among road segments (see Fig. 2)

3.3   Conformity Graph

As mentioned previously, installations for two (or more) similar lighting situations can be designed once. It reduces an overall preparation and cost assessment times. To achieve that, however, it is necessary to match all *similar* nodes, where similarity will be measured using functions from the $\mathcal{D}$ set introduced in Definition 2. To simplify that process we use a *conformity graph* defined below.

Definition 3. *Conformity graph* *(abbrev. CG), is a weighted similarity graph (see Definition 2) such that:*

1. $u, v \in V \implies type(u) = type(v)$,
2. $a, b \in \Sigma \implies attr(a) = attr(b)$,
3. $e \in E \implies type(e) = K \in \Gamma$ *and an edge weighting function w: (i) w : $E \longrightarrow [0, 1]$, (ii) $attr(K) = w$,*

where $V, E, \Sigma, \Gamma, K, type, attr$ were defined in Definitions 1 and 2.

In order to create a conformity graph we proceed following steps, starting from an initial similarity graph G_0, such as the one shown in Fig. 3, for instance.

Step 1 Select a desired node type **X** (say, **X** = S). Set a **similarity threshold** value, $\mathbf{t} \in [0, 1]$. Please note that an assumed value of $\mathbf{t}$ is a subject to an arbitrary decision, depending on a particular problem. For instance, in some cases we can admit even moderate disturbances in lamp spacings, what is reflected in a lower $\mathbf{t}$ value. It should be also noted that similarity between two nodes, say u and v, is tested against $\mathbf{t}$ using an $f(u, v)$ function value (see Definition 2).

Step 2 Remove all nodes of types other than **X** from G_0, together with incident edges. *Note*: At this moment we obtain a clique, G_1, with weighted edges (see Fig. 4).

Step 3 Find a maximum spanning tree, G_2, for the clique G_1. *Note*: That step, which can be performed using the modified Kruskal algorithm, is not deterministic, as several maximum spanning trees may exist for a single G_1.

Step 4 Remove all edges from G_2, for which $w(e) < \mathbf{t}$.

In Figs. 5a, 5b and 5c there are shown conformity graphs obtained for three similarity thresholds: $\mathbf{t}_1 = 0$ (no constraints are imposed on similarity), $\mathbf{t}_2 = 0.9$ and $\mathbf{t}_3 = 1$ (strict object similarity) respectively.

In the first of the above cases there are no limitations regarding similarity among segments. This implies that a lighting installation setup found for the segment $S1$ will be replicated to installations assigned to all other segments, namely, $S2, S3, S4$. Obviously, it may cause potential over- or under-lighting. The intermediate scenario, $\mathbf{t}_2 = 0.9$, represent the situation of the controlled replication of solutions among particular segments. The general rule is that reducing similarity threshold affects a confidence to the replicated solution quality but a benefit is the reduced preparation time. The trade-off between both determines a value of $\mathbf{t}$. In turn, the other extreme case, $\mathbf{t}_3 = 1$, reflects the situation when we reuse

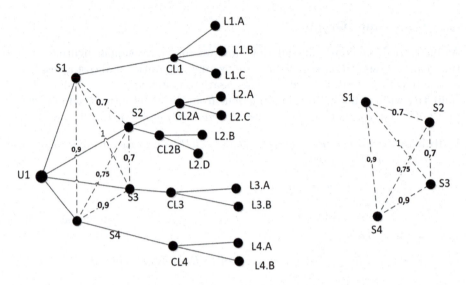

Fig. 4. Step 2 of the conformity graph creation process

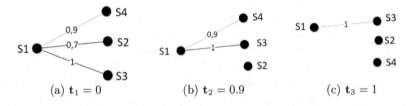

(a) $t_1 = 0$ (b) $t_2 = 0.9$ (c) $t_3 = 1$

Fig. 5. CGs with various similarity thresholds

existing solution if and only if there is a full conformity between two lighting situations (identical street geometries, lamp spacings, pole heights etc.). Although it does not allow to reduce significantly a computation time, unless there are multiple uniform lighting situations, it is guaranteed that solutions can be safely replicated among connected nodes.

4 Case Study

In this section application of similarity and conformity graphs in a real-life lighting installation retrofit process will be presented. The case we focus on is the investment carried out in the city of Cracow, Poland. The subject of modernization were 3,741 streetlights (approx. 5% of all streetlights in Cracow) located in the city center area (see Fig. 6). Its objective was replacing existing sodium fixtures with LED-based ones. The expected result was the power usage reduction which had reached about 72%. The achieved money savings were at the same level. The lighting project preparation took over two months for this investment.

Although the final result was satisfying, in terms of the power balance and financial goals, an investment's bottleneck was just the design process: note that only 5% of streetlights was modernized.

The analysis presented below gives an answer whether the process can be carried out faster. If we are able to shorten it, we will benefit from possibility of investigating several alternative setup variants, based on various combinations of fixtures, poles, arms and so on. Thus the final beneficiary is offered with a range of available options which can be selected dependently on actual business preferences and needs.

4.1 Optimization Process and Its Parameters

For the given investment 662 street segments (lighting situations) were considered. Figure 6 shows the investment scope. All segments are marked with individual colors.

Fig. 6. The investment scope

The goal of an optimization is selecting such values of particular lighting infrastructure parameters (see Table 3) that the resultant power usage is minimized.

In this case we also consider changing lamp positions (lamp spacings and setbacks), which is the extremely rare scenario in real-life retrofits.

Due to the financial constraints, the real-life investments are usually limited to changing fixtures and arms, sometimes the poles (lamp dimming and changing a mounting angle are obviously cost-free). The side effect of this limitation is less power usage reduction, compared to the full optimization. Performing a full optimization is much more complicated due to the time overhead but it offers a test bed for application the methods based on similarity and conformity graphs.

All parameters which were used in searching the optimal installation are summarized in Table 3. They produces the collection of 10,510,937,500 variants for a single segment only. It should be emphasized that the optimization engine used for calculations did not perform a brute force method but highly advanced methods and heuristics which are beyond the scope of this work. Finding the optimal setup for the entire considered investment area took about 8,220 min, on a single machine. The resultant power was 99.8 kW.

4.2   Application of Similarity and Conformity Graphs

Our goal, in this subsection, is application of similarity and conformity graphs, and thus reducing project preparation time. We also want to investigate how does a similarity threshold (see Subsection 3.3) value affects resultant installation power, solution quality and calculation time.

To answer those questions the light infrastructure was modeled by the base graphs. Each of 662 street segments had its BG representation, disjoint with other ones (it was assumed that neither lamp illuminates two lighting situations). Then similarity measures among segments were determined to obtain corresponding similarity graph. Such a SG was ready to generate CGs for subsequent threshold values between 0.50 and 1.00 (including).

Table 3. Optimization parameters

	From	To	Step	Number
Luminous flux dimming	1%	100%	1%	100
Fixture mounting angle	0°	30°	1°	31
Arm length	0 m	2 m	0.5 m	5
Lamp setback	0 m	2 m	0.5 m	5
Mounting height	6 m	12 m	1m	7
Lamp spacing	30 m	60 m	1 m	31
Fixtures types	n/a	n/a	n/a	625

After performing the series of calculations, we obtained the characteristics shown graphically in Fig. 7. Detailed results are presented in Table 4.

As shown previously, growing threshold value results in increasing number of connected components of a conformity graph.

When using the described method with the threshold value $t = 1$ we get the same final installation's power as for the standard approach, namely 99.8 kW. The calculation time, however, is 7,405 min. vs 8,220 min. for the standard

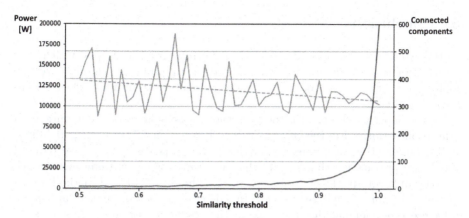

Fig. 7. Estimated resultant power of the installation and number of connected components of the conformity graph as a function of similarity threshold

Table 4. Detailed results using different conformity limit.

Similarity threshold t	Number of CG connected components	Estimated power [kW]	Power overhead*	Calculation time [min]
0.50	6	133.6	33.9%	74
0.60	6	134.3	34.6%	74
0.70	11	100.1	0.3%	135
0.80	17	108.2	8.4%	209
0.90	33	117.3	17.2%	406
0.95	65	107.0	7.2%	800
0.97	107	108.7	8.9%	1316
0.98	155	107.9	8.1%	1907
0.99	310	105.4	5.6%	3813
1.00	602	99.8	0.0%	7405

* The overhead relative to the optimal solution ($t = 1.00$)

method (i.e., its is 13.5 h shorter). This is because 60 of 662 initial segments had strictly similar neighbors (in the sense of CG vertex neighbors) with the same street geometry, so they inherited their solutions, without performing redundant optimization.

It should be noted that lowering the threshold by 5%, to $t = 0.95$, reduces the calculation time to 800 minutes (approx. 13 h 20 min) which is $\frac{1}{10}$ of the standard method calculation time, for the power overhead lower than 10%. This value is fully acceptable for estimating a final power of the installation.

5   Conclusions

In this work we present similarity and conformity graph concepts, which may be used for assessment purposes in soft computing problems, characterized by high complexity. One of the field of their application is outdoor lighting planning/retrofitting.

In the paper we also present the case study of the lighting infrastructure modernization performed in the city of Cracow, Poland, on 3,741 streetlights illuminating 662 lighting situations. By applying the proposed approach we reduced the design preparation time by 10, with only 10% worse power efficiency, which is acceptable rate in the considered context. The SG/CG-based method allows performing *what if* analyses as well. In this case, a decision maker can choose a fixture type yielding the best power usage reduction or to select streets for which a planned investment will give the best return on investment rate.

Analyzing test results we can see that although an obtained power efficiency can be worse compared to the optimal one, we get a result in hours rather than in days. The right choice of the similarity functions and thresholds is the important factor. Further analysis of these seems to be necessary.

The presented concept is also an outline for creating an agent system that would offer even faster estimation. The use of parallel processing might enable receiving initial estimates in a time comparable to the real time.

References

1. Bonini, N., Egidi, M.: Cognitive traps in individual and organizational behavior: some empirical evidence. RePEc Revue d économie industrielle (1999). https://doi.org/10.3406/rei.1999.1749
2. Escolar, S., Carretero, J., Marinescu, M.C., Chessa, S.: Estimating energy savings in smart street lighting by using an adaptive control system. Int. J. Distrib. Sens. Netw. (2014). https://doi.org/10.1155/2014/971587
3. Fujii, Y., Yoshiura, N., Takita, A., Ohta N.: Smart street light system with energy saving function based on the sensor network. In: International Conference on Future Energy Systems (e-Energy 2013) (2013)
4. Gómez-Lorente, D., Rabaza, O., Espín Estrella, A., Peña-García, A.: A new methodology for calculating roadway lighting design based on a multi-objective evolutionary algorithm. Expert Syst. Appl. **40**(6), 2156–2164 (2013)
5. International Energy Agency: Lights Labour's Lost, Policies for Energy-efficient Lighting (2006)
6. Komnata, K., Basiura, A., Kotulski, L.: Graph-based street similarity comparing method, theory and applications of dependable computer systems. In: DepCoS-RELCOMEX (2020). https://doi.org/10.1007/978-3-030-48256-5_36
7. Lima, GFM., Tavares, J., Peretta, IS., Yamanaka, K., Cardoso, A., Lamounier, E.: Optimization of lighting design using genetic algorithms. IEEEXplore (2010)
8. McKinsey&Company: Lighting the way: perspectives on the global lighting market (2011)
9. Nelson, M.A., Anderson, B.P., Cai, H.: Selection methods and procedure for evaluation of LED roadway luminaires. LEUKOS **13**, 159–175 (2017)

10. Peña-García, A., Gómez-Lorente, D., Espín, A., Rabaza, O.: New rules of thumb maximizing energy efficiency in street lighting with discharge lamps: the general equations for lighting design. Eng. Optim. **48**, 1080–1089 (2016)
11. PremiumLIGHTPro : LED Basics (2017). http://www.premiumlightpro. eu/fileadmin/user_upload/Education/PL-Pro-3-LED_Basics-3-Oct-2017.pptx. Accessed 4 Oct 2020
12. Rabaza, O., Gómez-Lorente, D., Pérez-Ocón, F., Peña-García, A.: A simple and accurate model for the design of public lighting with energy efficiency functions based on regression analysis. Energy **107**, 831–842 (2016)
13. Salata, F., et al.: Management optimization of the luminous flux regulation of a lighting system in road tunnels. A first approach to the exertion of predictive control systems. Sustainability **8**(11) (2016). https://doi.org/10.3390/su8111092. Art. no. 1092
14. Sędziwy, A.: Sustainable street lighting design supported by hyper-graph based computational model. Sustainability (2016)
15. Sędziwy, A., Basiura, A.: Energy reduction in roadway lighting achieved with novel design approach and LEDs. LEUKOS **14**(1), 45–51 (2018). https://doi.org/10. 1080/15502724.2017.1330155
16. Sędziwy, A., Basiura, A., Wojnicki, I.: Roadway lighting retrofit: environmental and economic impact of greenhouse gases footprint reduction, Sustainability Special Issue: Sustainable Lighting and Energy Saving (2018)
17. Sędziwy, A., Kotulski, L.: Graph-based optimization of energy efficiency of street lighting. In: Rutkowski, L., Korytkowski, M., Scherer, R., Tadeusiewicz, R., Zadeh, L.A., Zurada, J.M. (eds.) ICAISC 2015. LNCS (LNAI), vol. 9120, pp. 515–526. Springer, Cham (2015). https://doi.org/10.1007/978-3-319-19369-4_46
18. Sędziwy, A., Kotulski, L., Basiura, A.: Enhancing energy efficiency of adaptive lighting control. In: Nguyen, N.T., Tojo, S., Nguyen, L.M., Trawiński, B. (eds.) ACIIDS 2017. LNCS (LNAI), vol. 10192, pp. 487–496. Springer, Cham (2017). https://doi.org/10.1007/978-3-319-54430-4_47
19. Sędziwy, A., Kotulski, L., Basiura, A.: Agent aided lighting retrofit planning for large-scale lighting systems. In: 11th Asian Conference on Intelligent Information and Database Systems, ACIIDS (2019)
20. Tsao, J.Y., Waide, P.: The world's appetite for light: empirical data and trends spanning three centuries and six continents. LEUKOS **6**(4), 259–281 (2010)
21. Wojnicki, I., Ernst, S., Kotulski, L., Sędziwy, A.: Advanced street lighting control. Expert. Syst. **41**, 999–1005 (2014)
22. Wojnicki, I., Kotulski, L.: Empirical study of how traffic intensity detector parameters influence dynamic street lighting energy consumption: a case study in Krakow. Sustainability, Poland (2018)
23. Wojnicki, I., Kotulski, L., Ernst, S., Sedziwy, A., Strug, B.: A two-level agent environment for intelligent lighting control. Int. J. Mater. Prod. Technol. **53**, 187–201 (2015)

Pruned Simulation-Based Optimal Sailboat Path Search Using Micro HPC Systems

Roman Dębski$^{(\boxtimes)}$ and Bartlomiej Sniezynski

Institute of Computer Science, AGH University of Science and Technology,
Al. Mickiewicza 30, 30-059 Kraków, Poland
{rdebski,sniezyn}@agh.edu.pl

Abstract. Simulation-based optimal path search algorithms are often solved using dynamic programming, which is typically computationally expensive. This can be an issue in a number of cases including near-real-time autonomous robot or sailboat path planners. We show the solution to this problem which is both effective and (energy) efficient. Its three key elements – an accurate and efficient estimator of the performance measure, two-level pruning (which augments the estimator-based search space reduction with smart simulation and estimation techniques), and an OpenCL-based SPMD-parallelisation of the algorithm – are presented in detail. The included numerical results show the high accuracy of the estimator (the medians of relative estimation errors smaller than 0.003), the high efficacy of the two-level pruning (search space and computing time reduction from seventeen to twenty times), and the high parallel speedup (its maximum observed value was almost 40). Combining these effects gives (up to) 782 times faster execution. The proposed approach can be applied to various domains. It can be considered as an optimal path planing framework parametrised by a problem specific performance measure heuristic/estimator.

Keywords: Heterogeneous computing · SPMD-parallel processing · Trajectory optimisation · Dynamic programming

1 Introduction

Optimal path search – an important issue in robotics, aerospace engineering, and optimal control – in a number of cases has to be simulation based since the corresponding mathematical model is too complex for analytical methods. In such situations algorithms based on dynamic programming are often used. This approach usually leads to accurate results but is computationally expensive, because of the search space size and cost of simulation [8].

High computational costs can be an issue in a number of cases including (hard) real-time embedded systems but also near-real-time autonomous robot or sailboat path planners. Some of the time constraints can be met (at least

© Springer Nature Switzerland AG 2021
M. Paszynski et al. (Eds.): ICCS 2021, LNCS 12745, pp. 158–172, 2021.
https://doi.org/10.1007/978-3-030-77970-2_13

partially) by parallelising the algorithms and adapting them to contemporary CPU-GPU heterogeneous mobile/onboard computers, which are massively parallel micro HPC platforms. Although effective, this approach is often inefficient – the computation is accelerated but with no reduction of the algorithm computational cost. This can be an issue if energy consumption is of primary importance, which is the case in onboard/mobile computers (especially when at sea).

The aim of this paper, which is a significant extension of [8], is to present a simulation-based optimal sailboat path planning algorithm which is both effective and (energy) efficient. Its main contributions are:

- an estimator of the performance measure (i.e., sailing duration) that reflects its variational character and is accurate without being computationally complex (Sect. 4.2),
- the concept of two-level pruning, which augments the estimator-based search space reduction with smart simulation and estimation techniques (paragraph *External and internal pruning* in Sect. 4.3),
- the SPMD-parallel[1] algorithm for simulation-based optimal sailboat path search, based on the above two elements and adapted to on-board heterogeneous micro HPC systems (Sect. 4.3),
- numerical results which demonstrate three important aspects of the algorithm: the accuracy of the performance measure estimator, the efficacy of the two-level pruning, and the SPMD-parallelisation capabilities (Sect. 5).

The remainder of this paper is organised as follows. The next section presents related research. Following that, the search problem under consideration is defined and proposed algorithm is described. After that, experimental results are presented and discussed. The last section contains the conclusion of the study.

2   Related Research

The first scientific formulation of the problem of trajectory optimization[2] was proposed by Johan Bernoulli in 1696 as the brachistochrone problem (see [26] for discussion). For over two hundred years, the main approach for trajectory optimization was the calculus of variations (see, for instance [24]) based on Johan's brother Jakob's solution for the brachistochrone problem. This situation was changed when dynamic programming was introduced [1] after the development of the digital computer in the 1950's. Effective shortest path algorithms [2,9], the Pontryagin Maximum Principle [19] and non-linear programming (NLP) are foundations for many trajectory optimization algorithms. They are classified either as *direct*, to construct the best path step by step like our algorithm, or *indirect* in which the best path is a solution of some set of equations [15,25]. A subgroup of direct methods is a set of algorithms based on the shortest path algorithm [4,21].

---

[1] SPMD - Single Program Multiple Data.

[2] In this paper, the notions of *trajectory optimization* and *optimal path search* are used interchangeably.

An important part of optimal search problems is those having *black-box rep-resented* performance measures. In such cases the performance measure values are usually computed using computer simulation, and therefore classical opti-mization methods cannot be used directly. Algorithms for such problems are often based on heuristics, *soft-computing* and *AI* methods [20,27,28].

Another important direction of research is related to the parallelisation of both trajectory optimization and graph algorithms [6,13] which then can be executed using GPU[3] acceleration [10,12,17,22]. In our research we also utilized GPU. Although initially we applied machine learning (ML) algorithms with some success, even better results were obtained using a specially designed performance measure estimator allowing us to significantly limit the number of performance measure evaluations.

A different approach to speed up computations is pruning the search graph. In [11] uniform-cost grid environments are considered. They are simple but com-monly used in robotics and video games. The authors propose an algorithm find-ing optimal paths by expansion of selected nodes only. Pruning rules are defined to decide if a node should be skipped or expanded. In [29] an algorithm for path planning of a differential drive mobile robot is proposed. It is an extension of a Bi-directional Rapidly-exploring Random Tree (RRT) method [14]. It improves the performance of path planning by incorporating kinematic constraints and efficient branch pruning. In [30] another extension of RTT is proposed. An ini-tial tree covering the whole map is processed using branch pruning, reconnection, and regrowth operations. It allows for planning in a complex, dynamic environ-ments in which obstacles and the destination are moving.

Trajectory optimization is often used in robotics, as mentioned above, and also in other various domains, for example aerospace engineering [5]. However, the sailboat domain is also common [7,18,23,31].

3   Problem Formulation

Consider a sailboat going from point $A(q_A, y_A)$ to $B(q_B, y_B)$, where (q_i, y_i) are the coordinates of the corresponding point in either the Cartesian or polar sys-tem. We assume that the true wind can be expressed by the following vector field (see Fig. 1)

$$\boldsymbol{v}_t(q, y, t) = M(q, y, t)\, \hat{\boldsymbol{q}} + N(q, y, t)\, \hat{\boldsymbol{y}}, \tag{1}$$

where: $M(q, y, t)$, $N(q, y, t)$ are scalar functions, and $\hat{\boldsymbol{q}}$, $\hat{\boldsymbol{y}}$ are the unit vectors representing the axes of the corresponding coordinate system.

The set of admissible $\widetilde{AB}$ paths (i.e. the problem domain) consists of C^1-continuous curves which cover the given sailing area S_A (see Fig. 1) and do not violate the constraints embedded in a sailboat model (these constraints can be related to the state and/or control variables). This model is used to evaluate each path $(y^{(i)})$ through simulation, therefore

$$J[y^{(i)}] = \text{PerformSimulation}[y^{(i)}, \text{cfg}\,(\boldsymbol{v}_t, \dots)], \tag{2}$$

---

[3] Graphics Processing Unit.

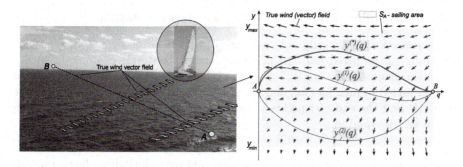

Fig. 1. Optimal sailboat path search problem: example admissible paths connecting points A and B, with $y^{(*)}(q)$ representing the optimal path.

where: J represents the given performance measure and $\mathrm{cfg}\,(\boldsymbol{v}_t, \dots)$ – the simulator configuration.

Problem Statement. The optimal sailboat path search problem under consideration can be defined as follows:

- find, among all admissible paths, the one with the best value of performance measure J;
- the values of J can be found only through simulation;
- only on-board, off-line computers can be used.

Remark 1. In the special case, when $J[y^{(i)}] = \Delta t[y^{(i)}]$, with Δt being the sailing duration, we get the *minimum-time problem*.

4   Solution

The approach we propose in this paper is a "pruning augmented" extension of the one introduced in [8]. It is based on the following two main steps:

1. transformation of the continuous optimisation problem into a (discrete) search problem over a specially constructed finite graph (*multi-spline*);
2. application of pruned dynamic programming to find the approximation of the optimal path represented as a C^1-continuous cubic Hermite spline.

These two steps repeated several times form an adaptive version of the algorithm in which subsequent grids are generated through mesh refinement making use of the best trajectory found so far. Key elements of the proposed algorithm are:

- multi-spline based solution space and the SPMD-parallel computational topology it generates;
- effective estimate of the performance measure (sail duration);
- two-level smart pruning that significantly reduces the time complexity of the reference algorithm.

They are discussed in the following sub-sections.

4.1   Multi-spline as the Solution Space Representation

A discretisation of the original continuous problem leads to a grid, G, which structure can be fitted into the problem domain. A simple example of such a grid is shown on the left of Fig. 2. The grid is based on equidistant nodes grouped in rows and columns: four regular rows plus two special ones – containing the start (A) and the end (B) points – and four columns. The number of nodes in such a grid is equal to

$$|G| = n_c \left(n_r - 2 \right) + 2 \tag{3}$$

where n_c and n_r are the numbers of columns and rows (including the two special ones), respectively.

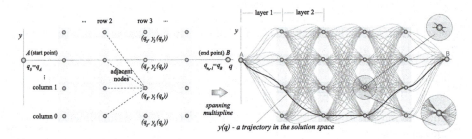

Fig. 2. Solution space representation: multi-spline built (spanned) on regular grid G_{ex}

Having assigned n_{ts} additional values (a vector of n_{ts} tangent slopes) to every node of grid G, we obtain a new structure, G_{ex}, that can store not only the coordinates of each node but also the n_{ts} slopes (angles) of path segments which start/end in that particular node (see the right part of Fig. 2).

Joining the nodes from subsequent rows of G_{ex} by using cubic Hermite spline segments, we get a *multi-spline* [8] which forms a discrete space of C^1-continuous functions (see Fig. 2). The properties of a multi-spline that are important from the point view of this paper are as follows [8]:

– when seen as a graph (knots are vertices, spline segments are edges) – it is directed (from A to B or vice-versa), acyclic and topologically sorted (i.e. the edges in layer l are followed by those from layer $l + 1$ and the vertices in row r are followed by those from row $r + 1$, see Fig. 2);
– it is built from $n_c n_{ts}^2 \left[\left(n_r - 3 \right) n_c + 2 \right]$ different spline segments;
– the discrete search space it spans represents $n_{ts}^{n_r} n_c^{n_r-2}$ different trajectories connecting points A and B; this value corresponds to the "inter-row complete" graph (i.e. the one in which all vertices from subsequent rows are connected);
– each of its internal layers (again, in the "inter-row complete" graph) consists of $n_c^2 n_{ts}^2$ spline segments.

Remark 2. Since the details of multi-spline auto-adaptation and formulae for Hermite spline segments are not of prime importance from the point of view of this paper, they have been omitted. If necessary, they can be found in [8].

4.2   Performance Measure Estimate

Simulation is the most complex and time-demanding phase of the optimal path search algorithm with the critical element being the computation of an instantaneous net-force[4], F, acting on a sailboat. Although the complete evaluation of a multi-spline spanned on a $G_{ex}\langle n_r, n_c, n_{ts}\rangle$ requires $n_c n_{ts}^2 [(n_r - 3) n_c + 2]$ simulations (see Sect. 4.1), and a significant number of them need more than a thousand computations of F, in a typical scenario, simulations for more than 80% of the multi-spline segments are not really necessary as their performance measures are significantly worse. Unfortunately, we do not know them up-front (i.e. before the simulation), and hence the need for a good estimator.

After some experiments with several ML-algorithms (both off-line and on-line) we have found an estimate that reflects the variational character of the performance measure and is accurate enough without being computationally complex. The solution was not at all obvious because of the circular-dependent nature of the problem: the performance measure of a path segment ($t_E = t_S + dt$) obviously depends on the time of reaching its start point (t_S), the initial velocity (v_S), and the segment length (dl). It also depends on F which, in turn, depends on the current position of the sailboat (see Eq. 1) and its instantaneous velocity, which depends (circularly) on F. The pseudo-code of the estimator is presented as Algorithm 1.

The proposed estimator is inspired by *the work-energy theorem* that states that the net-work done by the forces on an object (here the sailboat), $W_{AB} = F\, l_{AB}$, equals the change in its kinetic energy, $0.5\, m_s(v_B^2 - v_A^2)$, where m_s stands for the sailboat mass. The estimate is calculated simultaneously (as a kind of side-effect) with the Gaussian quadrature based computation of the segment length.

Remark 3. The estimation of path segments is also pruned (see line 11 in Algorithm 1).

4.3   Pruned Optimal Sailboat Path Search

The graph G_{ex}, on which the solution space (multi-spline) is spanned, is directed, acyclic (DAG) and has a layered structure. These properties can be naturally utilised in a parallel version of the dynamic programming based optimal path search algorithm.

---

[4] i.e., the sum of all forces acting on a sailboat; this can be found taking into account the wind vector field and the characteristics of the particular sailboat, see [16].

Algorithm 1: Path segment performance measure estimate

Input:
- s: the segment to be estimated,
- v_S: the initial velocity of the sailboat (at the start point of s),
- t_S: the time of reaching the start point of s (starting from A),
- t_{min}: the best estimated performance measure found so far,
- Δ_t: the penalty value (e.g., 10^3 seconds) used when the sailboat stops,
- E_M: the "safety factor" (e.g., 1.2) for turning-on the rough estimation mode

Output: the estimated value of the performance measure of s

1   **function** estimate(s; v_S, t_S, t_{min}, Δ_t, E_M):
2     $v_1 \leftarrow v_S$; $dt \leftarrow 0$
3   **foreach** x_i in Gauss nodes for segment s **do**
4       $dl_i \leftarrow$ the length of the i-th sub-segment of s
5       $F_i \leftarrow$ the net-force for the current position and velocity
6       $s_v \leftarrow v_1^2 + 2\,F_i\,dl_i\,(m_s)^{-1}$
7   **if** $s_v > 0$ **then** $v_2 \leftarrow \sqrt{s_v}$ **else** $v_2 \leftarrow 0$
8       $\bar{v} \leftarrow 0.5\,(v_1 + v_2)$
9   **if** $\bar{v} > 0$ **then** $dt \leftarrow dt + dl_i\,(\bar{v})^{-1}$ **else** $dt \leftarrow dt + \Delta_t$
10      $v_2 \leftarrow v_1$
11   **if** $(t_S + dt)\,t_{min}^{-1} > E_M$ **then return** (length (s) $(\sum_{k=0}^{i} dl_k)^{-1}\,dt$)
12   **return** $t_S + dt$

Principle of Optimality as the Algorithm Foundation. At the beginning of the search process the *cost matrix* is unknown – the performance measure of each path segment in the graph will be received from simulation using the *Principle of Optimality* [3]. This principle can be expressed for an example path A-$N_{r,c,s}$ (see Fig. 3) in the following way:

$$\widetilde{J}_A^{N_{r,c,s}} = \min_{c_j, s_k} \left(\widetilde{J}_A^{N_{r-1,c_j,s_k}} + J_{N_{r-1,c_j,s_k}}^{N_{r,c,s}} \right) \tag{4}$$

where: $c_j = (0, \ldots, n_c - 1)$, $s_k = (0, \ldots, n_{ts} - 1)$, $J_{N_s}^{N_e}$ is the cost corresponding to the path $N_s - N_e$ (N_s - start node, N_e - end node), $\widetilde{J}$ represents the optimal value of J and $N_{r,c,s}$ is the node of G_{ex} with "graph coordinates" $\langle row, column, tangent\_slope \rangle = \langle r, c, s \rangle$.

Figure 3 (a visualization of Eq. 4) presents the computation state in which the optimal costs of reaching all nodes in row $r - 1$ are known (they were calculated in previous stages of this multi-stage process). The optimal cost of path A-$N_{r,c,s}$ is calculated by performing simulations for all spline segments that join node $N_{r,c,s}$, which is located in row r, with nodes from the previous (i.e. $(r - 1)^{\text{th}}$) row.

Multi-spline Generated Computational Topology. The simulation-based multi-stage process can be visualized as a propagation of a "simulation-wave" presented in Fig. 4.

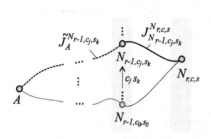

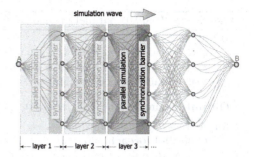

Fig. 3. Principle of optimality in dynamic programming (see Eq. 4).

Fig. 4. A sequence of parallel simulations for multi-spline segments from the same layer.

The computation begins from point A in layer 1, taking into account the corresponding initial conditions, and is continued (layer by layer) for the nodes in subsequent rows. On the completion of the simulations for the last layer (i.e. reaching the end node B), we get the optimal path and the corresponding value of the performance measure. The sequential component of the computation – presented in Fig. 4 as a synchronization barrier – is the result of the layer-on-layer dependence (to start simulation for a segment we have to know the corresponding initial conditions, see Eq. 4).

The Algorithm. The multi-spline generated computational topology is reflected in the SPMD structure of Algorithm 2. From its two sequential parts only the internal one – evaluating all segments which end in the same entry point (e_p) – needs some clarification. It has four important steps:

1. (lines 3–6) – estimation of all segments which end in entry point
2. (lines 7–10) – simulation for the k segments which have the best estimates (to verify the estimation accuracy; the value of k can be a constant or auto-adaptive variable);
3. (lines 11–13) – computation of the estimation accuracy measure Δt_{s_i};
4. (lines 14–17) – simulation for the rest of the "promising" segments (if there are any).

Pruning, discussed in the next paragraph, is applied in steps 1, 2, and 4.

External and Internal Pruning. Significant reduction of the average-case time complexity of the algorithm is important at least for two reasons. Firstly, we often need to know the solution as soon as possible (sometimes for safety reasons). Secondly, since we only use on-board computers, energy efficiency is critical while at sea. This is why the reference algorithm has been augmented with two-level pruning. The first pruning level (*external*) is related to the explicit reduction of the search space (lines 14–17 in Algorithm 2) using the performance measure estimate. The second level pruning mechanism (*internal*) is embedded both in the process of estimation (line 11 in Algorithm 1) and in simulation (lines 8–10

Algorithm 2: SPMD-parallel pruned search of the optimal path

Input:
- g_{AB}: initial (layered) grid with the start point, A, and the target point, B,
- $\boldsymbol{v}_t$: (true) wind vector field (see Eq.1),
- MODEL: sailboat model definition

Output: minimum time, t_{min}, sailboat path

1 **foreach** *layer in grid* g_{AB} **do**
2   **@parallel foreach** *entry point* e_p *of its nodes* **do**
3         $sgms\langle e_p \rangle \leftarrow$ select segments ending in the entry point e_p
4   **foreach** s_i *in* $sgms\langle e_p \rangle$ **do**
5             $t_{s_i}^{(est)} \leftarrow$ estimate $(s_i; v_{0i}, t_{0i}, t_{min}, \ldots)$
6   **if** $t_{s_i}^{(est)} < t_{min}$ **then** $t_{min} \leftarrow t_{s_i}^{(est)}$
7         $k\_best \leftarrow$ select k segments with best estimates
8   **foreach** s_i *in* k_best **do**
9             $t_{s_i}^{(sim)} \leftarrow$ pruned adaptive simulation for (s_i)
10   **if** $t_{s_i}^{(sim)} < t_{best}$ **then** $t_{best} \leftarrow t_{s_i}^{(sim)}$
11   **foreach** s_i *in* k_best **do**
12             $\Delta t_{s_i} \leftarrow |t_{s_i}^{(est)} - t_{best}| \, t_{best}^{-1}$
13   **if** $\Delta t_{s_i} > \Delta t_{max}^{(k)}$ **then** $\Delta t_{max}^{(k)} \leftarrow \Delta t_{s_i}$
14   **foreach** s_i *in* $\{ sgms\langle e_p \rangle \setminus k\_best \}$ **do**
15   **if** $|t_{s_i}^{(est)} - t_{best}| \, t_{best}^{-1} < C_M \, \Delta t_{max}^{(k)}$ **then**
16                 $t_{s_i}^{(sim)} \leftarrow$ pruned adaptive simulation for (s_i)
17   **if** $t_{s_i}^{(sim)} < t_{best}$ **then** $t_{best} \leftarrow t_{s_i}^{(sim)}$
18   save the best segment for e_p

and 14–17). It terminates the computation from "inside" that cannot lead to a better solution than the reference one.

Complexity Analysis. Algorithm 2 average-case *time complexity* is determined by the number of solution space refinements, n_i, the average number of net-force evaluations[5] for a single path segment, $\bar{n}_F$, and the number of such segments, $n_c n_{ts}^2 [(n_r - 3) n_c + 2]$ (see Sect. 4.1). For the sequential version of the algorithm it can be expressed as:

$$T_s = \Theta \left(n_i \, \bar{n}_F \, n_r \, n_c^2 \, n_{ts}^2 \right). \tag{5}$$

In the SPMD-parallel version of the algorithm, the evaluations for all nodes in a given row can be performed in parallel (using p processing units), thus:

$$T_p = \Theta \left(n_i \, \bar{n}_F \, n_r \, n_c \, n_{ts} \left\lceil \frac{n_c \, n_{ts}}{p} \right\rceil \right). \tag{6}$$

---

[5] Values of F are needed both in estimation and in simulation; Runge-Kutta-Fehlberg 4(5) method, used in the simulator, requires at each step six evaluations of F.

As for the reference algorithm [8], Algorithm 2 *space complexity* formula, $\Theta\left(n_r n_c n_{ts}\right)$, arises from the solution space representation.

Remark 4. The aim of pruning is to significantly reduce $\bar{n}_F$.

5   Experimental Verification

To demonstrate the effectiveness of the pruned optimal sailboat path search algorithm, a series of experiments was carried out using a MacBook Pro[6] with macOS 10.15.7 and OpenCL 1.2, having two (operational) OpenCL-capable devices: Intel Core i7-3740QM @ 2.7 GHz (the CPU) and Intel HD Graphic 4000 (the integrated GPU). The aim of the experiments was to investigate three important aspects of the algorithm: the accuracy of the performance measure estimator, the efficacy of different kinds of pruning, and the SPMD-parallelisation efficiency. The results are presented in the subsequent paragraphs.

Performance Measure Estimator Accuracy. This element has an explicit impact on the search space reduction since the more accurate the estimator is, the more segments can be omitted in the simulation phase (see lines 14–17 in Algorithm 2). The results of its experimental evaluation are given in the form of a violin plot in Fig. 5.

The plot shows the distributions of estimation relative errors (i.e., $|(t_{est} - t_{sim})/t_{sim}|$) for path segments from different search spaces.

Remark 5. In all cases the medians of relative estimation errors were smaller than 0.003 (i.e., 0.3%), which confirms the very high accuracy of the proposed estimator[7].

Pruning Efficacy. The application of different types of pruning is one of the two ways of lowering the total computation time. Its efficacy was verified using the reduction of the number of net-force computations, $(\bar{n}_F^{(base)} - \bar{n}_F^{(prun)})/\bar{n}_F^{(base)}$, as the measure. The corresponding experimental results are given in Table 1 and Fig. 6. To verify the accuracy of $\bar{n}_F$ as the measure of the algorithm time complexity (see Eqs. 5 and 6), the duration of sequential computations, t_{sim}, was also measured.

The results shown in Table 1 prove the very high accuracy since the Pearson correlation coefficient, $\rho(\bar{n}_F, t_{sim})$, is equal to 1 (up to 5 decimal places), which means (almost) perfect linear dependence of the two variables.

Remark 6. Different types of pruning working together reduced the sequential computation time from seventeen to twenty times (see Fig. 6).

---

[6] With 16 GB of DDR3 1600 MHz RAM; the laptop manufactured in 2013.

[7] The samples sizes (i.e., the number of segments) used to compute the distributions of errors were very large: from 29 760 for $n_c = 16$ to 950 528 for $n_c = 128$.

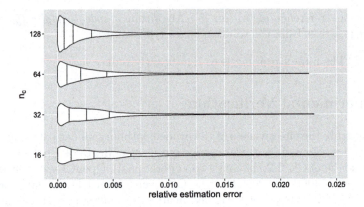

Fig. 5. Distributions of relative errors of the performance measure estimate, e, for grids with different n_c. In each case the calculation was based on the k best values (lines 7–10 in Algorithm 2) with outliers excluded (i.e., $e > \bar{e} + 3\,\sigma_e$).

Table 1. Efficacy of different kinds of pruning: average numbers of net-force evaluations, $\bar{n}_F$, and execution times, $\bar{t}_{sim}$ (in seconds), for different n_c. The solution space with $n_r = 32$, $n_{ts} = 8$; two refinements. *Auxiliary notation*: IPS - Internal Pruning in simulation, EP - External Pruning [+estimation], IPE - Internal Pruning in Estimation.

n_c	BASE		IPS		EP		IPS+EP		IPS+IPE+EP	
	$\bar{n}_F$	t_{sim}	$\bar{n}_F$	t_{sim}	$\bar{n}_F$	t_{sim}	$\bar{n}_F$	t_{sim}	$\bar{n}_F$	t_{sim}
16	748.8	206.0	165.5	44.3	172.1	15.0	53.7	14.1	44.1	12.2
32	735.8	812.4	142.8	154.9	170.5	59.9	51.6	55.0	42.6	47.3
64	731.4	3351.9	136.0	600.6	170.2	239.3	50.7	210.9	41.6	182.4
128	731.7	12950.2	137.6	2428.7	145.1	810.0	46.0	769.8	37.0	654.8

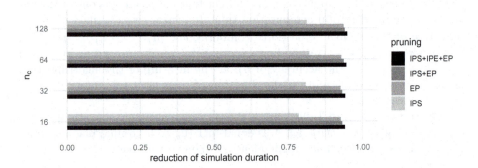

Fig. 6. Reduction of simulation duration for different kinds of pruning and different n_c. The auxiliary notation as in Table 1.

SPMD-*Parallelisation Efficiency.* Parallelisation is the second way of lowering the total computation time. Contemporary mobile/embedded computers are usually equipped with more than one type of processor, typically one CPU and one or two GPUs. OpenCL makes it possible to use these heterogeneous platforms effectively since the same code can be executed on any OpenCL-capable processor. The performance comparison of the two available processors is shown[8] in Table 2 and Fig. 7.

Table 2. SPMD-parallelisation efficiency: average execution times, $\bar{t}_{sim}$ (in seconds), and standard deviations, σ, from ten runs of the parallel version of the algorithm for different n_c. The solution space with $n_r = 32$, $n_{ts} = 8$; two refinements.

n_c	i7-3740QM (seq)		i7-3740QM+OCL		HD4000+OCL	
	$\bar{t}_{sim}$	σ	$\bar{t}_{sim}$	σ	$\bar{t}_{sim}$	σ
16	12.2	0.05	2.53	0.06	1.85	0.00
32	47.3	0.88	5.74	0.10	4.34	0.05
64	182.4	3.31	22.16	0.27	8.89	0.01
128	654.8	19.04	72.87	0.21	16.55	0.12

The results refer to the case of all three types of pruning being active with the pair of columns annotated with "seq" corresponding to the reference (sequential) variant of the algorithm. The two platforms differ significantly (see Fig. 7), and therefore there is definitely room for processor allocation plan optimisation done before or during code execution (run-time).

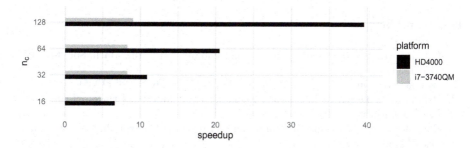

Fig. 7. SPMD-parallelisation efficiency: speed-ups, t_{seq}/t_{par}, for different Open-CL platforms and n_c ($t_{(.)}$ measured with $-Os$ flag). The remaining parameters as in Table 2.

---

[8] The results for $n_c < 16$ are omitted because the search spaces they define are too coarse-grained; it is worth noting, however, that for such cases the CPU was faster.

Remark 7. The maximum observed speedup was 39.56 (see Fig. 7). Setting the reference to the case with no pruning gives in total $19.78 \times 39.56 \approx 782.5$ times faster execution.

6   Conclusions

It has been shown that the simulation-based optimal sailboat path planning algorithm can be both effective and (energy) efficient. The three key elements in achieving this have been an accurate and efficient estimator of the performance measure (sailing duration), the two-level pruning (which augments the estimator-based search space reduction with smart simulation and estimation techniques), and the OpenCL-based SPMD-parallelisation of the algorithm.

The numerical results show the high accuracy of the estimator (the medians of relative estimation errors were smaller than 0.003, see Fig. 5), the high efficacy of the two-level pruning (search space and computing time reduction from seventeen to twenty times, see Table 1 and Fig. 6), and the high parallel speedup (its maximum observed value was 39.56, see Fig. 7). Combining these effects has given (up to) 782.5 times faster execution.

The proposed approach can be applied to various domains. It can be considered as an optimal path planing framework parametrised by a problem specific performance measure heuristic/estimator. Further exploration of this idea could be the first possible future research direction. Another could be the algorithm space complexity reduction.

Acknowledgement. Research presented in this paper was supported by the funds assigned to AGH University of Science and Technology by the Polish Ministry of Science and Higher Education.

References

1. Bellman, R.: The theory of dynamic programming. Bull. Am. Math. Soc. **60**, 503–515 (1954)
2. Bellman, R.: On a routing problem. Q. Appl. Math. **16**, 87–90 (1958)
3. Bellman, R., Dreyfus, S.: Applied Dynamic Programming. Princeton University Press, Princeton (1962)
4. Bertsekas, D.P.: Dynamic Programming and Optimal Control, 2nd edn. Belmont, Mass (2000)
5. Ceriotti, M., Vasile, M.: MGA trajectory planning with an ACO-inspired algorithm. Acta Astronaut. **67**(9–10), 1202–1217 (2010)
6. Crauser, A., Mehlhorn, K., Meyer, U., Sanders, P.: A parallelization of Dijkstra's shortest path algorithm. In: Brim, L., Gruska, J., Zlatuška, J. (eds.) MFCS 1998. LNCS, vol. 1450, pp. 722–731. Springer, Heidelberg (1998). https://doi.org/10.1007/BFb0055823
7. Dalang, R.C., Dumas, F., Sardy, S., Morgenthaler, S., Vila, J.: Stochastic optimization of sailing trajectories in an upwind regatta. J. Oper. Res. Soc. **66**, 807–821 (2014)

8. Dębski, R.: An adaptive multi-spline refinement algorithm in simulation based sailboat trajectory optimization using onboard multi-core computer systems. Int. J. Appl. Math. Comput. Sci. **26**(2), 351–365 (2016)
9. Dijkstra, E.W.: A note on two problems in connexion with graphs. Numer. Math. **1**, 269–271 (1959)
10. Dębski, R.: High-performance simulation-based algorithms for alpine ski racer's trajectory optimization in heterogeneous computer systems. Int. J. Appl. Math. Comput. Sci. **24**(3), 551–566 (2014)
11. Harabor, D., Grastien, A.: Online graph pruning for path finding on grid maps, vol. 2 (2011)
12. Harish, P., Narayanan, P.J.: Accelerating large graph algorithms on the GPU using CUDA. In: Aluru, S., Parashar, M., Badrinath, R., Prasanna, V.K. (eds.) HiPC 2007. LNCS, vol. 4873, pp. 197–208. Springer, Heidelberg (2007). https://doi.org/10.1007/978-3-540-77220-0_21
13. Jasika, N., Alispahic, N., Elma, A., Ilvana, K., Elma, L., Nosovic, N.: Dijkstra's shortest path algorithm serial and parallel execution performance analysis. In: MIPRO, 2012 Proceedings of the 35th International Convention, pp. 1811–1815. IEEE (2012)
14. Kuffner, J.J., LaValle, S.M.: Rrt-connect: an efficient approach to single-query path planning. In: Proceedings 2000 IEEE International Conference on Robotics and Automation, vol. 2, pp. 995–1001. IEEE (2000)
15. Lewis, R.M., Torczon, V., Trosset, M.W.: Direct search methods: then and now. J. Comput. Appl. Math. **124**, 191–207 (2000)
16. Marchaj, C.: Aero-hydrodynamics of Sailing. Adlard Coles Nautical (2000)
17. Park, C., Pan, J., Manocha, D.: Real-time optimization-based planning in dynamic environments using GPUs. In: 2013 IEEE International Conference on Robotics and Automation (ICRA), pp. 4090–4097. IEEE (2013)
18. Pêtres, C., Romero-Ramirez, M.A., Plumet, F.: Reactive path planning for autonomous sailboat. In: 2011 15th International Conference on Advanced Robotics (ICAR), pp. 112–117. IEEE (2011)
19. Pontryagin, L.S., Boltyanski, V.G., Gamkrelidze, R.V., Mischenko, E.F.: The Mathematical Theory of Optimal Processes. Interscience, NY (1962)
20. Pošík, P., Huyer, W., Pál, L.: A comparison of global search algorithms for continuous black box optimization. Evol. Comput. **20**, 509–541 (2012)
21. Rippel, E., Bar-Gill, A., Shimkin, N.: Fast graph-search algorithms for general-aviation flight trajectory generation. J. Guid. Control. Dyn. **28**(4), 801–811 (2005)
22. Singla, G., Tiwari, A., Singh, D.P.: New approach for graph algorithms on GPU using CUDA. Int. J. Comput. Appl. **72**(18), 38–42 (2013). Published by Foundation of Computer Science, New York, USA
23. Stelzer, R., Pröll, T.: Autonomous sailboat navigation for short course racing. Robot. Auton. Syst. **56**(7), 604–614 (2008)
24. Stillwell, J.: Mathematics and its History, 3rd edn. Springer, New York (2010). https://doi.org/10.1007/978-1-4419-6053-510.1007/978-1-4419-6053-5
25. von Stryk, O., Bulirsch, R.: Direct and indirect methods for trajectory optimization. Annals Oper. Res. **37**(1), 357–373 (1992)
26. Sussmann, H.J., Willems, J.C.: 300 years of optimal control: from the brachystochrone to the maximum principle. IEEE Control. Syst. **17**(3), 32–44 (1997)
27. Szłapczyński: Customized crossover in evolutionary sets of safe ship trajectories. Int. J. Appl. Math. Comput. Sci **22**(4), 999–1009 (2012)
28. Vasile, M., Locatelli, M.: A hybrid multiagent approach for global trajectory optimization. J. Global Optim. **44**(4), 461–479 (2009)

29. Wang, J., Li, B., Meng, M.Q.H.: Kinematic constrained bi-directional RRT with efficient branch pruning for robot path planning. Expert Syst. Appl. **170**, 114541 (2021)
30. Zhang, C., Zhou, L., Li, Y., Fan, Y.: A dynamic path planning method for social robots in the home environment. Electronics **9**, 1173 (2020)
31. Życzkowski, M.: Sailing route planning method considering various user categories. Polish Marit. Res. **27** (2020)

Two Stage Approach to Optimize Electricity Contract Capacity Problem for Commercial Customers

Rafik Nafkha⬤, Tomasz Ząbkowski⬤, and Krzysztof Gajowniczek⁽✉⁾⬤

Institute of Information Technology,
Warsaw University of Life Sciences SGGW, Warsaw, Poland
{rafik_nafkha,tomasz_zabkowski,
krzysztof_gajowniczek}@sggw.edu.pl

Abstract. The electricity tariffs available to Polish customers depend on the voltage level to which the customer is connected as well as contracted capacity in line with the user demand profile. Each consumer, before connecting to the power grid, declares the demand for maximum power which is considered a contracted capacity. Maximum power is the basis for calculating fixed charges for electricity consumption. Usually, the maximum power for the household user is controlled through a circuit breaker. For the industrial and business users the maximum power is controlled and metered through the peak meters. If the peak demand exceeds the contracted capacity, a penalty charge is applied to the exceeded amount which is up to ten times the basic rate. In this article, we present a solution for entrepreneurs which is based on the implementation of two stage approach to predict maximal load values and the moments of exceeding the contracted capacity in the short-term, i.e., up to one month ahead. The forecast is further used to optimize the capacity volume to be contracted in the following month to minimize network charge for exceeding the contracted level. As shown experimentally with two datasets, the application of multiple output forecast artificial neural network model and genetic algorithm for load optimization delivers significant benefits to the customers.

Keywords: Contracted capacity · Optimization · Genetic algorithm · Electricity load time series forecasting

1 Introduction

Energy storage and supply conditions are demanding and difficult on the electricity market when compared to other inputs of a typical production system. Therefore, forecasting the load demand is of high importance. To deal with those inconvenient conditions, energy producers propose different energy tariffs and contract options to their customers. Usually, voltage level and individual contracted capacity are the main factors to assign proper tariff. This strategy ensures that fluctuations in energy demand are controlled, what gives an insight for the energy quantity required to be generated and allows to transmit it to the customers.

© Springer Nature Switzerland AG 2021
M. Paszynski et al. (Eds.): ICCS 2021, LNCS 12745, pp. 173–184, 2021.
https://doi.org/10.1007/978-3-030-77970-2_14

The Polish demand side of the retail electricity market comprises end-users. In total, there are over 17.05 million of them, out of whom 90.3% (15.4 million) are the customers in G tariff group, with a great majority of household consumers (over 14.5 million) who purchase electricity for individual consumption [1]. The rest of end-users are industrial, business and institutional clients and they may belong to A, B or C tariff group [2]. There are three voltage levels distinguished in Poland: high voltage (110 kV and higher), medium voltage (higher than 1 kV but lower than 110 kV) and low voltage (less than 1 kV). Tariff groups A and B comprise customers supplied from the high and medium voltage grids, i.e., the so-called industrial customers, whereas group C is typical for the customers connected to the. low voltage grid consuming electricity for the purpose of small and medium business activity.

One of the main variables considered in the tariff structures is the capacity component so the users are charged for the availability to use the maximum power, in line with the connection agreement which is the maximum value of the averaged consumed power within the period of 15 min in an hour span [3]. In practice, households and small business are not obliged to monitor the level of power consumed on an ongoing basis. Customers who are using these tariffs are not charged for exceeding the contracted capacity. On the other hand, if the declared capacity quantity is exceeded by the consumer of tariff groups (C, B or A) penalty charge is levied. In line with the government's regulation with regards to the specific rules for the determination and calculation of the tariffs and billing in electricity industry [3], a fee is charged for exceeding the contractual capacity defined in the contract. The fee constitutes the product of the rate of capacity charge and the sum of ten largest quantities of the surplus consumed capacity over the contractual capacity, indicated by the measuring and clearing device or ten times the maximum amount of surplus of consumed capacity over the contracted capacity recorded during the reference period.

Large companies which are connected to medium or high voltage lines, usually have an energy specialist who takes care of the energy consumption parameters and supervises them on an ongoing basis. The level of power consumed is monitored by the metering systems, and in case of exceeding the contractual level, an additional fee related to exceeding value appears on the invoice. In practice, big companies are rarely unconscious about exceeding their contractual capacity. On the other hand, entrepreneurs operating in the field of services, production and processing, who are connected to low voltage with contracted capacities above 40 kW [3], do not always have the time, appropriate information, and knowledge to control their energy consumption parameters to ensure their optimal adjustment. Generally, in order to avoid over-contracted capacity amount, customers declare a level of contracted capacity that is much higher than their needs. On the other hand, those customers who are not using the planned capacity pay for the unused power.

Time Horizon Selection. Numerous papers consider load forecasting, but only few of them use the short-term load forecasting for tariff optimization. Most of the works are mainly related to long-term optimization of the electricity purchase and distribution process by suppliers and distributors. In general, load forecasting has been investigated by utilities and electricity suppliers where Long-Term Load Forecasts (LTLF) are used to predict the annual peak of the power system [4, 5] in order to manage future investments in terms of modernization and launching new units to maintain stability of nationwide electricity demand over time periods of up to 20 years [6]. The Medium-Term Load Forecasts (MTLF) use hourly loads to predict the weekly peak load for both, power and system operations planning [7]. The Short-Term Load Forecasts (STLF) usually aim to predict the load up to one-week ahead, while the Very Short-Term Load Forecasts (VSTLF) are used for a time-horizon of less than 24 h. Both, STLF and VSTLF have engaged the attention of most researchers, since they provide necessary information for the day-to-day utilities' operations [8]. These forecasts become also useful when dealing with smart grid, micro grids, peak load anticipation, and intelligent buildings [9, 10].

STLF Techniques. There are number of papers dedicated to short-term load forecasting of the commercial customers for the purpose of contract capacity optimization. Also, there are numerous approaches applied to load forecasting with good accuracy. Importantly, the quality of the forecast improves when the forecasting is applied to the higher aggregation levels (like group of customers, power stations or cities) and this can be achieved with quite low errors. In the earliest works, some classical techniques including auto regression (AR) models [11], linear regression models [12], dynamic linear and nonlinear models, general exponential smoothing models, spectral methods, and seasonal ARIMA models were used for forecasting [13–15]. Unfortunately, their capability to solve time series with complex seasonality and non-linear series is limited, in favor of artificial neural networks (ANN) techniques and expert systems [16–18]. Interestingly, load forecasting field is one of the most successful applications of ANN in power systems. Neural networks are able to deliver better performance when dealing with highly non-linear series resulting from e.g. the non-integer seasonality appearing as a result of averaging ordinary and leap years.

Optimization. Various types of hardware and software solutions. Load limiters [19] are currently used to prevent overruns. Another, more sophisticated option like Electric Power Distribution and Utility Monitoring System provided by [20] support large and medium-sized sites in Japan in terms of energy efficient management, preventive maintenance and capacity overruns. A measurement data screen is provided to display real-time measurement data which alarms if the predicted demand exceeds the present level. Then actions can be taken automatically or manually.

For small companies a simple and uncomplicated solutions using e.g. Excel are applicable to analyze the relation between the contracted capacity and the actual consumption. When contracted capacity levels are exceeded, they have then the ability to increase the level of contracted capacity or change the structure of energy consumption. As a result of the analysis, the client can reduce consumption in periods when the capacity is exceeded and increase consumption in periods when there is a capacity reserve. Unfortunately, these solutions are ineffective because they are based on the monthly averaged data from the electricity consumption invoices, while the excess of power is determined based on averaged data recorded over 15-min periods. More effective solutions for contracted capacity optimization can be achieved using different models including deep learning neural networks [21], Particle Swarm Optimization (PSO) [22], Genetic Algorithm (GA) [23] or Linear Programming (LP) optimization [24].

Motivated by aforementioned discussions, this paper presents solution for small and medium-sized enterprises from the C tariff group, concerning short-term load forecasts as a basis for calculating and optimizing the capacity required to avoid any additional fee related to exceeding contracted capacity level.

Specifically, Long Short-Term Memory (LSTM) Artificial Neural Network (ANN) is constructed to forecast the load values and the moments of exceeding the contracted capacity in the short-term horizon, i.e., up to one month ahead. The forecast is further used to optimize the capacity volume to be contracted in the following month for the commercial customer to minimize network charge for exceeding the contracted level.

Long Short-Term Memory networks belong to a complex area of deep learning methods. These are type of recurrent neural network (RNN) capable of learning order dependence in sequence prediction problems like time series. The reason for using recurrent networks is that these are different from traditional feed-forward neural networks and, in addition to the complexity and volatility in electricity time series, comes with the expectation to reveal new patterns and behaviors that the traditional methods cannot achieve [25, 26]. Standard RNNs often fail to learn correctly in the presence of time lags greater than 5–10 discrete time steps between the input events and target signals. The problem with disappearing error raises question whether standard RNNs can indeed provide significant practical advantages over time window-based feedforward networks. As provided in [27], LSTM model is not affected by this problem and it can learn to bridge minimal time lags in excess of 1000 discrete time steps by enforcing constant error flow through "constant error carrousels" (CECs) within special units, called cells.

The remainder of this paper is organized as follows. Sect. 2 proposes two stage approach to optimize electricity contract capacity problem. First stage presents multiple output strategy for forecasting supported by LSTM ANN model. The second stage uses genetic algorithm to optimize the electricity contract capacity. Sect. 3 provides a detailed description of the tariff structure in Poland. Section 4 applies the models to real datasets for two commercial customers in Poland. Section 5 concludes with the comments and provides directions for the future research.

2   Two Stage Approach to Optimize Electricity Contract Capacity Problem

2.1   Stage One – LSTM Electricity Load Time Series Forecasting

Time series forecasting is typically considered as one-step prediction. Due to the fact that electricity load forecasting is essential for both, the utility and the customer, it cannot be designed with one step prediction. Maximum power is used by the utility to provide the right amount of power for customers, whereas it is the basis for calculating, usually monthly, fixed charges for the industrial and business electricity users. Predicting multiple time steps is considered a multi-step forecasting and it includes prediction of the load values $[L + 1, \ldots, L + t]$ of historical load time series $[L1, \ldots, LN]$ composed of N load observations, where $t > 1$ denotes the forecasting horizon. In this paper we used multiple output strategy for forecasting which involves the development of a single model that is capable of predicting the entire forecast time horizon in a one-shot approach. Therefore, to predict load required for the next e.g. two data points, we would develop one model and use it to predict the next two data points as one operation. The model form would be as follows:

$$L\_prediction(t+1), L\_prediction(t+2) = model1(L(t-1), L(t-2), \ldots, L(t-n))$$

The model can learn the dependence structure between inputs and outputs as well as between outputs. Specifically, for this approach, the LSTM Artificial Neural Networks were constructed to forecast the load values and the moments of exceeding the contracted capacity in the short-term horizon, i.e., up to one month ahead and hour by hour. The forecast is further used to optimize the capacity volume to be contracted in the following month for the commercial customer to minimize network charge for exceeding the contracted level.

Additionally, the naive forecast was considered in the following manner: for the forecasting horizon, the values observed for the same hour and same day of the four previous week were averaged and taken as a forecast. However, the forecasts were far from the optimal and therefore, these were not further optimized.

2.2 Stage Two – Load Forecast Optimization

In this article, we consider peaks over contracted capacity in a given month. Most customers order the same amount of power for individual months of the year. If the peak demand does not exceed the contractual capacity, a fixed capacity charge will be levied. It constitutes the product of the fixed capacity rate R [PLN/kW], where PLN stands for Polish Zloty and contracted capacity demand for month Rt in kW. For exceeding the contractual capacity defined in the contract an additional surcharge for excess demand will be added. The annual fee per year can be, therefore, expressed as:

$$Cost_m = \begin{cases} R_m * d_m & d_m < d_m^c \\ R_m * d_m + R_m * (d_m - d_m^c) * n_m & d_m^c < d_m \end{cases} \tag{1}$$

$$Cost_{total} = \sum_{m=1}^{year} Cost_m \tag{2}$$

where

d_m^c – contracted capacity (kW) in month m;
d_m – maximum demand amount (kW) in month m;
n_m – the sum of up to ten largest amounts of surplus consumed capacity over the contractual capacity, indicated by the measuring.
R_m – rate of contractual capacity (PLN/kW) in month m.

In this work, since 12 months of data us available, we consider the total cost over 10 months, i.e., March–December 2016, due to the fact that January and February were considered for model training (including variable calculations with delays). The solution that minimizes the annual total contracted capacity cost and the penalties for excessive consumption over the fixed capacity amount can be solved using Particle Swarm Optimization, Genetic Algorithm (GA) or even the Excel's solver for linear programing.

However, in this paper we propose GA which can find multiple Pareto solutions for a multi-objective optimization problem in one run.

In principle, genetic algorithms are stochastic search algorithms inspired by biological evolution and natural selection processes. GAs simulate the evolution process where the fittest individuals dominate over the weaker ones, by reflecting the biological mechanisms of evolution, such as selection, crossover and mutation. For the experiments we used R package for GA as it provides a collection of various functions for optimization using genetic algorithms. The package includes a flexible set of tools for implementing genetic algorithms search in both the continuous and discrete case, whether constrained or not. Several genetic operators are available and can be combined to explore the best settings for the analyzed problem. Basically, the default parameters settings were used to maximize a fitness function using genetic algorithms in line with documentation [https://cran.r-project.org/web/packages/GA/GA.pdf].

3 Data Characteristics and Tariff Structure

There were two separate data sets used in the analysis. Each data set consists of data points at 15 min intervals gathered for medium-size commercial customers and covering time interval time between January 1st, 2016 and December 31st, 2016. In total, there are 35 136 observations in each data set. The customers belongs to C tariff group which is applicable to small and medium-size enterprises where the electricity is supplied with low voltage lines. The group includes C2x tariffs where contracted capacity is over 40 kW and the letter "x" designates the number of energy consumption zones per day. The following tariffs are available: C22a tariff with two-zones measurement (peak and off-peak), C22b tariff with two-zones measurement (day and night) and C23 tariff with three zones measurement per day.

The first data set contains details for the customer who belongs to C22a tariff. The customer is classified as a small pharmaceutical plant with a contracted capacity greater than 40 kW and who is mainly using electricity during the day hours. The contracted capacity for the customer is 51 kW. Figure 1a shows lower electricity consumption during morning and evening peak hours. Much higher consumption is observed between 10:00 and 16:00. The second data set contains details for the customer who belongs to C22b tariff. It is a confectionery plant which performs majority of its activities during the night. The contracted capacity for the customer is 80 kW. Figure 1b shows lower electricity consumption in the daytime zone, i.e., between 6:00 and 21:00 and higher consumption in the night time zone, between 21:00 and 6:00.

Most of the users within C2x tariff groups, do not possess detailed usage data to control energy consumption parameters and to ensure their optimal adjustment. As shown on Fig. 2. the contracted capacity is not adequately set as it is often being exceeded in the reality. On the other hand, average load consumption does not exceed 70% of the contracted capacity level, which translates into losses due to unused capacity. Therefore, it is crucial to determine the optimal contract capacity for each month so as to minimize the total cost of the electricity bills.

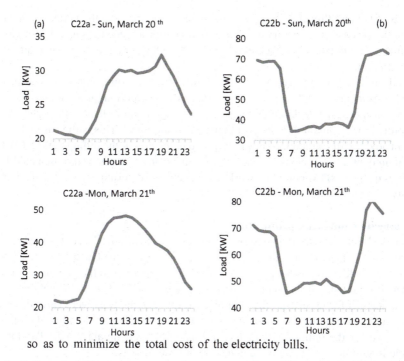

so as to minimize the total cost of the electricity bills.

Fig. 1. Daily and weekly energy consumption structures of two different users: (a) user C22a, (b) user C22b.

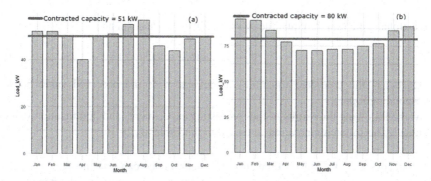

Fig. 2. Consumed load and contracted capacity (a) for customer C22a and (b) customer C22b.

4 Numerical Experiments

In this section, we use the multiple output forecast approach for electricity load forecasting as outlined in Sect. 2.1, and then, as the second stage, we apply genetic algorithm to optimize the user's contract capacity.

At the beginning, we start with hourly forecasts month by month through the entire year. Although the settlement with the power plant or electricity supplier is made on the basis of the monthly characteristics (including frequency and the volume of peaks), the hourly forecast is necessary for load optimization at second stage. These values constitute the input data to predict and then optimize the amount of capacity required in the next monthly period.

For the forecasting approach we determine the following components: (1) The quantities and costs incurred on the basis of the actual load consumptions and contract capacity, i.e., the constant value declared by the user at the beginning of the contract period; (2) The optimal load amount and the cost that user would incur. This is the case when we know, in advance, the amount of power required at the end of the billing period. (3) The optimal amount and the costs that user would incur on the basis of the predicted load quantities using a LSTM neural network. Ultimately, we determine the optimal contract capacity using a genetic algorithm.

The Capacity Contract Optimization

We used the Multiple output forecast strategy with ANN to predict the contract value with the maximum consumption values as well as the maximum load at specific hours and days of the week. These values were used further as the input to the genetic algorithm in order to establish such monthly contract capacity values that would help the user to avoid charges for exceeding the contracted level. The analysis were carried out for q100 quantile in order to check how large the maximum loads are and based on those, we searched for the optimal contract values using the genetic algorithm. The loss in Quantile Regression for an individual data point is defined as:

$$\mathbb{Q}(\varsigma_i | \alpha) = \begin{cases} \alpha \varsigma_i, & \text{if } \varsigma_i \geq 0 \\ (\alpha - 1)\varsigma_i, & \text{if } \varsigma_i < 0 \end{cases} \tag{3}$$

where alpha is the required quantile (a value between 0 and 1) and

$$\varsigma_i = y_i - f(x_i) \tag{4}$$

where $f(x)$ is the predicted (quantile) model and y is the observed value for the corresponding input x.

At the following the results of the forecasting experiments and optimization will be discussed. The following notations are used in the Tables 1, 2:

- Actual contract – the value of the customer's contracted capacity in kW;
- Actual cost – the customer's total cost of contracted capacity and the penalties of exceedances the contracted capacity level in PLN;
- Above actual contract – the number of loads consumed over the contracted level in kW;
- Opt contract capacity – the optimal amount of consumed capacity based on the historical usage in kW;
- Opt contract cost – the optimal cost of consumed capacity based on historical usage in PLN;
- Above opt contract – the number of loads over the contracted capacity based on historical usage;

- Opt contract capacity pred – the optimal contract based on the forecast obtained by neural network and optimized by GA in kW;
- Opt cost capacity pred – the total cost of optimal contract predicted by the network and optimized by GA in PLN;
- Above opt capacity pred – the number of loads over the contracted capacity based on the forecast obtained by neural network and optimized by GA.

Table 1. Multiple output forecast strategy with Q100 for C22a tariff user.

Month	Actual values			Optimal values based on historical usage			Optimal values based on predicted usage		
	Actual contract [kW]	Actual cost [PLN]	Above actual contract	Opt contract capacity [kW]	Opt contract cost [PLN]	Above Opt contract	Opt contract capacity pred [kW]	Opt cost capacity pred [kW]	Above opt capacity pred
Mar	51	510	0	50	506.74	1	53	530	0
Apr	51	510	0	47	471.39	1	49	490	0
May	51	510	0	50	501.72	1	51	510	0
Jun	51	522.89	2	52	520	0	52	520	0
Jul	51	950.3	122	56	560	0	56	560	0
Aug	51	1123.16	95	57	571.31	1	57	571.31	1
Sep	51	510	0	49	490	0	54	540	0
Oct	51	510	0	46	466.28	2	48	480	0
Nov	51	510	0	50	500	0	50	500	0
Dec	51	510	0	50	506.52	3	51	510	0
Total		6166.36			5093.99			5211.31	

Table 1 shows the results of the analysis for the customer who belongs to C22a tariff group, having a contracted capacity of 51 kW per month. During June-August period the customer consumed more capacity and therefore, the contracted level was exceeded several times in those months, even 122 times in July, what significantly increased the cost. In total, the actual cost for the customer between March and December was 6166.37 PLN. With a retrospective analysis, based on historical usage, one could see that the optimal values for contracted capacity would vary between 46 kW and 57 kW, depends on the month, as presented in the Table 1. Knowing that, the customer could benefit from lower bills, so the cost of the optimal contract would be 5093.99 PLN, which is 17.4% less than actual cost. Of course, for the customer it is difficult to specify correctly what would be the capacity required in the following months, therefore the optimal contract capacity should be forecasted. In our case we used multiple output forecast strategy with LSTM neural network to estimate maximum load for each hour and these values were used further as the input to the genetic algorithm the forecast for the optimal contract level so the total cost is minimized. As the result the forecasted capacity was between 48 kW and 57 kW, depends on the month. Importantly, only once, in August we would exceed the contracted capacity. This helped to keep the total cost very low, i.e., close to the optimal cost values. Eventually, the total cost of optimal contract predicted by the network and optimized by genetic algorithm was 5211.32 PLN which is very close to the optimal one (5093.99 PLN). In comparison to the actual cost, the benefit of the customer is quite material and amounts to 955.05 PLN (6166.37 – 5211.32) which is 15.5% of the actual bills.

In the similar manner the analysis for the second customer was prepared. Table 2 shows the results of the analysis for the customer who belongs to C22b tariff group, having a contracted capacity of 80 kW per month. During March, November and December the customer consumed more capacity and therefore, exceeded the contracted level many times, specifically even 266 times in December, what significantly impacted the actual bills. In total, the actual cost for the customer between March and December was 10789.34 PLN. With a retrospective analysis, based on historical usage, one could see that the optimal values for contracted capacity would vary between 72 kW and 90 kW, depends on the month, as presented in the Table 2. Knowing that, the customer could benefit from lower bills, so the cost of the optimal contract would be 7930.49 PLN, which is 26.5% less than actual cost. Once again, we used multiple output forecast strategy with ANN to estimate maximum load for each hour and these values were used further as the input to the genetic algorithm the forecast for the optimal contract level so the total cost is minimized. As the result the forecasted capacity was between 73 kW and 89 kW, depends on the month. There were instances where the usage would exceed the contracted capacity, e.g. 1 times in November and December. Finally, the total cost of optimal contract predicted by the network and optimized by genetic algorithm was 8068.66 PLN. In comparison to the actual cost, the benefit of the customer is also material, similarly to the previous customer, and amounts to 2720.68 PLN (10789.34 – 8068.66) which is 25.2% of the actual bills.

Table 2. Multiple output forecast strategy with Q100 for C22b tariff user.

	Actual values			Optimal values based on historical usage			Optimal values based on predicted usage		
Month	Actual contract [kW]	Actual cost [PLN]	Above actual contract	Opt contract capacity [kW]	Opt contract cost [PLN]	Above Opt contract	Opt contract capacity pred [kW]	Opt cost capacity pred [kW]	Above opt capacity pred
Mar	80	1402.71	62	86	860.27	1	88	880	0
Apr	80	800	0	78	786.63	1	85	850	0
May	80	800	0	72	722.82	1	77	770	0
Jun	80	800	0	72	724.28	1	73	730	0
Jul	80	800	0	73	737.81	2	74	740	0
Aug	80	800	0	74	740	0	74	740	0
Sep	80	800	0	76	760	0	76	760	0
Oct	80	800	0	78	780	0	78	780	0
Nov	80	1547.07	58	87	874.71	1	87	874.71	1
Dec	80	2239.55	266	90	943.96	1	89	943.96	1
Total		10789.34			7930.48			8086.66	

5 Conclusion

In this paper two stage approach is proposed to determine appropriate contract capacity amount that minimize financial losses in case of exceeding the amount of capacity defined in the contract. The LSTM neural network model was developed. The first stage was to forecast hourly capacity values as the basis for determining the monthly maximum capacity required. These maximum values were used to determine the optimal monthly capacity values at the second stage, so the values were provided as the input to the genetic algorithm in order to establish such monthly contract capacity level that would help the user to avoid charges for exceeding the contracted level.

As shown through the experiments, the application of multiple output forecast artificial neural network model and genetic algorithm for load optimization delivers significant benefits to the commercial customers. In comparison to the actual costs, the benefit for the customers, due to optimization, is material. Specifically, the benefit for the C22a customer is 15.5% of the actual bills while for the C22b customer it is 25.2%.

As a future work, we would continue the research towards fitting the models so these could potentially better deal with seasonality of the demand on the customers end. Although this research deals with Polish tariffs, we believe it can be applied to other electricity customers in capacity cost decision making.

References

1. Nafkha, R., Gajowniczek, K., Ząbkowski, T.: Do customers choose proper tariff? Empirical analysis based on polish data using unsupervised techniques. Energies **11**(3) (2018). Art. 514
2. ERO: National Report 2016 of the President of ERO in Poland (2016)
3. Regulation of the Minister of Economy of 27 April 2012 amending the regulation on the specific rules for the determination and calculation of tariffs and billing in electricity trading, Journal of Laws, No. 2, item 535 (2012)
4. Gajowniczek, K., Nafkha, R., Ząbkowski, T.: Electricity peak demand classification with artificial neural networks. Annals Comput. Sci. Inf. Syst. **11**, 307–315 (2017)
5. Rajakovic, N.L., Shiljkut, V.M.: Long-term forecasting of annual peak load considering effects of demand-side programs. J. Modern Power Syst. Clean Energy **6**(1), 145–157 (2017). https://doi.org/10.1007/s40565-017-0328-6
6. Mojica, J.L., Petersen, D., Hansen, B., Powell, K.M., Hedengren, J.D.: Optimal combined long-term facility design and short-term operational strategy for CHP capacity investments. Energy **118**, 97–115 (2017)
7. Torkzadeh, R., Mirzaei, A., Mirjalili, M.M., Anaraki, A.S., Sehhati, M.R., Behdad, F.: Medium term load forecasting in distribution systems based on multi linear regression and principal component analysis: a novel approach. In: Proceedings of the 19th Electrical Power Distribution Networks (EPDC), Tehran, Iran, pp. 66–70 (2014)
8. Hippert, H.S., Pedreira, C.E., Souza, R.C.: Neural networks forshort-term load forecasting: a review and evaluation. IEEE Trans. Power Syst **16**, 44–55 (2001)
9. Ryu, S., Noh, J., Kim, H.: Deep neural network based demand side short term load forecasting. Energies **10**, 3 (2017)
10. Gajowniczek K., Ząbkowski T.: Short term electricity forecasting based on user behavior using individual smart meter data. Intell. Fuzzy Syst. **30**(1), 223–234 (2015)
11. Zuhairi, B., Nidal K.: Autoregressive method in short term load forecast. In: 2nd IEEE International Conference on Power and Energy (PECon 2008), Johor Baharu, Malaysia (2008)

12. Karpio, K., Łukasiewicz, P., Nafkha, R.: Regression technique for electricity load modeling and outlined data points explanation. In: Pejaś, J., El Fray, I., Hyla, T., Kacprzyk, J. (eds.) ACS 2018. AISC, vol. 889, pp. 56–67. Springer, Cham (2019). https://doi.org/10.1007/978-3-030-03314-9_5
13. Alberg, D., Last, M.: Short-term load forecasting in smart meters with sliding window-based ARIMA algorithms. Vietnam J. Comput. Sci. **5**(3–4), 241–249 (2018). https://doi.org/10.1007/s40595-018-0119-7
14. Mohamed, N., Maizah H.A., Suhartono, S., Zuhaimy, I.: Improving short term load forecasting using double seasonal Arima model. World Appl. Sci. J. **15**, 223–231 (2010)
15. Ismit, M., Soeprijanto, A., Suhartono, S.: Applying of double seasonal ARIMA model for electrical power demand forecasting at PT. PLN Gresik Indonesia, Int. J. Electr. Comput. Eng. **8**, 4892–4901 (2018)
16. Kowm, D., Kim, M., Hong, C., Cho, S.: Artificial neural network based short term load forecasting. Int. J. Smart Home **8**, 145–150 (2014)
17. Shoeb M., Shahriar, Md, Khairul H., S., Refat, A.: an effective artificial neural network based power load prediction algorithm. Int. J. Comput. Appl. **178**, 35–41 (2019)
18. Charytoniuk, W., Chen, M.S.: Very short-term load forecasting using artificial neural networks. IEEE Trans. Power Syst. **15**, 263–268 (2000)
19. Haniffuddin, H., Mansor, W.: PLC based load limiter. In: 2011 IEEE Control and System Graduate Research Colloquium, Shah Alam, pp. 185–188 (2011)
20. Hayakawa, H., Watanabe, T., Kimura, T.: Electric power distribution and utility monitoring system for better energy visualization. Hitachi Rev. **63**(7), 60–64 (2014)
21. Bouktif, S., Fiaz, A., Ouni, A., Serhani, M.: Optimal deep learning LSTM model for electric load forecasting using feature selection and genetic algorithm: comparison with machine learning approaches. Energies **11**, 1636 (2018)
22. Niu, D., Dai, S.: A short-term load forecasting model with a modified particle swarm optimization algorithm and least squares support vector machine based on the denoising method of empirical mode decomposition and grey relational analysis. Energies **10**, 408 (2017)
23. Ali, S., Kim, D.-H.: Optimized power control methodology using genetic algorithm. Wireless Pers. Commun. **83**(1), 493–505 (2015). https://doi.org/10.1007/s11277-015-2405-3
24. Chiung-Yao, C., Ching-Jong, L.: A linear programming approach to the electricity contract capacity problem. Appl. Math. Model. **35**, 4077–4082 (2011)
25. Bengio, Y., Simard, P., Frasconi, P.: Learning long-term dependencies with gradient descent is difficult. IEEE Trans. Neural Netw. **5**(2), 157–166 (1994)
26. Gers, F.A., Schmidhuber, J., Cummins, F.: Learning to forget: continual prediction with LSTM. In: 1999 Ninth International Conference on Artificial Neural Networks ICANN 1999, (Conf. Publ. No. 470), Edinburgh, UK, vol. 2, pp. 850–855 (1999)
27. Sak, H., Senior, A., Beaufays, F.: Long short-term memory recurrent neural network architectures for large scale acoustic modeling, In: INTERSPEECH, pp. 338–342 (2014)

Improved Design Closure of Compact Microwave Circuits by Means of Performance Requirement Adaptation

Slawomir Koziel[1,2]([📧]) [iD], Anna Pietrenko-Dabrowska[2] [iD], and Leifur Leifsson[3] [iD]

[1] Engineering Optimization and Modeling Center, Department of Engineering, Reykjavík University, Menntavegur 1, 101 Reykjavík, Iceland
`koziel@ru.is`
[2] Faculty of Electronics Telecommunications and Informatics, Gdansk University of Technology, Narutowicza 11/12, 80-233 Gdansk, Poland
`anna.dabrowska@pg.edu.pl`
[3] Department of Aerospace Engineering, Iowa State University, Ames, IA 50011, USA
`leifur@iastate.edu`

Abstract. Numerical optimization procedures have been widely used in the design of microwave components and systems. Most often, optimization algorithms are applied at the later stages of the design process to tune the geometry and/or material parameter values. To ensure sufficient accuracy, parameter adjustment is realized at the level of full-wave electromagnetic (EM) analysis, which creates perhaps the most important bottleneck due to the entailed computational expenses. The cost issue hinders utilization of global search procedures, whereas local routines often fail when the initial design is of insufficient quality, especially in terms of the relationships between the current and the target operating frequencies. This paper proposes a procedure for automated adaptation of the performance requirements, which aims at improving the reliability of the parameter tuning process in the challenging situations as described above. The procedure temporarily relaxes the requirements to ensure that the existing solution can be improved, and gradually tightens them when close to terminating the optimization process. The amount and the timing of specification adjustment is governed by evaluating the design quality at the current design, and the convergence status of the algorithm. The proposed framework is validated using two examples of microstrip components (a coupler and a power divider), and shown to well handle design scenarios that turn infeasible for conventional approaches, in particular, when decent starting points are unavailable.

Keywords: Microwave design · Simulation-based optimization · Performance requirements · Design specification adaptation

1 Introduction

The involvement of numerical optimization techniques has been constantly growing in the area of high-frequency electronics [1, 2]. Parameter tuning through rigorous optimization replaces interactive methods, mostly involving parameter sweeping, due to

© Springer Nature Switzerland AG 2021
M. Paszynski et al. (Eds.): ICCS 2021, LNCS 12745, pp. 185–199, 2021.
https://doi.org/10.1007/978-3-030-77970-2_15

inadequacy of the latter. Topologically complex designs developed to meet the demands of emerging application areas (wireless sensing [3], Internet of Things (IoT) [4], 5G communications [5], autonomous vehicles [6]) require simultaneous adjustment of multiple variables and handling several performance figures and constraints. For reliability reasons, design closure is normally executed with the aid of full-wave electromagnetic (EM) analysis. EM-driven tuning is especially important for compact devices [7, 8], where common miniaturization means such as replacing conventional transmission lines by compact microstrip resonant cells (CMRCs) [9, 10], or incorporation of defected ground structures [11], lead to cross-coupling effects that cannot be accounted for by equivalent network models.

The major bottleneck of simulation-based microwave optimization is its high computational cost. It can be alleviated using various means such as utilization of adjoint sensitivities [12], sparse sensitivity updates [13], or surrogate-assisted procedures incorporating both data-driven [14], and physics-based replacement models [15]. Approximation models (e.g., kriging [16], support vector regression [17], polynomial chaos expansion [18], neural networks [19]) are fast and widely accessible but they are strongly affected by the curse of dimensionality. The methods of the second group (e.g., space mapping [20], response correction methods [21]) are more immune to dimensionality issues but require an underlying lower-fidelity model, which has to be customized for a given problem and application area. Another option are variable-fidelity methods such as co-kriging [22], or machine learning frameworks [23].

Computational efficiency is one of the essential aspects of the optimization routines. In many cases, the reliability, understood as the capability of the algorithm to deliver a satisfactory solution, is even more important. An example is the lack of good initial designs, which commonly occurs whenever the component at hand is to be re-designed to different operating frequencies. Another example is optimization of CMRC-based devices [9], where identification of a decent starting point may be difficult due to significant topology modifications incurred by the incorporation of the compact cells. Local search routines are prone to a failure under such circumstances, whereas utilization of global optimizers is associated with significant CPU expenses. The techniques outlined before [12–23] are mostly focused on achieving the computational speedup. Combining surrogate modelling methods with nature-inspired algorithms (e.g., [24]) may be a step towards reliability improvement, yet applicability of such methods is limited in microwave design due to high level of nonlinearity of the system outputs. As a matter of fact, only low-dimensional cases can be treated this way [25].

This paper proposes an alternative approach to improving the reliability of microwave design closure using local search procedures. The major component of our technique is design specification adaptation, which operates by adjusting the performance specifications on the grounds of the misalignment between the target and the actual operating frequencies and/or bandwidths of the device of interest. At the initial stages of the optimization process, the specifications are relocated to the vicinity of the actual operating frequencies, and gradually tightened to eventually reach their original values upon the convergence of the process. The major advantages of the proposed methodology include: (i) reducing the sensitivity of the search process to the starting point quality, (ii) enabling the possibility of system re-design within wide frequency ranges by means of local algorithms, (iii) eliminating the need for global procedures when handling challenging design scenarios. The presented framework is validated using two microstrip circuits, a branch-line coupler and a power divider. Successful optimization is demonstrated using inferior initial designs at which the operating conditions are severely misaligned with respect to the targets.

2 Microwave Design Closure with Adaptive Adjustment of Design Requirements

This section introduces the concept and the implementation of the adaptively adjusted performance requirements. For the sake of illustration, it is incorporated into the trust-region gradient-based algorithms, although the adaptation procedure itself is generic, i.e., can be combined with various iterative search routines.

2.1 Adaptive Adjustment of Performance Requirements. The Concept

The purpose of this work is to improve the reliability of EM-driven design closure of compact microwave circuits with the emphasis on handling poor initial designs. As mentioned before, a representative situation is dimension scaling of a given structure for the operating frequencies that are distant from those at the current design. The task is particularly challenging when the system at hand features narrow-band characteristics, and local optimization routines are prone to a failure. At the same time, globalized procedures (e.g., nature-inspired algorithms [26]) tend to be prohibitively expensive.

We begin by recalling the simulation-based parameter adjustment problem formulation. The assumption is that the intended operating frequencies of the circuit are f_k, $k = 1, \ldots, N$; here, N stands for the number of the operating bands. These frequencies are assembled into the target vector $\boldsymbol{F} = \left[f_1 \ldots f_N\right]^T$. The EM-evaluated system response will be denoted as $\boldsymbol{S}(\boldsymbol{x})$, where $\boldsymbol{x}$ is the vector of independent design variables. The design problem is defined as

$$\boldsymbol{x}^* = \arg \min_{\boldsymbol{x}} U(\boldsymbol{S}(\boldsymbol{x}), \boldsymbol{F}) \tag{1}$$

The function U in (1) is a scalar metric of the design utility. For the sake of clarification, let us consider a coupler circuit, which is to operate at the centre frequency f_0. The characteristics of interest (all complex functions of frequency) are matching S_{11}, transmission S_{21} and S_{31}, and port isolation S_{41}. The coupler is supposed to provide an equal power split, i.e., $|S_{21}| = |S_{31}|$ at f_0 (here, $|.|$ stands for the modulus of the complex number), and minimize both matching and isolation, also at f_0. Having this in mind, the objective function is formulated as follows

$$U(\boldsymbol{S}(\boldsymbol{x}), \boldsymbol{F}) = U([S_{11}(\boldsymbol{x}, f), S_{21}(\boldsymbol{x}, f), S_{41}(\boldsymbol{x}, f), S_{41}(\boldsymbol{x}, f)], [f_0])$$

$$= \max\{|S_{11}(\boldsymbol{x}, f_0)|, |S_{41}(\boldsymbol{x}, f_0)|\} + \beta\left[|S_{21}(\boldsymbol{x}, f_0)| - |S_{31}(\boldsymbol{x}, f_0)|\right]^2 \tag{2}$$

In (2), minimization of $|S_{11}|$ and $|S_{41}|$ is executed explicitly, whereas the equal power split condition is handled by means of a penalty function.

The typical challenges pertinent to parameter tuning have been illustrated in Fig. 1. There are two initial designs shown in the picture, one of which (marked black) is of sufficiently good quality to ensure that local optimization is capable of identifying the optimum at the target operating frequency. The second design (marked grey) is severely misaligned frequency-wise with respect to the target, and local search initiated from this point fails.

In this work, we aim at the conceptual development and implementation of the algorithmic framework capable of handling situations such as those illustrated in Fig. 1, while using local algorithms. This is realized by an adaptive adjustment of performance requirements, where the target operating frequencies are relocated having in mind the circuit characteristics at the current design. In particular, we strive to ensure attainability of the re-adjusted specifications at each iteration of the optimization algorithm. Clearly, the design requirements are to reach their original values upon the algorithm convergence.

The concept of design specifications adaptation has been graphically illustrated in Fig. 2. As the operating frequency at the initial design is away from the target, the specifications are relocated to about 1.1 GHz to become attainable by means of local search. The adjustments are subsequently made at each iteration of the process based on the operating frequency of the current design. Examples of intermediate stages are shown in Figs. 2(b) and 2(c). Towards the end of the optimization run, the target is allocated at its original value. It should be noted that performance requirement relocation and design update (e.g., within the iteration of a descent procedure) are interleaved. This allows for accomplishing the entire procedure within a single algorithm run. Section 2.2 will provide a rigorous formulation of the adaptation scheme.

2.2   Adaptive Adjustment of Performance Requirements. The Procedure

The proposed scheme for adaptive adjustment of performance requirements is developed having in mind the following factors:

- The amount of performance requirement adjustment should be computed using the actual operating frequencies of the current design;
- The attainability of the relocated specifications using local search should be ensured at each stage of the process;

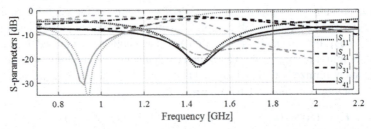

Fig. 1. S-parameters of a miniaturized microwave coupler (cf. Sect. 3.1). The circuit is supposed to operate at 1.8 GHz (marked with a vertical line). The target is attainable through local optimization from the design marked with the black lines. It is not attainable from the other (marked with the grey lines) because of severe misalignment of the operating bandwidth and the target.

- The aforementioned assessments should be accomplished at low computational expense, ideally without involving any additional full-wave EM simulations.

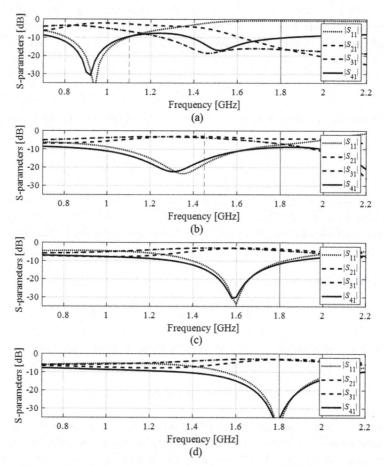

Fig. 2. Performance requirement adjustment concept explained using a microwave coupler (the initial design and the target, are as in Fig. 1): (a) target frequency moved towards the operating frequency at the initial design to ensure attainability of the current specifications (dashed line), (b) current design and specifications in the middle of the optimization run, (c) final stage: the target frequency converged to its original value, (d) final design optimized w.r.t. the original requirements.

The cost issue is particularly important from the perspective of practical utility of the optimization procedure. In this paper, the aim is to incorporate the specification adjustment procedure into gradient-based procedures, specifically, the trust-region framework briefly outlined in Sect. 2.3. Therein, the primary tool employed to render the candidate designs is the first-order (linear) expansion model of the circuit frequency characteristics. It is constructed using the Jacobian matrix, which needs to be estimated before each iteration of the trust-region algorithm as a part of its operation. Consequently, it is available at no extra cost.

For the purpose of subsequent considerations, we will denote the Jacobian matrix, representing the gradients of all relevant system outputs at the design x, as $J(x)$. The optimization algorithm is assumed to be iterative and to generate a series $x^{(i)}$, $i = 0, 1,$ ..., which are approximations of the optimum x^* of (1). The design $x^{(0)}$ stands for the starting point. Let

$$L^{(i)}(x) = S(x^{(i)}) + J(x^{(i)}) \cdot (x - x^{(i)}) \tag{3}$$

be the first-order model of $S(x)$ at the design $x^{(i)}$. We consider an optimization sub-problem

$$x^{tmp} = \arg \min_{||x - x^{(i)}|| \leq D} U(L^{(i)}(x), F) \tag{4}$$

where D is a search radius. The value of D is not critical, and will be set to unity in the numerical experiments of Sect. 3.

We use the following factors to adjust the design specifications in the course of the optimization process:

- The improvement factor F_r defined as

$$F_r = \left| U(L^{(i)}(x^{tmp}), F) - U(L^{(i)}(x^{(i)}), F) \right| \tag{5}$$

- The distance between the actual operating frequencies of the circuit at $x^{(i)}$, denoted as $F_c = [f_{c.1} \ldots f_{c.N}]^T$, and the target frequencies $F = [f_1 \ldots f_N]^T$, defined as

$$D_c = ||F_c - F|| \tag{6}$$

Herewith, F_r is used to evaluate the capability to improve the design when using $x^{(i)}$ as the starting point. D_c is employed to ensure that the current (re-adjusted) performance requirement are sufficiently close to the actual operating frequencies of the circuit, also at $x^{(i)}$. Furthermore, the two acceptance thresholds are defined, $F_{r.\min}$ and $D_{c.\max}$, and used to establish the conditions for performance requirement relocation. These are:

- $F_r < F_{r.\min}$, which indicates that the current design is unlikely to be improved when starting from $x^{(i)}$, or
- $D_c > D_{c.\max}$, which indicates that the operating frequencies at $x^{(i)}$ are too far away from the current targets.

Satisfaction of either of these conditions means that the current requirements may be too stringent, i.e., unlikely to be met through local search from $x^{(i)}$. Consequently, a relocation is necessary. The specific value of the aforementioned thresholds should take into account the shape of the system characteristics. A simple procedure for establishing both $F_{r.\min}$ and $D_{c.\max}$ is as follows:

1. Adjust $D_{c.\max}$ to approximately half of the system bandwidth(s) (frequency-wise) at the starting point $x^{(0)}$. With this value, satisfaction of the condition $D_c < D_{c.\max}$ corresponds to the target frequency (or frequencies) being allocated on the slopes of the frequency characteristics near the respective operating bandwidths;
2. Adjust the target operating frequencies to ensure $D_c = D_{c.\max}$;
3. Find x^{tmp} as in (4) using $D = 1$;
4. Set $F_{r.\min} = F_r$ with F_r computed using x^{tmp} obtained in Step 3.

The above procedure allows for setting up $F_{r.\min}$ so that it accounts for a typical objective function improvement under the assumption that the performance requirements have been altered to ensure $D_c < D_{c.\max}$. With this setup, the relocated specifications are attainable from the current design using local search.

For the purpose of subsequent considerations, the relocated target frequencies will be denoted as $F_{current}(a) = [f_{current.1}(a) \ldots f_{current.N}(a)]^T$. Here, a is a scalar (scaling) parameter, $0 \le a \le 1$. The entries of the vector $F_{current}$ are computed as.

$$f_{current.k}(a) = (1 - a)f_{c.k} + af_k \quad \text{for } k = 1, \ldots, N \tag{7}$$

where $f_{c.k}$ are the components of $F_c = [f_{c.1} \ldots f_{c.N}]^T$ (the vector of the operating frequencies at $x^{(i)}$). The key factor is an appropriate determination of a. It is set to be the maximum number within the interval $[0, 1]$, such that both $F_r \ge F_{r.\min}$ and $D_c \le D_{c.\max}$ at x^{tmp} obtained as

$$x^{tmp} = \arg \min_{||x - x^{(i)}|| \le 1} U(L^{(i)}(x), F_{current}(a)) \tag{8}$$

In practice, a is found by solving a separate optimization process, where it is lowered to make sure that both conditions, $F_r \ge F_{r.\min}$ and $D_c \le D_{c.\max}$, are satisfied for x^{tmp} produced by (8). The latter is an indication that the performance requirements have been relaxed to the extent that makes them attainable from the current design $x^{(i)}$.

It can be observed that $f_{current.k}$ is a convex combination of $f_{c.k}$ and the original target frequency f_k. Small values of a correspond to relaxed requirements. Upon convergence of the optimization algorithm, the conditions $F_r \ge F_{r.\min}$ and $D_c \le D_{c.\max}$ will hold for $a = 1$ (i.e., the original requirements). This is of course under the assumptions that the original specifications are attainable. If this is not the case, the optimization process will converge upon approaching the targets as closely as possible.

The above procedure describes the adaptation of the design requirements for one iteration of the optimization process. It is executed before each iteration, i.e., at all vectors $x(i)$, $i = 0, 1, \ldots$ As a result, the target frequencies are being continuously altered throughout the optimization run. An important feature of this technique is that it does not incur additional computational costs because the Jacobian matrix required to construct the linear model (cf. (3)) is already available. More specifically, it is generated as a part of the normal operation of the optimization algorithm.

2.3 Reference Algorithm: Trust-Region Gradient Search

The adaptive adjustment of performance requirements can be combined with various optimization algorithms. For the purpose of validation, it is incorporated into the trust-region (TR) framework [27], briefly recalled in the remaining part of this section. We aim at solving the problem (1). Toward this end, the TR procedure yields a series of vectors $x^{(i)}$, $i = 0, 1, \ldots$, being approximations to x^*. These are generated as

$$x^{(i+1)} = \arg \min_{||x-x^{(i)}|| \leq d^{(i)}} U(L^{(i)}(x), F_{current}) \tag{9}$$

In (9), the linear model $L^{(i)}$ is constructed using (3), whereas $F_{current}$ stands for the current performance specifications (target operating frequencies). The TR radius $d^{(i)}$ is updated at each iteration based on the gain ratio r, defined as

$$r = \frac{U(S(x^{(i+1)}), F_{current}) - U(S(x^{(i)}), F_{current})}{U(L^{(i)}(x^{(i+1)}), F_{current}) - U(L^{(i)}(x^{(i)}), F_{current})} \tag{10}$$

In particular, if $r > 0$, the candidate design $x^{(i+1)}$ is accepted. In addition, if r is sufficiently large (e.g., $r > 0.75$), $d^{(i+1)}$ is set to $2d^{(i)}$; otherwise, if r is too low (e.g., $r < 0.25$), $d^{(i+1)}$ is diminished to $d^{(i)}/3$. If no objective function improvement is observed, i.e., $r < 0$, the candidate design is rejected, and the iteration is repeated with a reduced trust region radius.

2.4 Optimization Framework

Figure 3 shows the pseudocode of the optimization procedure combining the trust-region algorithm and the procedure for adaptive adjustment of performance requirements discussed in Sects. 2.1 and 2.2. The termination condition is convergence in argument $||x^{(i+1)} - x^{(i)}|| < \varepsilon$, or reduction of the trust region radius below the same threshold, i.e., $d^{(i)} < \varepsilon$. In our numerical experiments (cf. Sect. 3), we set $\varepsilon = 10^{-3}$. More details about the TR algorithms can be found in the literature (e.g., [27]).

3 Verification Examples

This section demonstrates the operation and performance of the procedure for adaptive adjustment of performance specifications introduced in Sect. 2. We consider two test cases, a miniaturized branch-line coupler, and a dual-band power divider. In both cases, poor-quality initial designs are employed to emphasize the benefits of the proposed adjustment strategy, and to corroborate the efficacy of the method under challenging design scenarios.

<u>Input arguments</u>:
1. $x^{(0)}$ – initial design;
2. F – target operating frequency/band vector;
3. U_R – primary objective function encoding design goals;
<u>Algorithm operating flow</u>:
1. Set the iteration index $i = 0$;
2. Find the scalar a to determine current specification vector $F_{current}(a)$
   (cf. Section 2.2);
3. Perform TR iteration (9) to find the new iteration point $x^{(i+1)}$ according to $F_{current}$;
4. Update the TR radius $d^{(i)}$;
5. If the termination condition is satisfied, go to 7;
6. If $U(R(x^{(i+1)}),F_{current}) < U(R(x^{(i)}),F_{current})$
       Set $i = i + 1$;
       Go to 2;
   else
       Go to 3;
   end
7. END

Fig. 3. Operating flow of the trust-region optimization algorithm incorporating adaptive adjustment of performance specifications.

3.1 Compact Branch-Line Coupler

Consider a compact branch-line coupler (BLC) [28] shown in Fig. 4. The circuit is
implemented on RO4003 substrate ($\varepsilon_r = 3.5$, $h = 0.76$ mm) and described by ten
independent parameters $x = [g\ l_{1r}l_al_bw_1\ w_{2r}w_{3r}w_{4r}w_aw_b]^T$; we also have $L = 2dL +
L_s$, $L_s = 4w_1 + 4g + s + l_a + l_b$, $W = 2dL + W_s$, $W_s = 4w_1 + 4g + s + 2w_a$,
$l_1 = l_bl_{1r}$, $w_2 = w_aw_{2r}$, $w_3 = w_{3r}w_a$, and $w_4 = w_{4r}w_a$. The computational model
is evaluated using frequency-domain solver of CST Microwave Studio (~60,000 mesh
cells, simulation time 5 min).

The design objective is to make the circuit operate at the centre frequency of f_0
= 1 GHz. The matching $|S_{11}|$, and isolation $|S_{41}|$ should be minimized, and the circuit
should provide equal power split, i.e., $|S_{21}| = |S_{31}|$, all at f_0. We use the objective function
(2). The initial design and the final design rendered using the technique proposed in this
work are shown in Fig. 5 using grey and black lines, respectively. Note that the operating
frequency at $x^{(0)}$ is approximately 2.2 GHz. Consequently, re-designing the structure to
1 GHz using local procedures is virtually impossible. In particular, the conventional
TR routine fails to identify a satisfactory solution. In contrast to that, the approach
introduced in this paper works yields the design $x^* = [1.00\ 0.81\ 6.91\ 11.94\ 0.75\ 0.99
0.89\ 0.65\ 4.05\ 0.53]^T$, which satisfies the original requirements (cf. Fig. 5). Figure 6
shows the evolution of the performance specifications (target operating frequency). It
can be observed that the requirements had to be significantly altered at the early stages

of the optimization process, then gradually adjusted back to the original value of 1 GHz. Figure 7 shows selected intermediate designs obtained in the course of the optimization run.

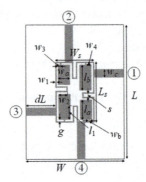

Fig. 4. Miniaturized branch-line coupler (BLC) [28]. The circuit ports marked using numbered circles.

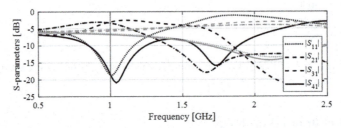

Fig. 5. Branch-line coupler: frequency characteristics of the circuit at the initial (grey) and the final design (black) obtained using the proposed procedure for adaptive adjustment of performance specifications. The target operating frequency is marked using the vertical line.

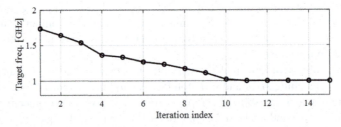

Fig. 6. Branch-line cupler: evolution of the target operating frquency verus iteration index of the optimization algorithm. Original target frequency marked using a horizontal line.

3.2 Dual-Band Power Divider

The second verification example is a dual-band power divider [29], shown in Fig. 8, implemented on AD250 substrate ($\varepsilon_r = 2.5$, $h = 0.81$ mm). The design parameters are

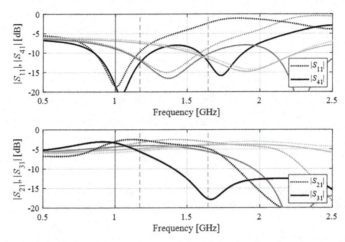

Fig. 7. Branch-line coupler: frequency characteristics of the circuit at three intermediate designs, marked using the light-grey, dark-grey, and black colors (optimum), along with the corresponding target frequencies. For the sake of clarity, the matching/isolation characteristics are shown separate from the transmission characteristics.

$x = [l_1 \; l_2 \; l_3 \; l_4 \; l_5 \; s \; w_2]^T$; $w_1 = 2.2$ and $g = 1$ are fixed (all dimensions in mm). The computational model is evaluated in the time-domain solver of CST Microwave Studio (~200,000 mesh cells, simulation time 2 min).

In this case, the circuit parameters are optimized to make the circuit operate at $f_1 =$ 2.4 GHz and $f_2 = 3.8$ GHz. More specifically, the input matching $|S_{11}|$, and the output matching $|S_{22}|$, $|S_{33}|$, as well as isolation $|S_{23}|$ are to be simultaneously minimized at both f_1 and f_2. Note that equal power split is ensured by the symmetry of the structure; therefore, it does not have to be explicitly handled within the objective function. Figure 9 shows the initial and the optimized designs (top and bottom panels, respectively). It can be observed that the operating frequencies at the initial design are severely misaligned with the targets. Conventional TR algorithm fails to identify the optimum design. On the other hand, TR procedure combined with adaptive adjustment of performance specifications produces the design $x^* = [26.86 \; 2.18 \; 21.92 \; 2.00 \; 3.82 \; 0.50 \; 4.56]^T$, which satisfied the design requirements (cf. Fig. 9). Figure 10 shows the evolution of the target operating frequencies, which is consistent with what was observed for the first example. Some of the intermediate designs produced in the course of the optimization run along with the corresponding target operating frequencies are shown in Fig. 11.

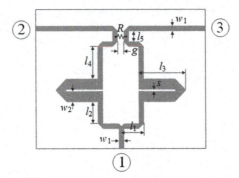

Fig. 8. Dual-band equal split power divider [29]: circuit topology; the ports marked using number in circles. The lumped resistor denoted as *R*.

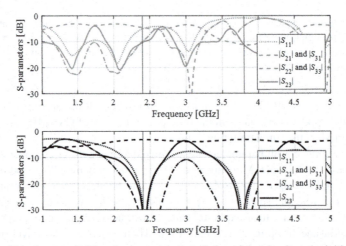

Fig. 9. Dual-band power divider: frequency characteristics of the circuit at the initial (top) and the final design (bottom) obtained using the proposed procedure for adaptive adjustment of performance sepcifications. The target operating frequencies are marked using the vertical lines.

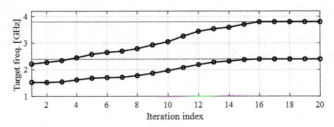

Fig. 10. Dual-band power divider: evolution of the target operating frequencies versus iteration index of the optimization alogorithm. Original target frequencies marked using horizontal lines.

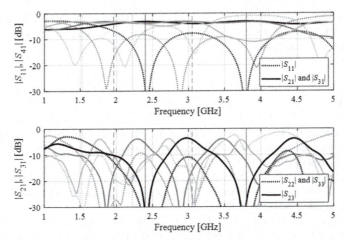

Fig. 11. Dual-band power divider: frequency characteristics of the circuit at the three intermediate designs, marked using the light-gray, dark-gray, and black colors (optimum), along with the corresponding target frequencies. For clarity, the input matching and transmission responses, as well as the output matching and isolation responses are shown in separate panels.

4 Conclusions

This paper proposed a procedure for adaptive adjustment of performance requirements as a tool for improving the reliability of simulation-based design closure of compact microwave circuits. The procedure is intended to work with local iterative search routines (in particular, the gradient-based ones). Its goal is to facilitate identification of the optimum design under demanding scenarios, especially in terms of the lack of quality starting points. Toward this end, the target operating frequencies are automatically relocated before each iteration of the optimization process to become attainable using local search. The necessary adjustment is quantified using rigorous criteria involving the analysis of the system response at the current design. Our methodology has been validated through the optimization of two microstrip circuits, a miniaturized branch-line coupler, and a dual-band power divider. In both cases, satisfactory designs were identified despite of poor starting points. The fundamental advantage of the adaptive adjustment of performance requirements is that it considerably reduces the sensitivity of the optimization algorithm to the initial design quality. Among others, this feature enables dimension scaling of microwave components within broad ranges of operating conditions using local procedures, but also eliminates the need for global optimization routines, normally required to handle more challenging design closure tasks. The future work will include generalization of the presented approach for other types of operating conditions (i.e., not necessarily operating frequencies), as well as its application to other classes of high-frequency structures (e.g., antenna systems).

Acknowledgement. The authors would like to thank Dassault Systemes, France, for making CST Microwave Studio available. This work is partially supported by the Icelandic Centre for

Research (RANNIS) Grant 206606 and by by Gdańsk University of Technology Grant DEC-41/2020/IDUB/I.3.3 under the Argentum Triggering Research Grants program - 'Excellence Initiative - Research University'.

References

1. Feng, F., Zhang, C., Na, W., Zhang, J., Zhang, W., Zhang, Q.: Adaptive feature zero assisted surrogate-based EM optimization for microwave filter design. IEEE Microwave Wirel. Comp. Lett. **29**(1), 2–4 (2019)
2. Bao, C., Wang, X., Ma, Z., Chen, C.P., Lu, G.: An optimization algorithm in ultrawideband bandpass Wilkinson power divider for controllable equal-ripple level. IEEE Microwave Wirel. Comp. Lett. **30**(9), 861–864 (2020)
3. Abdolrazzaghi, M., Daneshmand, M.: A phase-noise reduced microwave oscillator sensor with enhanced limit of detection using active filter . IEEE Microwave Wirel. Comp. Lett. **28**(9), 837–839 (2018)
4. Wang, Y., Zhang, J., Peng, F., Wu. S.: A glasses frame antenna for the applications in internet of things. IEEE Internet Things J. **6**(5), 8911–8918 (2019)
5. Ali, M.M.M., Sebak, A.: Compact printed ridge gap waveguide crossover for future 5G wireless communication system. IEEE Microwave Wirel. Comp. Lett. **28**(7), 549–551 (2018)
6. Kim, M., Kim, S.: Design and fabrication of 77-GHz radar absorbing materials using frequency-selective surfaces for autonomous vehicles application. IEEE Microwave Wirel. Comp. Lett. **29**(12), 779–782 (2019)
7. Gómez-García, R., Yang, L., Muñoz-Ferreras, J., Psychogiou, D.: Single/multi-band coupled-multi-line filtering section and its application to RF diplexers, bandpass/bandstop filters, and filtering couplers. IEEE Trans. Microwave Theory Tech. **67**(10), 3959–3972 (2019)
8. Tan, X., Sun, J., Lin, F.: A compact frequency-reconfigurable rat-race coupler. IEEE Microwave Wirel. Comp. Lett. **30**(7), 665–668 (2020)
9. Chen, S., Guo, M., Xu, K., Zhao, P., Dong, L., Wang, G.: A frequency synthesizer based microwave permittivity sensor using CMRC structure. IEEE Access **6**, 8556–8563 (2018)
10. Qin, W., Xue, Q.: Elliptic response bandpass filter based on complementary CMRC. Electr. Lett. **49**(15), 945–947 (2013)
11. Sen, S., Moyra, T.: Compact microstrip low-pass filtering power divider with wide harmonic suppression. IET Microwaves Ant. Propag. **13**(12), 2026–2031 (2019)
12. Sabbagh, M.A.E., Bakr, M.H., Bandler, J.W.: Adjoint higher order sensitivities for fast full-wave optimization of microwave filters. IEEE Trans. Microwave Theory Techn. **54**(8), 3339–3351 (2006)
13. Pietrenko-Dabrowska, A., Koziel, S.: Expedited antenna optimization with numerical derivatives and gradient change tracking. Eng. Comput. **37**(4), 1179–1193 (2019)
14. Zhang, Z., Cheng, Q.S., Chen, H., Jiang, F.: An efficient hybrid sampling method for neural network-based microwave component modeling and optimization. IEEE Microwave Wirel. Comp. Lett. **30**(7), 625–628 (2020)
15. Van Nechel, E., Ferranti, F., Rolain, Y., Lataire, J.: Model-driven design of microwave filters based on scalable circuit models. IEEE Trans. Microwave Theory Tech. **66**(10), 4390–4396 (2018)
16. Li, Y., Xiao, S., Rotaru, M., Sykulski, J.K.: A dual kriging approach with improved points selection algorithm for memory efficient surrogate optimization in electromagnetics. IEEE Trans. Magn. **52**(3), 1–4 (2016). Art 7000504

17. Cai, J., King, J., Yu, C., Liu, J., Sun, L.: Support vector regression-based behavioral modeling technique for RF power transistors. IEEE Microwave Wirel. Comp. Lett. **28**(5), 428–430 (2018)
18. Petrocchi, A., et al.: Measurement uncertainty propagation in transistor model parameters via polynomial chaos expansion. IEEE Microwave Wirel. Comp. Lett. **27**(6), 572–574 (2017)
19. Feng, F., et al.: Multifeature-assisted neuro-transfer function surrogate-based EM optimization exploiting trust-region algorithms for microwave filter design. IEEE Trans. Microwave Theory Tech. **68**(2), 531–542 (2020)
20. Li, S., Fan, X., Laforge, P.D., Cheng, Q.S.: Surrogate model-based space mapping in post-fabrication bandpass filters' tuning. IEEE Trans. Microwave Theory Tech. **68**(6), 2172–2182 (2020)
21. Koziel, S.: Shape-preserving response prediction for microwave design optimization. IEEE Trans. Microwave Theory Tech. **58**(11), 2829–2837 (2010)
22. Pietrenko-Dabrowska, A., Koziel, S.: Surrogate modeling of impedance matching transformers by means of variable-fidelity EM simulations and nested co-kriging. Int. J. RF Microwave CAE **30**(8), e22268 (2020)
23. Xiao, L., Shao, W., Ding, X., Wang, B., Joines, W.T., Liu, Q.H.: Parametric modeling of microwave components based on semi-supervised learning. IEEE Access **7**, 35890–35897 (2019)
24. Toktas, A., Ustun, D., Tekbas, M.: Multi-objective design of multi-layer radar absorber using surrogate-based optimization. IEEE Trans. Microwave Theory Tech. **67**(8), 3318–3329 (2019)
25. Lim, D.K., Yi, K.P., Jung, S.Y., Jung, H.K., Ro, J.S.: Optimal design of an interior permanent magnet synchronous motor by using a new surrogate-assisted multi-objective optimization. IEEE Trans. Magn. **51**(11), 8207504 (2015)
26. Luo, X., Yang, B., Qian, H.J.: Adaptive synthesis for resonator-coupled filters based on particle swarm optimization. IEEE Trans. Microwave Theory Tech. **67**(2), 712–725 (2019)
27. Conn, A.R., Gould, N.I.M., Toint, P.L.: Trust Region Methods, MPS-SIAM Series on Optimization (2000)
28. Tseng, C., Chang, C.: A rigorous design methodology for compact planar branch-line and rat-race couplers with asymmetrical T-structures. IEEE Trans. Microwave Theory Tech. **60**(7), 2085–2092 (2012)
29. Lin, Z., Chu, Q.X.: A novel approach to the design of dual-band power divider with variable power dividing ratio based on coupled-lines. Prog. Electromagn. Res. **103**, 271–284 (2010)

Graph-Grammar Based Longest-Edge Refinement Algorithm for Three-Dimensional Optimally p Refined Meshes with Tetrahedral Elements

Albert Mosiałek[1], Andrzej Szaflarski[1], Rafał Pych[1], Marek Kisiel-Dorohinicki[1], Maciej Paszyński[1(✉)], and Anna Paszyńska[2]

[1] Institute of Computer Science, AGH University of Science and Technology, Krakow, Poland
paszynsk@agh.edu.pl

[2] Faculty of Physics, Astronomy and Applied Computer Science, Jagiellonian University, Krakow, Poland

Abstract. Finite element method is a popular way of solving engineering problems in geoengineering. Three-dimensional grids employed for approximation the formation layers are often constructed from tetrahedral finite elements. The refinement algorithms that avoids hanging nodes are desired in order to avoid constrained approximation on broken edges and faces. We present a new mesh refinement algorithm for such the tetrahedral grids, with the following features (1) it is a two-level algorithm, refining the elements' faces first, followed by the refinement of the elements' interiors; (2) for the face refinements it employs the graph-grammar based version of the longest-edge refinement algorithm to avoid the hanging nodes; and (3) it allows for nearly perfect parallel execution of the second stage, refining the element interiors. We describe the algorithm using the graph-grammar based formalism. We verify the properties of the algorithm, by breaking 5,000 tetrahedral elements, and checking their angles and proportions. On the generated meshes without hanging nodes we span the polynomial basis functions of the optimal order, selected via metaheuristic optimization algorithm. We use them for the projection based interpolation of formation layers.

1 Introduction

The tetrahedral three-dimensional grids are commonly employed for representation of the formation layers in geophysics [1]. There are multiple applications of the three-dimensional tetrahedral grids representing the real earth models [2–6]. The quantities of interest in the geophysical domain can be better approximated on the grids refined towards them. The mesh refinements algorithms can generate hanging nodes on broken edges and faces when they refine grids by breaking tetrahedral elements into smaller ones. The hanging nodes are difficult to handle, and thus the mesh refinements algorithms avoiding hanging nodes are needed. On the other hand, the broken elements must keep their proportions, to

M. Paszynski et al. (Eds.): ICCS 2021, LNCS 12745, pp. 200–213, 2021.
https://doi.org/10.1007/978-3-030-77970-2_16

avoid approximations over elongated elements. Such badly shaped elements, i.e., with low angles, generate huge numerical errors during factorization. Uniform h-adaptation might cause significant computation cost with little improvement of approximation, especially when most of the error is caused by a few elements. For this reason, the meshing algorithm should be able to generate locally dense meshes, without hanging nodes, and keeping the proportions from the initial mesh elements. Various algorithms of tetrahedral mesh refinements, avoiding the hanging nodes, and preserving the proportions have been already proposed. Most of them suggest operations on tetrahedra like bisection [7–9] or refinement to multiple tetrahedra at once [10,11]. Algorithms involving erasing and remeshing regions with high error have also been proposed [12,13].

The state-of-the-art algorithm for refinements of the tetrahedral elements is the longest-edge refinement algorithm proposed by Rivara [14]. The algorithm generates the path of additional refinements required to remove the hanging nodes. The parallelization of the Rivara longest-edge refinement algorithm in three-dimensions is only possible by assigning the refinement paths resulting from breaking different elements at the same time to different threads, and avoiding conflicts. Recently, we proposed a graph-grammar based version of the longest-edge refinements algorithm in two-dimensions [15]. The algorithm benefits from the graph-grammar based implementaton by a better partitioning of the computational problem into basic undividable tasks that can be executed in parallel. Comparing to the classical two-dimensional Rivara algorithm [7] it has a better parallelization potential, as described in [15]. But the three-dimensional implementation of the graph-grammar model is needed.

In this paper we propose for the first time the graph-grammar based model of the three-dimensional longest-edge refinement algorithm. It starts with executing the two-dimensional graph-grammar based longest-edge refinement algorithm on faces of the tetrahedra, taking into account the three-dimensional topology of the connected faces. Later, it breaks all the interiors of the tetrahedra, with the faces already broken in the first stage. This method allows for both better concurrency of the first stage, as shown in [15], as well as ideally parallel execution of the breaking of tetrahedral interiors in the second stage.

The structure of this paper is the following. Firstly, we briefly discuss the longest-edge refinement algorithm. Then, the definition of the hypergraph grammar is introduced. Later, we pesent the graph-grammar based model for breaking the faces of tetrahedral elements (an extension of the "flat" 2D model described in [15]). Next, we describe the pseudo-code of the graph-transformation breaking the interior of a tetrahedron with its faces already broken in the previous step. Finally, we present some numerical experiments verifying the correctness of the proposed algorithm. Namely, we break 5,000 tetrahedral elements in the mesh approximating the geological formation layers, and we monitor the hanging nodes, the angles and the proportions of the newly created elements.

2   Mesh Refinement

In this section we present the definition of hypergraph grammar and all details concerning tetrahedral faces refinement.

2.1   Longest Edge Refinement in 2D

The longest edge refinement algorithm simply splits triangles by its longest edge. The division either creates a hanging node in an adjacent triangle or keeps the mesh conforming if the longest edge was on the edge of the mesh. Any triangles with hanging nodes are then refined again to their longest edge until there are no more triangles with hanging nodes. Note that if a triangle has a hanging node on an edge E, then it may only be refined to E or to edge longer than E. This means that hanging nodes may only be propagated to longer and longer edges. This ensures that the algorithm will eventually stop and the mesh will be conforming. However, in the worst-case scenario breaking edge E may require splitting all edges in the mesh longer than E.

2.2   Hypergraph Grammar Definition

We define a hypergraph as a system $G = (V, HE, t, l, a, v)$, where:

V - set of vertices
HE - set of hyperedges
t - function which maps hyperedge to sequence of vertices
l - function which maps hyperedge to label from label set C
a - function which maps vertex or hyperedge to set of attributes A
v - function which maps vertex or hyperedge attribute to value of that attribute

For our case we define $C = \{I, E, T\}$ and $A = \{x, y, z, HN, L, W, R\}$ where:

I - hyperedge label which represents interior of tetrahedron
E - hyperedge label which represents edge of tetrahedron
T - hyperedge label which represents wall of tetrahedron
$x \in \mathbb{R}$ - vertex attribute; x coordinate of vertex corner
$y \in \mathbb{R}$ - vertex attribute; y coordinate of vertex corner
$z \in \mathbb{R}$ - vertex attribute; z coordinate of vertex corner
$HN \in \mathbb{N}$ - vertex attribute; represents number of adjacent walls for which given vertex is a hanging node
$L \in \mathbb{R}$ - hyperedge attribute; represents length of the edge.
$W \in \mathbb{N}$ - hyperedge attribute; represents number of walls adjacent to given edge
$R \in \{TRUE, FALSE\}$ - hyperedge attribute; flag which denotes whether given triangle should be refined

We denote graph grammar production as $P = (LG, RG, VM, PR)$, where: LG and RG are hypergraphs
$VM : LG \ni l \rightarrow r \in RG$ is the function which maps some veritices from LG to RG
PR - predicate of applicability. We can apply production to graph G if there exist a subgraph of G isomorphic to LG and the predicate of applicability is fulfilled. Application of production is done by substituting with RG the subgraph of graph

G isomorphic with LG so that some vertices of subgraph isomorphic to LG are replaced with corresponding vertices from RG according to VM mapping.

Hypergraph grammar is a system $HG = (G_0, PS)$, where:
G_0 is the starting hypergraph, PS is a production set

An example of tetrahedral refinement represented as hypergraph production is presented in Fig. 1.

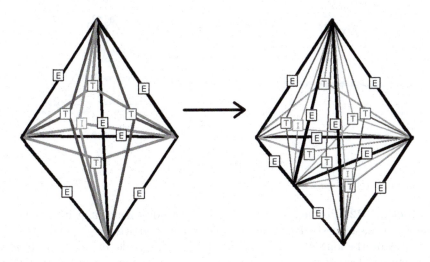

Fig. 1. Hypergraph production refining tetrahedron into two smaller tetrahedra.

2.3   Walls Refinement

In this section, we describe mesh refinement regarding walls and edges. Because of a 3D domain, our case has the following differences from two-dimensional mesh refinement:

1. There are no boundary edges. Each edge have two adjacent walls on the single tetrahedron.
2. Edge might be shared between multiple tetrahedra and their walls.
3. Each vertex might be a hanging node on multiple walls.

These properties of the graph require some adjustments in graph grammar. To keep track of hanging nodes, we introduce a hanging node counter (HN) in each vertex and adjacent wall counter in each hyperedge representing a wall.

Production P1. Production P1 from Fig. 2 refines element with no hanging nodes, marked with R=TRUE into two triangles. New vertex is created in the midpoint of the longest edge. It becomes a hanging node for all walls adjacent to split edge but one. Split edge keeps the number of adjacent walls. Both newly created triangles have R set to false because element is already refined. The new edge going through triangle has only two adjacent walls. Since this grammar does not support multiple hanging nodes on triangle edges, we need to ensure, that ends of the longest edge are not hanging nodes. The predicate for this production is as follows: $R(R1, L1, L2, L3, HN1, HN2) = R1$ AND $L2 \geq L1$ AND $L2 \geq L3$ AND $HN2 = 0$ AND $HN3 = 0$.

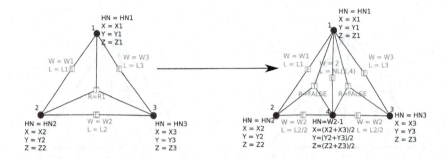

Fig. 2. Production P1.

Production P2. Production P2 from Fig. 3 considers elements with one hanging node on the longest edge of the triangle. No new vertices are created. Because the vertex on the longest edge is no longer a hanging node for this wall, its HN parameter is decremented. Because the algorithm refines all elements with a hanging node, the value of R parameter does not matter in this production. We do not create any new hanging node, so we can also omit $HN = 0$ conditions in the predicate:

$$R(L1, L2, L3, L4) = (L2 + L3) \geq L1 \text{ AND } (L2 + L3) \geq L4$$

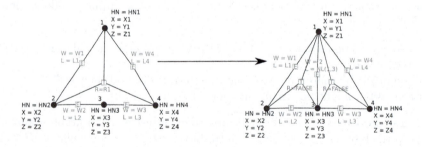

Fig. 3. Production P2.

Production P3. Production P3 from Fig. 4 refines the triangle with one hanging node on other than the longest edge. A hanging node might be left for the next productions. This production works similarly to P1. The only differences are that there is strict inequality between edge to be split and edge with hanging node and refinement does not depend on the R flag. If the edges were the same length, only P2 could be applied since we prioritize reducing hanging nodes number. Predicate for P3:

$$R(L1, L2, L3, L4, HN1, HN3) = L4 \geq L1 \text{ AND } L4 > (L2 + L3) \text{ AND } HN1 = 0 \text{ AND } HN4 = 0$$

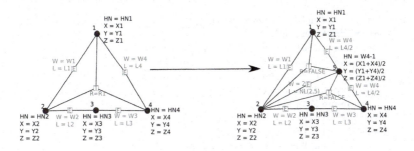

Fig. 4. Production P3.

Production P4. Production P4 from Fig. 5 refines the triangle with 2 hanging nodes. One of the hanging nodes is placed on the longest edge. Both predicate and values of parameters of new hyperedge are similar to these in P2. The predicate for P4: $R(L1, L2, L3, L4, L5) = (L4 + L5) \geq (L2 + L3) \text{ AND } (L4 + L5) \geq L1$.

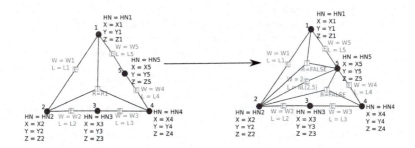

Fig. 5. Production P4.

Production P5. Production P5 from Fig. 6 refines the triangle with two hanging nodes on its shorter edges. As in P3, the length of the edge has to be strictly greater than the length of the edges with hanging nodes. The predicate for P5: $R(L1, L2, L3, L4, L5, HN1, HN3) = L5 > (L1+L2)$ AND $L5 > (L3+L4)$ AND $HN1 = 0$ AND $HN5 = 0$.

Production P6. Production P6 from Fig. 7 refines the triangle with hanging node on each edge. The predicate ensures that hanging node on the longest edge will be used to split the triangles. The predicate for P6: $R(L1, L2, L3, L4, L5, L6) = (L5 + L6) \geq (L1 + L2)$ AND $(L5 + L6) \geq (L3 + L4)$

```
1   procedure refineTetrahedron(IEdge t):
2       q := emptyQueue()
3       verticesInTetrahedron := emptyList()
4       tetrahedraToBeCreated := emptyList()
5       //Find all vertices in the tetrahedron
6       Vt := getVertices(t)
7       q.enqueue(Vt[0])
8       verticesInTetrahedron.add(Vt[0])
9       while q is not empty:
10          currentVertex := q.dequeue()
11          for each neighbor of currentVertex:
12              if (neighbor is inside t
13              AND neighbor is not in verticesInTetrahedron):
14                  q.enqueue(neighbor)
15                  verticesInTetrahedron.add(neighbor)
16      //Find all tetrahedra to be created
17      possibleTetrahedra :=
18      getAllFourElementsSubsetsOf(verticesInTetrahedron)
19      for each tetrahedronVertices in possibleTetrahedra:
20          allVerticesAreConnected := TRUE
21          for each subset in
22          getAllTwoElementsSubsetsOf(tetrahedronVertices):
23              if subset[0] has no edge 'E' to subset[1]:
24                  allVerticesAreConnected := FALSE
25          if allVerticesAreConnected:
26          tetrahedraToBeCreated.add(tetrahedronVertices)
27      //Add missing walls and tetrahedra
28      for each tetrahedronVertices in tetrahedraToBeCreated:
29          for each subset in
30          getAllThreeElementsSubsetsOf(tetrahedronVertices):
31              if subset elements have no common 'T' hyperedge:
32                  CreateTEdge(subset)
33          CreateIEdge(tetrahedronVertices)
34      RemoveIEdge(t)
```

Listing 1.1. Tetrahedron refinement based on split walls

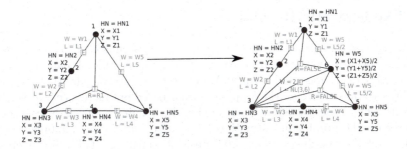

Fig. 6. Production P5.

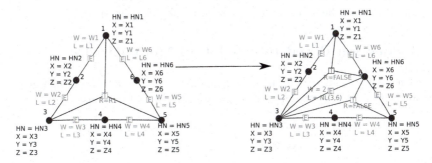

Fig. 7. Production P6.

2.4 Tetrahedron Refinement

Productions described in the previous section leave tetrahedra with split walls. In this section, we present all possible outcomes of 2D refinement as well as naive pseudocode for tetrahedron refinement. We have analyzed possible results of 2D refinement for cases with broken 1–4 walls. There are 20 different cases collected in Fig. 8. The graph-grammar model ensures that there are no hanging nodes. From such grids with refined faces, we can break tetrahedron in a deterministic way into smaller pieces by creating new tetrahedron between any fully connected four vertices. Therefore we may treat tetrahedra represented by nets on Fig. 8 as left sides of hypergraph grammar production while the right side would be multiple tetrahedra generated as described above.

In Listing 1.1 we present a procedure which refines tetrahedron which walls have been already split with 2D refinement. The only argument passed is the interior edge, which represents tetrahedron to be refined lines 8–17 performs Breadth-First Search (BFS) algorithm to find all vertices which are inside the tetrahedron. More precisely all new vertices have been created on tetrahedron's walls and edges, so it is sufficient to check whether the vertex is on any of the walls of the original tetrahedron. Lines 21–30 check all combinations of 4 vertices if they are connected with each other with 'E' hyperedges. If so, new tetrahedron will be created between the chosen vertices. All new tetrahedra will contain at least one wall inside the original one, so we need to create missing walls. In lines 33–39, we add relevant hyperedges representing missing walls and tetrahedra. Finally, the original tetrahedron is removed.

3   Experimental Results

In this section we verify our algorithm on the three-dimensional mesh refined towards the formation layers data. We start from the cube mesh partitioned into five tetrahedra, as presented in Fig. 9. We refine faces of tetrahedra that intersects two layers, and we execute the graph-grammar based productions expressing the longest-edge refinement algorithm removing the hanging nodes on faces. Finally, we execute the graph-transformation summarized in Listing 1.1 breaking the interiors of the tetrahedra. The final mesh obtained by 5,000 refinements are presented in Fig. 9. We monitor the minimal angles between edges of the created tetrahedra, as well as the ratios between the sum of the lengths of the three longest edges and the sum of all the edges, in Fig. 10 (Fig. 8).

4   Projection-Based Interpolation Metaheuristic

On each of the tetrahedral finite element we span the hierarchical basis functions following [16]. Namely, we introduce four master basis functions

$$\lambda_1(\xi_1, \xi_2, \xi_3) = 1 - \xi_1 - \xi_2\xi_3, \quad \lambda_2(\xi_1, \xi_2, \xi_3) = \xi_1$$
$$\lambda_3(\xi_1, \xi_2, \xi_3) = \xi_2, \quad \lambda_4(\xi_1, \xi_2, \xi_3) = \xi_3$$

and we define linear basis functions, one on each element vertex

$$\psi_1(\xi_1, \xi_2, \xi_3) = \lambda_1(\xi_1, \xi_2, \xi_3) \quad \psi_2(\xi_1, \xi_2, \xi_3) = \lambda_2(\xi_1, \xi_2, \xi_3)$$
$$\psi_3(\xi_1, \xi_2, \xi_3) = \lambda_2(\xi_1, \xi_2, \xi_3) \quad \psi_4(\xi_1, \xi_2, \xi_3) = \lambda_2(\xi_1, \xi_2, \xi_3)$$

We also span $p_i - 1$ polynomials of orders $2, ..., p_i$ over element edges, $(p_m - 1)(p_n - 1)$ polynomials of orders $(2, 2), ..., (p_m, p_n)$ on element faces, and $(p_a - 1)(p_b - 1)(p_c - 1)$ polynomials of order $(2, 2, 2), ..., (p_a, p_b, p_c)$ on element interiors. Exemplary second order polynomials on element edges, faces and interiors are obtain by multiplications of two, three or four master basis functions

$$\psi_5(\xi_1, \xi_2, \xi_3) = \lambda_1(\xi_1, \xi_2, \xi_3)\lambda_2(\xi_1, \xi_2, \xi_3)$$
$$\psi_6(\xi_1, \xi_2, \xi_3) = \lambda_2(\xi_1, \xi_2, \xi_3)\lambda_3(\xi_1, \xi_2, \xi_3)$$
$$\psi_7(\xi_1, \xi_2, \xi_3) = \lambda_3(\xi_1, \xi_2, \xi_3)\lambda_4(\xi_1, \xi_2, \xi_3)$$
$$\psi_8(\xi_1, \xi_2, \xi_3) = \lambda_4(\xi_1, \xi_2, \xi_3)\lambda_1(\xi_1, \xi_2, \xi_3)$$
$$\psi_9(\xi_1, \xi_2, \xi_3) = \lambda_1(\xi_1, \xi_2, \xi_3)\lambda_2(\xi_1, \xi_2, \xi_3)\lambda_3(\xi_1, \xi_2, \xi_3)$$
$$\psi_{10}(\xi_1, \xi_2, \xi_3) = \lambda_2(\xi_1, \xi_2, \xi_3)\lambda_3(\xi_1, \xi_2, \xi_3)\lambda_4(\xi_1, \xi_2, \xi_3)$$

$$\psi_{11}(\xi_1, \xi_2, \xi_3) = \lambda_2(\xi_1, \xi_2, \xi_3)\lambda_3(\xi_1, \xi_2, \xi_3)\lambda_4(\xi_1, \xi_2, \xi_3)$$
$$\psi_{12}(\xi_1, \xi_2, \xi_3) = \lambda_1(\xi_1, \xi_2, \xi_3)\lambda_3(\xi_1, \xi_2, \xi_3)\lambda_4(\xi_1, \xi_2, \xi_3)$$
$$\psi_{13}(\xi_1, \xi_2, \xi_3) = \lambda_1(\xi_1, \xi_2, \xi_3)\lambda_2(\xi_1, \xi_2, \xi_3)\lambda_4(\xi_1, \xi_2, \xi_3)$$
$$\psi_{14}(\xi_1, \xi_2, \xi_3) = \lambda_1(\xi_1, \xi_2, \xi_3)\lambda_2(\xi_1, \xi_2, \xi_3)\lambda_3(\xi_1, \xi_2, \xi_3)$$

Fig. 8. Possible outcomes of 2D refinements. Red lines represents manually broken walls. Blue lines represent additional refinements done by P1–P6 productions to keep the graph consistent (Color figure online)

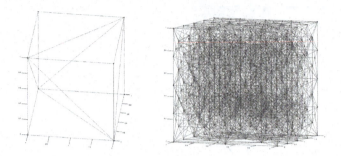

Fig. 9. Initial mesh and final mesh refined towards formation layers

$$\psi_{15}(\xi_1, \xi_2, \xi_3) = \lambda_1(\xi_1, \xi_2, \xi_3)\lambda_2(\xi_1, \xi_2, \xi_3)\lambda_3(\xi_1, \xi_2, \xi_3)\lambda_4(\xi_1, \xi_2, \xi_3)$$

For the full definition of the hierarchical basis functions we refer to [16].

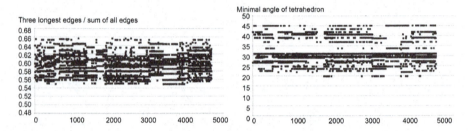

Fig. 10. Ratio between the sum of lengths of the three longest edges and the sum of all the lengths of all the edges for newly created tetrahedra (left panel), and the minimal angle between edges of the newly created tetrahedra (right panel).

Having the mesh refined towards the formation layers, we compute now the projection of the terrain data employing the projection-based interpolation algorithm. We compute coefficients for the four vertex functions at the vertices (x_i, y_i, z_i) for $i = 1, 2, 3, 4$

$$a_i = U(x_i) \quad i = 1, 2, 3, 4 \tag{1}$$

Next, we solve the L2 projection problem over each of six edges e_i

$$\left\| (U - \sum_{i=1,\ldots,4} a_i\psi_i) - a_j\psi_j \right\|_{L^2(e_j)} \to 0 \quad j = 5, 6, 7, 8, 9, 10 \tag{2}$$

L2 projection over four element faces f_j

$$\left\| (U - \sum_{i=1,\ldots,10} a_i\psi_i) - a_j\psi_j \right\|_{L^2(f_j)} \to 0 \quad j = 11, 12, 13, 14 \tag{3}$$

and finally L2 projection over element interior

$$||(U - \sum_{i=1,...,14} a_i\psi_i) - a_{15}\psi_{15}||_{L^2(K)} \to 0 \qquad (4)$$

We can select optimal polynomial orders of approximation at element interiors, element faces, and element edges, constructing a projection of the material data. We employ the metaheuristic algorithm presented in Listing 1.2, that is minimizing both numerical error and the computational cost. The algorithm is an iterative procedure. It first selects the minimal polynomial orders of approximation on finite element edges, faces and interiors. It solves the projection problem, and then it increases the polynomial order of approximation by one, in all directions, on the entire mesh. It solves the projection problem again on the finer mesh. Having the coarse $u(h,p)$ and the fine $u(h,p+1)$ mesh solutions, it considers different refinement strategies. We can increase the polynomial order on selected edges, selected faces and over element interiors. There are several possibilities, since we have different orders in different directions. All these possibilities are considered, and we project the resulting element solution from the fine mesh into the proposed configuration of basis functions related to the considered orders configuration, given the projected reference solution w. Each refinemenet comes with the prize to pay, expressed by $dnrdof$ the number of basis functions added to implement this refinement strategy. We compute the error decrease rate $rate(w) = |u(h,p+1) - u(h,p)| - |u(h,p+1) - w|)/dnrdof$ the ratio between the error decrease and the cost. Finally, we select such the refinement, that results in the maximum error decrease rate. We proceed with these computations element wise, and for the edges and faces shared between elements we do not change the optimal orders when we consider the other element.

```
1       procedure SelectPRefinement(Element K,
2           coarse solution u(h,p), fine solution u(h,p+1):
3       for mesh elements K:
4           for P(i,K) different polynomial configurations over K:
5               ratemin = infinity
6               Compute the projection based interpolant w
7                   of u(h,p+1) for polynomial order P(i,K)
8               Compute the error decrease rate
9               rate(w)=(|u(h,p+1)-u(h,p)|-|u(h,p+1)-w|)/dnrdof
10              if rate(w) < ratemin then
11                  ratemin = rate(w)
12                  Select P(opt,K) corresponding to ratemin
13                      as the optimal refinement for element K
```

Listing 1.2. Selection of optimal refinements over K

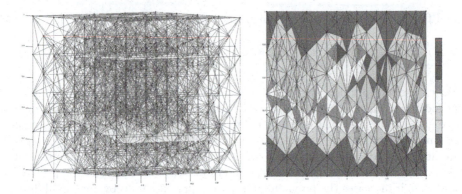

Fig. 11. Exemplary L2 projection of terrain data with three layers as well as the exemplary distribution of polynomial orders over element faces.

5   Conclusions

We proposed the graph-grammar based model of the three-dimensional longest-edge refinement algorithm for approximation of the geological formation layers. It starts with executing the two-dimensional graph-grammar based longest-edge refinement algorithm on faces of the tetrahedra, taking into account the three-dimensional topology of the computational mesh. In particular, it propagates the refinements between topologically adjacent faces. Next, it breaks all the interiors of the tetrahedra having some faces already broken by the first stage of the algorithm. We verified the graph-grammar based algorithm by a sequence of numerical experiments. Namely, we broke 5,000 tetrahedra towards the geological formation layers. We checked that the algorithm removed all the hanging nodes, and it preserved the proportions of the original tetrahedra. We also discussed the metaheuristic algorithm allowing for the selection of the optimal orders of approximation on element edges, faces and interiors.

Acknowledgement. This work was supported by National Science Centre, grant no. 2019/35/O/ST6/ 00571.

References

1. Farquharson, C.G., Lelièvre, P.G., Ansari, S., Jahandari, H.: Towards real earth models - computational geophysics on unstructured tetrahedral meshes? (2014)
2. Ansari, S., Farquharson, C.G.: Numerical modeling of geophysical electromagnetic inductive and galvanic phenomena, pp. 669–674 (2013)
3. Lelièvre, P.G., Farquharson, C.G.: Gradient and smoothness regularization operators for geophysical inversion on unstructured meshes. Geophys. J. Int. **195**(1), 330–341 (2013)
4. Lelièvre, P.G., Farquharson, C.G., Hurich, C.A.: Joint inversion of seismic travel times and gravity data on unstructured grids with application to mineral exploration. Geophysics **77**(1), K1–K15 (2012)

5. Puzyrev, V., Koldan, J., de la Puente, J., Houzeaux, G., Vázquez, M., Cela, J.M.: A parallel finite-element method for three-dimensional controlled-source electromagnetic forward modelling. Geophys. J. Int. **193**(2), 678–693 (2013)
6. Schwarzbach, C., Börner, R.-U., Spitzer, K.: Three-dimensional adaptive higher order finite element simulation for geo-electromagnetics–a marine CSEM example. Geophys. J. Int. **187**(1), 63–74 (2011)
7. Rivara, M.-C.: Mesh refinement processes based on the generalized bisection of simplices. SIAM J. Num. Anal. **21**(3), 604–613 (1984)
8. Arnold, D.N., Mukherjee, A., Pouly, L.: Locally adapted tetrahedral meshes using bisection. SIAM J. Sci. Comput. **22**(2), 431–448 (2000)
9. Balboa, F., Rodriguez-Moreno, P., Rivara, M.-C.: Terminal star operations algorithm for tetrahedral mesh improvement. In: Roca, X., Loseille, A. (eds.) IMR 2018. LNCSE, vol. 127, pp. 269–282. Springer, Cham (2019). https://doi.org/10.1007/978-3-030-13992-6_15
10. Bey, J.: Tetrahedral grid refinement. Computing **55**(4), 355–378 (1995)
11. Antepara, O., Balcázar, N., Oliva, A.: Tetrahedral adaptive mesh refinement for two-phase flows using conservative level-set method. Int. J. Num. Methods Fluids **93**(2), 481–503 (2020)
12. Marot, C., Pellerin, J., Remacle, J.-F.: One machine, one minute, three billion tetrahedra. Int. J. Num. Methods Eng. **117**(9), 967–990 (2019)
13. Guo, W., Nie, Y., Zhang, W.: Parallel adaptive mesh refinement method based on bubble-type local mesh generation. J. Parallel Distrib. Comput. **117**, 37–49 (2018)
14. Rivara, M.-C.: Local modification of meshes for adaptive and/or multigrid finite-element methods. J. Comput. Appl. Math. **36**(1), 79–89 (1991). Special Issue on Adaptive Methods
15. Podsiadło, K., et al.: Parallel graph-grammar-based algorithm for the longest-edge refinement of triangular meshes and the pollution simulations in lesser Poland area. Eng. Comput. **12** (2020)
16. Demkowicz, L., Kurtz, J., Pardo, D., Paszynski, M., Rachowicz, W., Zdunek, A.: Computing with HP-Adaptive Finite Element Method, Volume II Frontiers: Three Dimensional Elliptic and Maxwell Problems with Applications. Taylor & Francis, CRC Press, Boca Raton (2008)

Elitism in Multiobjective Hierarchical Strategy

Michał Idzik[(✉)] [iD], Radosław Łazarz[iD], and Aleksander Byrski[iD]

AGH University of Science and Technology, Krakow, Poland
{miidzik,lazarz,olekb}@agh.edu.pl

Abstract. The paper focuses on complex metaheuristic algorithms, namely multi-objective hierarchical strategy, which consists of a dynamically evolving tree of interdependent demes of individuals. The main contribution presented in this paper is the introduction of elitism in a form of an archive, locally into the demes and globally into the whole tree and developing necessary updates between them. The newly proposed algorithms (utilizing elitism) are compared with their previous versions as well as with the best state of the art multi-objective metaheuristics.

Keywords: Multi-objective optimization · Hierarchical metaheuristics · Elitist multi-objective evolutionary algorithms

1 Introduction

Contradictory objectives are present in the optimization area and such problems are much harder to solve than global-optimization ones. Their difficulty is even increased, as we have to seek not only one optimum but a whole set of Pareto-optimal solutions. We have to try to distribute those solutions evenly, make sure that we do not lose good solutions between the subsequent approximations of the Pareto-front, especially when the number of those functions becomes higher than several (many-objective problems). Natural weapons of choice to deal with such problems are metaheuristic algorithms.

One of the very interesting and successful metaheuristics is Hierarchic Genetic Search, introduced by Schaefer and Kolodziej [22]. It assumes a very strict structuralization of the sub-populations in a form of a tree. Particular demes are created when interesting solutions are found and the parameters of the search are adapted (the search grid is modified). The demes can sprout other demes. Of course, the demes can also be removed so the whole structure is dynamic.

HGS algorithm was successfully applied to solving global optimization problems, however, later it was adapted by Łazarz et al. [17] to multi-objective problems and further developed in this area. In this paper we want to present a subsequent, important development of the HGS-related multi-objective metaheuristics,

The research presented in this paper was financed by Polish National Science Centre PRELUDIUM project no. 2017/25/N/ST6/02841.

M. Paszynski et al. (Eds.): ICCS 2021, LNCS 12745, pp. 214–228, 2021.
https://doi.org/10.1007/978-3-030-77970-2_17

introducing the notion of elitism, by applying local and global archives into HO-mHGS, which step is necessary in to retain good solutions in the current frontier.

The next section of this paper shows the most important aspects of elitism in the area of multi-objective optimization, then the multi-objective hierarchical strategy is discussed along with the presentation of the aspects of local and global archives, then the experimental results are presented, showing not only the relation between newly introduced algorithms and its older versions but also referring to the most known state of the art algorithms, like NSGAII.

2   Elitism in Multi-objective Evolutionary Algorithms

Elitism, in a broad sense, is a mechanism enhancing evolutionary computation. It works by ensuring that the current best individuals have a higher chance of remaining in the population throughout the iterations (or being otherwise involved in the search process). As a result, it guarantees evolution towards the objective function improvement (or at least prevents deterioration to inferior solutions) [10]. On the other hand, it may decrease the population diverseness due to increased selection pressure.

There have been numerous attempts at implementing this paradigm in practice. One approach is to simply combine the population with its offspring and then apply the survival selection [20]. Its variation, known as multi-elitism [1], instead keeps in population the so-called residents—individuals occupying separate peaks of the fitness function. Another option, popular in the case of multi-objective problems, is to employ a relaxed version of the idea (termed controlled elitism [6]), where solutions are first divided into an ordered sequence of subsets (called fronts), and then each of them contributes to the next generation several individuals based on geometric distribution. Finally, δ-similar elimination [21] procedure can be used to filter the joint parent-offspring population and ensure that no two individuals are within a given distance from one another, thus giving an advantage to more exploratory solutions. All those variants have one thing in common: they strive to preserve the elitism benefits while simultaneously trying to counteract its deficiencies.

An alternative to those techniques is to store the chosen individuals, not in the working population, but in a separate set, usually referred to as archive [8]. Specific optimisation models tend to differ by both their organisation and how archived organisms are utilised during the run. PAES [14] keeps an archive of non-dominated solutions with additional crowding-based thinning. Its contents are used in two situations: to estimate the true dominance ranking in a pair of otherwise incomparable solutions, and to be returned as the final outcome. MOIPSO [26] modifies this scheme by making use of the distribution entropy as a crowding measure, introducing Gaussian mutation throw points as a way of improving the external archive uniformity, and applying it to a particle swarm optimisation technique. A similar archive system is also successfully adopted in the case of chaos optimisation algorithms, such as MPCOA [18]. ME-MOEA/D [9] (a decomposition-based method) identifies elite solutions based on a dominance index, fine-tunes them using a local search mechanism, and transfers them to the isolated external population afterwards. This supplementary population is then used as a potential substitute source of individuals during the crossover operation. Other notable examples of effective archive integration are the classical algorithms PESA-II [4] and SPEA2 [31].

Naturally, certain solutions elude being classified by such a simple dichotomy. For example, BBO [25] is an algorithm utilising a multi-population island-based computation scheme, with organisms representing solutions emigrating from and immigrating to the said islands. Its elitism is enforced by prohibiting migration for the top portion of individuals in each subset. Other researchers experiment with unbounded archives and propose novel frameworks for assembling the final result from the subset of all examined solutions [13]. Lastly, some studies explore the potential of methods that completely abstain from employing any forms of elitism whatsoever, demonstrating that they are still able to perform significantly well on particular types of problems [27].

3   Optimizing MO-mHGS with Archives

3.1   Multi-Objective Hierarchical Strategy

Multi-Objective Hierarchical Strategy (MO-HGS) [2] is a framework designed to adapt and run multi-objective evolutionary algorithms (MOEAs) in an environment split into several populations (called *demes*). MO-HGS dynamically creates population nodes and lays them out in a tree-like hierarchy. Each node runs an independent instance of MOEA (denoted as node's *driver*) and its search accuracy depends on its depth in the hierarchy. Nodes closer to the root perform a more chaotic search to find promising areas.

The process of MO-HGS consists of several steps, called *metaepochs*. In each metaepoch driver runs several epochs, progressing in an evolutionary process. Additionally, HGS-specific mechanics are applied. The most promising individuals (*delegates*) of each node have a chance to become seeds of next-level *child nodes* (*sprouting* procedure). A child node runs with reduced variance settings so that its population will mostly explore that region. To eliminate the risk of redundant exploration of independently evolving demes, two trimming procedures are performed. If the area is already explored by one of the other children, then sprouting is cancelled and the next candidate for sprouting is considered (*branch comparison*). Moreover, after each metaepoch MO-HGS checks if populations at the same level of the tree perform a search in the common landscape region or already explored regions and removes such nodes (*branch reduction*).

MO-HGS was inspired by a more general hierarchical scheme, Hierarchical Genetic Strategy [23], which was adapted into a multi-objective environment. Further improvements were made to the model, including the hypervolume-based sprouting procedure introduced in [17]. The result meta-model was denoted as Multiobjective Optimization Hierarchic Genetic Strategy with maturing (MO-mHGS).

With MO-mHGS we focus mainly on reducing the cost of fitness evaluation. We proved that this meta-model can be a perfect tool to solve problems with a high cost of the evaluation. It can be achieved by providing fitness operators with different accuracies bound to specific HGS tree levels. However, in the recent research [12] we have shown that even without this fitness adjusting mechanism, MOEA enforced by MO-mHGS can achieve better performance comparing to the situation when it's run alone. It is possible thanks to the natural parallel execution capabilities of the hierarchical model.

In this paper, we continue the exploration of MO-mHGS features as a generic tool improving the performance of its driver (preferably: any MOEA). This time we will consider two approaches to elitism in the MO-mHGS meta model: global fitness archive and local node-level archives. We expect it should lead to better convergence and more stable results, regardless of the incorporated driver.

3.2   Global Fitness Archive in MO-mHGS

In each metaepoch MO-mHGS delegates the work to its internal driver of the given tree node. As a result, it obtains a new state of all demes that can be used in further procedures (sprouting, reduction). These populations are completely independent which by design (combined with redundancy reduction procedures) leads to better coverage of multiple different areas. But, at the same time, we lose information about solutions found during the process, that might be valuable but were abandoned with simulation progress. Furthermore, we strongly rely on a driver ability to provide any elitism mechanics and compose the final MO-mHGS result as a union of final populations of all nodes. Such result population may be large in the case of broader HGS trees. It is unnecessary – there's no guarantee that a larger set of solutions is better in terms of individuals distribution over a Pareto front. To address these issues we modified our MO-mHGS model by adding a global fitness archive located in the nodes supervisor.

This MO-mHGS variant (denoted as *Multi-Objective Elitist Hierarchical Genetic Strategy*, MO-EHGS or EHGS) offers a new metaepoch procedure presented in Listing 1. Each alive (still progressing and not redundant) node invokes several steps of its driver and produces finalized population. How the population is finalized is driver-dependent – it may be similar to the delegates set described in the following section. Solutions from the population are inserted into the archive. Eventually, the archive is truncated according to the comparison operator. In our experiments, we use the crowding distance attribute from fast non-dominated sorting to compare results. Note that solutions from dead nodes can be still present in the final result if they are strong enough to prevail.

Additionally, in simulations driven by a budget of allowed fitness evaluations number, it is important to preserve fair assumptions – solutions are added to the archive only if node execution does not exceed the budget. Finally, fitness-based archive size can be easily constrained so we can now control the maximum size of MO-EHGS result front approximation. In our experiments, we set maximum archive size $|A|$ depending on a number of a problem's objectives k. According to the rules set in the CEC09 competition [29]: $|A| = 100$ for $k = 2$, $|A| = 150$ for $k = 3$ and $|A| = 800$ for $k = 5$.

Algorithm 1. MO-EHGS metaepoch procedure with elitism support

---

Require:
    hgsNodes, list of all HGS nodes
    archive, fitness based archive
    budget, maximum allowed evaluation cost
    totalCost < *budget*, current cost of HGS simulation

1: **for each** *node* in *hgsNodes* **do**:
2:     **if** IsAlive*node* **then**
3:         *totalCost* ← *totalCost* + RunMetaepoch(*node*)
4:         **if** IsBudgetMet(*budget, totalCost*) **then**
5:             **break**
6:         **end if**
7:         *nodePopulation* ← FinalPopulation(*node*)
8:         *archive* ← *archive* ∪ *nodePopulation*
9:     **end if**
10: **end for**
11: Truncate(*archive*, CrowdingDistanceOperator())

3.3 Local Archives

During our previous research, we observed OMOPSO [24] algorithm was the driver most significantly improved by MO-mHGS. OMOPSO is one of the particle swarm optimizers adjusted to solve multi-objective problems. This specific observation has lead us to the conclusion that characteristics of OMOPSO could be applied in the general scheme of the MO-HGS model to recreate these phenomena in other applications. The crucial part of OMOPSO algorithm is the way it handles elitism. There are two layers of best individual storages: swarm leaders archive and global ϵ-archive. The latter is based on the concept of ϵ-dominance [16]. A search space vector X_1 ϵ-dominates solution X_2 when: $f_i(X_1)/(1+\epsilon) \leq f_i(X_2), \forall_{i=1...k}$ and $f_i(X_1)/(1+\epsilon) < f_i(X_2), \exists i \in 1..k$, where $f_1,..f_k$ are objective functions of k-objective problem. The value of ϵ should be greater than 0 and is set arbitrary, depending on a solved problem.

An archive built on ϵ-dominance trimming operator can be used as an additional filter of best solutions, reducing outcomes size. This procedure may also supplement the MO-mHGS nodes sprouting process. After a metaepoch, each HGS node selects *delegates* – individuals that will seed new populations whenever sprouting is performed. Choosing the right delegates is specific to a driver, e.g. in NSGAII we take the first layer of non-dominance sorting results, while in SPEA2 we consider the whole result archive. More generically we could assume that the delegates set consists of the whole current node's population. It may lead to worse MO-mHGS performance, but sometimes it is hard to find a better adaptation that could take advantage of driver characteristics. However, by adding an additional layer in form of ϵ-archive we could improve this process and institute more robust independence from a used driver. Figure 1 shows how

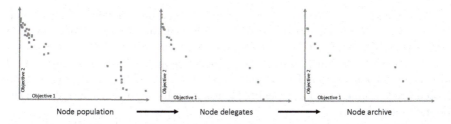

Node population ━━━━━▶ Node delegates ━━━━━▶ Node archive

Fig. 1. Presentation of MO-ϵ-EHGS node ϵ-archive update process for 2-objective problem example. Results of each metaepoch are filtered in two stages: as population delegates (specific to a driver) and as ϵ-archive truncation outcomes.

the final delegates nomination procedure works in another MO-mHGS modification (denoted as *Multi-Objective ϵ-Elitist Hierarchical Genetic Strategy*, MO-ϵ-EHGS or ϵ-EHGS). At the end of a metaepoch node produces a new population. Then we choose the best solutions using driver-specific mechanics. Finally, the delegates are inserted into ϵ-archive. At later stages, the sprouting procedure randomly selects new seeds from the archive.

It is also worth noting that each node possesses a separate (local) ϵ-archive. The state of each local archive is preserved between metaepochs thus it serves not only as a filtering operator but also as elitist storage. When MO-ϵ-EHGS gathers its final results from all nodes, it receives the content of these local archives (instead of finalized populations).

4   Experiments and Results

4.1   Preparing the Experiment's Environment

The main aim at this stage of experiments was to compare MO-EHGS and MO-ϵ-EHGS with specific MOEA-based driver against basic (single-deme) version of this MOEA. The previous (non-elitist) version of MO-mHGS was also included in the survey. We selected two drivers basing on two different MOEA approaches: NSGAII [5] and SPEA2 [31]. Additionally, we prepared a separate set of simulations for NSGAIII [7] driver, intended to solve many-objective problems.

We conducted several simulations that were set up according to rules inspired by CEC09 competition [29]: each algorithm had a maximum budget of 300000 fitness function evaluations and was run independently 30 times for each test problem to improve statistical accuracy. Simulations were run on benchmark problems from two widely known families: ZDT [30] and CEC09. We divided problems into 3 categories: basic 2-objective (ZDT1-ZDT6), more demanding 2-objective (UF1-UF7) and complex with more objectives. The last category included a 3-objective (UF8-UF10) and one 5-objective (UF11). Problems from the first two groups were tested with NSGAII and SPEA2 drivers and problems from the third group with NSGAIII.

At the end of each run (after reaching the maximum budget), result populations were stored and evaluated with 3 different quality indicators. Inverted Generational Distance (IGD) [3] computes the average size of the gap between the members of optimal Pareto front and the one approximated by an assessed algorithm. Hypervolume (HV) [28] measures the volume of the result space dominated by the found solution. Spacing [3] describes how uniformly distributed the points in solution are.

Metric values from all 30 runs were then statistically analyzed. We calculated minimum, average and maximum value with standard deviation. To ensure statistical significance we also performed Kruskal-Wallis [15] and Mann-Whitney U [19] tests. Basing on the tests results, we were able to create set of indifferent algorithms I_{α_i} for each result of an algorithm α_i. We define a result as a **global winner** if it outperforms other algorithms and $I_{\alpha_i} = \emptyset$. Because we are interested in analyzing impact of MO-mHGS in improving driver's performance we also introduce concept of **local winners**. An algorithm α_i is marked as strong if it outperforms other algorithms and:

$$
I_{\alpha_i} = \begin{cases} \emptyset \vee \{\alpha_j \mid \alpha_j \in A_{HGS} \wedge 0 \leq j \leq N \wedge j \neq i\} & \alpha_i \in A_{HGS} \\ \emptyset \vee \{\alpha_j \mid 0 \leq j \leq N \wedge j \neq i\} \wedge A_{HGS} \subsetneq I_{\alpha_i} & \text{otherwise} \end{cases}
$$

where A_{HGS} is set of all MO-mHGS algorithm variants and N number of all algorithms included in the simulation. In other words, MO-mHGS results have to be statistically different from single-deme algorithms, but they can be indifferent about other MO-mHGS variants. On the other hand, a single-deme algorithm should win over at least one MO-HGS variant to be a local winner.

For completeness, we also put in our summary number of **weak winners**, that is algorithms that outperformed other solutions regardless of significance test result.

All described simulations were performed with the use of kEMO[1], our new platform written in Kotlin. It is also a wrapper for popular Java library, MOEA Framework [11] containing implementations of state-of-the-art MOEAs and multi-objective problems. Therefore, all algorithm parameters were set with default values provided by MOEA Framework. That includes i. a. ϵ values used in ϵ-archives specific to each considered benchmark.

4.2   Improving MOEA Performance

Tables 1, 2 and 3 show comparison of MO-mHGS, MO-EHGS and MO-ϵ-EHGS with reference to NSGAII driver. In all problems but one MO-mHGS variants offer better performance than NSGAII in terms of Hypervolume and IGD. While final Hypervolume outcomes were close to each other in all cases (with notable MO-mHGS winners), in IGD indicator we can observe improvement up to 50–60% for simpler ZDT problems and 20–30% for UF family. Spacing of the outcomes also is the best in MO-mHGS variants, but it's less consistent – in simpler

---

[1] https://github.com/Soamid/kEMO.

Table 1. Summary of NSGAII with different variants of MO-mHGS after 300000 fitness evaluations (Hypervolume indicator, higher is better).

	Hypervolume															
	NSGAII				MO-mHGS+NSGAII				MO-EHGS+NSGAII				MO-ε-EHGS+NSGAII			
Problem	Min	Average	Max	Error	Min	Average	Max	Error	Min	Average	Max	Error	Min	Average	Max	Error
zdt1	0.65561	0.65678	0.65765	0.00041	0.66034	0.66086	0.66120	0.00019	**0.662015**	**0.662049**	**0.662073**	**0.000016**	0.66071	0.66112	0.661139	0.00020
zdt2	0.32265	0.32395	0.32486	0.00050	0.32706	0.32756	0.32785	0.00018	**0.328736**	**0.328764**	**0.328797**	**0.000016**	0.32765	0.32789	0.32814	0.00013
zdt3	0.51374	0.51412	0.51455	0.00020	0.515289	0.515455	0.515604	0.000070	**0.5159753**	**0.5160028**	**0.5160215**	**0.0000099**	0.51372	0.51468	0.51507	0.00033
zdt4	0.00	0.08	0.34	**0.10**	0.02	0.33	0.66	0.20	0.04	**0.34**	**0.66**	0.16	**0.04**	0.31	0.66	0.15
zdt6	0.39564	0.39632	0.39693	0.00035	0.39828	0.39964	0.40021	0.00038	**0.40050**	**0.40073**	**0.40105**	**0.00011**	0.39901	0.39957	0.39992	0.00022
UF1	0.470	0.558	0.609	0.033	0.515	0.565	0.612	0.028	0.545	**0.601**	**0.634**	**0.022**	**0.546**	0.599	0.631	0.022
UF2	0.6145	0.6222	0.6301	0.0039	0.6094	0.6251	0.6371	0.0063	0.6232	0.6320	**0.6389**	0.0047	**0.6252**	**0.6342**	0.6386	**0.0035**
UF3	0.353	0.452	0.525	0.051	0.369	0.458	0.524	**0.043**	0.375	**0.469**	**0.553**	0.052	0.351	0.454	0.544	0.050
UF4	**0.2518**	**0.2564**	**0.2591**	**0.0018**	0.2397	0.2488	0.2562	0.0042	0.2321	0.2426	0.2536	0.0053	0.2338	0.2445	0.2527	0.0049
UF5	0.055	0.169	0.297	0.069	0.152	0.235	0.303	0.031	0.179	**0.243**	0.304	**0.031**	0.133	0.232	**0.305**	0.036
UF6	0.000	0.193	0.324	0.084	0.193	**0.273**	**0.356**	**0.048**	0.060	0.231	0.349	0.072	0.088	0.247	0.352	0.074
UF7	0.10	0.34	0.46	0.11	0.192	0.364	0.461	0.099	0.193	0.383	**0.478**	0.095	**0.207**	**0.397**	0.477	**0.091**
Weak wins	1	1	1	2	1	1	1	2	6	8	9	6	4	2	1	2
Local wins	1	1	1	2	1	1	1	2	5	7	8	6	4	2	1	2
Global wins	1	1	1	2	0	0	0	2	4	4	5	6	1	1	0	2

Table 2. Summary of NSGAII with different variants of MO-mHGS after 300000 fitness evaluations (IGD indicator, lower is better).

	Inverted Generational Distance															
	NSGAII				MO-mHGS+NSGAII				MO-EHGS+NSGAII				MO-ε-EHGS+NSGAII			
Problem	Min	Average	Max	Error	Min	Average	Max	Error	Min	Average	Max	Error	Min	Average	Max	Error
zdt1	0.00693	0.00736	0.00814	0.00028	0.003863	0.004020	0.004190	0.000083	**0.003639**	**0.003683**	**0.003717**	**0.000018**	0.003795	0.003906	0.004091	0.000068
zdt2	0.00676	0.00751	0.00815	0.00033	0.00406	0.00423	0.00458	0.00011	**0.003739**	**0.003788**	**0.003869**	**0.000027**	0.00399	0.00421	0.00451	0.00012
zdt3	0.00483	0.00534	0.00589	0.00025	0.002848	0.002982	0.003098	0.000052	**0.002668**	**0.002713**	**0.002791**	**0.000026**	0.00326	0.00361	0.00450	0.00027
zdt4	0.26	0.73	1.60	0.34	0.00	0.29	0.68	0.21	**0.00**	**0.28**	**0.67**	0.17	**0.00**	0.30	0.67	**0.16**
zdt6	0.00498	0.00578	0.00891	0.00087	0.00308	0.00338	0.00423	0.00028	0.003126	**0.003308**	**0.003519**	**0.000084**	0.00356	0.00389	0.00440	0.00019
UF1	0.045	0.080	0.148	0.025	0.040	0.076	0.121	0.024	**0.020**	**0.044**	**0.081**	0.017	0.022	0.047	0.082	**0.017**
UF2	0.0258	0.0350	0.0545	0.0055	0.0212	0.0342	0.0604	0.0089	**0.0179**	0.0263	0.0461	0.0064	0.0190	**0.0249**	**0.0370**	**0.0043**
UF3	0.101	0.183	0.317	0.058	0.089	**0.162**	**0.231**	**0.044**	**0.083**	0.166	0.302	0.054	0.086	0.200	0.341	0.060
UF4	**0.0484**	**0.0501**	**0.0542**	**0.0013**	0.0499	0.0556	0.0641	0.0034	0.0525	0.0602	0.0703	0.0045	0.0529	0.0584	0.0668	0.0039
UF5	0.169	0.280	0.583	0.096	0.157	0.192	0.250	0.020	**0.146**	**0.186**	**0.234**	**0.020**	0.148	0.200	0.349	0.041
UF6	0.09	0.24	0.53	0.14	0.059	**0.139**	0.317	0.061	0.09	0.19	0.45	0.11	0.07	0.20	0.51	0.13
UF7	0.02	0.19	0.57	0.17	0.02	0.16	0.43	0.15	**0.01**	0.14	0.42	0.14	**0.01**	**0.12**	**0.40**	0.14
Weak wins	1	1	1	1	2	2	2	2	9	7	7	5	0	2	2	4
Local wins	1	1	1	1	2	1	1	2	8	7	7	5	0	2	2	4
Global wins	1	1	1	1	1	1	1	2	3	3	3	5	0	0	0	4

Table 3. Summary of NSGAII with different variants of MO-mHGS after 300000 fitness evaluations (Spacing indicator, lower is better).

	Spacing															
	NSGAII				MO-mHGS+NSGAII				MO-EHGS+NSGAII				MO-ε-EHGS+NSGAII			
Problem	Min	Average	Max	Error	Min	Average	Max	Error	Min	Average	Max	Error	Min	Average	Max	Error
zdt1	0.0074	0.0109	0.0133	0.0014	0.00229	0.00292	0.00386	0.00031	**0.00067**	**0.00095**	**0.00124**	**0.00017**	0.003	0.059	0.156	0.045
zdt2	0.00825	0.01066	0.01240	0.00088	0.00281	0.00352	0.00478	0.00035	**0.00067**	**0.00096**	**0.00160**	**0.00021**	0.00250	0.00325	0.00410	0.00044
zdt3	0.0090	0.0131	0.0166	0.0015	0.00342	0.00408	0.00485	0.00047	**0.00192**	**0.00249**	**0.00304**	**0.00032**	0.006	0.051	0.219	0.048
zdt4	0.0116	0.0145	**0.0195**	**0.0022**	0.004	0.026	0.080	0.028	**0.001**	**0.010**	0.073	0.017	0.00	0.16	1.25	0.28
zdt6	0.0070	0.0088	0.0110	0.0010	0.0032	0.0042	0.0091	0.0011	**0.00055**	**0.00095**	**0.00145**	**0.00022**	0.002	0.010	0.196	0.035
UF1	0.0017	0.0040	0.0098	0.0022	**0.0006**	**0.0024**	**0.0060**	**0.0015**	0.0018	0.0040	0.0099	0.0021	0.0012	0.0034	0.0084	0.0017
UF2	0.0073	0.0093	0.0116	**0.0010**	0.0037	0.0090	0.0225	0.0053	0.0033	**0.0051**	**0.0079**	0.0013	**0.0030**	0.0075	0.0267	0.0051
UF3	0.000	0.022	0.061	0.018	0.000	0.016	0.084	0.019	0.0001	0.0094	0.0360	0.0092	0.0001	**0.0061**	**0.0290**	**0.0084**
UF4	0.0097	0.0115	0.0137	**0.0010**	0.0039	0.0053	0.0093	0.0012	**0.0037**	**0.0053**	0.0129	0.0016	0.0039	0.0065	**0.0104**	**0.0014**
UF5	**0.000**	**0.030**	**0.117**	**0.031**	0.006	0.048	0.133	0.032	0.001	0.034	0.186	0.041	0.000	0.088	0.364	0.095
UF6	**0.000**	**0.026**	0.170	0.035	0.000	0.033	0.092	0.028	0.000	0.029	0.147	0.031	0.00	0.10	0.77	0.19
UF7	**0.0002**	0.0048	**0.0116**	0.0037	0.0002	0.0042	0.0141	0.0036	0.0008	0.0042	0.0129	0.0027	0.0005	**0.0034**	0.0137	0.0027
Weak wins	3	2	3	4	1	2	3	2	6	6	5	5	2	2	1	1
Local wins	1	1	2	4	1	2	2	2	6	6	5	5	2	1	1	1
Global wins	0	0	0	4	1	1	1	2	5	6	5	5	1	1	1	1

problems spacing gains even 90% improvement, but the last 3 UF problems reach slightly (but insignificantly) better results without MO-HGS.

Concerning elitist mechanics impact on MO-mHGS performance, it is also clear that MO-EHGS figures as the best variant in any considered metric. It wins in 6–8 out of 12 test cases with all competitors (**bold values** in the tables), almost

Table 4. Comparison of NSGAII and SPEA2 results with different variants of elitist MO-mHGS after 300000 fitness evaluations (Hypervolume indicator, higher is better).

	Hypervolume											
	NSGAII		EHGS+NSGAII		ε-EHGS+NSGAII		SPEA2		EHGS+SPEA2		ε-EHGS+SPEA2	
Problem	Average	Error	Average	Error	Average	Error	Average	Error	Average	Error	Average	Error
zdt1	0.65678	0.00041	**0.662049**	**0.000016**	0.66112	0.00020	0.65849	0.00021	0.661858	0.000058	0.66078	0.00016
zdt2	0.32395	0.00050	**0.328764**	**0.000016**	0.32789	0.00013	0.32548	0.00020	0.328655	0.000052	0.32782	0.00013
zdt3	0.51412	0.00020	**0.5160028**	**0.0000099**	0.51468	0.00033	0.51448	0.00013	0.515841	0.000042	0.51446	0.00012
zdt4	0.08	**0.10**	**0.34**	0.16	0.31	0.15	0.09	0.12	0.21	0.13	0.18	0.14
zdt6	0.39632	0.00035	**0.40073**	0.00011	0.39957	0.00022	0.39771	**0.00011**	0.40051	0.00022	0.39932	0.00024
UF1	0.558	0.033	0.601	0.022	0.599	0.022	0.553	0.021	0.618	0.021	**0.622**	**0.014**
UF2	0.6222	0.0039	0.6320	0.0047	0.6342	0.0035	0.6250	0.0085	**0.6351**	**0.0031**	0.6348	0.0033
UF3	0.452	0.051	0.469	0.052	0.454	0.050	0.433	**0.041**	0.488	0.048	**0.499**	0.043
UF4	0.2564	0.0018	0.2426	0.0053	0.2445	0.0049	**0.2626**	**0.0016**	0.2548	0.0036	0.2546	0.0025
UF5	0.169	0.069	0.243	**0.031**	0.232	0.036	0.155	0.088	**0.264**	0.037	0.261	0.048
UF6	0.193	0.084	0.231	0.072	0.247	0.074	0.232	0.083	0.305	0.054	**0.311**	**0.045**
UF7	0.34	0.11	0.383	0.095	0.397	0.091	0.32	0.11	0.404	0.089	**0.431**	**0.080**
Weak wins	0	1	5	4	0	0	1	3	2	1	4	3
Local wins	0	1	5	4	0	0	1	3	2	1	4	3
Global wins	0	1	4	4	0	0	1	3	0	1	0	3

always beating NSGAII significantly (in terms of local wins), but in several situations is also statistically different than other considered MO-mHGS variants (denoted with **red bolding** in the tables). MO-ε-EHGS is generally worse in this case, generating similar outcomes to the basic MO-mHGS algorithm.

Table 5. Comparison of NSGAII and SPEA2 results with different variants of elitist MO-mHGS after 300000 fitness evaluations (IGD indicator, lower is better).

	Inverted Generational Distance											
	NSGAII		EHGS+NSGAII		ε-EHGS+NSGAII		SPEA2		EHGS+SPEA2		ε-EHGS+SPEA2	
Problem	Average	Error	Average	Error	Average	Error	Average	Error	Average	Error	Average	Error
zdt1	0.00736	0.00028	**0.003683**	**0.000018**	0.003906	0.000068	0.00626	0.00013	0.003743	0.000044	0.004033	0.000072
zdt2	0.00751	0.00033	**0.003788**	**0.000027**	0.00421	0.00012	0.006224	0.000094	0.004006	0.000090	0.00430	0.00012
zdt3	0.00534	0.00025	**0.002713**	**0.000026**	0.00361	0.00027	0.00492	0.00013	0.002794	0.000040	0.00363	0.00014
zdt4	0.73	0.34	**0.28**	0.17	0.30	**0.16**	0.68	0.29	0.42	0.17	0.46	0.19
zdt6	0.00578	0.00087	**0.003308**	**0.000084**	0.00389	0.00019	0.00495	0.00026	0.00378	0.00020	0.00439	0.00031
UF1	0.080	0.025	0.044	0.017	0.047	0.017	0.087	0.017	0.033	0.015	**0.031**	**0.010**
UF2	0.0350	0.0055	0.0263	0.0064	0.0249	0.0043	0.036	0.016	**0.0226**	0.0042	0.0236	**0.0041**
UF3	0.183	0.058	0.166	0.054	0.200	0.060	0.202	0.047	**0.143**	0.046	0.144	**0.043**
UF4	0.0501	0.0013	0.0602	0.0045	0.0584	0.0039	**0.0469**	**0.0012**	0.0527	0.0027	0.0536	0.0019
UF5	0.280	0.096	0.186	**0.020**	0.200	0.041	0.290	0.094	**0.168**	0.029	0.172	0.031
UF6	0.24	0.14	0.19	0.11	0.20	0.13	0.22	0.13	**0.101**	**0.039**	0.101	0.066
UF7	0.19	0.17	0.14	0.14	0.12	0.14	0.22	0.16	0.12	0.13	**0.08**	**0.12**
Weak wins	0	0	5	5	0	1	1	1	4	1	2	4
Local wins	0	0	5	5	0	1	1	1	4	1	2	4
Global wins	0	0	5	5	0	1	1	1	0	1	0	4

In order to ensure if driver reinforced by elitist MO-mHGS can be still competitive when compared to another MOEA, we also show a comparison of NSGAII and SPEA2 (with their MO-mHGS versions) in Tables 4, 5 and 6. Again, basic single-deme versions of the algorithms are rarely seen as winners, with MO-EHGS+NSGAII remains the most profitable in ZDT family and MO-EHGS+SPEA2 in UF problems.

Table 6. Comparison of NSGAII and SPEA2 results with different variants of elitist MO-mHGS after 300000 fitness evaluations (Spacing indicator, lower is better).

Spacing												
	NSGAII		EHGS+NSGAII		ϵ-EHGS+NSGAII		SPEA2		EHGS+SPEA2		ϵ-EHGS+SPEA2	
Problem	Average	Error	Average	Error	Average	Error	Average	Error	Average	Error	Average	Error
zdt1	0.0109	0.0014	**0.00095**	**0.00017**	0.059	0.045	0.00519	0.00067	0.00198	0.00069	0.036	0.033
zdt2	0.01066	0.00088	**0.00096**	**0.00021**	0.00325	0.00044	0.00536	0.00062	0.00287	0.00072	0.00365	0.00058
zdt3	0.0131	0.0015	**0.00249**	**0.00032**	0.051	0.048	0.00662	0.00087	0.00355	0.00052	0.0142	0.0074
zdt4	0.0145	0.0022	0.010	0.017	0.16	0.28	**0.0068**	**0.0010**	0.025	0.032	0.101	0.089
zdt6	0.0088	0.0010	**0.00095**	**0.00022**	0.010	0.035	0.00377	0.00044	0.00170	0.00053	0.0055	0.0092
UF1	0.0040	0.0022	0.0040	0.0021	0.0034	0.0017	0.010	0.021	0.0040	0.0026	**0.0025**	**0.0012**
UF2	0.0093	**0.0010**	**0.0051**	0.0013	0.0075	0.0051	0.0062	0.0025	0.0057	0.0022	0.0069	0.0020
UF3	0.022	0.018	0.0094	0.0092	**0.0061**	**0.0084**	0.019	0.018	0.021	0.015	0.023	0.014
UF4	0.0115	0.0010	0.0053	0.0016	0.0065	0.0014	0.0067	0.0020	**0.00458**	**0.00040**	0.0053	0.0012
UF5	**0.030**	**0.031**	0.034	0.041	0.088	0.095	0.055	0.057	0.082	0.077	0.12	0.11
UF6	**0.026**	0.035	0.029	**0.031**	0.10	0.19	0.10	0.21	0.047	0.050	0.06	0.11
UF7	0.0048	0.0037	0.0042	0.0027	0.0034	0.0027	0.009	0.022	**0.0028**	**0.0014**	0.0050	0.0032
Weak wins	2	2	**5**	**5**	1	1	1	1	2	2	1	1
Local wins	1	2	**4**	**5**	1	1	1	1	1	2	0	1
Global wins	0	2	**4**	**5**	1	1	0	1	1	2	0	1

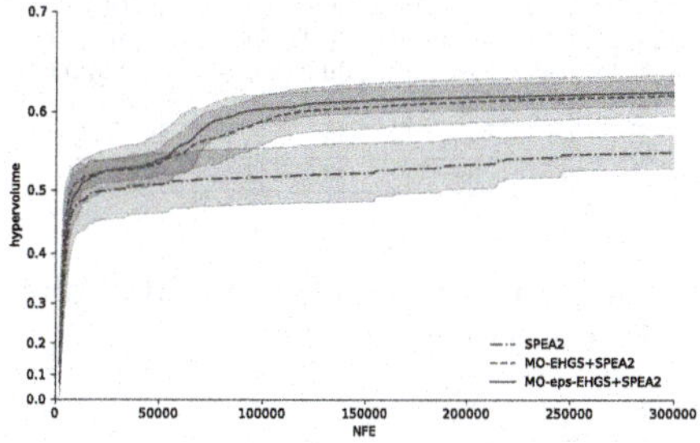

(a) Hypervolume

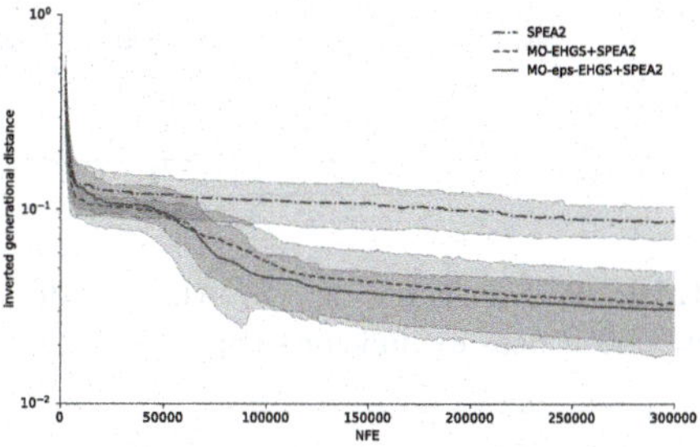

(b) IGD

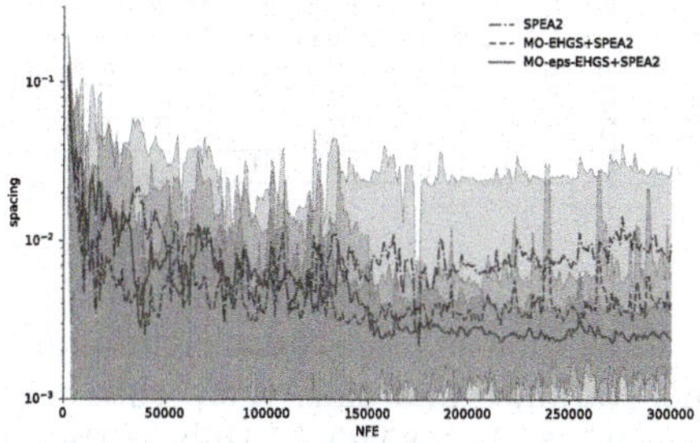

(c) Spacing

Fig. 2. Quality indicators values over number of fitness evaluations in UF1 (2-objective) problem (SPEA2 algorithm).

Figure 2 shows detailed simulation run of SPEA2 with elitist MO-mHGS in UF1 problem. While final results of MO-EHGS+SPEA2 and MO-ϵ-EHGS+SPEA2 does not differ significantly, they both quickly reach Hypervolume and IGD values far from single-deme algorithm. Better convergence of

Table 7. Summary of NSGAIII results with different variants of MO-mHGS after 300000 fitness evaluations (Hypervolume indicator, higher is better).

	Hypervolume															
	NSGAIII				MO-mHGS+NSGAIII				MO-EHGS+NSGAIII				MO-ε-EHGS+NSGAIII			
Problem	Min	Average	Max	Error	Min	Average	Max	Error	Min	Average	Max	Error	Min	Average	Max	Error
UF8	0.000	0.193	0.257	0.041	0.124	0.172	0.221	**0.023**	0.000	0.218	0.266	0.048	**0.207**	**0.243**	**0.307**	0.034
UF9	0.408	0.521	**0.618**	0.092	0.389	0.450	0.516	**0.035**	**0.461**	0.537	0.590	0.044	0.440	**0.562**	0.616	0.051
UF10	0.000	0.056	0.162	0.044	**0.013**	0.058	0.130	**0.029**	0.001	0.061	0.152	0.034	0.002	**0.072**	**0.185**	0.048
UF11	0.016	0.051	0.103	0.023	0.0000	0.0014	0.0089	**0.0020**	0.049	0.084	0.124	0.017	**0.167**	**0.195**	**0.229**	0.014
Weak wins	0	0	1	0	1	0	0	4	1	0	0	0	2	4	3	0
Local wins	0	0	1	0	0	0	0	4	0	0	0	0	2	2	2	0
Global wins	0	0	0	0	0	0	0	4	0	0	0	0	2	2	2	0

Table 8. Summary of NSGAIII results with different variants of MO-mHGS after 300000 fitness evaluations (IGD indicator, lower is better).

	Inverted Generational Distance															
	NSGAIII				MO-mHGS+NSGAIII				MO-EHGS+NSGAIII				MO-ε-EHGS+NSGAIII			
Problem	Min	Average	Max	Error	Min	Average	Max	Error	Min	Average	Max	Error	Min	Average	Max	Error
UF8	0.13	0.19	0.75	0.11	0.142	0.170	0.209	**0.018**	0.10	0.17	0.75	0.11	**0.094**	**0.150**	**0.207**	0.039
UF9	0.079	0.169	0.310	0.095	0.126	0.171	0.217	**0.024**	0.089	0.130	0.207	0.038	**0.070**	**0.109**	**0.186**	0.038
UF10	0.23	0.38	0.66	0.11	0.206	**0.261**	0.325	0.032	0.211	0.298	0.441	0.061	**0.19**	0.38	0.70	0.13
UF11	0.411	0.536	0.678	0.059	0.629	0.747	0.846	0.056	0.407	0.455	0.513	0.024	**0.313**	**0.342**	**0.385**	**0.016**
Weak wins	0	0	0	0	0	1	1	3	0	0	0	0	4	3	3	1
Local wins	0	0	0	0	0	1	1	3	0	0	0	0	3	3	3	1
Global wins	0	0	0	0	0	1	1	3	0	0	0	0	2	2	2	1

Table 9. Summary of NSGAIII results with different variants of MO-mHGS after 300000 fitness evaluations (Spacing indicator, lower is better).

	Spacing															
	NSGAIII				MO-mHGS+NSGAIII				MO-EHGS+NSGAIII				MO-ε-EHGS+NSGAIII			
Problem	Min	Average	Max	Error	Min	Average	Max	Error	Min	Average	Max	Error	Min	Average	Max	Error
UF8	**0.000**	0.098	0.181	0.037	0.038	0.077	0.304	0.047	0.010	**0.056**	0.072	**0.012**	0.040	0.057	0.102	0.013
UF9	**0.024**	0.057	0.110	0.021	0.042	0.070	0.152	0.027	0.0391	0.0546	0.0785	**0.0094**	0.0334	**0.0517**	**0.0732**	0.0099
UF10	0.033	0.125	0.410	0.075	0.091	0.145	0.268	0.048	0.038	0.112	**0.205**	**0.039**	**0.000**	**0.093**	0.216	0.057
UF11	0.157	0.228	0.313	0.038	0.216	0.255	0.305	0.023	0.1455	0.1601	0.1815	**0.0078**	0.126	**0.155**	0.184	0.013
Weak wins	2	0	0	0	0	0	0	0	0	1	3	4	2	3	1	0
Local wins	1	0	0	0	0	0	0	0	1	2	4		1	1	0	0
Global wins	1	0	0	0	0	0	0	0	0	1	4		1	1	0	0

MO-ε-EHGS+SPEA2 at earlier stages of simulation is also worth mentioning. Spacing metrics tend to have much higher variance and thus their outcomes are not always interpretable, but there's a visible trend of MO-ε-EHGS+SPEA2 minimizing both indicator value and error rate at later stages.

4.3 Evaluation of Many-Objective Problem

The last group of benchmarks were problems with higher objective dimensionality. In Tables 7, 8 and 9 we present outcomes of NSGAIII driver, whose basic form was designed to deal with many-objective problems. This time basic, non-elitist version of MO-mHGS was not able to improve MOEA – its performance was meaningfully worse (about 10–30% or even more in the hardest, 5-objective problem). At the same time, elitist variants provided the best possible solutions

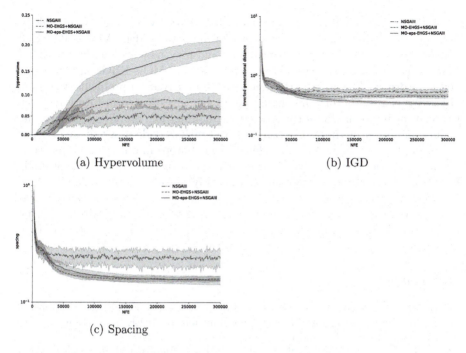

(a) Hypervolume   (b) IGD

(c) Spacing

Fig. 3. Quality indicators values over number of fitness evaluations in UF11 (5-objective) problem (NSGAIII algorithm).

regardless of metrics. Furthermore, even though MO-EHGS outperformed bare NSGAIII, it is MO-ϵ-EHGS that should be considered as a winner here. Local archives used in sprouting procedure improved delegates selection and in the end resulted in up to 2–3 times better improvement than MO-EHGS with only one, global archive. This behaviour is also visible in Fig. 3 of NSGAIII run solving the hardest, 5-objective problem. MO-ϵ-EHGS+NSGAIII reaches new regions in terms of Hypervolume and IGD while remaining stable and well distributed.

The reason why MO-ϵ-EHGS was considerably less impressive in previous experiments may be connected with our previous hypothesis about delegates nomination procedure impact. NSGAII and SPEA2 both incorporate simple elitism mechanics internally thus reducing their delegates set with $\epsilon - archive$ might not have been necessary. NSGAIII does not use an archive by itself and delegates nomination is based mainly on results of non-dominated sorting.

5   Conclusions

In this paper, we have shown the evolution of multi-objective hierarchical strategy, by introducing a notion of elitism by adding local and global archives into the search tree. We did our best to compare the proposed algorithms using a carefully selected set of benchmarks as well as the competitors.

Both proposed elitism upgrades of the MO-mHGS model were able to improve the performance of the hierarchical meta-model. While MO-EHGS stands out as a generic multi-purpose tool and may be paired with any MOEA in most of the considered situations, applications of MO-ϵ-EHGS are more specific to a driver. The good performance of MO-ϵ-EHGS in problems with more objectives is also notable and should be further investigated in future research. Correlation between MO-mHGS delegates nomination procedure and sprouts quality and other methods of controlling MO-mHGS node progress are also very promising topics that will be considered as next steps in our research project.

In our further research, we will also focus on the matters of concurrent implementation of the proposed algorithms, especially considering the scalability. Such hierarchical systems, constantly dynamically changing, should be implemented utilizing matters of concurrent processes, going towards parallelization, especially using HPC environments, which can help in speeding up the running of actually very complex metaheuristic algorithms.

References

1. Bellomo, D., Naso, D., Turchiano, B.: Improving genetic algorithms: an approach based on multi-elitism and Lamarckian mutation. In: IEEE International Conference on Systems, Man and Cybernetics, vol. 4, p. 6. IEEE (2002)
2. Ciepiela, E., Kocot, J., Siwik, L., Dreżewski, R.: Hierarchical approach to evolutionary multi-objective optimization. In: Bubak, M., van Albada, G.D., Dongarra, J., Sloot, P.M.A. (eds.) ICCS 2008. LNCS, vol. 5103, pp. 740–749. Springer, Heidelberg (2008). https://doi.org/10.1007/978-3-540-69389-5_82
3. Coello, C.A.C., Lamont, G.B., Van Veldhuizen, D.A., et al.: Evolutionary algorithms for solving multi-objective problems, vol. 5. Springer, Boston (2007). https://doi.org/10.1007/978-1-4757-5184-0
4. Corne, D.W., Jerram, N.R., Knowles, J.D., Oates, M.J.: PESA-II: region-based selection in evolutionary multiobjective optimization. In: Proceedings of the 3rd Annual Conference on Genetic and Evolutionary Computation, pp. 283–290 (2001)
5. Deb, K., Agrawal, S., Pratap, A., Meyarivan, T.: A fast elitist non-dominated sorting genetic algorithm for multi-objective optimization: NSGA-II. In: Schoenauer, M., et al. (eds.) PPSN 2000. LNCS, vol. 1917, pp. 849–858. Springer, Heidelberg (2000). https://doi.org/10.1007/3-540-45356-3_83
6. Deb, K., Goel, T.: Controlled elitist non-dominated sorting genetic algorithms for better convergence. In: Zitzler, E., Thiele, L., Deb, K., Coello Coello, C.A., Corne, D. (eds.) EMO 2001. LNCS, vol. 1993, pp. 67–81. Springer, Heidelberg (2001). https://doi.org/10.1007/3-540-44719-9_5
7. Deb, K., Jain, H.: An evolutionary many-objective optimization algorithm using reference-point-based non dominated sorting approach, part I: solving problems with box constraints. IEEE Trans. Evol. Comput. 18(4), 577–601 (2013)
8. Dulebenets, M.A.: Archived elitism in evolutionary computation: towards improving solution quality and population diversity. Int. J. Bio-Inspired Comput. 15(3), 135–146 (2020)
9. González-Almagro, G., Rosales-Pérez, A., Luengo, J., Cano, J.R., García, S.: Improving constrained clustering via decomposition-based multiobjective optimization with memetic elitism. In: Proceedings of the 2020 Genetic and Evolutionary Computation Conference, pp. 333–341 (2020)

10. Guariso, G., Sangiorgio, M.: Improving the performance of multiobjective genetic algorithms: An elitism-based approach. Information **11**(12), 587 (2020)
11. Hadka, D.: Beginner's guide to the MOEA framework (2016)
12. Idzik, M., Byrski, A., Turek, W., Kisiel-Dorohinicki, M.: Asynchronous actor-based approach to multiobjective hierarchical strategy. In: Krzhizhanovskaya, W., et al. (eds.) ICCS 2020. LNCS, vol. 12139, pp. 172–185. Springer, Cham (2020). https:// doi.org/10.1007/978-3-030-50420-5_13
13. Ishibuchi, H., Pang, L.M., Shang, K.: A new framework of evolutionary multi-objective algorithms with an unbounded external archive (2020)
14. Knowles, J.D., Corne, D.W.: Approximating the nondominated front using the Pareto archived evolution strategy. Evol. Comput. **8**(2), 149–172 (2000)
15. Kruskal, W.H., Wallis, W.A.: Use of ranks in one-criterion variance analysis. J. Am. Stat. Assoc. **47**(260), 583–621 (1952)
16. Laumanns, M., Thiele, L., Deb, K., Zitzler, E.: Combining convergence and diversity in evolutionary multiobjective optimization. Evol. Comput. **10**(3), 263–282 (2002)
17. Lazarz, R., Idzik, M., Gadek, K., Gajda-Zagorska, E.: Hierarchic genetic strategy with maturing as a generic tool for multiobjective optimization. J. Comput. Sci. **17**, 249–260 (2016)
18. Li, Q., Liu, L., Yuan, X.: Multiobjective parallel chaos optimization algorithm with crossover and merging operation. Math. Prob. Eng. **2020** (2020)
19. Mann, H.B., Whitney, D.R.: On a test of whether one of two random variables is stochastically larger than the other. Annals Math. Stat. 50–60 (1947)
20. Sano, R., Aguirre, H., Tanaka, K.: A closer look to elitism in ε-dominance many-objective optimization. In: 2017 IEEE Congress on Evolutionary Computation (CEC), pp. 2722–2729. IEEE (2017)
21. Sato, M., Aguirre, H.E., Tanaka, K.: Effects of δ-similar elimination and controlled elitism in the NSGA-II multiobjective evolutionary algorithm. In: 2006 IEEE International Conference on Evolutionary Computation, pp. 1164–1171. IEEE (2006)
22. Schaefer, R., Kolodziej, J.: Genetic search reinforced by the population hierarchy. In: Jong, K.A.D., Poli, R., Rowe, J.E. (eds.) Proceedings of the Seventh Workshop on Foundations of Genetic Algorithms, Torremolinos, Spain, 2–4 September 2002, pp. 383–400. Morgan Kaufmann (2002)
23. Schaefer, R., Kolodziej, J.: Genetic search reinforced by the population hierarchy. Found. Genet. Algorithms **7**, 383–401 (2002)
24. Sierra, M., Coello Coello, C.: Improving PSO-based multi-objective optimization using crowding, mutation and ϵ-dominance. In: Coello Coello, C., Hernández Aguirre, A., Zitzler, E. (eds.) Evolutionary Multi-Criterion Optimization. Lecture Notes in Computer Science, vol. 3410, pp. 505–519. Springer, Berlin Heidelberg (2005)
25. Simon, D., Ergezer, M., Du, D.: Population distributions in biogeography-based optimization algorithms with elitism. In: 2009 IEEE International Conference on Systems, Man and Cybernetics, pp. 991–996. IEEE (2009)
26. Sun, Y., Gao, Y.: A multi-objective particle swarm optimization algorithm based on Gaussian mutation and an improved learning strategy. Mathematics **7**(2), 148 (2019)
27. Tanabe, R., Ishibuchi, H.: Non-elitist evolutionary multi-objective optimizers revisited. In: Proceedings of the Genetic and Evolutionary Computation Conference, pp. 612–619 (2019)
28. While, L., Bradstreet, L., Barone, L.: A fast way of calculating exact hypervolumes. IEEE Trans. Evol. Comput. **16**(1), 86–95 (2011)

29. Zhang, Q., Zhou, A., Zhao, S., Suganthan, P.N., Liu, W., Tiwari, S.: Multiobjective optimization test instances for the CEC 2009 special session and competition (2008)
30. Zitzler, E., Deb, K., Thiele, L.: Comparison of multiobjective evolutionary algorithms: empirical results. Evol. Comput. $8(2)$, 173–195 (2000)
31. Zitzler, E., Laumanns, M., Thiele, L.: Spea 2: Improving the strength Pareto evolutionary algorithm. TIK-report 103 (2001)

Modelling and Forecasting Based on Recurrent Pseudoinverse Matrices

Christos K. Filelis-Papadopoulos[1]([✉]) [iD], Panagiotis E. Kyziropoulos[1] [iD],
John P. Morrison[1] [iD], and Philip O'Reilly[2] [iD]

[1] Department of Computer Science,
University College Cork, Western Gateway Building, Cork, Ireland
{christos.papadopoulos-filelis,j.morrison}@cs.ucc.ie,
panagiotis.kyziropoulos@ucc.ie
[2] University College Cork, Cork University Business School, O'Rahilly Building,
Cork, Ireland
philip.oreilly@ucc.ie

Abstract. Time series modelling and forecasting techniques have a wide spectrum of applications in several fields including economics, finance, engineering and computer science. Most available modelling and forecasting techniques are applicable to a specific underlying phenomenon and its properties and lack generality of application, while more general forecasting techniques require substantial computational time for training and application. Herewith, we present a general modelling framework based on a recursive Schur - complement technique, that utilizes a set of basis functions, either linear or non-linear, to form a model for a general time series. The basis functions need not be orthogonal and their number is determined adaptively based on fitting accuracy. Moreover, no assumptions are required for the input data. The coefficients for the basis functions are computed using a recursive pseudoinverse matrix, thus they can be recomputed for different input data. The case of sinusoidal basis functions is presented. Discussions around stability of the resulting model and choice of basis functions is also provided. Numerical results depicting the applicability and effectiveness of the proposed technique are given.

Keywords: Forecasting · Pseudoinverse matrix · Modelling

1 Introduction

General time series modelling and forecasting has become an essential tool in several scientific fields and business sectors, spanning from physics and engineering to workforce prediction and finance. Several methods have been proposed in the literature for modelling and forecasting time series based on either statistical or Machine Learning techniques. The most notable methods in the statistical methods class include Exponential Smoothing (ES), Auto-Regressive Integrated

The original version of this chapter was revised: chapter 18 made as open access and Acknowledgement section added. The correction to this chapter is available at https://doi.org/10.1007/978-3-030-77970-2_51

M. Paszynski et al. (Eds.): ICCS 2021, LNCS 12745, pp. 229–242, 2021.
https://doi.org/10.1007/978-3-030-77970-2_18

Moving Average (ARIMA), State Space and Structural models, Kalman filter, Nonlinear models and Generalized Auto-Regressive Conditionally Heteroscedastic (GARCH) models.

Exponential Smoothing (ES) was proposed in the1960 s by Brown, Holt and Winters [3,11,27], has been used extensively along with its variants due to its simplicity and adaptability to various scientific fields [14,25]. Auto-Regressive (AR), Auto-Regressive Moving Average (ARMA) and Auto-Regressive Integrated Moving Average (ARIMA) models were first proposed by Box and Jenkins [2] in the1970 s. Estimation techniques for the parameters of such methods have been discussed in [2]. Several univariate time series forecasting variants of such models have been proposed in the literature including ARARMA [18], Vector ARIMA (VARIMA) [21,23], Automatic univariate ARIMA type models [24] and seasonal approaches such as STL or $X - 11$ [4,5]. In the1980 s, state space models have been also used including Dynamic Linear Models (DLM) [9] and Basic Structural Models (BSM) [10] along with several variants [20]. In the quantitative finance domain Generalized Auto-Regressive Conditionally Heteroscedastic (GARCH) models [6] have been used. Nonlinear variants that handle asymmetric volatility have been also presented and studied [1,17].

Research around non-linear modelling techniques has also been conducted however not in such extent. This is primarily because of their increased complexity and the lack of closed form formulas. In this field notable contributions include the work of Wiener [26] and others [19]. Detailed overview on the advances of time series modelling and forecasting techniques over the last decades are given in [8] and references inside. Moreover, large scale benchmarking of popular modelling and forecasting techniques are given in [15].

Despite the popularity and wide use of the aforementioned techniques, they are mostly dedicated tools tuned specifically to a use case relying on restrictive assumptions for the form and nature of the data. General models such as Artificial Neural Networks or Support Vector Machines handle those limitations but they require substantial amounts of data, increased training and retraining times, as well as substantial tuning of the large number of hyperparameters. A different approach is followed by Fast Orthogonal Search (FOS) and its variants [12,13,16], which can form general models using combinations of linear and non-linear basis functions. These methods are based on Gram - Schmidt and Modified Gram - Schmidt orthogonalization as well as a Cholesky type factorization approaches. These methods build the model incrementally based on a re-orthogonalization approach. The time series is projected to this orthogonal set and until a prescribed modelling error is achieved. However, inherently parallel orthogonalization procedures have instabilities and stable variants are inherently sequential, thus cannot take advantage of novel multicore architectures. Furthermore, for models composed of a large number of trigonometric components, retraining the model requires re-computation of the coefficients performed by backward - forwards substitution, which is inherently sequential.

In this article a novel inherently parallel Schur complement based pseudoinverse matrix modelling and forecasting technique is proposed. The proposed scheme is based on a recursive pseudoinverse matrix to form a model based on a predefined set of linearly independent (linear or non-linear) basis functions. Basis functions are accumulated recursively into the model until a prescribed error is achieved. The stability of the model is ensured by enforcing positive definiteness of the dot product matrix of basis functions through monitoring of error reduction as well as monitoring the magnitude of the diagonal elements. Retraining of the produced model is limited to the pseudoinverse matrix by vector product. By exploiting orthogonality features, computing the residual time series is avoided, improving computational complexity. Discussions and implementation details regarding the computation of the pseudoinverse matrix and model formation are also provided. The case of sinusoidal basis functions is also discussed along with a technique to accurately determine frequencies based on the proposed approach. The efficiency and applicability of the proposed scheme is assessed in a variety of time series with different characteristics.

In Sect. 2, the recursive Schur complement based pseudoinverse matrix of basis functions is introduced. In Sect. 3, the Schur complement based pseudoinverse matrix is utilized in the design of the proposed modelling technique in order to computed the weights of the respective basis. Discussions on the stability of the scheme and implementation details are also provided. In Sect. 4, the case of sinusoidal basis functions is examined and discussed. In Sect. 5, numerical results are presented depicting the applicability and accuracy of the proposed scheme along with discussions for several time series.

2   Recursive Schur Complement Based Pseudoinverse Matrix of Basis Functions

Let us consider a matrix composed of a set of n basis functions:

$$X = \left[x_1(t) \; x_2(t) \; \ldots \; x_n(t) \right],\tag{1}$$

where $x_i(t), 1 \leq i \leq n$ are the basis functions and t is the time variable. We can expand a general time series y based on the $x_i(t)$:

$$y = a_1 x_1(t) + a_2 x_2(t) + \ldots + a_n x_n(t) + \epsilon(t),\tag{2}$$

where $\epsilon(t)$ is the error term and $a_i, 1 \leq i \leq n$ are the unknown coefficients that have to be determined. The Eqs. (2) can be written in block form:

$$y = Xa.\tag{3}$$

or equivalently,

$$a = X^+ y = (X^T X)^{-1} X^T y.\tag{4}$$

Thus, the coefficients can be determined by solving the least squares linear system (3) or through the pseudoinverse X^+ (4). However, in many cases the basis functions are not known *a priori* and are computed iteratively or recurrently up to prescribed tolerance. This requires formation and solution of multiple least squares linear systems, which increases substantially the computational work. To avoid such issues modelling techniques such as Fast Orthogonal Search (FOS) and variants [12,13,16] have been proposed, based on Gram - Schmidt and Modified Gram - Schmidt orthogonalization as well as Cholesky factorization type approaches. These methods build the model incrementally by adding basis functions and re-orthogonalize with the already computed ones in order to project the time series and reduce the modelling error.

In order to avoid computational costs and instabilities involved in using orthogonalization procedures, while retaining the flexibility of incrementally building the model, a novel approximate inverse scheme is proposed. Let as consider the matrix X_i containing up to the i-th base and an additional base F:

$$X_{i+1} = \begin{bmatrix} X_i \mid F \end{bmatrix} \tag{5}$$

or:

$$K_{i+1} = X_{i+1}^T X_{i+1} = \begin{bmatrix} X_i^T X_i & X_i^T F \\ F^T X_i & F^T F \end{bmatrix} = \begin{bmatrix} K_i & b \\ b^T & d \end{bmatrix}, \tag{6}$$

where $b = X_i^T F$ and $d = F^T F$. It should be noted that the matrix $X_{i+1}^T X_{i+1}$ is Symmetric Positive Definite and thus invertible, under the assumption that the basis function are linearly independent.

Computing the inverse of the matrix K_{i+1} enables formation of the pseudoinverse X_{i+1}^+ and consequently the computation of the coefficients a_i to form the model. The matrix K_{i+1} can be factored to enable easier inversion and avoid the update step for the elements of the inverse of the matrix K_i, which is essential in case of non-factored inverse matrices [7]:

$$K_{i+1} = \begin{bmatrix} G_i^{-T} & 0 \\ b^T K_i^{-1} & 1 \end{bmatrix} \begin{bmatrix} D_i & 0 \\ 0 & s_i \end{bmatrix} \begin{bmatrix} G_i^{-1} & K_i^{-1} b \\ 0 & 1 \end{bmatrix} \tag{7}$$

or equivalently:

$$K_{i+1}^{-1} = \begin{bmatrix} G_i & -K_i^{-1} b \\ 0 & 1 \end{bmatrix} \begin{bmatrix} D_i^{-1} & 0 \\ 0 & s_i^{-1} \end{bmatrix} \begin{bmatrix} G_i^T & 0 \\ -b^T K_i^{-1} & 1 \end{bmatrix} = G_{i+1} D_{i+1}^{-1} G_{i+1}^T \tag{8}$$

where the matrix D_i retains the Schur complements s_i corresponding to each addition of a basis functions. The Schur complements are of the form: $s_i = d - b^T G_i D_i^{-1} G_i^T b$. Due to the symmetric nature of the inverse K_{i+1}^{-1} only the factors D_{i+1} and G_{i+1} are required. Thus, we have:

$$G_{i+1} = \begin{bmatrix} G_i & -G_i D_i^{-1} z \\ 0 & 1 \end{bmatrix} \quad and \quad D_{i+1}^{-1} = \begin{bmatrix} D_i^{-1} & 0 \\ 0 & (d - z^T D_i^{-1} z)^{-1} \end{bmatrix} \tag{9}$$

where $z = G_i^T b$. Addition of a new basis is limited to simple "matrix times vector" operations, which can be efficiently performed in modern CPUs or accelerators such as GPUs. The storage requirements for the inverse are limited to the upper triangular matrix G_i, which retains $i(i+1)/2$ elements, and the diagonal matrix D_i retaining i elements.

In order to avoid breakdowns in case of weak linear independence between different basis functions or loss of numerical precision, a condition on the magnitude of the diagonal element (Schur Complement) should be imposed. If the diagonal element is close to machine precision, in practice less than $\sqrt{\epsilon_{mach}}$, it is substituted with $\sqrt{\epsilon_{mach}}$, thus avoiding breakdowns.

3   Schur Based Pseudoinverse Matrix Modelling and Forecasting

The choice of basis functions for creating a model can be arbitrary, since they are evaluated and stored in the columns of matrix X. However, the amount and type of functions chosen affects the accuracy of the model and computational work. Considering the basis functions of Eq. (1), the general time - series of Eq. (2) and the recursive inverse matrix of Eq. (8) we can obtain the coefficients a_i, after the addition of the $i+1$ basis function, as follows:

$$\begin{bmatrix} a_i \\ b \end{bmatrix} = \begin{bmatrix} G_i & -G_i D_i^{-1} G_i^T X_i^T F \\ 0 & 1 \end{bmatrix} \begin{bmatrix} D_i^{-1} & 0 \\ 0 & s_i^{-1} \end{bmatrix} \begin{bmatrix} G_i^T & 0 \\ -F^T X_i G_i D_i^{-1} G_i^T & 1 \end{bmatrix} \begin{bmatrix} X_i^T y \\ F^T y \end{bmatrix} \quad (10)$$

where $s_i = F^T F - F^T X_i G_i D_i^{-1} G_i^T X_i^T F$. The addition of a basis function involves updating of the values of already computed coefficients a_i corresponding to the i basis functions. Let $g_{i+1} = -G_i D_i^{-1} G_i^T X_i^T F$ then we have:

$$a_{i+1} = \begin{bmatrix} a_i^* \\ b \end{bmatrix} = \begin{bmatrix} a_i + g_{i+1} b \\ s_i^{-1}(F^T + g_{i+1}^T X_i^T)y \end{bmatrix} \quad (11)$$

where $a_i = G_i D_i^{-1} G_i^T X_i^T y$ and $s_i = F^T(F + X_i g_{i+1})$. The coefficients a_i^* are updated after the inclusion of basis function F, while matrix X_i retains all basis functions up to i-th. The residual time series update equation, with respect to the addition of a new basis function can be formed using Eq. (11) as follows:

$$\begin{bmatrix} X_i & F \end{bmatrix} a_{i+1} = \begin{bmatrix} X_i & F \end{bmatrix} \begin{bmatrix} a_i^* \\ b \end{bmatrix} \quad (12)$$

or:

$$r_{i+1} = y - \begin{bmatrix} X_i & F \end{bmatrix} \begin{bmatrix} a_i^* \\ b \end{bmatrix} = y - X_i a_i^* - Fb = r_i^* - Fb = r_i - (X_i g_{i+1} + F)b. \quad (13)$$

Using Eq. (13) the norm of the residual, after addition of a basis function, can be computed as follows:

$$r_{i+1} = r_i - (X_i g_{i+1} + F)b = r_i - (I - X_i G_i D_i^{-1} G_i^T X_i^T)Fb = r_i - (I - P_{X_i})Fb, \quad (14)$$

where P_{X_i} is an orthogonal projection operator onto the subspace spanned by the columns of X_i. The Eq. (14) implies also the following:

$$P_{X_i} r_i = 0 \quad or \quad r_i \perp span(X_i). \quad (15)$$

Thus, we have:

$$\|r_{i+1}\|_2^2 = \|r_i\|_2^2 - \|(X_i g_{i+1} + F)b\|_2^2 = \|r_i\|_2^2 - b^2 s_i, \qquad (16)$$

or

$$\|r_{i+1}\|_2 = \sqrt{\|r_i\|_2^2 - b^2 s_i}. \qquad (17)$$

Equation (17) can be used to assess progress of fitting, instead of computing the norm of the residual at each iteration, substantially improving performance especially in the case of increased number of basis functions. Moreover, the quantity $b^2 s$ is positive since matrix $X_i^T X_i$ is Symmetric Positive Definite, thus its Schur Complement is also Symmetric Positive Definite, leading to monotonic reduction of the norm of the residual r_{i+1}.

Algorithm 1. Pseudoinverse Matrix Modelling
$(a = pmm(y, F_i, \epsilon))$

1: **Let** y $(T \times 1)$ be the input time series of length T, X_i is the matrix retaining the up to the i-th basis function, F_{i+1} the $i+1$ basis function, M is the maximum number of basis functions, a_i the coefficients corresponding to the i-th basis function and ϵ is the prescribed tolerance.
2: $G_1 = 1$, $D_1 = (F_1^T F_1)$, $s = F_1^T F_1$
3: $b = s^{-1} F_1^T y$
4: $\rho = \|y\|_2^2$, $\rho_0 = \rho$
5: $a_1 = b$
6: $\rho = \rho - b^2 s$
7: $X_1 = F_1$
8: **for** $i = 1$ to $M - 1$ **do**
9:     $g = -G_i D_i^{-1} G_i^T X_i^T F_{i+1}$
10:     $F^* = F_{i+1} + X_i g$
11:     $s = F_{i+1}^T F^*$
12:     **if** $|s| < \sqrt{\epsilon_{mach}}$ **then**
13:         $s = \sqrt{\epsilon_{mach}}$
14:     **end if**
15:     $b = s^{-1}(F^*)^T y$
16:     $a_{1:i} = a_{1:i} + gb$, $a_{i+1} = b$
17:     $G_{i+1} = \begin{bmatrix} G_i & g \\ 0 & 1 \end{bmatrix}$
18:     $D_{i+1} = s$, $X_{i+1} = F_{i+1}$
19:     $\rho = \rho - b^2 s$
20:     **if** $\sqrt{\rho} < \epsilon \sqrt{\rho_0}$ **then**
21:         **return** $a_{1:i+1}$
22:     **end if**
23: **end for**

The initial conditions for Eqs. (11) and (13) are the following:

$$G_1 = 1 \qquad (18)$$

$$D_1^{-1} = (X_1^T X_1)^{-1} \tag{19}$$

and

$$a_1 = (X_1^T X_1)^{-1} X_1^T y. \tag{20}$$

The procedure for computing the coefficients a_i is described in the Algorithm 1. The chosen basis functions are evaluated for $t = 1, ..., T$ and stored in the columns of matrix X. Thus, the model can be formed by computing Xa. The formed model can be used to compute forecasts up to predefined horizon h. This can be achieved by initially progressing the set of basis function in time, e.g. evaluate them for $t^* = T+1, ..., T+h$ and store them to the columns of a matrix $X^* = [x_1(t^*) ... x_n(t^*)]$. Then, the forecasts can be computed by X^*a. In practice, the choice of basis functions can be arbitrary, e.g. linear or non-linear or combinations.

The stability of the model with respect to each addition of a basis function can be ensured by allowing basis function to be added only if the error reduction, caused by such an addition, is positive $(b^2 s \geq 0)$ and the reduction of the error does not render the error term negative $\rho < 0$ $(\rho \geq b^2 s)$. These conditions ensure invertibility of the matrix $X_i^T X_i$ and positive definiteness of the inverse matrix $G_i D_i^{-1} G_i^T$, by allowing inclusion of basis functions that are suitably linearly independent to the already selected ones.

4   The Case of Sinusoidal Basis Functions

Sinusoidal basis functions can be used to form a model for general time series. In the case of strong trigonometrical phenomena in a time series, such basis functions can be used to capture them. The sinusoidal basis functions are of the following form:

$$b_i(t) = A\cos(\omega_i t) + B\sin(\omega_i t), \tag{21}$$

where ω_i is the frequency. Estimation of frequencies can be performed using techniques such as the Fast Fourier Transform through the spectrum or the Quinn - Fernandes algorithm [22]. The proposed scheme allows for determining frequencies with arbitrary accuracy through frequency searching similar to the procedure described in [12]. In case of the proposed technique basis function of the form $F = \cos(\omega_i t)$ followed by a basis function of the form $F = \sin(\omega_i t)$ for various frequencies $\omega_i \in (0, pi]$ are fitted and the frequency that results in maximum error reduction is selected. The selected frequency becomes part of the model, is excluded from the search space, and the procedure continues until the error criterion is met. The search space $(0, pi)$ is sampled based on a prescribed sampling interval $\delta\omega$. The choice of $\delta\omega$ affects the accuracy in which the frequencies are determined. The advantages of this technique are that frequencies can be determined in parallel while the residual time series need not be computed explicitly.

In order to assess the accuracy of the technique the following example is provided. Let us consider the following:

$$y(t) = \cos(\omega_1 t) + 3\sin(\omega_2 t) + 2\cos(\omega_3 t) - \cos(\omega_4 t) + \sin(\omega_4 t) + \sigma(t) \tag{22}$$

where $\sigma(t)$ is following uniform random distribution with average equal to 0.5. The frequencies ω_i are: $\omega_1 = 0.0546$, $\omega_2 = 0.83120$, $\omega_3 = 1.87120$ and $\omega_4 = 1.91320$. The frequencies are estimated using Fast Fourier Transform (FFT), Quinn - Fernandes method based on FFT as initial guess coupled with Ordinary Least Squares (OLS) method and the proposed technique. The results are given in Table 1.

Table 1. Percentage errors for frequency estimation.

Frequency	FFT	QF-OLS	$\delta\omega = 10^{-4}$	$\delta\omega = 10^{-5}$
ω_1	3.5690	0.0250	0.1167	0.0592
ω_2	0.2189	0.0007	0.0079	0.0003
ω_3	0.0636	0.0091	0.0132	0.0048
ω_4	0.1626	0.0021	0.0016	0.0017

The choice of $\delta\omega$, in the proposed scheme, should be less than the sampling interval of the Fast Fourier Transform, e.g. $\delta\omega \leq \frac{2*pi}{N}$, in order to allow accurate determination of the frequencies and avoid undersampling.

The proposed technique can be used to estimate the frequency to improved accuracy compared to FFT or the QF-OLS method. The proposed scheme can be coupled also with either the FFT or QF-OLS method and hybrid schemes can be designed leveraging the advantages of those schemes. This will be studied in future work.

5   Numerical Results

In this section the applicability and accuracy of the proposed scheme is examined by applying the proposed technique to three time series. The two error measures used to assess the forecasting error was Mean Absolute Percentage Error (MAPE) and Mean Absolute Deviation (MAD):

$$MAPE = \frac{100}{T} \sum_{i=1}^{T} \frac{|y_i - \hat{y}_i|}{|y_i|} \text{ and } MAD = \frac{1}{T} \sum_{i=1}^{T} |y_i - \hat{y}_i|, \qquad (23)$$

where y_i are the actual values, $\hat{y}_i$ the forecasted values and T the length of the test set. The basis functions chosen to model the selected time series were:

$$F_1 = 1, F_2 = t, F_3 = e^t, F_i = Acos(\omega_i t) + Bsin(\omega_i t), i \geq 4, t \geq 0. \qquad (24)$$

The linear and exponential basis were added to automatically capture such trends in the data. It should be mentioned that the time variable t is scaled for the linear and exponential components to improve numerical behavior during inversion of the matrix of basis functions.

5.1   US Airline Passenger Volume

This US airline passenger volume dataset was extracted from R Studio and is composed of monthly total volumes of passengers spanning from January 1949 to December 1960 (144 samples). The training part was composed of 75% of the dataset, while the test part was composed of 25% of the dataset, specifically the training part included 108 samples and the test part included 36 samples, as presented in Fig. 1. The prescribed interval $\delta\omega$ for frequency search was set to 0.001 and the prescribed tolerance for fitting the model was set to $\epsilon = 0.01$. The forecasted values along with the actuals are given in Fig. 2. The MAPE and MAD of the forecasts were 9.3474 and 40.0410, respectively. From Fig. 2 we observe that the proposed scheme captured the exponential and linear tendency automatically as well as the underlying trigonometric phenomena, without requiring any pre-processing steps for the input data apart from maximum scaling.

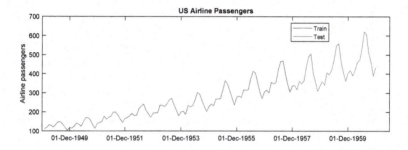

Fig. 1. Train and test parts for the US airline passenger volume dataset.

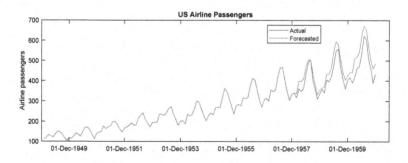

Fig. 2. Forecasted and actual values for the US airline passenger volume dataset.

With respect to the value of the coefficients comprising the model the time series exhibits a significant exponential component, a weak linear component along with a strong low frequency harmonic component, because of the yearly periodicity of the time series.

5.2   Monthly Expenditure on Eating Out in Australia

The monthly expenditure on eating out in Australia dataset was extracted from R Studio and is composed of the monthly expenditure on cafes, restaurants and takeaway food services in Australia in billion dollars. The dataset is composed of 426 samples spanning a period from April 1982 to September 2017. The training part was composed of $\approx 80\%$ of the dataset while the test part was composed of $\approx 20\%$ of the dataset, specifically the training part included 342 samples and the test part included 84 samples, as presented in Fig. 3. The prescribed interval $\delta\omega$ for frequency search was set to 0.001 and the prescribed tolerance for fitting the model was set to $\epsilon = 0.01$. The forecasted values along with the actuals are given in Fig. 4. The MAPE and MAD of the forecasts were 4.5292 and 0.1465, respectively.

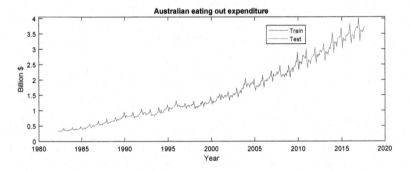

Fig. 3. Train and test parts for the monthly expenditure on eating out in Australia dataset.

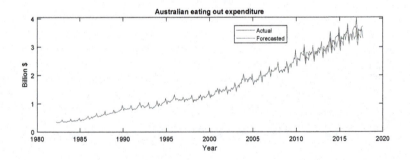

Fig. 4. Forecasted and actual values for the US airline passenger volume dataset.

With respect to the value of the coefficients comprising the model the time series exhibits a strong exponential component, a strong linear component along with a relatively significant medium frequency harmonic components.

5.3   Call Volume for a Large North American Bank

The call volume for a large North American bank dataset was extracted from R Studio and is composed of the volume of calls, per five minute intervals, spanning 164 d starting from 3 March 2003. The dataset is composed of 27716 samples. The training part was composed of $\approx 80\%$ of the dataset while the test part was composed of $\approx 20\%$ of the dataset, specifically the training part included 22325 samples and the test part included 5391 samples, as presented in Fig. 5. The prescribed interval $\delta\omega$ for frequency search was set to 10^{-5} and the prescribed tolerance for fitting the model was set to $\epsilon = 0.088$. The value of the tolerance ϵ is chosen as below that margin the rate of error reduction slows down significantly due to the presence of noise in the form of a large number of frequencies with the same magnitude in the spectrum. This issue can be overcome by increasing the samples of the spectrum, however this substantially increases the computational work, without significant improvement in the forecasting error. The forecasted values along with the actuals are given in Fig. 6. The MAPE and MAD of for the forecasts were 15.1580 and 25.4576, respectively.

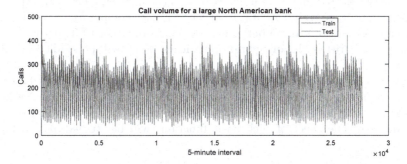

Fig. 5. Train and test parts for the call volume for a large North American bank dataset.

With respect to the value of the coefficients the model has a weak exponential component that is counteracted by a weak linear component. There are also strong low frequency components that contribute significantly in the reduction of the error.

5.4   Discussions

The proposed scheme was able to capture the dominant characteristics of the different time series. The choice of the basis functions substantially affects the estimation and the forecasting error. For example, for the model problem of Subsection 5.3 the linear and exponential basis do not contribute significantly to the accuracy of the model, while also increasing the computational complexity since the dimensions of the pseudoinverse matrix grow. However, to preserve

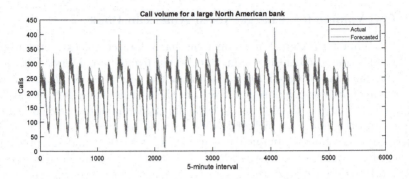

Fig. 6. Forecasted and actual values for call volume for a large North American bank dataset.

generality and wide applicability, a common set of basis functions was retained for all experiments.

Another important issue is the estimation of frequencies, which for the low value of the $\delta\omega$ parameter requires substantial computational work especially in the case of large training data. In order to reduce computational complexity, frequency estimation can be performed by means of either FFT or the Quinn - Fernandes algorithm [22] or hybrid approaches which will be studied in future research.

The generality of the approach allows the incorporation of basis functions based on nonlinear modelling techniques such as Artificial Neural Networks (ANN) and Support Vector Machines (SVM) trained by subsets of the available dataset. The effect of such basis functions will be studied also in future research.

6 Conclusion

A novel Schur complement based pseudoinverse matrix approach for modelling and forecasting general time series has been proposed. The proposed technique can incorporate linear and non-linear components during model formation, thus avoiding preprocessing and transformation of the time series or restrictive assumptions related to the statistical properties of the data. Stability of the model is ensured by enforcing positive definiteness of the dot product matrix of basis functions ($X^T X$) and its inverse, and monotonic reduction of the error. A frequency detection technique is also presented based on the proposed scheme. The proposed scheme does not require preprosessing of time series and is assessed by modelling several time series exhibiting combinations of exponential, linear, trigonometric and random characteristics. Moreover, the model relies on a single parameter and it is suitable for modelling general time series.

Future work is directed towards the design of a parallel approximate pseudoinverse matrix approach in order to reduce storage requirements especially in the case of large number of basis functions. Moreover, an adaptive approach for frequency estimation is under further research.

Acknowledgement. This publication has emanated from research conducted with the financial support of Science Foundation Ireland under Grant number [18/SPP/3459]. For the purpose of Open Access, the author has applied a CC BY public copyright licence to any Author Accepted Manuscript version arising from this submission.

References

1. Awartani, B.M., Corradi, V.: Predicting the volatility of the S&P-500 stock index via GARCH models: the role of asymmetries. Int. J. Forecast. **21**(1), 167–183 (2005)
2. Box, G.E.P., Jenkins, G.M.: Time Series Analysis Forecasting and Control. Holden Day, San Francisco (1976)
3. Brown, R.G.: Smoothing. Forecasting and prediction of discrete time series. Englewood Cliffs, NJ, Prentice Hall (1963)
4. Cleveland, R.B., Cleveland, W.S., McRae, J.E., Terpenning, I.: STL: a seasonal-trend decomposition procedure based on loess (with discussion). J. Official Stat. **6**, 3–73 (1990)
5. Dagum, E.B.: Revisions of time varying seasonal filters. J. Forecast. **1**(2), 173–187 (1982). https://doi.org/10.1002/for.3980010204
6. Engle, R.F.: Autoregressive conditional heteroscedasticity with estimates of the variance of united kingdom inflation. Econometrica **50**(4), 987–1007 (1982)
7. Filelis-Papadopoulos, C.K.: Incomplete inverse matrices. Numer. Linear Algebra Appl. (2021). https://doi.org/10.1002/nla.2380
8. Gooijer, J.G., Hyndman, R.: 25 years of IIF time series forecasting: a selective review. In: Monash Econometrics and Business Statistics Working Papers 12/05, Monash University, Department of Econometrics and Business Statistics (2005). https://econpapers.repec.org/paper/mshebswps/2005-12.htm
9. Harrison, P.J., Stevens, C.F.: Bayesian forecasting. J. R. Stat. Soc. Series B (Methodol.) **38**(3), 205–247 (1976)
10. Harvey, A.C.: Forecasting, Structural Time Series Models and the Kalman Filter. Cambridge University Press, Cambridge (1989)
11. Holt, C.C.: Forecasting seasonals and trends by exponentially weighted moving averages. Int. J. Forecast. **20**, 5–13 (2004)
12. Korenberg, M.J., Paarmann, L.D.: Orthogonal approaches to time-series analysis and system identification. IEEE Signal Process. Mag. **8**(3), 29–43 (1991)
13. Li, X., Tian, J., Wang, X., Dai, J., Ai, L.: Fast orthogonal search method for modeling nonlinear hemodynamic response in fMRI. In: Amini, A.A., Manduca, A. (eds.) Medical Imaging 2004: Physiology, Function, and Structure from Medical Images. International Society for Optics and Photonics, SPIE, vol. 5369, pp. 219–226 (2004). https://doi.org/10.1117/12.536165
14. Makridakis, S., Hibon, M.: Exponential smoothing: the effect of initial values and loss function on post-sample forecasting accuracy. Int. J. Forecast **7**, 317–330 (1991)

15. Makridakis, S., Spiliotis, E., Assimakopoulos, V.: The m4 competition: 100,000 time series and 61 forecasting methods. Int. J. Forecast. **36**(1), 54–74 (2020). https://doi.org/10.1016/j.ijforecast.2019.04.014, m4 Competition
16. Osman, A., et al.: Adaptive fast orthogonal search (fos) algorithm for forecasting streamflow. J. Hydrol. **586**, 124896 (2020).https://doi.org/10.1016/j.jhydrol.2020.124896, http://www.sciencedirect.com/science/article/pii/S0022169420303565
17. Pagan, A.: The econometrics of financial markets. J. Empirical Finance **3**(1), 15–102 (1996)
18. Parzen, E.: Ararma models for time series analysis and forecasting. J. Forecast. **1**(1), 67–82 (1982). https://doi.org/10.1002/for.3980010108
19. Poskitt, D., Tremayne, A.: The selection and use of linear and bilinear time series models. Int. J. Forecast. **2**(1), 101–114 (1986). https://doi.org/10.1016/0169-2070(86)90033-6
20. Proietti, T.: Comparing seasonal components for structural time series models. Int. J. Forecast. **16**(2), 247–260 (2000). https://doi.org/10.1016/S0169-2070(00)00037-6
21. Quenouille, M.H.: The Analysis of Multiple Time-Series. London, Griffin. 2nd edn. (1968)
22. Quinn, B.G., Fernandes, J.M.: A fast efficient technique for the estimation of frequency. Biometrika **78**(3), 489–497 (1991)
23. Riise, T., Tjozstheim, D.: Theory and practice of multivariate arma forecasting. J. Forecast. **3**(3), 309–317 (1984). https://doi.org/10.1002/for.3980030308
24. Tashman, L.J.: Out-of-sample tests of forecasting accuracy: an analysis and review. Int. J. Forecast. **16**(4), 437–450 (2000)
25. Taylor, J.W.: Exponential smoothing with a damped multiplicative trend. Int. J. Forecast. **19**, 273–289 (2003)
26. Wiener, N.: Non-linear Problems in Random Theory. Wiley, London (1958)
27. Winters, P.R.: Forecasting sales by exponentially weighted moving averages. Manage. Sci. **6**, 324–342 (1960)

Semi-analytical Monte Carlo Optimisation Method Applied to the Inverse Poisson Problem

Sławomir Milewski[✉] [iD]

Faculty of Civil Engineering, Chair for Computational Engineering,
Cracow University of Technology, Cracow, Poland
s.milewski@L5.pk.edu.pl
https://www.cce.pk.edu.pl/ slawek/

Abstract. The research is focused on the numerical analysis of the inverse Poisson problem, namely the identification of the unknown (input) load source function, being the right-hand side function of the second order differential equation. It is assumed that the additional measurement data of the solution (output) function are available at few isolated locations inside the problem domain. The problem may be formulated as the non-linear optimisation problem with inequality constrains.

The proposed solution approach is based upon the well-known Monte Carlo concept with a random walk technique, approximating the solution of the direct Poisson problem at selected point(s), using series of random simulations. However, since it may deliver the linear explicit relation between the input and the output at measurement locations only, the objective function may be analytically differentiated with the respect to unknown load parameters. Consequently, they may be determined by the solution of the small system of algebraic equations. Therefore, drawbacks of traditional optimization algorithms, computationally demanding, time-consuming and sensitive to their parameters, may be removed. The potential power of the proposed approach is demonstrated on selected benchmark problems with various levels of complexity.

Keywords: Inverse Poisson problem · Optimisation problem · Monte Carlo method · Meshless random walk

1 Introduction

Problems of computational mechanics and civil engineering may be classified as direct and inverse ones. In case of direct problems, the input data (geometry, material, load) are known and, therefore, the resulting initial-boundary problem remains well-posed and yields a unique output solution (displacement, temperature, flux, strain, stress), determined by means of rather numerical than analytical tools. However, in case selected input data are unknown, we deal with the inverse problem, either of topological optimisation or material/load identification nature. Additional information is required, for instance optimisation

© Springer Nature Switzerland AG 2021
M. Paszynski et al. (Eds.): ICCS 2021, LNCS 12745, pp. 243–256, 2021.
https://doi.org/10.1007/978-3-030-77970-2_19

criteria (e.g., minimal mass or maximal capacity) or measurement data of output functions at selected points of the problem domain and its boundary. The optimal number of measurements as well as their appropriate locations are substantial since ill-posed inverse problems, suffering from information shortage, are usually ill-conditioned (in Hadamard sense) and, therefore, extremely sensitive to small data modifications.

Load identification problems are considered in this research, being the important stage of the recovery process of the full current static/dynamic/thermal state of existing engineering constructions. This process is usually based upon non-destructive, contact (e.g., sensors attached to the construction) or non-contact (e.g., vision techniques) measurements, done at few selected locations (nodes, bars, surfaces). In this manner, the determination of the current state of the structure allows for the estimation of time of its safe operation and exploitation, which is a part of a wide range of SHM (Structural Health Monitoring) issues.

The attention is laid upon the Poisson differential equation, modelling a variety of processes occurring in nature, for instance the gravitational field potential in the presence of sources, selected linear elasticity problems (torsional deflection of a prismatic bar, plane stress/plane strain/axial symmetry), stationary heat flow, distribution of electric potential, or filtration through porous systems. In case of the inverse Poisson problem, the source load (input) function, being the right-hand side function of the Poisson equation, is considered as the additional unknown, along with the primary output function. The source function may be interpreted, for instance, as the intensity of an external live load, subjected to the construction, or as the intensity of a heat generation inside the domain. Since it cannot be directly measured, its comprehensive determination (values, support, localization, gradient) is crucial for the entire solution procedure.

The solution approach to inverse source problems strongly depends on the type of additional data. The simplest and rather not realistic cases, though allowing for the avoidance of the ill-conditioning, assume the existence of a continuous input/output solution, partially given on the boundary and/or inside the domain. Therefore, the original heterogeneous problem (with non-zero unknown source function) may be transformed into the auxiliary homogeneous problem, being solved using traditional computational tools, like Finite Element Method (FEM) [1], Boundary Element Method (BEM) [2], Finite Difference Method (FDM) or meshless methods [3,4], and then retransformed. The iterative algorithm with the source function, partially known on the boundary, is presented in [5]. Consideration of noisy data requires additional restoration algorithms, usually of probabilistic type, preceding an reverse transform [6,7]. On the other hand, more experimentally oriented approaches take real measurement data into account. Measurements, with assigned uncertainties, have fixed locations or may be arbitrarily scattered in the problem domain. Consequently, the non-linear optimisation problem is formulated and solved [8,9], incorporating aforementioned deterministic (FEM, FDM, meshless methods [10,11]) as well as stochastic solution approaches (genetic algorithms, neural networks [12,13]). Ill-conditioning of the original inverse problem is reduced by appropriate regularization techniques [10,14].

Although a variety of optimisation methods exists, most of them exhibit serious drawbacks, for instance time–consuming searching algorithms (non-gradient methods), computationally demanding multiple solutions of auxiliary differential problems for fixed source parameters, numerical evaluation of the objective function's gradient and/or Hessian, sensitiveness to initial guess solutions (for iterative procedures) as well as requirement for admissible solution intervals. The proposed Monte Carlo optimization method allows for the reduction of those disadvantages. It is based upon the old concept of the Monte Carlo (MC) method with a random walk technique for an approximate solution of the Laplace problem at the selected point of the rectangular grid of points [15]. Series of simulations (random walks) are performed, starting from the considered point and terminating at the boundary, where the solution is known. Sum of all boundary indication numbers, related to the total number of trials and scaled by boundary solution values, is an unbiased estimator of the Laplace problem at this point. Moreover, it is convergent to the corresponding finite difference (FD) solution of the same problem, providing the number of trials is large enough [16,17].

Its novel application to inverse source problems assumes the determination of indication numbers, separately for all measurement points, coinciding with approximation nodes, regardless of their distribution. Afterwards, resultant explicit relations between the input (source function parameters) and output (solution values at measurement points) may be substituted into the objective function, minimising the measurement error. This move allows for its analytical differentiation towards fulfilment of the necessary condition of the existence of its extreme. Eventually, one obtains small symmetric system of algebraic equations. Once the source function is recovered, the problem becomes a direct one, and the unknown output function along with its derivatives may be determined using standard computational tools. The entire solution procedure is a two-step one and it does not require any a-priori knowledge concerning the unknown solution and source function. The solution of the inverse problem is obtained in an explicit form, using semi-analytical transformations, based upon combined, stochastic (MC)–deterministic (FD) model.

The paper is organized in the following manner. Section 2 presents the formulation of the analysed inverse problem. Section 3 introduces the source function recovery, using either its global approximation or independent approximation mesh. Section 4 presents results of several benchmark problems. The paper is briefly concluded and directions of future work are indicated.

2 Formulation of the Inverse Poisson Problem

The following 2D Poisson equation is considered

$$\boldsymbol{\nabla}^2 F = f(\mathbf{x}) \quad \text{in} \quad \Omega \tag{1}$$

with essential boundary conditions

$$F = \bar{F}(\mathbf{x}) \quad \text{on} \quad \partial\Omega \tag{2}$$

where $\boldsymbol{\nabla}^2 = \boldsymbol{\Delta} = \dfrac{\partial^2}{\partial x_1^2} + \dfrac{\partial^2}{\partial x_2^2}$ is the Laplace operator, $\Omega = \{\mathbf{x} = \begin{bmatrix} x_1 & x_2 \end{bmatrix}\} \in \Re^2$ is the problem domain, $\partial\Omega$ – its boundary, $F : \mathbf{x} \in \Omega \to \Re$ is the unknown C^2 scalar output function, with given values $\bar{F}(\mathbf{x})$ at every boundary point $\mathbf{x} \in \partial\Omega$.

It is assumed that the right–hand side input function $f = f(\mathbf{x})$ is unknown and has to be determined on the basis of additional data

$$\hat{F}_k \pm \Delta F, \quad k = 1, 2, ..., m \tag{3}$$

being measurements $\hat{F}_k$ of the output function F with given tolerance ΔF at m measurement points $\hat{\mathbf{x}}_k = \begin{bmatrix} \hat{x}_1^{(k)} & \hat{x}_2^{(k)} \end{bmatrix}$. Furthermore, let us assume that the unknown function f is represented by the finite set of its n_f parameters (degrees of freedom), denoted as $\boldsymbol{\alpha} = \begin{bmatrix} \alpha_1 & \alpha_2 & ... & \alpha_{n_f} \end{bmatrix}$. Therefore, coefficients of $\boldsymbol{\alpha}$ constitute the set of primary unknowns to the considered inverse problem and $F = F(\mathbf{x}, \boldsymbol{\alpha})$, $f = f(\mathbf{x}, \boldsymbol{\alpha})$. The relevant non–linear optimisation problem may be formulated, namely find such optimal parameters

$$\boldsymbol{\alpha}_{\text{opt}} = \text{argmin}_{(\boldsymbol{\alpha})} J(\boldsymbol{\alpha}) \tag{4}$$

that minimise the following objective function

$$J(\boldsymbol{\alpha}) = \sqrt{\frac{1}{m} \sum_{k=1}^{m} \left(\frac{F(\hat{\mathbf{x}}_k, \boldsymbol{\alpha}) - \hat{F}_k}{\Delta F} \right)^2} \tag{5}$$

with inequality constraints

$$\left| F(\hat{\mathbf{x}}_k, \boldsymbol{\alpha}) - \hat{F}_k \right| < \Delta F, \quad k = 1, 2, ..., m \tag{6}$$

where $F(\hat{\mathbf{x}}_k, \boldsymbol{\alpha})$ are values of F at measurement points. The necessary condition of the existence of the extreme of (5) is

$$\boldsymbol{\nabla}_{\boldsymbol{\alpha}} J(\boldsymbol{\alpha}) = 0 \tag{7}$$

which leads to the algebraic system of n_f equations with n_f unknowns. After substitution of (5) to (7), one obtains the following expression

$$\boldsymbol{\nabla}_{\boldsymbol{\alpha}} J(\boldsymbol{\alpha}) = \sum_{k=1}^{m} \left(F(\hat{\mathbf{x}}_k, \boldsymbol{\alpha}) - \hat{F}_k \right) \boldsymbol{\nabla}_{\boldsymbol{\alpha}} F(\hat{\mathbf{x}}_k, \boldsymbol{\alpha}) \tag{8}$$

in which $\boldsymbol{\nabla}_{\boldsymbol{\alpha}} F$ is the $[n_f \times 1]$ gradient vector

$$\boldsymbol{\nabla}_{\boldsymbol{\alpha}} F(\mathbf{x}, \boldsymbol{\alpha}) = \left\{ \frac{\partial F}{\partial \alpha_l}, \quad l = 1, 2, ..., n_f \right\} \tag{9}$$

The sufficient condition of the existence of the minimum of (5) is the positive-definitiveness of the Hesse matrix (Hessian) of (5), defined as $\mathbf{H}(\boldsymbol{\alpha}) = \boldsymbol{\nabla}_{\boldsymbol{\alpha}}^2 J(\boldsymbol{\alpha})$, which corresponds to the convex objective function.

3   Recovery of the Source Function

Let us assume that the square grid Ω_h with boundary $\partial\Omega_h$, represented by one mesh modulus h and consisting of n_F nodes $\mathbf{x}_F \in \Omega_h$, has been generated. Therefore, each node $\mathbf{x}_F$ may be expressed using two indices (i,j), namely $\mathbf{x}_{i,j} = \left[x_1^{(i,j)} \ x_2^{(i,j)} \right]$ (Fig. 1a). Applying the finite difference solution approach to (1), we replace the differential operators with difference ones, using appropriate configurations of nodes (stars) as well as we generate difference equations at internal nodes using collocation technique, namely

$$\frac{F_{i-1,j} + F_{i,j+1} + F_{i+1,j} + F_{i,j-1} - 4F_{i,j}}{h^2} = f_{i,j} \tag{10}$$

By terms rearrangement, we obtain the relation between the central value of a star and the remaining ones

$$F_{i,j} = \frac{1}{4}F_{i-1,j} + \frac{1}{4}F_{i,j+1} + \frac{1}{4}F_{i+1,j} + \frac{1}{4}F_{i,j-1} - \frac{h^2}{4}f_{i,j} \tag{11}$$

which factors may be considered as probabilities of selection of a next move within each random walk component, equal for each sense of directions (top, right, bottom, left). The final Monte Carlo formula

$$F_{i,j} \approx \frac{1}{N} \left(\sum_{(r,s)\in\partial\Omega_h} \bar{F}_{r,s} \bar{N}_{(i,j),(r,s)} - \frac{h^2}{4} \sum_{(r,s)\in\Omega_h} f_{r,s} N_{(i,j),(r,s)} \right), \quad (i,j) \in \Omega_h \tag{12}$$

is the stochastic approximation of all a–priori known problem parameters. Here, N denotes the total number of all random walks, terminating at boundary nodes $\mathbf{x}_{r,s} = \left[x_1^{(r,s)} \ x_2^{(r,s)} \right] \in \partial\Omega_h$ with known solution values $\bar{F}_{r,s}$. Moreover, $\bar{N}_{(i,j),(r,s)}$ and $N_{(i,j),(r,s)}$ denote nodal indications, being the number of hits of each boundary node (r,s), and the number of visits of each internal node (r,s), respectively, for a random walk starting from the internal node (i,j). Error bounds of (12) may be estimated using the a–priori formula

$$e = \left| \frac{F_{i,j} - F_{i,j}^{fdm}}{F_{i,j}^{fdm}} \right| < \frac{1}{\sqrt{N}} \tag{13}$$

where $F_{i,j}^{fdm}$ is the corresponding finite difference solution of (1) at the internal node (i,j).

3.1   Simplest Constant Approximation

In case we expect the unknown source function to be a smooth one, namely values of its gradient are small, we may use the simplest constant approximation of f, in order to have the general impression concerning its basic features. Therefore,

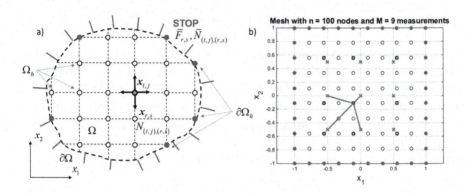

Fig. 1. Scheme of random walks for a regular grid of point (a) as well as exemplary regular distributions of approximation nodes $\mathbf{x}_F$ of F (red and white circles), measurements $\hat{\mathbf{x}}$ (blue squares) and approximation nodes $\mathbf{x}_f$ of f (magenta crosses) with a meshless star for a local approximation of f at $\mathbf{x}_F$ (b) (Color figure online)

we assume that the function f is ascribed by one unknown constant parameter $\alpha_1 = f = const$. In that case, the Monte Carlo relation (12) becomes

$$F_{i,j}(f) = \frac{1}{N}\left(\sum_{(r,s)\in\partial\Omega_h} \bar{F}_{r,s}\bar{N}_{(i,j),(r,s)} - \frac{h^2 f}{4}\sum_{(r,s)\in\Omega_h} N_{(i,j),(r,s)}\right) \qquad (14)$$

Condition (7) of existence of extreme solution $\dfrac{dJ}{df} = 0$ leads to the following equation

$$\sum_{i,j}^{m}\left(F_{i,j}(f) - \hat{F}_{i,j}\right)\sum_{(r,s)\in\Omega_h} N_{(i,j),(r,s)} = 0 \qquad (15)$$

which solution may be explicitly determined

$$f_{\text{opt}} = \frac{4\sum_{i,j}^{m}\left(\bar{S}_{i,j}S_{i,j} - N\hat{F}_{i,j}S_{i,j}\right)}{h^2\sum_{i,j}^{m}S_{i,j}^2}, \quad \begin{cases} \bar{S}_{i,j} = \displaystyle\sum_{(r,s)\in\partial\Omega_h} \bar{F}_{r,s}\bar{N}_{(i,j),(r,s)} \\ S_{i,j} = \displaystyle\sum_{(r,s)\in\Omega_h} N_{(i,j),(r,s)} \end{cases} \qquad (16)$$

in a stochastic, semi-analytical manner. Since f_{opt} corresponds to the analytical solution of (7), the inequality constraints (6) are a–priori satisfied.

3.2   Global Approximation of an Arbitrary Order

More general approximation of the unknown source function on the global level may incorporate the vector of n_f degrees of freedom $\boldsymbol{\alpha}$ as well as the vector of

n_f global basis functions $\mathbf{\Phi} = \{\Phi_l, \ l = 1, 2, ..., n_f\}$, namely

$$f(\mathbf{x}, \boldsymbol{\alpha}) = \sum_{l=1}^{n_f} \alpha_l \, \Phi_l(\mathbf{x}) = \boldsymbol{\alpha} \, \mathbf{\Phi}(\mathbf{x}) \tag{17}$$

In most approaches, degrees of freedom $\boldsymbol{\alpha}$ are values of f (physical degrees of freedom), providing basis functions are dimensionless (e.g., standard shape functions). However, since the initial values of $\boldsymbol{\alpha}$ are not required here, $\boldsymbol{\alpha}$ may be arbitrary mathematical degrees of freedom whereas basis functions may be simple monomials. On the basis of the Monte Carlo relation

$$F_{i,j}(\boldsymbol{\alpha}) = \frac{1}{N} \left(\sum_{(r,s) \in \partial \Omega_h} \bar{F}_{r,s} \bar{N}_{(i,j),(r,s)} - \frac{h^2}{4} \sum_{(r,s) \in \Omega_h} \boldsymbol{\alpha} \, \mathbf{\Phi}(\hat{\mathbf{x}}_{r,s}) N_{(i,j),(r,s)} \right) \tag{18}$$

as well as the necessary conditions

$$\frac{\partial J}{\partial \alpha_l} = \sum_{i,j}^{m} \left(F_{i,j}(\boldsymbol{\alpha}) - \hat{F}_{i,j} \right) \sum_{(r,s) \in \Omega_h} \Phi_l(\hat{\mathbf{x}}_{r,s}) N_{(i,j),(r,s)} = 0, \quad l = 1, 2, ..., n_f \tag{19}$$

we obtain the symmetric and positive–definite system of linear equations which solution may be expressed in the following matrix notation

$$\mathbf{A}\boldsymbol{\alpha}_{\mathrm{opt}} = \mathbf{B}, \quad \boldsymbol{\alpha}_{\mathrm{opt}} = \mathbf{A}^{-1}\mathbf{B} \tag{20}$$

where

$$A_{k,l} = \sum_{i,j}^{m} S_{i,j}^{(k)} S_{i,j}^{(l)}, \quad B_k = \frac{4}{h^2} \sum_{i,j}^{m} \left(\bar{S}_{i,j} - N\hat{F}_{i,j} \right) S_{i,j}^{(k)}, \quad k, l = 1, 2, ..., p$$

$$\bar{S}_{i,j} = \sum_{(r,s) \in \partial \Omega_h} \bar{F}_{r,s} \bar{N}_{(i,j),(r,s)}, \quad S_{i,j}^{\{(k),(l)\}} = \sum_{(r,s) \in \Omega_h} \Phi_{\{k,l\}}(\hat{\mathbf{x}}_{r,s}) N_{(i,j),(r,s)} \tag{21}$$

For instance, the linear approximation of f ($n_f = 3$) may be assumed as

$$f(\mathbf{x}, \boldsymbol{\alpha}) = \alpha_1 \Phi_1(\mathbf{x}) + \alpha_2 \Phi_2(\mathbf{x}) + \alpha_3 \Phi_3(\mathbf{x}) = \alpha_1 + \alpha_2 x_1 + \alpha_3 x_2 \tag{22}$$

whereas the optimal set of mathematical degrees of freedom $\boldsymbol{\alpha} = \begin{bmatrix} \alpha_1 & \alpha_2 & \alpha_3 \end{bmatrix}$ may be obtained from (20), namely

$$\begin{bmatrix} \sum_{i,j}^{m} \left(S_{i,j}^{(1)} \right)^2 & \sum_{i,j}^{m} S_{i,j}^{(2)} S_{i,j}^{(1)} & \sum_{i,j}^{m} S_{i,j}^{(3)} S_{i,j}^{(1)} \\ \sum_{i,j}^{m} S_{i,j}^{(1)} S_{i,j}^{(2)} & \sum_{i,j}^{m} \left(S_{i,j}^{(2)} \right)^2 & \sum_{i,j}^{m} S_{i,j}^{(3)} S_{i,j}^{(2)} \\ \sum_{i,j}^{m} S_{i,j}^{(1)} S_{i,j}^{(3)} & \sum_{i,j}^{m} S_{i,j}^{(2)} S_{i,j}^{(3)} & \sum_{i,j}^{m} \left(S_{i,j}^{(3)} \right)^2 \end{bmatrix} \begin{bmatrix} \alpha_1 \\ \alpha_2 \\ \alpha_3 \end{bmatrix} = \frac{4}{h^2} \begin{bmatrix} \sum_{i,j}^{m} \left(\bar{S}_{i,j} - N\hat{F}_{i,j} \right) S_{i,j}^{(1)} \\ \sum_{i,j}^{m} \left(\bar{S}_{i,j} - N\hat{F}_{i,j} \right) S_{i,j}^{(2)} \\ \sum_{i,j}^{m} \left(\bar{S}_{i,j} - N\hat{F}_{i,j} \right) S_{i,j}^{(3)} \end{bmatrix} \tag{23}$$

3.3   Local Approximation Using Independent Mesh

In case the unknown source function f is expected to be highly non–linear or localized, the global approximation (17) may provide the general information concerning its values and distribution. On the other hand, monomial terms, which order is higher than 2, may not have a significant influence on the final results. Therefore, a local approximation of f may be assumed, using physical degrees of freedom $\boldsymbol{\alpha} = \mathbf{f}$, based upon a mesh of n_f internal points $\Omega_f = \{\mathbf{x}_f \in \Omega\}$ (Fig. 1b). The mesh Ω_f may be totally independent from the approximation mesh $\Omega_h = \{\mathbf{x}_F\}$ for the primary function F, whereas its density may be directly related to the number (m) of measurement locations $\hat{\mathbf{x}}$. Therefore, the additional mapping, between both approximation meshes, namely Ω_f (of function f) and Ω_h (of function F), is required. It may be based upon the Moving Weighted Least Squares (MWLS) approximation technique [18], typical for meshless analysis, in which nodes may be distributed totally arbitrarily, without any imposed structure. Let us assume that the function f is defined by the finite set of its nodal values $\mathbf{f} = f(\mathbf{x}_f)$, given at n_f nodes $\mathbf{x}_f \in \Omega_f$. The function value f as well as its derivatives up to the p–th order are required at the arbitrary point $\mathbf{x}_F \in \Omega_h$. The configuration (called star or stencil) of $m_f < n_f$ nodes $\mathbf{S}_F = \{\mathbf{x}_f\}$, being neighbours of $\mathbf{x}_F$, is assigned to $\mathbf{x}_F$. The meshless star my be generated using various criteria, for instance, a distance criterion in which m_f nodes, closest to $\mathbf{x}$, are selected, or topology oriented ones, like cross or Voronoi neighbours criteria [19]. Afterwards, the local approximation of f is constructed, using the Taylor series expansion, namely

$$f(\mathbf{x}_F, \mathbf{x}) = \mathbf{p}(\mathbf{x}_F - \mathbf{x})\, \mathbf{D}_f(\mathbf{x}_F) \tag{24}$$

where $\mathbf{p}(\mathbf{x}_F - \mathbf{x}) = \begin{bmatrix} 1 & |\mathbf{x}_F - \mathbf{x}| & |\mathbf{x}_F - \mathbf{x}|^2 & \dots & |\mathbf{x}_F - \mathbf{x}|^p \end{bmatrix}$ is the vector of local interpolants, whereas $\mathbf{D}_f(\mathbf{x}_F) = \begin{bmatrix} f & \dfrac{\partial f}{\partial x_1} & \dfrac{\partial f}{\partial x_2} & \dfrac{\partial^2 f}{\partial x_1^2} & \dots & \dfrac{\partial^{(p)} f}{\partial x_2^{(p)}} \end{bmatrix}_{\mathbf{x}_F}$ is the vector of subsequent derivatives of f at $\mathbf{x}_F$. Fulfilling of (24) at all m_f node of the $\mathbf{S}_F$ star leads to the over–determined system of linear equations

$$\mathbf{P}(\mathbf{x}_F)\, \mathbf{D}_f(\mathbf{x}_F) = \mathbf{f}(\mathbf{x}_F) \tag{25}$$

where $\mathbf{P}(\mathbf{x}_F) = \mathbf{p}(\mathbf{x}_F - \mathbf{S}_F)$ and $\mathbf{f}(\mathbf{x}_F)$ is the vector of nodal values of f at star nodes. Its solution may be obtained by a minimisation of the weighted error function

$$I(\mathbf{x}_F) = (\mathbf{P}(\mathbf{x}_F)\mathbf{D}_f(\mathbf{x}_F) - \mathbf{f}(\mathbf{x}_F))^{\mathrm{T}}\, \mathbf{W}^2(\mathbf{x}_F)\, (\mathbf{P}(\mathbf{x}_F)\mathbf{D}_f(\mathbf{x}_F) - \mathbf{f}(\mathbf{x}_F)) \tag{26}$$

Here, $\mathbf{W}(\mathbf{x}_F)$ is the diagonal weighting matrix, with singular weights Ω assigned to each node of $\mathbf{S}_F$, according to the formula $\Omega(\mathbf{x}_F - \mathbf{x}) = \|\mathbf{x}_F - \mathbf{x}\|^{-p-1}$. Finally, we obtain the matrix of difference formulas

$$\mathbf{M}(\mathbf{x}_F) = \left(\mathbf{P}^{\mathrm{T}}(\mathbf{x}_F)\, \mathbf{W}^2(\mathbf{x}_F)\, \mathbf{P}(\mathbf{x}_F)\right)^{-1} \mathbf{P}^{\mathrm{T}}(\mathbf{x}_F)\, \mathbf{W}^2(\mathbf{x}_F) \tag{27}$$

rows of which correspond to difference coefficients of subsequent derivatives in $\mathbf{D}_f$, namely

$$\mathbf{D}_f(\mathbf{x}_F) = \mathbf{M}(\mathbf{x}_F)\mathbf{f}(\mathbf{x}_F) \tag{28}$$

Consequently, the approximation f at nodes $\mathbf{x}_F$ may be substituted into the Monte Carlo relation, yielding the following formula

$$F_{i,j}(\mathbf{f}) = \frac{1}{N}\left(\sum_{(r,s)\in\partial\Omega_h}\bar{F}_{r,s}\bar{N}_{(i,j),(r,s)} - \frac{h^2}{4}\sum_{(r,s)\in\Omega_h}\mathbf{M}^{(1)}(\hat{\mathbf{x}}_{r,s})\mathbf{f}(\hat{\mathbf{x}}_{r,s})N_{(i,j),(r,s)}\right) \tag{29}$$

where $\mathbf{M}^{(1)}$ denotes the first row of $\mathbf{M}$. The optimal parameters $\mathbf{f}_{\mathrm{opt}}$ are determined from (20) with (21) and modified terms

$$S_{i,j}^{\{(k),(l)\}} = \sum_{(r,s)\in\Omega_h}M_{1,\{k,l\}}(\hat{\mathbf{x}}_{r,s})N_{(i,j),(r,s)} \tag{30}$$

General principles of random walk strategy remain unmodified, namely the selection of four equally probable directions of each next move and its termination at boundary nodes. Moreover, it has to be stressed that for $n_f > m$ the matrix $\mathbf{A}$ from (20) is singular, for $n_f = m$ the matrix $\mathbf{A}$ is ill-conditioned, whereas its conditioning improves as n_f becomes smaller than m. Therefore, appropriate number of measurements m may be required in order to reproduce the source function with assumed accuracy.

4   Numerical Examples

Results of selected numerical experiments are presented in order to illustrate the effectiveness of the proposed approach. Since the research is in the preliminary state, only simulated measurement data are taken into account. The following strategy, based upon manufacturing solutions, is adopted

1. Geometry parameters, boundary conditions $\bar{F}$ as well as the right-hand side function f are assumed.
2. The regular mesh $\mathbf{x}_F \in \Omega_h$ is generated and the corresponding direct problem (with known $\bar{F}$ and f) is solved by means of the finite difference method (FDM), yielding the nodal solution $\mathbf{F}$ at $\mathbf{x}_F$.
3. Measurement locations $\hat{\mathbf{x}} \in \Omega_h$ are assumed and measurement data $\hat{\mathbf{F}}$ are generated on the basis of $\mathbf{F}$ values, randomly disturbed with the amplitude ΔF, corresponding to the measurement tolerance.
4. The selected approximation formula ((17) or (24)) for the unknown source function f is assumed, with the optional generation of additional mesh $\mathbf{x}_f \in \Omega_f$ and the mapping $\mathbf{x}_f \to \mathbf{x}_F$.
5. The first inverse solution step: nodal indications are determined at all m measurement locations using standard fixed random walk technique.
6. The second inverse solution step: unknown source parameters $\mathbf{f}$ are determined from the system of equations using appropriate Monte Carlo relations.

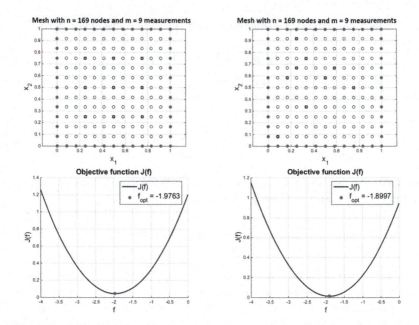

Fig. 2. Objective function and optimal solution for two sets of 9 measurement points

7. The source function f is recovered at all nodes $\mathbf{x}_F \in \Omega_h$ using previously introduced approximation.
8. The resulting direct problem, with recovered function f, is solved by FDM and the primary solution $\mathbf{F}$ is determined at $\mathbf{x}_F \in \Omega_h$.

Preliminary tests are performed for the problem with the constant source intensity $f = -2$, boundary values $\bar{F}(\mathbf{x}) = x_1^2 + x_2^2$, square domain $\Omega = \begin{bmatrix} 0 & 1 \end{bmatrix}^2$, nodes number $n = 13 \times 13$, the number of random walks $N = 1000$, measurements number $m = 3 \times 3$, as well as measurement tolerance $\Delta F = 0.2$ (up to the 10% of the original value). A–priori estimators allow to estimate the approximation error $e_{hp} = Ch^{p+1-k} < 0.0833C$ (with respect to the unknown exact solution), where C is the arbitrary constant, though independent from p and h, as well as the stochastic error (13), namely $e < 0.0316$ (with the respect to the FD solution). Figure 2 presents results (graph of the objective function and the optimal solution), obtained for fixed parameters, for two types of measurement locations, namely regular and randomly selected distributions. However, the final results of stochastic methods cannot be representative unless they are obtained after appropriate averaging of intermediate results. Therefore, histograms for 1000 various simulations of both random measurement locations and distributions of indication numbers are shown in Fig. 3. It may be observed that the dispersion of results is significant, especially for randomized locations. As a consequence, all following examples are executed in N–based series and properly averaged. Moreover, the influence of selected input parameters on f_{opt} is examined, using the regular grid of $m = 9$ measurement points. Figure 4 presents convergence of

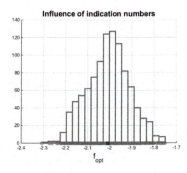
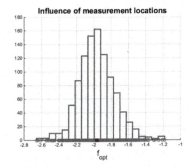

Fig. 3. Histograms of the optimal solution f_{opt}, for variability of indication numbers (left graph) and measurement locations (right graph), for 1000 simulations per 20 classes

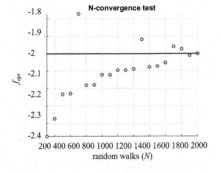
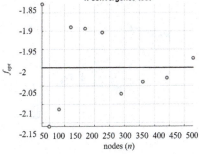

Fig. 4. Results of N-convergence (left graph) and h-convergence tests (right graph)

f_{opt} with respect to N (with fixed h, m and ΔF) and h (with fixed N, m and ΔF), whereas Fig. 5 shows convergence of f_{opt} with respect to m (with fixed N, h and ΔF) and Δ (with fixed N, h and m). Each time, results are stable and convergent to the original intensity value ($f = -2$). Afterwards, the global monomial approximation (17) of the full second order ($n_f = 6$) is applied for the recovery of the unknown source function. The simulated measurement data ($m = 9$, $\Delta F = 10\%$ of the original value) are generated on the basis of the FDM solution, obtained on the regular mesh with $n_F = 81$ nodes and corresponding to the original source function $f(\mathbf{x}) = -2 + x_1 + 3x_2 - 5x_1^2 + x_1 x_2 + 7x_2^2$. Nodal indications are determined by means of 10 series of $N = 1000$ random walks. Results are presented in Fig. 6. The formula of the recovered source function as well as relative source and solution errors, evaluated in L^2 and max norms, are given in graphs' titles. The maximum errors are approx. 2% (for f) and 1% for (F). Eventually, the local approximation of the first order as well as the independent mesh for the approximation of f are applied for the recovery of the source function, given by more complex formula of exponential type, namely $f(\mathbf{x}) = \exp\left(-5\left(x_1^2 + (x_2 - 9/10x_1)^2\right)\right)$. It may be computationally

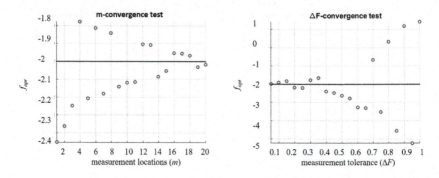

Fig. 5. Results of m-convergence (left graph) and ΔF-convergence tests (right graph)

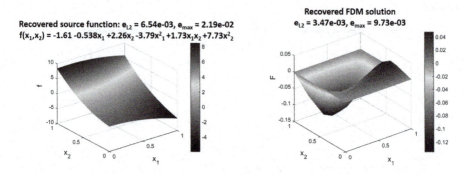

Fig. 6. Recovered source function f (left graph) and primary solution F (right graph) using global approximation and mathematical degrees of freedom

demanding even for the direct problem as it requires a dense discretization mesh and, therefore, a vast computational power. The applied model is based upon the approximation mesh with $n_F = 225$ nodes and $m = 49$ measurements, as well as the regular grid for the source function approximation with $n_f = 36$ physical degrees of freedom $\mathbf{f}$. Results are presented in Fig. 7. Relative source and solution errors, evaluated in L^2 and max norms, are given in graphs' titles. Similarly as in the previous example, the maximum errors are reasonably small, namely approx. 2.5% (for f) and 1% for (F). Furthermore, in all cases, the computational times are negligible (below several seconds). All results are obtained on 16 GB RAM and 1.80 GHz processor, using author's original software, written in Matlab R2014b.

5   Final Remarks

The semi–analytical approach for the numerical analysis of inverse Poisson problems is presented. In inverse Poisson problem, the source (load) function is unknown, whereas set of measurements of the primary function, being the

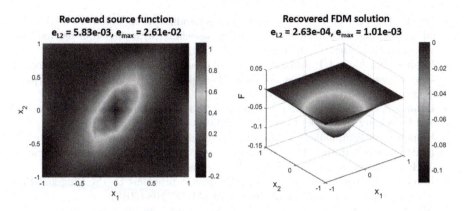

Fig. 7. Recovered source function f (left graph) and primary solution F (right graph) using local approximation and physical degrees of freedom

solution to this problem, may constitute the additional data. The problem may be formulated as the non–linear optimisation problem, in which the objective function is the average error between the measured and computed values whereas decision variables are source function parameters. The proposed approach is based upon the old and well–known concept of the Monte Carlo method as well as random simulations (random walks), performed on the regular mesh of points. It incorporates several features of the standard (selection probabilities) and meshless (mapping between independent meshes) finite difference methods. On the contrary to the standard optimisation methods, this approach requires neither an iterative procedure nor a searching of the admissible solution space. The optimal source functions parameters are determined from the system of linear equations, which source is the analytical differentiation of the objective function, approximated in the coupled stochastic-deterministic manner.

The future work may include the generalisation of the proposed approach for more complex geometries, for which unstructured meshes and arbitrarily irregular clouds of nodes are required as well as for problems with mixed boundary conditions. Selection directions and probabilities may be determined using meshless approximation techniques, similar to those already applied in the mapping between two independent approximation meshes. Moreover, the application of the approach to inverse problems of non–stationary thermal as well as to linear elastic types is planned.

References

1. Jiang, J., Mohamed, S.M., Seaid, M., Li, H.: Identifying the wavenumber for the inverse Helmholtz problem using an enriched finite element formulation. Comput. Methods Appl. Mech. Eng. **340**, 615–629 (2018)
2. Ohe, T., Ohnaka, K.: Boundary element approach for an inverse source problem of the Poisson equation with a one-point-mass like source. Appl. Math. Model. **18**(4), 216–223 (1994)

3. Gu, Y., Lei, J., Fan, C.M., He, X.Q.: The generalized finite difference method for an inverse time-dependent source problem associated with three-dimensional heat equation. Eng. Anal. Boundary Elem. **91**, 73–81 (2018)

4. Khan, M.N., Ahmad, I., Ahmad, H.: A radial basis function collocation method for space-dependent inverse heat problems. J. Appl. Comput. Mech. (2020). https://doi.org/10.22055/jacm.2020.32999.2123

5. Hamad, A., Tadi, M.: A numerical method for inverse source problems for Poisson and Helmholtz equations. Phys. Lett. A **380**(44), 3707–3716 (2016)

6. Koulouri, A., Rimpiläinen, V., Brookes, M., Kaipio, J.P.: Compensation of domain modelling errors in the inverse source problem of the Poisson equation: Application in electroencephalographic imaging. Appl. Numer. Math. **106**, 24–36 (2016)

7. Rond, A., Giryes, R., Elad, M.: Poisson inverse problems by the Plug-and-Play scheme. J. Vis. Commun. Image Represent. **41**, 96–108 (2016)

8. Alves, C., Martins, N., Roberty, N.: Full identification of acoustic sources with multiple frequencies and boundary measurements. Inverse Probl. Imaging **3**(2), 275–294 (2009)

9. Frackowiak, A., Wolfersdorf, J.V., Ciałkowski, M.: Solution of the inverse heat conduction problem described by the Poisson equation for a cooled gas-turbine blade. Int. J. Heat Mass Transf. **54**(5–6), 1236–1243 (2011)

10. Bergagio, M., Li, H., Anglart, H.: An iterative finite-element algorithm for solving two-dimensional nonlinear inverse heat conduction problems. Int. J. Heat Mass Transf. **126**(A), 281–292 (2018)

11. Wen, H., Yan, G., Zhang, C., He, X.: The generalized finite difference method for an inverse boundary value problem in three-dimensional thermo-elasticity. Adv. Eng. Softw. **131**, 1–11 (2019)

12. Wang, X., Li, H., He, L., Li, Z.: Evaluation of multi-objective inverse heat conduction problem based on particle swarm optimization algorithm, normal distribution and finite element method. Int. J. Heat Mass Transf. **127**(A), 1114–1127 (2018)

13. Milewski, S.: Determination of the truss static state by means of the combined FE/GA approach, on the basis of strain and displacement measurements. Inverse Probl. Sci. Eng. **27**(11), 1537–1558 (2019)

14. Liu, Ji-Chuan., Li, Xiao-Chen.: Reconstruction algorithms of an inverse source problem for the Helmholtz equation. Numer. Algorithms **84**(3), 909–933 (2019). https://doi.org/10.1007/s11075-019-00786-8

15. Reynolds, J.F.: A Proof of the Random-Walk Method for Solving Laplace's Equation in 2-D. Math. Gaz. **49**(370), 416–420 (1965)

16. Milewski, S.: Combination of the meshless finite difference approach with the Monte Carlo random walk technique for solution of elliptic problems. Comput. Math. Appl. **76**(4), 854–876 (2018)

17. Milewski, S.: Application of the Monte Carlo method with meshless random walk procedure to selected scalar elliptic problems. Arch. Mech. **71**(4–5), 337–375 (2019)

18. Lancaster, P., Salkauskas, K.: Curve and Surface Fitting. An Introduction, 1st edn. Academic Press Inc., London (1990). 280 pages

19. Orkisz, J.: Handbook of Computational Solid Mechanics: Finite Difference Method (Part III), 1st edn., pp. 336–431. Springer, Heidelberg (1998)

Modeling the Contribution of Agriculture Towards Soil Nitrogen Surplus in Iowa

Vishal Raul[1], Yen-Chen Liu[1], Leifur Leifsson[1]([✉]), and Amy Kaleita[2]

[1] Department of Aerospace Engineering, Iowa State University, Ames, IA 50011, USA
{vvssraul,clarkliu,leifur}@iastate.edu
[2] Departments of Agricultural and Biosystems Engineering, Iowa State University, Ames, IA 50011, USA
kaleita@iastate.edu

Abstract. The Midwest state of Iowa in the US is one of the major producers of corn, soybean, ethanol, and animal products, and has long been known as a significant contributor of nitrogen loads to the Mississippi river basin, supplying the nutrient-rich water to the Gulf of Mexico. Nitrogen is the principal contributor to the formation of the hypoxic zone in the northern Gulf of Mexico with a significant detrimental environmental impact. Agriculture, animal agriculture, and ethanol production are deeply connected to Iowa's economy. Thus, with increasing ethanol production, high yield agriculture practices, growing animal agriculture, and the related economy, there is a need to understand the interrelationship of Iowa's food-energy-water system to alleviate its impact on the environment and economy through improved policy and decision making. In this work, the Iowa food-energy-water (IFEW) system model is proposed that describes its interrelationship. Further, a macro-scale nitrogen export model of the agriculture and animal agriculture systems is developed. Global sensitivity analysis of the nitrogen export model reveals that the commercial nitrogen-based fertilizer application rate for corn production and corn yield are the two most influential factors affecting the surplus nitrogen in the soil.

Keywords: Food-energy-water nexus · Nitrogen export · Sensitivity analysis · Hypoxic zone · Interrelationship · System modeling

1 Introduction

The Gulf of Mexico dead zone is a hypoxic region of the water body that is caused by the nutrient-enriched water it receives from the Mississippi River. Nitrogen (N) is the principal nutrient that contributes to the formation of the hypoxic zone in the Gulf of Mexico [10], with obvious detrimental environmental, societal, and economic impacts. The Midwest agriculture region is a significant contributor of nitrogen loading in the form of nitrates (NO3) to the Mississippi river basin [4].

ⓒ Springer Nature Switzerland AG 2021
M. Paszynski et al. (Eds.): ICCS 2021, LNCS 12745, pp. 257–268, 2021.
https://doi.org/10.1007/978-3-030-77970-2_20

The state of Iowa, a Midwest state in the US, is the country's foremost producer of corn, soybean, ethanol, animal products and is known for disproportionately contributing nitrogen loads to the Mississippi river basin [8]. Iowa has significantly invested in developing a subsurface drainage system to handle a unique situation of too much water for improving the agriculture productivity of its abundant farmland that produces most corn in the US. Almost 57% of Iowa's produced corn is used for ethanol production [16]. The increased demand for food production and ethanol could increase the use of nitrogen-based fertilizers by farmers for maximizing crop yield. Additionally, there is a possibility that the increased demands of ethanol production could drive corn production by converting soybean areas to corn [5]. Such a situation could increase the rate of Nitrogen (a highly water-soluble nutrient) export from Iowa watersheds through the subsurface drainage system, ultimately exiting into the Mississippi River. Further, the rising demands in animal protein have increased and concentrated Iowa's animal agriculture industry. Animal manure, a rich source of nitrogen, is also used as fertilizer along with commercial nitrogen fertilizers, making animal agriculture one of the contributors of nitrogen export from Iowa [7].

Agriculture, animal agriculture, and ethanol production are integral to Iowa's economy. However, their combined operation increases the rate of Nitrogen carried through the water system, contaminating Iowa's drinking municipal water [7] and adversely impacting the ecosystem of the Gulf of Mexico. Thus, it is important to understand the interrelationship of the Iowa food-water-energy (IFEW) system for the generation of appropriate policies that could mitigate the adverse impacts on the environment and economy.

In this work, an IFEW system model is proposed that describes the interconnections of agriculture, animal agriculture, water, energy, and the weather system along with the exported Nitrogen. Further, a macro-scale nitrogen export computational model of the agriculture and animal agriculture systems is developed and a global sensitivity analysis of this model is conducted. The Sobol' indices [15] a global sensitivity analysis method is used that reveals the important parameters contributing to the nitrogen export from Iowa.

The remainder of this paper is structured as follows. The next section presents IFEW system interrelationship, nitrogen export computational model, and global sensitivity analysis with Sobol' indices. The following section presents the results of the sensitivity analysis of the macro-scale computational model. Finally, conclusions and suggestions for future work are described.

2   Methods

This section describes the proposed IFEW system model and involved interrelationship and the formulation of the macro-scale nitrogen export model of the agriculture system. Further, the global sensitivity analysis based on Sobol' indices is described.

2.1 IFEW System Interrelationship

The current IFEW system modeling approach is inspired from the Water, Energy, and Food security nexus Optimization model (WEFO) developed by Zhang et al. [20], which employs an integrated modeling approach for providing critical information to decision-makers and stakeholders for optimal management of food-energy-water (FEW) systems. In the current study, state-of-the-art multidisciplinary design optimization (MDO) [11] methodology is employed for the development of the IFEW system model to better understand FEW system interrelationship with the goal of providing critical information for efficient policy generation to mitigate the environmental and economic impact of the nitrogen export from Iowa.

The proposed IFEW model is developed such that it represents the major socioeconomic systems of Iowa that affect nitrogen export related to agricultural activity. Figure 1 shows the proposed IFEW system model, showing individual systems and their interrelationship. Further, the IFEW system is subjected to socioeconomic and environmental constraints. In particular, the IFEW system involves five distinct systems: weather, water, agriculture, animal agriculture, and energy. The weather system significantly impacts agriculture and water systems through different environmental parameters, such as temperature, precipitation, vapor pressure, and solar radiation. The weather system strongly influences agricultural output by directly affecting corn and soybean yields [19] whereas the amount of precipitation and snowfall affects surface and groundwater availability under the water system as well as the concurrent transport of excess nitrogen downstream. The water is consumed in the animal agriculture and energy systems. In the animal agriculture system, water is consumed as drinking and service water. In the energy system, water is consumed for ethanol and fertilizer production. The byproduct of ethanol production is the dried distillers grains with solubles (DDGS), which is a rich protein source used as animal feed. The commercially produced fertilizers and animal manure from animal agriculture are applied to crop fields under the agriculture system in Iowa to maximize crop yield, mainly for corn production, where a major portion of corn is used for ethanol production. Except for the weather, all other systems are responsible for meeting socioeconomic demands for corn, soybean, ethanol, water, and animal protein. Lastly, the water flowing through Iowa watersheds carries the surplus nitrogen from the soil in the form of nitrates that drains into the Mississippi river basin and further into the Gulf of Mexico.

2.2 Agriculture Nitrogen Export Computational Model

Almost 70% of Iowa's land is under high yield agriculture practices where nitrogen-based fertilizers are primarily used to enhance crop yield, especially for the corn production [7]. Further, the increasing demands of animal protein have increased animal agriculture operations in Iowa. The manure produced from animal agriculture is rich in nitrogen and have been used for soil fertilization [1]. Thus, animal agriculture mostly contributes to Iowa's nitrogen export

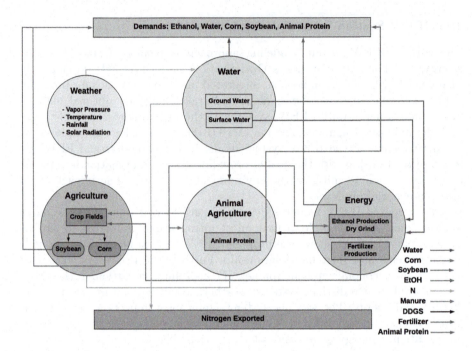

Fig. 1. A model of the interrelationship of the Iowa food-energy-water system.

through the agriculture system. With large agriculture operations, widespread availability of manure from animal agriculture, and high commercial nitrogen-based fertilizer application rates could create surplus nitrogen in the soil, which is then carried by the water through the subsurface drainage system. In this work, a nitrogen export model is developed for the state of Iowa based only on the agriculture and animal agriculture systems. The other systems are not accounted for in this simplified computational model. Figure 2 shows an extended design structure matrix of diagram [9] of the proposed nitrogen export model with the definitions of the parameters given in Tables 1 and 2.

The model yields a computation of the nitrogen surplus (N_s) based on the construction of a rough agronomic annual nitrogen budget [3] given as

$$N_s = CN + FN + MN - GN, \tag{1}$$

where CN represents the commercial nitrogen that the soil receives from the application of commercial nitrogen-based fertilizers, FN represents the nitrogen fixed in the soil by soybean crop, GN represents the nitrogen that is taken out from soil through the harvested grain, and MN represents the nitrogen generated from animal manure that is applied to the soil. Such nitrogen budgeting provides an insight into nitrate sources of Iowa.

The agriculture system receives four input parameters, which are the corn yield (x_1), soybean yield (x_2), rate of commercial nitrogen for corn (x_3), and

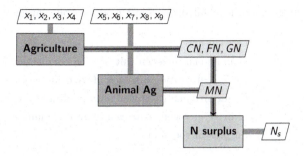

Fig. 2. An extended design structure matrix diagram of nitrogen export model considering only the agriculture and animal agriculture systems (Tables 1 and 2 give the parameter descriptions).

Table 1. Macro-level nitrogen export model input parameters.

Input parameters	Description
x_1	Corn yield
x_2	Soybean yield
x_3	Rate of commercial nitrogen for corn
x_4	Rate of commercial nitrogen for soybean
x_5	Hog/pigs population
x_6	Beef cattle population
x_7	Milk cows population
x_8	Other cattle population (heifers + slaughter cattle)
x_9	Chicken/hens population

the rate of commercial nitrogen for soybean (x_4). The output parameters of the agriculture system are CN, FN, and GN, respectively. The commercial nitrogen (CN) is computed as

$$CN = \frac{(x_3 A_{corn} + x_4 A_{soy})}{A}, \tag{2}$$

where A_{corn} and A_{soy} are Iowa corn and soybean acreage, and $A = A_{corn} + A_{soy}$ represents cumulative area under corn and soybean. In the current study, the corn and soybean acreages are obtained from USDA [18]. The biological fixation nitrogen by the soybean crop is computed using the relationship between soybean yield and FN provided by Barry et al. [2] given as

$$FN = \frac{(81.1x_2 - 98.5)A_{soy}}{A}, \tag{3}$$

where the soybean yield (x_2) is in tons per hectare to provide FN in kg per hectare. The nitrogen exported in the harvested grain from corn and soybean harvest is computed assuming 6.4% nitrogen in the soybean seed and 1.18% nitrogen in corn seed [3] given as

Table 2. Macro-level nitrogen export model output parameters.

Output parameters	Description
CN	Commercial nitrogen (nitrogen in commercial fertilizers)
FN	Biological fixation nitrogen of soybean crop
GN	Grain nitrogen (Nitrogen harvested in grain)
MN	Manure nitrogen (Nitrogen in animal manure)
N_s	Surplus nitrogen in soil

$$GN = \frac{x_1 \frac{1.18}{100} A_{corn} + x_2 \frac{6.4}{100} A_{soy}}{A}. \qquad (4)$$

In (2), (3), and (4), CN, FN and GN have the unit kg/ha.

Currently, Iowa holds the first rank in red meat, pork, and egg production in the U.S. [18]. Thus, for the manure nitrogen computation, the hogs/pigs, cattle, and chicken/hens populations are considered. The animal agriculture system model receives five input parameters representing the population of the following categories in Iowa, namely hogs/pigs (x_5), beef cattle (x_6), milk cows (x_7), total heifers and slaughter cattle (x_8), and layers chicken/hens (x_9). The annual manure nitrogen contribution from each animal category is given by [6]

$$MN_{Livestock\ group} = PN_m L_c, \qquad (5)$$

where P represent livestock population, N_m represents nitrogen present in animal manure, and L_c represents life cycle in days. Table 3 gives the numerical values of the parameters used in (5) for each livestock group.

The total MN contribution from animal agriculture is

$$MN = ((MN_{Hog/Pigs} + MN_{Beef\ cattle} + MN_{Milk\ cow}$$
$$+ MN_{other\ cattle} + MN_{Chicken/Hens}))A^{-1}, \qquad (6)$$

where the other cattle livestock group is composed of 50% heifers/steers population and 50% slaughter cattle population. In this study, the Iowa animal population data of 2012 [6] is used for the MN computation.

2.3   Global Sensitivity Analysis

For this study, global sensitivity analysis (GSA) of the nitrogen export model is performed to understand the contribution of each input parameter to the model output. In particular, Sobol' sensitivity analysis [15] is used in this work. Sobol' analysis provides a computation of the first-order and total-effect Sobol' sensitivity indices, which can be used to determine the sensitivities of individual parameters and their interactions with other input parameters on the model output.

Specifically, Sobol's method uses a variance decomposition to calculate the Sobol' indices. Consider the model response $y = f(\mathbf{x})$ as a function of vector $\mathbf{x}$ with n parameters. Then, the total variance $var(Y)$ in any model response Y can be decomposed as [13]

$$var(Y) = \sum_{i}^{n} V_i + \sum_{i}^{n} \sum_{i<j}^{n} V_{ij} + ... + V_{12...\, n}, \tag{7}$$

where V_i is $var(\mathbb{E}(Y|x_i))$ is the variance contribution of ith design parameter to the total variance $var(Y)$, V_{ij} is the variance contribution of ith and j^{th} parameter to $var(Y)$ and so on. The Sobol' indices are obtained by dividing (7) by the total variance $var(Y)$ to obtain

$$1 = \sum_{i}^{n} S_i + \sum_{i}^{n} \sum_{i<j}^{n} S_{ij} + ... + S_{12...\, n}, \tag{8}$$

where S_i represents the first-order Sobol' index given by [13]

$$S_i = \frac{V_i}{var(Y)}. \tag{9}$$

The total-effect Sobol' index is given as [13]

$$S_{T_i} = 1 - \frac{var(\mathbb{E}(Y|\mathbf{x}_{\sim i}))}{var(Y)}, \tag{10}$$

where $\mathbf{x}_{\sim i}$ represents the set of all parameters except x_i. Sobol' total-order index S_{T_i} indicates total contribution of ith parameter to total variation in Y including first-order S_i and interaction effects with other input parameters. The interaction effect is represented by difference between S_{T_i} and S_i ($S_{T_i} - S_i$). Zero interaction effect indicates that the particular parameter affects the model response only with first order-effect making $S_i = S_{T_i}$. Sobol' indices are typically computed using Monte Carlo-based numerical procedure. For this study, the numerical procedure provided by Saltelli et al. [13] is used.

Table 3. Nitrogen content in manure and life cycle for livestock groups used in manure nitrogen calculation [6]

Livestock group	Nitrogen in manure N_m (kg per animal per day)	Life cycle L_c (days per year)
Hog/pigs	0.027	365
Beef cattle	0.15	365
Milk cows	0.204	365
Heifer/steers (0.5 × other cattle)	0.1455	365
Slaughter cattle (0.5 × other cattle)	0.104	170
Chicken/Hens	0.0015	365

Table 4. Input parameter bounds.

Input parameters	Upper bound	Lower bound	Units
x_1	203	137	bushels/acre
x_2	45	60	bushels/acre
x_3	215	155	kg/hectare
x_4	30	5	kg/hectare
x_5	30,661,542	20,441,028	-
x_6	1,107,555	738,370	-
x_7	474,616	316,411	-
x_8	4,205,037	2,803,358	-
x_9	30,728,227	20,485,485	-

2.4 Parameter Sampling

The Latin hypercube sampling technique (LHS) [12] with uniform distributions
is used for the sample generation. The input parameter ranges for the agriculture system, and animal agriculture systems are given in Table 4. The corn and
soybean yield ranges are obtained from USDA report [18] and are based on
the maximum and minimum yield recorded during the 2008–2019 period. The
commercial nitrogen application rate for corn is obtained from the Iowa State
University extension guidelines [14] considering the average nitrogen rates for
corn following corn and corn following soybeans in Iowa. The commercial nitrogen rate of soybean are chosen based on the fertilizer use, and price data between
2008–2018 [17]. The acreage for corn and soybean is 13.5 and 9.2 million acres,
respectively, according to USDA 2019 statistics report [18]. The animal population headcount required for animal agriculture input parameters is acquired using
Iowa animal population data [6]. The lower bound shows the animal population
data from the year 2012 (see Table 4), whereas the upper bound is determined
by assuming a 50% increase from the lower bound for each livestock group.

3 Results

Figure 3 shows the sampling history of the Sobol' first-order and total-order
indices, indicating convergence with 10^5 samples. It should be noted that the
negative values of estimated Sobol' indices are due to numerical error and usually
occur when indices magnitudes are close to zero [13].

    The Sobol' computation is repeated 30 times to provide a statistical mean
and standard deviation for the Sobol' indices. Figure 4 shows the averaged Sobol'
indices of the nitrogen export model. The most influential parameters are corn
commercial nitrogen rate (x_3) and corn yield (x_1) based on their total-order Sobol'
index magnitudes. These two parameters chiefly affect the variation in nitrogen
surplus amount of soil. The soybean commercial nitrogen rate (x_4) and soybean

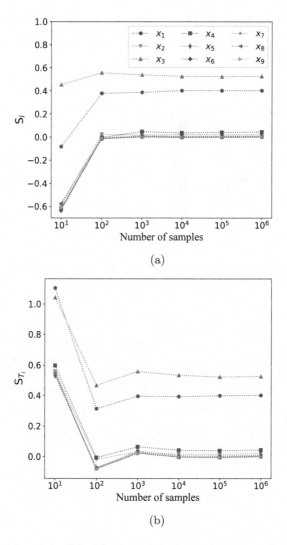

Fig. 3. Convergence of the Sobol' indices of the input parameters (cf. Table 1) in the nitrogen export model: (a) first-order indices, and (b) total-effect indices.

yield (x_2) slightly affects the nitrogen surplus amount while parameters (x_{5-9}) connected to animal agriculture has a negligible effect on soil nitrogen surplus amount. Further, it is observed that the interaction effects among input parameters are nonexistent. This is mainly due to the current limitation of the nitrogen export model, which employs only feed-forward design in its modeling approach.

The nitrogen export model is investigated to compute the average contribution of CN, FN, GN, and MN towards nitrogen surplus in soil. A total of 10^5 samples are used to compute average values. Figure 5 shows that CN, FN, and MN cumulatively contributes 214 kg/ha of nitrogen to the soil, whereas 171.8 kg/ha of

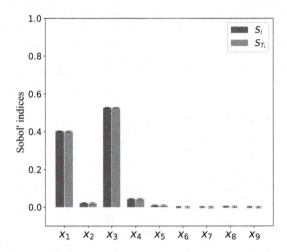

Fig. 4. Converged averages and standard deviations of the Sobol' indices for the input parameters (cf. Table 1) of the nitrogen export model.

nitrogen is removed from soil through harvested grains (corn and soybean), leaving on an average 42.2 kg/ha of nitrogen surplus in soil. Figure 5 also shows the contribution of corn and soybean fields in CN and GN computation. Additionally, MN contributes an average of 20.8 kg/ha of nitrogen to the soil, which is almost 9.7% of total nitrogen input to the soil ($CN + GN + MN$). The higher nitrogen surplus in soil subjected to high water flux could increase the rate of exported nitrogen to the Mississippi river basin. The current nitrogen export model could be used to reduce nitrogen surplus in the soil to mitigate nitrogen export.

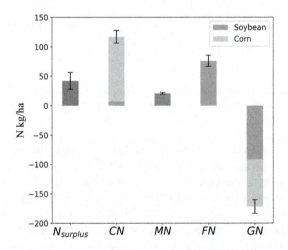

Fig. 5. Average contribution of commercial nitrogen (CN), fixation nitrogen (FN), and manure nitrogen (MN) towards the soil nitrogen surplus (N_s).

4 Conclusion

In this study, an Iowa food-energy-water (IFEW) systems model is proposed for understanding its interrelationship and to provide critical information for policy making to mitigate the environmental and economic impacts of the nitrogen export from Iowa. In particular, a macro-scale nitrogen export model of the agriculture and animal agriculture system is developed and a global sensitivity analysis is conducted to understand its influential parameters. It is observed that commercial nitrogen application rate for corn and corn yield are the two most influential parameters affecting soil nitrogen surplus. The parameters connected to soybean production and animal agriculture have a minimal effect on soil surplus; however, these parameters substantially contribute to soil nitrogen input and output. In future work, the IFEW system model will be further developed to include the water, weather, and energy systems to simulate different scenarios such as drought, flooding, or increased ethanol demand to regulate nitrogen export from Iowa.

Acknowledgements. This material is based upon work supported by the U.S. National Science Foundation under grant no. 1739551.

References

1. Bakhsh, A., Kanwar, R.S., Karlen, D.: Effects of liquid swine manure applications on no3-n leaching losses to subsurface drainage water from loamy soils in iowa. Agricult. Ecosyst. Environ. **109**(1–2), 118–128 (2005)
2. Barry, D., Goorahoo, D., Goss, M.J.: Estimation of nitrate concentrations in groundwater using a whole farm nitrogen budget. Technical report, Wiley Online Library (1993)
3. Blesh, J., Drinkwater, L.: The impact of nitrogen source and crop rotation on nitrogen mass balances in the mississippi river basin. Ecol. Appl. **23**(5), 1017–1035 (2013)
4. Burkart, M.R., James, D.E.: Agricultural-nitrogen contributions to hypoxia in the gulf of mexico. J. Environ. Qual. **28**(3), 850–859 (1999)
5. Donner, S.D., Kucharik, C.J.: Corn-based ethanol production compromises goal of reducing nitrogen export by the mississippi river. Proc. Natl. Acad. Sci. **105**(11), 4513–4518 (2008)
6. Gronberg, J.M., Arnold, T.L.: County-level estimates of nitrogen and phosphorus from animal manure for the conterminous united states, 2007 and 2012. Technical report, US Geological Survey (2017)
7. Jones, C.S., Drake, C.W., Hruby, C.E., Schilling, K.E., Wolter, C.F.: Livestock manure driving stream nitrate. Ambio **48**(10), 1143–1153 (2018). https://doi.org/10.1007/s13280-018-1137-5
8. Jones, C.S., Schilling, K.E.: Iowa statewide stream nitrate loading: 2017–2018 update. J. Iowa Acad. Sci. **126**(1), 6–12 (2019)
9. Lambe, A.B., Martins, J.R.: Extensions to the design structure matrix for the description of multidisciplinary design, analysis, and optimization processes. Struct. Multidiscip. Optim. **46**(2), 273–284 (2012)

10. Lu, C., et al.: Increased extreme precipitation challenges nitrogen load management to the gulf of Mexico. Commun. Earth Environ. **1**(1), 1–10 (2020)
11. Martins, J.R., Lambe, A.B.: Multidisciplinary design optimization: a survey of architectures. AIAA J. **51**(9), 2049–2075 (2013)
12. McKay, M.D., Beckman, R.J., Conover, W.J.: A comparison of three methods for selecting values of input variables in the analysis of output from a computer code. Technometrics **42**(1), 55–61 (2000)
13. Saltelli, A., et al.: Global Sensitivity Analysis: The Primer. Wiley, Chichester (2008). ISBN 978&0&470&05997&S
14. Sawyer, J.E.: Nitrogen use in Iowa corn production. Extension and Outreach Publications **107** (2015)
15. Sobol, I.M.: Global sensitivity indices for nonlinear mathematical models and their Monte Carlo estimates. Math. Comput. Simul. **55**(1–3), 271–280 (2001)
16. Urbanchuk, J.M.: Contribution of the renewable fuels industry to the economy of Iowa. Agricultural and Biofuels Consulting (2016)
17. USDA: Fertilizer use and price (2019). https://www.ers.usda.gov/data-products/fertilizer-use-and-price/
18. USDA: National agricultural statistics service quick stats (2020). https://quickstats.nass.usda.gov/
19. Westcott, P.C., Jewison, M.: Weather effects on expected corn and soybean yields. Technical report, USDA Economic Research Service, Washington DC (2013)
20. Zhang, X., Vesselinov, V.V.: Integrated modeling approach for optimal management of water, energy and food security nexus. Adv. Water Resour. **101**, 1–10 (2017)

An Attempt to Replace System Dynamics with Discrete Rate Modeling in Demographic Simulations

Jacek Zabawa$^{(\boxtimes)}$ ⓘ and Bożena Mielczarek ⓘ

Department of Operations Research and Business Intelligence, Wrocław University of Science and Technology, 50-370 Wrocław, Poland
{jacek.zabawa,bozena.mielczarek}@pwr.edu.pl

Abstract. The usefulness of simulation in demographic research has been repeatedly confirmed in the literature. The most common simulation approach to model population trends is system dynamic (SD). Difficulties in a reliable mapping of population changes with SD approach have been however reported by some authors. Another simulation approach, i.e. discrete rate modeling (DRM), had not yet been used in population dynamics modelling, despite examples of this approach being used in the modelling of processes with similar internal dynamics. The purpose of our research is to verify if DRM can compete with the SD approach in terms of accuracy in simulating population changes and the complexity of the model. The theoretical part of the work describes the principles of the DRM approach and provides an overview of the applications of the DRM approach versus other simulation methods. The experimental part permits the conclusion that DRM approach does not match the SD in terms of comprehensive accuracy in mapping the behavior of cohorts of the complex populations. We have been however able to identify criteria for population segmentation that may lead to better results of DRM simulation against SD.

Keyword: Computer simulation · Population modeling · Decision support · Cross-sectional analysis · Operations research · Healthcare demand modeling

1 Introduction

Reliable demographic projections are often a key and essential part of research on a wide range of socio-economic issues, such as forecasting health needs [1], designs of pension schemes [15], shaping of urban plans [14], natural interest rate forecasting [23] and others. Population analysis can be conducted, among others, using simulation methods, from which system dynamics approach (SD) is particularly often selected. This choice is justified by the specific characteristics of the SD, namely: causal loops, a macro-perspective focused on the entire population rather than on individual members of the population, and the form of mathematical relationships that facilitate the mapping of birth, death and migration processes. The usefulness of SD in demographic research has been repeatedly confirmed in the literature, starting with Forrester's work in the late

© Springer Nature Switzerland AG 2021
M. Paszynski et al. (Eds.): ICCS 2021, LNCS 12745, pp. 269–283, 2021.
https://doi.org/10.1007/978-3-030-77970-2_21

1960s and early 1970s [8]. However, it is reasonable to note there are some difficulties in a reliable mapping of population changes with this approach [16]. Eberlein and Thompson [6] discussed the so-called cohort blending problem of distortions in age structure: people entering a cohort at the same time "grow up" into the next cohort at different times. To overcome this problem, authors proposed the solution which assumes that the range of age-gender cohorts coincides with the simulation step. The disadvantage of this solution is the need to over-detail the model and to provide the model with data relating to one-year age groups. In our previous research [18] we proposed to modify the approach described in [6]. We suggested the use of so-called "cohort modeling", based simultaneously on complex (multi-annual) and elementary (annual) cohorts, in which the initial state and the intensity of the input and output stream can be taken individually. In the specific case, the parameters of the elementary cohorts that make up the compound cohort may be identical.

An analysis of the literature showed that another simulation approach, i.e. discrete rate modeling (DRM), had not yet been used in population dynamics modelling, despite examples of this approach being used in the modelling of processes with similar internal dynamics of change, cf. [5, 24] The DRM literature emphasizes the advantages of this approach, such as the quick reaction to changes in state variables, an avoidance of rounding and a speed of calculation. The purpose of our research is to verify if DRM can compete with the SD approach in terms of accuracy in simulating population changes and the complexity of the model measured by the number of blocks used. To ensure an unbiased comparison between the two approaches we extended the SD model, developed during our previous studies, with an additional module of an elementary cohort according to DRM approach. Therefore, we do not compare the time needed to get results using individual approaches.

2 Rationale for the Study

Our previous research focused on the application of a simulation approach to study healthcare services [17–20]. We used the SD approach in demographic modeling and studied the impact of demographic changes on the volume of demand for health services. As part of this research, we developed a model [18] that map 36 aggregate gender-age cohorts. Each cohort includes 5-year-old group of people, except for the oldest female and male cohort. The oldest cohort, i.e. 85+ is made up of 20-year-old group. Each aggregated cohort contains a number of (5 or 20) annual elementary cohorts. Thanks to the "cohort modeling" approach, we eliminated the so-called cohort blending problem described in [6]. The developed model has passed verification, but its disadvantage is the high complexity forced by the need to simultaneously control elementary and aggregate cohorts. Simplifying a model can be done in many ways, such as introducing blocks with a more versatile application, applying blocks tailored to a specific problem, or using a different simulation method. Our proposal is to apply the DRM approach in place of the SD approach inside of the hierarchical blocks. We will conduct all the research in the Extendsim environment.

Initial attempts to study the usefulness of the DRM have already been made by one of the authors. A comparison of DEVS (discrete event simulation) and DRM approaches was presented in the work [29]. It has been proved that DRM far exceeds DEVS performance, especially when dealing with high flow rate of objects. Moreover, duration of the simulation run in DRM approach is independent of source streams intensity and initial resource values.

3   The Concept of DRM Approach

3.1   Original Purpose of DRM

DRM is the approach for modelling the flow of uniform materials between different points. The flow is described by the actual flow rate (allowable and current one) and the initial and current state of the tanks with which the particular stream interacts. DRM can also be used to model systems whose dynamics are continuous or hybrid, i.e. continuous-discrete. The DRM approach can also be combined into a single, but compound, model with a discrete approach. For instance, discrete part may relate to road transport modelling and the DRM part may be related to the loading/unloading of bulk, liquid or even gaseous materials, i.e. when mapping processes of various and mixed materials, such as the production of foods [2, 24].

3.2   Principles of Calculations of DRM Simulation

The key concepts in the DRM approach are *tank* and *stream* (valve). The flow rate that fills or empties the tank is described by two values: the maximum (acceptable) flow rate in a given stream, and the actual (efficient) flow. The actual flow intensity depends on external conditions. For the incoming stream it depends on the availability of the material in the source tank and the capacity of the target tank. For the outcoming stream, the actual flow intensity depends on the availability of the flowing material in the source tank and the available capacity of the target tank. The above rule is called flow maximization [5] rule. The DRM approach implemented in Extendsim already provides possibilities for modeling discrete events related to the initiation or termination of a continuous process. This is the case when, for example, the vehicle gradually fills up the entire cargo space and it can drive away with the load. It is also possible to mix materials from different sources, when many source streams merge into a single output stream. In the output stream, it can be detected whether the mixture actually flows efficiently in it, or whether the intensity of one of the flows is zero. In conclusion, changes in the flow rate and the condition of the tank are mutually determined.

The uniqueness of the DRM [13] is also the ability to use SD concepts in DEVS environment. This is especially useful for modeling fast-changeable or large-scale processes that highlight flows, flow rate, events, constraints, storage capacity and routing. The bigger accuracy of simulations in the DRM approach is due to avoiding rounding errors that occur when time clocks are incompatible between continuous models and the discrete nature of events. Similarity to SD is due to the capturing of the reality by flow categories. The difference is that in SD we have fixed intervals between calculations and in DRM calculations occur when events occur. Similarity to DEVS is that model state changes occur only at discrete points over time. However, in DRM, calculations are performed only when the intensity of one of the flows changes. According to [13] it is possible to predict the occurrence of an event related to, for example, a system's state condition and reduce the number of calculations and the total error.

4 Literature Background

4.1 DRM Concept in Research

One of the early versions of the DRM library was called SD Industry. It was used by Siprelle and Phelps [24]. Authors pointed out that the DRM approach can be used in the food and pharmaceutical industries, when mass, continuous or semi-continuous materials are flowing at a fast rate. The alternative (DEVS) approach is not well suited for mapping these processes, as it focuses on individual parts or pieces of the flowing substance. Damiron and Nastasi [5] proved that in a situation where the main role is played by flows of things or fluids it is not appropriate to apply the DEVS or SD, and DRM is recommended. Additional advantages of the DRM approach include the ability to control the flow direction by prioritizing it and avoiding potential synchronization problems related to the creation of hybrid models that include discrete elements.

The literature contains studies comparing three basic simulation approaches, i.e. DEVS, SD and DRM. Muravjovs et al. [21] claimed that DRM resembles SD in terms of intensity and the usage of quantity/time units but the incorporation of randomness into the model is similar to DEVS. They also showed that results of simulation in the SD approach have a sharp step change, while the simulation results in the DEVS approach are gradual with minor changes of a step-by-step nature. The DRM approach produces more accurate results, represented by a fixed slope chart, without changes of a step-by-step nature. Similarly, Terlunen et al. [27] indicate that the DRM approach combines the key ideas and advantages of both DES and SD. They attempted to use DRM in supply chain modeling and identified two main problems. One problem is the need to perform conceptual abstractions to link supply chain elements and model elements. The second problem arises, because in the DRM approach an element of model can represent the flow of only one material type while real systems produce and transmit many different products and materials.

4.2 Demographic Simulations

We reviewed the literature on the use of simulation models in analysis of demographic processes. Our goal was to identify the target area of the demographic model and the type of simulation approach. In all models considered below, the following elements occur: resources, incoming streams, outgoing streams, initial states, and parameters. These are fundamental elements of any SD approach however implemented through different terms and technologies. Each of the following research problems does not solely concern the demographic model. The results of demographic simulations are used in target models, representing phenomena in areas such as finance, healthcare, construction and others.

Van Sonsbeek [25] presented a simulation model to analyze the impact of an ageing population on the burden of the pension system. He pointed out some shortcomings of the presented model, such as: too few age-gender cohorts, failure to consider different models of households and types of family life, inclusion in the model only state pensions. McGrattan et al. [15] studied risks to the pension system, healthcare and long-term care expenditure arising from the decline in the population and the lowered rate of working per pensioner in Japan. Because expenditures per person increase with age the authors stressed the need to take into account the heterogeneity of different cohorts. Giesecke and Meagher [9] studied the professional activity of individual social groups in Australia. They used a simulation model to predict the impact of population ageing on skills shortages and consumption levels. Han [10] used simulation model to predict the natural rate of interest that supports the economy at full employment while keeping an inflation at constant level, considering the ageing of Japan's population in current and future years. The natural interest rate in the light of demographic changes in the euro area was also addressed in [23]. In this model the overlapping generations were considered and the most important reason for the changes was the number of people at the working-age evaluated according to the age-dependent productivity over the number of people in the entire population. Colacelli and Corugedo [4] used demographic forecasts to study several scenarios for economic reform, such as monetary policy choice and debt ratio. Lauf et al. [14] built a city population model to study housing demand patterns. Feeding the model with data from demographic forecasts is also crucial in predicting future healthcare needs. Kingston et al. [12] predict that in the coming decades there will be a decrease in the need for assistance among younger older adults. Simulation was used in a number of studies to forecast the demand for long-term care [1, 7, 28]. Best et al. [3] based on death certificates, developed a model to analyze all-cause premature mortality and the commonest causes of premature death among different population groups. However, as indicated by the cancer incidence data provided by [26], morbidity rates for individual age cohorts may vary over time. Satyabudhi and Onggo [22] used a discrete parallel simulation approach in demographic research. They pointed out that it is more important to track the structure of the population than the size of the entire population.

In our opinion, in most of the examples discussed, the DRM approach can be used while retaining all the functionalities of models. We will use a multi-cohort model of regional population which was developed in accordance with the SD approach during our previous study [18]. We will suggest to replace the SD approach with a DRM approach. The use of the DRM approach should hypothetically allow for a more cross-cutting analysis of cohorts related to the size of residence, family situation, level of education or state of health. In models dealing with the impact of demographic change on the need for health services, the DRM approach should allow for the inclusion of healthcare facilities locations, thresholds for care qualifications, health status and etc.

5 Methods

5.1 Population Model

The reference population model describes two genders separately. Each age group is therefore represented by two cohorts: male and female. The model is based on blocks representing 1-year populations, equipped with an interface for growing from the pre-age group (if any) and up to the following age group (if any), taking into account births in age group 0, deaths in all groups, and migrations. The model provides an option of using two schemes to determine the initial values of the 1-year population status within aggregate cohorts. These are: assignment of uniform initial states and descending assignment according to triangular distribution. The selection of the scheme is carried out after the execution of test simulations. A total of 210 one-year-old cohorts, 34 five-year-old cohorts and two 20-year-old cohorts were used in the model.

The model requires the following data:

- initial states for all 5-year cohorts and 20-year cohort,
- annual migration volumes (in absolute terms) for all higher-level cohorts.
- the birth rate and the death rate, which is reported in proportion to the population status at the beginning of the year for all higher-level cohorts.

5.2 The Plan of Experiments

The research problem that we want to consider is as follows: will the replacement of the SD approach by DRM be efficient for a complex model which requires a wide range of input parameters and contains hundreds of parameters?

Each simulation experiment requires a range of tests to be performed to check for specific combinations of numeric values representing run parameters of the model. For example, the modeler needs to specify a time step for continuous simulation [11] to reconcile the batch order of objects in a DEVS simulation, as well as to agree on the order of calculations when performing parallel calculations in both approaches. Therefore it was decided that hierarchical blocks representing age-gender cohorts will be checked for different combinations of numerical values: initial state, birth rate and deaths rate. Figure 1 presents the SD and/or DRM population model at higher levels of hierarchy and Fig. 2 presents DRM population model (at low level).

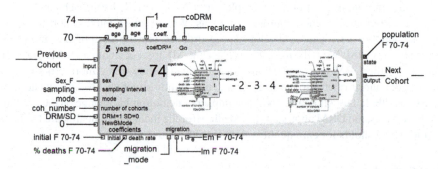

Fig. 1. The structure of one hierarchical (five-year) block representing a women's cohort aged 70–74. The block consists of five blocks (numbered 1 to 5) representing persons of a specific age range. It has 2 output connectors (cohort's population and rate of aging) and one main input connectors (rate of aging from younger cohort). Extendsim environment was used

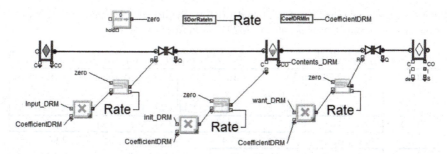

Fig. 2. Internal structure of a block representing persons of a specific age; fragment of DRM model. Extendsim environment was used

We want to verify if the use of the DRM approach will make it possible to obtain results comparable to the SD approach, taking into account the following criteria: absolute or relative deviations of simulation results from actual values and deviations of simulation results in the SD and DRM approach. Due to the models complexity, their modularity and the fact that observations take a matrix form (36 cohorts over many years), in part of our research we adopted extended criteria to consider the variation in deviations in each module. If the hypothesis cannot be considered true, we are interested in whether the DRM approach can be used to improve the SD approach, and when the DRM will be a better-rated approach than SD.

6 Results

6.1 Experiment 1: Comparing SD and DRM Results with Real Data

The first experiment was performed with a population model consisting of dozens of birth-population-death structures. This model is based on authors' previous study [18]. The model includes 36 main cohorts of the Wrocław, Poland region.

We take into account the size of the population by cohort: F0–4, F5–10,…, F85+ and M0–4, M5–10, …, M85+ from 2006 to 2019. We use the available actual data to verify the results of SD and DRM models. Calculations start in 2006. End-of-year simulation results are used to check the accuracy of calculations. The compatibility of the two models, i.e. the SD and DRM model, is compared using the X indicator (Formula 1).

$$X = \frac{real\ population - simulated\ population}{real\ population} \tag{1}$$

In the most favorable variant, the X will be equal to zero. A positive value of X occurs when the simulation result is less than the actual value. A negative value occurs when the simulation result is higher than the actual value. Because actual and simulated populations can only have positive values, the above claims are true. We start the experiment by performing simulations according to the SD (Table 1). Because there is no random factor in the model, a single run is sufficient to obtain simulation results.

The largest deviations occur for the M85+ cohort. This cohort is characterized by a small population size and diverse death rates. The number of deaths increases rapidly

Table 1. X-indicator values for the SD model (REAL-SD)/REAL (all values in percentages), separately for male and female population. MAD: Mean Absolute Deviation.

Year	M0-4	M5-9	M10-14	...	M80-84	M85+	F0-4	F5-9	F10-14	...	F80-84	F85+
2007	1.70	-2.21	-6.39	...	-1.84	-8.13	2.55	-2.50	-0.81		1.79	-6.04
...	...	...	...	...	...	...	...	...	...	...	...	...
2019	3.86	3.36	10.98		6.01	-22.54	3.41	3.11	12.15		5.72	-5.53
MAD	4.63	5.61	3.58		3.79	10.41	4.81	5.83	3.81		3.87	2.97

	Mean (Male)				3.4		Mean (Female)	2.96
Year	2007	2010	2013	2016	2019		Mean (All)	3.18
MAD	1.95	3.44	3.54	2.79	3.94			

as the age increases and the size of annual cohorts decreases, but these changes are not described by linear function. Moreover, there is a lack of the age structure data for the oldest cohorts. The relatively small mean deviation for the F85+ cohort can be explained by higher number of women than men of this age and perhaps an incidental but proper adjustment of the inner age structure of this cohort to the assumed variability in the model. Another explanation is that a uniform decrease in the size of one-year cohorts in both age-gender groups of the oldest population was assumed. Due to the lower life expectancy for men, probably after the age of 85 the number of individuals in this cohort decreases sharply, while in the F85+ cohort the number of women decreases gradually and is more evenly distributed over time.

We did a similar analysis for the DRM (Table 2) and we do not observe problems related to the M85+ cohort.

Table 2. The X values for the DRM model (REAL-DRM)/REAL (all values in percentages), separately for male and female population. MAD: Mean Absolute Deviation.

Year	M0-4	M5-9	...	M75-79	M80-84	M85+	F0-4	F5-9	F10-14	..	F40-44	..	F85+	MAD
2007	3.48	-3.09	...	3.05	2.89	10.71	4.24	-3.34	-4.07	...	-2.27	...	12.38	4.30
2008	7.4	-4.72	...	2.99	0.25	4.77	8.03	-4.53	-4.82	...	-3.95	...	6.14	4.40
2009	8.03	-4.53	...	2.49	-0.19	-0.48	8.29	-4.01	-6.23	...	-3.83	...	4.12	4.30
...	...	...	...	...	...	...	...	...	...	...	...	...	...	...
2017	2.12	6.13	...	0.59	1.43	4.42	2.40	6.64	8.52	...	0.52	...	5.86	3.52
2018	2.60	3.18	...	-2.52	5.81	-3.50	2.73	2.82	12.61	...	0.58	...	3.44	3.96
2019	1.27	2.47	...	-4.27	3.40	-1.17	0.80	2.23	13.45	...	0.07	...	3.64	3.93
MAD	4.34	6.26	...	2.44	2.47	3.78	4.45	6.42	5.67	...	2.12	...	6.31	

For almost all years the SD simulation results were above the actual values. In the DRM, simulation results are almost uniformly distributed relative to the course of actual values. In the DRM we can observe a higher mean deviation for the F85+ cohort than in SD, but it does not exceed 7%. We note that the mean deviation increases non-monotonically as the simulation time in the SD increases. In the DRM, the mean deviation is stable, however there are cohorts for which the deviation decreases non-monotonically over time - e.g. the F85+ and F10–14 cohorts.

In conclusion, the SD approach showed an advantage over DRM in terms of mean absolute deviation for male, female and both genders. However, there are cohorts for which the mean absolute deviation is lower in the DRM approach, e.g. M75–79, M80–84, M85+, F0–4, F40–44.

6.2 Experiment 2: Comparing SD and DRM According to the Lowest Absolute Deviation Criterion

Table 3 shows the results of the comparison between the SD and DRM models on the basis of absolute deviation criterion. In a small majority of cohorts (254 cells), the SD is a winning approach. The DRM approach resulted in better results in 214 cells. Note that for male cohorts, both approaches were found to be better exactly 117 times. For female cohorts, significantly better results are recorded for the SD approach. For each of the cohorts in the simulation, each approach won at least once.

Table 3. Comparison of simulation results according to DRM and SD approaches; The DRM or SD symbol indicates smaller relative differences between the simulation result for this approach and the actual data. For example, for the M0–4 cohort in 2019, the simulation result according to the DRM approach was more similar to historical data than the result for the SD approach.

Year	M0-4	M5-9	...	M60-64	...	M85+	F0-4	F5-9	...	F60-64	...	F85+	DRM	SD
2007	SD	SD	...	SD	...	SD	SD	SD	...	SD	...	SD	9	27
2008	SD	SD	...	SD	...	DRM	SD	SD	...	SD	...	SD	13	23
...	...	...	...	...	...	...	...	...	...	...	...	...	...	...
2018	DRM	DRM	...	DRM	...	DRM	DRM	DRM	...	DRM	...	DRM	18	18
2019	DRM	DRM	...	DRM	...	DRM	DRM	DRM	...	DRM	...	DRM	19	17
DRM	8	4	...	3	...	12	8	4	...	3	...	2		
SD	5	9	...	10	...	1	5	9	...	10	...	11		
DRM/SD	MALE		117/117				FEMALE		97/137					

6.3 Experiment 3: A Hybrid Approach

We combined the results (Table 4) of both approaches as follows: first we calculate average of the results for each cohort and then we use the weighted average formula (2):

$$Com = \alpha \times SD + (1 - \alpha) \times DRM \tag{2}$$

where *Com* means a common approach; α is a number between 0 and 1; *SD* is the average value for each cohort from SD model and *DRM* is the average value for each cohort from DRM model. We would like to find out which approach gives the closest real-world results in 13 simulated years for 36 cohorts: SD, DRM, or combined SD and

DRM results (Com). It turned out that in most cases the SD approach wins (206 out of 468), the second is the DRM approach (172 out of 468), while the combination of the two approaches won 90 times. However, the Common approach appeared to be a winner for at least one year for all cohorts.

Table 4. The comparison of DRM, SD and hybrid approach (Com). DRM, SD, or Com designation means the smallest relative differences between the simulation result of the given approach and actual data. Numbers mean how many times specified approach won. We chose $\alpha = 0.74$ because it provides the highest winning frequency by the common approach across the cohorts.

Year	M0-4	M5-9	...	M60-64	...	M85+	F0-4	F5-9	...	F60-64	...	F85+	DRM	SD	Com
2007	SD	SD	...	SD	...	Com	SD	SD	...	SD	...	Com	7	23	6
2008	SD	SD	...	SD	...	Com	SD	SD	...	SD	...	Com	10	19	7
...	...	...	...	...	...	...	...	...	...	...	...	...	...	...	...
2017	DRM	DRM	...	DRM	...	DRM	DRM	DRM	...	DRM	...	Com	13	11	12
2018	DRM	DRM	...	DRM	...	DRM	DRM	DRM	...	DRM	...	Com	14	15	7
2019	DRM	DRM	...	DRM	...	DRM	DRM	DRM	...	DRM	...	Com	15	13	8
DRM-M(F)	8	4	...	3	...	10	8	4	...	3	...	0			
SD-M(F)	5	9	...	10	...	0	5	9	...	10	...	7			
Com-M(F)	0	0	...	0	...	3	0	0	...	0	...	6			

MALE; DRM: 97/SD: 94/Com: 43   FEMALE; DRM: 75/SD: 112/Com: 47

6.4   Experiment 4: Analysis of Results by Gender

In experiments 1 to 3, we took into account each age-gender cohort separately. However, the different cohorts are not equinumerous. For example, the M85+ cohort had 3,361 individuals in 2006 and the F25–29 cohort had 53,727. Therefore we summed up the populations separately for each gender for each year of simulation and compared the deviations for aggregated cohorts (Table 5). It turns out that DRM almost always gives the best results and the combination of methods is never worse than SD.

The significant jump in the value of deviations in 2010 (Fig. 3) is generated by the significant increase in the population of M and F in reality. This deviation decreases over the years. Looking globally, these deviations are small and for the DRM the deviation never exceeds 1%.

Table 5. Population (in thousands) by gender; actual data, relative deviation values (in percentages) for SD, DRM, and COMMON approaches.

Year	SD M	SD F	DRM M	DRM F	Com M	Com F	REAL M	REAL F	DEVIATION SD. M/F		DRM.M/F		Com. M/F	
2007	559	610	559	611	559	611	559	611	0.04	0.13	-0.1	0.05	-0	0.09
2008	559	612	560	612	559	612	560	613	0.14	0.25	-0	0.13	0.07	0.19
...	...	...	...	...	...	...	...	...	...	...	...	...	...	...
2018	584	640	585	641	584	640	592	646	1.39	0.93	1.14	0.75	1.26	0.84
2019	588	643	589	645	589	644	596	650	1.37	0.99	1.14	0.8	1.26	0.89

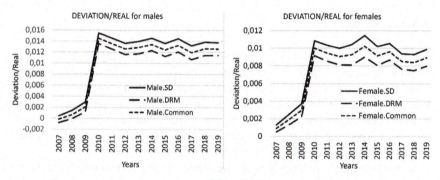

Fig. 3. Deviation graphs for SD, DRM, and COMMON relative to gender

6.5 Experiment 5: The Analysis of Results for Aggregated Age Cohorts

Another analysis is made for a population divided into two groups: youngers (0–59) and elders (60+), Fig. 4. It is obvious that 85+ cohorts are the least numerous of all cohorts and that their compliance with actual data is the most difficult to ensure. We see that for youngers by far the best is the DRM approach. However, when using other approaches, deviations are on the acceptable level equal to 1.6%. For the elders, the SD approach is the best choice and deviations do not exceed 5%. Omitting population 85+ results in a

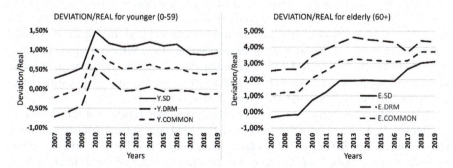

Fig. 4. Deviation graphs for SD, DRM, COMMON: population is divided relative to age

decreasing difference between the two approaches. If the 85+ population is taken into account, the advantage of the SD approach is stable.

6.6 Comparison of the Complexity of the Models

We define the complexity of two versions of the model by the additional number of blocks necessary for the proper operation of the elementary cohort. In the case of the DRM approach, we used 12 blocks. These are (Fig. 2): Tank (3), Valve (3), Multiply (3), Select Value In (3), Constant (1). In the case of the SD approach, we used 5 blocks: Holding Tank (2), Select Value In (2) and Constant (1). The presented data shows that we create a model with a simpler structure using the SD approach.

7 Conclusions

Continuous simulation modelling (SD) is widely used in demographic research. Population projections are, in turn, a key component of many socio-economic models created with discrete approaches. This leads to continuous-discrete hybrid models in which various technical and methodological problems arise. In industrial processes of material flows from one location to another, DRM method is used. This simulation approach combines certain aspects of a continuous and a discrete approach and outperforms DEVS for systems with high flow rates of objects in industrial processes.

The overall goal of our study was to investigate whether and under what conditions the SD approach could be replaced with DRM in demographic simulations, through the technique of modifying the elementary cohorts representing fragments of populations of the similar age and the same gender. The premises were as follows: the intent to test DRM hybridity in the sense of integration with modules implemented in accordance with the DEVS, literature reports on the usefulness of the DRM in non-socio-economic fields, the willingness to check the adequacy of demographic simulations using the DRM approach compared to results obtained from SD approach. Our goal was also to compare the complexity of models built using DRM and SD approaches.

The theoretical part of the work describes the principles of the DRM and provides an overview of the applications of the DRM versus two other simulation methods, i.e. SD and DEVS. The issue of simulation in demographic modeling is also discussed. For the research part, we developed a complex model consisting of dozens of feedback loops and 210 one-year-old cohorts, 34 five-year-old cohorts and two 20-year-old, and performed simulation experiments. We compared the results obtained from our simulation model with real data based on the example of the population of the Wrocław region in 2006–2019. We tested the SD, DRM and the combination of SD and DRM by performing the comparative analysis between all three approaches. We determined how many times (taking into account cohorts and years) specified approach won.

Taking into account the deviations between simulation results and actual data for individual cohorts, we come to the conclusion that DRM does not match the SD accuracy. The combination of SD and DRM has however in some cases shown an advantage over both components applied separately. If we consider the fact that the different cohorts differ significantly in number of individuals, we may conclude that DRM enables to achieve the accuracy level compared to SD. DRM approach does not match the SD in terms of comprehensive accuracy in mapping the behavior of cohorts of the complex populations. However, we have been able to identify criteria for population segmentation that may lead to better results of DRM simulation than SD. DRM results are more accurate compared to real data for younger populations (0–59 years) and for two genders. In some cases, it is also worth considering to use a hybrid method of weighted average of DRM and SD results.

The satisfactory results of our research entitle us to conclude that the research on using DRM in demographic simulations may help to overcome some of the limitations of the SD approach. One of them is the need to treat various factors affecting demographic dynamics as homogeneous. Since DRM enables mixing of input streams it is possible to model heterogeneous populations or subpopulations.

Acknowledgments. This project was financed by a grant from the National Science Centre Poland, titled *Simulation modeling of the demand for healthcare services.* It was awarded based on Decision 2015/17/B/HS4/00306.

ExtendSim blocks copyright © 1987–2021 Imagine That Inc. All rights reserved.

References

1. Ansah, J.P., Eberlein, R.L., Love, S.R., Bautista, M.A., et al.: Implications of long-term care capacity response policies for an aging population: a simulation analysis. Health Policy **116**(1), 105–113 (2014). https://doi.org/10.1016/j.healthpol.2014.01.006
2. Bechard, V., Cote, N.: Simulation of mixed discrete and continuous systems: an iron ore terminal example. In: Proceedings of the 2013 Winter Simulation Conference - Simulation: Making Decisions in a Complex World, WSC 2013, pp. 1167–1178 (2013)
3. Best, A.F., Haozous, E.A., Berrington de Gonzalez, A., Chernyavskiy, P., Freedman, N.D. et al.: Premature mortality projections in the USA through 2030: a modelling study. Lancet Public Heal. 3(8), e374–e384 (2018). https://doi.org/10.1016/S2468-2667(18)30114-2
4. Colacelli, M., Corugedo, E.F.: Macroeconomic effects of Japan's demographics: can structural reforms reverse them? IMF Working Paper (2018)
5. Damiron, C., Nastasi, A.: Discrete rate simulation using linear programming. In: Mason, S.J. et al. (eds.) Proceedings of the 40th Conference on Winter Simulation. pp. 740–749 Winter Simulation Conference (2008). https://doi.org/10.1109/WSC.2008.4736136
6. Eberlein, R.L., Thompson, J.P.: Precise modeling of aging populations. Syst. Dyn. Rev. **29**(2), 87–101 (2013). https://doi.org/10.1002/sdr.1497
7. Eggink, E., Ras, M., Woittiez, I.: Dutch long-term care use in an ageing population. J. Econ. Ageing. **9**, 63–70 (2017). https://doi.org/10.1016/j.jeoa.2016.08.001
8. Forrester, J.W.: Urban Dynamics. Productivity Press, Portland (1969)
9. Giesecke, J., Meagher, G.A.: Population ageing and structural adjustment. Aust. J. Labour Econ. **11**(3), 227–247 (2008)
10. Han, F.: Demographics and the Natural Rate of Interest in Japan. (2019). https://doi.org/10.5089/9781484396230.001
11. Keating, E.K.: Issues to Consider While Developing a System Dynamics Model. Technical report, Northwest. Univ. (1999)

12. Kingston, A., Comas-Herrera, A., Jagger, C.: Forecasting the care needs of the older population in England over the next 20 years: estimates from the Population Ageing and Care Simulation (PACSim) modelling study. Lancet Public Heal. **3**(9), e447–e455 (2018). https://doi.org/10.1016/S2468-2667(18)30118-X
13. Krahl, D.: ExtendSim advanced technology: discrete rate simulation. In: Rossetti, M.D., Hill, R.R., Johansson, B., Dunkin, A. R.G.I. (ed.) Proceedings - Winter Simulation Conference. pp. 333–338 IEEE (2009). https://doi.org/10.1109/WSC.2009.5429340
14. Lauf, S., Haase, D., Kleinschmit, B.: The effects of growth, shrinkage, population aging and preference shifts on urban development—a spatial scenario analysis of Berlin. Germany. Land Use Pol. **52**, 240–254 (2016). https://doi.org/10.1016/j.landusepol.2015.12.017
15. McGrattan, E., et al.: On financing retirement, health, and long-term care in Japan. IMF Work. Pap. **18**(249), 1–43 (2018). https://doi.org/10.5089/9781484384718.001
16. Meadows, D.L., et al.: Dynamics of Growth in a Finite World. Wright Allen Press, Cambridge (1974)
17. Mielczarek, B., Zabawa, J., Dobrowolski, W.: The impact of demographic trends on future hospital demand based on a hybrid simulation model. In: 2018 Winter Simulation Conference (WSC), pp. 1476–1487 (2018). https://doi.org/10.1109/WSC.2018.8632317
18. Mielczarek, B., Zabawa, J.: Modelling demographic changes using simulation: supportive analyses for socioeconomic studies. Socioecon. Plann. Sci. 100938 (2020). https://doi.org/10.1016/j.seps.2020.100938
19. Mielczarek, B., Zabawa, J.: Modelling population growth, shrinkage and aging using a hybrid simulation approach: application to healthcare. In: SIMULTECH 2016 - Proceedings of the 6th International Conference on Simulation and Modeling Methodologies, Technologies and Applications (2016)
20. Mielczarek, B., Zabawa, J.: Simulation model for studying impact of demographic, temporal, and geographic factors on hospital demand. In: Proceedings - Winter Simulation Conference, pp. 4498–4500 Institute of Electrical and Electronics Engineers Inc. (2017). https://doi.org/10.1109/WSC.2017.8248178
21. Muravjovs, A. et al.: The use of discrete rate simulation paradigm to build models of inventory control systems. In: Proceedings - 2nd International Symposium on Stochastic Models in Reliability Engineering, Life Science, and Operations Management, SMRLO 2016, pp. 650–655. IEEE (2016). https://doi.org/10.1109/SMRLO.2016.115
22. Onggo, B.S.S.: Parallel discrete-event simulation of population dynamics. In: Mason, S.J., Hill, R.R., Mönch, L., Rose, O., Jefferson, T., Fowler, J.W. (ed.) Proceedings - Winter Simulation Conference, Miami, FL, USA, pp. 1047–1054. IEEE Press, Piscataway (2008). https://doi.org/10.1109/WSC.2008.4736172
23. Papetti, A.: Demographics and the natural real interest rate: historical and projected paths for the euro area. ECB Work. Pap. 2258 (2019). https://doi.org/10.2866/865031
24. Siprelle, A.J., Phelps, R.A.: Simulation of bulk flow and high speed operations. In: Andradóttir, S., et al. (eds.) Winter Simulation Conference Proceedings, pp. 706–710, Atlanta, GE (1997). https://doi.org/10.1145/268437.268612
25. van Sonsbeek, J.M.: Micro simulations on the effects of ageing-related policy measures. Econ. Model. **27**(5), 968–979 (2010). https://doi.org/10.1016/j.econmod.2010.05.004
26. Sung, H., Siegel, R.L., Rosenberg, P.S., Jemal, A.: Emerging cancer trends among young adults in the USA: analysis of a population-based cancer registry. Lancet Public Heal. **4**(3), e137–e147 (2019). https://doi.org/10.1016/S2468-2667(18)30267-6
27. Terlunen, S,. et al.: Adaption of the discrete rate-based simulation paradigm for tactical supply chain decisions. In: A., T. et al. (eds.) Proceedings - Winter Simulation Conference, Savannah, GA, pp. 2060–2071. IEEE (2015). https://doi.org/10.1109/WSC.2014.7020051

28. Yu, W., Li, M., Ge, Y., Li, L. et al.: Transformation of potential medical demand in China: a system dynamics simulation model (2015). https://doi.org/10.1016/j.jbi.2015.08.015
29. Zabawa, J., Radosiński, E.: Comparison of discrete rate modeling and discrete event simulation. Methodological and performance aspects. In: Świątek, J., Wilimowska, Z., Borzemski, L., Grzech, A. (eds.) Information Systems Architecture and Technology: Proceedings of 37th International Conference on Information Systems Architecture and Technology—ISAT 2016—Part III. AISC, vol. 523, pp. 153–164. Springer, Cham (2017). https://doi.org/10.1007/978-3-319-46589-0_12

New On-line Algorithms for Modelling, Identification and Simulation of Dynamic Systems Using Modulating Functions and Non-asymptotic State Estimators: Case Study for a Chosen Physical Process

Witold Byrski[1][ID], Michał Drapała[1]([✉])[ID], and Jędrzej Byrski[2][ID]

[1] Department of Automatic Control and Robotics, AGH University of Science and Technology, Al. Mickiewicza 30, 30-059 Kraków, Poland
{wby,mdrapala}@agh.edu.pl
[2] Department of Applied Computer Science, AGH University of Science and Technology, Al. Mickiewicza 30, 30-059 Kraków, Poland
jbyrski@agh.edu.pl

Abstract. The paper presents an advanced application of computation methodology with complicated algorithms and calculation methods dedicated to optimal identification and simulation of dynamic processes. These models may have an unknown structure (the order of a differential equation) and unknown parameters. The presented methodology uses non-standard algorithms for identification of such continuous-time models that can represent linear and non-linear physical processes. Typical approaches, presented in the literature, most often utilize discrete-time models. However, for the case of continuous-time differential equation models, in which both, the parameters and the derivatives of the output variable are unknown, the solution is not easy. In the paper, for the solution of the identification task, the convolution transformation of the differential equation with a special Modulating Function will be used. Also, to be able to properly simulate the behaviour of the process based on the obtained model, the exact state integral observers with minimal norm will be used for the reconstruction of the exact value of the initial conditions (not their estimate). For multidimensional process case, with multiple control signals (many inputs), additional problems arise that make continuous identification and observation of the vector state (and hence simulation) impossible by the use of the standard methods. Application of the above-mentioned methods for solving this problem will be also presented. Both algorithms, for the parameter identification and the state observation, will be implemented on-line in two independent but cooperating windows that will simultaneously move along the time axis. The presented algorithms will be tested using data collected during the heat exchange process in an industrial glass melting installation.

Keywords: Complex algorithms · Multidimensional systems · Process identification · Modulating functions · State observers · Glass forehearth

© Springer Nature Switzerland AG 2021
M. Paszynski et al. (Eds.): ICCS 2021, LNCS 12745, pp. 284–297, 2021.
https://doi.org/10.1007/978-3-030-77970-2_22

1   Introduction

Many industrial processes can be locally approximated by linear models, de-scribed by ordinary differential equations. The parameters of these models are most often obtained based on performed identification experiments and because of process non-linearity, can be utilized only in a neighbourhood of specified operating points. However, performing active identification experiments is often impossible due to technological reasons, e.g. in many installations step changes of control signals, during identification experiments, could significantly deteriorate product quality. What is more, complete information about the process is not always available.

Passive identification methodology was widely discussed in the literature, especially for chemical processes, e.g. [1]. On-line implementation of such al-gorithms can be found in [2]. However, in most cases, discrete models are utilized, e.g. [3]. For the qualitative analysis of processes with distributed parameters, it is common to use the Computational Fluid Dynamics (CFD) approach. It enables precise simulation of glass melting phenomena, however requires a lot of compu-ting power. Hence, in works [4] and [5] procedures of model reduction based on the Proper Orthogonal Decomposition (POD) method are presented. Another, simpler approach, also using Partial Differential Equations (PDE), utilizes the Heat Transfer Equation modelling the glass conditioning process in forehearths, e.g. [6,7]. In engineering, linear time invariant (LTI) continuous models with lumped parameters are most popular. However, after reduction, in many cases the multidimensionality of such models should be taken into account, i.e. the fact that the process output depends on many inputs (many control signals) - Multi Input-Single Output model (MISO).

In some processes operating points are changed very often, that makes the necessity of changing the process model, which has to be once more re-identified. Hence, establishing of a passive identification algorithm for MISO models, that would be fast and accurate, as well as universal, is an important research goal. This research topic was considered in the previous works of the authors. In the paper [8] the special approach for the identification of MISO models was proposed by the assumption of the existence of several separate, internal, low order SISO sub-models, whose local outputs are unknown (only the main output is measured). The special iteration procedures for identification of each sub-model gave good results, but the algorithm occurred to be a bit complicated and computation time consuming.

In this paper, a new methodology for the identification of MISO models, with-out local sub-models, is presented. However, the high order of the main model has to be assumed. Identification of the parameters of high-order continuous dif-ferential equation is not easy, because only the output $y(t)$ is measured and the values of derivatives $y^{(i)}(t)$ are unknown. To solve this problem, an application of non-standard methods for parameter identification is proposed. A differential equation is transformed by its convolution with modulating functions.

Modulating Function Method (MFM) is the only identification method that leads to the optimal identification of parameters, without introducing any estimation or approximations at any stage of the differential equation transformation and during optimization calculations. Other methods used in the continuous system identification, e.g. by approximation of the input/output functions via orthogonal polynomials [9] introduce immediately methodological errors by definition. Additionally, the exact initial state observation is utilized to enable the model identification and hence the accurate simulation of the process output. The algorithm has been extended with the procedure of properly selecting new operating points, which gave very good modelling results and prediction of the process output values. The presented algorithms was tested using data collected during the glass conditioning process. The process takes place in a long channel, called glass forehearth, which is the final part of the glass melting installation.

The paper is organised as follows. In Sects. 2 and 3 theoretical basis of the Modulating Functions Method (MFM) and the Exact Integral State Observers are explained. Developed identification procedure is described in Sect. 4. Chosen industrial process of glass conditioning is briefly described in Sect. 5. Obtained simulation results are presented in Sect. 6. Section 7 draws conclusion.

2   MISO Model Optimal Identification

Linear Tine Invariant (LTI) Multi Input Single Output (MISO) system with K inputs is given as (1):

$$\sum_{i=0}^{n} a_i y^{(i)}(t) = \sum_{k=1}^{K}\sum_{j=0}^{m_k} b_{kj} u_k^{(j)}(t) = \sum_{j=0}^{m_1} b_{1j} u_1^{(j)}(t) + \ldots + \sum_{j=0}^{m_K} b_{Kj} u_K^{(j)}(t). \quad (1)$$

Functions $y^{(i)}$, $u_1^{(j)},\ldots,u_K^{(j)}$ are the inputs and output derivatives given on the interval $[t_0, T_{ID}]$. There are n output derivatives and m_k derivatives for the k-th input, where $m_k \leq n$, $\forall k = 1,\ldots,K$. Parameters a and b should be identified. The inputs u and the output y can be measured. A deep discussion about continuous-time systems identification can be found in the paper [10] and for MISO systems in [11].

2.1   Modulating Functions Method

Modulating Functions Method (MFM) was developed by M. Shinbrot [12]. Theoretical fundamentals of the method can be found in [13]. It utilizes the rule of integrating by parts. Left and right hand sides of (1) are convoluted with the known modulating function ϕ. This function should meet specified conditions:

- ϕ is supposed to have a compact support of width h (closed and bounded),
- $\phi \in C^{n-1}[0,h]$,
- $\phi^{(i)}(0) = \phi^{(i)}(h) = 0$ for $i = 0,\ldots,n-1$,
- $y * \phi = 0 \Rightarrow y = 0$ on the interval $[t_0 + h, T_{ID}]$.

In the described method, the Loeb and Cahen functions:

$$\phi(t) = t^N(h-t)^M \tag{2}$$

were used. Utilizing the properties:

$$y_i(t) = a_i \int_{-\infty}^{\infty} y^{(i)}(\tau)\phi(t-\tau)d\tau = a_i \int_0^h y(t-\tau)\phi^{(i)}(\tau)d\tau \overset{\text{def}}{=} a_i y_i(t), \tag{3}$$

$$u_{kj}(t) = b_j \int_{-\infty}^{\infty} u_k^{(j)}(\tau)\phi(t-\tau)d\tau = b_j \int_0^h u_k(t-\tau)\phi^{(j)}(\tau)d\tau \overset{\text{def}}{=} b_j u_{kj}(t), \tag{4}$$

the set of new known functions $y_i(t)$, $u_j(t)$ in the interval $[t_0+h, T_{ID}]$ is obtained. These functions should be stored in a computer memory. Then, the differential Eq. (1) can be transformed into the algebraic (5) with the same parameters:

$$\sum_{i=0}^{n} a_i y_i(t) = \sum_{j=0}^{m_1} b_{1j} u_{1j}(t) + \ldots + \sum_{j=0}^{m_K} b_{Kj} u_{Kj}(t) + \epsilon(t). \tag{5}$$

The term ϵ represents a difference between two sides of the equation. It can be treated as an identification performance index (6) in the Equation Error Method (EEM) for identification of the parameters in the Eqs. (1) and (5):

$$\epsilon(t) = \mathbf{c}^T(t)\boldsymbol{\theta} = [y_0(t), \ldots, y_n(t), -u_{10}(t), \ldots, -u_{1m_1}(t), \ldots, -u_{K0}(t),$$

$$\ldots, -u_{Km_K}(t)] \begin{bmatrix} \mathbf{a} \\ \mathbf{b}_1 \\ \vdots \\ \mathbf{b}_K \end{bmatrix}, \tag{6}$$

where $\mathbf{a}, \mathbf{b}_1, \ldots, \mathbf{b}_K$ are the column vectors of suitable dimensions $n+1, m_1+1, \ldots, m_K+1$, $\boldsymbol{\theta} \in R^{n+m_1+\ldots+m_K+K+1}$.

Minimization problem is typically solved using the least squares method. In [14] a different approach was proposed. Minimization problem is stated in the function space $L^2[t_0+h, T_{ID}]$ as:

$$\min_{\boldsymbol{\theta}} J^2 = \min \|\epsilon(t)\|_{L^2[t_0+h,T]}^2 = \min \|\mathbf{c}(t)^T\boldsymbol{\theta}\|_{L^2}^2. \tag{7}$$

The linear constraint $\boldsymbol{\eta}^T\boldsymbol{\theta} = 1$ is introduced to avoid the trivial solution. The norm in (7) can be written down as an inner product in the space L^2:

$$J^2 = \langle \mathbf{c}^T(t)\boldsymbol{\theta}, \mathbf{c}^T(t)\boldsymbol{\theta}\rangle_{L^2} = \boldsymbol{\theta}^T\langle \mathbf{c}(t), \mathbf{c}^T(t)\rangle\boldsymbol{\theta} = \boldsymbol{\theta}^T\mathbf{G}\boldsymbol{\theta}. \tag{8}$$

The square real and symmetric Gram matrix $\mathbf{G}$ is given as:

$$\mathbf{G} = \begin{bmatrix} \mathbf{YY} & \mathbf{YU}_1 & \ldots & \mathbf{YU}_K \\ \mathbf{U}_1\mathbf{Y} & \mathbf{U}_1\mathbf{U}_1 & \ldots & \mathbf{U}_1\mathbf{U}_K \\ \vdots & \vdots & \ddots & \vdots \\ \mathbf{U}_K\mathbf{Y} & \mathbf{U}_K\mathbf{U}_1 & \ldots & \mathbf{U}_K\mathbf{U}_K \end{bmatrix}, \tag{9}$$

where:

$YY(i,j) = \langle y_i, y_j \rangle$ and $i = 1 \ldots n$, $j = 1 \ldots n$,

$YU_k(i,j) = -\langle y_i, u_{kj} \rangle$ and $k = 1 \ldots K$, $i = 1 \ldots n$, $j = 1 \ldots m_k$,

$U_kY(i,j) = -\langle u_{ki}, y_j \rangle$ and $k = 1 \ldots K$, $i = 1 \ldots m_k$, $j = 1 \ldots n$,

$U_kU_l(i,j) = \langle u_{ki}, u_{lj} \rangle$ and $k = 1 \ldots K$, $l = 1 \ldots K$, $i = 1 \ldots m_k$, $j = 1 \ldots m_l$.

The matrix G is created by the inner products in L^2 of the $c(t)$ elements, e.g.:

$$\langle y_i, u_j \rangle = \int_{t_0+h}^{T_{ID}} y_i(\tau) u_j(\tau) d\tau. \tag{10}$$

The optimal vector $\boldsymbol{\theta}$, that minimizes the value of J, can be obtained using the Lagrange multiplier technique:

$$\min_{\boldsymbol{\theta}} J^2 = \min_{\boldsymbol{\theta}} L = \min_{\boldsymbol{\theta}}(\boldsymbol{\theta}^T G \boldsymbol{\theta} + \lambda[\boldsymbol{\eta}^T \boldsymbol{\theta} - 1]) \tag{11}$$

as:

$$\boldsymbol{\theta}^0 = \frac{G^{-1}\boldsymbol{\eta}}{\boldsymbol{\eta}^T G^{-1} \boldsymbol{\eta}}. \tag{12}$$

3   Exact Integral State Observers with Minimal Norm

In dynamic system theory, for every linear system (1) which describes input-output dependences for a MISO system, one can find corresponding to (1) description of the system in the state space, with the state variable $x(t)$ and the output variable $y(t)$:

$$\begin{aligned} \dot{x}(t) &= Ax(t) + Bu(t), \quad x(t_0) = x_0 \\ y(t) &= Cx(t), \end{aligned} \tag{13}$$

where:

$$\forall t \geq t_0 : x(t) \in R^n, u(t) \in R^r, y(t) \in R^l,$$

and the corresponding state, control and observation real matrices are: $A_{n \times n}$, $B_{n \times r}$, $C_{l \times n}$ and consist of parameters a_i, b_i from (1) or (5).

In many cases the state vector $x(t)$ is not measured. Only the system output $y(t)$ can be measured. The calculation of $x(t)$ is not easy because the matrix C is not square and hence it is not invertible. The reconstruction of the state vector $x(t)$ for a chosen time t, e.g. $t = t_0$ is well known problem in the control theory. The knowledge of $x(t_0)$ is very important for proper simulation of the real value of the output variable $y(t)$ for the given matrices A, B, C. To this end, one can use so called state estimators given by differential equations similar to (1), like Kalman Filter or Luenberger observers.

In contrast to differential estimators, the exact state observers have the structure of two integrals and can exactly reconstruct the state of a linear system. The described integral observers guarantee obtaining the real value of the observed state for the observation interval T_{OB}. The theory of optimal observers with minimal norm was described in [15].

In the paper, two types of observers are utilized. The initial state observer (14) allows obtaining the initial conditions for the identified models when a new operating point is determined. For $t_0 = 0$, the below equation is given:

$$x(0) = \int_0^{T_{OB}} \overline{G_1}(t)y(t)dt + \int_0^{T_{OB}} \overline{G_2}(t)u(t)dt, \tag{14}$$

where:

$$M_0 = \int_0^{T_{OB}} e^{A'\tau} C' C e^{A\tau} d\tau,$$

$$\overline{G_1}(t) = M_0^{-1} e^{A't} C',$$

$$\overline{G_2}(t) = M_0^{-1} \left[\int_t^{T_{OB}} e^{A'\tau} C' C e^{A\tau} d\tau \right] e^{-At} B.$$

The final state observer, given as (15), is used for simulation of the model output in the subsequent simulation intervals:

$$x(T_{OB}) = \int_0^{T_{OB}} G_1(t)y(t)dt + \int_0^{T_{OB}} G_2(t)u(t)dt, \tag{15}$$

where:

$$G_1(t) = e^{A T_{OB}} M_0^{-1} e^{A't} C',$$

$$G_2(t) = e^{A T_{OB}} M_0^{-1} \left[\int_0^t e^{A'\tau} C' C e^{A\tau} d\tau \right] e^{-At} B.$$

It allows to obtain the state value $x(t_j)$ at the end of j-th interval of width T_{OB}. The equation for the system state value at the end of each interval for the moving window version is given as (16):

$$x(t_j) = \int_{t_j - T_{OB}}^{t_j} G_1(T_{OB} - t_j + t)y(t)dt$$

$$+ \int_{t_j - T_{OB}}^{t_j} G_2(T_{OB} - t_j + t)u(t)dt, \tag{16}$$

where the successive time moments are:

$$t_j = p + T - (p \ \ modulo \ \ T) + (j-1) \cdot T, \quad j = 1, 2, 3, \ldots$$

and p is the current operating point.

4   Adaptive Identification Method

As it was mentioned previously, the described adaptive methodology assumes that the non-linear system can be linearised near a selected operating point.

The operating points p are defined for time moments in which input and output signals are almost unchanging functions of time, which can be verified by checking if the signal variance is small enough in the defined interval. After determining that the state of the process is steady and the operating point p_1 can be defined, the initial model of the process is identified for the n_{start} intervals assuming zero initial condition.

In case of finding another operating points, the new model is identified and a squared difference between the simulated output of this model and the real system output for the last $n_{reident}$ intervals is calculated and compared with the previous difference, based on the last valid model. If the obtained difference is less than the previous one, the model is updated. If not, then the current version of the model is upheld. The new value of the initial state, needed for the simulation procedure, is obtained with the use of the previously presented initial state observer, whereas the final value of the state in the last interval, essential for the future system output simulation, is determined using the final state observer. It is additionally assumed, that the correlation between the system inputs and output should be greater than the threshold value tr_{corr} for at least one interval among those used in the identification procedure. It prevents from obtaining inaccurate linear models.

The first identification window is significantly longer than the window used for the re-identification procedure. It results from the fact that changes of the input and output signals are rather small in the first intervals and longer signals are needed to obtain a sufficiently accurate model. In the next steps, the identification window is shorter, which allows to obtain the identified model faster, and at the same time causes that the prediction follows the real system output. The developed algorithm is presented in details in the form of Algorithm 1. The marker *empty* in the description means that the first model is still before the identification procedure.

It is worth noting that the MFM does not require zero initial condition. This advantage of the method is utilized, when the system signals preceding the new operating point are used in the identification algorithm. The state observers are used only for simulation purposes when the performance index (17), defined as an integral of the squared difference between the real system output and the simulated model output, is calculated:

$$E(t_0, t_{end}) = \int_{t_0}^{t_{end}} (y(t) - y_{sim}(t))^2 dt. \tag{17}$$

The state-space matrices, whose elements were obtained with the MFM, used in the simulation procedures have the form (18):

$$A = \begin{bmatrix} 0 \ldots 0 & -\frac{a_0}{a_n} \\ 1 & \ddots & \vdots \\ & \ddots & 0 & -\frac{a_{n-2}}{a_n} \\ 0 \ldots 1 & -\frac{a_{n-1}}{a_n} \end{bmatrix}_{(n \times n)}, \quad B = \begin{bmatrix} \frac{b_{10}}{a_n} & \cdots & \frac{b_{K0}}{a_n} \\ \vdots & \vdots & \vdots \\ \frac{b_{1n-1}}{a_n} & \cdots & \frac{b_{Kn-1}}{a_n} \end{bmatrix}_{(n \times K)},$$

$$C = \begin{bmatrix} 0 \dots 1 \end{bmatrix}, \quad D = \begin{bmatrix} 0 \dots 0 \end{bmatrix}. \qquad (18)$$
$$\underset{(1 \times n)}{} \qquad \underset{(1 \times K)}{}$$

Algorithm 1. Identification and output simulation procedure.

Step 1. Set the current interval counter $j = 1$.

Step 2.

    if operating point was found in the last n_{start} intervals **and** *empty*(current model) **then**

        Go to **Step 3**.

    else if operating point was found in the last $n_{reident}$ intervals **and** not *empty*(current model) **then**

        Go to **Step 4**.

    else

        Go to **Step 5**.

    end if

Step 3.

    if $j \geq n_{start}$ **and** input-output correlation $\geq tr_{corr}$ **then**

        Perform the identification procedure assuming zero initial condition for the n_{start} intervals to obtain the initial model.

    end if

Go to **Step 5**.

Step 4.

    if $j \geq n_{reident}$ **and** input-output correlation $\geq tr_{corr}$ **then**

        Perform the identification procedure for the $n_{reident}$ intervals to obtain the new model.

    end if

Calculate the performance index $E_{current}$ for the current model.

Calculate the performance index values for the new models (defined for different parameters) obtained in the last $n_{reident}$ intervals using the initial state observer.

Select the least value for the obtained models $E_{reident}$.

    if $E_{reident} < E_{current}$ **then**

        Update the current model parameters and save the new operating point.

    end if

Go to **Step 5**.

Step 5.

    if not *empty*(current model) **then**

        Calculate the current state value using the final state observer.

        Perform the future output simulation.

    end if

Increment the interval counter j.

Go to **Step 2**.

---

5   Process Description

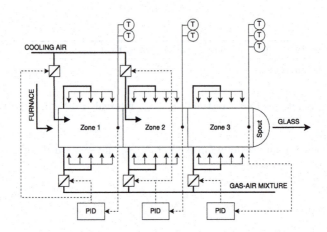

Fig. 1. Forehearth control system

The glass conditioning process involves phased in cooling of a molten glass smelt in order to obtain desired chemical and physical properties. The process is conducted in glass forehearths. Figure 1 presents an example glass forehearth installation. The forehearth is a long channel divided into several zones. In the first two of them, the glass can be cooled down or heated. The zone controllers adjust signals for cooling dampers and gas burners. In the last zone, there are only gas burners installed. Typically the desired temperature in each zone depends on the type of currently produced container and should be stabilised with an accuracy of at least 1 °C. Each forehearth zone is controlled regardless of neighbouring ones.

The on-line identification problem for this installation is associated with many difficulties. The lack of reliable information about the current glass pull rate (total weight of glass containers produced per day) seem to be the most problematic. In most glass factories this parameter is not measured. It can be evaluated only in steady states, when glass gobs fall into a forming machine and parameters of the machine are fixed. Transition between two steady states involves multiple glass pull rate changes, which causes disturbances visible as temperature fluctuations. In the following experiments, the dynamic model of the last zone will be identified. The linear MISO model have two inputs:

- temperature in the previous forehearth zone,
- gas-air mixture pressure,

so the corresponding state space matrix B has two columns. Both component models have the common matrix A, which follows directly from the algorithm presented in Subsect. 2.1.

6 Experimental Results

Simulation experiments were performed for two sets of data collected from the real glass forehearth installation. The glass pull rate and temperature set point values were noticeably changed for these sets.

The linear constraint vectors used in the MFM procedure have the suitable length depending on the rank of identified systems. The values of the parameters that were used during the experiments are presented in Table 1.

Table 1. Identification procedure coefficients.

Parameter	Description	Value
η	Linear constraint vector	[1...1]
T	Single interval width	250
n_{start}	Initial identification intervals	8
$n_{reident}$	Intervals for model re-identification	4
$T_{OB_{FIN}}$	Final state observer window width	500
$T_{OB_{INIT}}$	Initial state observer window width	1000
tr_{corr}	Input-output correlation threshold	0.5

The first experiment was performed for the glass pull rate changing from 82 t/24 h to 60 t/24 h. It caused that the model delay for the first input varied from 260 s to 319 s. In the second case, the glass pull rate was changed in the range 62 t/24 h to 55 t/24 h and the corresponding delay values fluctuated from 313 s to 337 s. The input signals are presented in Figs. 2 and 4 accordingly.

The experimental results are presented in Tables 3 and 5. Figures 3 and 5 show the predicted system output in both cases. Alternating blue and yellow backgrounds denotes the intervals in which subsequent models were applied for the system output prediction. The first green intervals concern the measurements needed for obtaining the initial model, when the system output could not be predicted. The same information are presented in Tables 2 and 4. Parameter *Ident. time* in Tables 2 and 4 concerns the time window used for the identification procedure, while the parameter *Sim. time* refers to the intervals when the identified model was used for the system output prediction. Subsequent operating points are depicted as dotted lines in Figs. 3 and 5.

Table 2. Model properties - 1. experiment

Model nr	Op. point	Ident. time [s]	Sim. time [s]	MFM parameters		
				N	M	$h[s]$
1	p_1	102–2000	2000–2250	3	4	50
2	p_2	1250–2250	2250–2500	3	4	50
3	p_2	1500–2500	2500–2750	3	4	50
4	p_2	1750–2750	2750–5750	3	4	100
5	p_3	4750–5750	5750–8000	3	4	100

Table 3. Identified model parameters - 1. experiment

Model nr	Parameters						
	a_0	a_1	a_2	b_{10}	b_{11}	b_{20}	b_{21}
1	62.86e−6	10.82e−3	1	141.61e−6	–	247.02e−6	–
2	92.77e−6	11.31e−3	1	119.82e−6	–	153.06e−6	–
3	469.26e−6	45.62e−3	1	694.11e−6	5.73e−3	1.52e−3	309.69e−6
4	321.73e−6	49.84e−3	1	337.46e−6	−3.54e−3	1.59e−3	9.01e−3
5	276.74e−6	44.73e−3	1	203.20e−6	16.20e−3	1.57e−3	11.11e−3

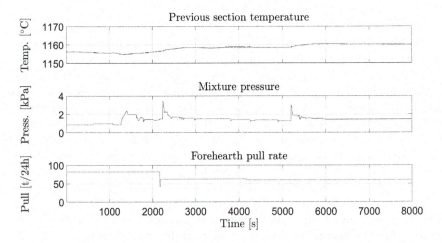

Fig. 2. System inputs and glass pull rate - 1. experiment

Table 4. Model properties - 2. experiment

Model nr	Op. point	Ident. time [s]	Sim. time [s]	MFM parameters		
				N	M	$h[s]$
1	p_1	102–2000	2000–4500	5	6	100
2	p_2	3500–4500	4500–4750	5	6	100
3	p_2	3750–4750	4750–9500	3	4	50

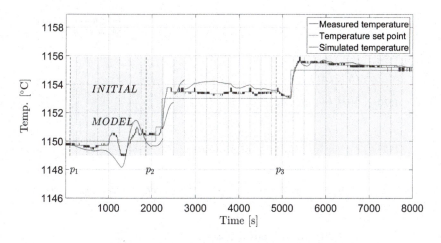

Fig. 3. Simulation results - 1. experiment

Table 5. Identified model parameters - 2. experiment

Model	Parameters							
nr	a_0	a_1	a_2	a_3	b_{10}	b_{11}	b_{20}	b_{21}
1	8.23e−6	2.6e−3	12.47e−3	1	141.61e−6	−	247.02e−6	−
2	6.13e−6	3.67e−3	15.93e−3	1	6.65e−6	−114.32e−6	110.12e−6	848.23e−6
3	168.80e−6	10.68e−3	1	−	65.79e−6	−	466.39e−6	−

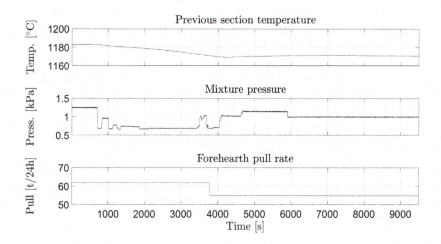

Fig. 4. System inputs and glass pull rate - 2. experiment

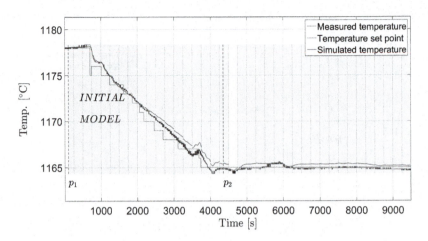

Fig. 5. Simulation results - 2. experiment

7   Conclusion

In the paper an original application of non-standard optimal method for identification of differential equation parameters was presented. For the purpose of checking the quality of the obtained models, an on-line simulation is performed. In order for the simulation to be correct and to guarantee that the models are accurate at different operating points, a non-standard precise method of identification (observation) of the initial and final states is used. The described parameters identification method using the MFM and the non asymptotic exact state observers allows to model the glass forehearth installation dynamics. Conducted experiments gave very satisfying results. The linear system output was very close to the real system one for both cases. The mean square error between the real system and the simulated output was equal 0.2841 for the first experiment and 0.2028 in the second case. The developed procedure in its current form can be used for wide variety of problems, e.g. PID controller tuning or feed forward control. For the purpose of implementation of the designed algorithms and their testing, an extensive programming environment with many modules was created. The packages were written in the Matlab language, suitable for rapid prototyping, and after automatic translation into C++, it would take tens of thousands of source code lines. It can be used in real computer control systems.

Acknowledgement. This work was supported by the scientific research funds from the Polish Ministry of Science and Higher Education and AGH UST Agreement no 16.16.120.773 and was also conducted within the research of EC Grant H2020-MSCA-RISE-2018/824046.

References

1. Rivas-Perez, R., Sotomayor-Moriano, J., Pérez-Zuñiga, G., Soto-Angles, M.E.: Real-time implementation of an expert model predictive controller in a pilot-scale reverse osmosis plant for brackish and seawater desalination. Appl. Sci. **9**(14), 2932 (2019). https://doi.org/10.3390/app9142932
2. Gough, B.P., Eng, P., Matovich, D.: Predictive-adaptive temperature control of molten glass. In: IEEE Industry Applications Society Dynamic Modeling Control Applications for Industry Workshop, Vancouver, BC, Canada, pp. 51–55 (1997). https://doi.org/10.1109/DMCA.1997.603511
3. Wang, Q., Chalaye, G., Thomas, G., Gilles, G.: Predictive control of a glass process. Control. Eng. Pract. **5**(2), 167–173 (1997). https://doi.org/10.1016/S0967-0661(97)00223-2
4. Huisman, L., Weiland, S.: Identification and model predictive control of an industrial glass-feeder. IFAC Proc. Vol. **36**(16), 1645–1649 (2003). https://doi.org/10.1016/S1474-6670(17)34996-0
5. Astrid, P.: Reduction of process simulation models: a proper orthogonal decomposition approach. Ph.D. thesis. Technische Universiteit Eindhoven (2004). https://doi.org/10.6100/IR581728
6. Kharitonov, A., Henkel, S., Savodny, O.: Two Degree of freedom control for a glass feeder. In: Proceedings of the European Control Conference 2007. Kos, Greece, July 2–5, 2007, pp. 4079–4086 (2007). https://doi.org/10.23919/ECC.2007.7068450
7. Malchow, F., Sawodny, O.: Model based feedforward control of an industrial glass feeder. Control. Eng. Pract. **20**(1), 62–68 (2012). https://doi.org/10.1016/J.CONENGPRAC.2011.09.004
8. Byrski, W., Drapała, M., Byrski, J.: An adaptive identification method based on the modulating functions technique and exact state observers for modeling and simulation of a nonlinear miso glass melting process. Int. J. Appl. Math. Comput. Sci. **29**(4), 739–757 (2019). https://doi.org/10.2478/amcs-2019-0055
9. Sinha, N.K., Rao, G.P.: Identification of Continuous-Time Systems. Springer, Heidelberg (1991). https://doi.org/10.1007/978-94-011-3558-0
10. Rao, G., Unbehauen, H.: Identification of continous-time systems. IEE Proc. Control Theory Appl. **153**(2), 185–220 (2006). https://doi.org/10.1049/ip-cta:20045250
11. Rao, G., Diekmann, K., Unbehauen, H.: Parameter estimation in large scale interconnected systems. IFAC Proc. Vol. **17**(2), 729–733 (1984). https://doi.org/10.1016/S1474-6670(17)61058-9
12. Shinbrot, M.: On the analysis of linear and nonlinear systems. Trans. Am. Soc. Mech. Eng. J. Basic Eng. **79**, 547–552 (1957)
13. Preisig, H., Rippin, D.: Theory and application of the modulating function method-I. Review and theory of the method and theory of the spline-type modulating functions. Comput. Chem. Eng. **17**(1), 1–16 (1993). https://doi.org/10.1016/0098-1354(93)80001-4
14. Byrski, W., Byrski, J.: The role of parameter constraints in EE and OE methods for optimal identification of continuous LTI models. Int. J. Appl. Math. Comput. Sci. **22**(2), 379–388 (2012). https://doi.org/10.2478/v10006-012-0028-3
15. Byrski, J., Byrski, W.: A double window state observer for detection and isolation of abrupt changes in parameters. Int. J. Appl. Math. Comput. Sci. **26**(3), 585–602 (2016). https://doi.org/10.1515/amcs-2016-0041

Iterative Global Sensitivity Analysis Algorithm with Neural Network Surrogate Modeling

Yen-Chen Liu[1]🆔, Jethro Nagawkar[1]🆔, Leifur Leifsson[1(✉)]🆔,
Slawomir Koziel[2,3]🆔, and Anna Pietrenko-Dabrowska[3]🆔

[1] Department of Aerospace Engineering, Iowa State University, Ames, IA 50011, USA
{clarkliu,jethro,leifur}@iastate.edu
[2] Engineering Modeling and Optimization Center, School of Science and Engineering,
Reykjavik University, Menntavegur 1, 101 Reykjavik, Iceland
koziel@ru.is
[3] Faculty of Electronics Telecommunications and Informatics,
Gdansk University of Technology, Narutowicza 11/12, 80-233 Gdansk, Poland
anna.dabrowska@pg.edu.pl

Abstract. Global sensitivity analysis (GSA) is a method to quantify the effect of the input parameters on outputs of physics-based systems. Performing GSA can be challenging due to the combined effect of the high computational cost of each individual physics-based model, a large number of input parameters, and the need to perform repetitive model evaluations. To reduce this cost, neural networks (NNs) are used to replace the expensive physics-based model in this work. This introduces the additional challenge of finding the minimum number of training data samples required to train the NNs accurately. In this work, a new method is introduced to accurately quantify the GSA values by iterating over both the number of samples required to train the NNs, terminated using an outer-loop sensitivity convergence criteria, and the number of model responses required to calculate the GSA, terminated with an inner-loop sensitivity convergence criteria. The iterative surrogate-based GSA guarantees converged values for the Sobol' indices and, at the same time, alleviates the specification of arbitrary accuracy metrics for the surrogate model. The proposed method is demonstrated in two cases, namely, an eight-variable borehole function and a three-variable nondestructive testing (NDT) case. For the borehole function, both the first- and total-order Sobol' indices required 200 and 10^5 data points to terminate on the outer- and inner-loop sensitivity convergence criteria, respectively. For the NDT case, these values were 100 for both first- and total-order indices for the outer-loop sensitivity convergence, and 10^6 and 10^3 for the inner-loop sensitivity convergence, respectively, for the first- and total-order indices, on the inner-loop sensitivity convergence. The differences of the proposed method with GSA on the true functions are less than 3% in the analytical case and less than 10% in the physics-based case (where the large error comes from small Sobol' indices).

© Springer Nature Switzerland AG 2021
M. Paszynski et al. (Eds.): ICCS 2021, LNCS 12745, pp. 298–311, 2021.
https://doi.org/10.1007/978-3-030-77970-2_23

Keywords: Global sensitivity analysis · Surrogate modeling · Neural networks · Sobol' indices · Termination criteria

1   Introduction

Sensitivity analysis (SA) [1,2] plays an important role in engineering, design, and analysis. SA quantifies the effects of individual input parameters, as well as combinations of input parameters, on the output model response [3,4]. Engineers and scientists can use SA in deciding which parameters are important while performing experimental or computational studies. SA can be classified as either local [5] or global [6] SA. In local SA, small perturbations in the inputs are used to quantify its effects on the output model response. In global SA, the variance of output model response due to the input variability is quantified. This work focuses on global variance-based SA with Sobol' indices [3,4].

Model-based SA often relies on using high-fidelity physics-based models. The use of such model-based SA can be challenging due to a variety of reasons, including (1) the physics-based models can be computationally costly to solve, (2) engineering problems can require a large number of variability parameters, and (3) SA requires multiple and repetitive physics-based model evaluations. The combination of these challenges may result in problems that are difficult to solve in a reasonable amount of time.

To reduce this computational burden, surrogate modeling methods [7] can be used. Surrogate models replace the high-fidelity physics-based models with a computationally efficient ones. Surrogate models (also called metamodels) can be broadly classified as either data-fit methods [7] or multifidelity methods [8]. In data-fit methods, a response surface is fitted through the responses of evaluated high-fidelity models. Examples include Kriging [9], neural networks (NN) [10], and support vector machines [11]. In multifidelity methods, data from multiple levels of fidelity are fused together to make predictions on the level of the high-fidelity model. Examples of multifidelity modeling methods include Cokriging [12] and manifold mapping [13]. This work utilizes data-fit surrogate modeling, namely NNs, in lieu of the high-fidelity physics-based models within a GSA framework.

In this paper, a new approach for surrogate-based GSA with Sobol' indices [3] is proposed. Specifically, in the proposed surrogate framework, the number of samples used to train the NNs is iteratively increased. The goal of the proposed algorithm is twofold: to minimize the training cost of the NNs while still yielding converged Sobol' indices and alleviating the specification of arbitrary surrogate modeling accuracy metrics. The latter is important because accuracy metrics for surrogate models do not guarantee that converged Sobol' indices are obtained.

The remainder of the paper is organized as follows. Next section gives the details of the proposed method, including the GSA framework with surrogate modeling, the procedure of iteratively improving sensitivity results, and the convergence criteria for termination. The following section describes the results of two numerical examples benchmarking the sensitivity results of the proposed method against the actual sensitivity. The last section concludes the paper and discusses potential future work.

2   Methods

This section describes a general SA problem and the proposed algorithm, including the iterative approach, the surrogate modeling, the Sobol' analysis, and the convergence of the algorithm.

2.1   Problem Statement

Quantifying the effects of the inputs on the output response of a system or model is essential for making design and engineering decisions. Global SA using Sobol' indices [3] is one such method of quantifying the variance of the model output due to variability in the inputs to the model. In this work, the physics-based models are considered as black box functions described as

$$y = f(\mathbf{x}), \tag{1}$$

where $\mathbf{x} \in \mathbb{R}^D$ is a set of D-dimensional input parameters and a single output y. The model output is perturbed due to probabilistically distributed random inputs, which imitate the uncertainties from different sources. This work proposes a method to determine the sensitivity of each input variability parameter on the model output. The following sections give a detailed explanation of the methodology used in this work.

2.2   Iterative Global Sensitivity Analysis Workflow

Figure 1 demonstrates the proposed method. The process starts by taking a small population of samples from the uncertainty input parameters. Samples are selected randomly from each probability distribution of the inputs via the Latin hypercube sampling (LHS) [14] method. The simulation model then generates the corresponding outputs and uses them as training data to construct a surrogate model that mimics the same input-output behavior of the simulation model. The surrogate modeling method used in this work is discussed in the next section. GSA is then performed with this surrogate model using Sobol' indices, and the local and global convergence of these indices are checked. In the inner-loop, the current surrogate model is used to calculate the Sobol' indices, and the number of samples is increased by one order of magnitude during each iteration until local convergence is achieved. The precision of the GSA is affected by the predicted outputs from the surrogate model, which is usually related to the number of physics-based model training samples. Therefore, to estimate the global convergence of the Sobol' indices, the above process is resampled with an increasing number of training data samples from the physics-based model in the outer-loop until convergence criteria are met. The outcome yields the converged Sobol' indices and the corresponding surrogate model.

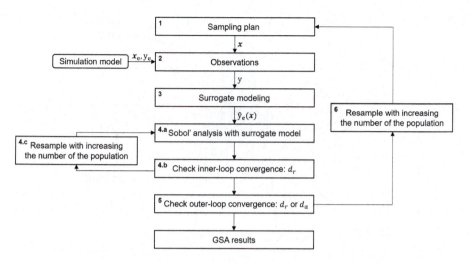

Fig. 1. flowchart of the surrogate-based iterative global sensitivity analysis algorithm.

2.3   Surrogate Modeling with Neural Networks

NNs are a subclass of surrogate models where any function can be approximated through a hierarchy of features [10]. Layers are steps in this hierarchy of features [15]. The layers in-between the input and output layers are called hidden layers [16]. An architecture of an NN is shown in Fig. 2. This NN has two hidden layers as well as an input and output layer. The input and output layers have six inputs and three outputs, respectively. The choice of the number of inputs and outputs is problem dependant, where this value is equal to the number of independent (input) variables and dependant (output) variables of the problem being solved. The output and hidden layers each consist of neurons. In a NN, neurons are fundamental units of computation [10]. Neurons output nonlinear transformations of a weighted sum of the outputs from a previous hidden or input layer[10]. This nonlinear transformation is termed activation [15]. Changing the number of hidden layers and the number of neurons in each hidden layer affects the complexity of the function being approximated [10].

The activation function in each neuron of a given hidden layer, L, is given by [10]

$$z_j^{(L)} = a\left(\sum_{i=1}^{N^{(L-1)}} \omega_{ji}^{(L)} z_i^{(L-1)} + b^{(L-1)}\right), \tag{2}$$

where $N^{(L-1)}$ refers to the number of neurons in the $L-1$ hidden layer, while $z_j^{(L-1)}$ and $z_i^{(L)}$ are the outputs of the jth and i^{th} neurons, respectively, in the L and $L-1$ hidden layers, respectively. The activation function is denoted by a, the weight between the ith neuron in the $L-1$ hidden layer and the jth neuron in the L hidden layer is denoted by $\omega_{ji}^{(L)}$, and the bias unit in the $L-1$ hidden

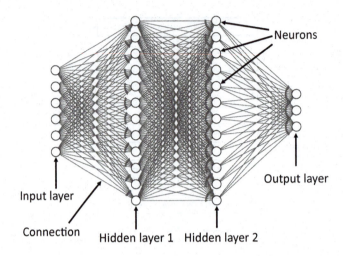

Input layer

Connection

Hidden layer 1   Hidden layer 2

Neurons

Output layer

Fig. 2. A schematic of a neural network.

layer is denoted by $b^{(L-1)}$. These weight and biases are termed parameters of the NN. To tune these parameters, an optimization problem is solved to reduce the loss function ($\mathcal{L}$) with respect to all the parameters of the NN. The Adaptive Moments (ADAM) [17] gradient-based optimizer [10] is used to find the values of these parameters, and the backpropagation algorithm [18] is used to find the gradients.

The loss function is defined to capture the mismatch between the training data observations, y, and the predicted, $\hat{y}$, of the NN. In this work, the mean squared error (MSE) along with a L_2 regularization term, averaged over all the training data, is used and is given by

$$\mathcal{L} = \frac{\sum_{l=1}^{N_{tr}} \sum_{m=1}^{N_o} (\hat{y}_m^{(l)} - y_m^{(l)})^2}{N_{tr} N_o} + \lambda \frac{\sum_{n=1}^{N_w} (W_n)^2}{2N_{tr}}, \tag{3}$$

where the first half of the equation represents the MSE, and the second half is the L_2 regularization term. The L_2 regularization term is an additional term added to prevent the NN from overfitting the training data [10]. N_o is the size of the output layer, N_{tr} is the number of training data sets, N_w is the total number of parameters (W) in the NN, l is the index of the training data-set, m is the index of the neuron in the output layer, and λ is the regularization constant. In practice, a subset of the training data, called mini-batch [10], is used to calculate the loss function.

To tune the hyperparameters of the NN, a testing set of $1,000$ samples is used. These hyperparameters are varied till both the training and testing loss values are of a similar order of magnitude and as low as possible. The common hyperparameter values for the cases selected in this study are the tangent hyperbolic function, a learning rate of 0.001, 100 neurons in each hidden layer, a mini-batch size of 16, and a maximum number of epochs of $2,000$. For the borehole case, the number of hidden layers was set to two, and the λ value was set to 0.1. For the NDE case, these values were set to one and 0.01, respectively. Definitions for these hyperparameters can be found in [10].

2.4   Sobol' Indices

In this study, the global variance-based sensitivity analysis with Sobol' indices [3,4] is used. It is used to quantify the effect of each individual inputs as well combination of inputs on the output model response. Given a model $y = f(\mathbf{x})$, where $\mathbf{x}$ is a set of D input parameters and y is the model output, it can be decomposed into the following form [4]

$$y = y_0 + \sum_{i=1}^{D} y_i + \sum_{i<j}^{D} y_{ij} + \dots + y_{1,2,\dots,D}, \tag{4}$$

where y_0 is a constant, y_i is the model output from varying individual x_i's, y_{ij} is the model output from varying x_i and x_j simultaneously, and so on. The total variance of y measures how far the uncertainties of the model inputs propagating to the model output, and the right-hand side of (4) becomes

$$\text{Var}(y) = \sum_{i=1}^{D} V_i + \sum_{i<j}^{D} V_{ij} + \dots + V_{12\dots D}, \tag{5}$$

where

$$V_i = \text{Var}_{x_i}(E_{\mathbf{x}_{\sim i}}(y|x_i)), \tag{6}$$

$$V_{ij} = \text{Var}_{x_{ij}}(E_{\mathbf{x}_{\sim ij}}(y|x_i, x_j)) - V_i - V_j, \tag{7}$$

and so on. $\mathbf{x}_{\sim i}$ represents all the variables except x_i. V_i represents the variance of the output due to individual x_i, while V_{ij} is the variance of the output due to interaction between x_i and x_j. Dividing Eq. (5) by $\text{Var}(y)$ results in

$$1 = \sum_{i=1}^{D} S_i + \sum_{i<j}^{D} S_{ij} + \dots + S_{12\dots D}, \tag{8}$$

where the main effect indices, also known as first-order Sobol's indices [4] are given by

$$S_i = \frac{V_i}{\text{Var}(y)} = \frac{\text{Var}_{x_i}(E_{\mathbf{x}_{\sim i}}(y|x_i))}{\text{Var}(y)}, \tag{9}$$

where S_i is the contribution of individual x_i on the output model variance. The total-effect Sobol' indices [4] are given by

$$S_{T,i} = 1 - \frac{\text{Var}_{\mathbf{x}_{\sim i}}(E_{x_i}(y|\mathbf{x}_{\sim i}))}{\text{Var}(y)}, \tag{10}$$

where $S_{T,i}$ measures the contribution of both individual x_i and the interaction between x_i and other model input parameters.

2.5   Convergence Criterion

The proposed iterative method includes an outer-loop that samples the physics-based model to train the surrogate NN models and an inner-loop that computes the Sobol' indices by sampling the trained NN surrogate model. Both the inner- and outer-loop are terminated based on the convergence of the Sobol' indices between successive iterations.

The inner-loop convergence is measured by the absolute relative change of Sobol' indices defined as

$$d_r[s_i] = \left| \frac{s_i^{(n)} - s_i^{(n-1)}}{s_i^{(1)}} \right|, \tag{11}$$

where n represents the current iteration index, i represents the index of input parameter, and s represents the value of the Sobol' indices and is calculated separately for first- and total-order indices. The inner-loop is terminated when $d_r[s_i] \leq \epsilon_r$ for all s_i. In this work, ϵ_r is set to 0.1.

Convergence of the outer-loop is measured by (11) and the absolute change of Sobol' indices, given by

$$d_a[s_i] = \left| s_i^{(m)} - s_i^{(m-1)} \right|, \tag{12}$$

where m represents the current iteration index. The outer-loop is terminated when $d_r[s_i] \leq \epsilon_r$ or $d_a[s_i] \leq \epsilon_a$ for all s_i. In this work, ϵ_a is set to 0.01.

3   Numerical Examples

This section presents two numerical problems; an analytical function and a physics-based system solved using the proposed method for GSA. Both cases involve multiple uncertainty input parameters in the computations, which usually require running numerous repetitive evaluations directly on the true function for GSA calculations.

3.1   Case 1: Analytical Function

The analytical function used in this study is the eight variable borehole function [19]. The borehole function is used to model the flow of water through a borehole and is given by

$$f_{\text{HF}} = \frac{2\pi T_u(H_u - H_l)}{ln(r/r_w)\left(1 + \frac{2LT_u}{ln(r/r_w)r_w^2 K_w} + \frac{T_u}{T_l}\right)}, \tag{13}$$

where r_w and r are the radius of the borehole and the radius of influence of the borehole, respectively, T_u and T_l are the transmissivity of the upper and lower aquifer, respectively, H_u and H_l are the potentiometric head of upper and lower aquifers, respectively, L is the length of the borehole, and K_w is the hydraulic conductivity of the borehole. Each of the variability parameters along with their units and distributions are given in Table 1.

Performing GSA with the proposed method starts by drawing 10 LHS samples and producing borehole function outputs to train the NN. Once trained, this NN model is then sampled to compute the Sobol' indices within the inner-loop. For this inner-loop, the NN is first sampled using 10^2 samples and these samples are increased by one order of magnitude during each iteration of the inner-loop until the convergence criterion of $d_r \leq \epsilon_r$ is met. This is done separately for both first- and total-order Sobol' indices and an example of the convergence plots are shown in Figs. 3(a) and 3(b), respectively. This process is continued by

Table 1. Variability parameters and their corresponding distribution of the borehole function [19].

Variability parameters	Distribution
Radius of borehole, $r_w(m)$	$N(0.1, 0.0161812^2)$
Radius of influence, $r(m)$	$LogN(7.71, 1.0056^2)$
Transmissivity of upper aquifer, $T_u(m^2/yr)$	$U(63070, 115600)$
Potentiometric head of upper aquifer, $H_u(m)$	$U(990, 1110)$
Transmissivity of lower aquifer, $T_l(m^2/yr)$	$U(63.1, 116)$
Potentiometric head of lower aquifer, $H_l(m)$	$U(700, 820)$
Length of borehole, $L(m)$	$U(1120, 1680)$
Hydraulic conductivity of borehole, $K_w(m/yr)$	$U(9855, 12045)$

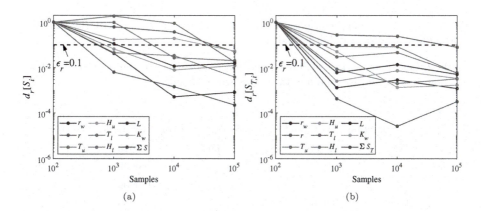

Fig. 3. Case 1 inner-loop convergence of s_i for the NN trained with 200 LHS samples: (a) first-order indices, and (b) total-order indices.

increasing the sample population to 50 in the global loop and by repeating the previous steps. The outer-loop termination criteria are checked between the converged Sobol' indices of the NN models trained with 10 and 50 samples and are shown in Figs. 4(a) and 4(b) for the first- and total-order Sobol' indices, respectively. The entire process is repeated by increasing the number of samples required to train the NN, until the outer-loop convergence criteria of either $d_r \leq \epsilon_r$ or $d_a \leq \epsilon_a$ is met.

For the borehole function, the NN needs to be trained with 200 samples until these criteria are met for both the first- and total-order indices. The converged Sobol' indices are displayed in Fig. 5. The GSA results from the proposed method are compared to the Sobol' indices computed by directly sampling the borehole function and is shown in Table 2. The radius of influence of the borehole, as well as the transmissivity of the upper and lower aquifers have negligible influence on

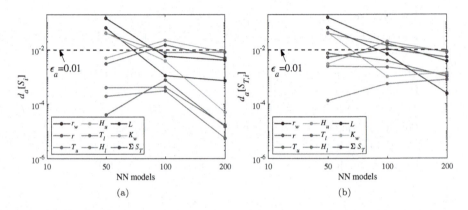

(a)   (b)

Fig. 4. Case 1 outer-loop convergences of s_i terminated on $d_a \leq \epsilon_a$ criteria: (a) first-order indices, and (b) total-order indices.

Table 2. Case 1 comparison of Sobol' index values calculated by sampling the borehole function directly and by sampling the converged NN model.

x	S_i			$S_{T,i}$		
	True function	NN model	% error	True function	NN model	% error
r_w	0.6638	0.6621	0.3%	0.6941	0.6910	0.4%
r	0	0	–	0	0.0002	–
T_u	0	0	–	0	0.0002	–
H_u	0.0949	0.0966	1.7%	0.1061	0.1088	2.5%
T_l	0	0	–	0	0	–
H_l	0.0948	0.0945	0.3%	0.1061	0.1051	0.9%
L	0.0907	0.0918	1.2%	0.1028	0.1041	1.3%
K_w	0.0219	0.0222	1.4%	0.0251	0.0256	2.0%

the output response (flow rate) of the borehole, while the radius of the borehole has the highest impact on the flow rate.

3.2   Case 2: Ultrasonic Nondestructive Testing

This study uses the spherically-void-defect under focused transducer ultrasonic (UT) nondestructive testing (NDT) benchmark case. This case was developed by the World Federal Nondestructive Evaluation Center [20]. The main goal of this case is to find the minimum number of training data required to accurately predict the Sobol' Indices using NN as well as quantify the contribution of each variability parameter on the output model response.

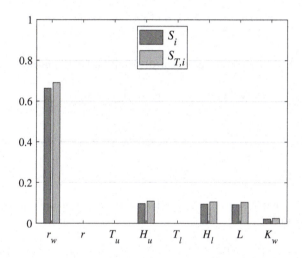

Fig. 5. Case 1 Sobol' index values of input parameters computed by the converged NN model.

Figure 6 shows the setup for the UT benchmark case used in this study. The variability parameters considered for this case are the probe angle (θ), the x location of the probe (x), and the F-number (F). The distributions of each of these parameters are shown in Table 3.

To predict the voltage waveforms at the receiver, the Thompson-Grey model [21] is used, while the velocity diffraction coefficient is calculated using the multi-Gaussian beam model [22]. Closed-form expressions of the scattering amplitude can then be calculated using the separation of variables [23]. In this study, a center frequency of 5 MHz is used for the transducer. The density of the fused quartz block is set to $2{,}000\,\mathrm{kg/m^3}$, while the longitudinal and shear wave speeds are set to $5{,}969.4\,\mathrm{m/s}$ and $3{,}774.1\,\mathrm{m/s}$, respectively. More information about this model can be found in Du et al. [24].

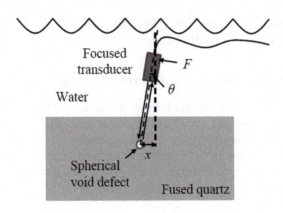

Fig. 6. Setup for the ultrasonic testing case.

Table 3. Variability parameters and their distribution for the ultrasonic testing case

Variability parameters	Distribution
θ (deg)	$N(0, 0.5^2)$
x (mm)	$U(0, 1)$
F	$U(13, 15)$

Table 4. Case 2 comparison of Sobol' index values between the true model and the converged NN model.

x	S_i			$S_{T,i}$		
	True model	NN model	% error	True model	NN model	% error
θ	0.8269	0.8253	0.2%	0.8367	0.8266	1.2%
F	0.0011	0.0010	9.1%	0.0014	0.0013	7.1%
x	0.1622	0.1616	0.4%	0.1732	0.1732	0%

A similar approach to the previous case is used for this case. The same convergence criteria were used in this case as in the previous case. The outer-loop iterated from 10 to 100 LHS samples, and the convergence plots for the first- and total-order indices are shown in Figs. 8(a) and 8(b), respectively. The convergence plots for the inner-loop of the first- and total-order indices of the trained 100 NN model are shown in Figs. 7(a) and 7(b), respectively. Both the first- and total-order indices require 100 samples each to terminate the outer-loop. The first-order indices require 10^6 samples to terminate the inner-loop, while the total-order indices require 10^3. Figure 9 shows that the F-number has a negligible effect on the output response, while the probe angle has the highest effect. Table 4 compares the Sobol' indices values from the proposed method to those from the true function, showing a good match.

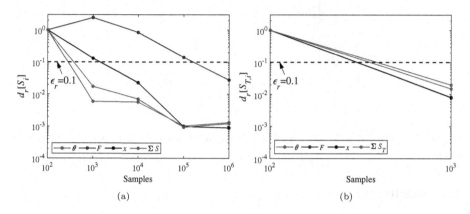

Fig. 7. Case 2 inner-loop convergence of s_i for the NN trained with 100 LHS samples: (a) first-order indices, and (b) total-order indices.

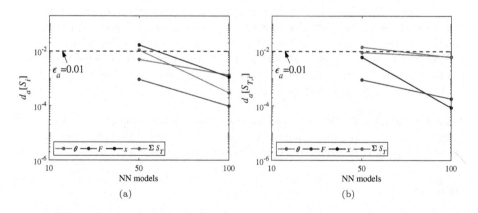

Fig. 8. Case 2 outer-loop convergence of s_i terminated on $d_a \leq \epsilon_a$ criteria: (a) first-order indices, and (b) total-order indices.

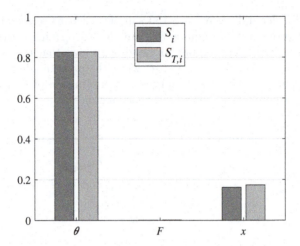

Fig. 9. Case 2 Sobol' index values of input parameters computed by the converged NN model.

4   Conclusion

This work has presented an algorithm for global sensitivity analysis (GSA) by evaluating Sobol' indices iteratively with surrogate modeling. The goal of the proposed approach is to obtain accurate GSA results while using few evaluations for the true model. Furthermore, the proposed method avoids the specification of arbitrary surrogate modeling accuracy metrics. The efficacy of the proposed algorithm is demonstrated using an analytical function and a physics-based model and comparing against the Sobol' indices obtained with the true functions. The results show that accurate and fully converged Sobol' indices can be achieved at a low computational cost. Future research will benchmark the computational cost and the accuracy of the proposed algorithm against other GSA and surrogate modeling techniques.

Acknowledgements. This material is based upon work supported by the U.S. National Science Foundation under grants no. 1739551 and 1846862, as well as by the Icelandic Centre for Research (RANNIS) grant no. 174573053.

References

1. Ferretti, F., Saltelli, A., Tarantola, S.: Trends in Sensitivity Analysis Practice in the Last Decades. Sci. Total Environ. **568**, 666–670 (2016). https://doi.org/10.1016/j.scitotenv.2016.02.133
2. Iooss, B., Saltelli, A.: Introduction to Sensitivity Analysis. Springer, Cham (2015)
3. Sobol', I., Kucherekoand, S.: Sensitivity estimates for nonlinear mathematical models. Math. Model. Comput. Exper. **1**, 407–414 (1993)
4. Sobol', I.: Global sensitivity indices for nonlinear mathematical models and their Monte Carlo estimates. Math. Comput. Simul. **55**, 271–280 (2001)

5. Zhou, X., Lin, H.: Local Sensitivity Analysis. Encyclopedia of GIS, pp. 1116–1119 (2017)
6. Homma, T., Saltelli, A.: Importance measures in global sensitivity analysis of nonlinear models. Reliab. Eng. Syst. Saf. **52**, 1–17 (1996)
7. Forrester, A.I.J., Keane, A.J.: Recent advances in surrogate-based optimization. Prog. Aerosp. Sci. **45**(1–3), 50–79 (2009). https://doi.org/10.1016/j.paerosci.2008.11.001
8. Peherstorfer, B., Willcox, K., Gunzburger, M.: Survey of multifidelity methods in uncertainty propagation, inference, and optimization. Soc. Ind. Appl. Math. **60**(3), 550–591 (2018). https://doi.org/10.1137/16M1082469
9. Krige, D.G.: A statistical approach to some basic mine valuation problems on the Witwatersrand. J. Chem. Metallurgical Min. Eng. Soc. South Africa **52**(6), 119–139 (1951)
10. Goodfellow, I., Bengio, Y., Courville, A.: Deep Learning. The MIT Press, Cambridge (2016)
11. Li, D., Wilson, P.A., Jiong, Z.: An improved support vector regression and its modelling of Manoeuvring performance in multidisciplinary ship design optimization. Int. J. Model. Simul. **35**, 122–128 (2015)
12. Kennedy, C.M., O'Hagan, A.: Predicting the output from a complex computer code when fast approximations are available. Biometrika **87**(1), 1–13 (2000). https://doi.org/10.1093/biomet/87.1.1
13. Echeverria, D., Hemker, P.: Manifold mapping: a two-level optimization technique. Comput. Vis. Sci. **11**, 193–206 (2008). https://doi.org/10.1007/s00791-008-0096-y
14. McKay, M.D., Beckman, R.J., Conover, W.J.: A comparison of three methods for selecting values of input variables in the analysis of output from a computer code. Technometrics **21**(2), 239–245 (1979)
15. Haykin, S.S.: Neural Networks and Learning Machines, 3rd edn. Pearson Education, Upper Saddle River (2009)
16. Schmidhuber, J.: Deep learning in neural networks: an overview. Neural Netw. **61**, 85–117 (2015). https://doi.org/10.1016/j.neunet.2014.09.003
17. Kingma, D.P., Ba, J.: Adam: a method for stochastic optimization. arXiv:1412.6980 (2014)
18. Chauvin, Y., Rumelhart, D.E.: Backpropagation: Theory, Architectures, and Applications. Psychology Press, Hillsdale (1995)
19. Harper, W.V., Gupta, K.S.: Sensitivity/uncertainty analysis of a borehole scenario comparing Latin Hypercube Sampling and deterministic sensitivity approaches. Technical report, Office of Nuclear Waste Isolation, Columbus, OH (1983)
20. Gurrala, P., Chen, K., Song, J., Roberts, R.: Full wave modeling of ultrasonic NDE benchmark problems using Nystrom method. Rev. Progress Quant. Nondestruct. Eval. **36**(1), 1–8 (2017)
21. Schmerr, L., Song, J.: Ultrasonic Nondestructive Evaluation Systems. Springer, New York (2007). https://doi.org/10.1007/978-0-387-49063-2
22. Wen, J.J., Breazeale, M.A.: A diffraction beam field expressed as the superposition of Gaussian beams. J. Acoust. Soc. Am. **83**, 1752–1756 (1988)
23. Schmerr, L.: Fundamentals of Ultrasonic Nondestructive Evaluation: A Modeling Approach. Springer, Boston (2013). https://doi.org/10.1007/978-1-4899-0142-2
24. Du, X., Leifsson, L., Meeker, W., Gurrala, P., Song, J., Roberts, R.: Efficient model-assisted probability of detection and sensitivity analysis for ultrasonic testing simulations using stochastic metamodeling. J. Nondestruct. Eval. Diagnost. Prognost. Eng. Syst. **2**(4), 041002(4) (2019). https://doi.org/10.1115/1.4044446

Forecasting Electricity Prices: Autoregressive Hybrid Nearest Neighbors (ARHNN) Method

Weronika Nitka⬤, Tomasz Serafin⬤, and Dimitrios Sotiros$^{(\boxtimes)}$⬤

Department of Operations Research and Business Intelligence,
Faculty of Computer Science and Management,
Wrocław University of Science and Technology, 50-370 Wrocław, Poland
`dimitrios.sotiros@pwr.edu.pl`

Abstract. The ongoing reshape of electricity markets has significantly stimulated electricity trading. Limitations in storing electricity as well as on-the-fly changes in demand and supply dynamics, have led price forecasts to be a fundamental aspect of traders' economic stability and growth. In this perspective, there is a broad literature that focuses on developing methods and techniques to forecast electricity prices. In this paper, we develop a new hybrid method, called ARHNN, for electricity price forecasting (EPF) in day-ahead markets. A well performing autoregressive model, with exogenous variables, is the main forecasting instrument in our method. Contrarily to the traditional statistical approaches, in which the calibration sample consists of the most recent and successive observations, we employ the k-nearest neighbors (k-NN) instance-based learning algorithm and we select the calibration sample based on a similarity (distance) measure over a subset of the autoregressive model's variables. The optimal levels of the k-NN parameter are identified during the validation period in a way that the forecasting error is minimized. We apply our method in the EPEX SPOT market in Germany. The results reveal a significant improvement in accuracy compared to commonly used approaches.

Keywords: Electricity price forecasting · Day-ahead market · ARX · k-nearest neighbors

1 Introduction

Electricity markets have witnessed significant changes over the last decades. Their deregulation, followed by the emergence of electrical power exchanges such as EPEX SPOT, OMIE and Nord Pool in Europe, or PJM in the USA, allowed for competitive electricity trading [23]. Electrical power exchanges usually consist of several markets. The market with the biggest volume of trade is the day-ahead (spot) market, which allows the traders to place bids and offers the day before the

---

The authors' names are ordered alphabetically.

M. Paszynski et al. (Eds.): ICCS 2021, LNCS 12745, pp. 312–325, 2021.
https://doi.org/10.1007/978-3-030-77970-2_24

physical delivery of electricity. The day-ahead market is usually supplemented by intraday and balancing markets, which allow trading until a few minutes before delivery and target at providing more accurate offers. However, this is often associated with paying significant balancing fees.

Notably, electricity market clearing prices, defined by the supply and the demand curve, are characterized by high volatility. The cost of storing electricity at large scales as well as the transition of power generation from conventional to renewable sources, permeated with uncertainty in the production levels, lead to fluctuations in the supply. On the other hand, demand may vary on an hourly (peak and off-peak hours) and daily (weekends, weekdays and festivities) basis. These factors, along with the requirement of supply and demand to be precisely balanced in the power grid, lead to highly volatile prices in the electricity markets which can undergo extreme changes within a span of a single day.

Traders, ideally, aim to maximize their profit as well as to minimize the financial risk by selecting the most appropriate strategy in an imperfect market, where there is incomplete information. Given the high level of price volatility and the limitations in storing electricity, the selection of a wrong strategy, based on price misinformation, may lead to economic losses or even bankruptcy. On the contrary, utilizing accurate forecasts may increase profits or reduce the risk of economic losses [9,14].

In this line of thought, there is a wide literature which focuses on providing accurate day-ahead electricity price forecasts. Extended literature reviews are provided in [1,21,23]. Two of the prominent broad classes of methods provided in the literature rely either on statistical approaches or on machine learning techniques. Statistical approaches utilize linear regression models or linear autoregressive models based on a set of variables related to observed prices and other exogenous variables (load, wind, solar, temperature) that may affect price levels. Differences in the implementation of the autoregressive models can be also identified in terms of the calibration window length, which can be predefined or estimated via more advanced econometric techniques [4,11,12]. However, in these cases, the calibration sample consists of the most recent and successive observations. Methods that rely on machine learning employ a variety of techniques such as artificial neural networks [24], support vector machines [27], clustering algorithms [22] or a combination of them [17]. It is worthy to mention that a hybrid approach that employs statistical and machine learning techniques has been also proposed in the literature. Specifically, in [18] three clustering algorithms and an autoregressive lag model were employed to predict consumers' energy consumption in a simulation suite. However, this approach was tested on simulated data.

In this paper we build on the bridge between the two aforementioned classes of methods and we propose a new hybrid method, called *autoregressive hybrid nearest neighbors* (ARHNN), for forecasting spot electricity prices. We generate one-day-ahead forecasts using a linear ARX (autoregressive with exogenous variables) model with parameters calibrated on samples selected with the k-nearest neighbors (k-NN) algorithm. ARX models are well-established in electricity price forecasting (EPF), as noted in [10,23]. The k-nearest neighbors algorithm has been found to be successful in the field of electricity market forecasts, mainly in forecasting electricity price and load [2,3,8,13,19,26] and renewable energy sources (RES) generation [25]. The proposed method is applied to the EPEX SPOT market in Germany. The results show a significant improvement in accuracy compared to commonly used benchmark approaches, while low increase in the computational load is ensured.

The rest of this paper unfolds as follows. Section 2 describes the most important features of the data used in this analysis. Section 3 provides an in-depth explanation of the proposed method. Section 4 illustrates the results of the proposed method applied to the EPEX SPOT data and provides comparison with commonly used benchmark models. Finally, conclusions are drawn in Sect. 5.

2   Data

To illustrate our method, we use data describing the day-ahead electricity prices in the EPEX SPOT market in Germany. As described in the Introduction, the day-ahead market is the most important market in terms of traded volume. The dataset, published by the transmission system operator (TSO), comprises four variables: the electricity price in EUR/MWh and the corresponding official TSO forecasts of total electrical load, wind energy generation and photovoltaic energy generation, expressed in GWh.

The dataset spans six full years, from January 2015 until December 2020, with hourly data (see Fig. 1). To evaluate the performance of the proposed algorithm, the data is divided into three periods with lengths of approximately two years each. The first 728-day period is reserved for the initial calibration window. Then, the middle period, of the same length, is utilized for validation and tuning the hyperparameters of the model as described in Subsect. 3.2. Finally, the procedure is tested on the last period with length equal to 736 days.

The time series of the price and the load forecasts, as well as the division into calibration, validation and testing periods, are depicted in Fig. 1. It can be seen that the spot prices are indeed highly volatile, with frequent upward and downward spikes multiple times greater in magnitude than the average price range. However, load is relatively predictable, exhibiting both weekly and yearly seasonality, which needs to be addressed by the predictive model.

3   Methods and Algorithms

As shown by numerous studies in the EPF [7,14], the selection of the calibration sample impacts the overall forecasting accuracy of the autoregressive model. While the majority of authors consider the longest possible portion of data for the model calibration, averaging predictions obtained from calibration samples of different lengths [16] or utilizing more sophisticated statistical methods [15] allows for the significant reduction of forecasting errors. In this paper, we propose a new method for the selection of the calibration sample, based on the k-nearest

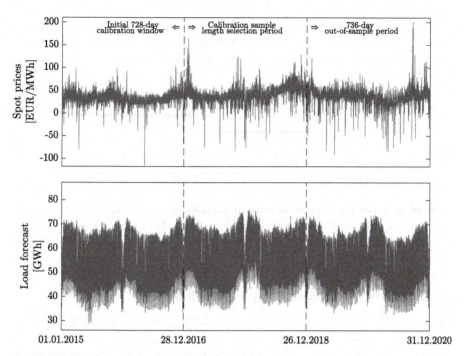

Fig. 1. Time series plot of the electricity spot prices (upper panel) and TSO load forecast (lower panel) from the EPEX SPOT market. Dashed lines indicate the split into calibration, validation and testing periods.

neighbors algorithm. The aforementioned methods rely on the time dimension to select the calibration sample, i.e. the most recent successive observations compose the calibration sample. On the contrary, in our method we define the calibration sample on the basis of a similarity measure over a set of features.

3.1   Predictive Model

To predict the spot prices in hour h of day $d + 1$ we use an expert ARX model with a specification well-established in the electricity price forecasting literature [7,20]. Due to the idiosyncratic nature of the electricity market, every hour of the day is treated as a distinct market product and separate forecasts are implemented for each hour, i.e. predicting the prices for the entire day $d + 1$ requires estimating 24 independent parameter sets. The models for every hour have an identical specification, incorporating an autoregressive structure with lags corresponding to two preceding days and a week, notated as $P_{d+1-p,h}$ where $p \in \{1, 2, 7\}$. The price dynamics are further captured by including the minimal and the maximal price from the previous day (respectively $P_{d,min}$ and $P_{d,max}$) as well as that day's price in hour 24 ($P_{d,24}$) – the previous day's last known price.

Finally, the model incorporates the publicly available forecasts of three exogenous variables relevant to the price levels: total electrical load $(\hat{L})$, wind energy generation $(\hat{W})$ and photovoltaic energy generation $(\hat{S})$. The complete model takes the form

$$P_{d+1,h} = \alpha_h D_{d+1} + \underbrace{\sum_{p\in\{1,2,7\}} \beta_{h,p} P_{d+1-p,h}}_{\text{AR component}} + \underbrace{\theta_{h,1} P_{d,min} + \theta_{h,2} P_{d,max}}_{\text{Daily statistics}}$$

$$+ \underbrace{\theta_{h,3} P_{d,24}}_{\text{Last known price}} + \underbrace{\theta_{h,4}\hat{L}_{d+1,h} + \theta_{h,5}\hat{W}_{d+1,h} + \theta_{h,6}\hat{S}_{d+1,h}}_{\text{Exogenous variables}} + \varepsilon_{d+1,h}, \qquad (1)$$

where D_{d+1} is the 1×7 vector of dummy variables representing days of the week and $\varepsilon_{d+1,h}$ is the Gaussian white noise. Henceforth, by $\hat{P}_{d+1,h}(\tau)$ we denote the prediction obtained from model (1) calibrated on a sample containing the τ most recent observations.

3.2   ARHNN Calibration Sample Selection

The k-Nearest Neighbors is an instance-based learning algorithm that can be used either for classification or regression. In the former case, an observation is assigned to the most common class label shared by its k-nearest neighbors. In the second case, the property value for an observation derives from the average of the k-nearest neighbors' values. In both cases, a neighbor weighting function can be employed [6].

To explain the applicability of the k-NN algorithm in our case study and the differentiation of our method, suppose that at day d we want to forecast the electricity price for the day ahead (day $d+1$). We denote by x_d the vector of the explanatory variables from model (1) for a given day d, after omitting dummy variables, 2-day and 7-day lagged prices and random noise. Within the matrix $X_{d+1} = (x_{d-726};\ldots;x_{d+1})$, it is evident that the most recent information we possess, x_{d+1}, provides the most accurate outlook at the current market state, i.e. prices from previous days as well as forecasts for day $d + 1$. Notably, the proposed statistical methods in the literature, rely on this assumption and they further extend it. Specifically, they assume that the most recent observations will provide the most accurate forecast and thus, they should compose the calibration sample. However, in case structural breaks exist among the last observations, the selected calibration sample will lead to a decreased forecasting accuracy. In addition, this approach relies exclusively on the last observations (in terms of time) and does not exploit information from other past data.

The main idea of our method is to identify past observations that resemble x_{d+1} as closely as possible and use them to estimate the parameters of the forecasting model. To this end, we employ the k-NN algorithm to select a calibration sample for the ARX model (1) consisting of the k-nearest neighbors of the point x_{d+1} (see Fig. 2), in terms of the Euclidean distance. In a sense, we invert the rationale of the k-NN method - instead of classifying the latest observation based on its neighboring points, we assume that the closest neighbors (in terms of the distance, not time) of x_{d+1} belong to the same market "regime".

Analogously to the notation in Sect. 3.1, we denote the price prediction for day $d+1$ and hour h, obtained by calibrating the forecasting model (1) on the sample consisting of k closest observations, by $\hat{P}^*_{d+1,h}(k)$. Note that for the clarity of notation, forecasts corresponding to the ARHNN method are marked with an asterisk.

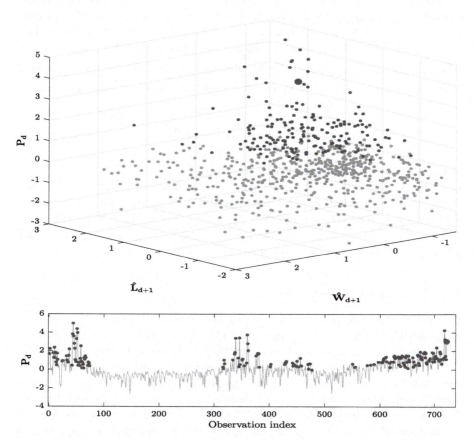

Fig. 2. The optimal (i.e. producing the lowest absolute prediction error) selection of the calibration sample ($\bar{k}_i = 181$) based on the matrix X_{d+1} for a specific day ($d+1$). The upper panel illustrates the sample selection, presented on three key variables, i.e. preceding day's price as well as forecasts of load and wind generation; while the lower panel depicts the corresponding selection in the time dimension. The most recent observation is marked with a red dot, while the observations selected for the model calibration are depicted with blue points. (Color figure online)

Obviously, the choice of the parameter k has a direct impact on the forecasting accuracy of the model. Disentangling its effects, is one of the main challenges that we address in the paper. As discussed in Sect. 2, in the validation period, we use the 728-day rolling window to identify the optimal values of the k parameter, which is responsible for the number of observations in the calibration sample. For each of the 728 days in the validation period, the procedure identifies (ex-post) the optimal value (i.e. the one that produced the lowest absolute prediction error for a certain day; see Figs. 2, 3) of the parameter, $\bar{k}_i$, $i = 1, \ldots, 728$. Next, in the evaluation (testing) procedure, instead of selecting only one value of k for each day, we consider 728 calibration samples, based on the set of past optimal values $(\bar{k}_1, \ldots, \bar{k}_{728})$. In such way, we obtain 728 price predictions for day $d + 1$, i.e. $\left(\hat{P}^*_{d+1,h}(\bar{k}_1), \ldots, \hat{P}^*_{d+1,h}(\bar{k}_{728}) \right)$. Eventually, inspired by [16], we obtain the final price prediction for day $d + 1$ and hour h from the average of these forecasts:

$$\hat{P}_{d+1,h} = \frac{1}{728} \sum_{i=1}^{728} \hat{P}^*_{d+1,h}(\bar{k}_i). \tag{2}$$

Notably, there may be cases where the values of $\bar{k}_i$, $i = 1, \ldots, 728$ coincide, i.e. $\bar{k}_i = \bar{k}_j$ for $i \neq j$. Therefore, the above expression is translated to the weighted average of forecasts calibrated to different samples, where the weight corresponding to a certain prediction $\hat{P}^*_{d+1,h}(\bar{k}_i)$ depicts the relative frequency of $\bar{k}_i$ in $(\bar{k}_1, \ldots, \bar{k}_{728})$. This can be interpreted as a weighting function which reflects the "relative significance" of the $\bar{k}_i$ values.

3.3 Benchmark Approaches

We evaluate the effectiveness of selecting the calibration period with the proposed ARHNN procedure by comparing it to a number of literature benchmarks. While all of them use Model (1) for computing the forecasts themselves, they differ in the selection of the calibration sample and in the forecasts post-processing. The first group of benchmark approaches provides forecasts obtained by using a single calibration window length throughout the entire test period. The calibration windows include from 56 to 728 days of the most recent data up to the moment of forecasting. The second group utilizes two additional approaches following [7]: the arithmetic mean of all the forecasts within the first group (673 predictions obtained from calibration windows of different lengths), and the average of forecasts from six hand-picked calibration windows: three short ones (56, 84, 112 days) and three long ones (714, 721, 728 days).

We assume the following convention to notate the aforementioned benchmark methods: the single-length windows with length τ are denoted as $\mathrm{Win}(\tau)$. The forecast averages are named using the MATLAB sequence convention, respectively becoming $\mathrm{Av}(56{:}728)$ and $\mathrm{Av}(56{:}28{:}112, 714{:}7{:}728)$.

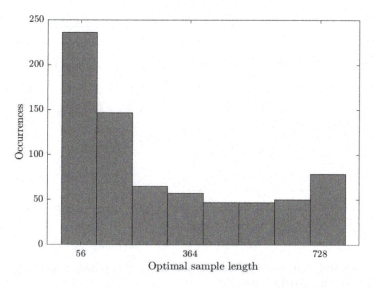

Fig. 3. Histogram of the optimal calibration sample lengths within the validation period (728 calibration sample lengths in total) for hour 18.

4   Results

We evaluate the accuracy of the forecasts obtained from different approaches with the use of the *root mean squared error* (RMSE). The reported error is calculated across all hours and days of the 736-day out-of-sample period. The results are presented in Fig. 4 and Table 1. The performance of single calibration window benchmarks (i.e. models trained on samples comprising a fixed amount of most recent observations) is presented with gray dots. In this approach, although the average error generally diminishes with the increase of the calibration window length τ and the longest window turns out to be the best choice, the decrease is not monotonic as we may expect. As shown by [16], for certain datasets, the error may even increase alongside with the calibration window length.

Table 1. The RMSE values of the selected benchmarks and the ARHNN method.

Method	RMSE
Win(364)	8.4443
Win(728)	8.2860
Av(56:728)	8.0584
Av(56:28:112, 714:7:728)	8.0286
ARHNN	7.8604

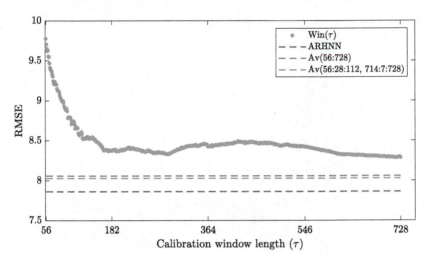

Fig. 4. The RMSE values as a function of calibration window length for the benchmark approaches and the ARHNN method.

As can be seen from Fig. 4, the ARHNN method as well as the averaging schemes outperform every approach based on a single, fixed calibration window length in terms of RMSE. Methods based on forecasts averaging, Av(56:728) and Av(56:28:112, 714:7:728), managed to outperform the predictive accuracy of the longest, 728-day calibration window, approximately by 3%. The forecasts obtained from the introduced ARHNN method exhibit over 5% lower error comparing to the best performing single calibration window length. The method also gains over 2% in terms of the forecasting accuracy compared to the well-performing literature benchmarks Av(56:728) and Av(56:28:112, 714:7:728). Since these results are not sufficient for determining the statistical significance of the difference between forecasts obtained from different approaches, we decided to use the Diebold and Mariano (DM) [5] test. First, for each pair of methods X and Y, we create a vector of errors for each day of the out-of-sample period. Here we consider two different perspectives - univariate and multivariate as classified by [28]. In the first one (multivariate), we consider 24-dimensional error vectors for each day:

$$\Delta_{X,Y,d} = ||\bar{\varepsilon}_{X,d}|| - ||\bar{\varepsilon}_{Y,d}||, \tag{3}$$

where $\bar{\varepsilon}_{X,d} = \sqrt{\frac{1}{24} \sum_{h=1}^{24} \varepsilon_{X,d,h}^2}$ and $\varepsilon_{X,d,h}$ is the error of forecasts obtained with method X for day d and hour h. In the second approach (univariate), instead of considering 24 h jointly, we are looking at each of them separately. More precisely:

$$\Delta_{X,Y,d,h} = |\varepsilon_{X,d,h}| - |\varepsilon_{X,d,h}|. \tag{4}$$

For each pair of approaches, we compute the p-value of the DM test with null hypothesis H_0: $\mathbb{E}(\Delta_{X,Y,d}) \leq 0$ (or H_0: $\mathbb{E}(\Delta_{X,Y,d,h}) \leq 0$ in case of the univariate approach) and additionally perform a complementary test with the reverse null hypothesis, $H_0^R\colon \mathbb{E}(\Delta_{X,Y,d}) \geq 0$ (or $H_0^R\colon \mathbb{E}(\Delta_{X,Y,d,h}) \geq 0$).

In Fig. 5 and Fig. 6, we present the p-values of the test. We use a heatmap to indicate the span of p-values. The closer they are to zero (dark green), the more significant is the difference between forecasts obtained with the approach from X-axis (superior) and predictions from the method in the Y-axis (inferior) [7,15,16]. The "chessboard" in Fig. 5 corresponds to the results of the multivariate approach, considering 24-dimensional error vectors (see Eq. 3). It turns out, that forecasts from the ARHNN method were able to significantly outperform predictions from nearly all benchmarks. The well-performing averaging scheme Av(56:28:112, 714:7:728) was neither significantly worse nor better than the proposed approach. Two "chessboards" in Fig. 6, correspond to the results of the univariate DM test for two exemplary hours. The selected Hour 9 and Hour 15 correspond to the worst and the best performance of the ARHNN method across all hours, respectively. For Hour 9, the forecasts based on the ARHNN approach were not able to statistically outperform predictions from any other method. Additionally, they are outperformed by the forecasts based on the Av(56:728) averaging scheme. When it comes to the results for Hour 15, the predictions from the proposed ARHNN method significantly outperform forecasts from all benchmarks, with p-values of the DM test close to zero. In general, the performance of the ARHNN approach across 24 h of the day is shown in Table 2. The columns, corresponding to 24 h are associated with six different performance classes, each of them representing a certain result of the DM test:

- **Class 1** (Hours 2, 6, 13, 14, 15, 16, 17) - forecasts from the ARHNN method significantly outperform predictions from all benchmarks and are not outperformed by any of them,
- **Class 2** (Hours 1, 3, 4, 5) - forecasts from the ARHNN method significantly outperform predictions from three out of four benchmarks and are not outperformed by any of them,
- **Class 3** (Hours 7, 8, 10, 11, 12, 18) - forecasts from the ARHNN method significantly outperform predictions from two out of four benchmarks and are not outperformed by any of them,
- **Class 4** (Hours 22, 23, 24) - forecasts from the ARHNN method significantly outperform predictions from two out of four benchmarks and are outperformed by one of them,
- **Class 5** (Hour 19) - forecasts from the ARHNN method do not significantly outperform predictions from any benchmark and are not outperformed by any of them,
- **Class 6** (Hours 9, 20, 21) - forecasts from the ARHNN method do not significantly outperform predictions from any benchmark and are outperformed by one of them.

Looking at the results of the Diebold-Mariano test it can be observed that forecasts from the ARHNN approach exhibit very satisfactory predictive accuracy compared to forecasts from the selected benchmarks. For eleven hours,

Table 2. Results of the statistical significance test between forecasts from the ARHNN approach and the selected benchmarks for all 24 h. Each class represents a certain result of the DM test.

Class	2	1	2	2	2	1	3	3	6	3	3	3	1	1	1	1	1	3	5	6	6	4	4	4
Hour	1	2	3	4	5	6	7	8	9	10	11	12	13	14	15	16	17	18	19	20	21	22	23	24

ARHNN forecasts were able to significantly outperform predictions from at least three out of four benchmarks and, for twenty hours, at least two out of four. Although the forecasts exhibit the worst performance for hours 9, 19, 20 and 21, they were significantly outperformed by at most one benchmark approach and, in the remaining twenty one hours of the day, by none of them.

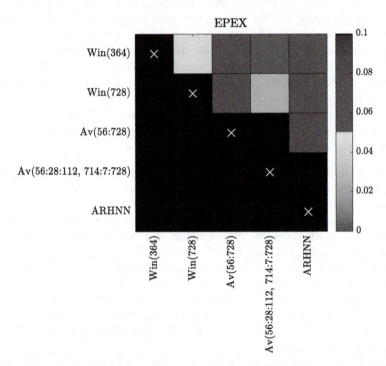

Fig. 5. Results of the multivariate approach to the pairwise Diebold-Mariano test between ARHNN method and the selected benchmarks. We illustrate the range of *p*-values using a heatmap: green squares indicate a statistically significant superiority of the forecasts from the method on the X-axis over the ones from the method on the Y-axis. (Color figure online)

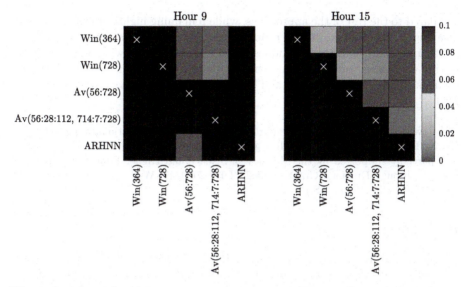

Fig. 6. Sample results of the univariate approach to the pairwise Diebold-Mariano test between ARHNN model and the selected benchmarks. We illustrate the range of *p*-values using a heatmap: green squares indicate a statistically significant superiority of the forecasts from the method on the X-axis over the ones from the method on the Y-axis. (Color figure online)

5   Conclusions and Discussion

In this paper we introduced a hybrid method for electricity price forecasting in day ahead markets. We employed a linear autoregressive model, with exogenous variables (total electrical load, wind energy generation and photovoltaic energy generation), as the underlying instrument for forecasts. Our novelty lies in the selection of the calibration sample which is achieved via a machine learning algorithm. Specifically, we utilized the *k*-NN instance-based learning algorithm to select the calibration sample based on a similarity (distance) measure between the most recent information and past observations, over a subset of the autoregressive model's variables. Our aim was to identify past observations that belong to the same "regime" with the latest available information.

The advantage of our method is therefore twofold. The selection of the calibration sample relies on a similarity measure over a set of variables rather than on the time dimension (i.e. to include only the most recent observations). With this type of selection, homogeneity within the calibration sample is secured and structural breaks are avoided. In addition, information from past observations is exploited and consequently, the selected calibration sample is expected to provide more accurate forecasts.

We applied our method on the EPEX SPOT market and we provided comparison with commonly used literature benchmarks. The results show that our proposed method achieves a statistically significant reduction in the forecasting error compared to the rest of the approaches, while remaining highly interpretable and meaningful. The accuracy of the proposed method in other markets, the adoption of other machine learning techniques as well as comparison with other methods relying exclusively on them, are subjects for future research. Nevertheless, our findings signify the importance and benefits of interdisciplinary research in this field.

Acknowledgments. This work is partially supported by the National Science Center (NCN, Poland) through MAESTRO grant no. 2018/30/A/HS4/00444 (D.S.). Also, it is partially supported by the Ministry of Science and Higher Education (MNiSW, Poland) through Diamond Grant no. 0027/DIA/2020/49 (W.N.) and Diamond Grant no. 0009/DIA/2020/49 (T.S.).

References

1. Aggarwal, S.K., Saini, L.M., Kumar, A.: Electricity price forecasting in deregulated markets: a review and evaluation. Int. J. Electr. Power Energy Syst. **31**(1), 13–22 (2009)
2. Ashfaq, T., Javaid, N.: Short-term electricity load and price forecasting using enhanced KNN. In: 2019 International Conference on Frontiers of Information Technology (FIT), pp. 266–2665 (2019)
3. Chaudhury, P., Tyagi, A., Shanmugam, P.K.: Comparison of various machine learning algorithms for predicting energy price in open electricity market. In: 2020 International Conference and Utility Exhibition on Energy, Environment and Climate Change (ICUE), pp. 1–7 (2020)
4. Chow, G.C.: Tests of equality between sets of coefficients in two linear regressions. Econometrica **28**(3), 591–605 (1960)
5. Diebold, F.X., Mariano, R.S.: Comparing predictive accuracy. J. Bus. Econ. Stat. **20**(1), 134–144 (2002)
6. Dudani, S.A.: The distance-weighted k-nearest-neighbor rule. IEEE Trans. Syst. Man Cybern. **SMC–6**(4), 325–327 (1976)
7. Hubicka, K., Marcjasz, G., Weron, R.: A note on averaging day-ahead electricity price forecasts across calibration windows. IEEE Trans. Sustain. Energy **10**(1), 321–323 (2019)
8. Jawad, M., et al.: Machine learning based cost effective electricity load forecasting model using correlated meteorological parameters. IEEE Access **8**, 146847–146864 (2020)
9. Kath, C., Nitka, W., Serafin, T., Weron, T., Zaleski, P., Weron, R.: Balancing generation from renewable energy sources: profitability of an energy trader. Energies **13**(1), 205 (2020)
10. Kiesel, R., Paraschiv, F.: Econometric analysis of 15-minute intraday electricity prices. Energy Econ. **64**, 77–90 (2017)
11. Killick, R., Fearnhead, P., Eckley, I.A.: Optimal detection of changepoints with a linear computational cost. J. Am. Stat. Assoc. **107**(500), 1590–1598 (2012)
12. Lavielle, M.: Using penalized contrasts for the change-point problem. Signal Process. **85**(8), 1501–1510 (2005)
13. Li, W., Kong, D., Wu, J.: A novel hybrid model based on extreme learning machine, k-nearest neighbor regression and wavelet denoising applied to short-term electric load forecasting. Energies **10**(5), 694 (2017)

14. Maciejowska, K., Nitka, W., Weron, T.: Day-ahead vs. intraday-forecasting the price spread to maximize economic benefits. Energies **12**(4), 631 (2019)
15. Maciejowska, K., Uniejewski, B., Serafin, T.: PCA forecast averaging—predicting day-ahead and intraday electricity prices. Energies **13**(14), 3530 (2020)
16. Marcjasz, G., Serafin, T., Weron, R.: Selection of calibration windows for day-ahead electricity price forecasting. Energies **11**(9), 2364 (2018)
17. de Marcos, R.A., Bunn, D.W., Bello, A., Reneses, J.: Short-term electricity price forecasting with recurrent regimes and structural breaks. Energies **13**(20), 5452 (2020)
18. Natividad, F., Folk, R.Y., Yeoh, W., Cao, H.: On the use of off-the-shelf machine learning techniques to predict energy demands of power TAC consumers. In: Ceppi, S., David, E., Hajaj, C., Robu, V., Vetsikas, I.A. (eds.) AMEC/TADA 2015-2016. LNBIP, vol. 271, pp. 112–126. Springer, Cham (2017). https://doi.org/10.1007/978-3-319-54229-4_8
19. Nawaz, M., et al.: An approximate forecasting of electricity load and price of a smart home using nearest neighbor. In: Barolli, L., Hussain, F.K., Ikeda, M. (eds.) CISIS 2019. AISC, vol. 993, pp. 521–533. Springer, Cham (2020). https://doi.org/10.1007/978-3-030-22354-0_46
20. Nowotarski, J., Raviv, E., Trück, S., Weron, R.: An empirical comparison of alternative schemes for combining electricity spot price forecasts. Energy Econ. **46**, 395–412 (2014)
21. Nowotarski, J., Weron, R.: Recent advances in electricity price forecasting: a review of probabilistic forecasting. Renew. Sustain. Energy Rev. **81**, 1548–1568 (2018)
22. Rocha, H.R.O., Honorato, I.H., Fiorotti, R., Celeste, W.C., Silvestre, L.J., Silva, J.A.L.: An Artificial Intelligence based scheduling algorithm for demand-side energy management in Smart Homes. Appl. Energy **282** (2021)
23. Weron, R.: Electricity price forecasting: a review of the state-of-the-art with a look into the future. Int. J. Forecast. **30**(4), 1030–1081 (2014)
24. Yamin, H.Y., Shahidehpour, S.M., Li, Z.: Adaptive short-term electricity price forecasting using artificial neural networks in the restructured power markets. Int. J. Electr. Power Energy Syst. **26**(8), 571–581 (2004)
25. Yesilbudak, M., Sagiroglu, S., Colak, I.: A novel implementation of kNN classifier based on multi-tupled meteorological input data for wind power prediction. Energy Convers. Manage. **135**, 434–444 (2017)
26. Zhang, R., Xu, Y., Dong, Z.Y., Kong, W., Wong, K.P.: A composite k-nearest neighbor model for day-ahead load forecasting with limited temperature forecasts. In: 2016 IEEE Power and Energy Society General Meeting (PESGM), pp. 1–5 (2016)
27. Zhao, J.H., Dong, Z.Y., Xu, Z., Wong, K.P.: A statistical approach for interval forecasting of the electricity price. IEEE Trans. Power Syst. **23**(2), 267–276 (2008)
28. Ziel, F., Weron, R.: Day-ahead electricity price forecasting with high-dimensional structures: Univariate vs. multivariate modeling frameworks. Energy Econ. **70**, 396–420 (2018)

Data-Driven Methods for Weather Forecast

Elias David Nino-Ruiz[✉][iD] and Felipe J. Acevedo García[iD]

Applied Math and Computer Science Lab, Computer Science Department,
Universidad del Norte, Colombia, Barranquilla, Colombia
{enino,fjacevedo}@uninorte.edu.co
https://sites.google.com/vt.edu/eliasn/home?authuser=1

Abstract. In this paper, we propose efficient and practical data-driven methods for weather forecasts. We exploit the information brought by historical weather datasets to build machine-learning-based models. These models are employed to produce numerical forecasts, which can be improved by injecting additional data via data assimilation. Our approaches' general idea is as follows: given a set of time snapshots of some dynamical system, we group the data by time across multiple days. These groups are employed to build first-order Markovian models that reproduce dynamics from time to time. Our numerical models' precision can be improved via sequential data assimilation. Experimental tests are performed by using the National-Centers-for-Environmental-Prediction Department-of-Energy Reanalysis II dataset. The results reveal that numerical forecasts can be obtained within reasonable error magnitudes in the L_2 norm sense, and even more, observations can improve forecasts by order of magnitudes, in some cases.

Keywords: Data assimilation · Markovian model · Machine learning

1 Introduction

Numerical weather forecasts are of extreme importance in different aspects of life, particularly in scenarios wherein human lives can be compromised (i.e., forecasts of storms, floods, hurricanes, and tsunamis) [10]. Numerical models are commonly employed to mimic the behavior of actual system dynamics, for instance, the ocean and/or the atmosphere [2, 6, 9]. Since numerical models are computationally demanding, high-performance-computing is a must to produce forecasts within reasonable computational times, especially for high-resolution grids. On the other hand, we can find data-driven models that can exploit decades of meteorological information to represent the future as some potential combination of the past. For instance, the National-Centers-for-Environmental-Prediction Department-of-Energy (NCEP-DOE) Reanalysis II [5] is a data set that holds meteorological information since 1979 onto global grids at varying resolutions. It is possible to use these sources of information to come up with statistical forecasts. We think

M. Paszynski et al. (Eds.): ICCS 2021, LNCS 12745, pp. 326–336, 2021.
https://doi.org/10.1007/978-3-030-77970-2_25

there is an opportunity to compute cheap models that produce forecasts with low computational effort, and even more, we can improve such forecasts by injecting real-time data via sequential data assimilation.

This paper is organized as follows: in Sect. 2 we discuss topics related to sequential data assimilation and machine learning models, Sect. 3 presents a sequential data assimilation method via machine learning models wherein numerical models are replaced by statistical ones, in Sect. 4 experiments are carried out by employing the NCEP-DOE Reanalysis II data set, and the conclusions of this research are stated in Sect. 5.

2 Preliminaries

In this section, we discuss some topics related to Machine Learning and Sequential Data Assimilation methods. These are necessary for the understanding of our proposed methods.

2.1 Machine Learning Models

In the context of Machine Learning (ML), parametric models can be seen as structures for the solution of the inverse problem [11]:

$$y = f(\boldsymbol{\beta}, \mathbf{z}) + \epsilon, \tag{1}$$

where y is an observation, $\boldsymbol{\beta} \in \mathbb{R}^{p \times 1}$ is a vector holding the parameters, p is the number of parameters, variables are stored in vector $\mathbf{z} \in \mathbb{R}^{v \times 1}$, v is the number of variables, $f : \mathbb{R}^{p \times 1} \times \mathbb{R}^{v \times 1} \to 1$, and ϵ can be described by some probability density function. The simplest model in which one can think is a linear one of the form:

$$y = \sum_{u=1}^{v} \beta_u \cdot z_u + \epsilon, \text{ with } \epsilon \sim \mathcal{N}\left(0, \sigma^2\right), \tag{2}$$

where $v = p$, β_u and z_u denote the u-th component of vectors $\boldsymbol{\beta}$ and $\mathbf{z}$, respectively. Since Gaussian assumptions can be easily broken on residuals in (9), local linear models can be built to preserve Gaussian shapes (i.e., by considering, local modes of error distributions). There are many manners to do this, the simpler, to weight each sample $\mathbf{x}_j \in \mathbb{R}^{v \times 1}$, for $1 \le j \le m$, with regard to its distance to the observations $\mathbf{y} \in \mathbb{R}^{m \times 1}$,

$$\mathbf{y} = \sum_{j=1}^{K<m} \beta_j \cdot \alpha_j(\mathbf{x}_j, \mathbf{y}) \cdot \mathbf{x}_j + \boldsymbol{\epsilon}, \text{ with } \boldsymbol{\epsilon} \sim \mathcal{N}\left(0, \sigma^2 \cdot \mathbf{I}\right), \tag{3a}$$

where m is the number of samples (observations), K is the number of closest points to $\mathbf{y}$ onto the hyperplane formed by the samples $\mathbf{x}_j$, $\boldsymbol{\epsilon}$ is a vector holding the residuals. Likewise, $\alpha_j(\mathbf{x}_s, \mathbf{y})$ is a weight/distance function, common choices are the Uniform distance

$$\alpha_j(\mathbf{x}_j, \mathbf{y})_\infty = \|\mathbf{x}_j - \mathbf{y}\|_\infty^{-1}, \tag{3b}$$

or the reciprocal of the Euclidean distance (L_2-norm):

$$\alpha_j(\mathbf{x}_j, \mathbf{y})_2 = \|\mathbf{x}_j - \mathbf{y}\|_2^{-1}. \tag{3c}$$

This kind of strategy is well-known in the statistical context as the K-Nearest-Neighbors (KNN) regression [12].

2.2   Sequential Data Assimilation

The ensemble Kalman filter (EnKF) is a well-established sequential Monte Carlo method for parameter and state estimation in highly non-linear models [3,4]. The EnKF describes the error statistics via an ensemble of model realizations:

$$\mathbf{X}^b = \left[\mathbf{x}^{b[1]}, \mathbf{x}^{b[2]}, \ldots, \mathbf{x}^{b[N]}\right] \in \mathbb{R}^{n \times N}, \tag{4}$$

where $\mathbf{x}^{b[e]}$, for $1 \le e \le N$, is the e-th ensemble member, N is the ensemble size, and n denotes the model dimension. The background ensemble (4) can be employed to estimate the moments of the background error distribution, this is, the background state $\mathbf{x}^b \in \mathbb{R}^{n \times 1}$

$$\mathbf{x}^b \approx b = \frac{1}{N} \cdot \sum_{e=1}^{N} \mathbf{x}^{b[e]} \in \mathbb{R}^{n \times 1}, \tag{5a}$$

and the background error covariance matrix $\mathbf{B} \in \mathbb{R}^{n \times n}$

$$\mathbf{B} \approx \mathbf{P}^b = \frac{1}{N-1} \cdot \boldsymbol{\Delta}\mathbf{X}^b \cdot \left[\boldsymbol{\Delta}\mathbf{X}^b\right]^T \in \mathbb{R}^{n \times n}, \tag{5b}$$

where b is the ensemble mean, and $\mathbf{P}^b$ is the ensemble covariance. Likewise, $\boldsymbol{\Delta}\mathbf{X}^b \in \mathbb{R}^{n \times N}$ stands for the matrix of member deviations $\boldsymbol{\Delta}\mathbf{X}^b = \mathbf{X}^b - b \cdot \mathbf{1}_N^T$ wherein $\mathbf{1}_N$ is an N-dimensional vector whose components are all ones. Observations are related to model states via the linear observation operator $\mathbf{H} \in \mathbb{R}^{m \times n}$

$$\mathbf{y} = \mathbf{H} \cdot \mathbf{x} + \boldsymbol{\varepsilon} \in \mathbb{R}^{m \times 1},$$

where $\mathbf{H}$ maps model states onto observation spaces, m is the number of observations, the white noise vector reads $\boldsymbol{\varepsilon} \sim \mathcal{N}(\mathbf{0}, \mathbf{R})$, and $\mathbf{R} \in \mathbb{R}^{m \times m}$ is the data error covariance matrix. By using Bayes' theorem, we can find the state that maximizes the posterior probability given an observation $\mathbf{y}$ as follows:

$$\mathbf{x}^a = \mathbf{x}^b + \mathbf{P}^b \cdot \mathbf{H}^T \cdot \left[\mathbf{R} + \mathbf{H} \cdot \mathbf{P}^b \cdot \mathbf{H}^T\right] \cdot \left[\mathbf{y} - \mathbf{H} \cdot \mathbf{x}^b\right] \in \mathbb{R}^{n \times 1}, \tag{6}$$

Since model realizations come at high computational costs, ensemble sizes are constrained by the hundreds while model resolutions range in the order of millions. This tigers spurious correlations between errors in distant model components. To mitigate this, better estimations of $\mathbf{B}$ are sought. For instance, sparse precision covariances of the form can be computed as follows:

$$\widehat{\mathbf{B}}^{-1} = \widehat{\mathbf{L}}^T \cdot \widehat{\mathbf{D}}^{-1} \cdot \widehat{\mathbf{L}} \in \mathbb{R}^{n \times n} \tag{7}$$

where $\widehat{\mathbf{L}} \in \mathbb{R}^{n \times n}$ is a sparse lower triangular matrix, and $\widehat{\mathbf{D}}^{-1}$ is a diagonal matrix [1]. By replacing (7) in (6) the EnKF via a modified Cholesky decomposition [7,8] can be obtained.

3   Proposed Method

Consider time snapshots of some physical variables onto a numerical mesh grid (i.e., the NCEP-DOE Reanalysis II data set). We group the snapshots (data) by time across different S days:

$$\mathbf{X}_\ell = \left[\mathbf{x}_\ell^{[1]}, \mathbf{x}_\ell^{[2]}, \ldots, \mathbf{x}_\ell^{[S]}\right] \in \mathbb{R}^{n \times S}, \tag{8}$$

where $\mathbf{X}_\ell$ is the ensemble holding all snapshots at time t_ℓ across different days, for $1 \leq \ell \leq L$, L is the number of snapshots in a single day, $\mathbf{x}_\ell^{[s]} \in \mathbb{R}^{n \times 1}$ is the snapshot of day s, for $1 \leq s \leq S$, at time t_ℓ, and S is the number of days (number of snapshots for time t_ℓ across different days). We consider evenly-spaced time snapshots, and we assume that the same number of snapshots are available for all ensembles $\mathbf{X}_\ell$. We then consider to fit models of the form:

$$\mathbf{X}_\ell = \mathbf{M}_{\ell,\ell-1} \cdot \mathbf{X}_{\ell-1} + \mathbf{E}_\ell, \tag{9}$$

where $\mathbf{E}_\ell \in \mathbb{R}^{n \times S}$ holds the residuals, and $\mathbf{M}_{\ell,\ell-1} \in \mathbb{R}^{n \times n}$ is a data-driven model which partially captures the evolution of system dynamics from time $t_{\ell-1}$ to t_ℓ. We then consider cost functions of the form:

$$\mathcal{J}\left(\mathbf{M}_{\ell,\ell-1}\right) = \frac{1}{2} \cdot \left\|\mathbf{X}_\ell - \mathbf{M}_{\ell,\ell-1} \cdot \mathbf{X}_{\ell-1}\right\|_2^2, \tag{10}$$

to estimate linear operators $\mathbf{M}_{\ell,\ell-1}$ which transport dynamics from time $t_{\ell-1}$ to t_ℓ, the optimization problem to solve reads:

$$\mathbf{M}_{\ell,\ell-1}^* = \arg \min_{\mathbf{M}_{\ell,\ell-1}} \mathcal{J}\left(\mathbf{M}_{\ell,\ell-1}\right). \tag{11}$$

It can be easily shown that the gradient of (10) reads

$$\nabla_{\mathbf{M}_{\ell,\ell-1}}\left(\mathcal{J}\left(\mathbf{M}_{\ell,\ell-1}\right)\right) = \mathbf{X}_\ell \cdot \left[\mathbf{X}_{\ell-1}\right]^T - \mathbf{M}_{\ell,\ell-1} \cdot \mathbf{X}_{\ell-1} \cdot \left[\mathbf{X}_{\ell-1}\right]^T, \tag{12}$$

from which the solution of (11) reads:

$$\mathbf{M}_{\ell,\ell-1}^* = \mathbf{X}_\ell \cdot \left[\mathbf{X}_{\ell-1}\right]^T \cdot \left[\mathbf{X}_{\ell-1} \cdot \left[\mathbf{X}_{\ell-1}\right]^T\right]^{-1}. \tag{13}$$

We now have a piecewise linear model $\left\{\mathbf{M}_{\ell,\ell-1}^*\right\}_{\ell=1}^L$ which mimics the behavior of the dynamical system in a single day. Note that, these set of models can be seen as first order Markovian models wherein all information needed to propagate dynamics from time $t_{\ell-1}$ to t_ℓ is condensed into $\mathbf{M}_{\ell-1,\ell}$.

3.1   Sequential Data Assimilation

Since linear models of the form (13) can partially capture the actual dynamics, we can improve their accuracies by using sequential data assimilation. For each time t_ℓ, we can build the precision matrix $\widehat{\mathbf{B}}_\ell^{-1}$ via the ensemble (8) and the modified Cholesky decomposition:

$$\widehat{\mathbf{B}}_\ell^{-1} = \widehat{\mathbf{L}}_\ell^T \cdot \widehat{\mathbf{D}}_\ell^{-1} \cdot \widehat{\mathbf{L}}_\ell \in \mathbb{R}^{n \times n}, \tag{14}$$

where the factors $\widehat{\mathbf{L}}_\ell$, and $\widehat{\mathbf{D}}_\ell^{-1}$ are computed as in Sect. 2. We model the uncertainties of any state at time t_ℓ by (14). Consider the analysis state $\mathbf{x}_{\ell-1}^a$, it is clear that:

$$\mathbf{x}_\ell^b = \mathbf{M}_{\ell,\ell-1}^* \cdot \mathbf{x}_{\ell-1}^a = \mathbf{X}_\ell \cdot [\mathbf{X}_{\ell-1}]^T \cdot \left[\mathbf{X}_{\ell-1} \cdot [\mathbf{X}_{\ell-1}]^T\right]^{-1} \cdot \mathbf{x}_{\ell-1}^a. \tag{15}$$

Consider the observation $\mathbf{y}_\ell \in \mathbb{R}^{m \times 1}$, at time t_ℓ, the analysis state can be computed as follows:

$$\mathbf{x}_\ell^a = \left[\widehat{\mathbf{B}}_\ell^{-1} + \mathbf{H}_\ell^T \cdot \mathbf{R}_\ell^{-1} \cdot \mathbf{H}_\ell\right]^{-1}$$
$$\cdot \left[\widehat{\mathbf{B}}_\ell^{-1} \cdot \mathbf{X}_\ell \cdot [\mathbf{X}_{\ell-1}]^T \cdot \left[\mathbf{X}_{\ell-1} \cdot [\mathbf{X}_{\ell-1}]^T\right]^{-1} \cdot \mathbf{x}_{\ell-1}^a + \mathbf{H}_\ell^T \cdot \mathbf{R}_\ell^{-1} \cdot \mathbf{y}_\ell\right] \tag{16}$$

The analysis state (16) is propagated until new observations are available.

3.2   Building Local Linear Models

In practice, model dynamics can be highly non-linear and therefore, Gaussian assumptions on residuals in (9) can be easily broken. To mitigate this, consider the ensemble $\mathbf{X}(\mathbf{x}_{\ell-1}^a, K) \in \mathbb{R}^{n \times K}$ formed by the K nearest states to $\mathbf{x}_{\ell-1}^a$ from the ensemble of snapshots $\mathbf{X}_{\ell-1}$ at time $t_{\ell-1}$: the ensemble $\mathbf{X}(\mathbf{x}_{\ell-1}^a, K)$ can be exploited to build local linear models

$$\mathbf{x}_\ell^b = \mathbf{M}\left(\mathbf{x}_{\ell-1}^a, K\right)_{\ell,\,\ell-1}^* \cdot \mathbf{x}_{\ell-1}^a, \tag{17a}$$

following a similar reasoning to that of (3a), (3b) and (3c). This is, by considering the reciprocal of Uniform or Euclidean distances. Besides, K can be employed as well to fit local Gaussian models for prior errors by using the forecast state (15), and its K nearest states (neighbors) in (8), $\mathbf{X}(\mathbf{x}_\ell^b, K) \in \mathbb{R}^{n \times K}$; from here we can obtain local approximations of precision matrices of the form (14):

$$\widehat{\mathbf{B}}^{-1}\left(\mathbf{x}_\ell^b, K\right)_\ell = \widehat{\mathbf{L}}\left(\mathbf{x}_\ell^b, K\right)_\ell^T \cdot \widehat{\mathbf{D}}\left(\mathbf{x}_\ell^b, K\right)_\ell^{-1} \cdot \widehat{\mathbf{L}}\left(\mathbf{x}_\ell^b, K\right)_\ell \in \mathbb{R}^{n \times n}, \tag{17b}$$

where $\widehat{\mathbf{L}}\left(\mathbf{x}_\ell^b, K\right)_\ell^T \in \mathbb{R}^{n \times n}$, and $\widehat{\mathbf{D}}\left(\mathbf{x}_\ell^b, K\right)_\ell^{-1} \in \mathbb{R}^{n \times n}$ are computed similar to those in (14). We decide to make use of the precision matrix $\widehat{\mathbf{B}}^{-1}$ in our formulation since it allows us to exploit the use of computational resources (the resulting estimator is sparse).

The background state (17a) and the precision matrix (17b) can be easily incorporated into the data assimilation framework (16), this is, we can produce forecasts via (17a), and analysis states:

$$\mathbf{x}_\ell^a(K) = \left[\widehat{\mathbf{B}}^{-1} \left(\mathbf{x}_\ell^b, K \right)_\ell + \mathbf{H}_\ell^T \cdot \mathbf{R}_\ell^{-1} \cdot \mathbf{H}_\ell \right]^{-1}$$
$$\cdot \left[\widehat{\mathbf{B}}^{-1} \left(\mathbf{x}_\ell^b, K \right)_\ell \mathbf{M} \left(\mathbf{x}_{\ell-1}^a, K \right)_{\ell, \ell-1}^* \cdot \mathbf{x}_{\ell-1}^a + \mathbf{H}_\ell^T \cdot \mathbf{R}_\ell^{-1} \cdot \mathbf{y}_\ell \right] \quad (17c)$$

Based on equation (17c), we denote by $\mathbf{x}_\ell^a(K)^\infty$ the analysis produced by using weights (3b) in the computation of (17a), and by $\mathbf{x}_\ell^a(K)^2$ the analysis obtained by employing weights (3c) in the computation of (17a).

Now, we are ready to test our proposed methods by using real-life meteorological data.

4   Experimental Results

To assess the accuracy of our method, we make use of the NCEP-DOE Reanalysis II dataset [5]. In this set, some physical variables such as air temperature T, relative humidity q, and wind components u and v are available four times daily, and those are the ones that we consider in our experiments. In this context, t_ℓ denotes the hour of the day in which snapshots are taken, $t_\ell \in \{0, 6, 12, 18\}$. Thus, we have four linear models:

- model $\mathbf{M}^*(\mathbf{x}_0, K)_{0,1}$ propagates a state $\mathbf{x}_0$ from hour 0 to hour 6,
- model $\mathbf{M}^*(\mathbf{x}_1, K)_{1,2}$ propagates a state $\mathbf{x}_1$ from hour 6 to hour 12,
- model $\mathbf{M}^*(\mathbf{x}_2, K)_{2,3}$ propagates a state $\mathbf{x}_2$ from hour 12 to hour 18, and
- model $\mathbf{M}^*(\mathbf{x}_3, K)_{3,0}$ propagates a state $\mathbf{x}_3$ from hour 18 to hour 0 (next day).

For each hour t_ℓ, by using the ensemble of snapshots $\mathbf{X}_\ell$, a $\widehat{\mathbf{B}}_\ell^{-1}$ matrix is estimated. A general structure of our piecewise linear model under these settings can be seen in Fig. 1.

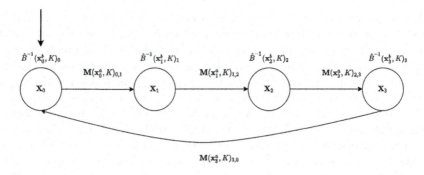

Fig. 1. Piecewise first order Markovian model for the NCEP-DOE Reanalysis II dataset.

Additional settings are described below:

– for all physical variables, we consider snapshots on the surface level,
– we consider the number of nearest states $K \in \{10, 20, 50\}$,
– local linear models are computed by using the weights (3b) and (3c),
– the local models are trained with snapshots from January 1st to July 31rd, 2020,
– the forecast and data assimilation process are carried out between august 24, 2020 to august 27, 2021,
– observations are taken every six hours (corresponding with the elapsed time between snapshots), these are simulated from the dataset by employing the following standard deviation for observation errors:
  • Temperature $1\,K°$.
  • Zonal Wind Component $1\,m/s$.
  • Meridional Wind Component $1\,m/s$.
  • Specific Humidity $10^{-3}\,kg/kg$.
– the number of observations is only 5% of model components, these are randomly placed during assimilation steps,
– as measures of accuracies, we consider the L_2-norm of errors

$$\epsilon_k = \|\mathbf{x}_k^* - \mathbf{x}_k\|_2\,, \tag{18}$$

where $\mathbf{x}_k^*$ and $\mathbf{x}_k$ are the reference state (actual snapshot) and its approximation at time t_ℓ, respectively. $\mathbf{x}_k$ can be obtained via pure forecasts or analysis states. The Root-Mean-Square-Error (RMSE) measures, in average, the performance of a method within an assimilation window with M time spaced set of observations $\{\mathbf{y}_k\}_{k=1}^{M}$:

$$\bar{\epsilon} = \sqrt{\frac{1}{M} \cdot \sum_{k=1}^{M} \epsilon_k}. \tag{19}$$

4.1   Time Evolution of Errors

In Fig. 3, the L_2-norm of errors are shown for pure forecasts as well as analysis states. The results are shown for different values of K and the uniform weights (3b) (U) and the Euclidean ones (3c) (E). We let dashed and solid lines represent forecasts and analysis errors, respectively. As can be seen, the proposed models (3a) can produce accurate forecasts. For instance, error levels remain almost constant for the entire assimilation window for all model variables. This can be explained as follows: since our general model is piecewise linear, linear models can properly capture actual dynamics in time periods of six hours. Moreover, by injecting "current" information into the system, such forecasts can be improved. For instance, errors in model variables such as T, u, and v can improve over a magnitude, in some cases. Improvements can be seen in the q variable as well for the overall time window. Note that forecast errors are sensitive to the choices

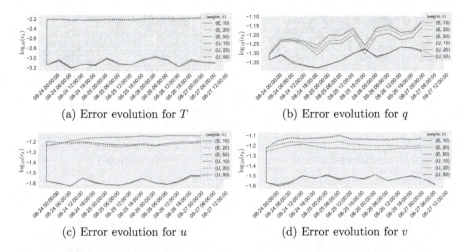

(a) Error evolution for T (b) Error evolution for q

(c) Error evolution for u (d) Error evolution for v

Fig. 2. Error evolution of model variables in the L_2 norm. Dashed lines denote pure forecasts while analysis solutions are shown in solid lines. Results are presented in the log-scale for easiness in reading.

of the number of neighbors K and the chosen distance (U or E). This is very common in the context of ML methods. Besides, those parameters are strictly related to the computation of (3a). For some variables, the differences in model trajectories for pure forecasts are evident. Nevertheless, the analysis solutions for all cases remain similar; their differences are imperceptible (recall that errors are in log-scale). This is very relevant since one of the current challenges in ML based methods is the tune of model parameters such as K. However, by using data assimilation (17c), the resulting states seem to be non-sensitive to the choices of K nor the weighing metric for distances. Of course, further research is needed to come up with a satisfactory answer to this (Fig. 2).

4.2   Mean of Errors for the Assimilation Window

In Fig. 3, we show the average of L_2-norm of errors for some model variables. As can be seen, the assimilation of observations can improve the quality of forecasts. For instance, in the T variable, many dense regions owing to the accumulation of errors can be dissipated by using the information brought by observations of the system. Furthermore, the use of local information for computing the precision matrices (17b) allows for the dissipation of spurious correlation in errors of distant model components (in space). The use of local samples (snapshots) can mitigate the impact of multi-modal prior error distributions (which are commonly in highly non-linear models). In this manner, samples that potentially belong to another mode of the prior distribution are neglected during the estimation of prior error moments. Similar behavior can be noted for other model variables such as v.

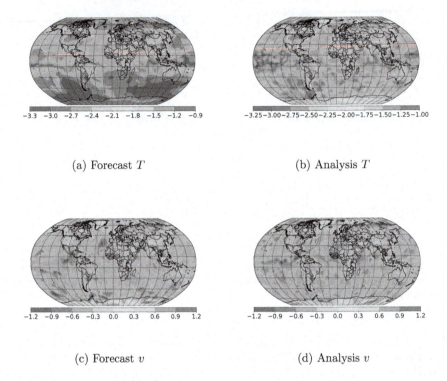

(a) Forecast T (b) Analysis T

(c) Forecast v (d) Analysis v

Fig. 3. Mean of L_2-norm of errors in the log-scale for the assimilation window for $K = 10$ and the weighting metric U.

In Fig. 4, we show the initial background, analysis state, and reference snapshot for the variable T. This step is of interest since no actual information of the system has been injected into the numerical model. We can see some spurious waves near the south pole for the forecast state. For instance, a lower level of temperature than those of the actual state is reported. We can see that, by using only 5% of observations, the accuracy can be drastically improved. For instance, the temperature levels are adjusted similarly to those of the actual snapshot. This mainly obeys the implicit background error correlations captured in $\hat{\mathbf{B}}_0^{-1}(\mathbf{x}_0{}^b, K)$: analysis increments are properly weighted as background error correlations are well estimated. This has a very important impact on the updating process of model components wherein no observations are available. In our settings, most of the Earth is unobserved (95%).

In the Table 1, we report the RMSE values for all model variables and all configurations. As we mentioned before, errors in forecasts are highly sensitive to ML parameters, as should be expected since these models rely on parameters such as K. Something that is very attractive is that we do not see high variations in errors, which is important since ML parameters can be hard to tune (i.e., parameter selection can vary from problem to problem). Hence, forecast errors are of similar magnitude regardless of the choice of K, for instance. On the other

230 240 250 260 270 280 290 300 310 320

240 248 256 264 272 280 288 296 304 312

210 225 240 255 270 285 300 315 330

(a) Actual Snapshot   (b) Forecast State   (c) Analysis State

Fig. 4. Snapshots in the globe for the T variable by using Euclidean weights and $K = 10$.

hand, we can see similar results of analysis errors for each configuration. Again, this obeys the fact that background error correlations are properly estimated during assimilation steps. Furthermore, the impact of multi-modal prior error distributions can be mitigated by considering similar snapshots to $\mathbf{x}_\ell^b$ during the estimation of $\widehat{\mathbf{B}}^{-1}(\mathbf{x}_\ell^b, K)_\ell$.

Table 1. Root-Mean-Square-Error values for the entire assimilation window, for all variables and different configuration of parameters.

Distance	K variables	Analysis			Forecast		
		10	20	50	10	20	50
E (3c)	T	0.0288	0.0287	0.0283	0.0790	0.0786	0.0802
	q	0.2190	0.2192	0.2187	0.2528	0.2482	0.2427
	u	0.1629	0.1626	0.1625	0.2479	0.2454	0.2624
	v	0.1713	0.1712	0.1705	0.2709	0.2540	0.2442
U (3b)	T	0.0288	0.0287	0.0283	0.0790	0.0789	0.0803
	q	0.2190	0.2190	0.2187	0.2513	0.2481	0.2425
	u	0.1629	0.1626	0.1625	0.2466	0.2448	0.2616
	v	0.1712	0.1712	0.1705	0.2692	0.2540	0.2439

5   Conclusions

In this paper, we propose a piecewise first-order Markovian model for the weather forecast. The proposed method employs linear models to mimic the behavior of the system in time intervals. For each time interval, a linear model is built to forecast system states. The accuracy of the forecast can be improved by using Machine Learning models such as the K-Nearest-Neighbor regression. In this context, we employ weighing distances to compute linear models: the reciprocal of the uniform and the euclidean distances. Besides, sequential data assimilation can be exploited to inject real-time information of the system into our Markovian model.

The estimation of background error correlation is performed by using a modified Cholesky decomposition and considering the K nearest snapshots to the forecast state. Experimental tests are performed by employing the National-Centers-for-Environmental-Prediction Department-of-Energy Reanalysis II dataset. The results reveal that numerical forecasts can be obtained within reasonable error magnitudes in the L_2 norm sense, they do not blow up, and even more, observations can improve forecasts by order of magnitudes, in some cases, for the entire assimilation window.

Acknowledgment. This work was supported by the Applied Math and Computer Science Laboratory (AML-CS) at Universidad del Norte, BAQ, COL.

References

1. Bickel, P.J., Levina, E., et al.: Covariance regularization by thresholding. Ann. Stat. **36**(6), 2577–2604 (2008)
2. Bouttier, F., Courtier, P.: Data Assimilation Concepts and Methods March 1999. Meteorological Training Course Lecture Series, vol. 718, p. 59. ECMWF (2002)
3. Evensen, G.: Data Assimilation: The Ensemble Kalman Filter. Springer, Heidelberg (2009). https://doi.org/10.1007/978-3-642-03711-5
4. Houtekamer, P.L., Mitchell, H.L.: Ensemble Kalman filtering. Q. J. Roy. Meteorol. Soc. A J. Atmos. Sci. Appl. Meteorol. Phys. Oceanogr. **131**(613), 3269–3289 (2005)
5. Kanamitsu, M., et al.: NCEP-DOE AMIP-II reanalysis (R-2). Bull. Am. Meteor. Soc. **83**(11), 1631–1644 (2002)
6. Law, K., Stuart, A., Zygalakis, K.: Data Assimilation, vol. 214. Springer, Cham (2015). https://doi.org/10.1007/978-3-319-20325-6
7. Nino-Ruiz, E.D., Sandu, A., Deng, X.: An ensemble Kalman filter implementation based on modified Cholesky decomposition for inverse covariance matrix estimation. SIAM J. Sci. Comput. **40**(2), A867–A886 (2018)
8. Nino-Ruiz, E.D., Sandu, A., Deng, X.: A parallel implementation of the ensemble Kalman filter based on modified Cholesky decomposition. J. Comput. Sci. **36**, 100654 (2019)
9. Reichle, R.H.: Data assimilation methods in the earth sciences. Adv. Water Resour. **31**(11), 1411–1418 (2008)
10. Richardson, L.F.: Weather Prediction by Numerical Process. Cambridge University Press, Cambridge (2007)
11. Tarantola, A.: Inverse Problem Theory and Methods for Model Parameter Estimation. SIAM, Philadelphia (2005)
12. Zhang, S., Li, X., Zong, M., Zhu, X., Cheng, D.: Learning k for KNN classification. ACM Trans. Intell. Syst. Technol. (TIST) **8**(3), 1–19 (2017)

Generic Case of Leap-Frog Algorithm for Optimal Knots Selection in Fitting Reduced Data

Ryszard Kozera[1,2]([✉])[ID], Lyle Noakes[2][ID], and Artur Wiliński[1][ID]

[1] Institute of Information Technology, Warsaw University of Life Sciences - SGGW, Ul. Nowoursynowska 159, 02-776 Warsaw, Poland
{ryszard_kozera,artur_wilinski}@sggw.edu.pl
[2] School of Physics, Mathematics and Computing, The University of Western Australia, 35 Stirling Highway, Crawley, Perth, WA 6009, Australia
lyle.noakes@uwa.edu.au

Abstract. The problem of fitting multidimensional reduced data $\mathcal{M}_n$ is discussed here. The unknown interpolation knots $\mathcal{T}$ are replaced by optimal knots which minimize a highly non-linear multivariable function $\mathcal{J}_0$. The numerical scheme called *Leap-Frog Algorithm* is used to compute such optimal knots for $\mathcal{J}_0$ via the iterative procedure based in each step on single variable optimization of $\mathcal{J}_0^{(k,i)}$. The discussion on conditions enforcing unimodality of each $\mathcal{J}_0^{(k,i)}$ is also supplemented by illustrative examples both referring to the generic case of *Leap-Frog*. The latter forms a new insight on fitting reduced data and modelling interpolants of $\mathcal{M}_n$.

Keywords: Interpolation · Optimization · Curve modelling

1 Introduction

In this work the problem of interpolating n points $\mathcal{M}_n = \{x_i\}_{i=0}^n$ in arbitrary Euclidean space $\mathbb{E}^m$ is addressed. The corresponding knots $\mathcal{T} = \{t_i\}_{i=1}^{n-1}$ are assumed to be unknown. The class of fitting functions (curves) $\mathcal{I}$ considered in this paper represents piecewise C^2 curves $\gamma : [0, T] \to \mathbb{E}^m$ satisfying $\gamma(t_i) = q_i$ and $\ddot{\gamma}(t_0) = \ddot{\gamma}(T) = \mathbf{0}$. It is also assumed that $\gamma \in \mathcal{I}$ is at least of class C^1 over $\mathcal{T}_{int} = \{t_i\}_{i=1}^{n-1}$ and extends to $C^2([t_i, t_{i+1}])$. Additionally, *the unknown internal knots $\mathcal{T}_{int}$ are allowed to vary* subject to $t_i < t_{i+1}$, for $i = 0, 1, \ldots, n-1$ (here $t_0 = 0$ and $t_n = T$). Such knots are called admissible and choosing them according to some adopted criterion permits *to control and model* the trajectory of γ. One of such criterion might focus on minimizing "average squared norm acceleration" of γ. In fact, for a given choice of fixed knots $\mathcal{T}$, the task of minimizing

$$\mathcal{J}_T(\gamma) = \sum_{i=0}^{n-1} \int_{t_i}^{t_{i+1}} \|\ddot{\gamma}(t)\|^2 dt , \tag{1}$$

(over $\mathcal{I}$) yields a unique optimal curve $\gamma_{opt} \in \mathcal{I}$ forming *a natural cubic spline* γ_{NS} - see [1] or [8]. Consequently, letting the internal knots $\mathcal{T}_{int}$ change, minimizing $\mathcal{J}_T$ over $\mathcal{I}$ reduces to searching for an optimal natural spline γ_{NS} with

© Springer Nature Switzerland AG 2021
M. Paszynski et al. (Eds.): ICCS 2021, LNCS 12745, pp. 337–350, 2021.
https://doi.org/10.1007/978-3-030-77970-2_26

$\mathcal{T}_{int}$ treated as free variables. Thus by [1], having recalled that γ_{NS} is uniquely determined by $\mathcal{T}$, minimizing $\mathcal{J}_{\mathcal{T}}$ amounts to optimizing a highly non-linear function J_0 in $n-1$ variables $\mathcal{T}_{int}$ satisfying $t_i < t_{i+1}$ (see [3]). Due to the high non-linearity of J_0 the majority of numerical schemes applied to optimize J_0 lead to numerical difficulties (see e.g. [3]). Similarly, the analysis of critical points of J_0 forms a complicated task. To alleviate the latter, *a Leap-Frog* can be applied to deal with J_0 - see [2] or [3]. This scheme minimizes J_0 with iterative sequence of single variable overlapping optimizations of $J_0^{(k,i)}$ subject to $t_i < t_{i+1}$.

The novelty of this work refers to *the generic* case of *Leap-Frog* (recursively applied over each internal snapshots). The analysis establishing sufficient conditions for *unimodality* of $J_0^{(k,i)}$ is conducted here. Numerical tests and illustrative examples supplement the latter. The discussion covers first a special case of data (see Sect. 4) extended next to its perturbation (see Sect. 5 and Theorem 1). More information on numerical performance of *Leap-Frog* and comparison tests with two standard numerical optimization schemes can be found in [2,3] or recently published [6]. Some applications of *Leap-Frog* optimization scheme used also as a modelling and simulation tool are discussed in [9,10] or [11].

2   Preliminaries

Recall (see [1]) that *a cubic spline interpolant* $\gamma_{\mathcal{T}}^{C_i} = \gamma_{\mathcal{T}}^C|_{[t_i,t_{i+1}]}$, for given admissible knots $\mathcal{T} = (t_0, t_1, \ldots, t_{n-1}, t_n)$ is defined as $\gamma_{\mathcal{T}}^{C_i}(t) = c_{1,i} + c_{2,i}(t - t_i) + c_{3,i}(t - t_i)^2 + c_{4,i}(t - t_i)^3$, (for $t \in [t_i, t_{i+2}]$) to satisfy (for $i = 0, 1, 2, \ldots, n-1$; $c_{j,i} \in \mathbb{R}^m$, where $j = 1, 2, 3, 4$) $\gamma_{\mathcal{T}}^{C_i}(t_{i+k}) = x_{i+k}$ and $\dot{\gamma}_{\mathcal{T}}^{C_i}(t_{i+k}) = v_{i+k}$, for $k = 0, 1$ with the velocities $v_0, v_1, \ldots, v_{n-1}, v_n \in \mathbb{R}^m$ assumed to be temporarily free parameters (*if unknown*). The coefficients $c_{j,i}$ read (with $\Delta t_i = t_{i+1} - t_i$):

$$c_{1,i} = x_i, \qquad\qquad c_{2,i} = v_i,$$

$$c_{4,i} = \frac{v_i + v_{i+1} - 2\frac{x_{i+1} - x_i}{\Delta t_i}}{(\Delta t_i)^2}, \qquad c_{3,i} = \frac{\frac{(x_{i+1} - x_i)}{\Delta t_i} - v_i}{\Delta t_i} - c_{4,i}\Delta t_i. \qquad (2)$$

The latter follows from Newton's divided differences formula (see e.g. [1, Chap. 1]). Adding $n-1$ constraints $\ddot{\gamma}_{\mathcal{T}}^{C_{i-1}}(t_i) = \ddot{\gamma}_{\mathcal{T}}^{C_i}(t_i)$ for continuity of $\ddot{\gamma}_{\mathcal{T}}^C$ at $x_1, \ldots, x_{n-1}$ (with $i = 1, 2, \ldots, n-1$) leads by (2) (for $\gamma_{\mathcal{T}}^{C_i}$) to the m tridiagonal linear systems (strictly diagonally dominant) of $n-1$ equations in $n+1$ vector unknowns representing velocities at $\mathcal{M}$ i.e. $v_0, v_1, v_2, \ldots, v_{n-1}, v_n \in \mathbb{R}^m$:

$$v_{i-1}\Delta t_i + 2v_i(\Delta t_{i-1} + \Delta t_i) + v_{i+1}\Delta t_{i-1} = b_i,$$

$$b_i = 3(\Delta t_i \frac{x_i - x_{i-1}}{\Delta t_{i-1}} + \Delta t_{i-1} \frac{x_{i+1} - x_i}{\Delta t_i}). \qquad (3)$$

(i) Both v_0 and v_n (*if unknown*) can be e.g. calculated from $a_0 = \ddot{\gamma}_{\mathcal{T}}^C(0) = a_n = \ddot{\gamma}_{\mathcal{T}}^C(T_c) = \mathbf{0}$ combined with (2) (this yields *a natural cubic spline interpolant* $\gamma_{\mathcal{T}}^{NS}$ - a special $\gamma_{\mathcal{T}}^C$) which supplements (3) with two missing vector linear equations:

$$2v_0 + v_1 = 3\frac{x_1 - x_0}{\Delta t_0} \ , \quad v_{n-1} + 2v_n = 3\frac{x_n - x_{n-1}}{\Delta t_{n-1}} \ . \tag{4}$$

The resulting m linear systems, each of size $(n + 1) \times (n + 1)$, (based on (3) and (4)) as strictly row diagonally dominant result in one vector solution $v_0, v_1, \ldots, v_{n-1}, v_n$ (solved e.g. by Gauss elimination without pivoting - see [1, Chap. 4]), which when fed into (2) determines explicitly *a natural cubic spline* γ_T^{NS} (with fixed $\mathcal{T}$). A similar approach follows for arbitrary a_0 and a_n.

(ii) If both v_0 and v_n are given then the so-called *complete spline* γ_T^{CS} can be found with $v_1, \ldots v_{n-1}$ determined solely by (3).

(iii) If one of v_0 or v_n is unknown it can be compensated by setting the respective terminal acceleration e.g. to $\mathbf{0}$. The above scheme relies on solving (3) with one equation from (4). Such splines are denoted here by $\gamma_T^{v_n}$ or $\gamma_T^{v_0}$. Two *non-generic* cases of *Leap-Frog* optimizations deal with the latter - omitted in this paper.

By (1) $\mathcal{J}_T(\gamma_T^{NS}) = 4\sum_{i=0}^{n-1}(\|c_{3,i}\|^2\Delta t_i + 3\|c_{4,i}\|^2(\Delta t_i)^3 + 3\langle c_{3,i}|c_{4,i}\rangle(\Delta t_i)^2)$, which ultimately reformulates into (see [2]):

$$\mathcal{J}_T(\gamma_T^{NS}) = 4\sum_{i=0}^{n-1}\Big(\frac{-1}{(\Delta t_i)^3}(-3\|x_{i+1} - x_i\|^2 + 3\langle v_i + v_{i+1}|x_{i+1} - x_i\rangle\Delta t_i$$

$$-(\|v_i\|^2 + \|v_{i+1}\|^2 + \langle v_i|v_{i+1}\rangle)(\Delta t_i)^2\Big) \ . \tag{5}$$

As mentioned before for fixed knots $\mathcal{T}$, the natural spline γ_T^{NS} minimizes (1) (see [1]). Thus upon relaxing the internal knots $\mathcal{T}_{int}$ the original infinite dimensional optimization (1) reduces to finding the corresponding *optimal knots* $(t_1^{opt}, t_2^{opt}, \ldots, t_{n-1}^{opt})$ for (5) (viewed from now on as a multivariable function $J_0(t_1, t_2, \ldots, t_{n-1})$) subject to $t_0 = 0 < t_1^{opt} < t_2^{opt} < \cdots < t_{n-1}^{opt} < t_n = T$. Such reformulated non-linear optimization task (5) transformed into minimizing $J_0(\mathcal{T}_{int})$ (here $t_0 = 0$ and $t_n = T$) forms a difficult task for critical points examination as well as for the numerical computations. The analysis addressing the non-linearity of J_0 and comparisons between different numerical methods used to optimize J_0 are discussed in [2,3] or [6]. One of the computationally feasible schemes handling (5) turns out to be *a Leap-Frog* (for its $2D$ analogue for image noise removal see also [11] or in other contexts see e.g. [9] or [10]). For optimizing J_0 this scheme is based on the sequence of single variable iterative optimization which in k-th iteration minimizes:

$$J_0^{(k,i)}(s) = \int_{t_{i-1}^k}^{t_{i+1}^{k-1}} \|\ddot{\gamma}_{k,i}^{CS}(s)\|^2 ds \tag{6}$$

over $I_i^{k-1} = [t_{i-1}^k, t_{i+1}^{k-1}]$. Here t_i is set to be a free variable s_i. The complete spline $\gamma_{k,i}^{CS} : I_i^{k-1} \to \mathbb{E}^m$ is determined by $\{t_{i-1}^k, s, t_{i+1}^{k-1}\}$, both velocities $\{v_{i-1}^k, v_{i+1}^{k-1}\}$ and the interpolation points $\{x_{i-1}, x_i, x_{i+1}\}$. Once s_i^{opt} is found one updates t_i^{k-1} with $t_i^k = s_i^{opt}$ and v_i^{k-1} with the $v_i^k = \dot{\gamma}_{k,i}^{CS}(s_i^{opt})$. Next we pass to the shifted overlapped sub-interval $I_{i+1}^k = [t_i^k, t_{i+2}^{k-1}]$ and repeat the previous step of updating t_{i+1}^{k-1}. Note that both cases $[0, t_2^{k-1}]$ and $[t_{n-2}^{k-1}, T]$ rely on splines discussed in *(iii)*, where the vanishing acceleration replaces one of the velocities v_0^{k-1} or v_n^{k-1}. Once t_{n-1}^{k-1} is changed over the last sub-interval $I_{n-1}^{k-1} = [t_{n-2}^k, T]$ the k-th iteration is terminated and the next local optimization over $I_1^k = [0, t_2^k]$ represents the beginning of the $(k+1)$-st iteration of *Leap-Frog Algorithm*. The initialization of $\mathcal{T}_{int}$ for *Leap-Frog* can follow normalized *cumulative chord parameterization* (see e.g. [8]) which sets $t_0^0 = 0, t_1^0, \ldots, t_{n-1}^k, t_n^0 = T$ according to $t_0^0 = 0$ and $t_{i+1}^0 = \|x_{i+1} - x_i\| \frac{T}{\hat{T}} + t_i^0$, for $i = 0, 1, \ldots, n-1$ and $\hat{T} = \sum_{i=0}^{n-1} \|x_{i+1} - x_i\|$.

3   Generic Middle Case: Initial and Last Velocities Given

Assume that for internal points $x_i, x_{i+1}, x_{i+2} \in \mathbb{E}^m$ (for $i = 1, 2, \ldots, n-3$ and $n > 3$) the interpolation knots t_i and t_{i+2} with the velocities $v_i, v_{i+2} \in \mathbb{R}^m$ are somehow given (e.g. by previous *Leap-Frog* iteration outlined in Sect. 2). We construct now a C^2 piecewise cubic (a complete spline - see Sect. 2), depending on varying $t_{i+1} \in (t_i, t_{i+2})$ (temporarily free variable). The curve $\gamma_i^c : [t_i, t_{i+2}] \to \mathbb{E}^m$ (i.e. a cubic on each $[t_i, t_{i+1}]$ and $[t_{i+1}, t_{i+2}]$) satisfies:

$$\gamma_i^c(t_{i+j}) = x_{i+j}, \quad j = 0, 1, 2; \quad \dot{\gamma}_i^c(t_{i+j}) = v_{i+j}, \quad j = 0, 2. \tag{7}$$

Letting $\phi_i : [t_i, t_{i+2}] \to [0, 1]$, $\phi_i(t) = (t - t_i)(t_{i+2} - t_i)^{-1} = s$ the curve $\tilde{\gamma}_i^c :$ $[0, 1] \to \mathbb{E}^m$ (with $\tilde{\gamma}_i^c = \gamma_i^c \circ \phi_i^{-1}$) by (7) satisfies, for $0 < s_{i+1} = \phi_i(t_{i+1}) < 1$:

$$\tilde{\gamma}_i^c(0) = x_i, \quad \tilde{\gamma}_i^c(s_{i+1}) = x_{i+1}, \quad \tilde{\gamma}_i^c(1) = x_{i+2}, \tag{8}$$

with the adjusted initial and the last velocities $\tilde{v}_i, \tilde{v}_{i+2} \in \mathbb{R}^m$ fulfilling:

$$\tilde{v}_i = \tilde{\gamma}_i^{c'}(0) = (t_{i+2} - t_i)v_i, \quad \tilde{v}_{i+2} = \tilde{\gamma}_i^{c'}(1) = (t_{i+2} - t_i)v_{i+2}. \tag{9}$$

To reformulate $\tilde{\mathcal{E}}_i$ define two cubics $\tilde{\gamma}_i^{lc}, \tilde{\gamma}_i^{rc}$ satisfying (with $s_{i+1} \in (0, 1)$) $\tilde{\gamma}_i^c = \tilde{\gamma}_i^{lc}$ (over $[0, s_{i+1}]$) and $\tilde{\gamma}_i^c = \tilde{\gamma}_i^{rc}$ (over $[s_{i+1}, 1]$) with $c_{ij}, d_{ij} \in \mathbb{E}^m$:

$$\tilde{\gamma}_i^{lc}(s) = c_{i0} + c_{i1}(s - s_{i+1}) + c_{i2}(s - s_{i+1})^2 + c_{i3}(s - s_{i+1})^3,$$
$$\tilde{\gamma}_i^{rc}(s) = d_{i0} + d_{i1}(s - s_{i+1}) + d_{i2}(s - s_{i+1})^2 + d_{i3}(s - s_{i+1})^3. \tag{10}$$

Since $\tilde{\gamma}_i^c$ is a complete spline the following constraints hold:

$$\tilde{\gamma}_i^{lc}(0) = x_i, \quad \tilde{\gamma}_i^{lc}(s_{i+1}) = \tilde{\gamma}_i^{rc}(s_{i+1}) = x_{i+1}, \quad \tilde{\gamma}_i^{rc}(1) = x_{i+2}, \tag{11}$$

$$\tilde{\gamma}_i^{lc'}(0) = \tilde{v}_i, \quad \tilde{\gamma}_i^{rc'}(1) = \tilde{v}_{i+2}, \tag{12}$$

together with two C^1 and C^2 smoothness constraints at $s = s_{i+1}$:

$$\tilde{\gamma}_i^{lc'}(s_{i+1}) = \tilde{\gamma}_i^{rc'}(s_{i+1}), \quad \tilde{\gamma}_i^{lc''}(s_{i+1}) = \tilde{\gamma}_i^{rc''}(s_{i+1}). \tag{13}$$

Upon shifting the coordinates origin in $\mathbb{E}^m$ to x_{i+1} we have for $\tilde{x}_{i+1} = \mathbf{0}$, $\tilde{x}_i = x_i - x_{i+1}$ and $\tilde{x}_{i+2} = x_{i+2} - x_{i+1}$ (by (11)):

$$\tilde{\gamma}_i^{lc}(0) = \tilde{x}_i, \quad \tilde{\gamma}_i^{lc}(s_{i+1}) = \tilde{\gamma}_i^{rc}(s_{i+1}) = \mathbf{0}, \quad \tilde{\gamma}_i^{rc}(1) = \tilde{x}_{i+2}. \tag{14}$$

Both (10) and $x_{i+1} = \mathbf{0}$ yield $c_{i0} = d_{i0} = \mathbf{0}$. Next (13) with $\tilde{\gamma}_i^{lc'}(s) = c_{i1} + 2c_{i2}(s - s_{i+1}) + 3c_{i3}(s - s_{i+1})^2$, $\tilde{\gamma}_i^{rc'}(s) = d_{i1} + 2d_{i2}(s - s_{i+1}) + 3d_{i3}(s - s_{i+1})^2$, $\tilde{\gamma}_i^{lc''}(s) = 2c_{i2} + 6c_{i3}(s - s_{i+1})$ and $\tilde{\gamma}_i^{lr''}(s) = 2d_{i2} + 6d_{i3}(s - s_{i+1})$, leads to $c_{i1} = d_{i1}$ and $c_{i2} = d_{i2}$. Hence one obtains:

$$\tilde{\gamma}_i^{lc}(s) = c_{i1}(s - s_{i+1}) + c_{i2}(s - s_{i+1})^2 + c_{i3}(s - s_{i+1})^3 ,$$
$$\tilde{\gamma}_i^{rc}(s) = c_{i1}(s - s_{i+1}) + c_{i2}(s - s_{i+1})^2 + d_{i3}(s - s_{i+1})^3 . \tag{15}$$

The unknown vectors $c_{i1}, c_{i2}, c_{i3}, d_{i3}$ in (15) follow from four linear vector equations obtained from (12) and (14) (i.e. with data $\tilde{\mathcal{M}}_i = \{\tilde{x}_i, \tilde{x}_{i+2}, \tilde{v}_i, \tilde{v}_{i+2}\}$):

$$\tilde{x}_i = -c_{i1}s_{i+1} + c_{i2}s_{i+1}^2 - c_{i3}s_{i+1}^3 ,$$
$$\tilde{x}_{i+2} = c_{i1}(1 - s_{i+1}) + c_{i2}(1 - s_{i+1})^2 + d_{i3}(1 - s_{i+1})^3 ,$$
$$\tilde{v}_i = c_{i1} - 2c_{i2}s_{i+1} + 3c_{i3}s_{i+1}^2 ,$$
$$\tilde{v}_{i+2} = c_{i1} + 2c_{i2}(1 - s_{i+1}) + 3d_{i3}(1 - s_{i+1})^2 . \tag{16}$$

Applying *Mathematica Solve* to (16) yields:

$$c_{i1} = -\frac{-s_{i+1}\tilde{v}_i + 2s_{i+1}^2\tilde{v}_i - s_{i+1}^3\tilde{v}_i - s_{i+1}^2\tilde{v}_{i+2} + s_{i+1}^3\tilde{v}_{i+2} - 3\tilde{x}_i + 6s_{i+1}\tilde{x}_i}{2(s_{i+1} - 1)s_{i+1}}$$
$$- \frac{-3s_{i+1}^2\tilde{x}_i + 3s_{i+1}^2\tilde{x}_{i+2}}{2(s_{i+1} - 1)s_{i+1}} ,$$

$$c_{i2} = -\frac{s_{i+1}\tilde{v}_i - s_{i+1}^2\tilde{v}_i - s_{i+1}\tilde{v}_{i+2} + s_{i+1}^2\tilde{v}_{i+2} + 3\tilde{x}_i - 3s_{i+1}\tilde{x}_i + 3s_{i+1}\tilde{x}_{i+2}}{(s_{i+1} - 1)s_{i+1}} ,$$

$$c_{i3} = -\frac{s_{i+1}(\tilde{v}_i + 2\tilde{x}_i) - s_{i+1}^3(\tilde{v}_i - \tilde{v}_{i+2}) - s_{i+1}^2(\tilde{v}_{i+2} + 3\tilde{x}_i - 3\tilde{x}_{i+2}) + \tilde{x}_i}{2(s_{i+1} - 1)s_{i+1}^3} ,$$

$$d_{i3} = -\frac{-s_{i+1}\tilde{v}_i + 2s_{i+1}^2\tilde{v}_i - s_{i+1}^3\tilde{v}_i + 2s_{i+1}\tilde{v}_{i+2} - 3s_{i+1}^2\tilde{v}_{i+2} + s_{i+1}^3\tilde{v}_{i+2} - 3\tilde{x}_i}{2(s_{i+1} - 1)^3 s_{i+1}}$$
$$- \frac{6s_{i+1}\tilde{x}_i - 3s_{i+1}^2\tilde{x}_i - 4s_{i+1}\tilde{x}_{i+2} + 3s_{i+1}^2\tilde{x}_{i+2}}{2(s_{i+1} - 1)^3 s_{i+1}} , \tag{17}$$

which satisfy (as functions in s_{i+1}) the system (16). Next, since $\|\gamma_i^{lc''}(s)\|^2 = 4\|c_{i2}\|^2 + 24\langle c_{i2}|c_{i3}\rangle(s - s_{i+1}) + 36\|c_{i3}\|^2(s - s_{i+1})^2$ and $\|\gamma_i^{rc''}(s)\|^2 = 4\|c_{i2}\|^2 + 24\langle c_{i2}|d_{i3}\rangle(s - s_{i+1}) + 36\|d_{i3}\|^2(s - s_{i+1})^2$ the formula for $\mathcal{E}_i$ reads as

$$\tilde{\mathcal{E}}_i(s_{i+1}) = \int_0^{s_{i+1}} \|\gamma_i^{lc''}(s)\|^2 ds + \int_{s_{i+1}}^1 \|\gamma_i^{rc''}(s)\|^2 ds = I_1 + I_2,$$

where $I_1 = 4(\|c_{i2}\|^2 s_{i+1} - 3\langle c_{i2}|c_{i3}\rangle s_{i+1}^2 + 3\|c_{i3}\|^2 s_{i+1}^3)$ and $I_2 = 4(\|c_{i2}\|^2(1 - s_{i+1}) + 3\langle c_{i2}|d_{i3}\rangle(1 - s_{i+1})^2 + 3\|d_{i3}\|^2(1 - s_{i+1})^3)$. Combining the latter with (17) (upon applying *NIntegrate* and *FullSimplify* from *Mathematica*) yields:

$$\tilde{\mathcal{E}}_i(s_{i+1}) =$$

$$\frac{1}{s_{i+1}^3(s_{i+1}-1)^3}(3\|\tilde{x}_i\|^2(s_{i+1}-1)^3(1+3s_{i+1})+s_{i+1}(-6\langle\tilde{v}_i|\tilde{x}_i\rangle$$

$$+s_{i+1}(\|\tilde{v}_{i+2}\|^2(s_{i+1}-4)(s_{i+1}-1)^2s_{i+1}+3\|\tilde{x}_{i+2}\|^2s_{i+1}(3s_{i+1}-4)$$

$$+\|\tilde{v}_i\|^2(s_{i+1}-1)^3(s_{i+1}+3)-2(s_{i+1}-1)^3s_{i+1}\langle\tilde{v}_i|\tilde{v}_{i+2}\rangle$$

$$+6(2+(s_{i+1}-2)s_{i+1}^2)\langle\tilde{v}_i|\tilde{x}_i\rangle-6(s_{i+1}-1)^2s_{i+1}\langle\tilde{v}_i|\tilde{x}_{i+2}\rangle-6(s_{i+1}-1)^3\langle\tilde{v}_{i+2}|\tilde{x}_i\rangle$$

$$+6(s_{i+1}-2)(s_{i+1}-1)s_{i+1}\langle\tilde{v}_{i+2}|\tilde{x}_{i+2}\rangle-18(s_{i+1}-1)^2\langle\tilde{x}_i|\tilde{x}_{i+2}\rangle)))\,. \tag{18}$$

Upon substituting for $\tilde{x}_{i+2}=x_{i+2}-x_{i+1}$ and $\tilde{x}_i=x_i-x_{i+1}$ one can reformulate (18) (and thus (19)) in terms of each data $x_i,x_{i+1},x_{i+2}\in\mathbb{E}^m$. *Mathematica* symbolic differentiation and *FullSimplify* applied to $\tilde{\mathcal{E}}_i$ yields:

$$\tilde{\mathcal{E}}_i'(s_{i+1}) =$$

$$\frac{-3}{(s_{i+1}-1)^4s_{i+1}^4}(3\|\tilde{x}_i\|^2(s_{i+1}-1)^4(1+2s_{i+1})+s_{i+1}(\|\tilde{v}_i\|^2(s_{i+1}-1)^4s_{i+1}$$

$$-\|\tilde{v}_{i+2}\|^2(s_{i+1}-1)^2s_{i+1}^3+3\|\tilde{x}_{i+2}\|^2s_{i+1}^3(2s_{i+1}-3)$$

$$+2(s_{i+1}-1)^4(2+s_{i+1})\langle\tilde{v}_i|\tilde{x}_i\rangle-2(s_{i+1}-1)^2s_{i+1}^3\langle\tilde{v}_i|\tilde{x}_{i+2}\rangle$$

$$-2(s_{i+1}-1)^4s_{i+1}\langle\tilde{v}_{i+2}|\tilde{x}_i\rangle+2(s_{i+1}-3)(s_{i+1}-1)s_{i+1}^3\langle\tilde{v}_{i+2}|\tilde{x}_{i+2}\rangle$$

$$-6(s_{i+1}-1)^2s_{i+1}(2s_{i+1}-1)\langle\tilde{x}_i|\tilde{x}_{i+2}\rangle))\,. \tag{19}$$

By (19) $\tilde{\mathcal{E}}_i'(s_{i+1})=(-1/((s_{i+1}-1)^4s_{i+1}^4))N_i(s_{i+1})$, where $N_i(s_{i+1})$ is a polynomial of degree 6 (use here e.g. *Mathematica* functions *Factor* and *CoefficientList*) $N_i(s_{i+1})=b_0^i+b_1^is_{i+1}+b_2^is_{i+1}^2+b_3^is_{i+1}^3+b_4^is_{i+1}^4+b_5^is_{i+1}^5+b_6^is_{i+1}^6$, where

$$\frac{b_0^i}{3}=3\|\tilde{x}_i\|^2\,, \qquad \frac{b_1^i}{3}=-6\|\tilde{x}_i\|^2+4\langle\tilde{v}_i|\tilde{x}_i\rangle\,,$$

$$\frac{b_2^i}{3}=\|\tilde{v}_i\|^2-6\|\tilde{x}_i\|^2-14\langle\tilde{v}_i|\tilde{x}_i\rangle-2\langle\tilde{v}_{i+2}|\tilde{x}_i\rangle+6\langle\tilde{x}_i|\tilde{x}_{i+2}\rangle\,,$$

$$\frac{b_3^i}{3}=-4\|\tilde{v}_i\|^2+24\|\tilde{x}_i\|^2+16\langle\tilde{v}_i|\tilde{x}_i\rangle+8\langle\tilde{v}_{i+2}|\tilde{x}_i\rangle-24\langle\tilde{x}_i|\tilde{x}_{i+2}\rangle\,,$$

$$\frac{b_4^i}{3}=6\|\tilde{v}_i\|^2-\|\tilde{v}_{i+2}\|^2-21\|\tilde{x}_i\|^2-9\|\tilde{x}_{i+2}\|^2-4\langle\tilde{v}_i|\tilde{x}_i\rangle-2\langle\tilde{v}_i|\tilde{x}_{i+2}\rangle$$

$$-12\langle\tilde{v}_{i+2}|\tilde{x}_i\rangle+6\langle\tilde{v}_{i+2}|\tilde{x}_{i+2}\rangle+30\langle\tilde{x}_i|x_{i+2}\rangle\,,$$

$$\frac{b_5^i}{3}=-4\|\tilde{v}_i\|^2+2\|\tilde{v}_{i+2}\|^2+6\|\tilde{x}_i\|^2+6\|\tilde{x}_{i+2}\|^2-4\langle\tilde{v}_i|\tilde{x}_i\rangle+4\langle\tilde{v}_i|\tilde{x}_{i+2}\rangle\rangle$$

$$+8\langle\tilde{v}_{i+2}|\tilde{x}_i\rangle-8\langle\tilde{v}_{i+2}|\tilde{x}_{i+2}\rangle-12\langle\tilde{x}_i|\tilde{x}_{i+2}\rangle\,,$$

$$\frac{b_6^i}{3}=\|\tilde{v}_i\|^2-\|\tilde{v}_{i+2}\|^2+2\langle\tilde{v}_i|\tilde{x}_i\rangle-2\langle\tilde{v}_i|\tilde{x}_{i+2}\rangle-2\langle\tilde{v}_{i+2}|\tilde{x}_i\rangle+2\langle\tilde{v}_{i+2}|\tilde{x}_{i+2}\rangle\,.$$

In a search for a global optimum of $\tilde{\mathcal{E}}_i$, instead of using any optimization scheme relying on initial guess, one can apply *Mathematica Solve* which finds all roots (real and complex). Indeed upon computing the roots of $N_i(s_{i+1})$ one selects only these from $(0,1)$. Next we evaluate $\tilde{\mathcal{E}}_i$ on each critical point $s_{i+1}^{crit}\in(0,1)$ and choose s_{i+1}^{crit} with minimal energy as optimal. This feature is particularly useful in implementation of Leap-Frog as opposed to the optimization of the initial energy (5) depending on $n-1$ unknown knots.

4   Special Conditions for Leap-Frog Generic Case

Assume $\tilde{x}_i, \tilde{x}_{i+1}, \tilde{x}_{i+2} \in \mathbb{E}^m$ with $\tilde{v}_i, \tilde{v}_{i+2} \in \mathbb{R}^m$ satisfy now the extra constraints:

$$\tilde{v}_i = \tilde{v}_{i+2}, \qquad \tilde{x}_{i+2} - \tilde{x}_i = \tilde{v}_i = \tilde{v}_{i+2}. \qquad (20)$$

By (20) we get $\|\tilde{v}_{i+2}\|^2 = \|\tilde{v}_i\|^2 = \langle \tilde{v}_i|\tilde{v}_{i+2}\rangle = \|\tilde{x}_{i+2}\|^2 + \|\tilde{x}_i\|^2 - 2\langle \tilde{x}_i|\tilde{x}_{i+2}\rangle$, $\langle \tilde{x}_i|\tilde{v}_i\rangle = \langle \tilde{x}_i|\tilde{v}_{i+2}\rangle = \langle \tilde{x}_i|\tilde{x}_{i+2}\rangle - \|\tilde{x}_i\|^2$ and $\langle \tilde{x}_{i+2}|\tilde{v}_i\rangle = \langle \tilde{x}_{i+2}|\tilde{v}_{i+2}\rangle = \|\tilde{x}_{i+2}\|^2 - \langle \tilde{x}_i|\tilde{x}_{i+2}\rangle$. Substituting the above into (19) (or into $\tilde{\mathcal{E}}_i^c$) yields $\tilde{\mathcal{E}}_i^c(s_{i+1}) =$

$$-\frac{3(\|\tilde{x}_i\|^2(s_{i+1}-1)^2 + s_{i+1}(\|\tilde{x}_{i+2}\|^2 s_{i+1} - 2(s_{i+1}-1)\langle \tilde{x}_i|\tilde{x}_{i+2}\rangle))}{(s_{i+1}-1)^3 s_{i+1}^3} \qquad (21)$$

and hence $\tilde{\mathcal{E}}_i^{c'}(s_{i+1}) =$

$$\frac{3}{(s_{i+1}-1)^4 s_{i+1}^4}(\|\tilde{x}_i\|^2(s_{i+1}-1)^2(4s_{i+1}-3)$$
$$+s_{i+1}(\|\tilde{x}_{i+2}\|^2 s_{i+1}(4s_{i+1}-1) - 4(1-3s_{i+1}+2s_{i+1}^2)\langle \tilde{x}_i|\tilde{x}_{i+2}\rangle)). \qquad (22)$$

The numerator of (22) forms now a polynomial of degree 3 (instead of degree 6 as in (19)) $N_i^c(s_{i+1}) = b_0^{ic} + b_1^{ic} s_{i+1} + b_2^{ic} s_{i+1}^2 + b_3^{ic} s_{i+1}^3$, where:

$$\frac{b_0^{ic}}{3} = -3\|\tilde{x}_i\|^2 < 0, \qquad \frac{b_1^{ic}}{3} = 2(5\|\tilde{x}_i\|^2 - 2\langle \tilde{x}_i|\tilde{x}_{i+2}\rangle),$$

$$\frac{b_2^{ic}}{3} = -11\|\tilde{x}_i\|^2 - \|\tilde{x}_{i+2}\|^2 + 12\langle \tilde{x}_i|\tilde{x}_{i+2}\rangle = 5(\|\tilde{x}_{i+2}\|^2 - \|\tilde{x}_i\|^2) - 6\|\tilde{x}_{i+2} - \tilde{x}_i\|^2,$$

$$\frac{b_3^{ic}}{3} = 4\|\tilde{x}_i\|^2 + 4\|\tilde{x}_{i+2}\|^2 - 8\langle \tilde{x}_i|\tilde{x}_{i+2}\rangle = 4\|\tilde{x}_{i+2} - \tilde{x}_i\|^2 \geq 0.$$

For $\tilde{\mathcal{E}}_i^c$ to be unimodal over $(0, 1)$ one needs $N_i^c(s_{i+1})$ with a single root in $(0, 1)$. *(i)* Note that if $\tilde{x}_{i+2} = \tilde{x}_i$ then $N_i^c(s_{i+1}) = -9\|\tilde{x}_i\|^2 + 18s_{i+1}\|\tilde{x}_i\|^2$ has exactly one root $\hat{s}_{i+1} = 1/2 \in (0, 1)$. By (20) we have $\tilde{v}_{i+2} = \tilde{v}_i = \mathbf{0}$. *(ii)* We assume now that $\tilde{x}_{i+2} \neq \tilde{x}_i$ then $N_i^c(s_{i+1})$ becomes *a cubic*. We find now the conditions for which N_i^c has exactly one root over $(0, 1)$. For the latter as $N_i^c(0) = -9\|\tilde{x}_i\|^2 < 0$ and $N_i^c(1) = 9\|\tilde{x}_{i+2}\|^2 > 0$ by Intermediate Value Th. it suffices to show that either $N_i^{c'}(s_{i+1}) = c_0^{ic} + c_1^{ic} s_{i+1} + c_2^{ic} s_{i+1}^2 > 0$ (over $(0, 1)$) or that the derivative $N_i^{c'}$ has exactly one root $\hat{u}_{i+1} \in (0, 1)$ (i.e. N_i^c has exactly one max/min/saddle at $\hat{u}_{i+1}$) and thus $N_i^c(s_{i+1}) = 0$ yields exactly single root $\hat{s}_{i+1} \in (0, 1)$ - note that if $\hat{s}_{i+1} = \hat{u}_{i+1}$ then $\hat{u}_{i+1}$ is a saddle point of N_i^c. Here *a quadratic* $N_i^{c'}(s_{i+1})$ (as $\tilde{x}_{i+2} \neq \tilde{x}_i$) has coefficients $(c_0^{ic}/6) = 5\|\tilde{x}_i\|^2 - 2\langle \tilde{x}_i|\tilde{x}_{i+2}\rangle$, $(c_1^{ic}/6) = 5(\|\tilde{x}_{i+2}\|^2 - \|\tilde{x}_i\|^2) - 6\|\tilde{x}_{i+2} - \tilde{x}_i\|^2$, and $(c_2^{ic}/6) = 6\|\tilde{x}_{i+2} - \tilde{x}_i\|^2 > 0$. The discriminant $\tilde{\Delta}$ of the quadratic $N_i^{c'}(s_{i+1})/6$ reads as:

$$\tilde{\Delta} = \|\tilde{x}_{i+2}\|^4 + \|\tilde{x}_i\|^4 - 98\|\tilde{x}_{i+2}\|^2\|\tilde{x}_i\|^2 + 24\langle \tilde{x}_i|\tilde{x}_{i+2}\rangle\|\tilde{x}_{i+2} + \tilde{x}_i\|^2. \qquad (23)$$

Define now *two auxiliary parameters* $(\lambda, \mu) \in \Omega = (\mathbb{R}_+ \times [-1, 1]) \setminus \{(1, 1)\}$:

$$\|\tilde{x}_i\| = \lambda\|\tilde{x}_{i+2}\|, \qquad \langle \tilde{x}_i|\tilde{x}_{i+2}\rangle = \mu\|\tilde{x}_i\|\|\tilde{x}_{i+2}\|. \qquad (24)$$

Here μ stands for $\cos(\alpha)$, where α is the angle between vectors $\tilde{x}_i$ and $\tilde{x}_{i+2}$ - hence $\mu = \lambda = 1$ is excluded as then $\tilde{x}_{i+2} = \tilde{x}_i$. Note, however that as analyzed in case *(i)* when $\tilde{x}_{i+2} = \tilde{x}_i$ there is only one optimal parameter $\hat{s}_{i+1} = 1/2$ - thus $(\mu, \lambda) = (1, 1)$ is also admissible. We examine various constraints on $(\mu, \lambda) \neq (1, 1)$ (with $\lambda > 0$ and $-1 \leq \mu \leq 1$) for the existence of either *no roots* or *one root* of $N_i^{c'} = 0$ over $[0, 1]$ *(yielding single critical point of $\tilde{\mathcal{E}}_i^c$ over $(0, 1)$)*.

1. $\tilde{\Delta} < 0$. Since $c_2^{ic} > 0$, clearly the following $N_i^{c'} > 0$ holds over $(0, 1)$. Substituting (24) into (23) yields (for $\Delta = (\tilde{\Delta}/\|\tilde{x}_{i+2}\|^4)$) $\Delta(\lambda, \mu) = \lambda^4 + 24\mu\lambda^3 + (48\mu^2 - 98)\lambda^2 + 24\mu\lambda + 1$. In order to decompose Ω into sub-regions Ω_- (with $\Delta < 0$), Ω_+ (with $\Delta > 0$) and Γ_0 (with $\Delta \equiv 0$) we resort to *Mathematica* functions *InequalityPlot*, *ImplicitPlot* and *Solve*. Figure 1(a) shows the resulting decomposition and Fig. 1(b) shows its magnification for λ small. The intersection points of Γ_0 and boundary $\partial\Omega$ (found by *Solve*) read: for $\mu = 1$ it is a point $(1, 1)$ (already excluded - see dotted point in Fig. 1) and for $\mu = -1$ we have two points $(-1, (1/(13 + 2\sqrt{42}))) \approx (-1, 0.0385186)$ or $(-1, 13 + 2\sqrt{42}) \approx (-1, 25.9615)$.

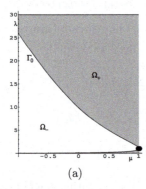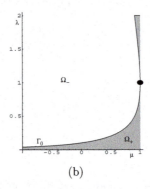

(a)   (b)

Fig. 1. Decomposition of Ω into sub-regions: (a) over which $\Delta > 0$ (i.e. Ω_+), $\Delta = 0$ (i.e. Γ_0) or $\Delta < 0$ (i.e. Ω_-),   (b) only for λ small.

The admissible subset $\Omega_{ok} \subset \Omega$ *of parameters* (μ, λ) *(for which there is one local minimum of $\tilde{\mathcal{E}}_i^c$) satisfies* $\Omega_- \subset \Omega_{ok}$. *The set to* $\Omega \setminus \Omega_-$ *is a potential exclusion zone* $\Omega_{ex} \subset \Omega \setminus \Omega_-$. Next we shrink an exclusion zone $\Omega_{ex} \subset \Omega$ (subset of shaded region in Fig. 1).

2. $\tilde{\Delta} = 0$. There is only one root $\hat{u}_{i+1}^0 \in \mathbb{R}$ for $N_i^{c'}(s_{i+1}) = 0$. As explained, irrespectively whether $\hat{u}_{i+1}^0 \in (0, 1)$ or $\hat{u}_{i+1}^0 \notin (0, 1)$ this results in exactly one root $\hat{s}_{i+1} \in (0, 1)$ of $N_i^c(s_{i+1}) = 0$, which in turn yields exactly one local (thus one global) minimum for $\tilde{\mathcal{E}}_i^c$. Hence $\Omega_- \cup \Gamma_0 \subset \Omega_{ok}$.

3. $\tilde{\Delta} > 0$. There are two different roots $\hat{u}_{i+1}^\pm \in \mathbb{R}$ of $N_i^{c'}(s_{i+1}) = 0$. Note that since $c_2^{ic} > 0$ we have $\hat{u}_{i+1}^- < \hat{u}_{i+1}^+$. They are either (in all cases we use Vieta's formulas):

(a) of opposite signs: i.e. $(c_0^{ic}/c_2^{ic}) < 0$ or

(b) non-positive: i.e. $(c_0^{ic}/c_2^{ic}) \geq 0$ and $(-c_1^{ic}/c_2^{ic}) < 0$ (as $\hat{u}_{i+1}^- < \hat{u}_{i+1}^+$) or

(c) *non-negative*: i.e. $(c_0^{ic}/c_2^{ic}) \geq 0$ and $(-c_1^{ic}/c_2^{ic}) > 0$ - split into:

(c1) $\hat{u}_{i+1}^+ \geq 1$: i.e.

(c2) $0 < \hat{u}_{i+1}^+ < 1$ (as here $\hat{u}_{i+1}^- < \hat{u}_{i+1}^+$).

Evidently for a), b) and c1) there is *up to one root* $\hat{u}_{i+1} \in (0,1)$ of $N_i^{c'}(s_{i+1}) = 0$. Therefore as already explained there is only one root $\hat{s}_{i+1} \in (0,1)$ of $N_i^c(s_{i+1}) = 0$, which is the unique critical point of $\tilde{\mathcal{E}}_i^c$ over $(0,1)$. We show now that the inequalities from a) or b) or c) extend (contract) the admissible (exclusion) zone Ω_{ok} (Ω_{ex}) of parameters $(\mu, \lambda) \in \Omega$. Indeed:

a) the constraint $(c_0^{ic}/c_2^{ic}) < 0$ upon using (24) reads (as $\lambda > 0$):

$$5\lambda^2 - 2\mu\lambda < 0 \quad \equiv \lambda < \frac{2\mu}{5} . \tag{25}$$

Figure 2 a) shows Ω_1 (over which (25) holds) cut out from the exclusion zone Ω_{ex} of parameters $(\mu, \lambda) \in \Omega$ (again *InequalityPlot* is used here).

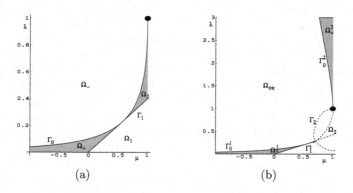

(a)   (b)

Fig. 2. Extension of admissible zone Ω_{ok} by cutting out from Ω_{ex}: (a) Ω_1, (b) Ω_2.

Thus $\Omega_- \cup \Gamma_0 \cup \Omega_1 \subset \Omega_{ok}$. The intersection $\Gamma_1 \cap \partial\Omega = \{(0,0),(1,0.4)\}$ (here $\Gamma_1 = \{(\mu,\lambda) \in \Omega : 5\lambda - 2\mu = 0\}$). Similarly the intersection $\Gamma_0 \cap \Gamma_1 = \{(5/(2\sqrt{19}), 1/\sqrt{19})\} \approx (0.573539, 0.229416) = p_1$.

b) the constraints $(c_0^{ic}/c_2^{ic}) \geq 0$ and $(-c_1^{ic}/c_2^{ic}) < 0$ combined with (24) yield:

$$\lambda \geq \frac{2\mu}{5} \quad \text{and} \quad 11\lambda^2 - 12\mu\lambda + 1 < 0 . \tag{26}$$

Using *ImplicitPlot* and *InequalityPlot* we find Ω_2 (cut out from Ω_{ex}) as the intersection of three sets defined by (26) and $\Delta > 0$ (for Ω_2 see Fig. 2 a–b)). Thus $\Omega_- \cup \Gamma_0 \cup \Omega_1 \cup \Omega_2 \subset \Omega_{ok}$ (see Fig. 2 b)). Note that for $\Gamma_2 = \{(\mu,\lambda) \in \Omega : 11\lambda^2 - 12\mu\lambda + 1 = 0\}$ the sets $\Gamma_0 \cap \Gamma_2 = \{(5/(2\sqrt{19}), 1/\sqrt{19}), (1,1)\}$, $\Gamma_1 \cap \Gamma_2 = \{(5/(2\sqrt{19}), 1/\sqrt{19})\}$, and intersection of Γ_2 with the boundary $\mu = 1$ yields $\{(1,1),(1,1/11)\}\}$ (use e.g. *Solve* in *Mathematica*).

c1) $(c_0^{ic}/c_2^{ic}) \geq 0$, $(-c_1^{ic}/c_2^{ic}) > 0$ and $u_{i+1}^+ \geq 1$ with (24) yield

$$\lambda \geq \frac{2\mu}{5}, \quad 11\lambda^2 - 12\mu\lambda + 1 > 0, \quad \sqrt{\Delta} \geq \lambda^2 - 12\mu\lambda + 11 . \tag{27}$$

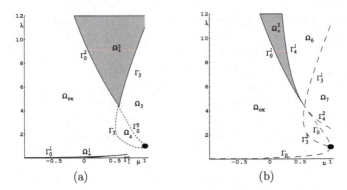

Fig. 3. Extension of admissible zone Ω_{ok} by cutting out from Ω_{ex}: (a) Ω_3, (b) Ω_4.

The last inequality in (27) is clearly satisfied for $\lambda^2 - 12\mu\lambda + 11 < 0$. This holds over $\Omega_5 = \Omega_3 \cup \Omega_4 \cup \Gamma_0^3$ which is the domain bounded by $\Gamma_3 = \{(\mu, \lambda) \in \Omega : \lambda^2 - 12\mu\lambda + 11 = 0\}$ and the boundary $\mu = 1$ (see Fig. 3 a)). Here $\Gamma_3 \cap \partial\Omega = \{(1,1), (1,11)\}$ and $\Gamma_3 \cap \Gamma_0 = \{(1,1), (5/(2\sqrt{19}), \sqrt{19})\} \approx \{(1,1), (0.573539, 4.3589)\}$ - again we resort here to *InequalityPlot*, *Implicit-Plot* and *Solve* functions in *Mathematica*. Intersecting Ω_5 with three subsets defined by the first two inequalities from (27) and $\Delta > 0$ yields cutting Ω_3 from the exclusion zone Ω_{ex} (see Fig. 3 a)), where Ω_3 is bounded by Γ_0^3, undashed Γ_3 and the boundary $\mu = 1$. Thus $\Omega_- \cup \Gamma_0 \cup \Omega_1 \cup \Omega_2 \cup \Omega_3 \subset \Omega_{ok}$. For the opposite case $\lambda^2 - 12\mu\lambda + 11 \geq 0$ (satisfied over $\Omega \setminus \Omega_5$) the last inequality from (27) yields $\Omega_8 = \Omega_6 \cup \Omega_7 \cup \Gamma_3^1$ with the bounding curve $\Gamma_4 = \{(\mu, \lambda) \in \Omega : \Delta - (\lambda^2 - 12\mu\lambda + 11)^2 = 0\}$ (see Fig. 3 b)) - here Ω_6 is bounded by Γ_4^1, Γ_3^1 and $\partial\Omega$ and Ω_7 is bounded by Γ_4^2, Γ_3^1 and $\partial\Omega$). The intersection of Γ_4 with boundary $\mu = 1$ yields single point $\{(1, 5/2)\}$. Since $\Gamma_0 \cap \Gamma_3 \cap \Gamma_4 = \{(5/(2\sqrt{19}), \sqrt{19})\} \approx \{(0.573539, 4.3589)\} = p_2$ the intersection of Ω_8 with the regions defined by first two inequalities in (27) (and by $\Delta > 0$ and $\lambda^2 - 12\mu\lambda + 11 \geq 0$) leads to the further cut out of $\Omega_6 \cup \Gamma_4^1 \cup \Gamma_3^1$ in the zone $\Omega_+^3 \subset \Omega_{ex}$. The inclusion $\Omega_- \cup \Gamma_0 \cup \Omega_1 \cup \Omega_2 \cup \Omega_3 \cup \Omega_6 \cup \Gamma_4^1 \cup \Gamma_3^1 \subset \Omega_{ok}$ follows.

5   Perturbed Special Case

Assume now that for data points $\{\tilde{x}_i, \tilde{x}_{i+1}, \tilde{x}_{i+2}\}$ and velocities $\{\tilde{v}_i, \tilde{v}_{i+2}\}$ condition (20) is not met. For the *perturbation vector* $\delta = (\delta_1, \delta_2) \in \mathbb{R}^{2m}$ we attempt to extend the results for (20) to its perturbed form (28). Indeed let (δ_1, δ_2):

$$\tilde{x}_{i+2} - \tilde{x}_i - \tilde{v}_{i+2} = \delta_1 , \qquad \tilde{v}_{i+2} - \tilde{v}_i = \delta_2 , \qquad (28)$$

with $\tilde{\mathcal{E}}_i^\delta$ derived as in (18). Of course, for $\delta_1 = \delta_2 = \mathbf{0} \in \mathbb{R}^m$ (28) collapses to (20) (i.e. with the notation $\tilde{\mathcal{E}}_i^0 = \bar{\mathcal{E}}_i^c$ derived for (20)). To obtain formulas for $\tilde{\mathcal{E}}_i^\delta$ and $\tilde{\mathcal{E}}_i^{\delta'}$ we resort to (by (28)):

$$\|\tilde{v}_{i+2}\|^2 = \|\tilde{x}_{i+2}\|^2 + \|\tilde{x}_i\|^2 - 2\langle\tilde{x}_i|\tilde{x}_{i+2}\rangle - 2\langle\tilde{x}_{i+2}|\delta_1\rangle + 2\langle\tilde{x}_i|\delta_1\rangle + \|\delta_1\|^2 \,,$$

$$\langle\tilde{v}_{i+2}|\tilde{x}_i\rangle = \langle\tilde{x}_i|\tilde{x}_{i+2}\rangle - \|\tilde{x}_i\|^2 - \langle\tilde{x}_i|\delta_1\rangle \,,$$

$$\langle\tilde{v}_{i+2}|\tilde{x}_{i+2}\rangle = \|\tilde{x}_{i+2}\|^2 - \langle\tilde{x}_i|\tilde{x}_{i+2}\rangle - \langle\tilde{x}_{i+2}|\delta_1\rangle \,,$$

$$\|\tilde{v}_i\|^2 = \|\tilde{x}_{i+2}\|^2 + \|\tilde{x}_i\|^2 - 2\langle\tilde{x}_i|\tilde{x}_{i+2}\rangle + \|\delta_1\|^2 + \|\delta_2\|^2 + 2\langle\delta_1|\delta_2\rangle$$
$$-2\langle\tilde{x}_{i+2}|\delta_1\rangle - 2\langle\tilde{x}_{i+2}|\delta_2\rangle + 2\langle\tilde{x}_i|\delta_1\rangle + 2\langle\tilde{x}_i|\delta_2\rangle \,,$$

$$\langle\tilde{v}_i|\tilde{v}_{i+2}\rangle = \|\tilde{x}_{i+2}\|^2 + \|\tilde{x}_i\|^2 - 2\langle\tilde{x}_i|\tilde{x}_{i+2}\rangle - 2\langle\tilde{x}_{i+2}|\delta_1\rangle + 2\langle\tilde{x}_i|\delta_1\rangle - \langle\tilde{x}_{i+2}|\delta_2\rangle$$
$$+\langle\tilde{x}_i|\delta_2\rangle + \|\delta_1\|^2 + \langle\delta_1|\delta_2\rangle \,,$$

$$\langle\tilde{x}_i|\tilde{v}_i\rangle = \langle\tilde{x}_i|\tilde{x}_{i+2}\rangle - \|\tilde{x}_i\|^2 - \langle\tilde{x}_i|\delta_1\rangle - \langle\tilde{x}_i|\delta_2\rangle \,,$$

$$\langle\tilde{x}_{i+2}|\tilde{v}_i\rangle = \|\tilde{x}_{i+2}\|^2 - \langle\tilde{x}_i|\tilde{x}_{i+2}\rangle - \langle\tilde{x}_{i+2}|\delta_1\rangle - \langle\tilde{x}_{i+2}|\delta_2\rangle \,,$$

leading by (18) to (with *FullSimplify*, *Factor* and *CoefficientList*): $\tilde{\mathcal{E}}_i^\delta(s_{i+1}) =$

$$\frac{1}{s_{i+1}^3(s_{i+1}-1)^3}(3\|\tilde{x}_i\|^2(s_{i+1}-1)^3(1+3s_{i+1}) + s_{i+1}(-6(\langle\tilde{x}_i|\tilde{x}_{i+2}\rangle$$
$$-\langle\tilde{x}_i|\delta_1\rangle - \langle\tilde{x}_i|\delta_2\rangle - \|\tilde{x}_i\|^2) + s_{i+1}(-18\langle\tilde{x}_i|\tilde{x}_{i+2}\rangle(s_{i+1}-1)^2$$
$$-6(\langle\tilde{x}_i|\tilde{x}_{i+2}\rangle - \langle\tilde{x}_i|\delta_1\rangle - \|\tilde{x}_i\|^2)(s_{i+1}-1)^3 + 6(-\langle\tilde{x}_i|\tilde{x}_{i+2}\rangle - \langle\tilde{x}_{i+2}|\delta_1\rangle$$
$$+\|\tilde{x}_{i+2}\|^2)(s_{i+1}-2)(s_{i+1}-1)s_{i+1} - 6(-\langle\tilde{x}_i|\tilde{x}_{i+2}\rangle - \langle\tilde{x}_{i+2}|\delta_1\rangle$$
$$-\langle\tilde{x}_{i+2}|\delta_2\rangle + \|\tilde{x}_{i+2}\|^2)(s_{+1}-1)^2 s_{i+1} + (-2\langle\tilde{x}_i|\tilde{x}_{i+2}\rangle + 2\langle\tilde{x}_i|\delta_1\rangle$$
$$-2\langle\tilde{x}_{i+2}|\delta_1\rangle + \|\tilde{x}_i\|^2 + \|\tilde{x}_{i+2}\|^2 + \|\delta_1\|^2)(s_{i+1}-4)(s_{i+1}-1)^2 s_{i+1}$$
$$-2(-2\langle\tilde{x}_i|\tilde{x}_{i+2}\rangle + 2\langle\tilde{x}_i|\delta_1\rangle + \langle\tilde{x}_i|\delta_2\rangle - 2\langle x_{i+2}|\delta_1\rangle - \langle\tilde{x}_{i+2}|\delta_1\rangle + \langle\delta_1|\delta_2\rangle$$
$$+\|\tilde{x}_i\|^2 + \|\tilde{x}_{i+2}\|^2 + \|\delta_1\|^2)(s_{i+1}-1)^3 s_{i+1} + (-2\langle\tilde{x}_i|\tilde{x}_{i+2}\rangle + 2\langle\tilde{x}_i|\delta_1\rangle$$
$$+2\langle\tilde{x}_i|\delta_2\rangle - 2\langle\tilde{x}_{i+2}|\delta_1\rangle - 2\langle\tilde{x}_{i+2}|\delta_2\rangle + 2\langle\delta_1|\delta_2\rangle + \|\tilde{x}_i\|^2 + \|\tilde{x}_{i+2}\|^2 + \|\delta_1\|^2$$
$$+\|\delta_2\|^2)(s_{i+1}-1)^3(3+s_{i+1}) + 3\|\tilde{x}_{i+2}\|^2 s_{i+1}(3s_{i+1}-4)$$
$$+6(\langle\tilde{x}_i|\tilde{x}_{i+2}\rangle - \langle\tilde{x}_i|\delta_1\rangle - \langle\tilde{x}_i|\delta_2\rangle - \|\tilde{x}_i\|^2)(2+(s_{i+1}-2)s_{i+1}^2)))) \tag{29}$$

yielding $\tilde{\mathcal{E}}_i^\delta(s_{i+1}) = M_i^\delta(s_{i+1})/(s_{i+1}^3(s_{i+1}-1)^3)$. Here $deg(M_i^\delta) = 6$ with the coefficients (using *Mathematica* functions *Factor* and *CoefficientList*): $a_0^{i,\delta} = -3\|\tilde{x}_i\|^2$, $a_1^{i,\delta} = -6\langle\tilde{x}_i|\tilde{x}_{i+2}\rangle + 6\langle\tilde{x}_i|\delta_1\rangle + 6\langle\tilde{x}_i|\delta_2\rangle + 6\|\tilde{x}_i\|^2$, $a_2^{i,\delta} = 6\langle\tilde{x}_i|\tilde{x}_{i+2}\rangle - 24\langle\tilde{x}_i|\delta_1\rangle - 18\langle\tilde{x}_i|\delta_2\rangle + 6\langle\tilde{x}_{i+2}|\delta_1\rangle + 6\langle\tilde{x}_{i+2}|\delta_2\rangle - 6\langle\delta_1|\delta_2\rangle - 3\|\tilde{x}_i\|^2 - 3\|\tilde{x}_{i+2}\|^2 - 3\|\delta_1\|^2 - 3\|\delta_2\|^2$, $a_3^{i,\delta} = 2(15\langle\tilde{x}_i|\delta_1\rangle + 9\langle\tilde{x}_i|\delta_2\rangle - 9\langle\tilde{x}_{i+2}|\delta_1\rangle - 6\langle\tilde{x}_{i+2}|\delta_2\rangle + 9\langle\delta_1|\delta_2\rangle + 3\|\delta_1\|^2 + 4\|\delta_2\|^2)$, $a_4^{i,\delta} = -12\langle\tilde{x}_i|\delta_1\rangle - 6\langle\tilde{x}_i|\delta_2\rangle + 12\langle\tilde{x}_{i+2}|\delta_1\rangle + 6\langle\tilde{x}_{i+2}|\delta_2\rangle - 18\langle\delta_1|\delta_2\rangle - 3\|\delta_1\|^2 - 6\|\delta_2\|^2$, $a_5^{i,\delta} = 6\langle\delta_1|\delta_2\rangle$ and $a_6^{i,\delta} = \|\delta_2\|^2$. The derivative of $\tilde{\mathcal{E}}_i^\delta(s_{i+1})$ reads as $\tilde{\mathcal{E}}_i^{\delta'}(s_{i+1}) = -N_i^\delta(s_{i+1})/(s_{i+1}^4(s_{i+1}-1)^4)$, where N_i^δ is the 6-th order polynomial in s_{i+1} with the coefficients (e.g. again upon using symbolic differentiation in *Mathematica* and functions *Factor* and *CoefficientList*): $b_0^{i,\delta} = -9\|\tilde{x}_i\|^2$, $b_1^{i,\delta} = -12\langle\tilde{x}_i|\tilde{x}_{i+2}\rangle + 12\langle\tilde{x}_i|\delta_1\rangle + 12\langle\tilde{x}_i|\delta_2\rangle + 30\|\tilde{x}_i\|^2$, $b_2^{i,\delta} = 3(12\langle\tilde{x}_i|\tilde{x}_{i+2}\rangle - 18\langle\tilde{x}_i|\delta_1\rangle - 16\langle\tilde{x}_i|\delta_2\rangle + 2\langle\tilde{x}_{i+2}|\delta_1\rangle + 2\langle\tilde{x}_{i+2}|\delta_2\rangle - 2\langle\delta_1|\delta_2\rangle - 11\|\tilde{x}_i\|^2 - \|\tilde{x}_{i+2}\|^2 - \|\delta_1\|^2 - \|\delta_2\|^2)$, $b_3^{i,\delta} = 3(-8\langle\tilde{x}_i|\tilde{x}_{i+2}\rangle + 32\langle\tilde{x}_i|\delta_1\rangle + 24\langle\tilde{x}_i|\delta_2\rangle - 8\langle\tilde{x}_{i+2}|\delta_1\rangle - 8\langle\tilde{x}_{i+2}|\delta_2\rangle + 8\langle\delta_1|\delta_2\rangle + 4\|\tilde{x}_i\|^2 + 4\|\tilde{x}_{i+2}\|^2 + 4\|\delta_1\|^2 + 4\|\delta_2\|^2)$, $b_4^{i,\delta} = 3(-26\langle\tilde{x}_i|\delta_1\rangle - 16\langle\tilde{x}_i|\delta_2\rangle + 14\langle\tilde{x}_{i+2}|\delta_1\rangle + 10\langle\tilde{x}_{i+2}|\delta_2\rangle - 12\langle\delta_1|\delta_2\rangle - 5\|\delta_1\|^2 - 6\|\delta_2\|^2)$,

$b_5^{i,\delta} = 3(8\langle\tilde{x}_i|\delta_1\rangle + 4\langle\tilde{x}_i|\delta_2\rangle - 8\langle\tilde{x}_{i+2}|\delta_1\rangle - 4\langle\tilde{x}_{i+2}|\delta_2\rangle + 8\langle\tilde{\delta}_1|\delta_2\rangle + 2\|\delta_1\|^2 + 4\|\delta_2\|^2)$
and $b_6^{i,\delta} = -6\langle\delta_1|\delta_2\rangle - 3\|\delta_2\|^2$.

The following result merging (20) with (28) holds (proof is omitted):

Theorem 1. *Assume that for unperturbed data (20) the corresponding energy $\tilde{\mathcal{E}}_i^0$ has exactly one critical point $\hat{s}_0 \in (0,1)$ with $\tilde{\mathcal{E}}_i^{0''}(\hat{s}_0) \neq 0$. Then there exists sufficiently small $\varepsilon_0 > 0$ such that for all $\|\delta\| < \varepsilon_0$ (where $\delta = (\delta_1, \delta_2) \in \mathbb{R}^{2m}$) the perturbed data (28) yield the energy $\tilde{\mathcal{E}}_i^\delta$ with exactly one critical point $\hat{s}_0^\delta \in (0,1)$ (a global minimum $\hat{s}_0^\delta$ of $\tilde{\mathcal{E}}_i^\delta$ is sufficiently close to $\hat{s}_0$).*

Example 1. Consider the planar points $\tilde{x}_i = (0,-1)$, $\tilde{x}_{i+1} = (0,0)$ and $\tilde{x}_{i+2} = (1,1)$ - we set here $i = 0$. Here cumulative chord parameterization yields $\hat{s}_1^{cc} = 1/(\sqrt{2}+1) \approx 0.414214$. Assume that given velocities $\tilde{v}_0, \tilde{v}_2$ (upon adjustment by some perturbation $\delta = (\bar{\delta}, \hat{\delta}) \in \mathbb{R}^4$) satisfy both constraints $\tilde{x}_2 - \tilde{x}_0 = \tilde{v}_2 + \bar{\delta}$ and $\tilde{v}_2 = \tilde{v}_0 + \hat{\delta}$. The above interpolation points $\{\tilde{x}_i, \tilde{x}_{i+1}, \tilde{x}_{i+2}\}$ for further testing in this example are assumed to be fixed. Here $\|\tilde{x}_0\|^2 = 1$, $\|\tilde{x}_2\|^2 = 2$, $\langle\tilde{x}_0|\tilde{x}_2\rangle = -1$ and $(\mu, \lambda) = (-1/\sqrt{2}, 1/\sqrt{2}) \approx (-0.707107, 0.707107) \in \Omega_{ok}$ (with $\delta = \mathbf{0}$). The unperturbed energy with $\tilde{v}_2 = \tilde{v}_0 = (1,2)$ (see also (21) or (29) with $\delta = \mathbf{0}$ and non-perturbed data satisfying (20)) amounts to: $\tilde{\mathcal{E}}_0^c(s) = -3(1 + s(5s - 4))((s - 1)^3 s^3)^{-1}$. Which yields a global minimum $\tilde{\mathcal{E}}_0^c(0.433436) = 41.6487$ (see Fig. 4). As here $(\mu, \lambda) = (-1/\sqrt{2}, 1/\sqrt{2}) \in \Omega_{ok}$ and thus $\tilde{\mathcal{E}}_0^0$ has exactly one critical point $\hat{s}_0 \in (0,1)$. One can show that $\tilde{\mathcal{E}}_0^{0''} \neq 0$ at any critical point $\hat{s}_0$ of $\tilde{\mathcal{E}}_0^0$. Hence the assumptions from Theorem 1 are clearly satisfied.

We add now the perturbation $\bar{\delta} = (2,-3)$ and $\hat{\delta} = (-1,2)$ (for $\tilde{v}_0 = (0,3)$ and $\tilde{v}_2 = (-1,5)$). The corresponding perturbed energy (see (29)) $\tilde{\mathcal{E}}_0^\delta(s) = (-3 + s(18 + s(-57 + s(34 + s(45 + s(5s - 48))))))((s - 1)^3 s^3)^{-1}$ is plotted in Fig. 5 with the optimal value $\hat{s}_0^\delta \approx 0.390407$ (close to $\hat{s}_1^{cc}$ as perturbation δ is sufficiently small - here $(\|\bar{\delta}\|, \|\hat{\delta}\|) = (\sqrt{13}, \sqrt{5})$) and $\tilde{\mathcal{E}}_\delta^c(\hat{s}_0^\delta) = 149.082 < \tilde{\mathcal{E}}_\delta^c(\hat{s}_1^{cc}) = 150.004$ - the convexity of $\tilde{\mathcal{E}}_0^c$ is visibly preserved by $\tilde{\mathcal{E}}_\delta^c$ (see Fig. 4 and Fig. 5).

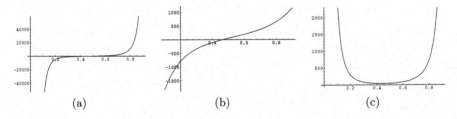

Fig. 4. The graph of $\tilde{\mathcal{E}}_0^{c'}$ for $\tilde{x}_0 = (0,-1)$, $\tilde{x}_2 = (1,1)$, $\tilde{v}_0 = \tilde{v}_2 = (1,2)$ (a) over $(0,1)$, (b) close to unique root $\hat{s}_0 \approx 0.433 \neq \hat{s}_1^{cc} = 1/(\sqrt{2}+1) \approx 0.414$, (c) the graph of $\tilde{\mathcal{E}}_0^c$.

For a large perturbation $\bar{\delta} = (16,7)$ and $\hat{\delta} = (-10,5)$ (for $\tilde{v}_0 = (-5,-10)$ and $\tilde{v}_2 = (-15,-5)$) the corresponding perturbed energy (see (29) and use *Simplify* in *Mathematica*) $\tilde{\mathcal{E}}_0^\delta(s) = (-3 + s(-60 + s(-189 + s(-74 + 5s(5s - 21)(5s - 9)))))((s - 1)^3 s^3)^{-1}$ is plotted in Fig. 6 a) with the unique optimal value

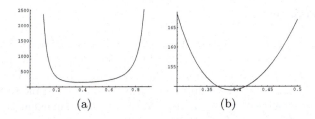

Fig. 5. The graph of $\tilde{\mathcal{E}}_\delta^c$ for $\tilde{x}_0 = (0, -1)$, $\tilde{x}_2 = (1, 1)$, $\tilde{v}_0 = (0, 3)$, $\tilde{v}_2 = (-1, 5)$, $\bar{\delta} = (2, -3)$ and $\hat{\delta} = (-1, 2)$ (a) over $(0, 1)$, (b) close to its unique min. $\hat{s}_0^\delta = 0.390 \neq s_1^{cc} = 1/(\sqrt{2} + 1) \approx 0.414$.

$\hat{s}_0^\delta \approx 0.432069$ for which $\tilde{\mathcal{E}}_\delta^c(\hat{s}_0^\delta) = 3229.81 < \tilde{\mathcal{E}}_\delta^c(\hat{s}_1^{cc}) = 3236.5$ - the convexity of $\tilde{\mathcal{E}}_0^c$ is here visibly also preserved by $\tilde{\mathcal{E}}_\delta^c$ (even for such a quite large perturbation δ - here $(\|\bar{\delta}\|, \|\hat{\delta}\|) = (125, 305)$). Note also that though cumulative chord $\hat{s}_1^{cc}$ is now farther away from a global minimum $\hat{s}_0^\delta$, it is still in its potential basin.

We add now very large $\bar{\delta} = (-25, -17)$ and $\hat{\delta} = (-6, 20)$ (for $\tilde{v}_0 = (32, -1)$ and $\tilde{v}_2 = (26, 19)$). The perturbed energy (see (29)) $\tilde{\mathcal{E}}_0^\delta(s) = (-3 + s(-6 + s(-3141 + 2s(3145 + s(-1221 - 570s + 218s^2)))))((s-1)^3 s^3)^{-1}$ is plotted in Fig. 6 b) with the optimal value $\hat{s}_0^\delta \approx 0.948503$ for which $\tilde{\mathcal{E}}_\delta^c(\hat{s}_0^\delta) = 11146 < \tilde{\mathcal{E}}_\delta^c(\hat{s}^{cc}) = 12667.7$ and another local minimum at $\hat{s}_0^1 \approx 0.563968$ for which $\tilde{\mathcal{E}}_\delta^c(\hat{s}_0^1) = 11781$. There is also a local maximum at $\hat{s}_{max} \approx 0.879929 > s_1^{cc} = 1/3$ - convexity of $\tilde{\mathcal{E}}_0^c$ is here clearly not preserved by $\tilde{\mathcal{E}}_\delta^c$ (δ is here too large for Theorem 1 to hold - here $(\|\bar{\delta}\|, \|\hat{\delta}\|) = (914, 436)$) - see also Fig. 4 and Fig. 6. Again the cumulative chord $\hat{s}_1^{cc} \approx 0.414214$ is here in the basin of $\hat{s}_0^1$ (not of $\hat{s}_0^\delta$) - see Fig. 6 b).   □

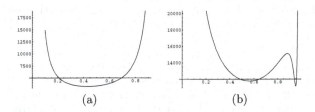

Fig. 6. The graph of $\tilde{\mathcal{E}}_\delta^c$ for $\tilde{x}_0 = (0, -1)$ and $\tilde{x}_2 = (1, 1)$ for (a) $\tilde{v}_0 = (-5, -10)$, $\tilde{v}_2 = (-15, -5)$ and a big $\bar{\delta} = (-16, 7)$ and $\bar{\delta} = (-10, 5)$ yielding global min. at $\hat{s}_0^\delta \approx 0.432 \neq s_1^{cc} \approx 0.414$, (b) $\tilde{v}_0 = (32, -1)$, $\tilde{v}_2 = (26, 19)$ and a very big $\bar{\delta} = (-25, -17)$ and $\bar{\delta} = (-6, 20)$ with global min. at $\hat{s}_0^\delta \approx 0.949$ and a local min. at $\hat{s}_0^1 = 0.564 \neq s_1^{cc} \approx 0.414$.

Example 1 suggests that δ in Theorem 1 can in fact be quite substantial. Thus a local character of Theorem 1 seems to be more a semi-global one.

6   Conclusions

We study the problem of finding optimal knots to fit reduced data. The optimization task (1) is reformulated into (5) (and (18)) to minimize a highly non-linear multivariable function $\mathcal{J}_0$ depending on knots $\mathcal{T}_{int}$. Leap-Frog is a feasible numerical scheme to handle (5). It minimizes iteratively single variable functions from (6). Generic case of Leap-Frog is addressed to establish sufficient conditions for unimodality of (18). First, its special case (20) is studied. Next a perturbed analogue (28) of the latter is addressed. The unimodality of (21) is shown to be preserved by large perturbations (28). The performance of Leap-Frog in minimizing (5) against Newton's and Secant Methods is discussed in [2,3] and [6]. Other contexts and applications of Leap-Frog can be found in [9,10] or [11]. For more work on fitting $\mathcal{M}_n$ (sparse or dense) see [4,5] or [7].

References

1. de Boor, C.: A Practical Guide to Splines, 2nd edn. Springer, New York (2001). https://www.springer.com/gp/book/9780387953663
2. Kozera, R., Noakes, L.: Optimal knots selection for sparse reduced data. In: Huang, F., Sugimoto, A. (eds.) PSIVT 2015. LNCS, vol. 9555, pp. 3–14. Springer, Cham (2016). https://doi.org/10.1007/978-3-319-30285-0_1
3. Kozera, R., Noakes, L.: Non-linearity and non-convexity in optimal knots selection for sparse reduced data. In: Gerdt, V.P., Koepf, W., Seiler, W.M., Vorozhtsov, E.V. (eds.) CASC 2017. LNCS, vol. 10490, pp. 257–271. Springer, Cham (2017). https://doi.org/10.1007/978-3-319-66320-3_19
4. Kozera, R., Noakes, L., Wilkołazka, M.: Parameterizations and Lagrange cubics for fitting multidimensional data. In: Krzhizhanovskaya, V.V., et al. (eds.) ICCS 2020. LNCS, vol. 12138, pp. 124–140. Springer, Cham (2020). https://doi.org/10.1007/978-3-030-50417-5_10
5. Kozera, R., Noakes L., Wilkołazka, M.: Exponential parameterization to fit reduced data. Appl. Math. Comput. **391**(C), 125645 (2021). https://doi.org/10.1016/j.amc.2020.125645
6. Kozera, R., Wiliński, A.: Fitting dense and sparse reduced data. In: Pejaś, J., El Fray, I., Hyla, T., Kacprzyk, J. (eds.) ACS 2018. AISC, vol. 889, pp. 3–17. Springer, Cham (2019). https://doi.org/10.1007/978-3-030-03314-9_1
7. Kuznetsov, E.B., Yakimovich A.Y.: The best parameterization for parametric interpolation. J. Comput. Appl. Math. **191**(2), 239–245 (2006). https://core.ac.uk/download/pdf/81959885.pdf
8. Kvasov, B.I.: Methods of Shape-Preserving Spline Approximation. World Scientific Pub., Singapore (2000). https://doi.org/10.1142/4172
9. Matebese, B., Withey, D., Banda, M.K.: Modified Newton's method in the Leapfrog method for mobile robot path planning. In: Dash, S.S., Naidu, P.C.B., Bayindir, R., Das, S. (eds.) Artificial Intelligence and Evolutionary Computations in Engineering Systems. AISC, vol. 668, pp. 71–78. Springer, Singapore (2018). https://doi.org/10.1007/978-981-10-7868-2_7
10. Noakes, L.: A global algorithm for geodesics. J. Aust. Math. Soc. Series A **65**(1), 37–50 (1998). https://doi.org/10.1017/S1446788700039380
11. Noakes, L., Kozera, R.: Nonlinearities and noise reduction in 3-source photometric stereo. J. Math. Imaging Vision **18**(2), 119–127 (2003). https://doi.org/10.1023/A:1022104332058

Intelligent Planning of Logistic Networks to Counteract Uncertainty Propagation

Przemysław Ignaciuk[ID] and Adam Dziomdziora[(✉)][ID]

Lodz University of Technology, 215 Wólczańska Street, 90-924 Łódź, Poland
przemyslaw.ignaciuk@p.lodz.pl, adam.dziomdziora@dokt.p.lodz.pl

Abstract. A major obstacle to stable and cost-efficient management of goods distribution systems is the bullwhip effect – reinforced demand uncertainty propagating among system nodes. In this work, by solving a formally established optimization problem, it is shown how one can mitigate the bullwhip effect, at the same minimizing transportation costs, in modern logistic networks with complex topologies. The flow of resources in the analyzed network is governed by the popular order-up-to inventory policy, which thrives to maintain sufficient stock at the nodes to answer *a priori* unknown, uncertain demand. The optimization objective is to decide how intensive a given transport channel should be used so that unnecessary goods relocation and the bullwhip effect are avoided while being able to fulfill demand requests. The computationally challenging optimization task is solved using a population-based evolutionary technique – Biogeography-Based Optimization. The results are verified in extensive simulations of a real-world transportation network.

Keywords: Transportation networks · Time-delay systems · Population-based optimization

1 Introduction

The bullwhip effect (BE) is a serious systemic distortion in logistic systems, manifesting itself as an enhanced variability of demand transmitted into the goods ordering signal. In addition to lowered earnings, it leads to unnecessary shipments, prolonged delays, and resource accumulation at subsidiary nodes. Thus far, its impact has been assessed primarily from local and chain-structure perspectives. In contrast, here, the BE formation and countermeasures are investigated in the context of modern networked systems, not restricted to specific, reduced topologies.

Forrester laid grounds for the BE examination in [1], with continued studies related to its formation within production-distribution environments reported later in [2–4]. The principal factors affecting the goods flow fluctuation in basic architectures were discussed by Lee et al. in [5] and [6], and in current settings in [7–9]. The essential BE triggers include: inaccurate demand prediction, production rate mismatch, non-negligible transportation time, batch arrangement, and price variations. A comprehensive classification

© The Author(s) 2021
M. Paszynski et al. (Eds.): ICCS 2021, LNCS 12745, pp. 351–364, 2021.
https://doi.org/10.1007/978-3-030-77970-2_27

of the BE causes in modern systems is given by Lin et al. in [10], with focalized treatment of erroneous stock level records in [11].

Besides seeking its origins, many scientists worked on techniques of decreasing the negative impact of the BE on supply system performance [12–15]. They emphasized statistical analysis and operations research methods. Another promising approach was to apply robust control techniques [16–19]. However, in practical installations, traditional methods are still preferred, e.g., order-up-to (OUT) policy. The BE formation in the systems organized in serial and arborescent configurations governed by the OUT policy has been examined in [20]. Preliminary treatment of mesh-type topologies has been given in [21]. A modified OUT policy, destined for centralized system management, was optimally tuned for holding and lost-sales costs reduction in [22].

In real-world logistic systems, the optimization typically targets delays and holding or transportation costs reduction. Finding the optimal solution for the considered objective functions, either analytically or numerically (e.g., through full search), is challenging. Therefore, non-weighted procedures, e.g., alternating, hierarchical, or phased optimization techniques, are applied. For example, to minimize both the whole-time cost of travelers and the number of essential transfers in a transit system, Arbex and da Cunha [23] introduced Alternating Objective Genetic Algorithm. As opposed to the traditional one [24, 25], it allowed them to use local search methods to deal with infeasibility of newly created individuals. Also, improved Simulated Annealing has recently been considered in the optimization of transportation networks [26]. However, the applied objectives overlook a fundamental problem that face modern systems: to work efficiently in a time-varying, perturbed environment. Hence, in this work, the OUT policy optimization explicitly targets reduction of a systemic distortion – the BE.

The considered class covers systems with an arbitrary configuration, with goods reflow subjected to non-negligible time delay. The popular OUT inventory policy governs lot sizing. The objective is to plan the network structure, i.e., to decide how intensively a given transportation route (channel) of goods distribution should be used to avoid the BE. As a result, a matrix of coefficients yielding reduced BE and transportation costs within a given time horizon is obtained. The coefficients may also be interpreted in terms of order splitting, i.e., which part of an order established by a controlled node is to be retrieved from a given supplier (a nearby controlled node or an external source). The optimization task is solved with one of recent evolutionary techniques – Biogeography-Based Optimization (BBO). The acquired tuning guidelines for the coefficient adjustment are straightforward in implementation and do not require considerable computational effort to calculate. As shown in the conducted research, the commonly exercised omission of the planning aspect through uniform lot partitioning among the transport channels is incorrect since it leads to reinforced perturbation and larger costs. The proposed intelligent planning technique enables one to place the goods distributor in a desirable situation with respect to the competition, reduces transportation costs, and throttles down the BE within the system.

2   Bullwhip Effect in Transportation Networks

An example supply chain is illustrated in Fig. 1, with S_1 – the external supplier, u_1, u_2, ..., u_i – ordering signals, and d_1, d_2, ..., d_i – imposed demands.

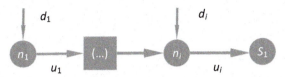

Fig. 1. Serial connection of n nodes. Resources are supplied by external source S_1. The arrows reflect the flow of information.

In [27], several indicators used to quantify the BE in serial structures, both in the time and the frequency domain, have been examined. One of the most popular ones is the order-to-demand variance ratio [28]. At node i, this bullwhip indicator (BI) is calculated as:

$$b_i = \frac{\text{var}[u_i]}{\text{var}[d_i]}, \tag{1}$$

where var$[\cdot]$ denotes variance. Value $b_i > 1$ means that at echelon i, the BE has been triggered.

In the nominal operating conditions, the OUT policy guarantees that for any demand pattern

$$b_1 = b_2 = \ldots = b_n = 1, \tag{2}$$

and the BE is absent [18].

Assuming that signals $d_1, \ldots, d_i$ are not correlated (given the knowledge about the demand imposed at a node, one should in principle not judge about the demand at other nodes), the BI can be measured having both internal and external demand incorporated as:

$$b_i = \frac{\text{var}[u_i]}{\text{var}[d_i + u_{i-1}]} = \frac{\text{var}[u_i]}{\text{var}[d_i] + \text{var}[u_{i-1}]}. \tag{3}$$

Nonetheless, to estimate the system propensity to the BE formation in current systems, one should examine more involving topologies than a serial chain. An example *networked* structure is illustrated in Fig. 2, where n_{1-3} denote controlled nodes, $S_{1,2}$ are external sources, and d_{1-3} is the exogenous demand imposed on the system.

In the system from Fig. 2, intuitively, the BE will be triggered when variance increase between the external replenishment signal $\mathbf{u} = [u_1\ u_2]^\text{T}$ and the imposed demand $\mathbf{d} = [d_1\ d_2\ d_3]^\text{T}$ is observed. Unlike the serial structure, the BI takes a matrix form – $\mathbf{B}$ – determined from the relation:

$$\begin{bmatrix} \text{var}[u_1] \\ \text{var}[u_2] \end{bmatrix} = \underbrace{\begin{bmatrix} b_{11}\ b_{12}\ b_{13} \\ b_{21}\ b_{22}\ b_{23} \end{bmatrix}}_{\mathbf{B}} \begin{bmatrix} \text{var}[d_1] \\ \text{var}[d_2] \\ \text{var}[d_3] \end{bmatrix}. \tag{4}$$

It is evident, that the application of indicators for serial connection is not sufficient to quantify the BE in a networked configuration, even in the basic setting from Fig. 2.

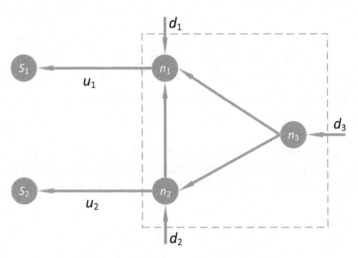

Fig. 2. A five-node transportation network.

Moreover, Eq. (2) does not hold in the networked case. Different measures are thus needed.

As opposed to the serial configuration, in which one may directly indicate the marginal nodes and use them to establish a BI within the transportation system, there is a limited possibility to adopt such approach in a networked environment. The absence of feasible measurement methods, providing the BE quantification in networked systems, motivates the search for alternatives [29]. The introduced indicator should allow the BE quantification, considering the networked topology as a holistic, multi-input multi-output entity. Hence, it shall relieve the complexity of determining each entry of matrix **B**. For the BE quantification in networked structures, a vector-based measure will be defined.

Instead of focusing on a particular node, all the demand and external replenishment signals will be considered. Within horizon of H periods, the record of replenishment signal placed by node i at an external supplier $u_i^H = [u_i(0) \, u_i(1) \, ... \, u_i(H-1)]^\mathrm{T}$. Similarly with respect to demand placed at node j one has $d_j^H = [d_j(0) \, d_j(1) \, ... \, d_j(H-1)]^\mathrm{T}$. The demand can be imposed on any node. Also, any node can generate a replenishment signal for an external supplier. The proposed BI, associated with the Euclidean distance, is calculated as:

$$\omega = \frac{\sqrt{\sum_{i\in\Omega_e} (\mathrm{var}[u_i^H])^2}}{\sqrt{\sum_{j\in\Omega_d} (\mathrm{var}[d_j^H])^2}}. \tag{5}$$

where Ω_e is the set of node indices that generate replenishment signals for the external suppliers, and Ω_d is the set of node indices at which the demand is placed.

With respect to the external actors, the controlled network is treated as a black-box entity. $\omega > 1$ implies an occurrence of the BE. The bigger the value of ω, the worse the BE.

3 System Model

3.1 Interconnection Structure

The considered class of system covers interaction between two types of actors:

- external sources – which supply the goods for the controlled network, yet are not affected by the customer demand, directly,
- controlled nodes – which serve both as intermediate suppliers for other controlled nodes and generate replenishment signals for the external sources to meet the demand.

The network encompasses N controlled nodes and S external sources, connected by unidirectional links. The links are characterized by:

- lot partitioning coefficient – to be determined in the optimization tasks – that says how intensively a given link (transportation channel) will be used,
- lead-time delay – the delay in order fulfillment, notably, the transport delay,
- transportation cost – related to the distance between the nodes.

For practical reasons, topologies with isolated nodes, i.e., having no connection to any supplier; or self-suppling nodes, are disregarded.

3.2 Node Dynamics

Let $k = 0, 1, 2, \ldots, H$ measure the duration of time. The stock level of goods accumulating at node i evolves according to

$$
x_i(k+1) = x_i(k) + \underbrace{\sum_{j=1}^{N+S} \alpha_{ji} u_i(k - \beta_{ji})}_{\text{incoming shipments}} - \underbrace{\sum_{j=1}^{N} \alpha_{ij} u_i(k)}_{\text{outgoing shipments}} - \underbrace{d_i(k)}_{\text{customer demand}} \tag{6}
$$

where:

- α_{ji} – the lot partitioning coefficient for the orders placed by node i at node j,
- β_{ji} – the lead-time delay of goods transferred from node j to i,
- $u_i(k)$ – the goods quantity requested by the node i in period k from its suppliers, both external sources and intermediate nodes,
- $d_i(k)$ – the external demand imposed on node i in period k. It exhibits arbitrary variations within $[0, d_i^{\max}]$, where $d_i^{\max}$ is the upper estimate.

The channel allocation matrix groups lot partitioning coefficients:

$$
\mathbf{A} =
\begin{bmatrix}
0 & \alpha_{12} & \cdots & \alpha_{1N} \\
\alpha_{21} & 0 & \cdots & \alpha_{2N} \\
\vdots & \vdots & \ddots & \vdots \\
\alpha_{N1} & \alpha_{N2} & \cdots & 0 \\
\vdots & \vdots & \cdots & \vdots \\
\alpha_{N+S,1} & \alpha_{N+S,2} & \cdots & \alpha_{N+S,N}
\end{bmatrix}_{N+S \times N} ,
\tag{7}
$$

$\alpha_{ji} \neq 0 \Rightarrow \alpha_{ij} = 0$ and for any i, j:

$$
0 \leq \alpha_{ji} \leq 1 \text{ and } \sum_{j=1}^{N+S} a_{ji} = 1.
\tag{8}
$$

3.3 OUT Inventory Policy

To manage the flow of resources in the network, the OUT inventory policy is applied. It is implemented in a distributed form, i.e., independently at the controlled nodes. According to the OUT policy, controlled node i generates the stock replenishment signal as

$$
u_i(k) = x_i^{ref} - x(k) - OR_i(k),
\tag{9}
$$

where x_i^{ref} is the reference level, e.g., that can be assigned to maximize sales [30], and $OR_i(k)$ is the open-order quantity, i.e., the goods in transit that have not yet reached the ordering node owing to lead-time delay.

3.4 Transportation Costs

The transportation costs are calculated by considering the length of the transportation trail and the quantity of the goods requested. Within the time horizon of H periods transportation cost

$$
\Psi = \sum_{k=0}^{H-1} \sum_{i=1}^{N} \sum_{j=1}^{N+S} \alpha_{ji} u_i(k) \varphi_{ji} \phi
\tag{10}
$$

where φ_{ij} is the transportation cost along the route $i-j$ determined as a product of a fixed unitary cost φ and the distance between the nodes.

3.5 Customer Satisfaction

A well-functioning goods distribution system is expected to ensure a high level of demand satisfaction. Denoting the satisfied demand at controlled node i in period k by $h_i(k)$, the customer satisfaction rate at that node is obtained as

$$\varepsilon_i = \frac{\sum_{k=0}^{H-1} h_i(k)}{\sum_{k=0}^{H-1} d_i(k)}, \tag{11}$$

An average satisfaction rate within the system can be calculated as

$$\vartheta = \frac{\sum_{i=1}^{N} \varepsilon_i}{N}. \tag{12}$$

4 Optimization Problem

The objective of the considered optimization problem is to establish a channel allocation matrix (7) so that the logistic system may satisfy the external demand with low transportation costs, yet avoiding the BE.

Formally, the optimization problem may be stated as follows:

$$\min J(\alpha_{ij}) = \Psi(k)\omega\vartheta^{-1}. \tag{13}$$

s.t. (6) – the stock level dynamics, (8) – the channel allocation constraint, and (9) – the method of replenishment signal computation. ω stands for the introduced BI for networked systems, Ψ is the transportation cost, and ϑ represents the mean customer satisfaction rate. Thus, one attempts to balance transportation costs and systemic distortion, at the same time maintaining a high customer service rate.

In the analyzed class of systems, the demand signal varies with time in an unpredictable way. Consequently, optimization problem (13) is not amenable to analytical treatment. It will be solved using a computational intelligence technique, specified in the next chapter.

5 Computational Framework

As a basis for constructing a computation framework to solve problem (13), a biogeography-based technique – BBO – will be adopted. BBO is an evolutionary algorithm that originates from the observations of the movement of species among separate areas called islands. It has been demonstrated to be a useful search procedure in optimization problems because it combines both examination and exploitation techniques based on migration [31]. Nowadays, it is one of the fastest growing in popularity algorithms, based on nature, used to tackle computationally-intensive optimization problems. In addition to numerous benefits, such as simplicity, flexibility, and efficiency, BBO does not demand computing derivatives of objective function. Its dynamic model has been described in [32]. The BBO algorithm is illustrated in Fig. 3.

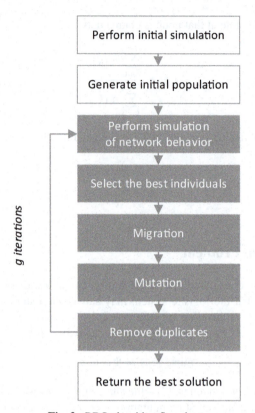

Fig. 3. BBO algorithm flowchart.

In evolutionary approaches, global recombination is applied to generate new solutions. However, in BBO it is migration that modifies existing solutions. The continuous domain of the search space allows for direct BBO application, i.e., without the classic translation to the binary form [24]. The matrix of lot partitioning coefficients reflects an individual, and an island (a set of individuals) corresponds to the set of a predefined size containing matrices of lot partitioning coefficients. A particular population comprises a set of islands. The component of an individual corresponds to a single lot partitioning coefficient – α_{ij} – in matrix **A**.

If a solution is intended for mutation, then a randomly chosen lot partitioning coefficient may be replaced with a newly generated one. The matrix of lot partitioning coefficients is created by randomly mutating the current columns, one-by-one, before going to a next algorithm iteration. It is performed by randomly increasing or decreasing each entry – α_{ij} – in the column. The value of the last entry in column j is calculated as

$$1 - \sum_{i=1}^{N+S-1} \alpha_{ji}.$$

6   Numerical Studies

The system considered in the numerical study represents the European distribution network of a firm from the premium-clothes fashion industry. The company root warehouses are located in Paris and Milan. The sales network extends through Central Europe, with the distribution centers in Brussels, Munich, Berlin, Warsaw, and Cracow, as illustrated in Fig. 4. The network graph representation is depicted in Fig. 5. The numbers displayed above the arrows indicate the lead-times and transportation costs associated with the routes.

Two cases (Network A and B) are given a closer examination. First, the star-network (Fig. 4), as the centralized architecture representation, is investigated. Afterward, a more complex topology, reflecting worldwide expansion, is investigated. In that case, additional sales points are introduced – with locations in Graz, Prague, and Budapest, as shown in Fig. 6. The graph representation of Network B is shown in Fig. 7.

The objective is to find the optimal lot partitioning coefficients for controlled nodes to minimize both the BE and transportation costs. Initially, the lot partitioning is distributed evenly among the connected nodes, as is customary in the literature. The simulation horizon is set as 10^3 periods. The demand, imposed on all the controlled nodes, exhibits stochastic variations generated according to the Poisson distribution with $\lambda = 0.6$. The unitary transportation cost equals $\phi = 0.04$ € per 10 km. Network A encompasses 7 nodes ($N = 5$, $S = 2$), which leads to 1.07×10^4 candidate solutions to perform a full-search with granularity 0.01. Network B comprised 10 nodes ($N = 8$, $S = 2$), with a search space of 6.54×10^{28} possible solutions, which is thus no longer viable for a full search. Therefore, the BBO method is applied. Each population in the BBO algorithm contains 10 individuals. The maximum number of generations is set as $g = 50$ epochs. Additional mutations are not applied in updating the emigration rates.

Fig. 4. Transportation network A. Red circles indicate external suppliers located in Paris and Milan. (Color figure online)

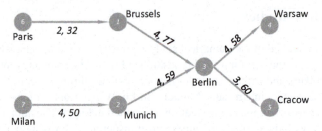

Fig. 5. Transportation network A–graph representation. The arrows indicate the goods flow direction. The numbers denote α_{ij} and φ_{ij}.

Network A shows an unstable behavior, reflected in the BE > 1. The lot partitioning coefficients modification in the considered distribution network has a minimal impact on the BE and transportation costs. The initial channel assignment and the best-obtained solution are

$$
\mathbf{A} = \begin{bmatrix} 0 & 0 & 0.5 & 0 & 0 \\ 0 & 0 & 0.5 & 0 & 0 \\ 0 & 0 & 0 & 1 & 1 \\ 0 & 0 & 0 & 0 & 0 \\ 0 & 0 & 0 & 0 & 0 \\ 1 & 0 & 0 & 0 & 0 \\ 0 & 1 & 0 & 0 & 0 \end{bmatrix} \rightarrow \begin{bmatrix} 0 & 0 & 0.23 & 0 & 0 \\ 0 & 0 & 0.77 & 0 & 0 \\ 0 & 0 & 0 & 1 & 1 \\ 0 & 0 & 0 & 0 & 0 \\ 0 & 0 & 0 & 0 & 0 \\ 1 & 0 & 0 & 0 & 0 \\ 0 & 1 & 0 & 0 & 0 \end{bmatrix} \tag{14}
$$

Before the optimization process, the BE is quantified as 2.58 and transportation costs as 2.05×10^5 €. Conducted optimization slightly improved overall performance by decreasing the BE to 2.57 and transportation costs value to 2.04×10^5 €, i.e., by 0.5%. Hence, the considered star-network leaves little room for optimization due to the limited number of lot partitioning coefficient modifications available in preordained interconnection structure.

The second investigated network, with a more complex topology, revealed factors that are non-negligible to the BE formation, i.e., the number of connections per node ($\vartheta = 2.3$) and the overall number of echelons ($\tau = 10$). The BBO algorithm significantly modified the lot partitioning coefficients for the controlled nodes. For Network B having 8 controlled nodes, the average cost function minimization of 24.48% enables one to bring down the BE by 24.86% and the transportation costs by 24.11%. Distribution networks with a denser topology (a bigger number of connections per node) give more space for improvement, both with respect to transportation costs and the BE. With even channel utilization, the BE is quantified as 1.73 and transportation costs as 5.06×10^5 €. The BBO allowed reducing the BE to 1.3 and transportation costs to 3.84×10^5 €.

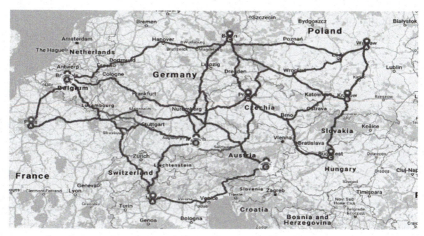

Fig. 6. Transportation network B. Red circles indicate the external suppliers located in Paris and Milan. (Color figure online)

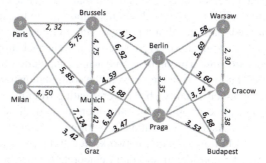

Fig. 7. Transportation network B – graph representation. The arrows indicate the goods flow direction. The numbers denote α_{ij} and φ_{ij}.

The initial channel assignment for network B

$$
\mathbf{A}_{init} =
\begin{bmatrix}
0 & 0.34 & 0.34 & 0 & 0 & 0 & 0.25 & 0 \\
0 & 0 & 0.33 & 0 & 0 & 0.34 & 0.25 & 0 \\
0 & 0 & 0 & 0.5 & 0.34 & 0 & 0.25 & 0.34 \\
0 & 0 & 0 & 0 & 0.33 & 0 & 0 & 0 \\
0 & 0 & 0 & 0 & 0 & 0 & 0 & 0.33 \\
0 & 0 & 0.33 & 0 & 0 & 0 & 0.25 & 0 \\
0 & 0 & 0 & 0.5 & 0.33 & 0 & 0 & 0.33 \\
0 & 0 & 0 & 0 & 0 & 0 & 0 & 0 \\
0.5 & 0.33 & 0 & 0 & 0 & 0.33 & 0 & 0 \\
0.5 & 0.33 & 0 & 0 & 0 & 0.33 & 0 & 0
\end{bmatrix}
\tag{15}
$$

and optimal one

$$
\mathbf{A}_{opt} = \begin{bmatrix}
0 & 0.01 & 0.30 & 0 & 0 & 0 & 0.64 & 0 \\
0 & 0 & 0.34 & 0 & 0 & 0.01 & 0.32 & 0 \\
0 & 0 & 0 & 0.17 & 0.25 & 0 & 0.01 & 0.08 \\
0 & 0 & 0 & 0 & 0.08 & 0 & 0 & 0 \\
0 & 0 & 0 & 0 & 0 & 0 & 0 & 0.43 \\
0 & 0 & 0.36 & 0 & 0 & 0 & 0.03 & 0 \\
0 & 0 & 0 & 0.83 & 0.67 & 0 & 0 & 0.49 \\
0 & 0 & 0 & 0 & 0 & 0 & 0 & 0 \\
0.99 & 0.01 & 0 & 0 & 0 & 0.01 & 0 & 0 \\
0.01 & 0.98 & 0 & 0 & 0 & 0.98 & 0 & 0
\end{bmatrix} \tag{16}
$$

7 Conclusions

The paper introduced a method of counteracting a major systemic distortion in distribution networks – the bullwhip effect – through an appropriate transportation channel assignment. A nontrivial multi-echelon topology, with arbitrary interconnection structure and time-delayed good relocation, is considered. The channel allocation is obtained via a formally stated optimization problem, solved using a population-based evolutionary technique – BBO. BBO allows one to circumvent the computational intricacy related to random demand and a dimensionality obstacle originating from the retarded argument in the network dynamical description. The allocation method allows for both the BE and transportation costs reduction. The validity is verified via numerical tests conducted for an example real-world transportation network. In a further study, other than OUT inventory policies will be considered and more elaborate tuning procedures covering sensitivity and robustness aspects.

Acknowledgment. This work has been completed while the second author was the Doctoral Candidate in the Interdisciplinary Doctoral School at the Lodz University of Technology, Poland.

References

1. Forrester, J.: Industrial dynamics: a major breakthrough for decision makers. Harv. Bus. Rev. **36**, 37–66 (1958)
2. Zabel, E.: Multiperiod monopoly under uncertainty. J. Econ. Theory **5**, 524–536 (1972)
3. Kahn, J.: Inventories and the volatility of production. Am. Econ. Rev. **77**(4), 667–679 (1987)
4. Baganha, M., Cohen, M.: The stabilizing effect of inventory in supply chains. Oper. Res. **36**(3), S72–S83 (1995)
5. Lee, H., Padmanabhan, P., Whang, S.: The bullwhip effect in supply chains. Sloan Manage. Rev. **38**(3), 93–102 (1997)
6. Lee, H., Padmanabhan, P., Whang, S.: Information distortion in a supply chain: the bullwhip effect. Manage. Sci. **43**(4), 546–558 (1997)

7. Croson, R., Donohue, K.: Behavioral causes of the bullwhip effect and the observed value of inventory information. Manage. Sci. **52**(3), 323–336 (2006)
8. Geary, S., Disney, S., Towill, D.: On bullwhip in supply chains - historical review, present practice and expected future impact. Int. J. Prod. Econ. **101**(1), 2–18 (2006)
9. Miragliotta, G.: Layers and mechanisms: a new taxonomy for the bullwhip effect. Int. J. Prod. Econ. **104**(2), 365–381 (2006)
10. Lin, W.-J., Jiang, Z.-B., Liu, R., Wang, L.: The bullwhip effect in hybrid supply chain. Int. J. Prod. Res. **52**(7), 2062–2084 (2014)
11. Ignaciuk, P., Dziomdziora, A.: Bullwhip effect – supply chain stability examination in the presence of demand uncertainty and delay. In: 24th International Conference on System Theory, Control, and Computing (ICSTCC), Sinaia, Romania, pp. 624–629 (2020)
12. Naim, M., Spiegler, L., Wikner, J., Towill, D.R.: Identifying the causes of the bullwhip effect by exploiting control block diagram manipulation with analogical reasoning. Eur. J. Oper. Res. **263**(1), 240–246 (2017)
13. Naim, M., Spiegler, L., Wikner, J., Towill, D.R.: Dynamic analysis and design of a semiconductor supply chain: a control engineering approach. Int. J. Prod. Res. **56**(13), 4585–4611 (2018)
14. Shaban, A., Shalaby, M.: Modeling and optimizing of variance amplification in supply chain using response surface methodology. Comput. Ind. Eng. **120**, 392–400 (2018)
15. Gupta, S., Saxena, A.: Classification of operational and financial variables affecting the bullwhip effect in Indian sectors: a machine learning approach. Recent Patents Comput. Sci. **12**(3), 171–179 (2019)
16. Boccadoro, M., Martinelli, F., Valigi, P.: Supply chain management by H-infinity control. IEEE Trans. Autom. Sci. Eng. **5**(4), 703–707 (2008)
17. Ignaciuk, P.: Nonlinear inventory control with discrete sliding modes in systems with uncertain delay. IEEE Trans. Industr. Inf. **10**(1), 559–568 (2014)
18. Ignaciuk, P.: Discrete inventory control in systems with perishable goods – a time-delay system perspective. IET Control Theory Appl. **8**(1), 11–21 (2014)
19. Bartoszewicz, A. and Latosiński, P.: Reaching law based discrete time sliding mode inventory management strategy. IEEE Access **4**, 10051–10058 (2016). Article ID 7762826
20. Dominguez, R., Cannella, S., Framinan, J.M.: The impact of the supply chain structure on bullwhip effect. Appl. Math. Model. **39**, 7309–7325 (2015)
21. Ignaciuk, P., Wieczorek, Ł: Continuous genetic algorithms in the optimization of logistic networks: applicability assessment and tuning. Appl. Sci. **10**(21), 7851 (2020)
22. Ignaciuk, P., Wieczorek, Ł: Minimum fuel resource distribution in multidimensional logistic networks governed by base-stock inventory policy. In: Bartoszewicz, A., Kabziński, J., Kacprzyk, J. (eds.) Advanced, Contemporary Control. AISC, vol. 1196, pp. 1141–1151. Springer, Cham (2020). https://doi.org/10.1007/978-3-030-50936-1_95
23. Arbex, R.O., da Cunha, C.B.: Efficient transit network design and frequencies setting multi-objective optimisation by alternating objective genetic algorithm. Transp. Res. Part B Methodol. **81**, part 2, 355–376 (2015)
24. Bielli, M., Caramia, M., Carotenuto, P.: Genetic algorithms in bus network optimization. Transp. Res. Part C **10**, 19–34 (2002)
25. Chew, J.S.C., Lee, L.S., Seow, H.V.: Genetic algorithm for biobjective urban transit routing problem. J. Appl. Math. **2013**, 15 p. (2013). Article ID 698645. https://doi.org/10.1155/2013/698645
26. Tikani, H., Setak, M.: Efficient solution algorithms for a time-critical reliable transportation problem in multigraph networks with FIFO property. Appl. Soft Comput. **74**, 504–528 (2019)
27. Dejonckheere, J., Disney, S.M., Lambrecht, M.R., Towill, D.R.: Measuring and avoiding the bullwhip effect: a control theoretic approach. Eur. J. Oper. Res. **147**(3), 567–590 (2003)

28. Chen, F., Drezner, Z., Ryan, J., Simchi-Levi, D.: Quantifying the bullwhip effect in a simple supply chain: the impact of forecasting, lead times, and information. Manage. Sci. **46**(3), 436–443 (2000)
29. Ignaciuk, P. and Dziomdziora, A.: Quantifying the bullwhip effect in networked structures with nontrivial topologies. In: ICBDM 2020: Proceedings of the 2020 International Conference on Big Data in Management, New York, USA, pp. 62–66 (2020)
30. Ignaciuk, P. and Wieczorek, Ł.: Networked base-stock inventory control in complex distribution systems. Math. Probl. Eng. **2019**, Article ID 3754367 (2019)
31. Simon, D.: Biogeography-based optimization. IEEE Trans. Evol. Comput. **12**(6), 702–713 (2008)
32. Simon, D.: A dynamic system model of biogeography-based optimization. Appl. Soft Comput. **11**(8), 5652–5661 (2011)

Modeling Traffic Forecasts with Probability in DWDM Optical Networks

Stanisław Kozdrowski[1]([✉]) [ID], Piotr Sliwka[2] [ID], and Sławomir Sujecki[3] [ID]

[1] Department of Computer Science, Faculty of Electronics, Warsaw University of Technology, Nowowiejska 15/19, 00-665 Warsaw, Poland
s.kozdrowski@elka.pw.edu.pl
[2] Department of Computer Science, Cardinal Wyszynski University, Woycickiego 1/3, 01-938 Warsaw, Poland
p.sliwka@uksw.edu.pl
[3] Telecommunications and Teleinformatics Department, Wroclaw University of Science and Technology, Wyb. Wyspianskiego 27, 50-370 Wroclaw, Poland
Slawomir.Sujecki@pwr.edu.pl

Abstract. Dense wavelength division multiplexed networks enable operators to use more efficiently the bandwidth offered by a single fiber pair and thus make significant savings, both in operational and capital expenditures. In this study traffic demands pattern forecasts (with probability) in subsequent years are calculated using statistical methods. Subject to results of statistical analysis numerical methods are used to calculate traffic intensity in edges of a dense wavelength division multiplexed network both in terms of the number of channels allocated and the total throughput expressed in gigabits per second. For the calculation of traffic intensity a model based on mixed integer programming is proposed, which includes a detailed description of optical network resources. The study is performed for a practically relevant network within selected scenarios determined by realistic traffic demand sets.

Keywords: DWDM system design · Optical network optimization · CDC-F technology · Optical node model · Network congestion · Mixed Integer Programming (MIP) · Forecasting with probability · Time series

1 Introduction

Optical networks based on Dense Wavelength Division Multiplexing (DWDM) technology form the backbone of today's telecommunications industry. To meet ever increasing traffic demands DWDM systems undergo constant development aimed at increasing the network capacity. For this purpose new concepts within DWDM network paradigm are proposed and studied, e.g. Ultra Wideband DWDM (UW-DWDM) systems or Space Division Multiplexing (SDM) systems.

© Springer Nature Switzerland AG 2021
M. Paszynski et al. (Eds.): ICCS 2021, LNCS 12745, pp. 365–378, 2021.
https://doi.org/10.1007/978-3-030-77970-2_28

From the network operator point of view, the ever increasing demand for high speed data services translates into the need to continually upgrade the networks to increase the data transmission rate per optical fiber. In currently deployed optical networks, which are based on single core fibers, the data transmission rate can be increased by using either a larger per-channel bit rate or by increasing the number of available channels [9].

Existing fiber optic networks that operate in C-band (1530 nm to 1565 nm) typically support up to 96 channels of 50 GHz bandwidth. The selection of the C-band for optical long-haul communications is low light attenuation in silica glass fiber and availability of high quality, low cost erbium ion doped fiber amplifiers (EDFAs). Additional capacity in C-band DWDM systems can be gained by using flexible grid, which enables provisioning flexible size DWDM channels with bandwidth as small as 12.5 GHz and the carrier wavelength step of 6.25 GHz. This allows for a more effective use of the available bandwidth within the C-band by reducing the guard bands. A significant proportion of operational expenditure for such networks is the cost of fiber lease which is usually quoted per fiber length unit [1].

During the last decade attention of the telecommunication community has been concentrated on Routing and Wavelength Assignment (RWA) and Routing and Spectrum Allocation (RSA) problems in static [3,6–8] and dynamic [4,17, 18] environment. This study concentrates on the analysis of DWDM networks, which not only pertains to the traffic present in the network but also allows for gaining insight into the future network development needed to meet the predicted demands matrix evolution.

In this article a statistical analysis of network traffic is performed based on the empirical data and used to calculate traffic forecasts. The application of statistical methods leads to calculation of the elements of the traffic demands matrix and their forecasts for the coming years. Subject to calculated values of the traffic demands matrix elements numerical methods are applied to calculate traffic intensity in the edges of a dense wavelength division multiplexed network both in terms of the number of channels allocated and the total throughput expressed in gigabits per second. For the calculation of traffic intensity a model based on integer programming is proposed, which includes a detailed description of optical network resources. Integer programming performs network cost optimisation subject to given data on traffic demand pattern predictions for subsequent time periods that was obtained using the aforementioned statistical methods. Network topology used in the simulations is realistic and representative for optical networks, and stems from [10].

The rest of the paper is organized as follows. In the second section, the mathematical background relevant to the applied optimizing procedures and statistical analysis is considered. In the third section, the usefulness of the proposed theoretical model is discussed in the context of the empirical data and is followed by a concise summary given in the last section.

2   Problem Formulation

As the case study a Polish network was selected (Fig. 1). Empirical data on the actual demand for transmission rates in the network between individual cities is not available. Real data on the volume of demand and the resulting transmission rates is usually proprietary to companies providing telecommunication services. Due to these constraints, the demand bandwidth for the access in the network nodes were determined assuming that the maximum demand for the bandwidth is a combination of demographic[1] (the percentage share of people aged over 14 in the cities of Poland considered) and the percentage of households with broadband access in 6 Polish macro-regions and separately for the Mazowieckie Voivodeship[2] in the period 2010–2019. Based on the above empirical data, optimal demands were determined, as well as their forecasts for 2020–2024.

Fig. 1. Schematic diagram of Polish national transmission optical network.

It is noted that the selection of the forecasting method depends on the phenomenon that one wants to forecast. The construction of long-term forecasts and the related methodology assumes that the time series of empirical data on the basis of which these forecasts are built meet at least stationary condition in wide-sense (the 1st moment and autocovariance do not depend on time t and the 2nd moment is finite for all t). Otherwise, short-term forecasts are made

---

[1] Based on Statistics Poland data.
[2] Based on Eurostat.

(for several periods ahead more about the use of short-term forecasting methods in [13–16]). Due to the very short time series of observations y_t available in this study, basic forecasting tools were used and short-term forecasts were made. Therefore, a linear or logarithmic trend depending on the nature of the empirical data was applied to model demand and objective function values. For the case studied the obtained estimates of the trend function parameters, from a statistical point of view, are significant (p-value for each estimated parameter of the corresponding trend model is <0.05, where 0.05 - a significance level) and the coefficient of determination R^2 is greater than 85%, reaching 98.5% in some cases, which means that the model is very well fit to empirical data.

One of the goals of this article is to determine the point forecast and the interval forecast with the appropriate probability. The $N(\mu, \sigma)$ distribution, bootstrap or resampling methods are usually used to determine the limits of the forecast confidence interval. In our case, due to the very small number of observations, the above methods cannot be used (the distribution of the e_t is not characterized by $N(\mu, \sigma)$). Therefore, a point forecast was determined for each edge, and then, taking into account the distribution of e_t residuals, as well as the naive forecast of e_t, the minimum and maximum forecast values were determined for this edge with an appropriate probability of its implementation (assumption: invariance of distribution of residuals e_t). For example, in the case of the Wroclaw-Lodz edge, the histogram of residuals e_{WrLo} for 3 classes gives the following distribution of values: $-0.0059, -0.0022, 0.0014, 0.0051$ with the respective probabilities: 0.2, 0.5 and 0.3 for the intervals: $\{-0.0059, -0.0022\}$, $\{-0.0022, 0.0014\}$, $\{0.0014, 0.0051\}$, where $e_{MIN_{WrLo}} = -0.0059$ is its left limit, and $e_{MAX_{WrLo}} = 0.0051$ right limit. Based on the predicted theoretical value of $\hat{y}_{WrLo_t}$ from the model and the dependence $y_t = e_t + \hat{y}_t$, the "real" value can be determined. The coefficients of the lower and upper matrix of the predicted values of $\hat{y}_t$ with the appropriate probability were determined in a similar way.

Knowing: a) the distribution of the residuals e_t over the period 2010–2019, b) the probability of meeting the maximum and minimum demand (in the figures as Max and Min, respectively) during the period 2010–2019 and c) point forecasts for 2020–2024, the thresholds have been set for the forecast between 2020 and 2024 to be realised with appropriate probability. Due to the problem of network congestion, which is taken into account here in particular, the main focus is on the top end of the $\hat{y}_{max}$ range, i.e. the maximum value of demand realised and probability of their occurrence.

Finally, one calculates the demand matrix elements, which provide the values of traffic flow between selected nodes expressed in Gbps. An example of a matrix of coefficients based on empirical data described in Sect. 2 only for level 0, and the corresponding demand matrix in 2020 for level 0 as well as additionally for the Min and Max levels with the probability of their occurrence (in parentheses) for the network in Fig. 1 are presented in Tables 1, 2, 3 and 4 respectively.

Once the elements of the demand matrix are determined the data traffic intensity in edges of the analysed network are calculated using Mixed Integer Programming (MIP). For this purpose the following sets are defined:

Table 1. An example of coefficient matrix: forecast, 2020, level 0.

	2	3	4	5	6	7	8	9	10	11	12
1	0.109	0.0758	0.1007	0.1624	0.1007	0.1514	0.1307	0.087	0.1144	0.3012	0.1457
2		0.061	0.0859	0.1476	0.0859	0.1366	0.1159	0.0722	0.0996	0.2864	0.1309
3			0.0527	0.1145	0.0528	0.1035	0.0827	0.039	0.0665	0.2532	0.0977
4				0.1393	0.0776	0.1283	0.1076	0.0639	0.0914	0.2781	0.1226
5					0.1393	0.19	0.1693	0.1256	0.1531	0.3398	0.1843
6						0.1283	0.1076	0.0639	0.0914	0.2781	0.1226
7							0.1583	0.1146	0.1421	0.3288	0.1733
8								0.0939	0.1213	0.3081	0.1526
9									0.0776	0.2644	0.1089
10										0.2919	0.1363
11											0.3231

Table 2. An example of demands matrix: forecast, 2020, level Min (probability).

	2	3	4	5	6	7	8	9	10	11	12
1	1293 (0.3)	898 (0.2)	1190 (0.2)	1917 (0.2)	1201 (0.3)	1799 (0.4)	1543 (0.2)	1040 (0.5)	1353 (0.2)	3585 (0.3)	1728 (0.1)
2		724 (0.3)	1016 (0.2)	1745 (0.2)	1019 (0.3)	1619 (0.4)	1369 (0.2)	859 (0.4)	1179 (0.2)	3412 (0.4)	1548 (0.2)
3			624 (0.2)	1354 (0.3)	619 (0.2)	1223 (0.3)	974 (0.2)	459 (0.2)	784 (0.2)	3021 (0.4)	1145 (0.2)
4				1652 (0.3)	913 (0.3)	1518 (0.5)	1269 (0.2)	753 (0.2)	1079 (0.2)	3318 (0.4)	1436 (0.2)
5					1651 (0.4)	2257 (0.4)	1997 (0.2)	1492 (0.4)	1808 (0.2)	4048 (0.4)	2179 (0.3)
6						1513 (0.3)	1271 (0.2)	752 (0.4)	1077 (0.2)	3313 (0.4)	1440 (0.2)
7							1872 (0.3)	1353 (0.3)	1680 (0.3)	3918 (0.4)	2053 (0.2)
8								1110 (0.2)	1430 (0.2)	3664 (0.2)	1814 (0.3)
9									917 (0.2)	3153 (0.4)	1275 (0.1)
10										3475 (0.3)	1616 (0.3)
11											3853 (0.3)

Table 3. An example of demands matrix: forecast, 2020, level 0.

	2	3	4	5	6	7	8	9	10	11	12
1	1307	910	1208	1949	1208	1817	1568	1044	1373	3614	1748
2		732	1030	1771	1031	1639	1390	866	1196	3437	1570
3			633	1374	633	1242	993	468	798	3039	1173
4				1672	931	1540	1291	767	1096	3337	1471
5					1672	2281	2032	1507	1837	4078	2212
6						1540	1291	767	1097	3337	1471
7							1900	1375	1705	3946	2080
8								1126	1456	3697	1831
9									932	3173	1306
10										3502	1636
11											3877

Table 4. An example of demands matrix: forecast, 2020, level Max (probability).

	2	3	4	5	6	7	8	9	10	11	12
1	1321	918	1220	1972	1221	1833	1577	1056	1383	3636	1760
	(0.2)	(0.3)	(0.3)	(0.2)	(0.2)	(0.3)	(0.4)	(0.2)	(0.4)	(0.3)	(0.3)
2		740	1042	1798	1040	1654	1400	875	1205	3461	1580
		(0.2)	(0.3)	(0.2)	(0.3)	(0.3)	(0.4)	(0.3)	(0.4)	(0.3)	(0.4)
3			646	1402	634	1257	997	469	801	3066	1173
			(0.2)	(0.2)	(0.6)	(0.4)	(0.5)	(0.6)	(0.5)	(0.3)	(0.6)
4				1716	943	1569	1302	780	1109	3379	1469
				(0.2)	(0.4)	(0.4)	(0.4)	(0.4)	(0.2)	(0.2)	(0.6)
5					1704	2329	2057	1540	1864	4135	2230
					(0.2)	(0.2)	(0.2)	(0.2)	(0.2)	(0.2)	(0.4)
6						1554	1296	773	1100	3364	1478
						(0.4)	(0.5)	(0.5)	(0.5)	(0.4)	(0.5)
7							1915	1391	1720	3990	2097
							(0.4)	(0.4)	(0.4)	(0.4)	(0.3)
8								1131	1462	3721	1846
								(0.4)	(0.5)	(0.3)	(0.4)
9									935	3201	1310
									(0.4)	(0.4)	(0.6)
10										3527	1646
										(0.3)	(0.4)
11											3897
											(0.3)

$\mathcal{N}$ set of all nodes
$\mathcal{T}$ set of transponders
$\mathcal{S}$ set of frequency slices
$\mathcal{E}$ set of edges

$\mathcal{P}_{(n,n')}$ set of paths between nodes $n, n' \in \mathcal{N}$; $p \subseteq \mathcal{E}$
$\mathcal{B}$ set of bands
$\mathcal{S}_b$ set of frequency slices used by band $b \in \mathcal{B}$; $\mathcal{S}_b \subseteq \mathcal{S}$; $\bigcup_{b \in \mathcal{B}} \mathcal{S}_b = \mathcal{S}$
$\mathcal{S}_t$ set of frequency slices that can be used as starting slices for transponder $t \in \mathcal{T}$; $\mathcal{S}_t \subseteq \mathcal{S}$

The following objective cost function is optimized using a MIP algorithm subject to the listed below constraints:

$$\min \left\{ \sum_{b \in \mathcal{B}} \left\{ \xi_1(b) \sum_{e \in \mathcal{E}} y_{be} + \sum_{t \in \mathcal{T}} \xi_2(t, b) \sum_{n,n' \in \mathcal{N}} \sum_{p \in \mathcal{P}_{(n,n')}} \sum_{s \in \mathcal{S}_t} x_{tnn'ps} \right\} \right\} \quad (1)$$

where, $\xi_1(b)$ is a cost of using band b at a single edge, y_{be} is a binary variable, equals 1 if band b is used on edge e and 0 otherwise, $\xi_2(t, b)$ is a cost of using a pair of transponders t in band b and $x_{tnn'ps}$ is a binary variable that equals 1 if transponders t are installed between node n and node n', routed on path p, and starting on frequency slice $s \in \mathcal{S}_t$ and 0 otherwise.

In the model the following three constraints have been included:

$$\sum_{t \in \mathcal{T}} \sum_{p \in \mathcal{P}_{(n,n')}} \sum_{s \in \mathcal{S}_t} v(t) x_{tnn'ps} \geq d(n, n') \quad \forall n, n' \in \mathcal{N} \quad (2)$$

where, $v(t)$ is a bitrate provided by transponder t and $d(n, n')$ is a bitrate demanded from node n to node n'.

$$x_{tnn'ps} h \nu(b) c(t) \Delta(t) \sum_{e \in \mathcal{E}} w(n, n', p, e) \cdot$$

$$\cdot \left(f(e) (e^{\frac{\lambda(s)l(e)}{1+f(e)}} + V - 2) + (e^{\frac{\lambda(s)l(e)}{1+f(e)}} + W - 2) \right) \leq P_0 \quad (3)$$

$$\forall t \in \mathcal{T}, \ \forall n, n' \in \mathcal{N}, \ \forall p \in \mathcal{P}_{(n,n')}, \ \forall b \in \mathcal{B}, \ \forall s \in \mathcal{S}_b$$

where, h is the Planck constant equal to $6.62607004 \cdot 10^{-34}$ m^2 kg/s, $\nu(b)$ is a frequency of band b, $c(t)$ is an Optical Signal to Noise Ratio (OSNR) of transponder t, which has been calculated using the standard formula, c.f. [2,5,11,12]. $\Delta(t)$ is the bandwidth used by a transponder t, $f(e)$ is a number of In-Line Amplifiers (ILAs) evenly distributed over edge e to re-amplify the signal in order to prevent OSNR from dropping to a very small value, $\lambda(s)$ is a loss per km using slice s, $l(e)$ is a length of edge e, V is the gain of ILA, W is the gain of fibre amplifier that compensates the nodal loss while the transmitter output power for a single WDM channel is assumed to equal to 1 mW and is represented by P_0. Finally, a constraint is added for avoiding duplicate allocation of the same wavelength in an edge:

$$\sum_{t \in \mathcal{T}} \sum_{n,n' \in \mathcal{N}} \sum_{p \in \mathcal{P}_{(n,n')}} \sum_{s \in \mathcal{S}_t} w(n, n', p, e) u(t, s, s') x_{tnn'ps} \leq y_{be}$$

$$\forall e \in \mathcal{E}, \ \forall b \in \mathcal{B}, \ \forall s' \in \mathcal{S}_b \quad (4)$$

where, $w(n, n', p, e)$ is a binary constant that equals 1 if a path p between nodes n and n' uses edges e and 0 otherwise, $u(t, s, s')$ is a binary constant that equals 1 if transponder t using bandwidth starting at frequency slice s also uses frequency slice s' and 0 otherwise.

The subject of minimization is the cost of installed amplifiers and transponders in (1). Constraints (2) ensure that all demands are satisfied. Constraints (3) ensure that all installed transponders are routed in such a way that their power budgets are not exceeded. Notice that these constraints can be precalculated and reduced to $x_{tnn'ps} = 0$ for some combinations of indices and removed for other combinations. Finally, (4) ensure that using a band results in installing appropriate amplifiers. Notice that these constraints also ensure that each frequency slice at each edge is not used more than once. It is noted that the constraints included do not allow for considering nonlinear interactions and resulting signal impairments.

Once the MIP problem is solved the number of the allocated channels and total throughput for each network edge is determined, which is the objective of this study. Section 3 provides illustrative examples for the network studied.

3   Results and Discussion

This section presents the results of computational experiments obtained by applying algorithms described in Sect. 2 to empirical data collected over the time span ranging from year 2010 until 2019. The objective of the simulations is first to forecast the elements of the demand matrix for the next years (2020–2024) and to calculate traffic intensity in the edges of a dense wavelength division multiplexed network both in terms of the number of channels allocated and the total throughput expressed in gigabits per second. Once the results are obtained the identification of the bottlenecks of the network studied is performed with intent to predict the time at which the capacity of a given network edge will achieve its limit and also to find network edges, which do not carry traffic.

Computational results were obtained for Polish national network, which topology is depicted in Fig. 1. The topology of the network was taken from [10]. The analyzed network consists of 12 nodes, 18 links and 66 traffic demands. The traffic demands (demand matrix elements) were calculated using statistical methods described in Sect. 2.

3.1   Simulation Parameters

Table 5a describes in detail the sets used by the optimisation procedures while Table 5b lists modelling parameters used for performing computations. Note, that the constants given in the first column of Table 5b are defined in Sect. 2.

The calculations were carried out using a linear solver engine of CPLEX 12.8.0.0 on a 2.1 GHz Xeon E7-4830 v.3 processor with 256 GB RAM running under Linux Debian operating system. The average calculation time for one particular result was approximately equal to 1800 s.

3.2   Results

Figure 2 presents the percentage of allocated channels in a network edge calculated for the empirical data, i.e. up until year 2019 whilst Figs. 3, 4 and 5 show analogous results calculated for the statistically estimated forecasts, i.e. years

Table 5. Sets and modelling parameters description.

Set	Set settings
$\mathcal{N}$	12
$\mathcal{E}$	18
$\mathcal{S}$	96 slots (opt. ch.)
$\mathcal{B}$	1 band
$\mathcal{T}$	3 transponders
$\mathcal{S}_b$	$\mathcal{S}_1 = \{1 \ldots 96\}$
$\mathcal{S}_t$	$\mathcal{S}_1 = \{1 \ldots 95\}$
	$\mathcal{S}_2 = \{1 \ldots 95\}$
	$\mathcal{S}_3 = \{1 \ldots 95\}$

(a) Set settings.

Constant	Constant settings
bitrate [Gbps]	$v(1) = 100$, $v(2) = 200$, $v(3) = 400$
OSNR[dB]	$c(1) = 12$, $c(2) = 15$, $c(3) = 22$
d(n,n')[Gbps]	an example in Tables $2 - 4$
$\xi_2(t, b)$	$\xi_2(1, 1) = 5$, $\xi_2(2, 1) = 7$, $\xi_2(3, 1) = 9$
$\Delta(t)$[GHz]	$\Delta(1) = \Delta(2) = \Delta(3) = 50$
$\nu(b)$[THz]	$\nu(1) = 193.8$
$\lambda(s)$[dB/km]	$\lambda(s) = 0.046$
W, V[dB]	15
P_0[W]	10^{-3}

(b) Constant settings.

2020–2024. The results shown in Figs. 2, 3, 4 and 5 are presented in a form of network maps, which helps fast identification of network edges, which are either unused or used to the full capacity.

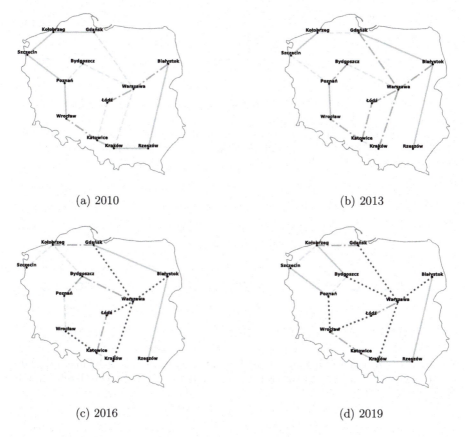

(a) 2010   (b) 2013

(c) 2016   (d) 2019

Fig. 2. The percentage of bandwidth used for all edges: 0%–70% - solid line, 71%–90% - dashed line, 91%–99% - dotted-dashed line and 100% - dotted line; empirical data, 2010–2019.

(a) Min   (b) 0   (c) Max

Fig. 3. The percentage of bandwidth used for all edges: 0%–70% - solid line, 71%–90% - dashed line, 91%–99% - dotted-dashed line and 100% - dotted line in levels: Min, 0, Max; forecast, 2020.

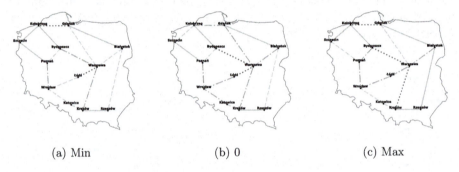

(a) Min   (b) 0   (c) Max

Fig. 4. The percentage of bandwidth used for all edges: 0%–70% - solid line, 71%–90% - dashed line, 91%–99% - dotted-dashed line and 100% - dotted line in levels: Min, 0, Max; forecast, 2022.

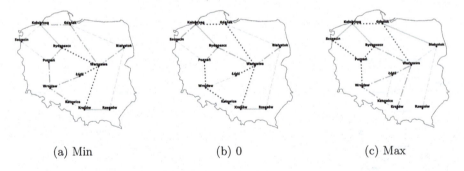

(a) Min   (b) 0   (c) Max

Fig. 5. The percentage of bandwidth used for all edges: 0%–70% - solid line, 71%–90% - dashed line, 91%–99% - dotted-dashed line and 100% - dotted line in levels: Min, 0, Max; forecast, 2024.

Figures 6 and 7 show the values for the number of the allocated channels and total throughput for given years calculated for the empirical data (Fig. 6, years 2010–2019) and forecasts (Fig. 7, years 2020–2024). The values of the allocated channels and total throughput were obtained by summing over all network edges. Based on the results presented in Figs. 2, 3, 4 and 5 one can make several observations that are potentially relevant to a network management team:

– some edges are not used at all, e.g. the edge Bialystok - Gdansk, Lodz - Wroclaw and Poznan - Bydgoszcz (Fig. 2a), or Gdansk - Bialystok and Bydgoszcz - Poznan (Fig. 2d),
– for the empirical data (Fig. 2), up to 16 edges were used between 2010 and 2019, out of the total of 18 possible edges. For the forecasts (Fig. 3a – 5b) 17 edges were used out of 18 possible,
– in the last year of empirical data (Fig. 2d), even though as many as 6 edges reached saturation, still 2 edges were not used at all. The utilisation of further edges takes place in the year 2020 (Fig. 3b), but instead some other edges are relieved,
– in the initial forecast period (2020), the network still large proportion of not allocated resources (Fig. 3b). In contrast, in 2022–2024 increasingly many edges reach saturation (Fig. 4b – 5b).

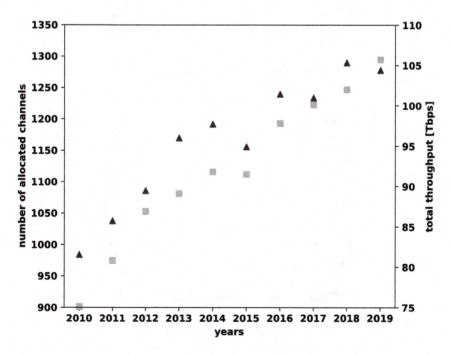

Fig. 6. Relationship between number of occupied channels (black triangles) and total capacity (grey squares) for empirical data in 2010–2019.

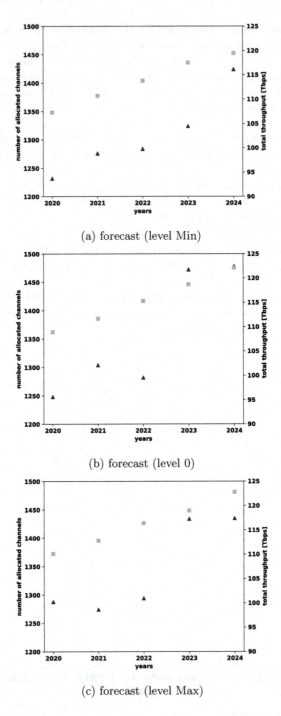

(a) forecast (level Min)

(b) forecast (level 0)

(c) forecast (level Max)

Fig. 7. Relationship between number of occupied channels (black triangles) and total capacity (grey squares) for forecast data.

Figures 6 and 7 show optical bandwidth utilisation (number of occupied optical channels) and total network capacity (throughput) for empirical and forecast data (in 2020–2024). Based on the presented results one can observe that:

- the increase in number of optical channels roughly corresponds to the increase in throughput for the empirical data,
- growth of total throughput for forecasts is nearly linear, while there is a large deviation from linearity for the number of optical channels.

4   Conclusion

The paper presents the analysis of fiber bandwidth utilization in DWDM optical network using statistical methods for the estimation of the demands matrix. The presented study provides methods for forecasting the matrix of traffic demands for the next years and for identification of both the bottlenecks in the network and network edges that are not used. Such an analysis is very useful to the telecommunication network operators, as it allows for optimal use of the allocated resources and aids the process of network expansion planning. This is because the results obtained allow assessing the need for additional investment into the DWDM network infrastructure. Hence, the developed model predicts the network edges that are most likely to be subjected to traffic congestion. This allows the network operator to plan in advance the network expansion and allocate appropriate means for the necessary capital expenditure.

To the best of the authors' knowledge, there are no publications in available literature on the problem considered in this contribution. Thus, this paper in a way initiates this topic and potentially can become a benchmark for future research.

Further research will be focused on improving the stochastic forecasting model to include the access and backbone network interfaces. In addition, modeling different network expansion scenarios taking into account existing network topologies and additional edge extensions. The cost function will include part of capital expenditures and operating expenditures. Additionally, we plan to use nature-inspired algorithms as additional heuristic methods to solve the problem.

References

1. CTC technology & energy: Dark fiber lease considerations. Technical report (2012). www.ctcnet.us/DarkFiberLease.pdf. Accessed 29 Aug 2019
2. Becker, P.M., Olsson, A.A., Simpson, J.R.: Erbium-Doped Fiber Amplifiers: Fundamentals and Technology. Elsevier, Amsterdam (1999)
3. Cai, A., Shen, G., Peng, L., Zukerman, M.: Novel node-arc model and multi-iteration heuristics for static routing and spectrum assignment in elastic optical networks. J. Lightwave Technol. **31**(21), 3402–3413 (2013). http://jlt.osa.org/abstract.cfm?URI=jlt-31-21-3402

4. Dallaglio, M., Giorgetti, A., Sambo, N., Velasco, L., Castoldi, P.: Routing, spectrum, and transponder assignment in elastic optical networks. J. Lightwave Technol. **33**(22), 4648–4658 (2015). https://doi.org/10.1109/JLT.2015.2477898
5. Desurvire, E., Bayart, D., Desthieux, B., Bigo, S.: Erbium-Doped Fiber Amplifiers: Device and System Developments. Wiley, Hoboken (2002)
6. Klinkowski, M., Walkowiak, K.: Routing and spectrum assignment in spectrum sliced elastic optical path network. IEEE Commun. Lett. **15**(8), 884–886 (2011). https://doi.org/10.1109/LCOMM.2011.060811.110281
7. Klinkowski, M., Żotkiewicz, M., Walkowiak, K., Pióro, M., Ruiz, M., Velasco, L.: Solving large instances of the RSA problem in flexgrid elastic optical networks. IEEE/OSA J. Opt. Commun. Netw. **8**(5), 320–330 (2016). https://doi.org/10.1364/JOCN.8.000320
8. Kozdrowski, S., Żotkiewicz, M., Sujecki, S.: Optimization of optical networks based on CDC-ROADM technology. Appl. Sci. **9**(3) (2019). https://doi.org/10.3390/app9030399, http://www.mdpi.com/2076-3417/9/3/399
9. Kozdrowski, S., Żotkiewicz, M., Sujecki, S.: Ultra-wideband WDM optical network optimization. Photonics **7**, 1 (2020). https://doi.org/10.3390/photonics7010016
10. Orlowski, S., Wessäly, R., Pióro, M., Tomaszewski, A.: Sndlib 1.0-survivable network design library. Networks **55**(3), 276–286 (2010). https://doi.org/10.1002/net.20371
11. Poggiolini, P., Bosco, G., Carena, A., Curri, V., Jiang, Y., Forghieri, F.: The GN-model of fiber non-linear propagation and its applications. J. Lightwave Technol. **32**(4), 694–721 (2014). https://doi.org/10.1109/JLT.2013.2295208
12. Shariati, B., Mastropaolo, A., Diamantopoulos, N., Rivas-Moscoso, J.M., Klonidis, D., Tomkos, I.: Physical-layer-aware performance evaluation of SDM networks based on SMF bundles, MCFs, and FMFs. IEEE/OSA J. Opt. Commun. Netw. **10**(9), 712–722 (2018). https://doi.org/10.1364/JOCN.10.000712
13. Sliwka, P.: Proposed methods for modeling the mortgage and reverse mortgage installment. In: Recent Trends In The Real Estate Market And Its Analysis. Oficyna Wydawnicza SGH, Warszawa, pp. 189–206 (2018)
14. Sliwka, P.: Application of the model with a non-Gaussian linear scalar filters to determine life expectancy, taking into account the cause of death. In: Rodrigues, J., et al. (eds.) ICCS 2019. LNCS, vol. 11538, pp. 435–449. Springer, Cham (2019). https://doi.org/10.1007/978-3-030-22744-9_34
15. Sliwka, P., Socha, L.: A proposition of generalized stochastic Milevsky-Promislov mortality models. Scand. Actuar. J. **2018**, 706–726 (2018)
16. Sliwka, P., Swistowska, A.: Economic forecasting methods with the R package. Wydawnictwo Naukowe UKSW (2019)
17. de Sousa, A., Monteiro, P., Lopes, C.B.: Lightpath admission control and rerouting in dynamic flex-grid optical transport networks. Networks **69**(1), 151–163 (2017). https://doi.org/10.1002/net.21715
18. Żotkiewicz, M., Ruiz, M., Klinkowski, M., Pióro, M., Velasco, L.: Reoptimization of dynamic flexgrid optical networks after link failure repairs. IEEE/OSA J. Opt. Commun. Netw. **7**(1), 49–61 (2015). https://doi.org/10.1364/JOCN.7.000049

Endogenous Factors Affecting the Cost of Large-Scale Geo-Stationary Satellite Systems

Nazareen Sikkandar Basha[1]📧, Leifur Leifsson[1]([✉])📧, and Christina Bloebaum[2]

[1] Iowa State University, Ames, IA 50011, USA
{nazareen,leifur}@iastate.edu
[2] Kent State University, Kent, OH 44240, USA
cbloebau@kent.edu

Abstract. This work proposes the use of model-based sensitivity analysis to determine important internal factors that affect the cost of a large-scale complex engineered systems (LSCES), such as geo-stationary communication satellites. A physics-based satellite simulation model and a parametric cost model are combined to model a real-world satellite program whose data is extracted from selected acquisitions reports. A variance-based global sensitivity analysis using Sobol' indices computationally aids in establishing internal factors. The internal factors in this work are associated with requirements of the program, operations and support, launch, ground equipment, personnel required to support and maintain the program. The results show that internal factors such as the system based requirements affect the cost of the program significantly. These important internal factors will be utilized to create a simulation-based framework that will aid in the design and development of future LSCES.

Keywords: Advanced high frequency satellite · Cost overrun · Cost modeling · Model-based sensitivity analysis · Endogenous factors

1 Introduction

A system associated with large-scale projects involving a vast number of stakeholders with high complexity is a large-scale complex engineered system (LSCES) [10]. Acquisition of an LSCES in the aerospace and defense industry undergo processes of design, engineering, construction, testing, deployment, sustaining, and disposal of the system [5,8,31]. Due to the high costs and risks involved, these systems are often acquired by large organizations, such as the National Aeronautics and Space Administration (NASA) and the Department of Defense (DoD), in the United States using the defense acquisition system (DAS) [1,28]. The costs involved during the acquisition process of LCSES are reported quarterly to the congress using the selected acquisition reports (SARs) [9].

Supported by U. S. National Science Foundation under grant no. 1662883.

M. Paszynski et al. (Eds.): ICCS 2021, LNCS 12745, pp. 379–391, 2021.
https://doi.org/10.1007/978-3-030-77970-2_29

Every year, the aerospace industry's cost overruns grow at least twice the estimated costs by the end of the program due to longer schedules and the system's high complexity [9]. Past research has shown that highly complex systems undergo costs and schedule overruns often exceeding 40% of their initial estimated costs [35]. SARs also demonstrates the trend of higher estimated costs for the space programs [35].

Cost overruns in LCSES may be due to the high complexity, size of the project, various stakeholders, organizations, political disruptions, and changes in requirements and scope [27]. The factors which affect the cost overrun are not clearly explained in the SARs and often described as to be caused due to underestimation of initial costs [10]. Although many factors go into the overall cost of a project, it is known that systems engineering efforts can reduce the costs [8]. But to use systems engineering frameworks, it is fundamentally essential to understand the factors affecting the cost overrun of the system [37]. The two main types of factors that affect the cost overrun are the exogenous and endogenous factors. Exogenous factors are factors not belonging to the system, and endogenous factors are within the program's realm [6]. Both factors play a significant role in the cost and schedule overrun and the DoD data can be used to validate it. Examples of endogenous factors are design errors, change in scope, and complexity of the system. In contrast, exogenous factors are changes due to natural disasters, political dynamics, warfare, and the scientific world [8, 29, 33]. Changing the factors within the system may or may not have a significant impact on the cost. The factors affecting the cost overrun are often identified by experts or using surveys, but they are often prone to error [34]. Global sensitivity analysis (GSA) has been used to identify critical factors in systems [7, 30].

In this work, a variance-based GSA is used to determine the effects of different factors on the program's cost [36]. Specifically, GSA is performed on a geostationary satellite system model by combining a physics-based simulation model with a cost model and evaluating the effects of its input parameters on the output parameter, in this case, the overall system cost. A geo-stationary communication satellite program is used as an example of a LCES program. In particular, the advanced extremely high frequency (AEHF) satellite program. A parametric cost-based model, the unmanned spacecraft cost model (USCM8), is used [40]. GSA with Sobol' indices is used to quantify how the internal factors affect the system's overall cost.

The remained of the paper is organized as follows. Next section introduces the AEHF satellite system and the data from SARs. The following section describes the methods used to construct the satellite physics model and the cost model as well as Sobol's method. The numerical results are presented in the following section. Lastly, the conclusions and suggestions for future work are presented.

2   The AEHF Satellite System

A geo-stationary communication satellites mission is to relay telecommunication signals using transponders between the satellite and different ground stations,

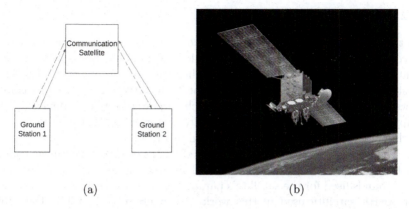

Fig. 1. A geo-stationary communication satellite: (a) block diagram of the system, and (b) the advanced extremely high frequency satellite (AEHF) [2].

as shown in Fig. 1(a). The objective is to transmit the signals from one ground station to another ground station efficiently and effectively. The example satellite program used in this work is the AEHF shown in Fig. 1(b).

The AEHF program consists of six geo-stationary communication satellites, and it is operated by the United States Air Force Space (USAF) command [3]. The expected lifetime of a satellite is 14 years [3]. The program baseline cost data is extracted from the selected acquisition reports (SARs) [1,3] and is shown in Fig. 2. The changes in the estimated costs during the development of AEHF are due to various internal and external factors to the system. In this work, a conceptual model of a satellite program is utilized to reveal how the internal factors affect the overall cost by utilizing GSA.

3   Methods

This section describes the development of a physics based model of a geo-stationary satellite, along with its cost model. Both the models have parameters that are internal to the system. A variance-based GSA model using Sobol' indices is constructed to study the effect of different internal factors on the cost of the system [36]. The workflow of the method is provided in the next subsection, followed by the satellite model, the cost model, and the variance based sensitivity analysis.

3.1   Workflow of the Model-Based Cost Sensitivity Analysis

A flowchart for exploring the effect of the systems cost due internal factors is shown in Fig. 3. The first step is to model the physics of the satellite. The mass of different subsystems is fed into the cost model to calculate the cost of the system. The satellite model and the cost model are then made to fit a real satellite program data from the SARs. This is followed by the variance-based GSA to determine the important internal parameters.

3.2   Satellite Model

In this work, the satellite system designed in this paper is conceptual. Generally, a communication satellite system includes a communication satellite, a launch station, and ground stations to accomplish the mission objective. The satellite system consists of the payload, including transmitting and receiving transponders for communication, subsystems such as power system, propulsion, ground support, and launch vehicle. The satellite system also follows a top-down hierarchical decomposition and this is shown in Fig. 4.

The satellite system is highly coupled with linear and nonlinear couplings. In this work, the physical satellite is defined by nine continuous parameters. The bounds used for the satellite's parameters are provided in Table 1. The data for AEHF satellite used in this work is also provided in table. Data for the AEHF satellite is classified and, hence, assumptions for those parameters are also provided in Table 1.

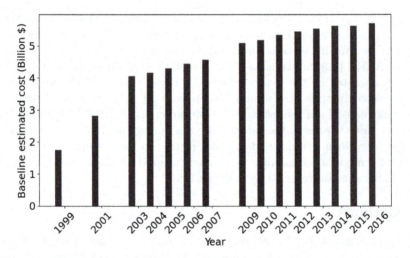

Fig. 2. The yearly baseline cost estimation of the advanced extremely high frequency satellite system (data obtained from [1,11–24,39]).

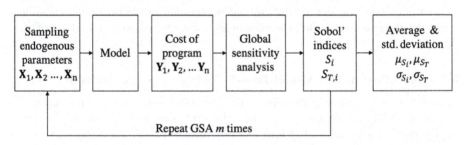

Fig. 3. Workflow of the model-based cost sensitivity analysis.

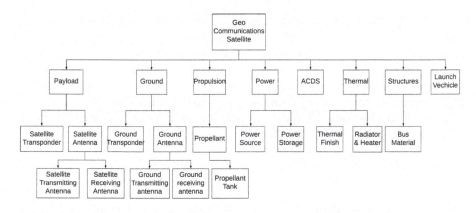

Fig. 4. System diagram of geo-stationary communication satellite.

The inherent couplings in the satellite system is represented using a design structure matrix (DSM) and this can be found in Kannan *et al.* [26] along with a detailed description of the parameters of the satellite system [25].

3.3   Cost Model

This section describes the cost model constructed with the inputs from the satellite system. Traditional cost models are based on the mass of the system. In this paper, a parametric cost estimation model, USCM8 developed by Telecote Research for the US Air Force, is used to calculate the cost of the satellite program [40]. USCM8 is mass-based cost and the mass of the system are provided from the previous satellite model. There are different costs involved in the design and development of a system. In this conceptual design, the cost involved are the cost of research and development, cost of the units, cost of operations and support and cost of launching the satellite. The total cost of the satellite system is

$$C_{total} = C_{r\&d} + C_{units} + C_{operations} + C_{launch}, \qquad (1)$$

Table 1. Geo-stationary communication satellite parameters

Parameters	Lower Bound	Upper Bound	AEHF Data
Downlink frequency (GHz)	1	100	20 [38]
Uplink frequency (GHz)	1	100	44 [38]
Satellite Transmitter power(Watts)	300	3,000	1,500 (assump)
Watts Ground Transmitter power (Watts)	300	30,000	15,000 (assump)
Satellite transmitting antenna diameter(m)	0.5	2.5	1.0 (assump)
Satellite receiving antenna diameter (m)	0.5	2.5	1.0 (assump)
Ground receiving antenna diameter (m)	2	20	0.3
Ground transmitting antenna diameter(m)	2	20	0.3
Energy density of the battery $(W - hr)/kg$	35	350	200

where C_{total} is the total cost of the satellite program, $C_{r\&d}$ is the cost of research and development of a satellite, C_{units} is the cost per unit of a satellite, $C_{operations}$ is the cost of operations and support of the program and C_{launch} is the cost of launching a satellite. The equations involved in calculating the cost of the research and development and the first unit of the satellite systems using USCM8 can be found in Wertz et $al.$ space mission analysis and design [40].

USCM8 involves the cost of the first unit and other units separately, the total cost of the satellite system is modified as

$$C_{total} = C_{r\&d} + C_{first-unit} + C_{other-units} + C_{operations} \qquad (2)$$
$$+ C_{launch} + C_{estimating},$$

where C_{units} is divided to $C_{first-unit}$ cost of first unit and $C_{other-units}$ is the cost of other remaining units.

The cost estimation of the research and development is

$$C_{r\&d} = \sum C_{subsystems} + C_{integration} + C_{Ground-Equipment}, \qquad (3)$$

where $\sum C_{subsystems}$ is the sum of all the subsystem costs, $C_{integration}$ is the cost of integration of the subsystems and $C_{Ground-Equipment}$ is the cost of ground equipment required for the program.

The cost estimation of the first unit is determined by

$$C_{first-unit} = C_{subsystem} + C_{integration}, \qquad (4)$$

where $C_{subsystem}$ is the cost of the subsystem within the satellite and $C_{integration}$ is the cost of integration of the subsystems.

The next step is to estimate the cost of other units. The cost of other units is extrapolated by using a learning curve representing the relationship between experience producing a good and efficiency of production learning curve is given by

$$L = N^{1-(log(1/S)/log(2))}, \qquad (5)$$

where N is the number of units and S is the learning curve percentage. In this work, the learning curve percentage is assumed to be 90%. The cost estimation of other units is

$$C_{unit-cost} = C_{first-unit} \cdot (L - 1), \qquad (6)$$

where L is the learning curve calculated from (5) and $C_{first-unit}$ is obtained from (4).

The launch costs are recurring and it depends on the launch vehicle and the number of the number of satellites. The launch cost is calculated as

$$C_{Launch} = C_{launch-vehicle} \cdot N, \qquad (7)$$

where N is the number of units. In this paper, the launch costs for Falcon 9 is utilized.

The operations and support cost are essential for a system to sustain its life cycle. These costs involve the personnel cost to operate the system, maintenance of the system as well as the facilities cost. The equations used in the estimation of operations cost is provided in the Space Mission Engineering textbook [40].

Estimating costs depends on various factors. In this paper, the estimating cost is obtained from the Technology Readiness Levels (TRL). These levels measure the maturity level of a particular technology. The estimating costs is calculated as follows

$$C_{estimating} = C_{units} \cdot Y_{Ef}, \tag{8}$$

where C_{Units} is the cost of a single unit and Y_{Ef} is the estimating factor which selected based on the TRL level. In this paper, the TRL level is assumed to be TRL 6 which has a multiplying factor of 1.

3.4 Sensitivity Analysis

A variance-based global sensitivity analysis, called Sobol' indices, is used to determine the internal parameters which affect the cost of the system [30,36]. Sobol' indices uses variance decomposition to find the sensitivity index. For a function $Y = f(\mathbf{X})$, with $\mathbf{X}$ as vector of n input parameters, the model response is provided as follows

$$y = f_0 + \sum_{i=1}^{n} f_i(\mathbf{X}_i) + \sum_{i<j}^{n} f_{i,j}(\mathbf{X}_i, \mathbf{X}_j) + \dots + f_{1,2,\dots,m}(\mathbf{X}_1, \mathbf{X}_2, \dots, \mathbf{X}_n), \tag{9}$$

where f_0 is a constant, n are first order function (f_i), $\sum_{i<j}^{n} f_{i,j}(\mathbf{X}_i, \mathbf{X}_j)$ second order functions and so on. All the decomposed terms are orthogonal which can be further decomposed in terms of conditional expected values

$$f_0 = E(f(\mathbf{X})), \tag{10}$$

$$f_i(\mathbf{X}_i) = E(f(\mathbf{X}|\mathbf{X}_i) - f_0, \tag{11}$$

and

$$f_{i,j}(\mathbf{X}_i, \mathbf{X}_j) = E(f|\mathbf{X}_i, \mathbf{X}_j) - f_0 - f_i(\mathbf{X}_i) - f_j(\mathbf{X}_j). \tag{12}$$

The variance of (9) is

$$Var(f(\mathbf{X})) = \sum_{i=1}^{n} V_i + \sum_{i<j}^{n} V_{i,j} + \dots + V_{1,2,\dots,m}, \tag{13}$$

where V_i is $Var[E(f(\mathbf{X}|\mathbf{X}_i)]$ and $V(f(\mathbf{X}))$ represents the total variance. The Sobol' indices are obtained by dividing (13) by $V(f(\mathbf{X}))$ to obtain

$$1 = \sum_{i}^{n} S_i + \sum_{i<j}^{n} S_{i,j} + \dots + S_{1,2,\dots n} \tag{14}$$

where S_i represents the first-order Sobol' indices as

$$S_i = \frac{V_i}{Var(f(\mathbf{X}))}. \tag{15}$$

The total-effect Sobol' indices are given as

$$S_{T_i} = 1 - \frac{Var_{\mathbf{X}_{\sim i}}(E(f(\mathbf{X})|\mathbf{X}_{\sim i}))}{Var(f(\mathbf{X}))}, \tag{16}$$

where $\mathbf{X}_{\sim i}$ gives the set of all the parameters except $\mathbf{X}_i$. By using the Sobol' indices S_i and S_T, the effect of $\mathbf{X}_i$ can be computed, thus, providing added information about the parameter interactions.

3.5   Parameter Sampling

Model-based GSA involves the sampling of the input parameters and then performing the corresponding evaluations of the simulation model. The sampling needs to be performed for several combinations to capture the model response trend. In this work, the internal factors are sampled from the physics model as well the cost model. Eleven parameters are used to identify which factor affects the cost of the system. Table 2 includes the parameters and their bounds (assuming a uniform distribution) which are used in the GSA using Sobol' indices. Generation of the samples for this study is performed using random samples from a uniform distribution within the bounds provided in Table 2 [4]. A Monte Carlo-based numerical procedure is used to compute the Sobol' indices [32].

4   Results

The calculated cost of the AEHF along with the estimated costs from SARs data [11–24,39] of the program are provided in Fig. 5. The calculated costs of AEHF match well in general with the data. From Fig. 5, it is seen that there is a mismatch for some years. This is due to the various assumptions made in the cost model (cf. Sec. 3.3) of the satellite program. Changing the assumptions to match the costs for the years 2003 to 2007 will change the costs in other years. Therefore, for the simplicity of calculations, the assumptions are retained and the model is considered accurate enough for performing the GSA.

The cost model is then used to determine the different internal factors which have an effect on the total cost by using sensitivity analysis. The GSA needs $n = 10^6$ samples to reach convergence on the Sobol' indices (Figs. 6 and 7). The GSA is repeated $m = 10$ times to obtain the mean and standard deviation of the Sobol' indices estimates (Fig. 8).

Table 2. Internal parameters used in GSA

Parameters	Description	Lower Bound	Upper Bound
X_1	Number of transponders (units)	2	20
X_2	Power of the satellite (Watts)	2	30,000
X_3	Salary of the engineers and technicians ($)	100×10^3	300×10^3
X_4	Number of engineers (units)	2	20
X_5	Number of technicians (units)	2	20
X_6	Years of operations (number)	5	30
X_7	Estimating cost (TRL) (number)	0.1	4
X_8	Learning curve (number)	0.6	0.95
X_9	Percentage of Hardware operations (%)	1	40
X_{10}	Percentage of PMSE (%)	1	20
X_{11}	Number of units	1	10

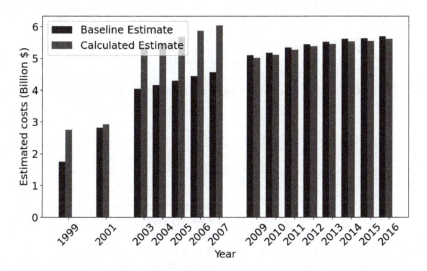

Fig. 5. Calculated yearly cost estimate and baseline estimated costs of AEHF satellite (data obtained from [1, 11–24, 39]).

From Fig. 8 it is seen that the parameters X_1 (number of transponders), X_2 (power of the satellite), X_7 (technology readiness level(estimating cost), X_8 (learning curve) and X_{11} (number of units) have significant effect on the cost of the program. Other internal parameters used in this program have negligible impact on the cost. It is also seen that X_1 (number of transponders) and X_2 (power of the satellite) have the highest impact on the cost and these factors are a part of the physics-based satellite model.

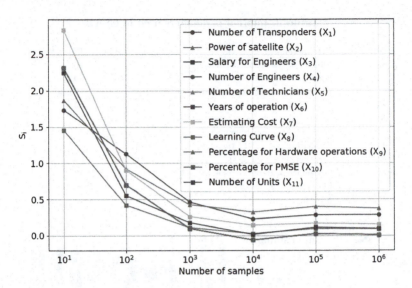

Fig. 6. Sobol' indices of 1st-order of the endogenous parameters.

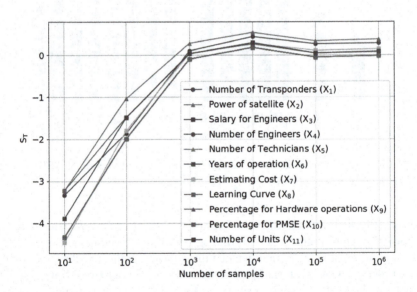

Fig. 7. Total-order Sobol' indices of the endogenous parameters.

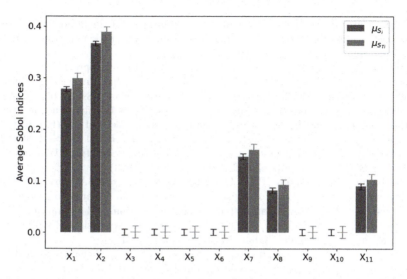

Fig. 8. Average and standard deviation of the first- and total-order Sobol' indices.

5   Conclusion

In this work, model-based global sensitivity analysis (GSA) is used to determine the effects of large-scale satellite program's, the advanced extremely high frequency (AEHF) satellite, endogenous (internal) parameters on the program overall cost. A physics-based satellite model and a cost model were been used to estimate the cost of AEHF.

Sobol' analysis performed on eleven internal factors shows that parameters such as the number of transponders and power of the satellite have the most significant impact on the satellite's cost. This study shows that GSA can be used to determine the system's internal factors and helps in determine which internal parameters affect the cost the most. These results will be used in future work to determine the effects of both internal and external parameters on the program's actual cost.

Acknowledgements. This material is based upon work supported by the U. S. National Science Foundation under grant no. 1662883. The authors would like to thank Hanumanthrao Kannan for providing the physics-based satellite model.

References

1. Selected acquisition reports. https://www.law.cornell.edu/uscode/text/10/2432
2. Advanced extremely high frequency system (2021). https://www.afspc.af.mil/News/Photos/igphoto/2001722287/
3. Defense Acquisition Management Information Retrieval: Selected acquisition report - advanced extremely high frequency (2011)

4. Beaurepaire, P., Broggi, M., Patelli, E.: Computation of the Sobol' indices using importance sampling. In: Vulnerability, Uncertainty, and Risk: Quantification, Mitigation, and Management, pp. 2115–2124 (2014)
5. Bhatia, G.V., Kannan, H., Bloebaum, C.L.: A game theory approach to bargaining over attributes of complex systems in the context of value-driven design. In: 54th AIAA Aerospace Sciences Meeting, San Diego, California, USA, p. 0972 (2016)
6. Bhatia, G.V.: A game theory approach to negotiations in defense acquisitions in the context of value-driven design: An aircraft system case study. MsT, p. 63 (2016)
7. Bloebaum, C., Sobieszczanski-Sobieski, J.: Sensitivity based coupling strengths in complex engineering systems. In: 34th Structures, Structural Dynamics and Materials Conference, La Jolla, CA, USA, p. 1472 (1993)
8. Bloebaum, C.L., Collopy, P.D., Hazelrigg, G.A.: NSF/NASA workshop on the design of large-scale complex engineered systems—from research to product realization. In: 12th AIAA Aviation Technology, Integration, and Operations (ATIO) Conference and 14th AIAA/ISSMO Multidisciplinary Analysis and Optimization Conference, Indianapolis, Indiana, USA, vol. 5572 (2012)
9. Chaplain, C.T.: Space acquisitions: DOD continues to face challenges of delayed delivery of critical space capabilities and fragmented leadership, statement of Cristina T. Chaplain, director, acquisition and sourcing management, testimony before the subcommittee on strategic forced, committee on armed services, US Senate. United States. Government Accountability Office. No. GAO-17-619T, United States. Government Accountability Office (2017)
10. Deshmukh, A., Collopy, P.: Fundamental research into the design of large-scale complex systems. In: 13th AIAA/ISSMO Multidisciplinary Analysis Optimization Conference, Fort Worth, Texas, USA, p. 9320 (2010)
11. Dodaro, G.L.: Assessments of major weapon programs. Technical report, GAO-04-476. General Accounting Office, Washington, DC (2004)
12. Dodaro, G.L.: Assessments of major weapon programs. Technical report, GAO-05-476. General Accounting Office, Washington, DC (2005)
13. Dodaro, G.L.: Assessments of major weapon programs. Technical report, GAO-06-391. General Accounting Office, Washington, DC (2006)
14. Dodaro, G.L.: Assessments of major weapon programs. Technical report, GAO-07-406. General Accounting Office, Washington, DC (2007)
15. Dodaro, G.L.: Assessments of major weapon programs. Technical report, GAO-08-467. General Accounting Office, Washington, DC (2008)
16. Dodaro, G.L.: Assessments of major weapon programs. Technical report, GAO-09-326. General Accounting Office, Washington, DC (2009)
17. Dodaro, G.L.: Assessments of major weapon programs. Technical report, GAO-10-388. General Accounting Office, Washington, DC (2010)
18. Dodaro, G.L.: Defense acquisitions: assessments of selected weapon programs. GAO-11-233. General Accounting Office, Washington, DC (2011)
19. Dodaro, G.L.: Assessments of major weapon programs. Technical report, GAO12-400. General Accounting Office, Washington, DC (2012)
20. Dodaro, G.L.: Assessments of major weapon programs. Technical report, GAO-13-294. General Accounting Office, Washington, DC (2013)
21. Dodaro, G.L.: Assessments of major weapon programs. Technical report, GAO-14-340. General Accounting Office, Washington, DC (2014)
22. Dodaro, G.L.: Assessments of major weapon programs. Technical report, GAO-15-342. General Accounting Office, Washington, DC (2015)
23. Dodaro, G.L.: Assessments of major weapon programs. Technical report, GAO-16-329. General Accounting Office, Washington, DC (2016)

24. Dodaro, G.L.: Assessments of major weapon programs. Technical report, GAO-17-333. General Accounting Office, Washington, DC (2017)
25. Kannan, H.: An MDO augmented value-based systems engineering approach to holistic design decision-making: a satellite system case study. Ph.D. thesis, Iowa State University (2015)
26. Kannan, H., Bloebaum, C.L., Mesmer, B.: Incorporation of coupling strength models in decomposition strategies for value-based MDO. In: 15th AIAA/ISSMO Multidisciplinary Analysis and Optimization Conference, Atlanta, GA, USA, p. 2430 (2014)
27. Kwasa, B., Kannan, H., Mesmer, B., Bloebaum, C.: Capturing trust as organizational uncertainty in a value-based systems engineering framework. In: International Annual Conference of the American Society for Engineering Management, Huntsville, AL (2017)
28. Lewis, K.E., Collopy, P.D.: The role of engineering design in large-scale complex systems. In: 12th AIAA Aviation Technology, Integration, and Operations (ATIO) Conference and 14th AIAA/ISSMO Multidisciplinary Analysis and Optimization Conference, Indianapolis, Indiana, USA, vol. 5573 (2012)
29. Lightsey, B.: Systems engineering fundamentals. Technical report, Defense Acquisition University, Fort Belvoir, VA (2001)
30. Lilburne, L., Tarantola, S.: Sensitivity analysis of spatial models. Int. J. Geogr. Inf. Sci. **23**(2), 151–168 (2009)
31. McBride, S., Paret, A.: Cost estimating in the department of defense and areas for improvement. Technical report, Naval Postgraduate School (2010)
32. Saltelli, A.: Making best use of model evaluations to compute sensitivity indices. Comput. Phys. Commun. **145**(2), 280–297 (2002)
33. Schwartz, M.: Defense acquisitions: how DOD acquires weapon systems and recent efforts to reform the process. Library of Congress Washington Dc Congressional Research Service (2010)
34. Shao, A., Wertz, J.R., Koltz, E.A.: Quantifying the cost reduction potential for earth observation satellites. Proceedings of the 12th Reinventing Space Conference, pp. 199–210. Springer, Cham (2017). https://doi.org/10.1007/978-3-319-34024-1_16
35. Sikkandar Basha, N., Bloebaum, C., Kwasa, B.: Study of cost overrun and delays of DoD's space acquisition program. In: Proceedings of the International Annual Conference of the American Society of Engineering Management, Coeur d'Alene, Idaho, USA (2018)
36. Sobol, I.M.: Global sensitivity indices for nonlinear mathematical models and their Monte Carlo estimates. Math. Comput. Simul. **55**(1–3), 271–280 (2001)
37. Spero, E., Bloebaum, C.L., German, B.J., Pyster, A., Ross, A.M.: A research agenda for tradespace exploration and analysis of engineered resilient systems. Procedia Comput. Sci. **28**, 763–772 (2014). https://doi.org/10.1016/j.procs.2014.03.091. 2014 Conference on Systems Engineering Research
38. About Us Fact Sheets: Advanced extremely high frequency system (2021). https://www.afspc.af.mil/About-Us/Fact-Sheets/Display/Article/249024/advanced-extremely-high-frequency-system/
39. Walker, D.M.: Assessments of major weapon programs. Technical report, GAO-03-476. General Accounting Office, Washington, DC (2003)
40. Wertz, J.R., Everett, D.F., Puschell, J.J.: Space Mission Engineering: The New SMAD. Microcosm Press, Portland (2011)

Description of Electricity Consumption by Using Leading Hours Intra-day Model

Krzysztof Karpio [iD], Piotr Łukasiewicz [iD], Rafik Nafkha[(✉)] [iD],
and Arkadiusz Orłowski [iD]

Institute of Information Technology,
Warsaw University of Life Sciences SGGW, Warsaw, Poland
{Krzysztof_karpio,piotr_lukasiewicz,rafik_nafkha,
arkadiusz_orlowski}@sggw.edu.pl

Abstract. This paper focuses on parametrization of one-day time series of electricity consumption. In order to parametrize such time series data mining technique was elaborated. The technique is based on the multivariate linear regression and is self-configurable, in other words a user does not need to set any model parameters upfront. The model finds the most essential data points whose values allow to model the electricity consumptions for remaining hours in the same day. The number of data points required to describe the whole time series depends on the demanded precision which is up to the user. We showed that the model with only four describing variables, describes 20 remaining hours very well, exhibiting dominant relative error about 1.5%. It is characterized by a high precision and allows finding non-typical days from the electricity demand point of view.

Keywords: Time series · Data mining technique · Electricity consumption modeling · Multivariate linear regression

1 Introduction

The load curve of the power system can be considered as a random function of a random variable of its load. The power system load curve is formed as a result of the superposition of consumers individual load curves and is the sum of the components of random functions. It should be emphasized that these are functions that have only one implementation during the considered time interval, and they are also correlated and are subject to an increase during the year, which is also a random variable. Functions of the load of component consumers of the system, and thus functions of the system load, are influenced by non-accidental factors [1–3]: (1) Location of the considered area (climate, changes in the angle of sunlight beams, changes in the times of sunrise and sunset). (2) Features of the power system, the most important of which are: the structure of consumers, the dynamics of economic development, the statutory number of working hours within a day and the system of shifts binding in a given country. The above-mentioned factors of a non-accidental nature constitute a set of features that determine the course of the system load in terms of the process average.

© Springer Nature Switzerland AG 2021
M. Paszynski et al. (Eds.): ICCS 2021, LNCS 12745, pp. 392–404, 2021.
https://doi.org/10.1007/978-3-030-77970-2_30

The load curve of the power system is also influenced by random factors, such as changes in temperature [4–6], rainfall, wind speed [7], errors of receiving devices, changes in the structure of electricity consumers, etc. The life rhythm of the population, its connection to the variability of seasons, traditions and customs, and the nature of the work performed by people to regulate the rhythm of the working and non-working days, become enhanced in the variation of the load curve of the power system and are the main cause of the phenomenon called load variability.

Electricity consumers load variation have always been the basis of modeling electricity demand. Initially, planning and forecasting electricity consumption was mainly related to the assurance of electricity delivery. Due to the specific nature of electricity, as it cannot be permanently stored, means that its demand and supply should be equal. The first electricity consumption models were, therefore, oriented towards forecasting electricity demand. Depending on the purpose of model preparation, many types and classes of models can be distinguished. The models used in the electricity management can be divided into two basic groups. A group of analytical and forecasting models and a group of forecasting models. The most intensively developed and studied models are prognostic models [8]. Currently, models based on artificial neural networks [9–12] and genetic algorithms [13] are most often used in forecasting, due to the fact that prognostic effects obtained with their use provide best results [14, 15]. However, the disadvantage of these types of models, is that they do not have information about reasons of the problems. Based on that, it is difficult to determine which of the factors has the strongest impact on the shape of the load.

In the case of electricity consumption analysis and modeling, the most commonly used models are: regression [16, 17], economic processes and time series [17–19], which, however, inform about certain tendencies, but not about the causality of the phenomenon. In such situation, econometric models that describe phenomena while taking into account many external factors are useful.

Time-series and regression have been extensively used in modelling for decades. In electrical engineering, regression techniques are used to model the electric load as a function of load consumption in relationship with different dynamics like seasonal patterns, meteorological conditions, day type, customer social class etc. Different regression techniques (linear, multiple linear, quadratic, and exponential regression models) with the hour-by-hour load data based upon a specific day, and considered temperature as the variant parameter were used to perform the hour by hour model for a Grid Station in Pakistan [20]. A multiple regression model was performed for monthly electricity demand of New South Wales [21]. The climatic variables such as temperature, humidity, and rainy days appear as the most affective the electricity demand consumption in this case. The multivariate linear regression is used to model electricity consumption of Jordian Industrial sector as function of variables such as number of establishments, number of employees, electricity tariff, prevailing fuel prices, and production outputs [22].

Undoubtedly, the involvement of many variables provides better explanations which of the factors most strongly influences the shape of the load. This, however, requires collecting a lot of data, tedious work with processing it and introduction of increasingly powerful processors. According to the econometric methodology, there is a gap between the best (meaning the best compatibility) model and the most economical approach. The constructed model should be as simple as possible, i.e., contain the smallest possible number of variables, and explain the fluctuations of the analyzed dependent variable as precise as possible. Usually, the first choice for modeling is a linear regression model estimated by the least squares' method. This method enables deterministic dependencies modeling. The ability to describe a given phenomenon in terms of deterministic components lets us predict most of the variability. In this paper, we present a method that finds a subset of the points that are enough to recreate the entire electricity demand curve for a given time allocation. It turns out that it is enough to know 4 appropriate points to parameterize e.g. one day time series of electricity consumption with very good accuracy. With 5 points, the accuracy is already super good. The method uses the multivariate linear regression technique.

2 Motivation

Electricity consumption depends on many factors like day/night, weekdays/weekends, working days/holidays. One must also keep in mind the seasons and environmental conditions like temperature and humidity. Let us note that hour is the important factor too. The Fig. 1a contains all possible electricity consumptions at each hour (the data is described in the following chapter). The consumption varies between 0.98×10^4 and 2.57×10^4 MWh, thus, having the range of 1.56×10^4. However, ranges for the single hours are smaller for night hours, they are only about 0.9×10^4 MWh. It is due to two groups of reasons: (1) not all combinations of factors' values exist and (2) behaviors of people are typical. First, there is a night at 7 o'clock only during autumn or winter, very low temperatures can possibly exist during summer, but they are extremely unlikely. Secondly, activities of people, business and public transportation are very limited during the night. Those behaviors lead to the relatively low electricity consumption regardless of current temperature or humidity. Moreover, the electricity demand does not change rapidly between hours. Even, if a number of people start a job at the same time they get up at different hours, travel to work for various times and arrive to the workplace at similar but still different hours. All of those make changes of electricity demands to spread for a few hours even the factors change rapidly, what is visible in Fig. 1a. The lowest electricity consumptions are concentrated around 6:00 o'clock and it takes about 5 additional hours to reach the first local maximum. We can point out that the electricity consumptions at different hours are not independent on one another. The dependence between electricity consumptions at various hours is visible in Fig. 1b. One showed there the differences between current electricity demands and electricity consumption at 12:00 o'clock. We have a zero value at 12:00 o'clock but the widths of ranges are not growing rapidly when going away of 12:00. Moreover, these ranges are not symmetrical around zero. They exhibit bigger values after the 12:00 and significantly smaller before. The base hour has been chosen arbitrary but in general we observe the following behavior. If the electricity consumption is x [MWh] at the hour n, then one observes the expected electricity consumptions at hours $n-1$ and $n+1$ in some range around x [MWh]. This range will be significantly narrower than the total dispersion of electricity consumptions at those hours. Such an „inertia" of electricity consumption provides the idea that we can describe time series within the whole day by the consumptions at a relatively small number of hours regardless of other factors. In our earlier work [23] we studied many possible factors potentially influencing electricity consumption. Now, the only independent variable which we take into account is hourly electricity consumption.

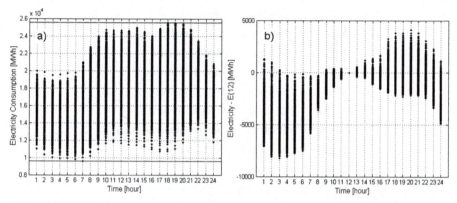

Fig. 1. a) Electricity consumption vs hour. Two horizontal lines indicate maximum and minimum electricity consumption; b) Difference between the current electricity demand and the electricity consumption at 12 o'clock vs hour

3 The Data

This study was carried out based on the historical data representing total electricity consumption in Polish power system [24]. The consumption is denoted on the hourly basis covering the time span between January 1st 2008 and June 23rd 2015, what corresponds to 2,731 days. Each consumption is accompanied by additional attributes listed in Table 1.

The electricity consumption is accompanied by the additional 10 attributes, including Hour. The majority of the attributes: Year, Month, Day as well as nDay, and nWeek are derived from dates (originally in the data set). The date was also complemented by the Holiday attribute, which is similar to nDay but additionally to weekends it takes into account also holidays. Some of the hours are during sunlight some are not. The Night_Day attribute reflects the presence of the sunlight. There are hours when this attribute changes its value during the year. The Night_Day is also influenced by time changes between winter and summer. The last two attributes are continuous: temperature and humidity. The data set contains whole days, and it holds 55,536 data points. The power system load is presented as the time series in Fig. 2a. The vertical line on May 3rd 2014 represents the split of the data into the learning and test data subsets. The learning data subset was used during the model construction and tuning. The other data subset was used for testing and during model evaluation. The first data part contains 2,314 days and the second one consists of 416 days.

The individual days and weeks are not visible but annual periodicity is clear. The high power demands take place during winters, low – during summers. There is a visible weak rising trend, but the data was not correct for the trend. The data is smeared vertically due to

Table 1. List of all features/variables used in the model

Symbol	Type	Description
Ener	Continuous	Electricity [MWh]
Hour	Discrete	The hour
Year	Discrete	Year's number
Month	Discrete	Month's number
Day	Discrete	Day of the month
nDay	Discrete	Day of the week
nWeek	Discrete	Week of the year
Holiday	Boolean	Holiday or working day
Night_Day	Boolean	Night or day
Tem	Continuous	Temperature
Hum	Continuous	Humidity

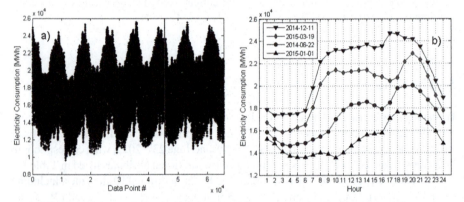

Fig. 2. a) The whole data series; b) Electricity consumption during selected days.

various days of the week, holidays, temperature, humidity and so on as was mentioned earlier. The Fig. 2b contains sample daily time series of electricity consumption for selected four days. The attributes of presented days are listed in Table 2.

The highest power demand is on (1) 2014-12-11, which is weekday in winter (December). It is a regular working day. Next, (2) 2015-03-19 is also a regular working day but during March, month almost always significantly warmer than December. Slightly lower demand is experienced on (3) 2014-06-25, the day in June, during summer. Probably, because it is significantly warmer day than two previous ones, electricity consumption is significantly lower. In this case there is no second maximum at evening hours. Even lower electricity demand is on (4) 2015-01-01 which is the New Year. The electricity consumption starts rising very late and its maximum is at about 18 o'clock. Although, we are looking at time series for various seasons, days of the week, etc. we can still see

Table 2. The attributes of the selected days

No	Data	nDay	Working Day	Month
1	2014-12-11	Thursday	Yes	December
2	2015-03-19	Thursday	Yes	March
3	2014-06-25	Wednesday	Yes	June
4	2015-01-01	Thursday	No	January

common features. There is a local minimum during night hours followed by a rising power demand. Next, we can observe small local maximum, wide plateau or slow rise. Afterwards, the power consumptions drop a little bit or at least stop rising and then rise again towards local maximum. At the end of a day there is a significant drop in the power consumption. Described common features and visible smoothness of daily time series lead to the idea of the model which is presented below.

4 Model Construction

4.1 The Model

The model presented herein is based on a linear multidimensional regression and is self-configurable – no model parameters have to be set up front. Its purpose is to find the most essential data points (hours) whose values allow to model the electricity consumptions for remaining hours within the same day. The whole data set consists of the time series of hourly electricity consumption.

In order to build and configure a model, we use a learning subset of data (see Fig. 2a). The remaining data is used for testing purposes of the final model. The described variables are hourly electricity consumptions $E(h)$, $h = 1, 2, \ldots, 24$. The construction of the model is performed in successive steps. Each step expands the model by changing the type of one variable from described to describing. During each step another hour is chosen till one reaches a demanded accuracy. The procedure starts with 24 described variables corresponding to 24 h in step zero. One calculates a root of the mean error squared (standard deviation) for each variable:

$$s(h) = \sqrt{\frac{1}{N} \sum_{i=1}^{N} \left(E_i(h) - \overline{E(h)} \right)^2} \tag{1}$$

where, $E_i(h)$ – electricity consumption at hour h in i-th day, $\overline{E(h)}$ – average electricity consumption at hour h, N – total number of analyzed days. The mean electricity consumption and standard deviation are plotted vs hour in Fig. 3.

Now, the first step starts and the hour for which $s(h)$ is the biggest is chosen, in our case 18 o'clock. This hour is denoted by h1 and an electricity consumption at this hour by E_i (h1). The chosen variable is used to construct 23 linear regressions of the form:

$$E_i(h) = a_{1h} E_i(h_1) + a_{0h} + \varepsilon_i \tag{2}$$

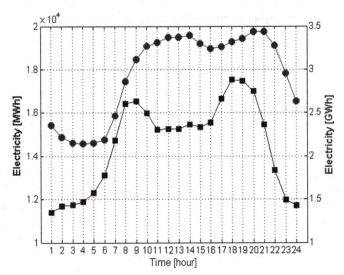

Fig. 3. Mean energy consumption (blue circles) and standard deviation (green squares) vs hour. (Color figure online)

where $h = 1, 2, ..., 24$ and $h \neq h1$, $i = 1, 2, ..., N$ and a_{1h}, a_{0h} are model parameters. For each linear model root of the mean squared error (3) is calculated.

$$\sigma(h) = \sqrt{\frac{1}{N} \sum_{i=1}^{N} \left(E_i(h) - \hat{E}_i(h) \right)^2} \qquad (3)$$

where the error $E_i(h) - \hat{E}_i(h)$ is the difference between theoretical and empirical value. In the second step the hour corresponding to the biggest $\sigma(h)$ – the most poorly described variable is chosen: $h2 = 9$ (see Fig. 4). Now, there are 2 describing and 22 described variables. The 22 linear regressions of the form (4)

$$E_i(h) = a_{2h}E_i(h_2) + a_{1h}E_i(h_1) + a_{0h} + \varepsilon_i \qquad (4)$$

are built and another set of $\sigma(h)$ is calculated. During the third step, the third describing variable is chosen and 21 models are calculated. The procedure is repeated till only one described variable will remain or when demanded accuracy will be reached. The values of $S(h)$ and $\sigma(h)$ are presented in Fig. 4. One showed values for steps: 0 and 1 as well as another sample steps: 4 and 8. One can notice that the values of $\sigma(h)$ for the first step are significantly smaller than those obtained during the previous step. In the following steps the values of $\sigma(h)$ became smaller. During each step more accurate models are constructed, each with one more describing variable but for the price of dropping the number of described variables.

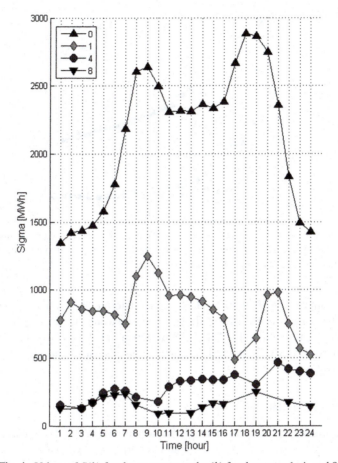

Fig. 4. Values of $S(h)$ for the step zero and $\sigma(h)$ for the steps: 1, 4, and 8.

The procedure finishes with one linear regression with 23 describing variables and only one described variable. The biggest, mean and the smallest $\sigma(h)$ for modeled hours vs step number are plotted in Fig. 5.

The biggest $\sigma(h)$, which are taken into account during each step are also plotted in logarithmic scale. The progress during the first steps is very big and became smaller afterward. For example, having 4 describing variables the most poorly described hour has the root of the mean squared error below 500. That means a $\sigma(h)$ dropped almost 6 times in comparison to the starting step when it was almost 3000. We want the error to be as small as possible. On the other hand, we want to have as few describing variables as possible possessing a model with big support. In the Fig. 5 we see that, starting from 4 variables the maximum sigma drops systematically according to the exponential law. It seems that the optimal number of steps is 4, when we get high accuracy while having a small number of describing variables and big support (20 described variables). One should keep in mind we take into account the biggest $\sigma(h)$, while taking into account mean $\sigma(h)$ we would get even better results – they would drop from about 2200 to

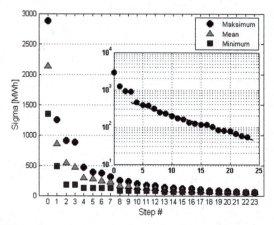

Fig. 5. Values of $S(h)$ for the step zero and $\sigma(h)$ vs step number. There are maximum, mean, and minimum values for each step. Inserted plot shows maximal values vs step number in log scale. Line represents linear fit to data points starting with step 4.

about 250 so approximately 9 times. However, the interpretation of the maximal $\sigma(h)$ is clearer: a time series is described with the accuracy equal or better than a given sigma. For comparison, we present $\sigma(h)$ for 4-th and 8-th step in Fig. 4. Having 4 and 8 describing variables we would have the poorest described variables with sigma equal to about 500 and 250 respectively.

4.2 Testing the Model

The last stage is the test of the model with 4 describing variables. The model is based on variables related to the electricity consumptions at: $h_1 = 18$, $h_2 = 9$, $h_3 = 2$, and $h4 = 20$. They were selected at first four successive steps. The model is described by the formula:

$$E_i(h) = a_{4h}E_i(h_4) + a_{3h}E_i(h_3) + a_{2h}E_i(h_2) + a_{1h}E_i(h_1) + a_{0h} + \varepsilon_i \tag{5}$$

We use test data set – the last 9 984 h that corresponds to 416 days. We compared empirical time series of electricity consumptions for one day with those provided by the model. Results for sample days are presented in Fig. 6. The red markers indicate empirical, black – theoretical data, and open circles indicate describing variables. Attributes of the selected days were listed in Table 2.

Although presented plots are for different days of weeks, seasons and environmental conditions (thus having various shapes), the model describes data well. However, some discrepancies are visible. The shapes of the empirical and theoretical plots are similar to each other. Let note that the majority of the differences are between -1% and 1%. The biggest differences are for New Year 2015-01-01 10:00 and 14:00 o'clock (see Fig. 6). In order to investigate the model quality in the complex manner, we defined the model quality measure given by the formula (6):

$$Q(i) = \sqrt{\frac{1}{20} \sum_{h=1}^{24} \left(E_{ih} - \hat{E}_{ih}\right)^2} \Big/ \frac{1}{20} \sum_{h=1}^{24} \hat{E}_{ih}, \tag{6}$$

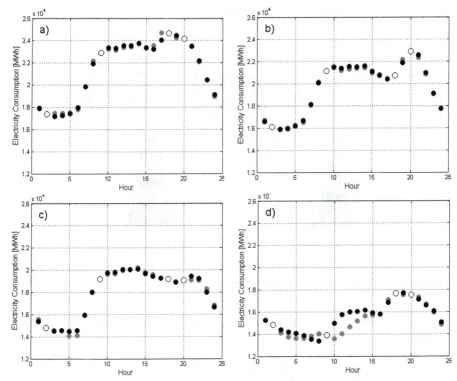

Fig. 6. Empirical (red dots) and theoretical (black dots) electricity consumptions for selected days: (a) 2014-12-11, (b) 2015-03-19, (c) 2014-06-25, (d) 2015-01-01. Theoretical values are provided by the model (5) with 4 describing variables indicated by open circles. (Color figure online)

where, $h \neq h1, h2, h3, h4$ and $i = 1, 2, \ldots N$. Measure (6) is a relative root of the mean squared error. Figure 7a contains values of the quality measure $Q(i)$ for the full range of the test data set. Presented results indicate a very high quality of the model in the entire data range. Moreover, there is no visible trend of values of $Q(i)$. The distribution of the measure values is presented in Fig. 7b, while the distribution of the log-values is in Fig. 7c. One can notice that the distribution of the measure values has a shape similar to a log-normal. The distribution starts at about 0.5%, has the maximum at 1.5% and falls rapidly thereafter. Almost all measure values (95%) are within the range of 0.5%–2.8%, however there are a few high spikes, the highest reaches almost 7%. We investigated attributes of days related to the highest six values of the quality measure: 7%, and three around 5% and two about 3.5%. They are listed in the Table 3.

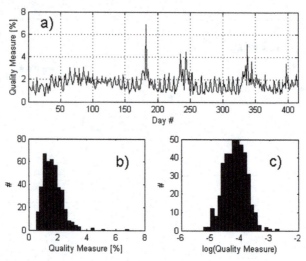

Fig. 7. a) Quality measure $Q(i)$ for each day in test data set; b) Distribution of quality measure values; c) Distribution of logarithm of quality measure values.

Table 3. Days with the highest values of the quality measure. All the days are holydays (names in brackets)

Data	Week Day	$Q(i)$ [%]
2014-11-01	Saturday (All Saints' Day)	6.89
2014-12-24	Wednesday (Christmas)	4.31
2015-01-01	Thursday (Labour Day)	4.43
2015-04-05	Sunday (Easter)	5.11
2015-04-12	Sunday (Orthodox Easter)	3.59
2015-06-04	Thursday (Corpus Christi)	3.44

If the days with the highest values of the measure are not typical, that would explain the differences between the model and data. The first spike is on the first of November – a holiday. All Saints' Day is celebrated solemnly in Poland. It is a bank holiday during which people travel a lot to visit family graves. Additionally, in 2014 this day was on Saturday, which favored trips. That's why the spike in this case was so strong. We deal with not a typical day. The second spike is in Christmas Eve which occurred on Wednesday. It is worth to point out that during Christmas, next two days are non-working days in Poland, thus, there was so called long-weekend: Thursday-Sunday. The next spike was related to the New Year, which was on Thursday. There is a common practice in Poland to take the next day off and have a long weekend. The fourth spike occurred during Easter Sunday, while the next day is also a bank holiday in Poland. The last two spikes are lower than previous ones. The first one corresponds to Orthodox Easter, holiday celebrated by the biggest religious minority in Poland. The next day is also celebrated by them. The last spike occurs on Thursday (Corpus Christi bank holiday) and it makes a long weekend possible. All the most deviated days are related to the bank holidays which are the first days of groups of non-working days. They surely cannot be regarded as typical. To conclude, the elaborated model can be used to distinguish non typical days from the electricity consumption point of view.

5 Conclusion

In this work, the hourly electricity consumption was investigated. In order to describe shapes of time series during each day, the data mining model was elaborated. The model uses the multivariate linear regression technique. Its aim is to find points in data series which will describe remaining data points in the same day. The construction of the model is performed in steps and every step improves the accuracy but lowers support. It is up to the user when to stop a construction of the model and, thus, obtaining required precision. The accuracy was measured as a maximum root of the mean squared error. Based on the Polish data, we showed that the model with 4 describing variables, describes 20 h remaining very well, exhibiting dominant relative error about 1.5%. The model describes data very well, independently on season, temperature time series, humidity, or day of the week. It is characterized by a high precision and allows finding non-typical days from the electricity demand point of view.

During future studies this model will be used for classification of the daily electricity demand profiles as well as for finding non-typical electricity consumptions. The model will be also used for identification factors influencing electricity consumption at different hours of day. We also expect to obtain good results during a prognosis of the electricity demand.

References

1. Pérez-Arriaga, I.J. (ed.): Regulation of the Power Sector. Springer, London (2013). https://doi.org/10.1007/978-1-4471-5034-3
2. Khatoon, S., Ibraheem, Arunesh, Kr.S., Priti: Effects of various factors on electric load forecasting: an overview. In: 6th IEEE Power India International Conference (PIICON), Delhi, pp. 1–5 (2014)
3. Kremers, E.: Modelling and Simulation of Electrical Energy Systems through a Complex Systems Approach using Agent-Based Models. KIT Scientific Publishing, Karlsruhe (2013)
4. Neng, S., Jingjing, Z., Wenjie, Z.: Energy efficiency measures and convergence in China. Taking into account the effects of environmental and random factors. Pol. J. Environ. Stud. **24**(1), 257–267 (2015)
5. Li, J., Yang, L., Long, H.: Climatic impacts on energy consumption: Intensive and extensive margins. Energy Econ. **71**, 332–343 (2018)
6. McCulloch, J., Ignatieva, K.: Forecasting high frequency intra-day electricity demand using temperature. SSRN Electron. J. (2017)
7. Xie, J., Hong, T.: Wind speed for load forecasting models. Sustainability **9**(5), 795 (2017)
8. Kalimoldayev, M., Drozdenko, A., Koplyk, I., Abdildayeva, M.A., Zhukabayeva, T.: Analysis of modern approaches for the prediction of electric energy consumption. Open Eng. **10**, 350–361 (2020)

9. Manowska, A.: Using the LSTM Network to forecast the demand for Electricity in Poland. Appl. Sci. **10**, 8455 (2020)
10. Del Real, A.J., Dorado, F., Duran, J.: Energy demand forecasting using deep learning: applications for the French grid. Energies **13**, 2242 (2020)
11. Kang, T., Lim, D.Y., Tayara, H., Chong, K.T.: Forecasting of power demands using deep learning. Appl. Sci. **20**, 7241 (2020)
12. Ciechulski, T., Osowski, S.: Deep learning approach to power demand forecasting in Polish power system. Energies **13**, 6154 (2020)
13. Elias Barrón, I., et al.: Genetic algorithm with radial basis mapping network for the electricity consumption modeling. Appl. Sci. **10**, 4239 (2020)
14. Gajowniczek, K., Nafkha, R., Ząbkowski, T.: Seasonal peak demand classification with machine learning techniques. In: 2018 International Conference on Applied Mathematics & Computer Science (ICAMCS), pp. 101–1014 (2018)
15. Dudek, G.: Neural networks for pattern-based short-term load forecasting: a comparative study. Neurocomputing **205**, 64–74 (2016)
16. Dudek, G.: Pattern-based local linear regression models for short-term load forecasting. Electr. Power Syst. Res. **130**, 139–147 (2016)
17. Fan, G.F., Qing, S., Wang, H., Hong, W.C., Li, H.J.: Support vector regression model based on empirical mode decomposition and auto regression for electric load forecasting. Energies **6**, 1887–1901 (2013)
18. Divina, F., Torres, M.G., Vela, F.A.G., Noguera, J.L.V.: A comparative study of time series forecasting methods for short term electric energy consumption prediction in smart buildings. Energies **12**, 1934 (2019)
19. De Gooijer, J.G., Hyndman, R.J.: 25 years of time series forecasting. Int. J. Forecast. **22**, 443–473 (2006)
20. Halepoto, I.A., Uqaili, M.A., Chowdhry, B.S.: Least square regression based integrated multi-parameteric demand modeling for short term load forecasting. Mehran Univ. Res. J. Eng. Technol. **33**(2), 215–226 (2014)
21. Vu, D.H., Muttaqi, K.M., Agalgaonkar, A.P.: A variance inflation factor and backward elimination based robust regression model for forecasting monthly electricity demand using climatic variables. Appl. Energy **140**, 385–394 (2015)
22. Ghandoor, A., Samhouri, M.: Electricity consumption in the industrial sector of Jordan: application of multivariate linear regression and adaptive neuro-fuzzy techniques. Jordan J. Mech. Ind. Eng. **3**(1), 69–76 (2009)
23. Karpio, K., Łukasiewicz, P., Nafkha, R.: Regression technique for electricity load modeling and outlined data points explanation. In: Pejaś, J., El Fray, I., Hyla, T., Kacprzyk, J. (eds.) ACS 2018. AISC, vol. 889, pp. 56–67. Springer, Cham (2019). https://doi.org/10.1007/978-3-030-03314-9_5
24. Polish power system dataset. http://www.pse.pl/index.php?dzid=77. Accessed 12 Aug 2017

The Problem of Tasks Scheduling with Due Dates in a Flexible Multi-machine Production Cell

Wojciech Bożejko[1]([envelope]) [iD], Piotr Nadybski[2] [iD], Paweł Rajba[3] [iD],
and Mieczysław Wodecki[4] [iD]

[1] Department of Automatics, Mechatronics and Control Systems, Wrocław University of Technology, Janiszewskiego 11-17, 50-372 Wrocław, Poland
wojciech.bozejko@pwr.edu.pl

[2] Witelon State University of Applied Science in Legnica, Sejmowa 5A, 59-220 Legnica, Poland
nadybskip@pwsz.legnica.edu.pl

[3] Institute of Computer Science, University of Wrocław, Joliot-Curie 15, 50-383 Wrocław, Poland
pawel@cs.uni.wroc.pl

[4] Department of Telecommunications and Teleinformatics, Wrocław University of Science and Technology, Wybrzeże Wyspiańskiego 27, 50-370 Wrocław, Poland
mieczyslaw.wodecki@pwr.edu.pl

Abstract. In the paper we consider an NP-hard problem of tasks scheduling with due dates and penalties for the delay in a flexible production cell. Each task should be assigned to one of the cell's machines and the order of their execution on machines should be determined. The sum of penalties for tardiness of tasks execution should be minimized. We propose to use the tabu search algorithm to solve the problem. Neighborhoods are generated by moves based on changing the order of tasks on the machine and changing the machine on which the task will be performed. We prove properties of moves that significantly accelerate the search of the neighborhoods and shorten the time of the algorithm execution and in result significantly improves the efficiency of the algorithm compared to the version that does not use these properties.

1 Introduction

Optimizing the production process relies on designating of such a schedule for the execution of the elements that gives optimal effects measured by the value of a certain criterion (e.g. execution time, sum of penalties for delays, etc.). This usually comes down to the formulation of a combinatorial problem in which an optimal element (sequence or permutation) should be determined from a finite set, usually very large, of feasible solutions.

In Just-in-Time manufacturing systems there is a requirement to complete each element before the expiration of the due date, otherwise a penalty is being

© Springer Nature Switzerland AG 2021
M. Paszynski et al. (Eds.): ICCS 2021, LNCS 12745, pp. 405–419, 2021.
https://doi.org/10.1007/978-3-030-77970-2_31

calculated for exceeding it (for the delay or being tardy). In the problem considered here a set of tasks and a set of machines are given. Machines of the same type (parallel machines), i.e. with the same functional properties, form a flexible production cell. Each task must be performed on a machine of the cell. The data includes the times of completing tasks on each of machines, and their due dates. The penalty for being tardy depends on the tardiness value and the penalty factor, different for each task. The problem is to assign tasks to machines and arrange their execution on each machine to minimize the sum of penalties (costs). In short, we will denote this problem by **MTC**.

If the cell consists of only one machine then the **MTC** problem comes down to the single machine scheduling problem. In the literature it is denoted by $1||\sum w_i T_i$ and it belongs to the class of NP-hard problems (see [7]; of course it means, that **MTC** problem is also NP-hard). Its exact description, specific properties and a very effective tabu search algorithm can be found in Bożejko et al. [2]. The parallel exact algorithm described Wodecki [14]. In contrast, a generalization of the **MTC** problem is the flexible task shop problem in which a set of machines is partitioned into cells. According to the relationship of the technological order, each task passes through many cells. In the literature, it is mainly considered with the $C_{\max}$ criterion. Unfortunately, models and problems of scheduling production with other cost criteria, despite significant practical needs, are relatively rarely considered. This is mainly due to the irregularity of the criterion functions and the lack of specific properties of problems leading to a limitation of the set of solutions. The most important, containing main theoretical results, and with good algorithms are papers written by: Karabulut [5], Liu [9], Kayvanfar et al. [6], Ojstersek, Buchmeister [11], Bulfin, Hallah [4], Bożejko et al. [3], Park et al. [10], Tocovicha et al. [13].

In this work, we will present a description of the problem, prove some of its properties and provide constructive and tabu search algorithms. In the design of algorithms, the neighborhood generated by the moves based on changing the order of execution of tasks on the machine and changing the machine on which the task will be performed is used. The properties of the problem have been proven here to enable the elimination of worse solutions from the neighborhood. As a result, the number of elements of the neighborhood is significantly reduced, thus, the time needed for search also decreases. Due to the lack of bigger instances of the considered problem, computational experiments were performed on randomly generated examples.

2   Problem Description

There is a set $\mathcal{J} = \{1, 2, \ldots, n\}$ of n tasks given to be performed on the m machines from the set $\mathcal{M} = \{1, 2, \ldots, m\}$. For the task $i \in \mathcal{J}$, there are the following notions introduced:

$p_{i,j}$ - execution time , $j \in \mathcal{M}$,

d_i - required completion date ,

w_i - weight of the penalty function (cost of delay).

In the **MTC** problem, the tasks are to be assigned to machines and there must be determined the order of their execution on each machine to minimize the sum of the tardinesses costs. If the assignment of tasks to machines and the order in which they are performed is determined, then for a task $i \in \mathcal{J}$ the following designations are introduced:

C_i - *completion date* ,

$T_i = \max\{0, C_i - d_i\}$ - *delay,*

$w_i \cdot T_i$ - *penalty (cost) of the delay.*

Therefore, the problem consists in designating such an allocation of tasks to machines and the order of their execution to minimize *the sum of penalties (costs of tardinesses)*, i.e. $\sum_{i=1}^{n} w_i \cdot T_i$, wherein the following constraints must be met:

(i) the task must be performed on exactly one machine,
(ii) the task execution cannot be interrupted,
(iii) at any given time, the machine can perform only one task.

Let a sequence of tasks' sets

$$A = (A_1, A_2, \ldots, A_m),$$

such that $A_i \subset \mathcal{J}$, $A_i \cap A_j = \emptyset$, $i \neq j$, $i, j \in \mathcal{M}$ and $\sum_{i=1}^{m} A_i = \mathcal{J}$, is called *assignment of tasks to machines.* By $\mathcal{A}$ we denote the *set of all such assignments.*

For assignment $A = (A_1, A_2, \ldots, A_m)$, $A \in \mathcal{A}$, let $\pi = (\pi_1, \pi_2, \ldots, \pi_m)$ will be a sequence of permutations such that π_i is an n_i-elementary $(n_i = |A_i|)$ permutation (order of execution) of tasks from the set A_k, $k \in \mathcal{M}$, i.e. performed by a k-th machine. By $\mathcal{P}(A)$ we denote the set of all such sequences of permutations (in short these sequences will be called *task permutations*).

The set of solutions of the **MTC** problem of performing tasks on machines can be defined as follows:

$$\mathcal{AP} = \{(A, \pi) \colon A \in \mathcal{A}, \pi \in \mathcal{P}(A)\}. \tag{1}$$

Let the solution $\mathbf{S} = (A, \pi) \in \mathcal{AP}$, where $A = (A_1, A_2, \ldots, A_m)$ and $\pi = (\pi_1, \pi_2, \ldots, \pi_m)$. If the task $i \in \mathcal{J}$ is executed on a machine $k \in \mathcal{M}$ (i.e. $i \in A_k$) as the l-th in the order $(\pi_k(l) = i)$, then the earliest moment of its completion $C_i = \sum_{j=1}^{l} p_{\pi_k(j)}$, tardiness $T_i = \max\{0, C_i - d_i\}$, and a penalty for the tardiness $f_i(\mathbf{S}) = w_i \cdot T_i$. In turn

$$T(\mathbf{S}) = \sum_{j=1}^{n} w_{\pi(i)} T_{\pi(i)} = \sum_{j=1}^{n} f_j(\mathbf{S}), \tag{2}$$

is the cost of the execution of all tasks (the sum of the penalties for delays).

The problem of task execution on **MTC** machines considered in the work is to determine the optimal solution $\mathbf{S}^* \in \mathcal{AP}$. This problem is NP-hard because for the number of machines $m = 1$, we get NP-hard single-machine problem, as it was written in the Introduction.

In the case where the number of machines $m = 2$ the cardinality of the set of all assignments $|\mathcal{A}| = 2^n$. For each assignment, the number of all possible sequences of tasks is $O(n!)$. In this case all possible solutions to the problem are $O(2^n \cdot n!)$.

3   Solution Method

Scheduling operations in a flexible production cell, solving the **MTC** problem, requires simultaneous decision making on two levels:

1. assigning tasks to machines,
2. determining the order of performing tasks on each machine.

Therefore, an idea of the problem solving method can be presented in the form of the following algorithm:

ATMC algorithm

Let $\mathbf{S}$, $\mathbf{S}^*$ be solutions, $\mathbf{S} = (A, \pi)$, $\mathbf{S} \in \mathcal{AP}$ and $\mathbf{S}^* := \mathbf{S}$.
repeat
      Step 1: generate from $\mathbf{S}$ a new assignment $A' \in \mathcal{A}$;
      Step 2: for assignment A' generate permutation
             $\pi' \in \mathcal{P}(A')$ - of the task execution order;
             ($\mathbf{S}' = (A', \pi')$ is a new solution);
                if $\mathcal{T}(\mathbf{S}') < \mathcal{T}(\mathbf{S}^*)$ then $\mathbf{S}^* := \mathbf{S}'$;
      $\mathbf{S} := \mathbf{S}'$;
until {*Stop condition*}

$\mathbf{S}^*$ is the solution determined by the algorithm (output).

Step 1 can be accomplished in many ways (e.g. random selection, constructive algorithm, etc.). It should only be emphasized that for two machines the number of possible assignments is 2^n. In turn the implementation of Step 2 requires, for each machine, determining the sequence of performing the assigned tasks. Determining the optimal order (i.e. minimizing the sum $\sum w_i T_i$) is an NP-hard problem. It boils down to solving the NP-hard one-machine task scheduling problem $1 || \sum w_i T_i$. Therefore, to solve the **MTC** problem under consideration there will be the approximate algorithm based on the tabu search method used. Its essential elements are neighborhoods, the subsets of the set of feasible solutions generated from the current solution by transformations called moves. When browsing the neighborhood, we select the element with the lowest value of the criterion function, which we adopt as new, current solutions in the next iteration of the algorithm. From a fixed solution, another solution can be generated by executing the move (**Step 1**) consisting of:

1. changing the order of execution of tasks on a certain machine, or
2. *transfer* of the task from one machine to another.

Both of these moves will be described in details further.

For a solution $\mathbf{S} \in \mathcal{AP}$ $(\mathbf{S} = (A, \pi))$ by

$$T_k(\mathbf{S}) = \sum_{j \in A_k} f_j(\mathbf{S}), \tag{3}$$

we denote the cost of performing tasks by k-th machine. This value will be denoted in short k-th cost. Therefore, the cost of performing all tasks (2) is equal to the sum of the costs of individual machines, i.e. $T(\mathbf{S}) = \sum_{k=1}^{m} T_k(\mathbf{S})$.

3.1 Changing the Order of Tasks on the Machine

To change the order of execution of tasks on a machine in a solution $(\mathbf{S} = (A, \pi))$ there will be an *insert*-type of move used. It consists in shifting a task into a different position. Let π_k be n_k-elementary permutation – an order of tasks on k-th machine. If $1 \le s, t \le n_k$, then *insert*-type of move consists in swapping the element from position s (of task $\pi_k(s)$) to position t in permutation π_k. Two cases will be considered.

Case 1. Let us assume that $t \ge s$.

Then this move will be denoted by with $\overrightarrow{\eta_t^s}$, and generated permutation $\overrightarrow{\eta_t^s}(\pi_k) = \pi_k'$. Then:

$$\pi_k'(i) = \begin{cases} \pi_k(i), & \text{if } i < s \vee i > t, \\ \pi_k(i+1), & \text{if } s \le i < t, \\ \pi_k(s), & \text{if } i = t. \end{cases} \tag{4}$$

where $(\mathbf{S}' = (A, \pi'))$. Having executed the move $\overrightarrow{\eta_t^s}$, in order π_k'

$$T_k'(\mathbf{S}') = \sum_{i=1}^{s-1} f_{\pi_k'(i)}(\mathbf{S}) + f_{\pi_k'(s)}(\mathbf{S}) + \sum_{i=s+1}^{t-1} f_{\pi_k'(i)}(\mathbf{S}) + f_{\pi_k'(t)}(\mathbf{S}) + \sum_{i=t+1}^{n_k} f_{\pi_k'(i)}(\mathbf{S}).$$

It follows from definition (4) that

$$f_{\pi_k'(t)}(\mathbf{S}) = \max\{0, C_{\pi_k(t)} - d_{\pi_k(t)}\}, \tag{5}$$

$$\sum_{i=1}^{s-1} f_{\pi'(i)}(\mathbf{S}) = \sum_{i=1}^{s-1} f_{\pi,(i)}(\mathbf{S}), \text{ and } \sum_{i=t+1}^{n_k} f_{\pi(i)}(\mathbf{S}) = \sum_{i=t+1}^{n_k} f_{\pi'(i)}(\mathbf{S}). \tag{6}$$

Since for the task completion times: $\pi_k'(s), \pi_k'(s+1), \ldots, \pi_k'(t)$

$$C_{\pi_k'(t)} = C_{\pi_k(t)} \text{ and } C_{\pi_k'(i)} = C_{\pi_k(i)} - p_{\pi_k(s)} + p_{\pi_k(s+1)}, \text{ for } i = s, s+1, \ldots, t-1,$$

where the cost of the task execution:

$$f_{\pi_k'(i)}(\mathbf{S}) = \max\{0, C_{\pi_k(i)} - p_{\pi_k(s)} + p_{\pi_k(s+1)} - d_{\pi_k(i)}\}. \tag{7}$$

For *insert*-type of move $\overrightarrow{\eta_t^s}$ $(t \geq s)$, generating from π_k permutation π_k', let

$$\delta_k(\overrightarrow{\eta_t^s}) = \sum_{i=s}^{t-1} f_{\pi_k'(i)}(\mathbf{S}) + f_{\pi_k'(t)}(\mathbf{S}) - \sum_{i=s}^{t} f_{\pi_k(i)}(\mathbf{S}). \tag{8}$$

Theorem 1. *If permutation π_k' was generated from π_k by execution of insert-type of move $\overrightarrow{\eta_t^s}$ $(t \geq s)$, then*

$$T_k'(\mathbf{S}') = T_k(\mathbf{S}) + \delta_k(\overrightarrow{i_t^s}). \tag{9}$$

Proof. To prove of the theorem the following equality should be used (5)–(8).

Case 2. $t < s$. Let us assume that $t < s$. Then the *insert*-type of move generating from π_k a new permutation π_k'' by swapping the task from position s to position t will be denoted by $\overleftarrow{\eta_t^s}$. In this case a generated permutation

$$\pi_k''(i) = \begin{cases} \pi_k(i), & \text{if } i < s \vee i > t, \\ \pi_k(i+1), & \text{if } s \leq i < t, \\ \pi_k(s), & \text{if } i = t. \end{cases} \tag{10}$$

where $(\mathbf{S}'' = (A, \pi''))$. Similarly as in **Case 1** it is possible to determine equality similar to (5)–(7). Next, let

$$\delta_k(\overleftarrow{\eta_t^s}) = \sum_{i=s}^{t-1} f_{\pi_k''(i)}(\mathbf{S}) + f_{\pi_k''(t)}(\mathbf{S}) - \sum_{i=s}^{t} f_{\pi_k(i)}(\mathbf{S}). \tag{11}$$

Theorem 2. *If permutation π_k'' was generated from π_k by insert- type of move $\overleftarrow{\eta_t^s}$ $(t < s)$, then*

$$T_k''(\mathbf{S}'') = T_k(\mathbf{S}) + \delta_k(\overleftarrow{\eta_t^s}). \tag{12}$$

Proof. The proof of equality (12) is similar to the proof of Theorem 1. It follows from the Theorem 1 that if $\delta_k(\overrightarrow{\eta_t^s}) < 0$, then the execution of the *insert*-type of move generates solution with lower cost of execution of tasks (improvement of the solution). It is the same with the move $\overleftarrow{\eta_t^s}$ (Theorem 2). Each of these moves can therefore be a move improving the current solution.

3.2   Transferring of a Task to Another Machine

Let the solution $\mathbf{S} = (A, \pi)$, where $A = (A_1, A_2, \ldots, A_m)$, $\pi = (\pi_1, \pi_2, \ldots, \pi_m)$. We are considering two machines k and l $(k \neq l, \ k, l \in \mathcal{M})$. From a solution $\mathbf{S}$ we generate $\mathbf{S}' = (A', \pi')$ by transferring a single task from a machine k to l. Let s be a position in permutation π_k, and t position in permutation π_l. The *transfer*-type of moves (τ - *move*) transfers the task from the position s on machine k

to position t on machine l generating in this way a new solution $\mathbf{S}'$ (in short we will designate it by $\mathbf{S}' = \tau_l^k(s,t)(\mathbf{S})$). In the generated solution $\mathbf{S}' = (A', \pi')$

$$A' = (A_1', A_2', \ldots, A_m'), \text{ and } \pi' = (\pi_1', \pi_2', \ldots, \pi_m'), \text{ where}$$

$$n_k' = n_k - 1, \ n_l' = n_l + 1, \ A_i' = A_i \text{ and } \pi_i' = \pi_i, \text{ for } i \neq k, l \ (k, l \in \mathcal{M}), \quad (13)$$

$$A_k' = A_k \setminus \{\pi_k(s)\}, \ A_l' = A_l \cup \{\pi_k(s)\},$$

$$\pi_k' = (\pi_k(1), \pi_k(2), \ldots, \pi_k(s-1), \pi_k(s+1), \ldots, \pi_k(n_k')),$$

$$\pi_l' = (\pi_l(1) \ldots, \pi_l(t-1), \pi_k(s), \pi_l(t), \pi_l(t+1) \ldots, \pi_l(n_l')).$$

It is easy to check that the determined in this way solution $\mathbf{S}' = (A', \pi')$ is feasible for the **MTC** problem. Then we determine the cost of performing tasks for both machines.

After executing the move $\tau_l^k(s,t)(\mathbf{S})$ (transferring the task from machine k to machine l) The cost of executing tasks on machine k

$$\mathcal{T}_k(\mathbf{S}') = \sum_{i=1}^{n_k'} f_{\pi_k'(i)}(\mathbf{S}') = \sum_{i=1}^{s-1} f_{\pi_k'(i)}(\mathbf{S}') + \sum_{i=s}^{n_k'} f_{\pi_k'(i)}(\mathbf{S}'). \quad (14)$$

From definition of permutation π_k' the first sum $\sum_{i=1}^{s-1} f_{\pi_k'(i)}(\mathbf{S}') = \sum_{i=1}^{s-1} f_{\pi_k(i)}(\mathbf{S})$. Since permutation π_k' is created from π_k by removing the task $\pi_k(s)$, then the execution by the k-th machine the tasks $\pi_k'(s), \pi_k'(s+1), \ldots, \pi_k'(n_k')$ in solution $\mathbf{S}'$ is exceeded by the $p_{\pi_k(s)}$ (this is $\pi_k(s)$). Therefore, the moment of completion of tasks execution $C_{\pi_k'(i)} = C_{\pi_k(i+1)} - p_{\pi_k(s)}$, and the cost

$$f_{\pi_k'(i)}(\mathbf{S}') = w_{\pi_k(i+1)} \cdot \max\{0, C_{\pi_k(i+1)} - p_{\pi_k(s)} - d_{\pi_k(i+1)}\}, \quad (15)$$

for $i = s, s+1, \ldots, n_k'$. Ultimately, using (15) the cost of execution of tasks by k-th machine in the order π_k'

$$\mathcal{T}_k(\mathbf{S}') = \sum_{i=1}^{s-1} f_{\pi_k(i)}(\mathbf{S}) + \sum_{i=s+1}^{n_k-1} w_{\pi_k(i+1)} \max\{0, C_{\pi_k(i+1)} - p_{\pi_k(s)} - d_{\pi_k(i+1)}\}. \quad (16)$$

Similarly, for the l-th machine, the cost of execution of tasks on machine l

$$\mathcal{T}_l(\mathbf{S}') = \sum_{i=1}^{n_l'} f_{\pi_l'(i)}(\mathbf{S}') = \sum_{i=1}^{t-1} f_{\pi_l'(i)}(\mathbf{S}') + f_{\pi_l'(t)}(\mathbf{S}') + \sum_{i=t+1}^{n_l'} f_{\pi_l'(i)}(\mathbf{S}'). \quad (17)$$

It follows from definition of permutation π_l' that $\sum_{i=1}^{t=1} f_{\pi_l'(i)}(\mathbf{S}') = \sum_{i=1}^{t=1} f_{\pi_l(i)}(\mathbf{S})$, and $f_{\pi_l'(t)}(\mathbf{S}') = w_{\pi_k(s)} \cdot \max\{0, C_{\pi_l(t-1)} + p_{\pi_k(s)} - d_{\pi_k(s)}\}$. The moment of completing each task $\pi'_l(t+1), \pi'_l(t+2), \ldots, \pi'_l(n'_l)$ is transferred (relative to its execution time in π_k) by the time of the duration of the task $\pi_k(s)$, namely by $p_{\pi_k(s)}$. Therefore, similarly as above

$$\sum_{i=t+1}^{n'_l} f_{\pi'_l(i)}(\mathbf{S}') = \sum_{i=t}^{n_l} w_{\pi_k(i)} \cdot \max\{0, C_{\pi_l(i)} + p_{\pi_k(s)} - d_{\pi_l(i)}\}.$$

Ultimately

$$\mathcal{T}_l(\mathbf{S}') = \sum_{i=1}^{t=1} f_{\pi_l(i)}(\mathbf{S}) + w_{\pi_k(s)} \cdot \max\{0, C_{\pi_l(t-1)} + p_{\pi_k(s)} - d_{\pi_k(s}\}$$

$$+ \sum_{i=t}^{n_l} w_{\pi_k(i)} \cdot \max\{0, C_{\pi_l(i)} + p_{\pi_k(s)} - d_{\pi_l(i)}\}. \tag{18}$$

For the solution $\mathbf{S} \in \mathcal{QP}$, let

$$\delta_{\mathbf{S}}^- = \sum_{i=s+1}^{n_k-1} w_{\pi_k(i)} \cdot \max\{0, C_{\pi_k(i+1)} - p_{\pi_k(s)}$$

$$- d_{\pi_k(i+1)}\} - f_{\pi_k(s)}(\mathbf{S}) + \sum_{i=s+1}^{n_k} f_{\pi_k(i)}(\mathbf{S}). \tag{19}$$

Lemma 1. *If the solution $\mathbf{S}' = \tau_l^k(s, t)(\mathbf{S})$, then the cost (14) of tasks execution on k-th machine*

$$\mathcal{T}_k(\mathbf{S}') = \mathcal{F}_k(\mathbf{S}) + \delta_{\mathbf{S}}^-,$$

Proof. From definition the cost of tasks execution

$$\mathcal{T}_k(\mathbf{S}) = \sum_{i=1}^{s-1} f_{\pi_k(i)}(\mathbf{S}) + f_{\pi_k(s)} + \sum_{i=s+1}^{n_k} f_{\pi_k(i)}(\mathbf{S}).$$

$$\mathcal{T}_k(\mathbf{S}') = \mathcal{T}_k(\mathbf{S}) + \delta_{\mathbf{S}}^- = \sum_{i=1}^{s-1} f_{\pi_k(i)}(\mathbf{S}) + f_{\pi_k(s)} + \sum_{i=s+1}^{n_k} f_{\pi_k(i)}(\mathbf{S}) +$$

$$+ \sum_{i=s+1}^{n_k-1} w_{\pi_k(i)} \cdot \max\{0, C_{\pi_k(i+1)} - p_{\pi_k(s)} - d_{\pi_k(i+1)}\} - (f_{\pi_k(s)}(\mathbf{S}) + \sum_{i=s+1}^{n_k} f_{\pi_k(i)}(\mathbf{S}))$$

$$= \sum_{i=1}^{s-1} f_{\pi_k(i)}(\mathbf{S}) + \sum_{i=s+1}^{n_k-1} w_{\pi_k(i)} \cdot \max\{0, C_{\pi_k(i+1)} - p_{\pi_k(s)} - d_{\pi_k(i+1)}\}.$$

The last equality follows from the formula (16). A similar lemma will be proven for *l*-th machine. Let

$$\delta_{\mathbf{S}}^+ = w_{\pi_{\pi_l(s)}} \cdot \max\{0, C_{\pi_l(t-1)} + p_{\pi_k(s)} - d_{\pi_k(s}\}$$

$$+ \sum_{i=t}^{n_l} w_{\pi_k(i)} \cdot \max\{0, C_{\pi_l(i)} + p_{\pi_k(s)} - d_{\pi_l(i)}\} - \sum_{i=t}^{n_l} f_{\pi_l(i)}(\mathbf{S}). \tag{20}$$

Lemma 2. *If the solution* $\mathbf{S}' = \tau_l^k(s,t)(\mathbf{S})$, *then the cost (17) of tasks execution by l-th machine*

$$T_l(\mathbf{S}') = T_l(\mathbf{S}) + \delta_{\mathbf{S}}^+$$

Proof. Similarly as in the case of the proof of the Lemma 1

$$T_l(\mathbf{S}) = \sum_{i=1}^{t-1} f_{\pi_k(i)}(\mathbf{S}) + \sum_{i=t}^{n_k} f_{\pi_k(i)}(\mathbf{S}).$$

Then, using equality (18) and definition $\delta_{\mathbf{S}}^+$ we obtain

$$T_l(\mathbf{S}') = T_l(\mathbf{S}) + \delta_{\mathbf{S}}^+ = \sum_{i=1}^{t-1} f_{\pi_l(i)}(\mathbf{S}) + \sum_{i=t}^{n_l} f_{\pi_t(i)}(\mathbf{S})+$$

$$+ w_{\pi_{\pi_l(s)}} \cdot \max\{0, C_{\pi_l(t-1)} + p_{\pi_k(s)} - d_{\pi_k(s)}\}+$$

$$+ \sum_{i=t}^{n_l} w_{\pi_k(i)} \cdot \max\{0, C_{\pi_l(i)} + p_{\pi_k(s)} - d_{\pi_l(i)}\} - \sum_{i=t}^{n_l} f_{\pi_l(i)}(\mathbf{S}).$$

Let

$$\Delta_l^k(s,t)(\mathbf{S}) = \delta_{\mathbf{S}}^- + \delta_{\mathbf{S}}^+. \tag{21}$$

The value of this expression can be determined in time $O(n)$. It will be used to estimate the effectiveness (change of the criterion value) of the *transfer*-type move. It will allow us to determine the best solution from the neighborhood.

Theorem 3. *If* $\mathbf{S}$ *is a solution to the* **MTC** *problem and* $\mathbf{S}'$ *was generated from* $\mathbf{S}$ *by the transfer-type move* $\tau_l^k(s,t)$, *then the cost of tasks execution*

$$T(\mathbf{S}') = T(\mathbf{S}) + \Delta_l^k(s,t)(\mathbf{S}). \tag{22}$$

Proof. The proof results directly from the Lemma 1, 2 and the definition (22).

The value of the expression (22) can be computed in time $O(n)$. It will be used to quickly compute the value of the solution criteria generated by the *transfer*-type of move.

4 Elimination of Solutions

Here we will present some properties of the **MTC** problem that allow one to eliminate 'the worse' solutions from the neighborhood. This will reduce the number of elements in the neighborhood, and thus the time it takes to search.

Let $\mathbf{S}$ be some solution, and l some machine. For task v ($v \notin A_l$) transferred to machine l

$$\lambda_l(v) = \max\{j : C_{\pi_i(j-1)} + p_v > d_v, 1 \le j \le n_i\} - 1 \tag{23}$$

is the maximum number of positions in the permutation π_l on which the task v was transferred if not delayed (its penalty equals to 0). In definition (23) there is an assumption that $C_{\pi_i(0)} = 0$. For a fixed machine l and task $v \notin A_l$ parameter (23) can be determined in time $O(n)$. We will prove the theorem enabling the elimination of certain solutions from the neighborhood generated by transfer-type of moves.

Theorem 4. *Let* $\mathbf{S}(A, \pi)$ *be some solution to* **MTC** *problem. For any pair of moves* $\tau_l^k(r, s)$ *and* $\tau_l^k(r, t)$, *where* $k \neq l$, $k, l \in \mathcal{M}$, $1 \leq s < t \leq \lambda_k(\pi_k(r))$ *there is*

$$T(\tau_l^k(r, s)(\mathbf{S})) \geq T(\tau_l^k(r, t)(\mathbf{S})). \tag{24}$$

Proof. In order to simplify the notation, in the proof of the theorem we assume that the order of execution of tasks on the l-th machine is h element identity permutation $\pi = (1, 2, \ldots, h)$, and $T(\pi)$ is the cost of tasks execution on l-th machine in the order π (this is equivalent to the definition (3)). Further, by π^i ($1 \leq i \leq h$) we denote the permutation resulting from π by inserting into position i tasks $v = \pi_k(r)$, this is

$$\pi^i = (1, 2, \ldots, i = 1, v, i, i + 1, \ldots, h, \eta), \text{ where } \eta = h + 1.$$

We are considering two positions s and t ($1 \leq s < t \leq \lambda_l(v)$) in permutation π. By inserting the task v to permutation π respectively to position s and t we generate two permutations (order of tasks execution by machine l) π^s and π^t.

We are now calculating the cost of tasks execution by the machine l respectively for the order π^s and π^t.

$$T(\pi^s) = \sum_{i=1}^{s-1} f_i(C_{\pi^s(i)}) + f_s(C_{\pi^s(s)}) + \sum_{i=s+1}^{t} f_i(C_{\pi^s(i)}) + \sum_{i=t+1}^{\eta} f_i(C_{\pi^s(i)}). \tag{25}$$

Similarly, for permutation π^t

$$T(\pi^t) = \sum_{i=1}^{s-1} f_i(C_{\pi^t(i)}) + + \sum_{i=s+1}^{t-1} f_i(C_{\pi^t(i)}) + f_t(C_{\pi^t(t)}) + \sum_{i=t+1}^{\eta} f_i(C_{\pi^t(i)}). \tag{26}$$

From definition of both permutations

$$\pi^s(i) = \pi^t(i) \text{ and } C_s(i) = C_t(i), \text{ for } i = 1, 2, \ldots, s-1, t+1, t+2, \ldots, \eta.$$

Therefore, in expressions (25) and (26) we have the equality of components:

$$\sum_{i=1}^{s-1} f_i(C_{\pi^s(i)}) = \sum_{i=1}^{s-1} f_i(C_{\pi^t(i)}) \text{ and } \sum_{i=t+1}^{\eta} f_i(C_{\pi^s(i)}) = \sum_{i=t+1}^{\eta} f_i(C_{\pi^t(i)}.$$

Since $\pi^s(s) = v$ and the task v is not delayed (by assumption of the theorem $s \leq \lambda_l(v)$), thus $f_s(C_{\pi^s(s)}) = 0$. Similarly $f_t(C_{\pi^t(t)}) = 0$. The proof of the theorem thesis comes down to showing that $\sum_{i=s+1}^{t} f_i(C_{\pi^s(i)}) \geq \sum_{i=s}^{t-1} f_i(C_{\pi^t(i)})$. The tasks $s, s + 1, \ldots, t - 2, t - 1$ are placed in permutation π^t to positions $s, s+1, \ldots, t-2, t-1$. In turn, in permutation π^s they are on positions $s+1, s+2, \ldots, t-1, t$. This shift to the right by one position is due to the fact that in π^s is preceded by the task v.

Since $C_{\pi^s(i)} = C_{\pi^t(i-1)} + p_{v,}$, it's easy to show that

$$f_i(C_{\pi^s(i)}) \geq f_i(C_{\pi^t(i-1)}) \text{ for } i = s + 1, s + 2, \ldots, t.$$

Summing up these inequalities in sides, we finally get $\sum_{i=s+1}^{t} f_i(C_{\pi^s(i)}) \geq \sum_{i=s}^{t-1} f_i(C_{\pi^t(i)})$, which ends the proof of the theorem.

Using the above theorem, it is easy to prove the properties that allow one to eliminate certain transfer-type moves generating worse solutions from the neighborhood.

Property 1. (Elimination of solutions). If **S** is solution to **MTC** problem, k and l are machine numbers, and $\pi_k(s)$ a task transferred from the machine k to l by transfer-type move, then

$$\mathcal{T}(\tau_l^k(s,1)(\mathbf{S})) \geq \mathcal{T}(\tau_l^k(s,2)(\mathbf{S})) \geq, \ldots, \geq \mathcal{T}(\tau_l^k(s, \lambda_l(s))(\mathbf{S})).$$

Therefore, by designating the elements of the neighborhood for each task $\pi_k(s) \notin A_k$ it is possible to omit the moves from the set

$$\mathcal{R}_l^k(s) = \{\tau_l^k(s,1)(\mathbf{S}), \tau_l^k(s,2)(\mathbf{S}), \ldots, \tau_l^k(s, \lambda_l(s) - 1)(\mathbf{S})\}. \tag{27}$$

They generate solutions no better (of not lower value than the penalty function) than the move $\tau_l^k(s, \pi_l(\lambda_l(s))(\mathbf{S})$. This move will be called a *representative* of the moves from the set (27). Then

$$\mathcal{R}(\mathbf{S}) = \bigcup_{i=1}^{m} \bigcup_{\substack{j=1, \\ j \neq i}}^{m} \bigcup_{s=1}^{n_j} \mathcal{R}_j^i(\mathbf{S}). \tag{28}$$

Is the set of all the *transfer*-type of moves, which can be omitted.

5   Solution Method

In order to solve the problem considered at work there was the tabu search (**TS**) algorithm used for solving the single-machine problem $1||\sum w_i T_i$, adopted from the work of Bożejko et al. [2]. The essential component of the algorithm based on this method is a neighborhood – subset of the set of all solutions. In each iteration of the algorithm, the element with the minimum value of the criterion function is determined from the neighborhood. Generation procedure and searching the neighborhood have a decisive influence on the values of the determined solutions, the pace of convergence and time calculations. Below we describe the method of generating and searching the neighborhood in algorithm solving the **MTC** problem.

Let $\mathcal{I}(\mathbf{S})$ and $\Theta(\mathbf{S})$ be respectively, the set of all insert and transfer-type of moves for a certain solution $\mathbf{S} \in \mathcal{AP}$. Neighborhood $\mathbf{S}$ is the set of solutions

$$\mathcal{N}(\mathbf{S}) = \{\lambda(\mathbf{S}) : \lambda \in \mathcal{I}(\mathbf{S}) \cup \Theta(\mathbf{S})\}. \tag{29}$$

If by generating the neighborhood $\mathcal{N}(\mathbf{S})$ 'the worse' moves will be removed, i.e. moves from the set $\mathcal{R}(\mathbf{S})$ then we obtain a subneighborhood $\mathcal{N}_{sub}(\mathbf{S})$. When determine an element from the neighborhood, we also omit moves whose attributes are on the so-called tabu list.

For the $\mathbf{S} \in \mathcal{AP}$ solution, generating and searching the neighborhood is performed in two steps:

Step 1. Designate the best solution from the set generated by *insert*-type of moves, i.e. from the set of moves $\mathcal{I}(\mathbf{S})$.

Step 2. Designate the best solution from the set generated by *transfer*-type of moves, i.e. from the set of moves $\Theta(\mathbf{S})$.

If the solution in Step 1 is better than the $\mathbf{S}$ solution, then Step 2 is omitted.

In the description of the algorithm $\mathbf{S}$ is a starting solution, and $\mathbf{S}^*$ is the best designated solution.

Tabu Search (TS) algorithm

$\mathbf{S}^*:=\mathbf{S}$;
repeat
      Determine solution $\mathbf{S}'$ such that $\mathcal{T}(\mathbf{S}') = \min\{\mathcal{T}(\lambda(\mathbf{S})) : \lambda \in \mathcal{I}(\mathbf{S})\}$
      omitting moves whose attributes are on the tabu list;
      if ($\mathcal{T}(\mathbf{S}') < \mathcal{T}(\mathbf{S}^*)$) then $\mathbf{S}^* := \mathbf{S}'$; $\mathbf{S} := \mathbf{S}'$; else
      begin
        Determine solution $\mathbf{S}''$ such that $\mathcal{T}(\mathbf{S}'') = \min\{\mathcal{T}(\lambda(\mathbf{S})) : \lambda \in \Theta(\mathbf{S})\}$
        omitting moves whose attributes are on the tabu list;
        if ($\mathcal{T}(\mathbf{S}'') < \mathcal{T}(\mathbf{S}^*)$) then $\mathbf{S}^* := \mathbf{S}''$; $\mathbf{S} := \mathbf{S}''$;
      end{*else*}
      Update move's attributes on a tabu list;
until (*stop_condition*).

The above algorithm with subneighborhood $\mathcal{N}_{sub}(\mathbf{S})$ will be denoted by $\mathbf{TS}_{sub}$.

6   Computational Experiments

Variations of the problem are considered in the literature, e.g. [1,8], but the sizes of test data used do not correspond to the contemporary requirements of industrial practice ($n = 18$, $m = 4$ in [8] (B&B), $n = 500$, $m = 10$ in [1] (heuristics only). Therefore, we decided to propose new test examples of large sizes ($n \leq 1000$, $m \leq 20$). Therefore the test examples were generated similarly to those described in the work by Potts and Van Wassenhove [12]: parameters of each the tasks $i \in \mathcal{J}$ are the implementation of random variables of uniform distribution. The execution time p_i - on the interval $[1, 100]$, the weight of the penalty function w_i – on the interval $[1, 10]$, and the latest completion date d_i – on the interval $[P(1-TF-RDD/2), P(1-TF+RDD/2)]$, $P = \lceil (\sum_{i=1}^{n} p_i)/m + 1 \rceil$, RDD, $TF \in \{0.2, 0.4, 0.6, 0.8, 1.0\}$.

For each of 25 pairs of RDD and TF parameter values there were 5 examples generated. So, for each pair n ($n = 50, 100, 500, 1000$) and m ($m = 2, 5, 10, 20$), there were 125 examples of varying difficulty determined - a total of 2,000 examples was reached.

The percentage relative deviation (PRD) was used to evaluate the algorithms: $\delta = \frac{T_{ref}-T_{alg}}{T_{ref}} \cdot 100\%$, where: T_{ref} – value of the reference solution, T_{TS} – the value of the solution designated by tested algorithm. There are no reference data in the literature. Therefore, there will be a comparison of the solutions determined by the **TS** algorithm and the results of a constructive algorithm.

Constructive GRC algorithm

$n_k = 0$; $\pi_k := ()$, $k = 1, 2, \ldots, m$;
Number tasks in such a way that $p_1/w_1 \geq p_2/w_2 \geq, \ldots, \geq p_n/w_n$;
for $l := 1$ to n do begin
       Determine machine k such that $T(\pi_k) = \min\{T(\pi_i) : i = 1, 2, \ldots, m\}$;
       Insert the task l on position in permutation π_k so that the penalty
           for the tasks execution tardiness on machine k was minimal;
   end

The computational complexity of the algorithm is $O(mn^2)$.

Algorithms **GRC** and **TS** were programmed in C ++ language and run on a personal computer with an Intel processor Core i7 3.5 GHz. First, the effectiveness of the **GRC** algorithm was tested, the solutions of which will be the reference when evaluating the results of the **TS** algorithm. In the work of Bożejko et al. [2] presents the tabu search algorithm for solving the $1||\sum w_i T_i$ problem and the results of computational experiments. On the set of these examples, the mean relative error of the **GRC** algorithm is 13.7%.

On the basis of the preliminary calculations of the **TS** algorithm, the following parameter values were determined: the length of the tabu list $L_{LT} = 11$, the maximum number of iterations $L_{iter} = 2n$. The main calculations were made on the examples described at the beginning of this section. The mean relative deviation for each group of examples is presented in Table 1.

This error (the average relative improvement of the solutions determined by the **GRC** algorithm) for all computed examples is 10.4%. The lowest value 7.2% is for $n = 20$ and $m = 2$ and it grows with increasing number of tasks and machines. In general, the average errors do not vary much. This is due to the fact that the P parameter, on which the sizes of the intervals from which

Table 1. The mean relative deviation of the **TS** algorithm in reference to **GRC**.

n	$m = 2$	$m = 5$	$m = 10$	$m = 20$
20	7.2	7.8	9.4	12.7
50	8.0	7.9	8.6	12.6
100	7.9	8.2	9.8	11.3
500	9.3	9.1	1.5	14.1
1000	11.4	12.0	11.6	16.5
Average	8.7	9.0	10.3	13.4

the critical lines d_i are drawn, depends on the number of machines. The total computation time for all 2000 examples was 35.5 min.

The mean relative error of the $\mathbf{TS}_{sub}$ algorithm is 10.3%. Hence, it is almost identical to the **TS** algorithm. However, the total computation time decreased significantly by 21.5%. Thus, the use of elimination properties in the construction of the algorithm resulted in a significant improvement.

7   Comments and Conclusions

In the paper there was an NP-hard task scheduling problem considered with the earliest requested completion dates. The tasks should be assigned to machines and the order of their execution should be determined so that the sum of penalties for delays is minimal. Quick methods of calculating the value of the criterion function and properties have been introduced to enable the elimination of 'worse' solutions. The conducted computational experiments have shown that the use of elimination properties in the construction of the tabu search algorithm significantly reduces the computation time.

Acknowledgments. The paper was partially supported by the National Science Centre of Poland, grant OPUS no. 2017/25/B/ST7/02181.

References

1. Alidaee, B., Rosa, D.: Scheduling parallel machines to minimize total weighted and unweighted tardiness. Comput. Oper. Res. **24**(8), 775–788 (1997)
2. Bożejko, W., Grabowski, J., Wodecki, M.: Block approach tabu search algorithm for single machine total weighted tardiness problem. Comput. Ind. Eng. **50**(1–2), 1–14 (2006)
3. Bożejko, W., Uchroński, M., Wodecki, M.: Blocks for two-machines total weighted tardiness flow shop scheduling problem. Bull. Pol. Acad. Sci. Tech. Sci. **68**(1), 31–41 (2020)
4. Bulfin, R.L., Hallah, R.: Minimizing the weighted number of tardy tasks on two-machine flow shop. Comput. Oper. Res. **30**, 1887–1900 (2003)

5. Karabulut, K.: A hybrid iterated greedy algorithm for total tardiness minimization in permutation flowshops. Comput. Ind. Eng. **98**, 300–307 (2016)
6. Kayvanfar, V., Komaki, G.H.M., Aalaei, A., Zandieh, M.: Minimizing total tardiness and earliness on unrelated parallel machines with controllable processing times. Comput. Oper. Res. **41**, 31–43 (2014)
7. Lenstra, J.K., Rinnooy Kan, A.G.H., Brucker, P.: Complexity of machine scheduling problems. Ann. Discrete Math. **1**, 343–362 (1977)
8. Liaw, C.-F., Lin, Y.-K., Cheng, C.-Y., Chen, M.: Scheduling unrelated parallel machines to minimize total weighted tardiness. Comput. Oper. Res. **30**(12), 1777–1789 (2003)
9. Liu, C.: A hybrid genetic algorithm to minimize total tardiness for unrelated parallel machine scheduling with precedence constraints. Math. Probl. Eng. (2013). ID 537127
10. Park, J.M., Choi, B.C., Min, Y., Kim, K.M.: Two-machine ordered flow shop scheduling with generalized due dates. Asia-Pac. J. Oper. Res. **37**(01), 1950032 (2020)
11. Ojstersek, R., Tang, M., Buchmeister, B.: Due date optimization in multi-objective scheduling of flexible job shop production. Adv. Prod. Eng. Manag. **15**(4), 481–492 (2020)
12. Potts, C.N., Van Wassenhove, L.N.: A decomposition algorithm for the single machine total tardiness problem. Oper. Res. Lett. **1**, 177–181 (1982)
13. Toncovicha, A., Rossit, D., Frutos, M., Rossit, D.G., Daniel, A.: Solving a multi-objective manufacturing cell scheduling problem with the consideration of warehouses using a simulated annealing based procedure. Int. J. Ind. Eng. Comput. **10**, 1–16 (2019)
14. Wodecki, M.: A block branch-and-bound parallel algorithm for single-machine total weighted tardiness problem. Adv. Manuf. Technol. **37**, 996–1004 (2008)

Discovering the Influence of Interruptions in Cycling Training: A Data Science Study

Alen Rajšp[(✉)] and Iztok Fister Jr.

Faculty of Electrical Engineering and Computer Science, University of Maribor,
2000 Maribor, Slovenia
{alen.rajsp,iztok.fister1}@um.si

Abstract. The usage of wearables in different sports has resulted in the potential of recording vast amounts of data that allow us to dive even deeper into sports training. This paper provides a novel approach to classifying stoppage events in cycling, and shows an analysis of interruptions in training that are caused when a cyclist encounters a road intersection where he/she must stop while cycling on the track. From 2,629 recorded cycling training sessions 3,731 viable intersection events were identified on which analysis was performed of heart-rate and speed data. It was discovered that individual intersections took an average of 4.08 s, affecting the speed and heart-rate of the cyclist before and after the event. We've also discovered that, after the intersection disruptions, the speed of the cyclist decreased and his heart-rate increased in comparison to his pre intersection event values.

Keywords: Artificial sport trainer · Data science · Smart sports training · Cycling · Wearables

1 Introduction

Cycling is considered as one of the most pleasant ways of doing recreation. It is definitely popular all over the world. People are cycling for transportation, commuting and maintaining a healthy lifestyle on the one hand, while some others are also cycling competitively, participating in various cycling race competitions. The professional sports scene has become more and more competitive in recent years, and various new training regimes and tactics have been discovered incorporating digital technology [17]. This is also, in part, due to the introduction of the Internet of Things (IoT) devices to the Sports Training domain [7]. The data generated by such devices are, nowadays, used for building Machine Learning models, which can be used to optimize training and actual competition performance in sports [13]. A recent survey [13] identified that many sports have already undergone the new smart training revolution. Cycling is definitely one of them, where researchers developed various applications for tackling the process of sport training using more data-driven approaches.

© Springer Nature Switzerland AG 2021
M. Paszynski et al. (Eds.): ICCS 2021, LNCS 12745, pp. 420–432, 2021.
https://doi.org/10.1007/978-3-030-77970-2_32

Current research indicates that the focus on the researched domain has focused on planning fitness exercises [5], generating training plans [14], planning match day strategies [10] and generating eating plans [2].

In this paper, we go a step forward, and approach the process of sport training from another perspective. We focus on the role of interruptions within a sports activity. The characteristic of carrying out cycling training or activity is that it can take a very long time (up to 7 h or even more), and during this time interruptions are inevitable.

Therefore, it is important to investigate how cyclists approach the interruptions and what are the side effects of those interruptions. Usually, interruptions are not planned in advance, but they appear spontaneously.

We investigated the actual interruptions during a training exercise. In fact, studies investigating training interruption in other sports were only related to the interruption of training routines over prolonged periods of missed training days [1,8,9,15,16] and not interruption during the training itself.

The events which interrupt the standard training procedure certainly influence the training performance and results of the training conducted. Since the interruptions in training have not been studied thoroughly yet, their frequency is not yet known. The interruptions are certainly an important factor in cycling since, due to the outdoor nature of training and the fact that the training happens on public roads and paths, the performance of the cyclist also depends on his highly volatile and changing environment in contrast with other sports where the environment can be controlled fully (e.g. swimming in indoor pools, training sports in gyms), or at least partially (e.g. soccer training in a closed stadium).

In a nutshell, the purpose of this paper is to propose a methodology for discovering the influence of interruptions in cycling training's on the athletes' performances. The main contributions of this paper are summarized as follows:

- proposal of taxonomy of cycling training interruptions,
- a novel algorithm proposal for detection of stoppage interruption events, and
- data analysis of effects the interruption events have on the training session.

2   Interruptions in Cycling Training

For the purpose of this paper, interruptions in cycling training relate to interruptions during the actual training which were not anticipated in advance, and not the interruptions [8] which are sometimes defined as missed training days. Since no previous works on interruption detection in cycling training have been identified, a taxonomy of interruptions was constructed, based on the general cause for the interruption. As such, we divided the most probable interruptions which can impact the cyclists' training into four main categories: **Environmental factors**, which are factors over which the cyclist has no influence, or his influence was limited to the planning part of the exercise; **Biological factors**, which are factors related to the physical condition and current state of the athlete; **Equipment factors** that can be semi-controlled by proper equipment maintenance and using quality equipment, and **Other factors** which are all the other factors which may or may not be controlled and predicted. The aforementioned factors and their sub-factors are shown in Fig. 1.

Environmental factors are related to: <u>Intersections</u>*[1] which are one of the most common causes for cycling interruptions. This means that traffic lights, priority roads and roundabouts may force the cyclist to reduce his speed or come to a complete stop; <u>Poor track conditions</u> resulting from unexpected potholes, obstacles and traffic on the road; <u>Weather factors</u>, unfavorable weather conditions may interrupt the training. Weather conditions may impact visibility (e.g. fog), decrease the stability of the cyclist on the track (e.g. snow, rain, wind, road ice), or increase the workload of the cyclist (e.g. high humidity, high temperatures, low temperatures) and also unfavorable vision conditions (e.g. sunset, sunrise) which may block the view of the cyclist.

Biological factors which are factors that can be split into **Need fulfilment** which is the result of an unfulfilled need, such as tiredness, thirst, hunger and when nature calls; and **Health factors** that can interrupt the training, and may be a result of an injury sustained during the training, feeling unwell and falling, when the cyclist loses control of his bicycle.

Equipment factors which relate mostly to **Defects and failures**, that may be caused by a flat tire or failure of the bicycle's mechanical parts (e.g. gearbox failure, detachment of the chain).

Other factors which can be grouped into **Social factors** resulting mostly from calling and receiving phone calls, meeting a known person while training and stopping to greet them, but may also be caused when a cyclist is stopped

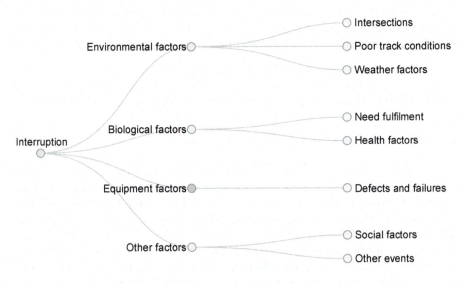

Fig. 1. Taxonomy of interruption events

---

[1] Interruption factors marked with asterisk (*) were the ones monitored in our research.

by the authorities for a traffic control. There are countless *Other events*, that may interrupt the training, such as issues with motivation, forgetting something at home (e.g. wallet with documents).

3   Materials and Methods

3.1   OpenStreetMap

OpenStreetMap [12] is a project that creates and distributes free geographic data for the world. The map data are saved and distributed in a collection of XML documents which consist of four main elements [11], namely **nodes**, **ways**, **relations** and **tags**. **Nodes** represent specific points (locations) on planet Earth. Each node has an id, latitude and longitude. **Ways** represent ordered lists between at least 2 and up to 2,000 nodes defining a poly-line. Ways are used to represent rivers, roads and areas (e.g. forest, municipality, building, etc.) If something is described by a boundary of more than 2,000 nodes then multiple ways are combined with relations. **Relations** are used to describe a relationship between two or more data elements. Tags describe the purpose of the relation. These relations may be a route relation (on which the roads are defined in a way), turn restriction, direction restriction, etc. **Tags** are data elements which can be included in all of the other elements, and are used to give meaning to the attached elements.

3.2   Method for Interruption Detection

All the methods described refer to detection and classification of interruption events in cycling sports training data. The detected interruption events are classified as events where speed dropped below 2 km/h, or the cyclist came to a complete stop. The data used were in the form of TCX [3] and GPX [4] training records.

The initial data pre-processing was divided in two parts. In the Sect. 3.3 the algorithm for detecting interruptions from exercise data is described, and in the Sect. 3.4 the algorithm is described for detecting the nearest intersections from the interruption locations. A visualization of the event data is seen in Fig. 2, indexes 1–60 represent the speed and heart-rate at 1 to 60 s pre and post the

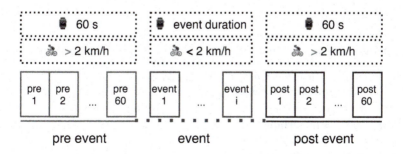

Fig. 2. Combined event data.

event. For the event, the i represents the duration of the event, since the events lasted different amounts of time. The pre and post event data points all span 60 s before and after the event. In each of those points the speed is above 2 km/h, while the length of the event itself varies, since stoppage by each intersection event varied, and some demanded more time off than others. It should be noted that when comparing values between pre and post event data, a pre_1 point can be compared to a $post_{60}$ point, since they both span the same amount of time from the actual interruption event.

3.3 Algorithm for Interruption Detection

The goal of this algorithm was to detect interruption events where speed dropped below 2 km/h. The input of the algorithm is a list of lines between different GPX/TCX points, and the output of the algorithm is the list of detected stoppage events. Each event consists of the pre-event (what happened before the

---

Algorithm 1: Algorithm for detecting exercise stoppage events

---

Input: List of lines between GPX/TCX points (two succeeding track points are connected to calculate their average speed and heart-rate)

$$L = (l_{x_1 x_2}, l_{x_2 x_3}, ..., l_{x_{n-1} x_{n-1}}), min\_speed, time\_interval$$

Output: List of detected stoppage events $E = (e_1, e_2, ...e_n)$,

$$e_n = (en_{pre}, en_{mid}, en_{post})$$

1 **for** $i=1$ **to** $i=n-1$ **do**
2      $event = false$
3      $i = 1$
4   **while** $L[i]_{speed} \leq SPEED\_MIN$ **and** $i \leq n$ **do**
         // Add to main event (actual stoppage)
5          $event = true$
6          $en_{mid} = insertToEn_{mid}(l[i])$
7          $i = i + 1$
8      $i_{postEvent} = i$
9      $timeStamp = time(L[i])$
10   **while** $i_{postEvent} \leq n - 1$ **and** $timeStamp \leq time(l[i]) + timeInterval$ **do**
         // Add to post event event (what happens after the stoppage)
11          $en_{post} = insertToEn_{post}(l[i_{postEvent}])$
12          $timeStamp = time(L[i_{postEvent}])$
13          $i_{postEvent} = i_{postEvent} + 1$
14      $timeStampStart = time(i_{preEven})$
15   **while** $i_{preEvent} \geq 1$ **and** $timeStamp \geq timeStampStart - timeInterval$ **do**
         // Add to pre event event (what happens before the stoppage)
16          $en_{mid} = InsertToStartEn_{pre}(l[i_{preEvent}])$
         $timeStamp = time(L[i_{preEvent}])$
17          $i_{preEvent} = i_{preEvent} - 1$
18   **if** $event == true$ **then**
19          $e = (en_{pre}, en_{mid}, en_{post})$
20          $addToEventList(e)$

---

event), event (the time period during which speed was below 2 km/h), and post event (what happened after the event). The boundaries of pre and post event are defined by the *time_interval* variable, which defines the maximum difference between the time of the event and the surrounding data. The algorithm goes through each point in the training data, and detects and groups those where the cyclists dropped below the determined minimum speed (*min_speed* variable), after such an event was detected. After the event is detected in the first sub-loop (lines 1–6), in the following two sub loops (9–17) pre and post events are identified, it is necessary to compare the time stamps, since not all recording equipment records data at the same frequency.

3.4 Algorithm for Detecting Nearby Road Intersections

The input of the algorithm are the identified events of the previous algorithm. For each event, thelocations of said event points are inspected. In addition to that, OpenStreetMap data from Geofabrik [6], described in Sect. 3.1, was needed to give meaning to the coordinates, *maximum_distance* defined the maximum distance between the intersection and an event location to still be identified as such.

Algorithm 2: Algorithm for detecting nearby road intersections

---

Input: List of events with locations $E = (e_1, e_2, e_3...e_n)$, OpenStreetMap data, max_distance

Output: List of events with nearby intersections $E = e_{1*}, e_{2*}, e_{3*}...e_{n*}$

1   **for** *i=1 to i=n-1* **do**

2     $nearbyIntersection = false$ // each location has altitude and longitude

3     $location = getLocation(e_i)$

4     $topBorderLongitude = location.longitude + 0.005$

5     $bottomBorderLongitude = location.longitude - 0.005$

6     $topBorderAltitude = location.altitude + 0.005$

7     $bottomBorderAltitude = location.altitude - 0.005$

    // get intersections in box is a query to the openstreet map to discover all intersections in the box with maximum and minimums of longitudes and altitudes

8     $intersectionsList =$ $getIntersectionsInBox(topLong, bottomLong, topAlt, bottomAlt)$

9     $closestDistance = \infty$

10    $closestIntersection = null$

11   **for** *intersection* **in** *intersectionsList* **do**

12      $distance = calculateGeodesic(location, intersection.location)$

13   **if** $distance \leq closestDistance$ **and** $distance < max\_distance$ **then**

14        $closestDistance = distance$

15        $closestIntersection = intersection$

16    $e_{i*} = e_i \; e_{i*}.intersection = closestIntersection$

The output of this algorithm is a list of all events of type intersection, with their corresponding intersections.

The output was also visualized on the training route visualization, and its display is shown in Fig. 3, where the event is marked with the green box, and the identified closest intersection, which was identified by the algorithm, is shown by a blue circle. The direction of the route is shown by semi-transparent blue arrows. Each other color coded (in reference to speed) circle represents a recorded point of the exercise. It can be seen that the speed when approaching the intersection drops rapidly, and reaches the lowest point at the intersection, where the cyclist must take extra precautions, and increases rapidly again after leaving the intersection.

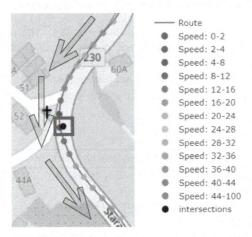

Fig. 3. Visualization of an stoppage event with a known intersection cause (Color figure online)

The algorithm receives the list of all identified events, and then checks for each of the events if there are any intersections in the near vicinity (*latitude* ± 0.005, *longitude*±0.005). It then goes through all the identified intersections and calculates the distance between the event and the intersection.

After each distance is calculated it is compared if it is smaller than the previous closest distance and the maximum distance. If both checks are returned true, this becomes the new closest intersection. After all the intersections are checked, and if any with the matching conditions were found, the closest intersection is added to the event.

3.5   Algorithm for Filtering Viable Exercise Stoppage Events

Further filtering of viable events among intersection events was conducted to determine the actual effects the event had on the pre-event and post-event performances. Only the events which had no additional stoppage events in the surrounding 60 s before and after the actual interruption were considered, which means that we eliminated any events that were less than 60 s apart themselves. The input of this filtering algorithm was a list of all previously detected intersection events, and the output were the viable events. In a loop each event had it's pre-event and post-event points checked. Had no drops of speed (*stoppage events*) occurred, the event was considered viable and added to the list for further analysis.

4   Experiments and Results

The purpose of the experiments was to discover the effects the interruptions had on the heart rate and pace (speed) of the cyclist before and after the event happened. We wanted to explore if the cyclist speed and heart-rate changed after the event compared to the value beforehand.

A total of 2629 training records were examined from 7 different athletes, where each athlete had a minimum of 148 training sessions recorded. Only data from athletes training exclusively in the Cycling and Road Cycling training category were used, and Mountain biking, Triathlon and Multi-sport athlete training sessions were excluded. A total of 3,914 intersection events were identified. This does not mean that the cyclist rode only past 3,914 intersections, but that not all intersections resulted in a drop of speed below 2 km/h. The events were then filtered further, so that only events where, in the preceding and succeeding minutes no other stoppage events occurred; this led to a total of 3,731 events which were then analysed further. Analysis of intersection events showed that an average intersection event lasted an average of 4.08 s. The analysis of viable events is presented in this Section. The values[2] were analyzed in reference to changes in heart-rate and speed as a result of an intersection event. Special attention was focused on how to represent data properly. Because of heart-rate and cycling speed differences between athletes and individual intersections events, the data of each individual event were first standardized by calculating the arithmetic mean speed and heart-rate of each individual intersection event. After that, all the values were divided by their respective means, so that speed and heart-rate in reference to average speed were presented, as shown in Eq. 1.

$$a_{standardized} = \frac{a_{actual}}{Mean} \tag{1}$$

where:

$a_{standardized}$ = standardized speed ratio with respect to mean speed recorded
a_{actual} = actual speed at a point
a_{Mean} = calculated mean speed of the pre and post event points.

These standardized values were than compared at 20, 40 and 60 s pre/post intervals, so that we could investigate how the intersection event influenced the cyclist. The difference between these values represented a change in value in

---

[2] When referring to values the same calculation was done for heart-rate and speed.

respect to the calculated means (e.g. so a difference between two points in respect to the average value). The Eq. 2 describes the calculated delta values, which were then used in individual Sect. 4.1 Heart-rate analysis and Sect. 4.2 Speed analysis subsections. The t used in post-speed value was deducted from 60, since we wanted to get the values at the same interval from the pre-post events.

$$\Delta a = a_{post\_event}(t) - a_{pre\_event}(60 - t) \tag{2}$$

where:

$$t = 0\text{–}60\,\text{s, point of time at the recorded event}$$
$$a_{post\_event}(60 - t) = \text{calculated mean value at the } t \text{ time of the post event}$$
$$a_{pre\_event}(t) = \text{calculated mean value at the } t \text{ time of the pre event}$$

For the resulting values, the statistics are presented in Table 1. The values are presented to an accuracy of 3 significant digits. The calculated values can be higher than 1, because they represent differences between average values and not percentage ratios, so as not to skew the values in any direction.

4.1   Speed Analysis

The aim of the speed analysis subsection was to discover what the relationship between interruptions and speed was, and if the speed changed in any meaningful way after the event compared to the previous baseline. What was discovered was that, in 54.8% of all cases (20 s - 54.7, 40 s - 54.2, 60 s - 54.8), the speed at the end of the post-event was **lower** than the speed at the start of the pre-event. This may mean that either (1) The athlete needs more than one minute to accelerate to the previous speed, or that (2) The extra work caused to accelerate from the stoppage event tired out the cyclist temporarily, resulting in his lag in speed. We found some evidence pointing out that the real cause is the explanation offered in the 2nd scenario, as explained in Subsect. 4.2. It is also interesting to observe the speed of the cyclist and at which point it started decreasing, as shown in Fig. 4. The pre and post event values presented on the chart are shown for median, the first quartile (q1) and the third quartile (q3), and the values are normalized with reference to average speeds. It can be seen that the majority of the steepest speed drop happened in the last 20 s before the event.

4.2   Heart-Rate Analysis

The aim of the heart-rate analysis subsection was to discover what the relationship between interruptions and heart-rate was, and if the stoppage event influenced the heart-rate compared to the previous baseline. We discovered that, in 51.4% of all events, the heart rate at the end of the post-event was **higher** than the heart rate at the start of the pre-event, for the 60 s comparison interval. It was, however, higher in only 46.9% of cases for the 20 and 48.2% of cases for the 40 s interval. This further points towards the proposed explanation offered in Subsect. 4.1. From the perspective of actions performed, the heart rate was lower

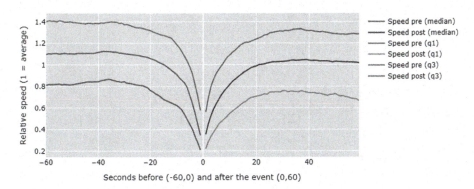

Fig. 4. Line plot speed

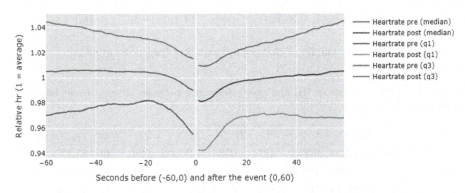

Fig. 5. Line plot of normalized heart-rate averages

immediately after the event, because the interruption (stoppage) also served as a very short rest for the cyclist, lowering his heart rate as seen on Fig. 5.

The heart rate increased, however, as the cyclist had to spend energy to accelerate, and at the end of the event was higher than before, while the speed at the end was, on average, still lower than at the start. This can also be seen if heart rate values from all events are categorized by seconds before and after the event, and their first quartile (q1), median and third quartile (q3) values are plotted as shown in Fig. 5, and the HR appears to be slightly higher at the median and third quartile values at the end of the event.

It can also be seen again that the interruption slightly lowers (notice the gap between the end of the pre-event line and the start of the post-event line) the heart-rate, as it provides a short break for the cyclist.

The slight increase of heart rate we are discussing, however, is only true for an average case of interruption and not when individual heart-rates are plotted on an line chart, as seen in Fig. 6. We deliberately chose some events where the heart-rate was lower after the conclusion of the post event than at the start of the pre event. But what can be seen is that, after some events, the heart rate

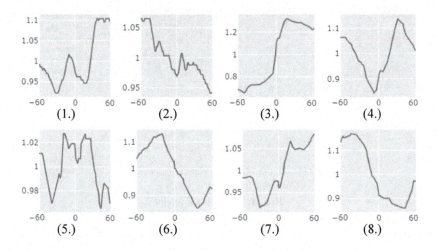

Fig. 6. Examples of changes in heart-rate (pre/post) event

Table 1. Statistical data of HR and speed difference between pre-post event based on standardised values

		Mean	σ	min	Q_1	Q_2	Q_3	max	$x_{post} < x_{pre}$
Δhr	20 s	−.00720	.0963	−.487	−.0566	−.00600	.0393	.568	53.1%
	40 s	−.00348	.123	−.508	−.0623	−.00397	.0517	.563	51.8%
	60 s	.00246	.138	−.650	−.0653	0	.0628	.597	48.4%
$\Delta speed$	20 s	−.0789	.729	−2.51	−.522	−.0706	.387	2.39	54.7%
	40 s	−.0776	.738	−2.22	−.532	−.0646	.370	2.55	54.2%
	60 s	−.0952	.801	−3.13	−.568	−.0640	.346	3.72	54.8%

might actually decrease (cases 2, 4, 5, 6, 8). This is completely normal, and might happen for a number of cases that depend on the environment (e.g. the cyclist was riding on a slight downhill slope; the cycle track entered a road in a forest and the surrounding temperature dropped). Over a large number of cases these favorable or unfavorable environments even out, and the slightly increased value can still be observed.

5   Conclusion

Nowadays, many athletes use various sport trackers for tracking the sport activities. During the activity, many parameters are monitored, which can later be analyzed in order to improve the performance of an athlete.

In this paper, the data monitored by sport trackers were used for the investigation of the interruptions that appear during the sport training resulting from intersections on the cyclist training track. A new method was proposed for identification of stoppage events and discovering intersection events during cycling training. Our research showed that an average intersection provided 4.08 s of stoppage time, which meant that, on average, these were only short stops, where the cyclist stopped gradually before the intersection, possibly due to safety concerns, and shortly resumed with his cycling training.

We've also discovered that in 54.8% of all intersection events (20 s - 54.7, 40 s - 54.2, 60 s - 54.8) the speed at the same interval, i.e. 20 s before and after the event, was **lower**, which shows that intersection interruptions result in lower speed of the cyclist after the event has concluded. We also discovered that immediately after the event the heart-rate was higher in only 46.9% of cases for the 20 and 48.2% of cases for the 40 s pre versus post interval. This ratio, however, changed at the end of the one minute interval, where, in 51.4% of all intersection events, the post event heart rate was **higher** than the heart rate at the start of the pre event. This implies that, while the short rest induced by the intersection interruption at first reduced the heart rate of the cyclist, the extra effort and work to try and attain the previous cycling speed resulted in a higher ending heart rate than before the interruption.

In the future, our approach will also be applied for the analysis of interruptions that appear in other sports, i.e. running. By the same token, we are also planning to conduct the study for analyzing the interruptions that are the result of poor track conditions, and also analysis of the effect weather conditions pose on the individual training sessions.

References

1. Comyns, T.M., Harrison, A.J., Hennessy, L.K., Jensen, R.L.: The optimal complex training rest interval for athletes from anaerobic sports. J. Strength Conditioning Res. **20**(3), 471–476 (2006). https://doi.org/10.1519/00124278-200608000-00003
2. Fister, D., Fister, I., Rauter, S.: Generating eating plans for athletes using the particle swarm optimization. In: CINTI 2016–Proceedings of the 17th IEEE International Symposium on Computational Intelligence and Informatics, pp. 193–198. Institute of Electrical and Electronics Engineers Inc., February 2017. https://doi.org/10.1109/CINTI.2016.7846402
3. Fister, I., Rauter, S., Fister, D.: A collection of sport activity datasets for data analysis and data mining 2017a. Technical report, Faculty of Electrical Engineering and Computer Science (UM FERI) (2017). http://www.academictorrents.com
4. Fister Jr., I., Rauter, S., Fister, D., Fister, I.: A collection of sport activity datasets with an emphasis on powermeter data
5. Fister Jr, I., Rauter, S., Ljubič Fister, K., Fister, D., Fister, I.: Planning fitness training sessions using the bat algorithm. In: 15th Conference on ITAT 2015. CEUR Workshop Proceedings, vol. 1422. pp. 121–126 (2015). ISSN 1613-0073
6. Geofabrik GmbH: Geofabrik Download Server (2020). https://download.geofabrik.de/
7. Goasduff, L.: Professional Sports Going Digital by Embracing ICT - Smarter With Gartner (2016). https://www.gartner.com/smarterwithgartner/professional-sports-going-digital-by-embracing-ict/

8. Hedrick, A.: Learning from each other: missed training days. Strength Condition-ing J. **27**(6), 87–89 (2015). https://insights.ovid.com/strength-conditioning/scjr/2005/12/000/learning-missed-training-days/15/00126548

9. Kovacs, M.S., Pritchett, R., Wickwire, P.J., Green, J.M., Bishop, P.: Physi-cal performance changes after unsupervised training during the autumn/spring semester break in competitive tennis players. Br. J. Sports Med. **41**(11), 705–710 (2007). https://doi.org/10.1136/bjsm.2007.035436. https://www.ncbi.nlm.nih.gov/pmc/articles/PMC2465299/

10. Ofoghi, B., Zeleznikow, J., MacMahon, C., Dwyer, D.: Supporting athlete selection and strategic planning in track cycling omnium: a statistical and machine learning approach. Inf. Sci. **233**, 200–213 (2013). https://doi.org/10.1016/j.ins.2012.12.050. http://www.sciencedirect.com/science/article/pii/S0020025513000431

11. OpenStreetMap Community: Elements - OpenStreetMap Wiki (2020). https://wiki.openstreetmap.org/wiki/Elements

12. OpenStreetMap Community: OpenStreetMap Wiki (2020). https://wiki.openstreetmap.org/wiki/Main_Page

13. Rajšp, A., Fister, I.: A systematic literature review of intelligent data analysis methods for smart sport training. Appl. Sci. **10**(9), 3013 (2020). https://doi.org/10.3390/app10093013. https://www.mdpi.com/2076-3417/10/9/3013

14. Silacci, A., Khaled, O.A., Mugellini, E., Caon, M.: Designing an e-coach to tailor training plans for road cyclists. In: Ahram, T., Karwowski, W., Pickl, S., Taiar, R. (eds.) IHSED 2019. AISC, vol. 1026, pp. 671–677. Springer, Cham (2020). https://doi.org/10.1007/978-3-030-27928-8_102

15. Stokes, K.A., et al.: Returning to play after prolonged training restrictions in pro-fessional collision sports. Int. J. Sports Med. (2020). https://doi.org/10.1055/a-1180-3692

16. Tremblay, A., Nadeau, A., Fournier, G., Bouchard, C.: Effect of a three-day inter-ruption of exercise-training on resting metabolic rate and glucose-induced ther-mogenesis in training individuals. Int. J. Obes. **12**(2), 163–168 (1988). https://europepmc.org/article/med/3290133

17. Xiao, X., et al.: Sports digitalization: an overview and a research agenda. In: 38th International Conference on Information Systems, ICIS 2017, December 2017

Analysis of Complex Partial Seizure Using Non-linear Duffing Van der Pol Oscillator Model

Beata Szuflitowska$^{(\boxtimes)}$ and Przemyslaw Orlowski

West Pomeranian University of Technology in Szczecin, Sikorskiego 37,
70-313 Szczecin, Poland
bszuflitowska@zut.edu.pl

Abstract. Complex partial seizures belong to the most common type of epileptic seizures. The main purpose of the case study is the application of the Van der Pol model oscillator to study brain activity during temporal left lobe seizures. The oscillator is characterized by three pairs of parameters: linear and two nonlinear, cubic and Van der Pol damping. The optimization based on the normalized power spectra of model output and real EEG signal is performed using a genetic algorithm. The results suggest that the estimated parameter values change during the course of the seizure, according to changes in brain waves generation. In the article, based on values of sensitivity factor of parameters, and, sample entropy non-stationary of considered seizure phases are analyzed. The onset of the seizure and the tangled stage belongs to strongly non-stationary processes.

Keywords: Van der Pol oscillator · EEG · Parameter estimation · Biological signal

1 Introduction

Electroencephalography (EEG) reflects the averaged electrical activity of neurons associated with different neural processing placed in different brain regions and structures [1, 2]. The International Federation of Societies for Electroencephalography and Clinical Neurophysiology considers EEG as non-invasive, safe for human health and easily controlled by clinicians technique [2]. The EEG signals belong to non-stationary and quasi-rhythmic signals that produce contained oscillations [3–5]. The main purpose of EEG measurement is the diagnosis of epilepsy [6, 7]. The epilepsy is theoretically characterized by abnormal synchronization between brain regions. In 2017, the International League Against Epilepsy (ILAE) released a new classification of seizure types, including focal motor and non-motor onset, generalized motor, and absences, unknown motor and non-motor onset and the last unclassified types [8]. Time series during some epileptic seizures or Parkinson's disease are much more ordered oscillatory than in healthy records [2, 5, 7, 9]. EEG signals have been studied in literature as a Bag-of-Words modelas a random and the back propagation (BP) neural networks, and coupled oscillators [10–12]. Interesting results related to modelling of a biological control system using a system composed of two coupled internal van der Pol oscillators [13]. A classical van

© Springer Nature Switzerland AG 2021
M. Paszynski et al. (Eds.): ICCS 2021, LNCS 12745, pp. 433–440, 2021.
https://doi.org/10.1007/978-3-030-77970-2_33

der Pol attractors are applied to distinguish between chaotic and stochastic behaviors of stationary EEG recorded from five normal subjects. The modification of Van der Pol oscillators,i.e. a generalized Van der Pol equation with fractional-order derivative and parametric excitation is derived from the Fitz–Hugh–Nagumo equations or the Wilson and Cowan model was considered as might be an efficient tool to control the dynamics of the action potentials [14, 15]. Even more popular in the analysis of EEG time series is duffing oscillator alone or in combination with Holmes and Lorenz oscillators [13]. Based on the results presented in the literature using these oscillators Ghorbanian at all proposed a coupled duffing Van der Pol Oscillator Model to distinguish two states healthy and Alzheimer's disease [2, 16]. While analyzing the literature, we noticed the possibility of using the couple duffing system in the analysis of epileptic seizures. The duffing equations lead to show different phase states between normal and epileptic signals [17]. Therefore we used the proposed by Ghorbanian deterministic duffing Van der Pol Oscillator Model to modelled pre-,ictal and post-ictal signals for the first time [18]. The paper is an attempt to analyze the possibility of using the model in the detection of stages of an epileptic seizure occurring in one patient, i.e. the onset of the seizure, during patient's moves, movement automatics, tangled stage and the end of the seizure. To the best of our knowledge, this model has not been used to analyse ictal carefully extracted phases yet. The estimated values of model parameters are determined for each considered phase. The parameters have been obtained using cost function L in the form mean square of normalized power spectrum of real and corresponding generated EEG signals. Additionally, the non-stationary character of the individual ictal stage is studied and compared using sensitive optimal values u, of model parameters, and sample entropy. The signal analysis and model oscillator are presented in Sect. 2. Sections 3 and 4 contain results and discussion, respectively. In Sect. 5, the main conclusions have been collected.

2   Materials and Methods

2.1   EEG Signals

EEG signals presented in this paper were recorded from right-handed 55 aged female who takes Phenytoin, at Temple University Hospital and is seizure-free since 7 months. This patient is selected for analysis for few reasons. Primarily, the description of records have been performed very carefully by the doctor. Patient behaviour has been associated with changes in the brain wave patterns. At the beginning of the EEG, the patient was calm and relaxed. The seizure begins at the end of hyperventilation when the patient's resting comfortably [19]. Digital video EEG is performed in the lab using standard 10–20 system of electrode placement with 1 channel of EKG. We considered sequences 10 s (length of samples N = 2500) registered by electrode T3. The sequences, according to clinical description include the onset of seizure without symptoms, behavior changes in the form shaking and moving leg, the movement automatisms, the confusion and the end of the epileptic seizure.

The medical equipment records the signal in a discretized form of time. To determine the number of the sequence we introduced the parameter d, where $d = 1, 2, \ldots , 12$. Therefore, the signal will be marked hereinafter as $x^d(n)$. Before calculating the discrete

Fourier transform of each sample of EEG sequence has been multiplied by the appropriate Blackman's window coefficient:

$$x_w^d(n) = x^d(n)w(n) \tag{1}$$

where:

$$w(n) = 0.42 - 0.5\cos(2\pi n/N - 1) + 0.08\cos(4\pi n/N - 1)$$

The discrete Fourier transform (DFT) takes the form:

$$X^d(k) = \sum_{n=0}^{N-1} x_w^d(n)\omega_N(n, k) \tag{2}$$

where:
$\omega_N = \exp(-j2\pi nk/N)$ is the Nth root of unity.

Next, the amplitude of DFT of the signal is normalized in the range of $[0, 1]$ according to the following formula:

$$\hat{X}^d(k) = \frac{|X^d(k)|}{\max|X^d(k)|} \tag{3}$$

The power P_b^d of normalized DFT amplitude sequences in five major frequency bands are calculated according to the formula:

$$P_b^d = \frac{1}{|S_b|} \sum_{k \in S_b} \left(\hat{X}^d(k)\right)^2 \tag{4}$$

where: $b = 1, \ldots, 5$ is the number of frequency band, S_b - set of discrete frequencies, corresponding to five major frequency bands [2]: delta (δ, 1–4 Hz, $b = 1$), theta (θ, 4–8 Hz, $b = 2$), alpha (α, 8–13 Hz, $b = 3$), beta (β, 13–30 Hz, $b = 4$) and gamma (γ, 30–60 Hz, $b = 5$).

2.2 Duffing Van der Pol Oscillator

The coupled system of duffing Van der Pol oscillators analyzed in this section was proposed by Ghorbanian and all to distinguish healthy and Alzheimer's disease signals [1, 5]. A four state equations representing coupled duffing Van der Pol oscillators model can be written as:

$$\begin{aligned}
\dot{x}_1^m &= x_3^m \\
\dot{x}_2^m &= x_4^m \\
\dot{x}_3^m &= -(\varsigma_1 + \varsigma_2)x_1^m + \varsigma_2 x_2^m - \varsigma_1\left(x_1^m\right)^3 - \rho_2\left(x_1^m - x_2^m\right)^3 + \varepsilon_1 x_3^m\left(1 - x_1^m\right) \\
\dot{x}_3^m &= \varsigma_2 x_1^m - \varsigma_2\left(x_1^m - x_2^m\right)^3 + \varepsilon_2 x_4^m\left(1 - \left(x_2^m\right)^2\right)
\end{aligned} \tag{5}$$

where superscript m indicates that the signal is generated by the model, ς is the linear stiffness coefficient, ρ is the nonlinear stiffness coefficient, which indicates the strength of the duffing nonlinearity resulting in multiple resonant frequencies, ε is the Van der Pol damping coefficient which determines the strength of van der Pol nonlinearity. Parameters $\varsigma_1, \rho_1, \varepsilon_1$ and $\varsigma_2, \rho_2, \varepsilon_2$ belong to the first and second oscillator, respectively. The output may be selected as any combination of the positions and velocities to mimic an EEG signal. In this study, the velocity of the second oscillator is selected as the model output. The initial conditions are equal to:

$$x_1^m(0) = 0, \; x_2^m(0) = 1, \; x_3^m(0) = 0, \; x_4^m(0) = 0$$

Runge-Kutta iterative method is selected from standard numerical integration methods to solve these dynamic equations. The amplitude of the generated signal is discretized and normalized according to the formula 2–3. The components of the power spectra of the signal generated by the oscillator model in five considered frequency bands were also needed to build the objective function. The genetic algorithm is adapted to perform the optimization procedure. In the optimization process, we set the population size as 2000. The cost function is chosen as a root mean square of the errors in the power spectra of signal EEG P_b^d and generated $\breve{P}(\varsigma_1, \varsigma_2, \rho_1, \rho_2, \varepsilon_1, \varepsilon_2)$ in each selected brain frequency band, as shown in [3]. The functions L can be formally written as (7):

$$L(\mathbf{\Omega}, d) = \sum_{v=1}^{5} \left(P_b^d - \breve{P}(\varsigma_1, \; \varsigma_2, \; \rho_1, \; \rho_2, \; \varepsilon_1, \; \varepsilon_2) \right)^2$$

$$\mathbf{\Omega} = [\varsigma_1, \; \varsigma_2, \; \rho_1, \; \rho_2, \; \varepsilon_1, \; \varepsilon_2] \tag{6}$$

where: L is the cost function, $\mathbf{\Omega}$ is the vector of design model variables, $\varsigma_1, \varsigma_2, \rho_1, \rho_2, \varepsilon_1,$ and ε_2 are the decision variables of the optimization. The initial guesses for the optimization search were randomly generated within the bounds defined as:

$$0 \le \zeta_{1,2} \le 200, 0 \le \rho_{1,2} \le 100, 0 \le \varepsilon_{1,2}$$

The optimization goal is error minimization:

$$\min_{\varsigma_1, \varsigma_2, \rho_1, \rho_2, \varepsilon_1, \varepsilon_2} L(\varsigma_1, \varsigma_2, \rho_1, \rho_2, \varepsilon_1, \varepsilon_2) \tag{7}$$

The solution set of decision variables for each sequence d will be denoted as:

$$\varsigma_1^d, \; \varsigma_2^d, \; \rho_1^d, \; \rho_2^d, \; \varepsilon_1^d, \; \varepsilon_2^d$$

3 Results

The analysis based on the optimal values of parameters, relative sensitivity factor of them and the objective function is performed taking into account types of sequences. Table 1 shows the results obtained in highlighted phases of the epileptic seizure.

Table 1. The model's parameters, the sensitivity factor u, and the cost function L

Phase		Estimated Values			Sensitivity factor u		
		Seq.1	Seq. 2	Seq. 3	Seq. 1	Seq. 2	Seq. 3
Onset	ς_1	10.79	11.79	10.97	4567.00	424.82	−215.60
Leg movement		131.88	88.99	54.00	4764.00	1218.00	−123.70
Movement automatics		164.00	120.34	11.32	0.25	44.85	9.85
Entanglement		98.81	53.56	69.69	164.40	330.19	−98.18
End		12.81	12.74	8.33	0.05	8.07	0.87
Cost function L		0.06	0.03	0.06			
Onset	ς_2	59.33	47.33	40.50	4567.00	424.82	−215.60
Leg movement		164.00	120.34	13.24	0.56	1.43	−1.08
Movement automatics		35.84	45.54	37.03	8.11	23.07	2.01
Entanglement		139.63	115.00	72.39	−72.84	3.40	139.63
End		18.34	17.63	22.22	0.07	1.10	0.28
Cost function L		0.18	0.22	0.06			
Onset	ρ_1	68.61	48.90	35.67	4715.00	22.69	−58.90
Leg movement		28.21	31.54	31.89	0.34	0.77	0.07
Movement automatics		54.20	47.23	45.02	8.27	70.91	2.44
Entanglement		33.97	43.00	59.59	172.82	−495.01	−98.18
End		6.72	6.45	7.87	0.09	6.44	1.27
Cost function L		0.06	0.19	0.06			
Onset	ρ_2	70.52	54.20	48.45	4875.00	481.49	−87.60
Leg movement		49.91	52.13	35.00	0.34	0.60	−0.79
Movement automatics		51.00	49.91	64.82	10.75	69.74	0.88
Entanglement		99.55	81.53	56.99	164.40	330.19	99.55
End		93.45	110.32	84.45	0.49	13.89	1.73
Cost function L		0.08	0.03	0.06			
Onset	ε_1	7.53	5.13	3.43	2416.00	51.10	52.55

(*continued*)

Table 1. (*continued*)

Phase		Estimated Values			Sensitivity factor u		
		Seq.1	Seq. 2	Seq. 3	Seq. 1	Seq. 2	Seq. 3
Leg movement		5.59	6.34	17.16	0.44	17.16	0.50
Movement automatics		2.78	4.29	4.72	2.21	0.07	−1.17
Entanglement		19.06	13.66	16.02	0.60	3.75	19.06
End		16.75	15.60	23.03	0.60	3.34	0.95
Cost function L		0.17	0.21	0.07			
Onset	ε_2	15.35	23.34	24.67	−6287.0	173.81	149.97
Leg movement		14.35	13.45	20.00	0.07	20.00	−0.07
Movement automatics		12.34	25.10	14.86	44.60	62.34	−5.38
Entanglement		22.59	28.47	19.17	0.49	−265.84	25.59
End		21.34	20.97	22.63	0.49	13.89	1.73
Cost function L		0.08	0.03	0.06			

It can be clearly seen, that high estimated values of linear stiffness parameters ς_1 and ς_2 are obtained for sequences associated with changes in the patient's behavior: during leg movement, movement automatics, and entanglement. The initial seizure is accompanied by high values of non-linear stiffness coefficients ρ_1 and ρ_2. A high value of ρ_2 is also observed in the final stage of the seizure. The small values of cost function L indicate obtaining a similar power spectrum of the real and generated signal. The obtained values of sensitivity coefficients for the onset are very high. In order to assess the application of sensitivity factor of optimal parameter values in the determination of range non-stationary nature of the signal, the sample entropy is calculated for each considered sequence. The results are collected in Table 2. According to the value of the sensitivity factor, the initial stages of the seizure are characterized by the highest entropy values.

Table 2. Values of the sample entropy calculated for the each sequence.

Phase	Sample entropy		
	Seq. 1	Seg. 2	Seq. 3
Onset	0.24	0.12	0.18
Leg movement	0.22	0.16	0.08
Movement automatics	0.004	0.0004	0.05
Entanglement	0.03	0.13	0.08
End	0.08	0.03	0.08

4 Discussion

The seizures described in the work are characterized by a sudden onset occurring within the neural network in the left hemisphere [19]. The results show that the onset epileptic seizure is a strongly non-stationary process. In the early stages of the seizure, according to the description, in the record high-amplitude left temporal spike and slow-wave complexes. The female began to move her left leg. The increase of amplitude that occurs here can be due to an increase in the force acting on the springs, which is represented by linear stiffness parameters ς_1 and ς_2. Next, the patient has motor automatisms. In the EEG recording can be seen slow frontal delta waves, which, combined with the characteristic of the waveform of the temporal lobe, associated with high, similar values of ς_2, ρ_1, and ρ_2. In this phase, occur the greatest changes in the value of the sensitivity factor of the model parameters (from 0.07 for ε_2 to 70.91 for ρ_1). During the seizure, the patient experiences further behavioral changes, including confusion. According to the doctor describing the study, the EEG record during cis difficult to interpret at this moment. High values of ς_1, ς_2, ρ_1, and ρ_2 changing the kinematics of the seizure, which allows the spread of discharges to the other hemisphere (two excitation networks). Under the influence of damping, the force decreases. At the end of the seizure, in the EEG recording bilateral slow brain wave synchronization with sharp waves is observed. The stationary character dominates until the seizure has completely ceased. The low values ς_1 and ρ_1 suggest, that the first oscillator no longer stimulates the system to further vibrate. The seizures initially persist due to higher values of the second oscillator.

5 Conclusions

Having some facts about temporal lobe epilepsy, including individual phases of seizure propagation, the deterministic coupled duffing Van der Pol oscillator model is proposed to model the brain activity of epileptic patients. From the results obtained from an individual patient, it is shown, that the proposed model explains the problem of the generation of different rhythms of the EEG ictal signal. The model allows determining the EEG changes that occur during the seizure. An increase in linear parameter values is seen during slow delta and sharp wave detection,and the generation of fast spikes and high-amplitude is associated with decrease values of liner parameters and increase values of nonlinear cubic parameters.

References

1. Verma, A., Radtke, R.: EEG of Partial Seizures. J. Clin. Neurophysiol. **4**(23), 333–339 (2006)
2. Ghorbanian, P.: Non-stationary time series analysis and stochastic modeling of EEG and its application to Alzheimer's Disease. Ph.D. dissertation, Villanova University (2014)
3. Sornmo, L., Laguna, P.: Bioelectrical Signal Processing in Cardiac and Neurological Applications. Elsevier Academic Press, Burlington. MA (2005)
4. Fisher, R.S. Cross, J.H. French, J.A., et al.: Operational classification of seizure types by the International League Against Epilepsy: Position Paper of the ILAE Commission for Classification and Terminology. Epilepsia **4**(58), 522–530 (2017)
5. Ghorbanian, P., Ramakrishnan, S., Ashrafiuon, H.: Stochastic non-linear oscillator models of EEG: the Alzheimer's disease case. Front. Comput. Neurosci. **9**(48) (2015). https://doi.org/10.3389/fncom.2015.00048
6. Acharya, U.R., Molinari, F., Vinitha, S., Chattopadhyay, S.: Automated diagnosis of epileptic EEG using entropies. Biomed. Signal Process. Control **4**(7), 401–408 (2012)
7. Soltesz, I. Staley, K. Computational Neuroscience in Epilepsy. Elsevier Science, Amsterdam (2011)
8. Fisher, R.S., et al.: Z. Epileptol. **31**(4), 282–295 (2018). https://doi.org/10.1007/s10309-018-0217-7
9. Botcharova, M.: Modelling and analysis of amplitude, phase and synchrony in human brain activity patterns. Ph.D. dissertation, University College London (2014)
10. Hussein, R., Elgendi, M., Wang, J.Z., Ward, R.K.: Robust detection of epileptic seizures based on L1-penalized robust regression of EEG signals. Expert Syst. Appl. **104**, 153–167 (2018)
11. Ghorbanian, P., Ramakrishnan, S., Ashrafioun, H.: Nonlinear dynamic analysis of EEG using a stochastic duffing-van der pol oscillator model. In: Proceedings of the ASME 2014 Dynamic Systems and Control Conference. https://doi.org/10.1115/DSCC2014-5854
12. Liu, L.: Recognition and analysis of motor imagery EEG signal based on improved BP neural network. IEEE Access **7**, 47794–47803 (2019). https://doi.org/10.1109/ACCESS.2019.2910191
13. Ohsuga, M., Jamaguchi, J., Schimuzi, H.: Entrainment of two coupled van der pol oscillators by an external oscillation. Biol. Cybernet. **51**, 325–333(1985)
14. Kawahara, T.: Coupled Van der Pol oscillators-A model of excitatory ad inhibitory neural interactions. Biol. Cybern. **39**, 37–43 (1980)
15. Tabi, C.B.: Dynamical analysis of the FitzHugh-Nagumo oscillatons through a modified Van der Pol equation with fractional-order derivative term. Int. J. Non-Linear Mech. **105**, 173–178 (2018)
16. Meijer, H.G.E., Eissa, T.L., Kiewiet, B., et al.: Modeling focal epileptic activity in the Wilson–Cowan model with depolarization block. J. Math. Neurosci. **5**(7) (2015)
17. Rakshit, S., Bera, B., Majhi, S., et al.: Basin stability measure of different steady states in coupled oscillators. Sci. Rep. **7**, 45909 (2017). https://doi.org/10.1038/srep45909
18. Szuflitowska B., Orlowski P.: Statistical and physiologically analysis of using a Duffing-van der Pol oscillator to modeled ictal signals. In: ICARCV Shenzhen, China, pp. 1137–1142 (2020). https://doi.org/10.1109/ICARCV50220.2020.9305339
19. Obeid, I., Picone, J., Harabagiu, S.: Automatic discovery and processing of EEG Cohorts from clinical records. In: Big Data to Knowledge All Hands Grantee Meeting, p. 1. National Institutes of Health, Bethesda (2016). www.isip.piconepress.com/publications/conference_presentations/2016/nih_bd2k/cohort/

Computational Science in IoT
and Smart Systems

A Review on Visual Programming for Distributed Computation in IoT

Margarida Silva[1], João Pedro Dias[1,2(✉)], André Restivo[1,3],
and Hugo Sereno Ferreira[1,2]

[1] DEI, Faculty of Engineering, University of Porto, Porto, Portugal
{ana.margarida.silva,jpmdias,arestivo,hugo.sereno}@fe.up.pt
[2] INESC TEC, Porto, Portugal
[3] LIACC, Porto, Portugal

Abstract. Internet-of-Things (IoT) systems are considered one of the most notable examples of complex, large-scale systems. Some authors have proposed visual programming (VP) approaches to address part of their inherent complexity. However, in most of these approaches, the orchestration of devices and system components is still dependent on a centralized unit, preventing higher degrees of dependability. In this work, we perform a systematic literature review (SLR) of the current approaches that provide visual and decentralized orchestration to define and operate IoT systems, reflecting upon a total of 29 proposals. We provide an in-depth discussion of these works and find out that only four of them attempt to tackle this issue as a whole, although still leaving a set of open research challenges. Finally, we argue that addressing these challenges could make IoT systems more fault-tolerant, with an impact on their dependability, performance, and scalability.

Keywords: Internet-of-Things · Orchestration · Visual programming · Decentralized computation · Large-scale systems

1 Introduction

The Internet-of-Things (IoT) comprises many devices with a wide range of capabilities, directly or indirectly connected to the Internet. This allows them to transfer, integrate, analyze and act according to data generated among themselves [11]. IoT systems use devices at an unprecedented scale, with applications ranging from mission-critical to entertainment and commodity solutions [14].

The widespread usage of IoT led to a mostly uncontrollable and ever-growing heterogeneity of devices, differing in computational power, protocols, and architectures, comprising a large-scale and distributed (geographically and logically) system of systems. These characteristics raise many development challenges in guaranteeing their scalability, maintainability, security, and dependability [56]. Consequently, the active pursuit of reducing the complexity and technical knowledge needed to configure and adapt such systems to their needs (from manufacturing floor automation to *smart home* system customization) eventually led to

© Springer Nature Switzerland AG 2021
M. Paszynski et al. (Eds.): ICCS 2021, LNCS 12745, pp. 443–457, 2021.
https://doi.org/10.1007/978-3-030-77970-2_34

the exploration of *low-code* [10,20] and conversational approaches [36]. Visual programming (VP) approaches (*model-* or *mashup*-based) provide such means via the arrangement of visual elements, which are then automatically translated into executable artifacts, by leveraging some kind of a Visual Programming Language [13,20,45]. One of the most popular approaches is Node-RED [34,39], which provides both a visual editor and a run-time environment for IoT systems.

Most VP solutions (Node-RED included) provide a centralized approach (which can be either *on-premises* or cloud-based) where the main component transforms and processes most of the computation on data provided by edge and fog devices. Consequences of this approach are well-known: (1) single point of failures, (2) violation of data boundaries (private, technological, and political), and (3) unused edge computational power. Recent research effort put in *Fog and Edge Computing* [20,54] focus on solving these by leveraging the resources available in lower-tier devices to improve overall dependability [22–24,46], performance [44], scalability, observability [52], and reproducibility [19,21].

In this paper we present a systematic literature review (SLR) on VP approaches for IoT, focusing on those related to orchestration of multiple components. Our initial search yielded 2698 results across three different scientific databases. We refined this selection with *inclusion* and *exclusion* criteria, resulting in 21 papers. Through snowballing and taking into account previous (non-systematic) surveys, we found 8 new works, resulting in 28 approaches across 22 papers. We compared their characteristics, including scope, architecture, scalability, and VP paradigms. We then carried out an in-depth analysis on the subset that provided mechanisms for decentralized computation.

The remainder of this paper is structured as follows: Sect. 2 presents the methodology used in this research, Sect. 3 presents the results of the literature review, followed by an in-depth analysis of the current alternative for visual IoT decentralized orchestration in Sect. 4. An overview of the current issues and research challenges, and some final remarks, is given in Sect. 5.

2   Literature Review Methodology

This work follows a Systematic Literature Review (SLR) methodology to gather information on the state of the art of VP applied to the IoT paradigm, with a particular emphasis on orchestration concerns. The goal of a systematic literature review is to synthesize evidence with emphasis on its quality [43]. We started by defining the research questions to be answered and choosing data sources to search for publications. We outlined the following research questions (RQ):

RQ1. *What are the relevant VP approaches applied to distributed computation and orchestration in IoT?* Using VPs to make IoT development easier for the end-user is a common go-to approach. However, we argue that there is a scarcity of those that provide decentralized approaches;

RQ2. *What architectures and tiers characterize the approaches found in* **RQ1**? IoT systems can target one or more tiers, as well as be implemented in a

Table 1. Inclusion and exclusion criteria.

	ID	Criterion
Exclusion	EC1	Not written in English
	EC2	Presents just ideas, tutorials, integration experimentation, magazine publications, interviews, or discussion papers
	EC3	Does not address multiple devices' orchestration
	EC4	Has less than two (non-self) citations when more than five years old
	EC5	Duplicated articles
	EC6	Articles in a format other than camera-ready (PDF)
Inclusion	IC1	Must be on the topic of VP in IoT
	IC2	Contributions, challenges, and limitations are detailed
	IC3	Research findings include sufficient explanation on how the approach works
	IC4	Publication year in the range between 2008 and 2019

centralized or decentralized architecture. A VP tool applied to IoT can facilitate the development of systems that operate and distribute computing tasks among the available tiers. Each tier and type of architecture offers advantages and disadvantages; understanding these characteristics is essential to understand how they can be used;

RQ3. *What was the evolution of VP approaches applied to IoT over the years focusing on its decentralized operation?* To understand the field of VP applied to IoT, more specifically, its visually-defined decentralized operation, it is essential to perceive its evolution.

```
((vpl OR visual programming OR visual-programming) OR
(node-red OR node red OR nodered) OR (data-flow OR
dataflow)) AND (iot OR internet-of-things)
```

Listing 1.1. Search query for relevant literature on *IEEEXplore, ACM DL* and *Scopus.*

Answering these questions will provide valuable insights for both practitioners (in terms of summarizing the current practices on the usage of VP methodologies for IoT orchestration are) and researchers (showing current challenges and issues that can be further researched). We follow the criteria detailed in Table 1 and outlined in Fig. 1. Three popular and reputable scientific databases were used, namely *IEEE Xplore, ACM Digital Library*, and *Scopus*. Listing 1.1 shows the query we used to convey the most probable keywords to appear in our target candidates, including popular variants. This search was performed in October 2019, and the results can be seen in the first step of Fig. 1. The evaluation process of the publications then followed eight steps:

1. **Automatic Search**: Run the query string in the different scientific databases and gather results;
2. **Filtering** (*EC1*, *IC4*, and *EC6*): Publications are selected regarding its (1) language, being limited to the ones written in the English language, (2) publication date, being limited to the ones published between 2008 and 2019, and (3) publication status, being selected only the ones that are published in their final versions (camera-ready PDF format);
3. **Filtering to remove duplicates** (*EC5*): The selected papers are filtered to remove duplicated entries;
4. **Filtering by *Title* and *Abstract*** (*EC2–EC4*, and *IC1–IC3*): Selected papers are revised by taking into account their *Title* and *Abstract*, by observing the (1) stage of the research, only selecting papers that present approaches with sufficient explanation, some experimental results, and discussion on the paper contributions, challenges, and limitations, (2) contextualization with recent literature, filtering papers that have less than two (non-self) citations when more than five years old, and (3) leverages the use of visual notations for orchestrating and operating multi-device systems;
5. **Filtering by *Introduction* and *Conclusions*** (*EC2–EC4*, and *IC1–IC3*): The same procedure of the previous point is followed but taking into consideration the *Introduction* and *Conclusion* sections of the papers;
6. **Selected Papers Analysis**: Selected papers are grouped, and surveys are separated; their content is analyzed in detail;
7. **Survey Expansion**: For surveys found, the enumerated approaches are analyzed and filtered, taking into account their scope and checking if they are not duplicates of the currently selected papers;
8. **Wrapping**: Works gathered from the *Selected Papers Analysis* (individual papers) and the *Survey Expansion* are presented and discussed.

The total number of publications was 2698, 22 of each were selected (*cf.* Fig. 1). One was a survey [47] (pointing to 8 new works), and the others presented approaches relevant for our RQs.

3   Results

From the 22 publications, 28 different approaches were analyzed and distributed among categories, according to several characteristics:

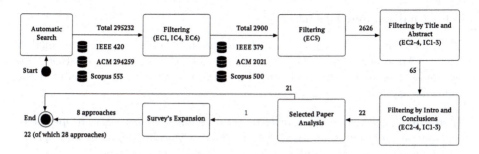

Fig. 1. Pipeline overview of the SLR Protocol.

1. **Scope.** Some approaches have specific use cases in mind. Therefore, knowledge of a tool's scope helps assess if it solves a problem or fills a specific gap in the literature. Example values consist of *Smart Cities, Home Automation, Education, Industry* or *Several* if there is more than one;
2. **Architecture.** VP approaches applied to IoT can have a centralized or decentralized architecture, based on their use of Cloud, Fog, or Edge Computing architecture. Possible values are *Centralized, Decentralized*, and *Mixed*;
3. **License.** The license of software or tool is essential in terms of its usability. Normally, an open-source software reaches a bigger user base, allowing everyone to expand and contribute to it. Possible values are the name of the tool license or N/A if it does not have one;
4. **Tier.** IoT systems are composed of three tiers—*Cloud, Fog*, and *Edge*. A tool can interact in several of these, shaping its features and how it is built;
5. **Scalability.** Defines how the tool or framework scales. It can be calculated based on metrics used to test the performance of the system. We considered scalability in terms of number and different types of devices supported. Possible values are *low, medium, high*, or N/A if information not sufficient;
6. **Programming.** According to Downes and Boshernitsan [9], VPs can be classified in five (possibly overlapping) categories: (1) Purely visual languages, (2) Hybrid text and visual systems, (3) Programming-by-example systems, (4) Constraint-oriented systems and (5) Form-based systems. It is important to know which type so that it might be possible to assess the type of experience the tool provides to the user and its architecture;
7. **Web-based.** Defines if the VP and/or environment can be used in a browser. It is useful in terms of the usability of the tool.

The resulting categorization is depicted in Table 2. Some key takeaways are easily observable, namely: (1) most approaches use a centralized architecture, (2) the hybrid visual-textual programming paradigm is predominant, and (3) most approaches are web-based. The extended findings and their categorization is presented in Table 3, following the same previously defined categories.

3.1   Analysis and Discussion

The approaches presented in this SLR passed the evaluation process defined in Sect. 2. Approaches supporting only one device or extending an existent VPL by applying it to IoT were left out. From the resulting ones, we analyzed the following aspects:

Domain. The surveyed approaches target different domains: six were specific to home automation, four to education, three to specific domains, and one for the industry; the remainder 14 had a wide range of use cases;

Architecture. Sixteen have a centralized architecture, three are decentralized, and the remaining nine do not present enough information on this topic;

License. Most did not mention a license; those that did were mostly open-source (*e.g.*, GPL v2, GPL v3, Apache v2 and LGPL v3);

Scalability. The majority do not consider scalability (*e.g.* the number of devices they were tested with); those that do, claim high scalability;

Table 2. VP approaches applied to IoT and their characteristics. Small circles (•) mean *yes*, hyphens (-) means *no information available*, empty means *no*.

Tool	Scope[h]	Centralized	License	Tier	Scalability	Programming	Web-based
Belsa et al. [5]	*	•	-	Cloud	High	Hybrid	•
Ivy [25]	*	•	-	Cloud	Medium[g]	Purely visual	
Ghiani et al. [29]	HA	•	-	Cloud	-	Form-based	•
ViSiT [2]	*	•	-	Cloud	High	Hybrid	•
Valsamakis et al. [53]	AAL	•	-	Cloud	-	Hybrid	•
WireMe [42]	EDU, HA	•	-	Cloud	-	Hybrid	
VIPLE [17]	EDU	•	-	Cloud	-	Hybrid	
Smart Block [4]	HA	•	-	Cloud	-	Hybrid	•
PWCT [28]	*	•	GPL v2	-[a]	High	Hybrid	
DDF [31]	-		Apache v2	Fog	High	Hybrid	•
GIMLE [51]	IND	•	-	Cloud	High	Hybrid	•
DDFlow [41]	SEC		-	Fog/Edge	-	Hybrid	•
Kefalakis et al. [35]	-	•	LGPL v3[c]	Cloud	-	Hybrid	
Eterovic et al. [26]	HA	-[d]	-	-	-	Hybrid	-
FRED [8]	*	•	-[e]	Cloud	High	Hybrid	•
WoTFlow [7]	-		-	Fog/Edge	-	Hybrid	•
Besari et al. [6] [49]	EDU	•	-	Cloud	-	Hybrid	•
CharIoT [50]	HA	•[f]	-	Cloud/Edge[f]	High[f]	Form-based	•
Desolda et al. [18]	SM	-	-	-	-	Hybrid	
Eun et al. [27]	HA	•	-	-	-	Form-based	•

[a] Used for several purposes, did not specify the tier it is located in regarding IoT.
[b] Since it uses Node-RED, this information was based on its architecture.
[c] Under the same license of OpenIoT.
[d] No information *w.r.t* the architecture of the environment created, only the VPL.
[e] No information about the license is given, but further research discovered that it had paid plans and no source code available.
[f] CharIoT uses the Giotto stack [1] from where we retrieved this information.
[g] Certainty regarding this information is low.
[h] Several (*), Home Automation (HA), Ambient Assisted Living (AAL), Education (EDU), Industry (IND), Security (SEC), and Smart Museums (SM).

Programming. 22 employ a hybrid text and visual system VP paradigm, while three use a purely visual, and the other three a form-based one;

Web-based. The majority of analyzed approaches are web-based. One tool did not specify the environment, only mentioned being a VPL.

The following paragraphs present an evolution-over-time analysis and attempt to give an answer to the aforementioned research questions.

Evolution Analysis. To understand the evolution of VP approaches applied to IoT, we analyzed the years when the selected papers were published and the surveyed approaches launched. Figure 2 clearly display an increased trend in research during the last years.

Research Questions. The research questions presented in Sect. 2 served to direct this SLR and obtain answers to relevant questions regarding the available visual programming approaches for IoT. These answers are:

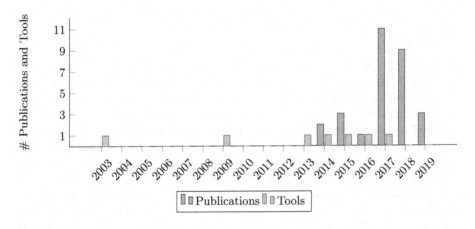

Fig. 2. Evolution of publications and number of visual approaches per year.

RQ1. *What are the relevant VP approaches applied to distributed computation and orchestration in IoT?* From the analyzed approaches in Section 3, we found 28 that share these concerns in IoT-scope;

RQ2. *What architectures and tiers characterize the approaches found in* **RQ1**? Tables 2 and 3 give an overview of the surveyed approaches characteristics. Our analysis (Sect. 3.1) concludes most of them have a centralized architecture and work in the Cloud tier;

RQ3. *What was the evolution of VP approaches applied to IoT over the years focusing on its decentralized operation?* As seen in Sect. 3.1 and Fig. 2, some approaches share this concern since 2003, though 2017–2018 saw a significant increase in publications focusing on it.

4   Visually-Defined Distributed Computing

From the analyzed approaches, we found four trying to tackle visual and decentralized orchestration in IoT. We discuss them in the following subsections.

Table 3. Characterization of the visual programming approaches for IoT [47].

Tool	Scope[c]	Centralized	License	Tier	Scalability	Programming	Web-based
Node-Red [39]	*	Yes	Apache v2	Cloud/Edge	High	Hybrid	Yes
NETLab Toolkit [38]	N/A	N/A	GPL	Edge[b]	N/A	Hybrid	Yes
NooDL [40]	*	N/A	NooDL[a]	Cloud[b]	N/A	Hybrid	No
DGLux5 [40]	*	N/A	DGLux	Cloud/Fog[b]	High[b]	Purely visual	No
AT& T Flow Designer [3]	*	N/A	GPL v3	Cloud[b]	High[b]	Hybrid	Yes
GraspIO [32]	EDU	N/A	BSD	Cloud[b]	N/A	Purely visual	No
Wyliodrin [55]	*	N/A	GPL v3	All[b]	N/A	Hybrid	Yes
Zenodys [12]	*	N/A	GPL v3	Cloud[b]	High[b]	Hybrid	Yes

[a] Available at https://www.noodl.net/eula
[b] Certainty regarding this information is low.
[c] Several (*), Education (EDU), and Not Available (N/A).

4.1   DDF

WoTFlow [7], DDF [41], and subsequent works [30,31] are systems extending Node-RED and focusing on Smart Cities. Their goal is to make it more suitable for developing fog-based applications that are context-dependent on edge devices where they operate. DDF starts by implementing D-NR (Distributed Node-RED), which contains processes that can run across devices in local networks and servers in the Cloud. The application, called *flow*, is built with a VP environment, running in a development server. All the other devices running D-NR subscribe to an MQTT topic that contains the status of the flow. When a flow is deployed, all devices running D-NR are notified and subsequently analyze the given flow. Based on a set of constraints, they decide which nodes they may need to deploy locally and which sub-flow (parts of a flow) must be shared with other devices. Each device has characteristics, from its computational resources, such as bandwidth and available storage, to its location. The developer can insert constraints into the flow by specifying which device a sub-flow must be deployed in or the computational resources needed. Further, each device must be inserted manually into the system by a technician.

Subsequent work focused on support for the Smart Cities domain, including the deployment of multiple instances of devices running the same sub-flow and the support for more complex deployment constraints of the application flow [31]. The developer can specify requirements for each node on device identification, computing resources needed (CPU and memory), and physical location. In addition to these improvements, the coordination between nodes in the fog was tackled by introducing a coordinator node. This node is responsible for synchronizing the device's context with the one given by the centralized coordinator. Recent work [30], support for CPSCN (Cyber-Physical Social Computing and

Networking) was implemented, making it possible to facilitate the development of large scale CPSCN applications. Additionally, to make this possible, the contextual data and application data were separated so that the application data is only used for computation activities. The contextual data is used to coordinate the communication between those activities.

4.2   UFlow and FogFlow

Szydlo et al. [48] focused on the transformation and decomposition of data flow. Parts of the flow can be translated into executable parts, such as Lua. Their contribution includes data flow transformation concepts, a new portable run-time environment (uFlow) targetting resource-constrained embedded devices, and its integration with Node-RED. Their solution transforms a given data flow by allowing the developer to choose the computing operations run on the devices. These operations are implemented using uFlow. The communication between the devices requires a Cloud layer, without support for peer-to-device communication. The results are promising, showing a decrease in the number of measurements made by the sensors. However, there is room for improvement *w.r.t.* the automatic decomposition and partitioning of the initial flow, and detecting current conditions in deciding when to move computations between fog and cloud. Later, the authors proposed FogFlow [48], which enables the decomposition into heterogeneous IoT environments according to a chosen decomposition schema. To achieve a certain level of decentralization and heterogeneity, they abstract the application definition from its architecture and rely on graph representations to provide an unambiguous, well-defined computation model. The application definition is infrastructure-independent and only contains data processing logic, and its execution should be possible on different sets of devices with different capabilities. Several algorithms for flow decomposition are mentioned [33,37], but none were explored/provided results.

4.3   FogFlow (Yet Another)

A different tool with the same name FogFlow by Cheng et al. [15,16] proposes a standards-based programming model for Fog Computing and scalable context management. The authors start by extending the dataflow programming model with hints to facilitate the development of fog applications. The scalable context management introduces a distributed approach, which allows overcoming the limits in a centralized context, achieving much better performance in throughput, response time, and scalability. The FogFlow framework focuses on a Smart City Platform use case, separated into three areas: (1) Service Management, typically hosted in the Cloud, (2) Data Processing, present in cloud and edge devices, and (3) Context Management, which is separated in a device discovery unit hosted in the Cloud and IoT brokers scattered in Edge and Cloud. This was later improved to empower infrastructure providers with an environment that allows them to build decentralized IoT systems faster, with increased stability and scalability. Dynamic data representing the IoT system flows are orchestrated between sensors (Producers) and actuators (Consumers). An application is first designed using the `FogFlow` Task Designer (a hybrid text and VP environment), which outputs an abstraction called *Service Template*. This abstraction contains details about the resources needed for each part of the system. Once the Service Template is submitted, the framework determines how to instantiate it using the context data available. Each task is associated with an operator (a Docker image), and its assignment is based on (1) how many resources are available on each edge node, (2) the location of data sources, and (3) the prediction of workload. Edge nodes are autonomous since they can make their own decisions based on their local context without relying on the central Cloud. Obviously, the dependency in Docker completely discards constrained devices.

4.4   DDFlow

`DDFlow` [41], presents another distributed approach by extending Node-RED with a system run-time that supports dynamic scaling and adaption of application deployments. The distributed system coordinator maintains the state and assigns tasks to available devices, minimizing end-to-end latency. Dataflow notions of *node* and *wire* are expanded, with a *node* in `DDFlow` representing an instantiation of a task deployed in a device, receiving inputs and generating outputs. *Nodes* can be constrained in their assignment by optional parameters, *Device*, and *Region*, inserted by the developer. A *wire* connects two or more nodes and can have three types: *Stream* (one-to-one), *Broadcast* (one-to-many), and *Unite* (many-to-one).

Table 4. IoT decentralized visual programming approaches and their characteristics.

Tool	Leveraging edge devices	Communication capabilities	Open-source	Computation decomposition	Run-time adaptation
DDF [30]	Limited[a]	Yes	Yes	Limited[b]	Yes
uFlow [48]	Yes	Limited[c]	No	Limited[b]	Limited[c]
FogFlow [15]	Yes	N/A	Yes	Limited[b]	Yes
DDFlow [41]	Limited[d]	Yes	No	Limited[b]	Yes

[a]Assumes all devices run Node-RED (doesn't apply to constrained devices).
[b]Does not specify the algorithm used.
[c]Communication between devices is made through the cloud (Internet-dependent).
[d]Assumes all devices have a list of specific services they can provide.

In a `DDFlow` system, each device has a set of capabilities and a list of services that correspond to an implementation of a *Node*. The devices communicate this information through their Device Manager or a proxy if it is a constrained device. The coordinator is a web server responsible for managing the `DDFlow` applications. It is composed of: (1) a VP environment where `DDFlow` application are built, (2) a Deployment Manager that communicates with the Device Managers of the devices, and (3) a Placement Solver, responsible for decomposing and assigning tasks to the available devices. When an application is deployed, a network topology graph and a task graph are constructed based on the real-time information retrieved from the devices. The coordinator proceeds with mapping tasks to devices by minimizing the task graph's end-to-end latency of the longest path. Dynamic adaptation is supported by monitoring the system; if changes in the network are detected, such as the failure or disconnection of a device, adjustments in the assignment of tasks are made. The coordinator can also be replicated into many devices to improve the system's reliability and fault-tolerance. They show DDFlow recovering from network degradation or device overload, whereas in a centralized system this would likely cause its (total) failure.

5 Conclusion

The mentioned approaches were characterized based on their mentions or support for the following features and characteristics:

Leveraging edge devices. A decentralized architecture takes advantage of the computational power of the devices in the network, assigning them tasks. However, some approaches have limitations on the type of supported devices or only focus on the Fog tier and not Edge;

Communication capabilities. The orchestrator must know each device's capabilities so that it can make informed decisions regarding the decomposition and assignment of tasks;

Open-source. The license of software or tool is essential in terms of its usability. Open-source allows access to the code, making it possible for its analysis, improvement, and reuse;

Computation decomposition. To implement a decentralized architecture, it is important to decompose the computation of the system into independent and logical tasks that can be assigned to devices. This is made using algorithms, which can be specified or mentioned;

Run-time adaptation. A system needs to adapt to run-time changes, such as non-availability of devices or even network failure. The system notices these events and can take action to circumvent the problems and keep functioning;

From the analysis of Table 4, we can conclude that the current research for visual programming approaches that leverage the decentralized nature of IoT is incomplete. All the surveyed approaches leverage the devices in the network but in a different way. DFF assumes that all devices run Node-RED, limiting the types of devices used. FogFlow and uFlow are the only ones that specify how they truly leverage constrained devices, with the transformation of sub-flows into Lua code. DDFlow assumes that all devices have a list of specific services they can provide that should match the node assigned to them. Regarding the method used to decompose and assign computations to the available devices, DDFlow describes the process using the longest path algorithm focused on reducing end-to-end latency between devices. FogFlow and uFlow mention several algorithms that could be used but do not specify which one was implemented. Both DDF and `FogFlow` do not specify the algorithm used besides some constraints but are the only ones with accessible source code and an open-source license. All the surveyed approaches claim to support run-time adaptation to changes in the system, such as device failures.

Acknowledgement. This work is financed by National Funds through the Portuguese funding agency, FCT - Fundação para a Ciência e a Tecnologia, within project UIDB/50014/2020.

References

1. Agarwal, Y., Dey, A.K.: Toward building a safe, secure, and easy-to-use internet of things infrastructure. IEEE Comput. **49**(4), 88–91 (2016)
2. Akiki, P.A., Bandara, A.K., Yu, Y.: Visual simple transformations: empowering end-users to wire internet of things objects. ACM Trans. Comput.-Hum. Interact. **24**(2), 1–43 (2017). https://doi.org/10.1145/3057857
3. AT&T Mobility LLC: AT&T Flow Designer. https://flow.att.com. Accessed 2020
4. Bak, N., Chang, B.M., Choi, K.: Smart block: a visual programming environment for SmartThings. In: International Computer Software and Applications Conference, vol. 2, pp. 32–37 (2018). https://doi.org/10.1109/COMPSAC.2018.10199
5. Belsa, A., Sarabia-Jacome, D., Palau, C.E., Esteve, M.: Flow-based programming interoperability solution for IoT platform applications. In: IEEE International Conference on Cloud Engineering, IC2E 2018, pp. 304–309 (2018)
6. Besari, A.R.A., Wobowo, I.K., Sukaridhoto, S., Setiawan, R., Rizqullah, M.R.: Preliminary design of mobile visual programming apps for Internet of Things applications based on Raspberry Pi 3 platform. In: International Electronics Symposium on Knowledge Creation and Intelligent Computing, pp. 50–54 (2017)
7. Blackstock, M., Lea, R.: Toward a distributed data flow platform for the Web of Things (Distributed Node-RED). In: ACM International Conference Proceeding Series, vol. 08, pp. 34–39 (2014)
8. Blackstock, M., Lea, R.: FRED: a hosted data flow platform for the IoT. In: 1st International Workshop on Mashups of Things and APIs (2016)
9. Boshernitsan, M., Downes, M.S.: Visual programming languages: a survey. Technical report UCB/CSD-04-1368, EECS Department, University of California, Berkeley, December 2004. http://www2.eecs.berkeley.edu/Pubs/TechRpts/2004/6201.html
10. Burnett, M., Kulesza, T.: End-user development in Internet of Things: we the people. In: International Reports on Socio-Informatics (IRSI), Proceedings of the CHI 2015 - Workshop on End User Development in the Internet of Things Era, vol. 12, no. 2, pp. 81–86 (2015)
11. Buyya, R., Dastjerdi, A.V.: Internet of Things: Principles and Paradigms. Elsevier (2016)
12. Zenodys B.V.: Zenodys. https://www.zenodys.com/. Accessed 2020
13. Chang, S.: Handbook of Software Engineering and Knowledge Engineering. World Scientific Publishing Co. (2002)
14. Chen, S., Xu, H., Liu, D., Hu, B., Wang, H.: A vision of IoT: applications, challenges, and opportunities with China Perspective. IEEE Internet Things J. **1**(4), 349–359 (2014)

15. Cheng, B., Kovacs, E., Kitazawa, A., Terasawa, K., Hada, T., Takeuchi, M.: Fogflow: orchestrating IoT services over cloud and edges. NEC Techn. J. **13**, 48–53 (2018)
16. Cheng, B., Solmaz, G., Cirillo, F., Kovacs, E., Terasawa, K., Kitazawa, A.: FogFlow: easy programming of IoT services over cloud and edges for smart cities. IEEE Internet Things J. **PP**, 1–1 (2017)
17. De Luca, G., Li, Z., Mian, S., Chen, Y.: Visual programming language environment for different IoT and robotics platforms in computer science education. CAAI Trans. Intell. Technol. **3**(2), 119–130 (2018)
18. Desolda, G., Malizia, A., Turchi, T.: A tangible-programming technology supporting end-user development of smart-environments. In: Proceedings of the Workshop on Advanced Visual Interfaces, pp. 59:1–59:3. ACM, New York (2018)
19. Dias, J.P., Couto, F., Paiva, A.C.R., Ferreira, H.S.: A brief overview of existing tools for testing the Internet-of-Things. In: IEEE International Conference on Software Testing, Verification and Validation Workshops (ICSTW), pp. 104–109 (2018)
20. Dias, J.P., Faria, J.P., Ferreira, H.S.: A reactive and model-based approach for developing Internet-of-Things systems. In: 11th International Conference on the Quality of Information and Communications Technology, pp. 276–281 (2018)
21. Dias, J.P., Ferreira, H.S., Sousa, T.B.: Testing and deployment patterns for the Internet-of-Things. In: Proceedings of the 24th European Conference on Pattern Languages of Programs. EuroPLop 2019. Association for Computing Machinery, New York (2019). https://doi.org/10.1145/3361149.3361165
22. Dias, J.P., Restivo, A., Ferreira, H.S.: Empowering visual Internet-of-Things mashups with self-healing capabilities. In: 2021 IEEE/ACM 2nd International Workshop on Software Engineering Research Practices for the Internet of Things (SERP4IoT) (2021)
23. Dias, J.P., Sousa, T.B., Restivo, A., Ferreira, H.S.: A pattern-language for self-healing Internet-of-Things systems. In: Proceedings of the 25th European Conference on Pattern Languages of Programs. EuroPLop 2020. Association for Computing Machinery, New York (2020). https://doi.org/10.1145/3361149.3361165
24. Dias, J.P., Lima, B., Faria, J.P., Restivo, A., Ferreira, H.S.: Visual self-healing modelling for reliable Internet-of-Things systems. In: Proceedings of the 20th International Conference on Computational Science, pp. 27–36. Springer (2020)
25. Ens, B., Anderson, F., Grossman, T., Annett, M., Irani, P., Fitzmaurice, G.: Ivy: exploring spatially situated visual programming for authoring and understanding intelligent environments. In: Proceedings - Graphics Interface, pp. 156–163 (2017)
26. Eterovic, T., Kaljic, E., Donko, D., Salihbegovic, A., Ribic, S.: An Internet of Things visual domain specific modeling language based on UML. In: Proceedings of the 25th International Conference on Information, Communication and Automation Technologies (2015)
27. Eun, S., Jung, J., Yun, Y.S., So, S.S., Heo, J., Min, H.: An end user development platform based on dataflow approach for IoT devices. J. Intell. Fuzzy Syst. **35**(6), 6125–6131 (2018). https://doi.org/10.3233/JIFS-169852
28. Fayed, M.S., Al-Qurishi, M., Alamri, A., Al-Daraiseh, A.A.: PWCT: visual language for IoT and cloud computing applications and systems. In: ACM International Conference Proceeding Series (2017)
29. Ghiani, G., Manca, M., Paterno, F., Santoro, C.: Personalization of context-dependent applications through trigger-action rules. ACM Trans. Comput. Hum. Interact. **24**(2), 14:1–14:33 (2017)

30. Giang, N.K., Lea, R., Leung, V.C.M.: Exogenous coordination for building fog-based cyber physical social computing and networking systems. IEEE Access **6**, 31740–31749 (2018)
31. Giang, N.K., Blackstock, M., Lea, R., Leung, V.C.: Developing IoT applications in the Fog: a distributed dataflow approach. In: Proceedings of the 5th International Conference on the Internet of Things, pp. 155–162 (2015)
32. Grasp IO Innovations Pvt. Ltd.: GraspIO. https://www.grasp.io/. Accessed 2020
33. Gupta, H., Vahid Dastjerdi, A., Ghosh, S.K., Buyya, R.: iFogSim: A toolkit for modeling and simulation of resource management techniques in the Internet of Things, Edge and Fog computing environments. Softw. Pract. Exp. **47**(9), 1275–1296 (2017)
34. Ihirwe, F., Di Ruscio, D., Mazzini, S., Pierini, P., Pierantonio, A.: Low-code engineering for internet of things: a state of research. In: Proceedings of the 23rd ACM/IEEE International Conference on Model Driven Engineering Languages and Systems: Companion Proceedings. MODELS 2020, USA (2020)
35. Kefalakis, N., Soldatos, J., Anagnostopoulos, A., Dimitropoulos, P.: A visual paradigm for IoT solutions development. In: Interoperability and Open-Source Solutions for the Internet of Things, vol. 9001, pp. 26–45. Springer, Cham (2015)
36. Lago, A.S., Dias, J.P., Ferreira, H.S.: Managing non-trivial Internet-of-Things systems with conversational assistants: a prototype and a feasibility experiment. J. Comput. Sci. **51**, 101324 (2021)
37. NAAS, M.I., Lemarchand, L., Boukhobza, J., Raipin, P.: A graph partitioning-based heuristic for runtime IoT data placement strategies in a fog infrastructure. In: 33rd Annual ACM Symposium on Applied Computing, pp. 767–774 (2018)
38. NETLabTK: Tools for Tangible Design. www.netlabtoolkit.org/. Accessed 2020
39. Node-RED. https://nodered.org/. Accessed 2020
40. NooDL. https://classic.getnoodl.com/. Accessed 2020
41. Noor, J., Tseng, H.Y., Garcia, L., Srivastava, M.: DDFlow: visualized declarative programming for heterogeneous IoT networks. In: Proceedings of the 2019 Internet of Things Design and Implementation, pp. 172–177. ACM (2019)
42. Pathirana, D., Sonnadara, S., Hettiarachchi, M., Siriwardana, H., Silva, C.: WireMe - IoT development platform for everyone. In: 3rd International Moratuwa Engineering Research Conference, MERCon 2017, pp. 93–98 (2017)
43. Petersen, K., Vakkalanka, S., Kuzniarz, L.: Guidelines for conducting systematic mapping studies in software engineering: an update. Inf. Softw. Technol. **64**, 1–18 (2015)
44. Pinto, D., Dias, J.P., Sereno Ferreira, H.: Dynamic allocation of serverless functions in IoT environments. In: 2018 IEEE 16th International Conference on Embedded and Ubiquitous Computing (EUC), pp. 1–8, October 2018
45. Prehofer, C., Chiarabini, L.: From IoT mashups to model-based IoT. In: W3C Workshop on the Web of Things (2013)
46. Ramadas, A., Domingues, G., Dias, J.P., Aguiar, A., Ferreira, H.S.: Patterns for things that fail. In: Proceedings of the 24th Conference on Pattern Languages of Programs, PLoP 2017. ACM - Association for Computing Machinery (2017)
47. Ray, P.P.: A survey on visual programming languages in Internet of Things. Sci. Program. **2017**, 1–6 (2017)
48. Sendorek, J., Szydlo, T., Windak, M., Brzoza-Woch, R.: Fogflow - computation organization for heterogeneous fog computing environments. In: Computational Science - ICCS 2019. pp. 634–647. Springer, Cham (2019)

49. Setiawan, R., Anom Besari, A.R., Wibowo, I.K., Rizqullah, M.R., Agata, D.: Mobile visual programming apps for Internet of Things applications based on raspberry Pi 3 platform. In: International Electronics Symposium on Knowledge Creation and Intelligent Computing, pp. 199–204, October 2019
50. Tomlein, M., Boovaraghavan, S., Agarwal, Y., Dey, A.K.: CharIoT: an end-user programming environment for the IoT. In: ACM International Conference Proceeding Series (2017). https://doi.org/10.1145/3131542.3140261
51. Tomlein, M., Grønbæk, K.: A visual programming approach based on domain ontologies for configuring industrial IoT installations. In: ACM International Conference Proceeding Series (2017). https://doi.org/10.1145/3131542.3131552
52. Torres, D., Dias, J.P., Restivo, A., Ferreira, H.S.: Real-time feedback in node-red for IoT development: an empirical study. In: IEEE/ACM 24th International Symposium on Distributed Simulation and Real Time Applications, pp. 1–8 (2020)
53. Valsamakis, Y., Savidis, A.: Visual end-user programming of personalized AAL in the Internet of Things. In: Lecture Notes in Computer Science, LNCS, vol. 10217, pp. 159–174 (2017). https://doi.org/10.1007/978-3-319-56997-0_13
54. Varshney, P., Simmhan, Y.: Demystifying fog computing: characterizing architectures, applications and abstractions. In: 2017 IEEE 1st International Conference on Fog and Edge Computing (ICFEC), pp. 115–124. IEEE (2017)
55. Wyliodrin. https://wyliodrin.com/. Accessed 2020
56. Zhang, K., Han, D., Feng, H.: Research on the complexity in Internet of Things. In: IET Conference Publications 2010 (571 CP), pp. 395–398 (2010)

Data Preprocessing, Aggregation and Clustering for Agile Manufacturing Based on Automated Guided Vehicles

Rafal Cupek[1]([envelope]), Marek Drewniak[2], and Tomasz Steclik[3]

[1] Silesian University of Technology, Gliwice, Poland
rcupek@polsl.pl
[2] AIUT Sp. z o.o. (LTD), Gliwice, Poland
mdrewniak@aiut.com.pl
[3] Institute of Innovative Technologies EMAG, Katowice, Poland
tsteclik@ibemag.pl

Abstract. Automated Guided Vehicles (AGVs) have become an indispensable component of Flexible Manufacturing Systems. AGVs are also a huge source of information that can be utilised by the data mining algorithms that support the new generation of manufacturing. This paper focuses on data preprocessing, aggregation and clustering in the new generation of manufacturing systems that use the agile manufacturing paradigm and utilise AGVs. The proposed methodology can be used as the initial step for production optimisation, predictive maintenance activities, production technology verification or as a source of models for the simulation tools that are used in virtual factories.

Keywords: Autonomous guided vehicles (AGV) · AI-driven analytics · Data clustering · Pattern searching

1 Introduction

Automated Guided Vehicles are one of major symbols of the fourth industrial revolution [1]. Their importance can be compared to the conveyor belt that was invented for producing the Ford Model T automobiles, which became the symbol of the second industrial revolution. However, the conveyor belt was not the reason for the changes in manufacturing, on the contrary, it was one of the results that came from the technological and social challenges that were visible at the beginning of the 20th century [2]. Similarly, the widespread introduction of AGVs should not be considered to be a new breakthrough technology (AGVs have been used by industry for more than twenty years), but should rather be considered to be an answer to the new technological challenges and a visible sign of the ongoing changes in manufacturing.

The new generation of manufacturing systems is characterised by a high degree of flexibility, which is required in order to cope with frequently changing customers' orders, low material buffers and agile production technologies. Moreover, production is often performed by robotic production stations that can execute many variants of the technological operations [3]. Contemporary manufacturing has to be supported by data mining mechanisms, artificial intelligence and cloud computing. AGVs are not stand-alone technological solutions because they have to cooperate and become a part of the highly advanced information services that are performed during the successive steps in the production chain. This means that AGVs have to cooperate with the advanced informatics tools that support the local optimisation of the ongoing production tasks and also have to support the automatic cooperation with production stands that are based on Machine-to-Machine communication. AGVs also have to be equipped with self-diagnostic tools, which are necessary for their predictive maintenance activities and have to support the new information architectures that are characterised by a high degree of autonomy and the distribution of the decision-making processes in manufacturing [4].

This research is focused on processing and managing the data that are collected by AGVs. The main research topics are data aggregation and clustering. The main research challenges were (i) to discover an unknown number of variants of transport orders performed by AGVs; (ii) to assign the subsequent, realised transport orders flowing in stream to the appropriate variant and (iii) to identify new, previously unknown variants of the transport orders.

The authors used two well-known clustering algorithms, K-means and DB-Scan, and used communication middleware that is based on the OPC UA standard. The authors consciously selected these "nonstreaming" clustering algorithms despite the fact that there are solutions that are dedicated to data that is flowing in a stream, for example, CluStream or Den-Stream, which are available in the MOA project [17]. The "streaming clustering algorithms" are primarily based on the concept of two steps (online and offline) and microclustering, which most often means high-speed computation. However, when the duration of a single transport cycle that is performed by an AGV (seconds, even minutes) is considered, the speed of a calculation is not crucial. The main reasons for selecting the above-mentioned "classic" clustering algorithms were (i) the authors were more interested in the possibility to immediately compare the incoming data (about the transport cycle being performed) with the existing reference clusters and anomaly detection [19] and (ii) "streaming clustering algorithms" enable clusters to "shift" in space when fluctuations occur in the incoming data values.

The main research contributions of the paper are (i) the proposed architecture for a data collection and aggregation system for AGV, which is presented in chapter two and (ii) the methodology for AGV data clustering and the experimental research that verified the proposed solutions, which is presented in chapter four. The presented results are part of the industrial research on the new generation of AGVs that are being designed and produced by AIUT LTD.

2 Aggregating Production Records for an AGV

The operating principle for the control systems that are used in manufacturing can be based on the event-triggered mode, which is typical for high-level control (e.g. state machine based Sequential Function Chart SFC, which is defined under the IEC 61131–3 standard or the distributed control architecture that is defined under the IEC 61499 standard) [5] or can follow a time-triggered periodic processing, which is typical for low-level control systems (e.g. a PLC running a Ladder or Function block diagram, which cyclically sample information from the sensors, execute the control logic and update the actuators) [6]. In both cases, the sources of information provide data streams (subsequent events or cyclically updated samples) that are quite difficult for data mining tools to directly analyse [7].

AGVs, which and use both control principles, are used to perform internal logistic tasks [8]. Time-triggered periodic data processing is used for the continuous control of an AGV's movement while the event-triggered mode is typical for communication with MES (Manufacturing Execution Systems) or high-level navigation support. Both types of data processing can be a source of valuable information for a data mining analysis. Each transportation task has a beginning and an end and can be characterised by one production record. Although from a technological point of view, the tasks that are performed by AGVs can be separated (one delivery can be transported by more than one AGV) or combined (several deliveries can be transported simultaneously) [x1], separating and joining transportation cycles was omitted for this research study. It was assumed that each transportation cycle included all of the operations that an AGV performs between the loading and unloading point.

In order to make the machine learning process effective and efficient, the streams of data that come from different sources have to be joined and grouped according to their relevance to the production steps being analysed. Next, the collected data have to be aggregated in order to create new information that reflects the statistically important features. This step combines the external engineering knowledge that can be expressed by the data models with the current information that is being collected from the production system. In this way, the streams of data from AGVs can be transformed into the discreet production records that are created by the aggregation functions.

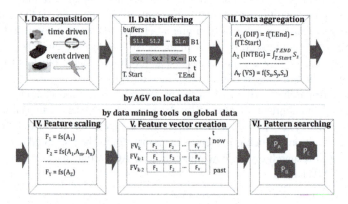

Fig. 1. AGV data processing methodology

The authors assumed that each aggregation function produces one scalar value according to the data model that is defined for a given AGV type. On the one hand, the information that is collected in one production record comes from different sources (e.g. sensors, control signals, machine states, messages transmitted from the parent system, independent security systems etc.), while on the other hand, each information source can be used many times by the different aggregation functions (different points of view on the same data source). The fusion of the aggregates forms a production record, which is an input vector for the data mining tools and characterises a single transportation cycle. Because this vector has the same structure for all of the transportation cycles (for a given data model), it is a convenient source of information for the data mining algorithms.

The proposed methodology (Fig. 1) for AGV data processing is composed of six steps: (i) data acquisition – sensors, control systems, supervisory systems, external events and other system provide a number of information streams. Data is exchanged based on a time-driven paradigm (where the samples are delivered cyclically according to a defined time interval) or by an event-driven paradigm (where the information is sent by its source in the event that a predefined event occurs); (ii) data buffering – the information has to be temporarily stored (buffered) in order to make its aggregation possible. The local storing is for a specific time period that is determined by the beginning and the end of each transportation cycle; (iii) data aggregation – each information buffer has to be transformed into a single scalar value that characterises the collected samples. The data stream from a single sensor can be processed by different aggregation functions (in order to detect the different features of a given signal) or an aggregate can be built on the data from multiple sensors (data fusion). Examples of the aggregates that were used in this research are DIF – the difference between the values at the beginning and at the end of a transportation cycle (e.g. for the energy consumption meters), INTEG – the integral of a parameter that is limited for a production cycle (e.g. the rotational speed of the drive wheels, which are integrated over the duration of the transportation cycle) or VS – the virtual sensors that generate new information that is based on one or more information sources and that includes any additional knowledge about the production technology; (iv) feature scaling, which is designed to change the values of the aggregates that are measured in the engineering units to a new artificial scale that reflects the technological importance of a given information source. This process differs from the classic normalisation that is used in data mining because the range of values cannot be predicted in advance (e.g. the duration of an operation may be extended many times due to the occurrence of a production error). Moreover, the standard normalisation procedure does not permit the technological significance of individual parameters to be taken into account; (v) feature vector creation, which is the step during which the data from the different aggregates are combined into the input feature vector and (vi) pattern searching, which on the one hand, is the output from the proposed aggregating methodology, while on the other hand, the database of the patterns that represent the repetitive transportation use cases can be used as the input for the subsequent stages of data mining.

The implementation should permit agile manufacturing including the use of big data analytics, cyber-physical systems and prediction technologies [10]. This can be divided into two parts: steps (i) – (v) can be performed in a distributed manner, while step (vi) should be performed on the global data using the data mining tools. However, the second part can also be parallelised when specific models of the aggregates are used by a number of the data mining tools that are used to simultaneously search for different transportation patterns. Another challenge is integrating the proposed approach with an actual manufacturing system. The analytical module has to combine the production data that comes from different AGVs that has actually been collected with the simulation data that has been created based on the relevant models. Additionally, the new models should be dynamically created and based on the collected data. This goal leads to the proposed system architecture that is shown in Fig. 2. The authors' research is based on the OPC UA communication middleware [11], which seems to be one of key enabling technologies for the new generation of manufacturing systems.

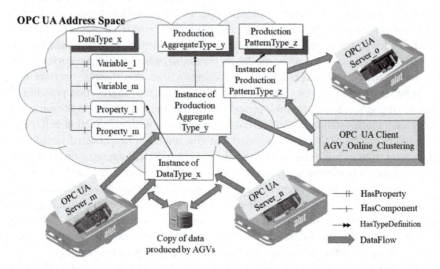

Fig. 2. Proposed system architecture.

The proposed architecture can be described by the following features: (i) the information models for the aggregated production records are transparent and are common for all of the AGVs. The definition is visible in the OPC UA address space and the object models that are used are the same for all of the OPC UA servers; (ii) many different models can be created and processed for the aggregates. Each record is identified according to the object-oriented approach by a class type that defines its content. New aggregates can be added based on the requirements that are generated by the data mining modules; (iii) the Historical Data Access mechanism, which is performed by OPC UA servers enables a later aggregation, which includes any new patterns that emerge in the data mining applications. It can also be used for simulations in virtual factories; (iv) the patterns for the transportation cycles are also visible in the OPC UA address space. These enable new fields of the data mining applications that support manufacturing decisions and (v) the data collected by the OPC UA servers can be accessed by many OPC UA clients in the parallel access mode. The parallel processing can be more efficient, e.g. in the event that the various patterns that are associated with the different data models have to be analysed.

3 The AGV that Was Used for the Experiments

The AGV that was used for the research was a Formica-1, which is designed and manufactured by AIUT LTD. The unit was designed to perform transportation tasks including both the typical intra-logistics operations such as delivering resources to the production stations or to be used as a transportation carrier for the manufactured products. The Formica-1 is presented in Fig. 3.

Fig. 3. The Formica-1 AGV that was used in the experiments

An AGV can be equipped with various types of navigation methods. One of the most common is natural navigation, which is based on two onboard laser scanners that provide measurement data about the surroundings of a unit. The measurements are used to create a map of the production environment and later to compare what the AGV is currently observing in relation to this map. The techniques that are used to calculate the travelled distance and the heading of the machine are based on odometry. In order to provide the complete diagnostic information about the performed transportation tasks as well as the operation of all of the onboard devices, all of the signals from the automation control and industrial IT systems are passed through the unit controller. This controller gathers the data, adds timestamps and transmits it to third party diagnostic systems cyclically using a TCP/IP connection.

Depending on the purpose of the analysis of this data, different values are available: the process memories and statuses of work, the signals of the activity of specific devices, the measured energy and time, the distance travelled, velocities, directions, payload control, sensor indications etc. The data are grouped into functional structures, marked with specific identifiers and transmitted in the TCP frames. For this study, all of the data associated to the work statuses, diagnostic information, the activity of the devices, energy measurements and cycle determination were received and entered into the OPC UA server. The AGV was used in the testing area in AIUT LTD, where units are functionally tested before they are sent to and set up in the target production environments. The testing area has space for many production scenarios and therefore, the experiments were combined with preliminary customer tests, in which the AGV performed the transportation tasks in several variants. The data, which was recorded and stored for further data mining approach, was also gathered during the actual operation of the Formica-1. These data included the duration of a cycle; the energy that was required to implement it; the duration of the forward and backward movements; the rotational speed of the left and right drive wheels, which were integrated over the transportation cycle; the times when the LED lighting was activated (red, blue and green); the duration of the blinker signals, which were aggregated as virtual sensors and the downtime. The units that were used to measure these data were milliseconds, kilowatts and RPM x s.

Communication between an AGV and the data mining system is based on OPC UA, which utilises Webservices to communicate with the enterprise management systems and for the TCP-based communication with the control systems. In terms of security, OPC UA uses the solutions that are appropriate for a given family of protocols. The security of the information exchange is managed by a Secure Channel Service Set, which defines a long-running logical connection between an OPC UA client and a server. OPC UA is a scalable, reliable and safe middleware that can not only be used as a data connector but also as a translator for information models that can be used to match different information systems [12].

4   Pattern Searching

The presented research was performed in a near-real environment during the preliminary tests of the Formica-1 before it was deployed at the customer's facility. During the course of the experiment, more than two hundred transport cycles of the AGV were logged, recorded and clustered. The pattern searching analytical module performed the clustering of production records that were collected from the AGV. The result was information about the clusters of transportation patterns and the assignment of the observed production cycles to these clusters. The main objectives of clustering were (i) to continuously monitor the work being performed by the AGV; (ii) to monitor the current condition of the AGV regarding it efficiency and the wear on its components (physical, power, executive); (iii) to detect any failures in the AGV; (iv) to detect any "problem spots" that could result in downtime and stoppages in realizing an order; (v) to detect any anomalies when realizing an order and (iv) to detect any new types or ways of outperforming the transportation cycles.

The work of the analytical module was divided into two phases: initialisation and production. The first phase was the one in which the analytical module was fed the data about the clean and undisturbed transport cycles that were considered to be correct in a given production scenarios – the observations that represented the assumed and possible states of the monitored object. The analytical module attempted to conclude how these cycles could be grouped (divided into clusters). It is worth mentioning that for the person analysing the work of the monitored object, the number of different states of the object remained unknown.

The production phase, which was next, was when the data flowed (as a stream) into the analytical module – the data that represented the observations that were actually recorded during the AGV's operation. The analytical module attempted to use the knowledge that was obtained in the initialisation phase and to qualify the obtained production records (data on the work that was performed by the AGV) into one of the previously discovered clusters. The analytical module enabled an analysis using one of the two tested methods that were (i) based on the K-means algorithm [13, 14] and (ii) based on the DB-SCAN algorithm [16].

4.1   Normalisation and Rescaling of the Values

Because the analytical module is intended for continuous operation and future transport cycles are not known, the authors could not normalise the source data values. The only process that the values were subjected to was rescaling – the problem was the different orders of magnitude that were obtained from the server. A single transportation cycle lasted for several dozen to several hundred seconds, while the values for the durations were presented in milliseconds. Similarly, the drive rotations were expressed in the number of rotations per minute, which led to poor results in the clustering. After the initial experiments, it was decided to use a single rescaling of the values function, where the basis of the time was a second. Therefore, all of the times were divided by 1000 and the RPM x s by 60. It was decided that the energy consumption should remain in kilowatts.

4.2 Initialisation Phase

During the initialisation phase of the experiment, more than 37 transportation cycles of the AGV were logged and recorded. Because the logging was done during the functional tests of three transportation variants, authors expected that three different types of production scenarios would be obtained from the input data. However, this information and the detailed information about the exact number of registered transportation cycles remained invisible for the clustering system.

The method that was based on the K-means algorithm was divided into two stages: (i) when searching for the most optimal K (number of clusters), the most optimal K was selected using the elbow method (Fig. 4) [15] and (ii) dividing the observations into the K-discovered clusters and remembering the information about the centroids and standard deviation for each of them.

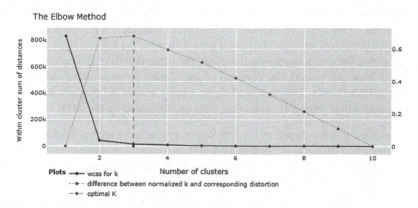

Fig. 4. The elbow method

The method that was based on the DB-Scan algorithm was also divided into two stages: (i) determining the maximum distance between the observations based on the average distance between N nearest neighbours (it was assumed that the maximum average distance on the histogram of the distances for the five nearest neighbours would be found, see Fig. 5) and (ii) dividing the observations into clusters taking into account the maximum distance between observations that was determined.

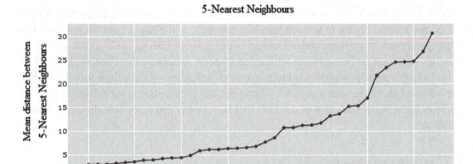

Fig. 5. Histogram of distances for the five nearest neighbours

Obtained Results of the Initialisation Phase. According to the initial suppositions, both methods divided the observations into identical clusters: 19 transportation cycles were assigned to the type 1 production pattern, 10 cycles to the type 2 production pattern and 9 cycles to the type 3 production pattern. The results were then discussed with the person who had ordered the transport cycles. It was found that all of the cycles were recognised, and that the orders were grouped correctly. Because an input vector that consisted of seventeen features was selected for analysis, in order to be able to present the graphical division into the clusters for the purposes of visualisation, the number of dimensions was reduced to three [18]. In Fig. 6, the authors present this division but only in the space of two dimensions ("cycle time" and "cycle energy"). These two features were selected as the most representative. The other parameters were the duration of the forward and backward movements, the rotational speed of the left and right drive wheels integrated over the entire transportation cycle, the times that the LED lighting was activated (red, blue and green), the duration of the blinker signals (aggregated as virtual sensors) and downtime. The presented figure clearly shows the clusters that represent the various assumed transport cycles that were implemented by the AGV.

The method that was based on the DB-Scan required more attention when determining the maximum distance in combination with the minimum number of observations that were included in a cluster. This results directly from the way in which the algorithm operates, which treats any observations that cannot be categorised (as a core/border point) as noise. The research showed that several attempts usually led to the expected effect (the correct number of clusters, the appropriate distance and the minimum number of observations that formed a cluster).

4.3 Production Phase

In the production phase, the analytical module used the previously acquired knowledge about the clusters for the new, received data and then attempted to assign them into one of them. For the K-means-based method, the process was as follows: using the information about the number of clusters and centroids to assign each of the new observations (received in a stream after each subsequent completed cycle) to the nearest one

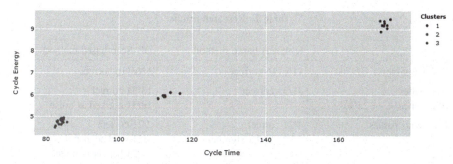

Fig. 6. Clustering results in two dimensions

and compare the distance with the value of the standard deviation for that cluster. For the second method (DB-Scan-based method), the process was as follows: assign new, received observations into clusters by comparing their distance to every core points with the maximum distance and radius that was assumed during the initialisation phase.

Obtained Results of the Production Phase. During this part of the experiment, there were 168 transport cycles, streamed and fetched during the AGV's operation. For the transport cycles whose realisation was not disturbed in any way (116 transport cycles), both methods divided those observations into the correct clusters. For the observations in which the values for some of the features deviated from the standard values (there were 31 anomalies with differences of between 10%−25%), both methods behaved differently. The first method required the observations to be assigned to the nearest cluster, regardless of whether it was the correct one or not. Because the differences in the values for the individual parameters for the transport cycles belonging to different clusters were significant (e.g. the difference in the duration of the order execution was greater than 30%), all of the anomalies were correctly assigned. For the second method, 18 of those observations were marked as noise. A detailed analysis of these observations showed that this procedure was correct because the values of the features for these observations differed from the standard ones to such an extent that the distances exceeded the radius that had been declared in the initialisation phase.

During the research, additional erroneous transport cycles were also discovered (there were 21 such cycles), which could not be qualified for any of the clusters. These were the situations in which the AGV was in a state of a safety failure that was caused by a poorly planned route or a downtime of tens of minutes. For the first method, four cycles were clustered correctly. For the second method, two were assigned correctly, some were classified as noise (5 of the 21) and the rest to were assigned bad clusters.

The research results are summarised and presented in Table 1.

Discussion of Results. Using the method based on the K-means algorithm permitted the analysis of the individual incoming observations. The advantage of this method is its speed. However, its disadvantage is that each observation must be assigned to one of the clusters. This can lead to incorrect assignments that are difficult to identify quickly and is possible if the distances and the standard deviations for individual features are

Table 1. Research results

	K-means-based method	DB-Scan-based method
Correct passes	116/116 correct	116/116 correct
Anomalies (difference < = 25%)	31/31 correct	13/31 correct 18/31 marked as noise
Erroneous passes	4/21 correct 17/21 bad assignment	2/21 correct 5/21 marked as noise 14/21 bad assignment

analysed (keeping in mind the differences in the values for the individual parameters for the transport cycles belonging to different clusters).

Using this method to discover new clusters (e.g. by starting the implementation of new types of transport cycles, which were unknown in the earlier stages) might not be an easy task and requires restarting the initialisation phase with all the data that had already been collected (point 4.2). This process can be performed at any time when: (i) there are several cycles for which the distance from the centroid is significant (a significant change in the standard deviation value) or (ii) the fact that a new type of transport cycle has occurred has been confirmed by the person ordering the work.

Using the method based on the DB-Scan algorithm also was determined to be relatively easy. As was the case of the previous method, all of the production cycles that did not have any anomalies or errors were assigned to the appropriate clusters. Although the behaviour of this algorithm was definitely different for the disturbed passes, it cannot be said that identifying them was easy. In some cases, the anomalies were assigned as noise, and in these cases, they were immediately identified. At the same time, some of the anomalies were assigned incorrectly and only a small proportion was correctly assigned. In a situation in which the observations that were classified as noise were individual cases and at the same time the values for the individual features did not differ significantly from the others, this could indicate the need to increase the distance that defines the cluster boundary (due to the fluctuations in the values). On the other hand, when the observations begin to resemble clusters, it could mean that it is a seed of a new cluster.

During the research, it was observed that the DB-Scan-based method was quite sensitive to changes in the values of the features, even if they were part of the observations that should belong to the same cluster, which made it possible, for example, to discover any incorrect AGV settings for several passes. An example could be a situation in which an AGV carries out a transport order, but instead of moving forward, part of the route was reversed by 180 degrees. A similar situation occurred for the routes in which the steering angles for some of the turnings were slightly larger, which was associated with the activation of the turn signals by the AGV.

Finally, it could be difficult for the end user to specify two parameters for the method using the DB-Scan: how many nearest neighbours should the smallest cluster consist of and what should the maximum distance between the observations be, which means that it OR they belong to a given cluster. This is especially important when analysing the observations for which there is no reference point.

However, with this method, it is possible to observe the formation of new clusters – if the subsequent observations (transport cycles) begin to form a new cluster and their number is sufficient to create it. However, correctly recognising a new cluster and assigning transport cycles to it using this method should begin with restarting the initialisation phase with all of the previously registered observations.

5 Conclusions

In this paper, the authors present industrial research on the methodology and system architecture that can support gathering, preprocessing, aggregating and clustering data for AGV-based internal logistics. The methods, which are based on K-means and DB-Scan, for clustering production records were verified in a near-real test environment. This work and the obtained results show that it is possible to use classical clustering algorithms for data that is flowing in a stream. It seems that using them would be easier than using the stream algorithms, especially in systems where the processing time is not a critical parameter. The proposed methods make it possible to detect an unknown number of clusters, correctly assign incoming observations to the appropriate groups and observe the formation of new clusters. Both of the proposed methods are divided into two phases, where the first phase is creates the reference point and the second is the operational phase. For both methods, it is possible to run the first phase at any time that is selected by the user, which results in the creation of a new reference model. It is important to emphasise that the results of the conducted experiments show that the two proposed methods behave differently in the case of any anomalies. OPC UA-based communication middleware makes it possible to separate the machine learning process from the data gathering and preprocessing. The knowledge about the AGV sensors and its technological meaning is stored in and expanded on by the OPC UA address space. This knowledge can be easily shared between AGVs and is also available for other systems, in particular, for data mining applications. The production patterns that are discovered by the data mining part can be used for production optimisation, predictive maintenance activities or as a source of models for the simulation tools.

Acknowledgements. The research leading to these results received funding from the Norway Grants 2014–2021, which is operated by the National Centre for Research and Development under the project "Automated Guided Vehicles integrated with Collaborative Robots for Smart Industry Perspective" (Project Contract no.: NOR/POLNOR/CoBotAGV/0027/2019 -00) and from the project "Hybrid systems of automated internal logistics supporting adaptive manufacturing" (grant agreement no POIR.01.01.01–00-0460/19–01) realised as Operation 1.1.1.: "Industrial research and development work implemented by enterprises" of the Smart Growth Operational Programme from 2014–2020 co-financed by the European Regional Development Fund.

References

1. Bechtsis, D., Tsolakis, N., Vlachos, D., Iakovou, E.: Sustainable supply chain management in the digitalisation era: the impact of Automated Guided Vehicles. J. Clean. Prod. **142**, 3970–3984 (2017)

2. Womack J.P., Jones D. T., Roos, D.: The machine that changed the world: the story of lean production—Toyota's secret weapon in the global car wars that is now revolutionizing world industry. In: Rawson Associates Macmillan Publishing Company, pp. 48–70 (2007)
3. Maskell B.: The age of agile manufacturing. Supply Chain Manag. Int. J. **6**(1), 5–11 (2001)
4. Cupek, R., Ziebinski, A., Drewniak, M., Fojcik, M.: Knowledge integration via the fusion of the data models used in automotive production systems. Enterprise Inf. Syst. **13**(7–8), 1094–1119 (2018)
5. Zhabelova, G., Vyatkin, V., Dubinin, V.N.: Toward industrially usable agent technology for smart grid automation. IEEE Trans. Ind. Electron. **62**(4), 2629–2641 (2014)
6. Stouffer, K., Falco, J., Scarfone, K.: Guide to industrial control systems (ICS) security. NIST Special Publ. **800**(82), 2_1–2_14 (2014)
7. Fei, X., et al.: CPS data streams analytics based on machine learning for cloud and fog computing: a survey. Futur. Gener. Comput. Syst. **90**, 435–450 (2019)
8. Yoshitake, H., Kamoshida, R., Nagashima, Y.: New automated guided vehicle system using real-time holonic scheduling for warehouse picking. IEEE Robot. Autom. Lett. **4**(2), 1045–1052 (2019)
9. Digani, V., Sabattini, L., Secchi, C.: A probabilistic eulerian traffic model for the coordination of multiple AGVs in automatic warehouses. IEEE Robot. Autom. Lett. **1**(1), 26–32 (2016)
10. Lin, Y.C., Hung, M.H., Huang, H.C., Chen, C.C., Yang, H.C., Hsieh, Y.S., Cheng, F.T.: Development of advanced manufacturing cloud of things (AMCoT)—a smart manufacturing platform. IEEE Robot. Autom. Lett. **2**(3), 1809–1816 (2017)
11. Lange, J., Iwanitz, F., Burke, T.J.: OPC – From Data Access to Unified Architecture. VDE Verlag, pp. 111–130 (2010)
12. Cupek, R., Folkert, K., Fojcik, M., Klopot, T., Polaków, G.: Performance evaluation of redundant OPC UA architecture for process control. Trans. Inst. Meas. Control. **39**(3), 334–343 (2017)
13. Linde, Y., Buzo, A., Gray, R.: An algorithm for vector quantizer design. IEEE Trans. Commun. **28**, 84–95 (1980)
14. David, A., Vassilvitskii, S.: k-means++: the advantages of careful seeding. In: Proceedings of the eighteenth annual ACM-SIAM symposium on Discrete algorithms, Society for Industrial and Applied Mathematics (2006)
15. Syakur, M.A., Khotimah, B.K., Rochman, E.M.S., Satoto, B.D.: Integration k-means clustering method and elbow method for identification of the best customer profile cluster. IOP Conf. Series: Mater. Sci. Eng. **336**, 1–7 (2017)
16. Ester, M., Kriegel, H.P., Sander, J., Xu, X.: A density-based algorithm for discovering clusters in large spatial databases with noise. In: Proceedings of the 2nd International Conference on Knowledge Discovery and Data Mining, Portland, OR, AAAI Press, pp. 226–231 (1996)
17. Bifet, A., Gavalda, R., Holmes, G., Pfahringer, B.: Machine learning for data streams with Practical Examples in MOA. Massachusetts Institute of Technology (2017). ISBN: 978-0-262-03779-2
18. Abdi, H., Williams, L.J.: Principal component analysis. Wiley Interdisc. Rev. Comput. Stat. **2**(4), 433–459 (2010). arXiv:1108.4372. https://doi.org/10.1002/wics.101
19. Gomes, H.M., Read, J., Bifet, A., Barddal, J.P., Gama, J.: Machine learning for streaming data: state of the art, challenges, and opportunities. ACM SIGKDD Explorations Newsl. **21**(2), 6–22 (2019)

Comparison of Speech Recognition and Natural Language Understanding Frameworks for Detection of Dangers with Smart Wearables

Dariusz Mrozek[1]([✉])(iD), Szymon Kwaśnicki[1], Vaidy Sunderam[3](iD),
Bożena Małysiak-Mrozek[2](iD), Krzysztof Tokarz[2], and Stanisław Kozielski[1]

[1] Department of Applied Informatics, Silesian University of Technology,
Akademicka 16, 44-100 Gliwice, Poland
dariusz.mrozek@polsl.pl
[2] Department of Graphics, Computer Vision and Digital Systems,
Silesian University of Technology, Akademicka 16, 44-100 Gliwice, Poland
krzysztof.tokarz@polsl.pl
[3] Department of Computer Science, Emory University, Atlanta, GA 30322, USA
vss@emory.edu

Abstract. Wearable IoT devices that can register and transmit human voice can be invaluable in personal situations, such as summoning assistance in emergency healthcare situations. Such applications would benefit greatly from automated voice analysis to detect and classify voice signals. In this paper, we compare selected Speech Recognition (SR) and Natural Language Understanding (NLU) frameworks for Cloud-based detection of voice-based assistance calls. We experimentally test several services for speech-to-text transcription and intention recognition available on selected large Cloud platforms. Finally, we evaluate the influence of the manner of speaking and ambient noise on the quality of recognition of emergency calls. Our results show that many services can correctly translate voice to text and provide a correct interpretation of caller intent. Still, speech artifacts (tone, accent, diction), which can differ even for each individual in various situations, significantly influences the performance of speech recognition.

Keywords: Internet of Things · Cloud computing · Natural language processing · Wearable sensors · Intention recognition · Speech recognition · Older adults

This work was supported by pro-quality grant for highly scored publications or issued patents (grant No 02/100/RGJ21/0009), the professorship grant (02/020/RGP19/0184) of the Rector of the Silesian University of Technology, Gliwice, Poland, and partially, by Statutory Research funds of Department of Applied Informatics, Silesian University of Technology, Gliwice, Poland (grant No BK-221/RAu7/2021).

M. Paszynski et al. (Eds.): ICCS 2021, LNCS 12745, pp. 471–484, 2021.
https://doi.org/10.1007/978-3-030-77970-2_36

1   Introduction

In many developed countries, the population is aging, due partly to longer life expectancy, which in turn is partly due to better medical care [1]. Among the elderly, human independence decreases, and the risk of disability increases. Despite family support and geriatric care, the elderly increasingly stay alone at home most of the time [26]. In the event of a sudden deterioration in health or an accident, they are often unable to call for help. In the event of an immediately life-threatening condition, e.g., myocardial infarction, stroke, or hypoglycemia, self-call for help is often impossible, and every minute of delay in implementing appropriate treatment carries irreversible consequences, including the patient's death. Smart wearable devices can help seniors (and others) in such situations by giving them a simple, technology-assisted, way to call their loved ones or request medical assistance.

The Internet of Things (IoT) has become an essential technology that allows people to monitor themselves during daily activities. In particular, there is growing interest in using IoT devices for personalized healthcare, monitoring older people and children, and detecting potential dangers in their lives [24,28]. Smart wearable devices, like smart bands and smart bracelets, can gather information about vital health parameters, including pulse, ECG, body temperature, blood pressure, and oxygen saturation. Smartwatches and mobile phones can detect falls and notify caregivers automatically when fall is detected [23]. External sensors and cameras can monitor a person constantly and raise the alarm when the situation is dangerous. However, sometimes it is much easier for the monitored senior to call for help than to monitor all possible life parameters and vital signs with such wearable devices and external sensors. A smart solution for this purpose can be based on an IoT personal device (like a pendant, necklace, or smart band) that can register the voice of the monitored person, analyze it or transmit it for analysis after connecting to a data center (Fig. 1). Such technologies can be especially useful in (a) providing voluntary additional information to physiological measurements, and (b) distinguishing between false and true alarms.

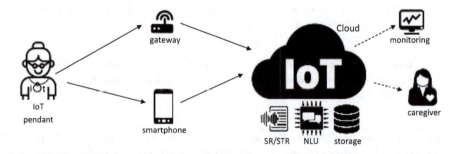

Fig. 1. Schematic diagram of monitoring an older adult with an IoT pendant registering voice and transmitting it for the analysis to a data center in the Cloud.

In this paper, we analyze the capabilities of various speech recognition and natural language understanding frameworks in the automatic detection of situations when a monitored person is requesting help. We rely on the architecture of the experimental environment, where the detection of attention-requiring situations occurs in the central Cloud-based telemedicine system for monitoring and data processing.

2   Related Works

The most natural way of communication between humans is speech, which has led to a focus on spoken language as an important human-machine interface [8,13]. Computer technology can understand phrases and whole sentences using spoken language understanding (SLU) systems. Such systems typically consist of a pipeline of two main modules – automatic speech recognition (ASR) followed by natural language understanding (NLU) [4].

Automatic speech recognition (ASR), also called speech to text recognition (STR), or speech to text analysis is an active area of research for many years. Currently available systems can recognize words in different noise conditions, spoken by different people in many languages, with high accuracy. Although their level of recognition does not achieve 100% they have been implemented in many solutions for everyday life usage, i.e., as the user interface to computers or smartphones, in smart homes [21], in telecommunication [18], and in the health care sector for automatic generation of medical documentation [11] or recording medical interview data [17]. STR systems give accessibility options for people with disabilities [18]. For example, Dimauro et al. presented research on using STR in measuring speech impairment in Parkinson's disease [11].

Natural Language Understanding (NLU) is one of the biggest challenges in computer science. Currently, recognizing separate words is quite simple but understanding their meaning while spoken in conjunction with other words is a complex task for computers [5]. To prevent misunderstandings, the Controlled Natural Language (CNL) approach can be used. Strong CNL has well-structured semantics, and words have a single meaning; it can be equivalent to formal language [25]. Weak CNL allows more than one word meaning but specifies some restrictions on the construction of sentences to avoid ambiguity [5]. It must be noted that in emergency situations, patients may not be able to utter a sentence according to the rules, placing CNL effectiveness in doubt. Although properly-recognized sentences can be highly informative, additional information can be obtained with emotion detection [9]. This can help when patients cannot speak a complete sentence, but utter emotionally short phrases or separate words.

NLU systems are implemented using deep neural networks (DNN) [10] or recurrent neural networks [12]. The continuing development of cloud and networking technology has driven the growth of a number of cloud-based services with artificial intelligence services, including such DNN-based services, and has allowed moving the voice recognition process to the Cloud. The first commercially available voice assistant (VA) was Apple's Siri launched in 2011 [3]. Subsequently, competitors created their own systems, with the Microsoft Cortana

(2014), Amazon Alexa (2014), and Google Assistant (2016) being the most popular. Cloud NLP is more effective than voice recognition technology built into end-devices [3] for several reasons: the possibility of installing custom applications; constant learning of new words and phrases from many users; and of course, computing power.

A comparison of the most popular VAs by Lopez et al. [19] shows that it is a non-trivial task to obtain both naturalness and correct operation at the same time. According to their research, the most natural but less accurate is Google Assistant, while Apple's Siri is the most accurate but the least natural [19]. Ammari et al. [2] surveyed everyday usage of voice assistant systems. Analyzing results, they found that the commands most often used with voice assistant systems are used for playing music, hands-free searching, and controlling IoT devices.

In the literature, there are many interesting examples of successful voice-controlled systems implementation. Many of them are created with open-source tools, including Node-RED, IFTTT, and well-known communication protocols like MQTT [16]. An example of smart home implementation based on Android is presented in [14]. In [3] Austerjost et al., a system to control real laboratory equipment with voice commands is presented. Using their system, it is possible to read measurements from sensors, report status, laboratory equipment parameters, and read operating procedures. In Mitrevski [22], the author describes events where the conversational interface used by a four-year-old child helped save his mother's life. The current research focuses mainly on methods of natural language processing (NLP) that allow recognizing not only simple commands but also conditional or personalized actions [20], and complex phrases with the ability to ask clarifying questions.

In our project, we adopt a different outlook that complements the descried approaches – we propose to build a system that acts as a personal assistant similar to home assistants, but focuses on the aspect of medical care and calling for help. The designed system has several functionalities, which are described in the next chapter.

3   Personal Assistant with Cloud Voice Analysis

A personal voice assistant is an IoT device that detects a dangerous event by registering the voice of a person that utters a call for help sentence or phrase. The assistant can be a band, pendant, or another type of device that an older person can easily carry. This mobile device directly or indirectly accesses the Internet to connect to a data center located in the Cloud. The main task of this device is to listen to the surrounding environment in standby mode. After detecting sounds from the environment, the sounds are analyzed to determine if they contain a speech recording or just irrelevant background noise. If the recording is classified as a voice, it is processed to highlight the desired features and remove unnecessary silence and surrounding sounds. Then the recording is converted to a fixed file type with a certain quality (kbps) and uploaded to the cloud data center for further analysis, as shown in Fig. 2.

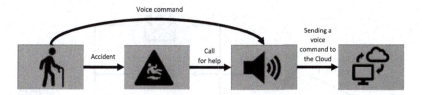

Fig. 2. Main operational phases in a call for help system with a data center for voice analysis located in the Cloud.

After receiving the speech recording by a dedicated application in the cloud data center, specific system modules begin intelligent processing of the voice command to determine if it contains a call for help. The intelligent processing of voice recordings consists of two steps (Fig. 3):

- the transcription of a voice command into text,
- appropriate analysis of the command represented in text, including intention recognition.

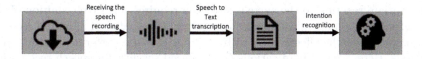

Fig. 3. Steps of voice command analysis.eps

Among the command recognition techniques, we use a combination of automatic speech recognition and information extraction using intent and entity (key elements in a sentence) recognition approaches. Among the systems found in the literature, we did not find any that use assistant mechanisms. Moreover, research has shown that older people have problems with commands that require strict construction [21]. Their statements differ from the standard ones by interwoven pauses and the addition of polite phrases (e.g., *Please find* ...). Therefore, the method used must have some flexibility.

The last step is the appropriate reaction to the recognized intention (Fig. 4). If the intention is to call for help, the caregivers mentioned in the command are notified by sending a specific message type. Moreover, in case of critical level messages indicating a life-threatening event, medical assistance can be called.

Both the positive (call for help cases) and negative results of the situation classification are recorded in the database for iterative training of the Natural Language Understanding models.

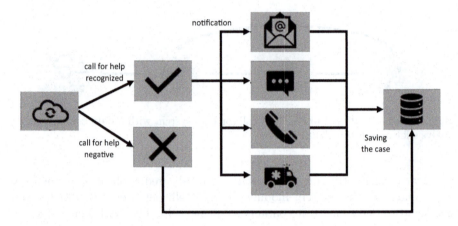

Fig. 4. Notification and saving the information on the recognized intention.

4   Experimental Results

Appropriate reaction to the current situation requires correct recognition of the intention that is hidden in the voice command. The quality of intention recognition depends on the effectiveness of both steps that are performed during voice analysis. Here, we compare various Cloud-based services for speech recognition and natural language understanding in terms of their capabilities to serve the purpose of the designed system.

4.1   Quality Assessment for Speech to Text Transcription

Relying on integrated platforms with well-tested services is one of the most common ways developers build their solutions that cover many loosely coupled components (including IoT devices for personal assistance and their software). When designing our Cloud-based systems to call for help and monitor older adults, we also followed this pattern. Therefore, when testing the speech-to-text transcription services, we mainly looked at large Cloud platforms that provide the possibility to connect many personal IoT devices and integrate them within a coherent system. For the speech to text transcription, we tested the following services:

– Google Cloud Speech-to-Text,
– IBM Watson Speech to Text,

– Microsoft Azure Speech to Text,
– Amazon Transcribe.

The research was carried out on a VoxForge data set[1] consisting of 551 files in WAV format with different sampling rates (16 kHz or 44 kHz to 48 kHz). This collection was selected from speech recordings in the English language. Speech recordings are of various dialects: British English, European and American. The length of the recordings was usually from 4 to 10 s, which was adapted to the length of the danger notification commands. The test data set consisted of recordings (WAV, waveform audio format) and text files containing the correct transcripts corresponding to the given WAV file. After obtaining a result in the form of a text sentence for a given cloud service, it was compared with the pattern (correct transcript, a ground truth).

One of the most popular ways to measure the quality of speech recognition is to evaluate effectiveness using the *Word error rate* (WER) [15]. WER is a simple index, which is the quotient of the sum of substituted S, omitted D and inserted I words by their number in a given sentence N. One of the variants of this metric is to use a weight of 0.5 for the removed and inserted words:

$$WER = \frac{S + D + I}{N}. \tag{1}$$

The second metric used to examine the entire group of results of a given service is *Sentence Error Rate* (SER) – determining the sentence recognition error rate. It is the number of incorrectly recognized sentences F in a given research group N:

$$SER = \frac{F}{N}. \tag{2}$$

Values of the *Word error rate* (WER) achieved in our experiments for various speech-to-text cloud services are presented in Fig. 5. Values of the WER for all tested services are low, indicating the possible usefulness of each of the services for the task being performed. The lowest value of $WER = 8.40$ was achieved with the Amazon Transcribe service. The largest value of $WER = 13.80$ was achieved for the IBM Watson Speech to Text.

Values of the *Sentence Error Rate* (SER) achieved in our experiments for various speech-to-text cloud services are presented in Fig. 6. Values of the SER for all tested services are also low, which confirms the possible usefulness of each of the services for the task being performed. The lowest value of $SER = 4.05$ was achieved with the Microsoft Azure Speech to Text service.

---

[1] VoxForge open speech dataset with transcribed speech: http://www.voxforge.org/home/downloads/speech/english.

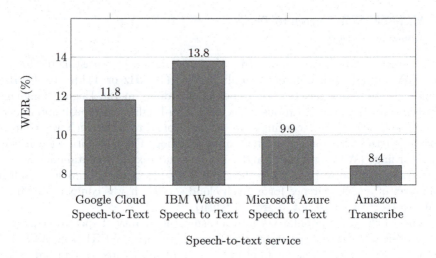

Fig. 5. Values of the *Word error rate* (WER) achieved for various speech-to-text cloud services.

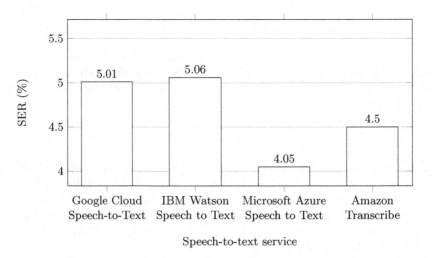

Fig. 6. Values of the *Sentence Error Rate* (SER) achieved for various speech-to-text cloud services.

4.2   Experimental Evaluation of Natural Language Understanding Services

In the second series of experiments, we conducted tests of the capabilities of Cloud services to understand natural language. We investigated Google DialogFlow, IBM Watson Assistant, Amazon Lex (the same deep learning technologies that power Amazon Alexa), Microsoft LUIS, and the free Rasa NLU library for this purpose. Within these experiments, we wanted to:

- verify the effectiveness of speech understanding depending on the service used,
- study the impact of the size of the training set on the effectiveness of the speech understanding services in use.

As in the case of speech recognition, we could not find any test data related to the Emergency Command issues. Alternatively, for our studies, we used the SNIPS dataset [7], consisting of 7 intentions:

- GetWeather – related to weather conditions,
- BookRestaurant – related to booking a meal in a given restaurant,
- PlayMusic – related to playing an artist, album or music track,
- AddToPlaylist – related to adding a music track to a playlist,
- RateBook - related to book reviews,
- SearchScreeningEvent – related to searching for film events,
- SearchCreativeWork – related to searching for creative activities.

The full data set used in our experiments contained 13,784 sentences (approximately 2,000 sentences for each intention), and besides, each file with intentions contained between 3,419 and 6,418 entities. We used two data sets for the training process: (1) a limited data set that contained 300 samples per intention, and (2) full data set with approximately 2,000 sentences per intention. For the limited training set, the number of entities ranged from 533 to 1,146. The test set consisted of 700 sentences (100 for each intention). Importantly, the test set for each intention was already distinguished. Therefore, no cross-validation or other modifications of the training set were required. Table 1 contains results of effectiveness evaluation for various NLU services trained with up to 300 samples per intention. Table 2 contains results of effectiveness evaluation for various NLU services trained with around 2000 samples per intention.

The largest number of properly-identified intentions was achieved with Microsoft LUIS service - 99.14% and 98.57% - and Rasa NLU - 98.57% for both data sets. For most of the services, we noticed an increase in effectiveness measures for the larger data set used for the training phase. This was expected. However, it is interesting that despite the increase in the number of training data, Microsoft LUIS was the only one that experienced a decrease in effectiveness. In the prediction with models trained with the limited data set, Microsoft LUIS made only two mistakes in the intention recognition for the testing sentences (Table 3). For the prediction based on training with the full data set, Microsoft LUIS made 10 wrong predictions (Table 4).

Table 1. Effectiveness of NLU for various services trained with the limited data set of 300 samples per intention.

NLU Service	Recognized intentions (%)	F1	Precision	Recall	Recognized entities (%)	Mean time (ms)
Amazon Lex	93.71	0.5582	0.5876	0.5741	50.01	277.88
Google DialogFlow	96.71	0.6022	0.7522	0.5266	55.17	328.95
Microsoft LUIS	99.14	0.4409	0.4626	0.4739	56.45	202.92
Rasa NLU	98.57	0.6526	0.6705	0.6446	80.64	9.50
IBM Watson NLU	98.14	0.4987	0.5130	0.5369	56.71	289.70

Table 2. Effectiveness of NLU for various services trained with the full data set of 2000 samples per intention.

NLU Service	Recognized intentions (%)	F1	Precision	Recall	Recognized entities (%)	Mean time (ms)
Amazon Lex	94.14	0.6089	0.6207	0.6347	56.83	286.87
Google DialogFlow	98.14	0.6888	0.7797	0.6393	67.94	337.76
Microsoft LUIS	98.57	0.5594	0.5434	0.6135	77.57	113.89
Rasa NLU	98.57	0.6794	0.6871	0.6766	86.42	9.56
IBM Watson NLU	98.43	0.5152	0.4791	0.6225	69.32	280.34

Table 3. Sentences with wrongly assigned intentions by Microsoft LUIS trained with the limited data set.

Sentence	Real intention	Predicted intention
When is sunrise for AR	GetWeather	SearchScreeningEvent
I want to eat in Ramona	BookRestaurant	GetWeather

Table 4. Sentences with wrongly assigned intentions by Microsoft LUIS trained with the full data set.

Sentence	Real intention	Predicted intention
When is sunrise for AR	GetWeather	SearchScreeningEvent
I want to eat in Ramona	BookRestaurant	GetWeather
Live In L.a Joseph Meyer please	PlayMusic	BookRestaurant
Where is Belgium located	GetWeather	SearchCreativeWork
Please tune into Chieko Ochi 's good music	PlayMusic	AddToPlaylist
I want to see JLA Adventures: Trapped In Time	SearchScreeningEvent	SearchCreativeWork
Where can I see The Prime Ministers: The Pioneers	SearchScreeningEvent	SearchCreativeWork
I want to see Medal for the General	SearchScreeningEvent	SearchCreativeWork
I want to see Fear Chamber	SearchScreeningEvent	SearchCreativeWork
I want to see Outcast	SearchScreeningEvent	SearchCreativeWork

4.3   Influence of the Manner of Speaking and Noise on the Quality of Recognition of Emergency Calls

The third part of the experiments covered investigations of the influence of the manner of speaking and the impact of noise on the quality of recognition of emergency calls. For this purpose, we recorded 15 sentences (including 13 calls for help and two random commands) spoken by each person. The study aimed to check how the emergency commands detection system reacts in case of danger and how susceptible it is to the change in the way of speaking and the surrounding noise. Tests covered four manners of pronouncing calls for help: normal speech, whispering, slurred speech, and calling for help with background noise. We performed the tests with the Microsoft LUIS service, which achieved the best results in the previous series of intention recognition tests. Results are presented in Fig. 7.

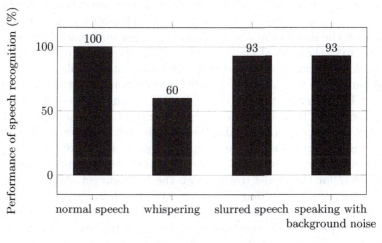

Fig. 7. Effectiveness (%) in recognizing intentions in uttered calls for help depending on the manner of speaking. Obtained for Microsoft LUIS service.

As can be observed from these experiments, the best results were obtained for normal speaking – the effectiveness reached 100%. Very good results were also obtained for slurred speech and speaking with a background noise – both 93%. The worst results were obtained for whispering, for which we achieved only 60% of properly-recognized intentions. This is an important observation, since after an accident older people may not be able to speak normally. This leaves room for future research on the development of algorithms for NLU for these types of situations.

5   Discussion and Conclusions

In the conducted study, we compared the capabilities of speech recognition and natural language understanding services. In the evaluation of speech recognition services, two services turned out to be the best – Amazon Transcribe with $WER = 8.4\%$ and $SER = 4.50\%$ and Microsoft Azure Speech to Text with $WER = 9.90\%$ and $SER = 4.05\%$. This shows that both cloud services are among the leaders in this field. The obtained values of the WER error are several times lower than in related works [27], which may indicate that we used a simpler test set in our experiments. The speech recognition execution time is at the level of a few seconds, which allows using these methods (that require Cloud communication) in emergency call applications.

In the second part of the experiments, we compared the results of speech understanding for Microsoft LUIS, Google DialogFlow, Amazon Lex, IBM Watson Assistant, and Rasa NLU services to recognize the intention of the speech. When trained with the full data set (about 2,000 sentences per intention), Microsoft LUIS and Rasa NLU turned out to return the best quality results. For services trained with the limited data set (300 sentences per intention), Microsoft LUIS turned out to be the most effective. These results and the order of services in terms of top-performing ones correspond well with related studies [6].

Comparing the effectiveness of entity recognition (intention discovery) by using the F1 indicator, we can conclude that among the tested services, the most efficient was Rasa NLU for the model trained with the limited data set – $F1 = 0.6526$ and Google DialogFlow for the full data set - $F1 = 0.6888$. Considering the percentage of the recognized intentions, Microsoft LUIS proved to be the most effective, with 99.14% for the model trained with the limited data set and 98.57 for the model trained with the full data set. Anyway, all tested services achieved a recognition rate above 93%, which is a good result.

In summary, it is worth mentioning that Rasa NLU achieved excellent results compared to paid Cloud-based services, being in the lead when recognizing speech intention in each of the categories. Among Cloud-based services, excellent results were achieved by services offered by Microsoft, which allow for the inclusion of these services in the construction of emergency call applications. In the case of speech understanding, comparing our results to related works, we can conclude that the key is to properly adjust the training data. For each intention, statements should be thoughtful and as unique as possible compared to other intentions. For example, in detecting dangers in the health state of older people, it would be reasonable to train the people to start the emergency call with some predefined expression, like 'Emergency, I need help.' It would also protect against an unintentional analysis of a speech related to everyday life and human conversations (i.e., reduce the amount of data that undergoes the analysis). This is also important from the viewpoint of older adults' privacy and information security since older people want to be sure that they are not overheard by IoT devices, other people, and unauthorized third parties. Therefore, data security and peoples' privacy are important issues when using IoT solutions that serve people.

References

1. World Health Organization: Global health and aging. Tech. Rep. 11–7737, NIH Publication (2011)
2. Ammari, T., Kaye, J., Tsai, J.Y., Bentley, F.: Music, search, and IoT: how people (really) use voice assistants. ACM Trans. Comput.-Hum. Interact. **26**(3), 17 (2019)
3. Austerjost, J., et al.: Introducing a virtual assistant to the lab: a voice user interface for the intuitive control of laboratory instruments. SLAS Technol. Translating Life Sci. Innov. **23**(5), 476–482 (2018)
4. Bhosale, S., Sheikh, I., Dumpala, S.H., Kopparapu, S.K.: Transfer learning for low resource spoken language understanding without speech-to-text. In: 2019 IEEE Bombay Section Signature Conference (IBSSC), pp. 1–5 (2019)
5. Braines, D., O'Leary, N., Thomas, A., Harborne, D., Preece, A.D., Webberley, W.M.: Conversational homes: a uniform natural language approach for collaboration among humans and devices. Int. J. Intell. Syst. **10**(3), 223–237 (2017)
6. Braun, D., Hernandez Mendez, A., Matthes, F., Langen, M.: Evaluating natural language understanding services for conversational question answering systems. In: Proceedings of the 18th Annual SIGdial Meeting on Discourse and Dialogue, pp. 174–185. Association for Computational Linguistics, Saarbrücken, Germany (2017)
7. Coucke, A., et al.: Snips voice platform: an embedded spoken language understanding system for private-by-design voice interfaces. ArXiv abs/1805.10190 (2018)
8. Cupek, R., et al.: Autonomous guided vehicles for smart industries - the state-of-the-art and research challenges. In: Krzhizhanovskaya, V.V., et al. (eds.) Computational Science - ICCS 2020, pp. 330–343. Springer International Publishing, Cham (2020)
9. de Velasco, M., Justo, R., Antón, J., Carrilero, M., Torres, M.I.: Emotion detection from speech and text. Proc. IberSPEECH **2018**, 68–71 (2018)
10. Deng, L., et al.: Recent advances in deep learning for speech research at microsoft. In: 2013 IEEE International Conference on Acoustics, Speech and Signal Processing, pp. 8604–8608 (2013)
11. Dimauro, G., Di Nicola, V., Bevilacqua, V., Caivano, D., Girardi, F.: Assessment of speech intelligibility in Parkinson's disease using a speech-to-text system. IEEE Access **5**, 22199–22208 (2017). https://doi.org/10.1109/ACCESS.2017.2762475
12. Graves, A., Fernández, S., Gomez, F., Schmidhuber, J.: Connectionist temporal classification: labelling unsegmented sequence data with recurrent neural networks. In: Proceedings of the 23rd International Conference on Machine Learning, pp. 369–376. ICML 2006. Association for Computing Machinery, New York, NY, USA (2006). https://doi.org/10.1145/1143844.1143891
13. Grzechca, D., Ziebinski, A., Rybka, P.: Enhanced reliability of ADAS sensors based on the observation of the power supply current and neural network application. In: Nguyen, N.T., Papadopoulos, G.A., Jedrzejowicz, P., Trawiński, B., Vossen, G. (eds.) Computational Collective Intelligence, pp. 215–226. Springer International Publishing, Cham (2017)
14. Kishore Kodali, R., Rajanarayanan, S.C., Boppana, L., Sharma, S., Kumar, A.: Low cost smart home automation system using smart phone. In: 2019 IEEE R10 Humanitarian Technology Conference (R10-HTC)(47129), pp. 120–125 (2019)
15. Klakow, D., Peters, J.: Testing the correlation of word error rate and perplexity. Speech Commun. **38**(1), 19–28 (2002)

16. Lago, A.S., Dias, J.P., Ferreira, H.S.: Conversational interface for managing non-trivial internet-of-things systems. In: Krzhizhanovskaya, V.V., et al. (eds.) Computational Science - ICCS 2020, pp. 384–397. Springer International Publishing, Cham (2020)
17. Laksono, T.P., Hidayatullah, A.F., Ratnasari, C.I.: Speech to text of patient complaints for bahasa Indonesia. In: 2018 International Conference on Asian Language Processing (IALP), pp. 79–84 (2018). https://doi.org/10.1109/IALP.2018.8629161
18. Lero, R.D., Exton, C., Le Gear, A.: Communications using a speech-to-text-to-speech pipeline. In: 2019 International Conference on Wireless and Mobile Computing, Networking and Communications (WiMob), pp. 1–6 (2019)
19. López, G., Quesada, L., Guerrero, L.A.: Alexa vs. Siri vs. Cortana vs. Google assistant: a comparison of speech-based natural user interfaces. In: Nunes, I.L. (ed.) Advances in Human Factors and Systems Interaction, pp. 241–250. Springer International Publishing, Cham (2018)
20. Mehrabani, M., Bangalore, S., Stern, B.: Personalized speech recognition for Internet of things. In: 2015 IEEE 2nd World Forum on Internet of Things (WF-IoT), pp. 369–374 (2015). https://doi.org/10.1109/WF-IoT.2015.7389082
21. Mishakova, A., Portet, F., Desot, T., Vacher, M.: Learning natural language understanding systems from unaligned labels for voice command in smart homes. In: 2019 IEEE International Conference on Pervasive Computing and Communications Workshops (PerCom Workshops), pp. 832–837 (2019)
22. Mitrevski, M.: Conversational interface challenges. In: Developing Conversational Interfaces for iOS, pp. 217–228. Apress, Berkeley, CA (2018). https://doi.org/10.1007/978-1-4842-3396-2_8
23. Mrozek, D., Koczur, A., Małysiak-Mrozek, B.: Fall detection in older adults with mobile IoT devices and machine learning in the cloud and on the edge. Inf. Sci. **537**, 132–147 (2020)
24. Mrozek, D., Milik, M., Małysiak-Mrozek, B., Tokarz, K., Duszenko, A., Kozielski, S.: Fuzzy intelligence in monitoring older adults with wearables. In: Krzhizhanovskaya, V.V., et al. (eds.) Computational Science - ICCS 2020, pp. 288–301. Springer International Publishing, Cham (2020)
25. Schwitter, R.: Controlled natural languages for knowledge representation. In: Coling 2010: Posters, vol. 2, pp. 1113–1121 (2010)
26. Sovariova Soosova, M.: Determinants of quality of life in the elderly. Central Euro. J. Nurs. Midwifery **7**(3), 484–493 (2016)
27. Vyas, M.: A Gaussian mixture model based speech recognition system using Matlab. Sign. Image Process. **4**(4), 109–118 (2013)
28. Wan, J., et al.: Wearable IoT enabled real-time health monitoring system. EURASIP J. Wirel. Commun. Netw. (1), 298 (2018)

A Decision Support System Based on Augmented Reality for the Safe Preparation of Chemotherapy Drugs

Sarah Ben Othman[1]([✉]), Hayfa Zgaya[1], Michèle Vasseur[2], Bertrand Décaudin[3], Pascal Odou[3], and Slim Hammadi[1]

[1] Univ. Lille, CNRS, Centrale Lille, UMR 9189 CRIStAL, 59000 Lille, France
Sara.ben-othman@centralelille.fr
[2] Lille University Hospital-Institut de Pharmacie, 59000 Lille, France
[3] Univ. Lille, CHU Lille ULR, 7365 – GRITA, 59000 Lille, France

Abstract. The preparation of chemotherapy drugs has always presented complex issues and challenges given the nature of the demand on the one hand, and the criticality of the treatments on the other. Chemotherapy involves handling special drugs that require specific precautions. These drugs are toxic and potentially harmful for people handling them. Their preparation entails therefore particular and complex procedures including preparation and control. The relevant control methods are often limited to the double visual control. The search for optimization and safety of pharmaco-technical processes leads to the use of new technologies with the main aim of improving patient care. In this respect, Augmented Reality (AR) technology can be an effective solution to support the control of chemotherapy preparations. It can be easily adapted to the chemotherapy drugs preparation environment. This paper introduces SmartPrep, an innovative decision support system (DSS) for the monitoring of chemotherapy drugs preparation. The proposed DSS uses the AR technology, through smart glasses, to facilitate and secure the preparation of these drugs. Controlling the preparation process is done with the help of the voice since hands are busy. SmartPrep was co-developed by the research laboratory CRISTAL, GRITA research group and the software publisher Computer Engineering.

Keywords: Chemotherapy drugs preparation · Decision support system · Augmented Reality · Voice control

1 Introduction

Since 2004, cancer has been the main cause of premature mortality in the world, in front of cardiovascular diseases. According to the National Cancer Institute (NCaI), in France, the number of new cancer types was about 382 000 and the number of cancer deaths has reached 157 400 in 2018 [1]. The mortality rate has been decreasing steadily for the past 25 years. However, cancers with a poor prognosis (with a 5-year survival rate of less than 33%) represents 31% of cancers in males and 17% in females. Research in

© Springer Nature Switzerland AG 2021
M. Paszynski et al. (Eds.): ICCS 2021, LNCS 12745, pp. 485–499, 2021.
https://doi.org/10.1007/978-3-030-77970-2_37

the field of oncology is extremely active. The objective is to accelerate the emergence of innovation for the benefit of patients [2–4].

University Hospitals are facing an increase in the number of preparations of chemotherapy both in and out-of-clinical trial. Preparations are becoming more and more complex. The demand for chemotherapeutic drugs is daily. Each day, different preparations are requested depending on the treatments prescribed for the patients. What makes the task even more difficult is that each patient requires a specific treatment, depending on his or her weight, the protocol prescribed, the stage of treatment, etc. In addition, the drugs have to be prepared with great precision, and any mistake could present a great risk on the patient's life. So, pharmacists are confronted every day with dozens of (sometimes very complex) recipes that require great vigilance and double checks during their preparation. This implies the need for an effective additional support for preparation and control processes [5–8]. Thus, in order to cope with this increase in activity, the chemotherapy preparation units are searching for an effective solution for the preparation and control process support [9, 10]. This solution should be able to make patient management as safe as possible, reduce the time of the whole preparation process of and meet the proponent requirements.

The main aim of this paper is to present the design, the development and evaluation of SmartPrep, our decision support system using AR, which is intended to improve safety in the preparation of chemotherapy drugs in health organizations and specifically in a hospital pharmacy. Our approach is based on the introduction of a free-hands appliance to assist the operators and record their acts. SmartPrep makes complex operations safe by ensuring their conformity to pre-established protocols and so reduce the incidence of errors. Our software solution is integrated into AR glasses: from a preparation/administration protocol, it is possible to give a step-by-step guidance to the operator carrying out the preparation/administration, to photo-record (triggered by the operator himself) the different steps, to obtain on-line control and to archive photos in the drug-batch file compiled for any possible posteriori control. SmartPrep gives also the operator the possibility to interact with the environment through a voice command. Our solution is studied, developed, implemented and tested in collaboration with the pharmacy of the Hospital University Center (HUC) of Lille and connected to the software CHIMIO®[1] of Computer Engineering company, containing all information about chemotherapy circuit including preparation protocols.

2   State of the Art

Technology has always been a key factor in logistics. AR has taken off since the emergence of Smartphones. It has been appearing everywhere and in various fields. It is seen as a revolutionary alternative to solutions, which have many limitations, especially in terms of functionality and comfort. AR can be defined as an interface between virtual data and the real world [11]. It is a technique that overlays to reality its digital representation updated in real time. It thus offers the user real-time interaction possibilities [12]. This technology is actively growing in the healthcare field. For example, it is used

---

[1] https://www.computer-engineering.fr/applications/chimio/.

to model patients' organs and blood vessels in 3D, thus facilitating the learning and work of surgeons, particularly to create precise 3D reconstructions of tumors [13]. The AccuVein® scanner is another example. It can illuminate the veins in a patient's skin, helping nurses and doctors locate them before inserting a needle [14].

For the traceability and control of operations, several solutions are available: a double control by a second operator, a gravimetric control during preparation with the help of scales integrated into the workstation [15], discharge control with identification and quantification of the drug in the finished preparation using a standard analytical method (CLHP-UV for example) or dedicated equipment (UV-Raman QCPrep + spectrometry for example), video assistance and control by camera [16]. These currently available solutions have several limitations. For example, a double control by a second operator implies considerable time consumption in terms of human resources from day to day, problems of reliability and an absence of traceability of the controlled elements. Finally, assistance and video control imply a high cost and the need to install equipment in the production zone as well as the presence of operators during its setting-up. A comparison of these different methods based on a SWOT analysis is available in a recent article dealing with current controls in the pharmacy context [17].

The proposed solution in this paper is to use AR glasses which display and reel off in the user's visual field information useful for making preparations as well as identifying the actions accomplished and the products used by the operator with the aim of preventing errors in real time and establishing the traceability of the completed operations. The interfaces present in the glasses help guiding the operators in their preparation tasks and warning them if errors are detected while at the same time limiting the workload. Information must therefore be presented in the simplest way and the best adapted to each user. The proposed DSS allows for an extension of the existing adaptation systems for AR, as the existing solutions only deal with WIMP interfaces (i.e., Windows, Icons, Menus, Pointing). Moreover, interaction is bidirectionally effective: it is not simply a question of displaying information (interface output) adapted to the user and his context of use but also of enabling him to interact with the system (interface input) especially in the case of error. This is an original approach in as far as a search in the Medline database for articles linked to «drug related problem and AR» and «medication error and AR» comes across only two articles linked to using an application for mobile phone for the purpose of training prescribers.

3   Current Practices at Lille University Hospital Center (UHC)

At the UHC of Lille, the clinical trial circuit is presented in Fig. 1. This circuit is composed of three main inter-connected blocs: administration and prescription (steps 1 and 8), verification and allocation (step 2) and preparation (steps 3 to 7). The circuit begins with the prescription of chemotherapy by the doctor, which is carried out in 2 formats:

- A paper format consisting of a pre-filled prescription specific to each clinical trial.
- A digital format based on CHIMIO® software for products prepared by the Centralized Cytotoxic Preparation Unit (CCPU) of the pharmacy.

In the Clinical Trials sector, after verification of its conformity, the printed prescription allows the allocation and dispensing of the vials needed for the preparation. The vials are then transferred and received after microbiological decontamination in the CCPU production area. They are later stored and packaged by patient and by test until pharmaceutical validation of the preparation. At the pharmaceutical validation station, the pharmacist/pharmacy intern carries out the computer data entry in storage of the vials and the pharmaceutical validation. The pharmacy assistant prepares the basket following the instructions on the paper preparation sheet and initials the preparation sheet.

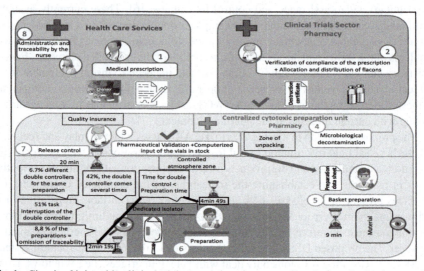

Fig. 1. Circuit of injectable clinical trial preparations in cancerology before the implementation of SmartPrep

Each basket is double-checked by a third person, who initials the production sheet to ensure that all the necessary equipment and the appropriate vials are present. Sterilization is then performed to allow the basket to be transferred to the isolator dedicated to clinical trial preparations. After transfer into the isolator, the pharmacy assistant carries out the preparation following the instructions on the preparation sheet. The preparation steps are controlled by a second person with a traceability (initials) on the preparation sheet. The preparation is afterwards transferred to the Quality Insurance Room (QIR) with the associated preparation sheet and the vials used for control. After verification of the concordance between the preparation, the identity of the patient, the good conservation and the stability, the Registered Nurse administers the preparation. The evaluation of the ordinary chemotherapy drugs preparation circuit is presented in the following section.

4 Evaluation of the Ordinary Preparation Process for Injectable Chemotherapy Clinical Trials at Lille HUC

4.1 Materials and Methods

In order to evaluate the ordinary preparation and control circuit for injectable chemotherapy clinical trials (Fig. 1), a data collection has been carried out with multiple objectives. The first purpose is to highlight any missing traceability of the preparation and control steps on the paper preparation sheets. A "traceability anomaly collection sheet" has been created to trace over a given period, all the traceability omissions detected on the preparation sheets at the time of pharmaceutical release. The second objective is to highlight the parameters that influence the preparation time and to achieve a time impact for the double visual inspection. For this data collection, clinical trials were selected to have a representative panel of preparations according to the presence or not of a reconstitution, the nature of the final packaging and thus the complexity of the operating procedure and the presence of the observer. The third objective is to highlight the interruptions of tasks in the preparation area related to the double visual control through the same data collection. In order to collect data on the functioning of the chemotherapy preparation unit, we rely on three sources of information:

- The "Sterilization Start Time Collection Sheet": This sheet is placed at the entrance to the isolators and as soon as a clinical trial chemotherapy preparation is introduced into the isolator for sterilization, the sheet is completed. Each line corresponds to a preparation with its order number, the time of the start of sterilization, the sterilization workstation used, and the number of diverse preparations placed in this workstation.
- The "Clinical Trial Preparations Observation Sheet": It tracks the progress of the chemotherapy preparation in the isolator. This form is completed on an observational basis by a person who acts as an external observer and is not involved in the preparation process or in the control of the preparation.
- The CHIMIO® software that provides information about:

- the type of preparation: packaging, volume of active ingredient, whether or not a reconstitution step is necessary.
- the time of the beginning and the time of the end of the preparation.
- the number of preparations carried out during the day.

4.2 Results

In order to track and optimize the different phases of the preparation (basket preparation, preparation under isolator, double visual control and release control) five preparation times have been defined:

- Total processing time (TPT): the total time required to carry out the preparation from the reception of the vials to the validation of the release control.
- The basket preparation time (BPT): the time needed to prepare and double check the basket from the paper preparation sheet.

- The real preparation time (RPT): the time necessary for the operator to carry out the preparation in the isolator.
- The double-control time (DCT): the time during which a second person is called in to carry out the double visual control of the preparation. The second person can control several times the same preparation.
- The release time (RT): the time required for the finished preparation to be checked, released and made available.

At the same time, a collection of all preparation anomalies or non-conformities are carried out. All this data is entered in a spreadsheet, whose schedule number serves as a link between the different sources of information. Analysis is handled by the PROSER-PINE platform of the Faculty of Pharmacy of Lille University and are performed with SAS version 9.4 software (SAS Institute Inc., Cary, NC, USA). The data are presented as mean standard deviation or median [interquartile range] for continuous variables according to the asymmetry of their distribution, and as numbers (proportions) for categorical variables. The search for factors predictive of preparation and double control times was carried out with mixed linear regression models, with the variable ln-transformed (application of the Nerian logarithmic function) and including the manipulator and/or double controller as a random effects factor. The multivariate models are constructed in including all covariates, regardless of their degree of significance in univariate analyses. The selection of covariates is then done manually. The regression assumptions underlying the regression for the final multivariate model are verified graphically using residuals. For all analysis, p-value = 0.05.

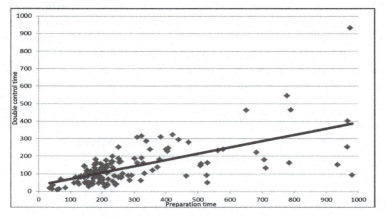

Fig. 2. Double-control time (DCT) according to the real preparation time (RPT)

Figure 2 shows that the DCT is proportional to the RPT: the more the real preparation time increases, the more the double-checking time increases. We can easily explain that the lowest DCTs are associated with placebo preparations (average $\bar{X} = 26$ s and standard deviation SD $= 21$ s) and the longest with so-called "complex" preparations ($\bar{X} = 421$ s and SD $= 256$ s). We have also made a projection over one year of the time devoted to the double visual control of the preparations in clinical trials. For this purpose, we relied on the number of clinical trials carried out in 2018 (3673 preparations) and on the average time for double visual control (DCT). In total, it represents 145 h/year equivalent to 8% of activity. This analysis shows the deficiencies of this preparation and control process. Therefore, in order to secure chemotherapy drugs preparation process and save time, further improvements are required.

5   The New Circuit of Preparation of Clinical Trials in Chemotherapy: AR Technology Integration

5.1   Implementation of the New Version of the Software CHIMIO®

The aim here is to improve the safety of the entire preparation process, including chemotherapy clinical trials, and to prepare for the implementation of AR glasses as a tool to assist in preparation and control. It includes the implementation of a new version of CHIMIO® called "New Preparation", which targets to secure the preparation process by tracing each step and identifying each component of the preparation (vials, active ingredients, solvent, etc.), and to dematerialize the preparation spreadsheets.

Each preparation sheet is substituted by a digital form that is displayed on each workstation. The edition in paper format is reduced to the printing of the preparation label only (Fig. 3) with a bar code corresponding to the order number. This latter allows to make an order between steps and to identify the content and the operating mode of the preparation.

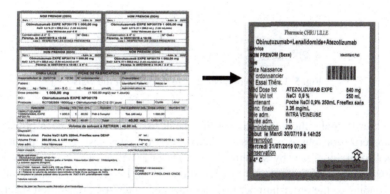

Fig. 3. Evolution to the preparation label in the new chemotherapy drugs preparation circuit

5.2   Design of the New Clinical Trial Preparation Circuit

According to the usual preparation circuit of clinical trials in injectable chemotherapy presented in Sect. 3 (Fig. 1: steps 3 to 7), we have made several modifications in order to be able to meet the requirements of AR integration (Fig. 4). The improvements require the presence of data matrix (two dimensional square or rectangular pattern code) on the vials and on the solvent bags. As clinical trial vials do not have data matrix, a prior relabelling step is required. So, when the vials are entered into the computerized inventory by the pharmacist/pharmacy intern, a vial label edition is performed (Fig. 5). The vial labels are then sent to the operator who re-labels the vials. Instead of the preparation sheets, preparation labels (Fig. 3) are edited. As regards the preparation of the basket, it is based on a digital check list obtained from the bar code on the preparation label. This checklist corresponds to the vials and the material necessary to carry out the preparation in the isolator.

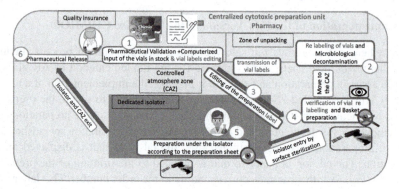

Fig. 4. Improvements in the preparation circuit for chemotherapy clinical trials with the implementation of the "New Preparation Process".

Fig. 5 Example of label after re-labelling of the vials

In order to make the generation of the check list automatic for each preparation, a parameterization phase is necessary beforehand for each test. During preparation, the operator is identified by scanning a barcode on a badge, which ensures computerized traceability of the operator at each stage. Then, he scans the preparation label corresponding to the order number to access the computerized preparation sheet. During preparation, the control steps are modified: the pre-filled vials are checked by data matrix scanning. Only the control of volumes (reconstitution, purging and volume of active ingredient) remains carried out by a double visual control by a third person. In order to eliminate any omissions in traceability, the steps of the double visual control are traced at the IT level (identification of the double controller by name badge scanning). This step has been made obligatory for the validation of the end of the preparation. At the release control level, a control of the preparation is carried out via an access to the computerized preparation sheet. The steps of the double visual control appear on a release checklist. As regards the number of steps, those removed are the double visual control of the basket and at the preparation level the control of the vial(s) and the final packaging. As for the new steps added, they concern the relabelling of the pre-filled vials with the label edition.

5.3 Description of the Process of a Chemotherapy Drug Preparation with AR Glasses

The preparation process is as follows: first of all, the operator wears glasses and is identified by scanning a personal data matrix. The glasses used are Optinvent ORA-2 Smart Glasses (Fig. 6) operating with Android 4.4. The proposed solution is transposable to any other Android AR technology.

Fig. 6. The AR glasses used in SmartPrep

The identification of the preparation is done by scanning the specific barcode on the preparation label in the isolator. The operator can then see the protocol operating mode, step by step on the glasses in a comfortable way without obstructing his field of vision. The protocol is given by the CHIMIO® software.

Fig. 7. Lille UHC Staff testing SmartPrep

To avoid the use of generic barcodes, the glasses are equipped with voice recognition (Sect. 7). At the end of the preparation, all the photos taken are saved as PDF files corresponding to the batch file of the preparation and archived at the level of the CHIMIO® software batch file. This file allows a posteriori control of the preparation thanks to the photos taken during the preparation. The different stages of the preparation and the photos taken can be configured before the preparation in the CHIMIO® software. This system allows a computerized identification of the operator, the realization of the preparation according to a step-by-step mode (which secures the preparation), a traceability of the stages of preparation and a suppression of the interruptions of tasks.

6 Image Processing

Integrating image processing features enables an automated control of some preparation stages (e.g., control of the sample volume by analysing the photo of the syringe). Our objective is to integrate into operators 'work environment information enabling them to reduce the risk of errors which may occur. According to the current preparation circuit analysis, the predominant factor was individual error (75.4%) followed by distraction or interruption (3.5%), insufficient training (2.9%), work overload (2.5%) and understaffing (2.3%). The access to this information is without disrupting completed tasks and without handling, mainly for reasons of hygiene and convenience. In order to automate the monitoring of the quality of each preparation phase, we have chosen to use the camera of the AR glasses to take pictures of each sample and to use a program to measure directly on the image the volume of the sample.

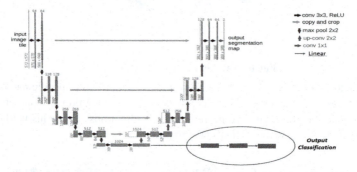

Fig. 8. Contour recognition of syringes: Output classification

The validation is made afterwards with the double control. The algorithm used involves machine learning via the UNet model implemented with the Python Pytorch module. This program allows, in particular, to learn an artificial intelligence how to recognize particular outlines (Fig. 8). In our situation, the program is trained to recognize the outlines of the syringes: the outline corresponding to the volume of the drug and the outline corresponding to the total volume of the syringe. Thus, the program recognizes the type of syringe used and the ratio between drug volume and total volume. From an image (input), it makes both a segmentation (output contour) and a classification (output type of syringe).

For the database, in order to have a reasonable amount of data (images of syringes with a certain amount of liquid), we took 2-min videos at the pharmacy for each type of syringe. These videos were then sequenced and cut into images to be processed by the program. For each of these images, we traced the outlines corresponding to the volume of the drug and the total volume of the syringe (Fig. 9).

The ratio is the one between the surface area corresponding to the drug and the surface area corresponding to the capacity of the syringe. To optimize the speed of learning, we used an online GPU which allows to save time; learning with this GPU took only a few hours. The results of the training on our database are represented by:

XXXX_Y_Z where XXXX is the image number, Y is the recognized syringe type, Z is the ratio. Figure 10 shows that the result is relatively good: it is a syringe type 5; the ratio is between 0.83 and 0.73. The histogram of the Fig. 10 gives the distribution of the measurement values given by the program on a set of images from our database. Ideally, this histogram should be as close as possible to a Gaussian curve centred around the real value.

Fig. 9. Photos of a syringe taken by the glasses

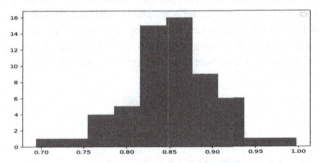

Fig. 10. Distribution of the measured values given by the program on a set of images from our database

The closer the histogram is to a Gaussian curve with a small standard deviation, the better the program performs. To improve accuracy, the database must be provided with relevant images. The program containing the learning phase is coded in Python. It uses modules specific to this language and is consistent from a storage viewpoint. This is why the learning program is not directly implemented in the Android program; a connection is then established between the glasses and an external computer server. An Android program function sends the photos taken to the image processing program which, once the processing is done, returns the measured value that will be compared to the set value.

7   Preparation Voice Control

Recently, many companies like Google and Android have introduced voice control applications to facilitate the phone navigation to people with reduced mobility [18–20], or even to control applications using Siri, Google Assistant, or Cortana [21, 22]. Some researchers have studied real-time control of critical operations of mobile devices [23] and even of the entire Android operating system [24]. In the case of SmartPrep, we use the library Droid Speech which supports continuous voice recognition and can be also used offline if the speech package is installed in the device. We have implemented recognition of the following commands: (1) "next" – validates the current step and passes to the following one; (2) "previous" – goes back to the previous step; (3) "camera" – turns on the camera; (4) "pause" – interrupts the preparation in the case of preparation with a pause time after reconstitution; "photo" – takes a photo; (5) "back" – turns off the camera and goes back to the preparation steps.

7.1   Connection of the SmartPrep to CHIMIO® Software

Developed by computer engineering, CHIMIO® software is a safe and easy tool for the implementation of healthcare protocols. Thanks to this tool, the pharmacy team has access to a dematerialization sheet which will allow to guide them in their activities. It is used everywhere in France and thus by the Pharmacy of Lille UHC as well. In summary, all information related to different parts of the chemotherapy circuit (patient, doctor, pharmacist, treatment…) is stored and managed by CHIMIO®. For the integration of the developed decision support system SmartPrep, we established a connector between CHIMIO® and the glasses, in order to retrieve automatically the processing steps. Since this information is stored on a web server, a web service is required to retrieve it (Fig. 11).

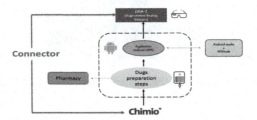

Fig. 11. The connector functioning between CHIMIO® and SmartPrep

A web service is a technology that allows applications to dialogue remotely via Internet, and this independently of the platforms and languages on which they are based. This communication is based on the request and response concept, carried out notably with http, XML messages. According to Computer Engineering, a web service was implemented in 2017 to facilitate the "exchange of data between the CHIMIO® software and external preparation systems". Therefore, it is used to retrieve all information concerning the preparations (operating procedures). The chosen web service is "Geocode" developed by Google, allowing to have geolocation coordinates from a specific postal address. We chose this web service in particular because it is of the same architecture (REST) and returns the same type of file (JSON, XML) as the web service corresponding to the CHIMIO® software. Tests were performed on Android Studio and were successful. Then we get the response, and we swap it with the example operating mode to carry out the following treatments as the steps of separation and display.

8   Evaluation

The objective of the evaluation protocol is to qualify the whole SmartPrep system which is used in an environment with which it will interact, through an Interface, and from which a performance is expected. One of the requirements for the implementation of AR glasses is the wireless network and the implementation of the CHIMIO® New preparation version. SmartPrep has been implemented in the pharmacy of Lille UHC on July 22, 2019. As after each implementation, a learning phase is necessary. During this period, data collection cannot be carried out in order not to introduce a bias in our data which would make the results unusable. Nevertheless, initial tests have been carried out. AR glasses were used in the preparation area on fictive preparations (Fig. 7). These tests helped us to validate the voice commands, the way to display of the protocol steps and the lecture of the protocol. After the tests, we could also be sure that the use of the AR glasses for the preparation under isolator (with the glass) is not problematic. The first tests carried out have proved the different concepts including the possibility of use in a complex environment (Wifi, Controlled Atmosphere Zone, through an isolator…), the integration of information (preparation protocol to be followed step by step), the traceability of the key steps of the preparation with by taking photos and the connexion with the software CHIMIO®. AR glasses improve the security of clinical trials chemotherapy preparation. In fact, the preparation according to a step-by-step mode obliges the operator to respect a chronology. This is particularly interesting for clinical trials because for each study, preparations are not carried out on a daily basis. The traceability of the realization of the critical steps is also interesting by taking and storing photos. During tests, we observed that operators were becoming more and more at ease with the system. All the feedback we received from the operators has been very positive. They showed a real interest in the system. We distributed a satisfaction questionnaire after each test in order to evaluate the evolution of the preparers' satisfaction. The opinions and remarks are very positive and constructive.

9   Conclusion and Perspectives

This work has enabled us to highlight the weaknesses of the current preparation circuit for injectable chemotherapy in clinical trials in the pharmacy of Lille HUC, particularly with regard to the double visual control itself and its impact on the preparation environment. AR glasses are an innovative solution to improve the preparation and control of chemotherapy preparations in clinical trials. This technology can be easily adapted to the preparation environment while securing preparations from the simplest to the most complex. SmartPrep can be used as a control solution for other types of preparations such as advanced therapy drugs or for clinical trial preparations beyond chemotherapy. It could also be used as a decision support system for any kind of drug preparation in health care services. Finally, in order to perfect the system and thus better satisfy user requirements, we are currently studying possible improvements to this decision-support system such as the appropriateness of adding more control steps and enhancing the learning system.

References

1. L'Institut publie L'essentiel des faits et chiffres des cancers en France (édition 2019) - Actu-alités, 7 Juill 2019. https://www.e-cancer.fr/Actualites-et-evenements/Actualites/L-Institut-publie-L-essentiel-des-faits-et-chiffres-des-cancers-en-France-edition-2019
2. La Food and Drug Administration a accordé un nombre record d'autorisations de mise sur le marché de médicaments sur l'année 2018 - Mission pour la Science et la Technologie de l'Ambassade de France aux Etats-Unis, 7 Juill 2019. https://www.france-science.org/La-Food-and-Drug-Administration-a.html
3. INCa. Plan cancer 2014-2019. 2019 mars
4. Le registre des essais cliniques - Le registre des essais cliniques, 13 Juill 2019, https://www.e-cancer.fr/Professionnels-de-la-recherche/Recherche-clinique/Le-registre-des-essais-cliniq ues/Le-registre-des-essais-cliniques
5. Dictionnaire français de l'erreur médicamenteuse. - Résultats de votre recherche - Banque de données en santé publique, 7 Juill 2019. https://bdspehesp.inist.fr/vibad/index.php?action = getRecordDetail&idt = 358157
6. ANSM. Bonnes pratiques de préparation (2007)
7. Anthony, K., Wiencek, C., Bauer, C., Daly, B., Anthony, M.K.: No interruptions please: impact of a No Interruption zone on medication safety in intensive care units. Crit. Care Nurse. **30**(3), 21–29 (2010)
8. Bateman, R., Donyai, P.: Errors associated with the preparation of aseptic products in UK hospital pharmacies: lessons from the national aseptic error reporting scheme. Qual Saf Health Care. **19**(5), (2010)
9. Marchiori, M., Menegazzo, C., Pedrocca, N., Pregnolato, S., Rossetti, C., Tomaselli, G., et al.: A gravimetric method for the quali-quantitative control of anticancer drugs. Eur. J. Hosp. Pharm. **21**(2), 87–90 (2014)
10. Lecordier, J., Heluin, Y., Plivard, C., Bureau, A., Mouawad, C., Chaillot, B., et al.: Safety in the preparation of cytotoxic drugs: How to integrate gravimetric control in the quality assurance policy? Biomed. Pharmacother. févr. **65**(1), 17–21 (2011)
11. Définition réalité augmentée – Qu'est-ce que la RA ? [Internet]. Réalité- Virtuelle.com, 7 Juill 2019. https://www.realite-virtuelle.com/definition-realite-augmentee
12. Abdallah, S.B., Ajmi, F., Ben Othman, S., Vermandel, S., Hammadi, S.: Augmented Reality for Real-Time Navigation Assistance to Wheelchair Users with Obstacles' Management. In: Rodrigues, J.M.F., et al. (eds.) ICCS 2019. LNCS, vol. 11540, pp. 528–534. Springer, Cham (2019). https://doi.org/10.1007/978-3-030-22750-0_47
13. Réalité augmentée : la chirurgie du futur [Internet]. E-Santé, 7 Juill 2019. https://www.e-sante.fr/realite-augmentee-chirurgie-futur/actualite/194715
14. AccuVein Vein Visualization-Improves IV 1-Stick Success 3.5x [Internet]. AccuVein, 7 Juill 2019. https://www.accuvein.com/.
15. http://www.cato.eu/fr/home.HTML
16. http://www.eurekam.fr/fr/drugcam
17. Bazin, C., Cassard, B., Caudron, E., Prognon, P., Havard, L.: Comparative analysis of methods for real-time analytical control of chemotherapies preparations. Int. J. Pharm. **494**(1), 329–336 (2015)
18. Carman, A.: Google launches a voice control app to help people with limited mobility navigate their phones. The Verge, Google (2018)
19. Hesse, B.: How to Use Google's Voice Access App to Control Your Android Device Hands-Free, in LifeHacker (2018)
20. Vertelney, L., Arent, M., Lieberman, H.: Two disciplines in search of an interface. Art Hum.-Comput. Interface, 45–55 (1990)

21. Hartt, M.: Adding voice control to your projects. in MEDIUM. Hackers at Cambridge (2018). https://medium.com/hackers-atcambridge/adding-voice-control-to-your-projects-7096fdee7c45. Accessed 10 April 2019
22. Matthews, K.: How to Make Any App Work With Voice Commands on Android, in Makeuseof. Android (2018). https://www.makeuseof.com/tag/make-any-app-work-voice-commands-android/. Accessed 10 April 2019
23. Omyonga, K., Shibwabo, K.B.: The Application of real-time voice recognition to control critical mobile device operations (2015)
24. Zhong, Y., Raman, T.V., Burkhardt, C., Biadsy, F., Bigham, J.F.: JustSpeak: enabling universal voice control on android. In: Proceedings of the 11th Web for All Conference W4A 2014, 7–9 April, Seoul, Korea (2014)

Metagenomic Analysis at the Edge with Jetson Xavier NX

Piotr Grzesik and Dariusz Mrozek[✉]

Department of Applied Informatics, Silesian University of Technology,
ul. Akademicka 16, 44-100 Gliwice, Poland
`dariusz.mrozek@polsl.pl`

Abstract. Nanopore sequencing technologies and devices such as Min-ION Nanopore enable cost-effective and portable metagenomic analysis. However, performing mobile metagenomics analysis in secluded areas requires computationally and energetically efficient Edge devices capable of running the whole analysis workflow without access to extensive computing infrastructure. This paper presents a study on using Edge devices such as Jetson Xavier NX as a platform for running real-time analysis. In the experiments, we evaluate it both from a performance and energy efficiency standpoint. For the purposes of this article, we developed a sample workflow, where raw nanopore reads are basecalled and later classified with Guppy and Kraken2 software. To provide an overview of the capabilities of Jetson Xavier NX, we conducted experiments in various scenarios and for all available power modes. The results of the study confirm that Jetson Xavier NX can serve as an energy-efficient, performant, and portable device for running real-time metagenomic experiments, especially in places with limited network connectivity, as it supports fully offline workflows. We also noticed that a lot of tools are not optimized to run on such Edge devices, and we see a great opportunity for future development in that area.

Keywords: Nanopore sequencing · Edge computing · Edge analytics · Bioinformatics · Jetson Xavier NX · Cloud computing · Metagenomics

1 Introduction

In recent years, we've observed the rapid development of third-generation sequencing technologies that allow for cost-effective and portable metagenomic analysis with devices such as MinION. It is a miniaturized sequencing device developed by Oxford Nanopore Technologies (ONT). Its small dimensions, weight of 90 g, cost of 1000$, and ability to carry out sequencing experiments in less than an hour make it a perfect choice for use in mobile laboratories for various in-field applications such as real-time monitoring. So far, it has been successfully used for monitoring Zika outbreak in Brazil [15], Ebola virus in Kenya [14], Lassa virus outbreak in Nigeria [16], in freshwater [24] or sewage monitoring systems [6], for early detection of plant viruses in Africa [7] and African swine

© Springer Nature Switzerland AG 2021
M. Paszynski et al. (Eds.): ICCS 2021, LNCS 12745, pp. 500–511, 2021.
https://doi.org/10.1007/978-3-030-77970-2_38

fever characterization [19]. It also has been tested in extreme conditions such as on International Space Station [9], and for studying microbial communities during an ice cap traverse expedition [13].

The portable metagenomics analysis, however, has to overcome a few challenges. The major one is having access to limited computational power at edge devices, where most of the bioinformatic tools require significant computing capabilities and are usually designed to run in cloud environments on multi-node clusters. Another challenge is the amount of data that has to be processed. According to official documentation [4], MinION can produce up to 15 GB of data per day, making it impossible to upload it to the cloud for further processing in scenarios where network connectivity is either not available or network throughput is very slow. The next important thing is energy consumption. For in-field applications in secluded areas, it might be challenging to have access to a reliable power supply, so such devices should operate on battery power or alternative energy sources such as solar panels. For that reason, it's important to use devices that offer a good ratio of computational power to energy consumption.

This paper aims to evaluate the feasibility, performance, and energy consumption of metagenomics analysis in the context of carrying it out on a low-cost constrained edge device in the form of Jetson Xavier NX. The paper is organized as follows. In Sect. 2, we review the related works in the area. In Sect. 3, we describe the considered analysis workflow along with bioinformatics tools used. Section 4 describes the testing environment with a focus on selected hardware. Section 5 contains a description of the testing methodology along with performance experiments that we carried out for selected scenarios. Finally, Sect. 6 concludes the results of the paper.

2   Related Works

In the literature, there is little research concerning portable metagenomics analysis. In [17], Ko et al. describe the applications and challenges related to real-time mobile DNA analysis. The authors highlight time-sensitive and location-sensitive sequencing as two important areas for portable sequencing applications. They also cite energy management, data management, network management, and consumables management as the biggest challenges for portable DNA analysis. The authors also suggest potential ideas for addressing the aforementioned challenges.

Oliva et al. [20] focused on porting and benchmarking bioinformatics tools on an Android smartphone to evaluate the possibility of performing portable analytics on regular smartphones. In their evaluation, they managed to port and run 11 out of 23 considered tools successfully. In most cases, failures were caused by implementations relying on instruction sets unsupported on ARM architectures. They also conducted performance experiments. However, the paper does not include benchmark details for basecalling with Nanocall [10], which was the only basecaller that was successfully ported. They conclude the paper with the suggestion that in order to support real-time portable analytics on commodity smartphones, the development of optimized and dedicated tools will be needed.

In his research [22], Parker benchmarked a cluster of Raspberry Pi single-board computers for the purposes of portable, scalable real-time analysis in comparison to a consumer laptop and a high-performance server. Evaluated analysis pipeline included tools such as Guppy, BLASTN, and Kraken. The author identified Raspberry Pi as a good candidate to run tasks such as BLASTN, where Guppy offered significantly lower performance and Kraken could only support tiny databases due to memory constraints of Raspberry Pi. The author concludes the paper with the observation that such computing clusters can be the right, inexpensive choice for carrying out several bioinformatics tasks.

D'Agostino et al. in their article [11], propose an architecture that combines edge and cloud computing for low-power and cost-effective metagenomics analysis. They evaluate the metagenomics workflow of basecalling and classification with Deepnano basecaller and Kraken software running on Intel System-on-Chip boards. They also consider various compression algorithms to reduce the amount of data that needs to be transferred to the cloud environment for further processing. The authors conclude the paper with the suggestion that while it is possible to run metagenomics sequencing on such devices directly at the edge, none of the tested SoCs could process data in real-time. For such applications, they suggest using fog-based architecture, with more computing nodes closer to the edge device.

Verderame et al. [25] proposed an edge-cloud architecture for metagenomics analysis. Authors in the paper focus on security aspects, providing an overview of security mechanisms handling authentication of devices and ensuring the confidentiality of processed and transmitted data. They evaluated the SSL overhead for MQTT protocol, which turned out to be around 8% of transmission time over using plain MQTT protocol.

Merelli et al. presented research [18] concerning the evaluation of the low-power portable devices for metagenomics analysis in fog computing architecture, based on Intel System-on-Chip boards. They evaluated the performance of various boards during basecalling operations, both from a throughput and energy consumption perspective. In the tested scenario, they concluded that the system that would support real-time workflows could not be battery-powered and would require multiple SoCs to be able to process estimated output from a single MinION device.

In [12], Gamaarachchi provided an overview of a real-time, portable DNA sequence analysis using a System-on-Chip computer. The author highlights the challenges related to portable metagenomic analysis and presents optimizations to tools like minimap2, and introduces a replacement for Nanopolish in the form of f5c. The paper also includes a successful evaluation of the workflow on devices such as Jetson TX2 or Jetson Nano. However, the whole workflow does not include the basecalling step.

Palatnick et al. in [21], introduce a comprehensive metagenomic analysis tool dedicated to mobile devices running the iOS operating system. It offers the ability to align reads, perform variant calling and visualize results directly on an iPhone or an iPad. However, currently, it's impossible to perform basecalling directly on an iOS device, which makes it impossible to run the whole metage-nomics workflow without having access to a separate device for sequencing with MinION. The author suggests that as soon as it's possible to connect MinION directly to these devices, iGenomics will be adjusted to execute the whole end-to-end analysis pipeline.

In [23], Samarakoon et al. presents a corresponding metagenomic analysis tool but dedicated to Android devices. It has integrated multiple known tools (such as minimap2, Samtools, f5c, and Bedtools, to mention just a few) and allows building customizable analytical workflows. Currently, similarly to the iGenomics, it does not support basecalling directly on an Android smartphone and requires a separate device for basecalling and sequencing with MinION.

Based on the above, it can be concluded that there is a lot of interest in being able to perform metagenomic analysis workflow in a portable manner. However, based on research from D'Agostino [11] and Merelli [18], evaluated approaches and tools were not sufficient to run such workflows in a real-time manner. This paper aims to expand knowledge in that area and evaluate the capabilities of low-powered boards such as Jetson Xavier NX in the context of running such workflows on a constrained edge device in a real-time scenario.

3   Analysis Workflow and Software Used

For the evaluation, we selected a workflow where after nanopore sequencing, reads in the form of FAST5 files are immediately basecalled to FASTQ format in real-time, which are further processed by classification software to identify microorganisms present in the tested sample. Figure 1 presents the analyzed workflow.

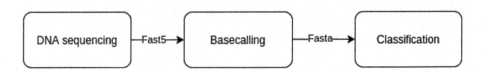

Fig. 1. Metagenomic analysis workflow

For the basecalling step, several tools were considered. The first one was Deepnano-blitz [8], an open-source basecaller developed by Boža, V, based on bidirectional recurrent neural networks. Unfortunately, after preliminary test-ing on Jetson Xavier NX and Jetson Nano, we couldn't evaluate it since the implementation heavily relies on an instruction set not supported by proces-sors powering testing boards. Next considered basecaller was Causalcall [28], another open-source basecaller, based on temporal convolutional network and CTC decoder. Unfortunately, similarly to Deepnano-blitz, we weren't success-ful in running it on Jetson Xavier NX. Next considered basecaller was Guppy [26], which is a closed-source, state-of-the-art basecaller developed by Oxford

Nanopore Technologies. It offers support for GPU acceleration, including GPU on Jetson Xavier NX, and supports multiple basecalling models (fast and high accuracy). The last considered basecaller is Bonito [1], an open-source basecaller developed by Oxford Nanopore Technologies. Likewise the Deeepnano-blitz, it also uses a recurrent neural network. Even though the basecaller is open-sourced, some of its dependencies are not, and they were not available for the ARM architecture, which made running and evaluating Bonito impossible on Jetson Xavier NX.

For classification, we selected Kraken2 [27], which is a taxonomic classification tool that relies on a pre-built database of k-mers associated with the lowest common ancestor of all genomes that contain that k-mer. Based on that association, the classification algorithm then classifies each read by querying the database with each k-mer from a sequence. Then, based on the results, it constructs the classification for the given input. Kraken2 is an improved version of Kraken, offering smaller database sizes, faster classification speed, and faster database build times. It is important to note that the default Kraken2 database requires more than 30 GB of memory, while the minified version of the default database requires 8 GB.

4 Testing Environment

The testing environment was based on a group of edge devices, where one of them served as a source of previously prepared MinION Nanopore reads in the form of FAST5 files. The other one served as a main analytical device that was responsible for carrying out the analysis workflow. It was also connected to an external SSD drive that served as storage for raw and basecalled reads. The main device was also powered through a power supply with an integrated current measurement circuit based on INA219 [2] current sensor connected to Raspberry Pi 4. Figure 2 presents the diagram of the described system.

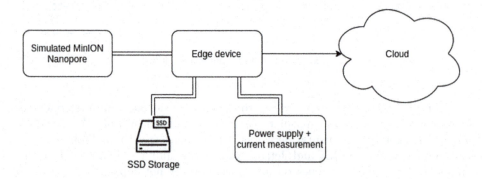

Fig. 2. Diagram of the testing environment.

In this research, we focused on the performance evaluation of the analysis workflow of the presented system. Its central piece is the edge computing device, which was selected to be Jetson Xavier NX, due to offering enough RAM for classification with Kraken 2 and having access to integrated GPU that supports Guppy basecaller acceleration. One of the most important features of Jetson Xavier NX from the perspective of using it in the field is its energy efficiency. It offers five power consumption modes, 10 W with either 2 or 4 CPU cores enabled and 15 W with 2, 4 or 6 CPU cores enabled, which makes it possible to power it with a portable powerbank or by using solar panels. The full technical specification of Jetson Xavier NX is presented below [3]:

- CPU - 6-core NVIDIA Carmel ARM®v8.2 64-bit CPU 6 MB L2 + 4 MB L3
- GPU - NVIDIA Volta™ architecture with 384 NVIDIA® CUDA® cores and 48 Tensor cores
- Memory - 8 GB 128-bit LPDDR4 × 51.2 GB/s
- Storage - SDHC card (32 GB, class 10)
- OS - Ubuntu 18.04.5 LTS with kernel version 4.19.140-tegra

We also considered other low-powered single board computers. However, they were excluded after preliminary testing due to their limitations. While having access to GPU, Jetson Nano was not compatible with the GPU version of Guppy basecaller due to its lower compute capability. Its limited amount of RAM additionally makes running classification with Kraken2 more challenging. Raspberry Pi 4 was also considered, but its lack of GPU acceleration turned out to be a limiting factor for basecalling purposes.

5   Performance Experiments

During experiments, we decided to measure the basecalling and classification capabilities of the Jetson Xavier NX board with different power modes set. For each experiment scenario, we recorded the number of bases or signals processed per second, as well as average power consumption, to assess if these processes could be carried out in real-time while offering energy-efficiency that allows the system to be battery-powered and usable in a portable manner. For evaluating power consumption, we used a dedicated INA219 sensor with Raspberry Pi to collect data, but after evaluating its results against outputs from tegrastats [5] module available on Jetson Xavier NX, we decided to use data from tegrastats to simplify the testing environment, as the values reported were very similar. We measured power every second during the experiment and computed the average to obtain the final value. To carry out experiments, we used a subset of the Klebsiella pneumoniae reads dataset that was used for benchmarking in [26]. During the experiments, we used Guppy in the 4.0.14 version and Kraken2 in the 2.1.1 version.

5.1   Basecalling

During the basecalling evaluation step, we considered two separate Guppy models, fast and high accuracy. Both of them were dedicated to basecalling reads obtained with flow cell using chemistry R9.4.1. We observed that the fast model offers up to 10 times higher throughput, which can be seen on Fig. 3 and Fig. 4, while also recording lower average power consumption, presented on Fig. 5 and Fig. 6. We did not observe significant differences between all three 15 W power modes when using GPU acceleration. However, we observed that from an energy-efficiency standpoint, it might be better to use a 10 W power mode, as the ratio of samples per second to average power is higher in that scenario. We also observed that basecalling without GPU acceleration is not feasible on Jetson Xavier NX, with throughput being lower in that scenario more than 50 times.

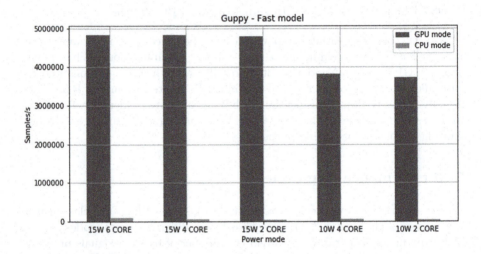

Fig. 3. Samples per second basecalled with Guppy fast model.

5.2   Classification

During classification experiments, in order to accommodate for the fact that Jetson Xavier NX offers only 8 GB of memory, we built a custom Kraken2 database that takes only up to 6 GB of RAM, which made running the experiments on Jetson Xavier NX feasible. As input data, we used the results obtained during the basecalling step in the form of FASTQ files. We observed an almost linear change in basepairs processed per second with changing the number of available cores in different power modes, which can be seen in Fig. 7, with 42,4 Mbp processed per second with 15 W 6 CORE mode. However, as can be seen in Fig. 8,

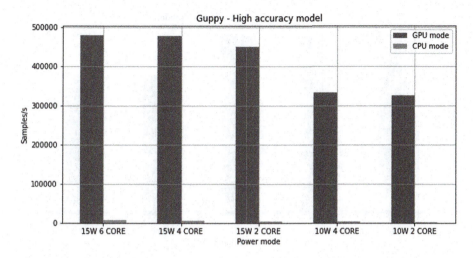

Fig. 4. Samples per second basecalled with Guppy high accuracy model.

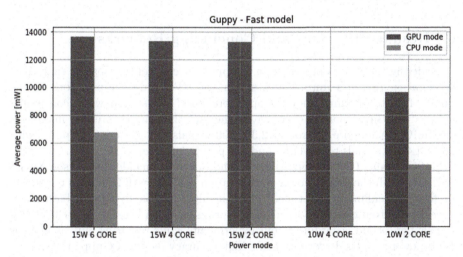

Fig. 5. Average power during basecalling with Guppy fast model.

average power does not follow similar linear change, and the most efficient from an energy standpoint is 15 W 6 CORE mode. As Kraken2 does not support GPU acceleration, we could not evaluate the potential benefits of taking advantage of Jetson Xavier NX's GPU.

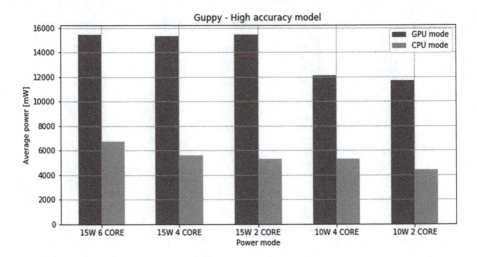

Fig. 6. Average power during basecalling with Guppy high accuracy model.

6   Results Summary and Concluding Remarks

Considering the results presented in the previous chapter, we observe that GPU acceleration provides improvements over the CPU only computations in the range of 45 to 70 times more samples processed per second while consuming only 2 to 3 times more energy. Given the fact that the theoretical maximum of MinION Nanopore is around 2,300,000 signals per second (512 pores with around 4,500 signals read per second per pore), with real-world scenarios resulting in less than 2,000,000 signals per second, we can conclude that Jetson Xavier NX, even with the lowest power configuration of 10 W with 2 enabled cores, can perform basecalling in real-time. In fact, it might be able to support basecalling in real-time from two MinION sequencing runs as it can basecall up to 3,826,260 signals per second according to our experiments while requiring only as little as 10 W power supply. However, when a high accuracy model of Guppy is required, then even on the highest power mode 15 W 6 CORE, Jetson Xavier NX can only process up to 480,000 samples per second, which means that it cannot keep up with real-time sequencing, but can still be useful if only few sequencing experiments are performed during the day. We also observed that basecalling is a more computationally expensive process than classification. In our experiment, 1,000,000 of signals processed translated to roughly 89,525 bases, which means that the classification process is orders of magnitude faster than basecalling and is not a limiting factor for real-time metagenomic analysis at the edge. The classification process can be run successfully on such edge devices, as long as the database that we're using for classification can fit into just 8 GB of memory available on Jetson Xavier NX.

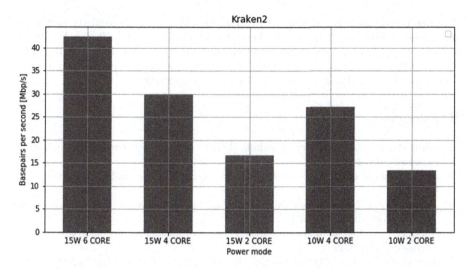

Fig. 7. Basepairs processed per second during classification with Kraken2.

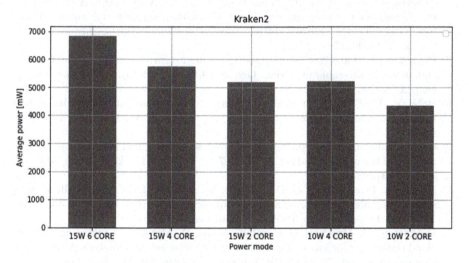

Fig. 8. Average power during classification with Kraken2.

Given the above results, we conclude that computing capabilities, small size, and low energy consumption make Jetson Xavier NX a suitable edge device for portable metagenomics analysis, especially in places with limited network connectivity as whole analysis can be done fully offline. It is important to note that it is still challenging to run alternative open-source basecallers. Many bioinformatics tools are designed for high-performance clusters in cloud environments, which suggests that there is still a lot of room for improvement and development in that area.

Acknowledgments. The research was supported by the Polish Ministry of Science and Higher Education as a part of the CyPhiS program at the Silesian University of Technology, Gliwice, Poland (Contract No. POWR.03.02.00-00-I007/17-00) and by Statutory Research funds of Department of Applied Informatics, Silesian University of Technology, Gliwice, Poland (grant No BK-221/RAu7/2021).

References

1. Bonito basecaller repository on github. (https://github.com/nanoporetech/bonito. Accessed 5 Fe 2021

2. INA219 specification. https://www.ti.com/lit/ds/symlink/ina219.pdf. Accessed 5 Feb 2021

3. Jetson Xavier NX specification. https://developer.nvidia.com/embedded/jetson-xavier-nx-devkit. Accessed 5 Feb 2021

4. Nanopore product comparison. https://nanoporetech.com/products/comparison. Accessed 5 Feb 2021

5. Tegrastats utility. https://docs.nvidia.com/jetson/l4t/index.html#page/Tegra. Accessed 5 Feb 2021

6. Acharya, K., Blackburn, A., Mohammed, J., Haile, A.T., Hiruy, A.M., Werner, D.: Metagenomic water quality monitoring with a portable laboratory. Water Res. **184**, 116112 (2020). https://www.sciencedirect.com/science/article/pii/S0043135420306497

7. Boykin, L.M., et al.: Tree lab: portable genomics for early detection of plant viruses and pests in sub-saharan Africa. Genes, **10**(9), 632 (2019). https://www.mdpi.com/2073-4425/10/9/632

8. Boža, V., Perešíni, P., Brejová, B., Vinař, T.: Deepnano-blitz: a fast base caller for minion nanopore sequencers. Bioinformatics (Oxford, England), **36**, 4191–4192 (2020)

9. Castro-Wallace, S.L., et al.: Nanopore DNA sequencing and genome assembly on the international space station. Sci. Rep. **7**(1), 18022 (2017). DOI: https://doi.org/10.1038/s41598-017-18364-0

10. David, M., Dursi, L.J., Yao, D., Boutros, P.C., Simpson, J.T.: Nanocall: an open source basecaller for Oxford Nanopore sequencing data. Bioinformatics, **33**(1), 49–55 (2016). https://doi.org/10.1093/bioinformatics/btw569

11. D'Agostino, D., Morganti, L., Corni, E., Cesini, D., Merelli, I.: Combining edge and cloud computing for low-power, cost-effective metagenomics analysis. Future Gen. Comput. Syst. **90**, 79–85 (2019). https://www.sciencedirect.com/science/article/pii/S0167739X18300293

12. Gamaarachchi, H., Smith, M.A., supervisor, Parameswaran, S.: Esweek: G: Real-time, portable and lightweight nanopore DNA sequence analysis using system-on-chip (2020)

13. Gowers, G.O.F., Vince, O., Charles, J.H., Klarenberg, I., Ellis, T., Edwards, A.: Entirely off-grid and solar-powered DNA sequencing of microbial communities during an ice cap traverse expedition. Genes **10**(11) (2019). https://www.mdpi.com/2073-4425/10/11/902

14. Hoenen, T., et al.: Nanopore sequencing as a rapidly deployable Ebola outbreak tool. Emerg. Infect. Dis. **22**(2), 331–334 (2016). https://pubmed.ncbi.nlm.nih.gov/26812583, 26812583[pmid]

15. Jesus, J., Giovanetti, M., Faria, N., Alcantara, L.: Acute vector-borne viral infection: Zika and minion surveillance. Microbiol. Spect. **7**(4) (2019)

16. Kafetzopoulou, L.E., et al.: Metagenomic sequencing at the epicenter of the Nigeria 2018 lassa fever outbreak. Science, **363**(6422), 74–77 (2019). https://science.sciencemag.org/content/363/6422/74

17. Ko, S.Y., Sassoubre, L., Zola, J.: Applications and challenges of real-time mobile DNA analysis. In: Proceedings of the 19th International Workshop on Mobile Computing Systems and Applications, pp. 1–6. HotMobile 2018, Association for Computing Machinery, New York, NY, USA (2018). https://doi.org/10.1145/3177102.3177114

18. Merelli, I., et al.: Low-power portable devices for metagenomics analysis: fog computing makes bioinformatics ready for the Internet of things. Future Gen. Comput. Syst. **88**, 467–478 (2018). https://www.sciencedirect.com/science/article/pii/S0167739X17324123

19. O'Donnell, V.K., Grau, F.R., Mayr, G.A., Sturgill Samayoa, T.L., Dodd, K.A., Barrette, R.W.: Rapid sequence-based characterization of African swine fever virus by use of the oxford nanopore minion sequence sensing device and a companion analysis software tool. J. Clin. Microbiol. **58**(1), e01104–19 (2019). https://pubmed.ncbi.nlm.nih.gov/31694969, 31694969[pmid]

20. Oliva, M., Milicchio, F., King, K., Benson, G., Boucher, C., Prosperi, M.: Portable nanopore analytics: are we there yet? Bioinformatics, **36**(16), 4399–4405 (2020). https://doi.org/10.1093/bioinformatics/btaa237

21. Palatnick, A., Zhou, B., Ghedin, E., Schatz, M.C.: iGenomics: comprehensive DNA sequence analysis on your Smartphone. GigaScience, **9**(12) (2020). https://doi.org/10.1093/gigascience/giaa138, giaa138

22. Parker, J.: Lightweight bioinformatics: evaluating the utility of single board computer (SBC) clusters for portable, scalable real-time bioinformatics in fieldwork environments via benchmarking. bioRxiv (2018). https://www.biorxiv.org/content/early/2018/06/02/337212

23. Samarakoon, H., et al.: Genopo: a nanopore sequencing analysis toolkit for portable android devices. Commun. Biol. **3**(1), 1–5 (2020)

24. Urban, L., et al.: Freshwater monitoring by nanopore sequencing. eLife Sci. **10**, 1–27 (2021)

25. Verderame, L., et al.: A secure cloud-edge computing architecture for metagenomics analysis. Future Gen. Comput. Syst. 111 (2019)

26. Wick, R.R., Judd, L.M., Holt, K.E.: Performance of neural network basecalling tools for oxford nanopore sequencing. Genome Biol. **20**(1), 129 (2019). https://doi.org/10.1186/s13059-019-1727-y

27. Wood, D.E., Lu, J., Langmead, B.: Improved metagenomic analysis with kraken 2. Genome Biol. **20**(1), 257 (2019). https://doi.org/10.1186/s13059-019-1891-0

28. Zeng, J., Cai, H., Peng, H., Wang, H., Zhang, Y., Akutsu, T.: Causalcall: nanopore basecalling using a temporal convolutional network. Front. Genet. **10**, 1332 (2020). https://www.frontiersin.org/article/10.3389/fgene.2019.01332

Programming IoT-Spaces: A User-Survey on Home Automation Rules

Danny Soares[1], João Pedro Dias[1,2(✉)], André Restivo[1,3],
and Hugo Sereno Ferreira[1,2]

[1] Faculty of Engineering, DEI, University of Porto, Porto, Portugal
{up201505509,jpmdias,arestivo,hugo.sereno}@fe.up.pt
[2] INESC TEC, Porto, Portugal
[3] LIACC, Porto, Portugal

Abstract. The Internet-of-Things (IoT) has transformed everyday manual tasks into digital and automatable ones, giving way to the birth of several end-user development solutions that attempt to ease the task of configuring and automating IoT systems without requiring prior technical knowledge. While some studies reflect on the automation rules that end-users choose to program into their spaces, they are limited by the number of devices and possible rules that the tool under study supports. There is a lack of systematic research on (1) the automation rules that users wish to configure on their homes, (2) the different ways users state their intents, and (3) the complexity of the rules themselves—without the limitations imposed by specific IoT devices systems and end-user development tools. This paper surveyed twenty participants about home automation rules given a standard house model and device's list, without limiting their creativity and resulting automation complexity. We analyzed and systematized the collected 177 scenarios into seven different interaction categories, representing the most common smart home interactions.

Keywords: Internet-of-Things · Home automation · Trigger-action programming · End-user development

1 Introduction

The Internet-of-Things (IoT) has been converting traditionally analog interactions with digital-based ones, opening doors to an increase of automation in IoT-enabled spaces, leveraging their sensing and actuating capabilities.

Smart homes are a primary example of an IoT-space, although its adoption has been slower than expected [28]. The installation of sensors and actuators in houses can improve the lives of the people who live in them in several ways, including providing more comfort and reducing costs. For example, having heaters, fans, or A/C devices monitored and controlled by an IoT system can improve temperature management efficiency, making houses more comfortable for residents and saving energy [28,30].

© Springer Nature Switzerland AG 2021
M. Paszynski et al. (Eds.): ICCS 2021, LNCS 12745, pp. 512–525, 2021.
https://doi.org/10.1007/978-3-030-77970-2_39

Fig. 1. Trigger-action programming (TAP) example, using the nomenclature typically used in the literature [32]. Using the IFTTT online service nomenclature [17], in a *if A then B* rule, *A* is a trigger, *B* is an action, and the entire rule is an applet.

Automating IoT systems, including smart homes, is not without its challenges, especially when most end-users do not have any technical background [16]. The heterogeneity and a large number of devices, platforms, and services used in IoT, together with the need for end-users to be able to configure and automate them, requires a different approach. While traditional programming (using code editors and integrated development environments) has been the go-to solution for developers and other technical individuals, as the number of IoT application scenarios, environments, and non-technical users increased, it became necessary to build abstractions of sensors, actuators, and whole devices, with additional supporting solutions as a way to reduce the complexity of developing and managing them [3,16]. This lead to the (re)birth of several low-code programming strategies for end-user development (EUD), which include trigger-action programming (TAP—see Fig. 1) [24], programming by demonstration [20], visual programming [25], and domain-specific languages [15]. Most of the programming solutions that use these strategies also differ on the means of programming, leveraging visual notations [9,18,25,26,31], natural language processing tools, and voice assistants [1,19] as a way for users to configure (*viz.*, program) their systems.

While several authors [1,18] state that these low-code solutions for end-user development still have considerable limitations, they also point out their growth in the variety and number of users. Thus, it becomes of paramount importance to understand what end-users wish to automate, how they state their intents, and grasp into the users' programming mental models. Knowing this can provide valuable information to future research, allowing researchers and industry-alike to model their own systems, and making it more easy find their limitations. This also provides ground for the development of test scenarios, using both testbeds and simulations [8,10]. As far as we could find, there is a lack on the literature of a systematic study on the concrete rules that users would define for smart home automation given a base set of devices and a minimal but realistic definition of a home (*i.e.*, akin to a house floor plan).

Some studies already reflect on the automation rules that end-users program into their spaces [1,22,32]. However, these studies are limited by the number of devices and ways of interaction that the development tool under study supports (which, in most cases, is limited to the IFTTT online service [17] due to the easy access to the applet dataset). Earlier works [4,6] attempted to survey automation rules and their complexity, but due to the continuous and rapid evolution of the IoT ecosystem, they fall short to represent the current spectrum of automation possibilities.

We surveyed 20 participants for home automation rules given a standard house model and devices in this work. The study intended to gather as many and as varied home automation scenarios as possible from individuals with different backgrounds and technical know-how while maintaining a certain level of similarity with real-world scenarios and not limiting their creativity and resulting automation complexity.

We proceeded to split the gathered scenarios into categories according to similarities in their structure and types of conditional statements. This survey also added knowledge on how users typically describe their home automation scenarios using text, allowing us to understand if different individuals use different phrases to describe the same scenarios.

The remainder of this paper is structured as follows: Sect. 2 presents an overview of the relevant literature. Section 3 presents the devised model of a smart home, the methodology of our survey, and its results, which are further analyzed and discussed in Sect. 5. Finally, Sect. 6 presents an overview of our study's threats, and Sect. 7 gives some final remarks.

2   Related Work

Dey *et al.* [6], circa 2006, gathered 371 automation scenario descriptions from 20 participants. Almost all of the participants (95%) stated their automation rules in a *if-then* fashion, and around 23.5% of the rules used explicit Boolean logic (*e.g.*, use of **and** or **or** statements). They also categorize the rules accordingly to their complexity: 78.6% fell into the *simple if-then* category, 7% mentioned temporal constraints, 7% mentioned spatial constraints, 6.5% mentioned personal relationships, and, less than 1% focused on environmental personalization (which depends on knowing the inhabitants preferences). Around 14% of the rules mentioned some kind of pre-defined user preference (*e.g.*, preferred ambient temperature).

Brush *et al.* [4], circa 2010, conducted an *in-situ* study of the home automation, by doing semi-structured home visits to 14 households and carried both an analyse of the in-place system and several interviews with the inhabitants. They found out that most of the automation's were of two levels of automation: *user controlled*—where the household explicitly performs an action that triggers one or more actuators—and *rule-based*—where actions happen based on events or at certain times (TAP-like).

Ur *et al.* [32] carried three studies to understand how users use TAP on smart home scenarios. In the first study, they asked 318 workers on Amazon's Mechanical Turk (MTurk) to list five things that they would want a smart home system to do, concluding that most of them fit into four categories, namely: (1) programming, *e.g.*, "automatically turning on the lights when it is dark outside"; (2) self-regulation, *e.g.*, "adjust the house to my preferred temperature at all times"; (3) remote control, "hitting a button on my phone to turn on the lights"; and (4) specialized functionality, *e.g.*, "a breakfast-making machine". To further check the ability to model the workers' intents, they tried to fit them into the TAP model, finding that 62.6% of the submitted answers fit the model. In a second study [32], the authors downloaded a dataset of 67169 recipes (TAP rules) from IFTTT [17], focusing only on the recipes related to smart home automation, which corresponded to a total of 1107 (2,1%), concluding that 513 recipes (0.8%) use physical devices as triggers and 594 (1.3%) use physical devices as actions. A third-study required that a sample of 226 MTurk workers complete pre-defined automation tasks using IFTTT, concluding that 80% or more of the participants successfully implement the presented automation cases. However, the authors' study on the diversity and complexity of the rules that end-users want to configure is too open and vague. There is no common base of devices to automate nor sample building schematic. Also, there is no dataset provided with the first study's collected cases (which is very relevant for this work).

Ammari *et al.* [1] study on how people use voice assistants, which included 82 Amazon Alexa and 88 Google Home users, concluded that IoT-related commands are one of the three most uses of these assistants. They conclude that 85% of the Amazon Alexa and Google Assistant IoT commands involved switching lights on and off. In the case of Alexa, about 10% involved adjustments in light's color and temperature, and a small percentage involved adjusting the temperature of different parts of the house. For Google Assistant, around 10% of the queries were related to changing light colors, dimming lights, and changing fan speeds. By their work, only five study participants created trigger-action rules using Alexa, and the authors do not discuss the intricacies of these rules.

Mi *et al.* [22] also carried a survey on IFTTT, including an analysis of over 408 services (third-party services such as Amazon Alexa), 1490 triggers, 957 actions, 320000 applets. Although their work is not IoT-focused, they carry an analysis on the IoT-related subset of the dataset. In their study, they conclude that the majority of the entries by users are trigger-action ones (*e.g.*, "turn on the light"); thus, the resulting applets are, in their majority, relatively simple, mostly due to the limited and simple interfaces exposed by most IoT devices. The authors also add that this is due to "the fact that most tasks (in the smart home context) we want to automate are indeed simple". While we agree that the limitations posed by the devices limit what end-users can program them to do and that the majority of the rules are indeed simple by nature, as the number of inhabitants and devices increases, the resulting operational context can be complex to model and reason about [21].

3 Survey

A survey was envisioned as the most effective way to gather as many automation scenarios as possible in a timely fashion. The methodology was based on the one presented by Molléri et al. [23].

3.1   Smart Home Model

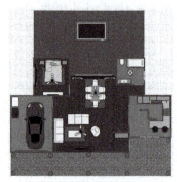 

Fig. 2. 2D and 3D floor plan of the smart house used for the survey.

To have a pre-defined, common foundation from where the participants could base themselves to draft their own automation scenarios, we designed a house floor plan and 2D/3D models of it, as shown in Fig. 2. The house has a total of 8 *spaces*: (a) a *garage*, (b) a *front patio*, (c) a *pool*, (d) a *garden*, (e) a *living room*, (f) a *kitchen*, (g) one *bedroom*, and (h) a *bathroom*.

Along with the home model, we provide a list of smart devices containing various types of sensors and actuators for the participants to use. Namely, across all home divisions, there are the following IoT devices: (1) motion, temperature, humidity, smoke, and air quality sensors, (2) security cameras, (3) controllable lights, (4) controllable windows and blinds, (5) A/C system, (6) robot vacuum cleaner, and (7) sound system.

The (a) *garage* has (a.1) controllable outside and inside doors, (a.2) washing machine, and (a.3) a dryer machine. The (b) *front patio* has only a (b.1) controllable entry door. The (c) *pool* has a (c.1) automated pool cover, (c.2) cleaning system, (c.3) water temperature sensor, and (c.4) water heating system. The (d) *garden* has a (d.1) water sprinkler system, (d.2) soil moisture sensor, and (d.3) robot lawnmower. The (e) *living room* has a (e.1) smart TV. The (f) kitchen has a (f.1) stove, (f.2) oven, (f.3) exhaust hood, (f.4) dishwasher, and (f.5) coffee machine. The (g) *bedroom* has a (g.1) a smart TV, and (g.2) controllable bedside lamps. Lastly, the (h) *bathroom* has a (h.1) heated towel rack.

We also allowed and instigated participants to include other devices in their home automation scenarios as long they were available as off-the-shelf IoT solutions. No limitations on the interoperability of the IoT system parts nor in the end-user programming interface were defined nor presented.

3.2   Methodology

An online form[1] was picked as a data collection method, given the study's motivation to gather as many automation scenarios as possible while attempting to reducing any bias on the respondent population. The form presented the smart home model, namely, the house floor plan and the list of available devices. Users had only one question where they could insert as many scenarios as they wished to, without any limitation in size or form.

The survey was then disseminated among 20 participants from different educational fields and ages. All the answers were collected in a spreadsheet, anonymized, and the individual scenarios identified, allowing further analysis.

4   Results

The survey resulted in 177 scenarios grouped into categories according to their structure and format similarity. This allowed us to filter duplicated entries, keeping only the mostly-unique and representative ones. The list of categories and a brief description and a sample of three (when available) representative scenarios extracted from the dataset are given in the following paragraphs.

Sensors and Actuators. Scenarios that only use sensors and actuators, where the sensors trigger the actuators, *e.g.*:
- "When there is movement in the garage, turn on the garage lights";
- "Adjust pool water temperature according to the outside temperature if someone is using it";
- "When smoke detectors are activated, alert through the sound system, alert the house owner via SMS, and warn the fire department".

Actuators on Schedule. Scenarios where actuators are triggered on a fixed schedule, *e.g.*:
- "When time is 7:30 am, turn on the coffee machine, the hot water system, and the kitchen lights";
- "Every Saturday at 3 pm, turn on the robot vacuum cleaner, and the robot lawn mower";
- "When time is 7:30 am, turn on the coffee machine, the hot water system and the kitchen lights (if they were turned off)".

Actuators on Time Interval, with Sensors. Scenarios that combine sensors information and time intervals to trigger actuators, *e.g.*:
- "During the night, when there is motion in one room, light that room at 200 brightness";
- "When a TV series starts, and I am home, turn on the TV on that channel and prepare popcorn. If I am not home, record it";
- "When turning on the dishwasher (or washing machine) after 20h, wait for 00h to start".

Sensors with Timers. Scenarios that combine sensors information and time intervals to trigger actuators, *e.g.*:

---

[1] Google Forms, https://forms.google.com.

- "During the night, when there is motion in one room, light that room at 200 brightness";
- " When a TV series starts, and I am home, turn on the TV on that channel and prepare popcorn. If I am not home, record it";
- "When turning on the dishwasher (or washing machine) after 20h, wait for 00h to start".

Actuators with Timers. Scenarios where the actuators are triggered when the status of the sensor does not change for some time, *e.g.*:

- "When it is 23:00, turn on the garden watering system for 10 min";

External Services. Scenarios that depend on external services to trigger the actuators, *e.g.*:

- "When the sun is expected during the following hours, turn off the heating system";
- "If it will rain in the next hour, inform that drying outside is not the best plan";
- "With a solar panel; and weather information; schedule a machine to run sometime during the day (*e.g.*, washing machine). It is expected that the system automatically schedules it to the time of day that is most likely to be sunny for power saving."

One-time Actions. Scenarios that are meant to happen only once, instead of being recurring, *e.g.*:

- "Shut all unnecessary devices";
- "When it is 7:00 today, turn on the bedroom lights".

The full dataset and the model house floor plan and device list are available on Zenodo [29] to ease the study's replication and allow further analysis.

5   Analysis

In this section, an analysis of the submitted 177 scenarios is done, attempting to extract insights from the different rules submitted given the suggested smart home model. We considered all the submitted entries valid smart home automation scenarios, and we were able to categorize all the submitted scenarios in one of the seven defined categories. The scenarios differ (1) in the granularity of application (*e.g.*, with some of them being specific to a house part or domestic appliance), (2) in complexity (rules range from direct triggers to one device to triggers of multiple devices depending on several conditional statements) and (3) in writing fashion (with most of them being close to a conditional programming logic).

Table 1 details the distribution of scenarios through the categories, showing the absolute and relative frequency of submissions per category as identified in Sect. 4.

Responses such as *"Intensity of lights based on the available natural light"*, *"Blinds inclination system based on outside light"*, *"On schedule turn on the coffee machine"* would need modifications to be closer to a programming-like

Table 1. Categories of submitted scenarios, along with the absolute and relative frequency of submissions for each category.

Category	Absolute frequency	Relative frequency
Sensors and actuators	103	0.58
Actuators on schedule	42	0.24
Actuators on time interval, with sensors	15	0.08
Sensors with timers	9	0.05
External services	5	0.03
One-time actions	2	0.01
Actuators with timers	1	0.01
Total	**177**	**1.00**

format to be possible to implement them in usual end-user programming solutions. For example, these should look more like *"When the luminosity in the living room is below $value, then increase the light's intensity by $increment"*, or *"When time is 7:00, then turn on the coffee machine"*. These responses (and respective scenarios) are still valid because the information portrayed is enough to understand their meaning and to which category they belong.

With the information collected, the house plan, and devices provided to the participants, we created a **resulting ecosystem** that represents the house with all the devices that the participants used. This ecosystem is represented in Fig. 3, showing the house plan with all the devices used by the participants. Some respondents also mentioned using an external weather forecast API and wearables (which are not represented in the isometric visualization).

The most common way of specifying scenarios is by using the structure "when *condition*, then *action*" or "*action* when *condition*" (with the recurrent use of "if" instead of "when"). This is close to the representation commonly used by TAP approaches. However, some scenarios are depicted differently, mainly for scheduled actions, such as "*action* at *time*", or "everyday at *time*, *action*".

Looking at the dataset, we can see that there are home areas more frequently identified. In contrast, others are almost unseen, as depicted in the chart of Fig. 4. We consider direct mentions all the mentions to specific rooms in the submitted scenarios, *e.g.*, garage or bedroom. All the indirect mentions consist of remarks about certain things that are, typically, only present in certain rooms, namely: washing machine mentions are part of the garage (by the given device list); entrance is considered front patio; lawnmower references considered as part of the garden; kitchen includes mentions to the dishwasher, coffee, and oven; alarm clock, waking up, sleep are all related to the bedroom; and all mentions to shower and toilet are considered part of the bathroom.

Users also tend to specify similar (or equal) scenarios using different expressions, granularity and forms. This was expected since the participants' background (*e.g.*, educational level, previous experience with IoT, age, and the way

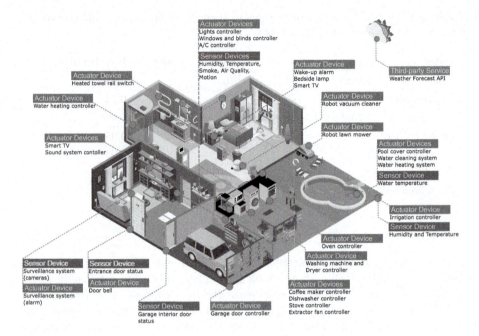

Fig. 3. Isometric visualization of the resulting ecosystem based on the survey responses.

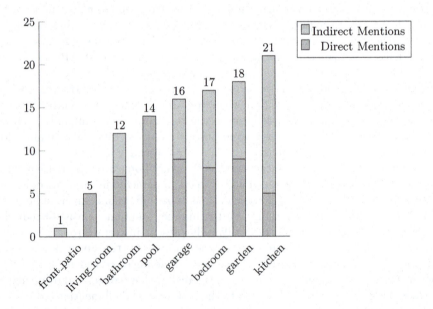

Fig. 4. The number of mentions to specific home parts in the submitted scenarios. Direct mentions consider situations where the surveyed participants directly mentioned a given house part and the indirect mentions consider references to certain things that are, typically, only present in certain rooms.

of expressing ideas) was homogeneous. As an example, "Turn the heating system on, set to the preferred temperature, on schedule.", "Maintain room temperature in between a specified permissible range.", and "Turn on the AC when the temperature is higher than a given value." transmit the same rule—adjusting the house temperature according to a preference value—in different fashions, either in format or precision.

Approx. 28% of the scenarios mention "turn on" actions, and there are 28 direct mentions to *lights*, 23 to *water*, 21 to *blinds*, and 20 to *temperature*.

It is noted the use of pre-conditionals in some scenarios, more specifically, defining some condition that should be met before enforcing the rule, *e.g.*, "With a solar thermal collector (for heating water); when the sun is expected during the following hours, turn off traditional water heating system". The use of macros that aggregate a set of tasks and sub-rules is also visible, *e.g.*, "*Holiday mode*: when any device is used/triggered notify the owner" and "*Garden automation*: stable soil moisture level, temperature stabilization in adverse weather".

Although integration with external services is only mentioned once, there are several rules that, when implemented, would depend on some information provider. For example, we can consider "If the hot water system is based on electricity, heat when electricity is cheaper" would depend on knowing the market prices of electricity. Further, voice control, *e.g.*, "Voice control over coffee machine/blinds/lights/etc.", is typically accomplished by integrating with a third-party voice assistant such as Amazon Alexa, Apple Siri, or Google Assistant [1].

Finally, it is also noticeable that some rules are too generic, *e.g.*, "Shut down all unnecessary devices", which would require that the IoT system had some degree of contextual awareness to be able to execute them. There is also some concern about the failure of the system parts and the use of IoT to detect them, *e.g.*, "Send SMS alert if faulty freezer/fridge".

6   Threats to Validity

For this survey, we have identified some threats that may affect the validity of the results attained.

We asked participants for home automation scenarios and did not give them any **structure for the phrases**, to understand how they would write the scenarios, which resulted in many scenarios being just a brief description and not specific enough. Perhaps, having requested the participants to provide more detailed scenarios would have resulted in more concrete scenarios. However, constraining participants to use a specific format or having certain degrees of detail for the scenario descriptions would not have allowed us to evaluate whether users tend to follow a pattern or typical structure.

Although we let participants use any off-the-shelf device, we provided them with an initial list of devices to pick from. This, together with the specified house plant, might have introduced a bias into the chosen scenarios.

The **sample size** for this survey was not very large since we were only capable of gathering 20 participants. Having a larger sample could have resulted in more varied scenarios. This could enrich our analysis and provide more insights into the typical way that users express their automation rules and what these rules typically consist of.

The **level of expertise** of the participants could impact the scenarios provided by them. For example, participants with more experience with IoT should provide more complex and realistic scenarios than participants with no experience in that field. To tackle this threat, we attempted to choose participants with different levels of expertise with low-coding programming solutions and IoT. This resulted in having scenarios from participants whose experience ranged from never having thought of a home automation scenario to participants that had already implemented IoT systems and worked with Node-RED extensively.

Even though the participants had different levels of expertise in home automation, and some even had a lot of experience, the results show **little variety in categories** for the scenarios. After collecting 177 scenarios, we only identified seven categories, and 82% of the submissions belonged to two categories (*i.e.*, 58% to Sensors and Actuators category and 24% to Actuators on schedule category). Increasing the number of participants in the survey would probably, have resulted in more varied scenarios and more categories.

7   Conclusions

This paper presented a survey conducted with 20 participants to collect home automation scenarios, which resulted in 177 scenarios to be categorized and analyzed. We consider that all the scenarios fit in one of the seven defined categories, representing different types of automation.

The most common pattern used by users to define their automation scenarios shares a similar structure close to conditional programming—"when *condition, then action*", or "*action*, when *condition*"—which is compatible with the trigger-action programming model used by several market solutions including IFTTT. This shows that it is intuitive for regular users to describe home automation scenarios in a mostly-structured fashion, easily transposed to a conditional programming fashion. Besides, the users tend to use (or mention) macros and/or aliases that represent more than one device (*e.g.*, a group of lights) or more than one action (*e.g.*, garden control).

Taking into account the available solutions in the market for end-user programming and their programming strategies, we can consider—taking into account relevant literature such as the one presented in Sect. 1—that while most of the scenarios could be easily mapped into TAP rules, the rules that do not follow such model appear as a challenge which is mostly ignored by existing solutions, especially the ones that focus users with little to none technical knowledge. In this case, voice assistants can become of utmost importance, allowing users to create automation rules in a conversation, adding complexity by steps instead of specifying everything in one statement or by a diagram [1,13,19].

Contextual awareness also adds value to low-coding solutions [16], since with such contextualization, they can use information about the system and their parts, the system surroundings, and the current defined behaviors and rules to provide user information, insights, and alerts when new rules (or changes) create conflicts with already defined rules, can lead to malfunctions, or nefarious effects—*e.g.*, avoiding turning off the CO_2 sensors by mistake by alerting the user to the changes that will occur.

Finally, as the number of devices being Internet-connected increases rapidly, depending on end-users to control and manage all of them appears to be too complex of a challenge. Strategies that hinder this complexity by the end-users seem to be the *natural direction* for IoT systems by enabling devices to manage themselves in terms of software and hardware configurations [2], optimizations in resource utilization (*e.g.*, energy) [27] and usage requirements (continuous adaptation by learning from the environment) [5], and in prevention and recovery of failures or other issues [11,12,14].

The data collected in this study[2], including the definition of typical automation categories, can be used as a foundation for follow-up studies, ranging from human-computer interaction (*e.g.*, low-code and end-user programming) to IoT systems design and implementation (*e.g.*, what are the current sensing and actuating needs that are not meet by the existent systems which are of the end-user interest). This data allows the definition of system prototypes by researcher and industry communities to meet current user needs, and be validated against a known dataset of user-based interaction and automation scenarios. In previous work, we already used this data partially for defining research directions and build validation testbeds and scenarios (*e.g.*, definition of a *SmartLab*[3]) [14,19].

Acknowledgement. This work is financed by National Funds through the Portuguese funding agency, FCT - Fundação para a Ciência e a Tecnologia, within project UIDB/50014/2020.

References

1. Ammari, T., Kaye, J., Tsai, J.Y., Bentley, F.: Music, search, and iot: howpeople (really) use voice assistants. ACM Trans. Comput.-Hum. Interact. **26**(3), 17 (2019)
2. Athreya, A.P., DeBruhl, B., Tague, P.: Designing for self-configuration and self-adaptation in the Internet of Things. In: Proceedings of the 9th IEEE International Conference on Collaborative Computing: Networking, Applications and Worksharing, COLLABORATECOM 2013, pp. 585–592 (2013). https://doi.org/10.4108/icst.collaboratecom.2013.254091
3. Baresi, L., Ghezzi, C.: The disappearing boundary between development-time and run-time. In: Proceedings of the FSE/SDP Workshop on Future of Software Engineering Research, pp. 17–22. FoSER 2010, Association for Computing Machinery, New York, NY, USA (2010)
4. Brush, A.B., Lee, B., Mahajan, R., Agarwal, S., Saroiu, S., Dixon, C.: Home automation in the wild: challenges and opportunities. In: proceedings of the SIGCHI Conference on Human Factors in Computing Systems, pp. 2115–2124 (2011)

---

[2] The replication package of the user study can be found on Zenodo [29].
[3] The replication package for the *SmartLab* can be found on Zenodo [7].

5. Chatzigiannakis, I., et al.: True self-configuration for the IoT. Proceedings of 2012 International Conference on the Internet of Things, IOT 2012, pp. 9–15 (2012). https://doi.org/10.1109/IOT.2012.6402298

6. Dey, A.K., Sohn, T., Streng, S., Kodama, J.: icap: interactive prototyping of context-aware applications. In: Fishkin, K.P., Schiele, B., Nixon, P., Quigley, A. (eds.) Pervasive Computing, pp. 254–271. Springer, Berlin Heidelberg, Berlin, Heidelberg (2006)

7. Dias, J.P.: jpdias/smartlab: Replication package for smartlab (2021). https://doi.org/10.5281/zenodo.4657647

8. Dias, J.P., Couto, F., Paiva, A.C.R., Ferreira, H.S.: A brief overview of existing tools for testing the Internet-of-things. In: 2018 IEEE International Conference on Software Testing, Verification and Validation Workshops (ICSTW), pp. 104–109 (2018). https://doi.org/10.1109/ICSTW.2018.00035

9. Dias, J.P., Faria, J.P., Ferreira, H.S.: A reactive and model-based approach for developing internet-of-things systems. In: 2018 11th International Conference on the Quality of Information and Communications Technology (QUATIC), pp. 276–281 (2018). https://doi.org/10.1109/QUATIC.2018.00049

10. Dias, J.P., Ferreira, H.S., Sousa, T.B.: Testing and deployment patterns for the Internet-of-things. In: Proceedings of the 24th European Conference on Pattern Languages of Programs. EuroPLoP 2019. Association for Computing Machinery, New York, NY, USA (2019). https://doi.org/10.1145/3361149.3361165

11. Dias, J.P., Restivo, A., Ferreira, H.S.: Empowering visual Internet-of-Things mashups with self-healing capabilities. In: 2021 IEEE/ACM 2nd International Workshop on Software Engineering Research Practices for the Internet of Things (SERP4IoT) (2021)

12. Dias, J.P., Sousa, T.B., Restivo, A., Ferreira, H.S.: A pattern-language for self-healing Internet-of-Things systems. In: Proceedings of the 25th European Conference on Pattern Languages of Programs. EuroPLop 2020. Association for Computing Machinery, New York, NY, USA (2020). https://doi.org/10.1145/3361149.3361165

13. Lago, A.S., Dias, J.P., Ferreira, H.S.: Conversational interface for managing non-trivial Internet-of-Things systems. In: Krzhizhanovskaya, V.V., et al. (eds.) ICCS 2020. LNCS, vol. 12141, pp. 384–397. Springer, Cham (2020). https://doi.org/10.1007/978-3-030-50426-7_29

14. Dias, J.P., Lima, B., Faria, J.P., Restivo, A., Ferreira, H.S.: Visual self-healing modelling for reliable Internet-of-Things systems. In: Krzhizhanovskaya, V., et al. (eds.) ICCS 2020. LNCS, vol. 12141, pp. 357–370. Springer, Cham (2020). https://doi.org/10.1007/978-3-030-50426-7_27

15. Einarsson, A.F., Patreksson, P., Hamdaqa, M., Hamou-Lhadj, A.: Smarthomeml: towards a domain-specific modeling language for creating smart home applications. In: 2017 IEEE International Congress on Internet of Things (ICIOT), pp. 82–88 (2017)

16. Ghiani, G., Manca, M., Paternò, F., Santoro, C.: Personalization of context-dependent applications through trigger-action rules. ACM Trans. Comput.-Hum. Interact. **24**(2), 1–33 (2017). https://doi.org/10.1145/3057861

17. IFTTT: Ifttt helps your apps and devices work together (2019). https://ifttt.com

18. Ihirwe, F., Di Ruscio, D., Mazzini, S., Pierini, P., Pierantonio, A.: Low-code engineering for internet of things: a state of research. In: Proceedings of the 23rd ACM/IEEE International Conference on Model Driven Engineering Languages and Systems: Companion Proceedings. MODELS 2020. Association for Computing Machinery, New York, NY, USA (2020)

19. Lago, A.S., Dias, J.P., Ferreira, H.S.: Managing non-trivial Internet-of-Things systems with conversational assistants: a prototype and a feasibility experiment. J. Comput. Sci. **51**, 101324 (2021)
20. Li, T.J.J., Li, Y., Chen, F., Myers, B.A.: Programming IoT devices by demonstration using mobile apps. In: Barbosa, S., Markopoulos, P., Paternò, F., Stumpf, S., Valtolina, S. (eds.) End-User Development, pp. 3–17. Springer International Publishing, Cham (2017)
21. Manca, M., Fabio, Paternò, Santoro, C., Corcella, L.: Supporting end-user debugging of trigger-action rules for IoT applications. Int. J. Hum.-Comput. Stud. **123**, 56–69 (2019)
22. Mi, X., Qian, F., Zhang, Y., Wang, X.: An empirical characterization of IFTTT: ecosystem, usage, and performance. In: Proceedings of the 2017 Internet Measurement Conference, pp. 398–404 (2017)
23. Molléri, J.S., Petersen, K., Mendes, E.: An empirically evaluated checklistfor surveys in software engineering. Inf. Softw. Technol.**119**, 106240 (2020)
24. Rahmati, A., Fernandes, E., Jung, J., Prakash, A.: IFTTT vs. zapier: A comparative study of trigger-action programming frameworks. arXiv abs/1709.02788 (2017)
25. Ray, P.P.: A survey on visual programming languages in Internet of Things. Sci. Program. **2017**, 1–6 (2017)
26. Reiss, S.P.: IoT end user programming models. In: 2019 IEEE/ACM 1st International Workshop on Software Engineering Research & Practices for the Internet of Things (SERP4IoT), pp. 1–8. IEEE (2019)
27. Seo, J., Kim, W.H., Baek, W., Nam, B., Noh, S.H.: Optimally Self-Healing IoT Choreographies. In: International Conference on Architectural Support for Programming Languages and Operating Systems - ASPLOS Part F1271(1), 91–104 (2017)
28. Shin, J., Park, Y., Lee, D.: Who will be smart home users? an analysis of adoption and diffusion of smart homes. Technol. Forecast. Soc. Change **134**, 246–253 (2018)
29. Soares, D., Dias, J.P., Restivo, A., Ferreira, H.S.: Smart home automation survey (2021). https://doi.org/10.5281/zenodo.4531395
30. Sundmaeker, H., Guillemin, P., Friess, P., Woelfflé, S.: Vision and Challenges for Realising the Internet of Things for the Information society, E.C.D.G., Media. Publications Office of the European Union (2010)
31. Torres, D., Dias, J.P., Restivo, A., Ferreira, H.S.: Real-time feedback in node-red for IoT development: an empirical study. In: 2020 IEEE/ACM 24th International Symposium on Distributed Simulation and Real Time Applications (DS-RT), pp. 1–8 (2020)
32. Ur, B., McManus, E., Pak Yong Ho, M., Littman, M.L.: Practical trigger-action programming in the smart home. In: Proceedings of the SIGCHI Conference on Human Factors in Computing Systems, pp. 803–812. CHI 2014. Association for Computing Machinery, New York, NY, USA (2014)

Application of the Ant Colony Algorithm for Routing in Next Generation Programmable Networks

Stanisław Kozdrowski[1]([✉]) [iD], Magdalena Banaszek[1], Bartosz Jedrzejczak[1],
Mateusz Żotkiewicz[2] [iD], and Zbigniew Kopertowski[3] [iD]

[1] Computer Science Institute, Warsaw University of Technology, Nowowiejska 15/19,
00-665 Warsaw, Poland
s.kozdrowski@elka.pw.edu.pl,
{magdalena.banaszek.stud,bartosz.jedrzejczak.stud}@pw.edu.pl
[2] Institute of Telecommunications, Warsaw University of Technology,
Nowowiejska 15/19, 00-665 Warsaw, Poland
mzotkiew@tele.pw.edu.pl
[3] Orange Labs Polska, Orange Polska Obrzeżna 7, 02-691 Warszawa, Poland
Zbigniew.Kopertowski@orange.com

Abstract. New generation 5G technology provides mechanisms for network resources management to efficiently control dynamic bandwidth allocation and assure the Quality of Service (QoS) in terms of KPIs (Key Performance Indicators) that is important for delay or loss sensitive Internet of Things (IoT) services. To meet such application requirements, network resource management in Software Defined Networking (SDN), supported by Artificial Intelligence (AI) algorithms, comes with the solution. In our approach, we propose the solution where AI is responsible for controlling intent-based routing in the SDN network. The paper focuses on algorithms inspired by biology, i.e., the ant algorithm for selecting the best routes in a network with an appropriately defined objective function and constraints. The proposed algorithm is compared with the Mixed Integer Programming (MIP) based algorithm and a greedy algorithm. Performance of the above algorithms is tested and compared in several network topologies. The obtained results confirm that the ant colony algorithm is a viable alternative to the MIP and greedy algorithms and provide the base for further enhanced research for its effective application to programmable networks.

Keywords: Heuristics · Artificial intelligence · Ant colony · Internet of Things · Software defined networking · Mixed integer programming · Programmable networks.

The work on this paper was done in FlexNet project in EUREKA CELTIC-NEXT Cluster for next-generation communications under partial funding of The National Centre for Research and Development in Poland.

M. Paszynski et al. (Eds.): ICCS 2021, LNCS 12745, pp. 526–539, 2021.
https://doi.org/10.1007/978-3-030-77970-2_40

1   Introduction

In the paper, we present a problem solution studied in the FlexNet project and related to efficient, dynamical, flexible Software Defined Networking (SDN) resource allocation for Internet of Things (IoT) applications with different Quality of Service (QoS) requirements. In the FlexNet project [5,8], the Artificial Intelligence (AI) based developed solution is responsible for controlling the routing of intents in SDN. Typically, network is managed statically using commands and scripts, so the efficiency of resource provisioning is low and practically without automatization. Over the last few years we can observe new network solutions with improved management, where most advanced is the SDN solution [17,28]. In IETEF RFC7149 [12] SDN is defined as a set of mechanisms and techniques used to build network services in deterministic, dynamic, and scalable methodology, suitable for use in 5G technology [24]. SDN Controller allows for adaptive, dynamic resources provisioning by applying management rules to traffic flows in the network [6]. On the other hand, also the traffic in the network becomes more complex, especially in IoT applications [2,21,22,30], with big data volume generated to the network, and requires more flexibility and scalability [14,25]. New applications and appearing different types of devices generate different patterns of network traffic. Therefore, current network solutions often do not meet the arising needs and their management is not effective. Another disadvantage of legacy networks is complex architecture in case of introducing QoS and security policies [23].

1.1   Motivation

Therefore, it is envisioned to use for future networks the SDN solutions with their capability of programmable flexible control of network resources and dynamic on demand configuration according to application requirements [27]. Such approach allows for flexible creation of new services and applications that are installed over the network controller while no changes in the actual network devices are needed.

Artificial Intelligence (AI) has seen a surge of interest in the networking community [32]. Recent contributions include data-driven flow control for wide-area networks, job scheduling, and network congestion control [18]. A particularly promising domain is the network management. Researchers have used Machine Learning (ML) to address a range of network tasks such as network resource control, routing, and traffic optimization [1,4,15]. In IoT networks we expect network conditions to vary over time and space. Time varying conditions may be long term (seasonal) or short term resulting in a significant impact on network performance [20]. ML techniques will be developed in order to detect such changes and signal them to the SDN layer for timely action to be taken to improve the overall network performance. Another solution is a knowledge-based network (KDN) which also is a next step on the path towards an implementation of a self-driving network [19]. KDN is a complementary solution for SDN that brings reasoning processes and ML techniques into the network control plane to enable autonomous and fast operation and minimization of operational costs.

1.2   AI Challenges and FlexNet AI Concept

In the FlexNet platform, SDN network orchestration will be supported by AI for solving selected problems of network resource control. The concept of AI application covers such capabilities as: flexible traffic control in the network, flexible adaptation to the conditions of the system, using appropriate learning algorithms based on the state-of-the-art approaches to the problem, i.e., Reinforcement Learning (RL), [31], reaction in real time. The sole goal of the AI will be to distribute network resources in the way to maximize a global objective function related to QoS, e.g., minimize buffer occupancy sizes in the nodes in order to minimize network characteristics like packet losses or packet delays.

FlexNet AI uses the concept of Off-Platform Application (OPA). The application consists of two main blocks, i.e., Path Generator and Maintainer, and communicates directly with ONOS controller [26] and their two build-in applications, i.e., Intent Forwarding (IFWD) and Intent Monitoring and Rerouting (IMR). Also, it communicates directly with switches on a network using ifstat external application [13] or sFlow-RT monitoring tool [29] (see Fig. 1).

The first block of FlexNet AI, i.e., Path Generator, is responsible for generating a pair of paths for each intent being registered. The paths are selected in a way to balance efficiency and available capacity providing one fast and one spacious path.

The second block, i.e., Maintainer, is responsible for switching intents between paths selected by the Path Generator in a way to maximize efficiency. The data used by FlexNet AI to select paths and maintain intents consists of static and dynamic entries. In the first group, entries deal with the

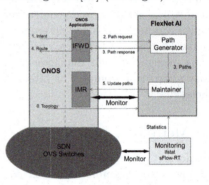

Fig. 1. Flexnet AI architecture.

topology of a network. They are obtained directly from ONOS during start-up. In the second group the dynamic data are collected, constantly updated using the feedback generated by the ONOS IMR build-in application and the external ifstat application.

The first task of FlexNet AI is to route intents. The IFWD application contacts the Path Generator informing FlexNet AI that the intent is present. The Path Generator computes two paths based on the current network state and previously routed intents. One of the computed paths is selected by the Path Generator and sent back to the IFWD application, where it is forwarded to ONOS controller to establish a route for the intent. The main goal is to find two paths, which are edge-disjoint and have the best cost function related to QoS parameters, i.e., delay and/or loss. For this purpose, we used a nature-inspired Ant Colony (AC) algorithm [7,10]. The novelty of our approach is to use the

AC algorithm to find two paths simultaneously. In general, until now, the AC algorithms have been used to find only one path at a time. [3,11,16].

The article is organized as follows: in Sect. 2 we briefly describe the problem and algorithms we use in the contribution. Section 3 presents the results of the experimental study. Contributions of this work are summarized and future work directions are discussed in Sect. 4.

2   Problem Formulation and Algorithms

In this contribution the goal of the AI is to rationally control the routes of intentions in an SDN network. To this end, for each intention we choose a pair of potential paths that are disjoint as much as possible. This pair is computed in a way that minimises the weighted average cost of the paths by assuming that the cost of one path is given by its length and the cost of the other by its occupancy. In this way, we obtain a pair of paths, one of which minimises the transmission time and the other minimises the probability of loss. We find the sought pair using mainly the Ant Colony (AC) algorithm. Our AC algorithm is designed specifically for this problem and is compared to commonly known algorithms (MIP, Mixed Integer Programming) that guarantee an optimal solution and to greedy algorithms.

2.1   Problem Formulation

The problem we are considering in this article belongs to the class of NP-complete problems [9] and is presented as follows:

Data. Input data consist of the following items:

- demand between s and d nodes,
- network with capacities,
- actual traffic on arcs.

They can be formally presented using the following sets and constants.

Sets

$\mathcal{V}$ vertices
$\mathcal{A}$ arcs
$\delta^+(v)$ set of arcs entering vertex $v \in \mathcal{V}$
$\delta^-(v)$ set of arcs leaving vertex $v \in \mathcal{V}$
$\mathcal{I}$ set of possible lengths of the first path; $|\mathcal{I}| = |\mathcal{A}|$

Constants

s source; $s \in \mathcal{V}$
d destination; $d \in \mathcal{V}$
$b(a)$ used fraction of bandwidth on arc $a \in \mathcal{A}$
ξ' weight of the first path
ξ'' weight of the second path
L_X maximum tolerable used fraction of bandwidth for the first path
L_Y maximum tolerable used fraction of bandwidth for the second path

Assumptions. In the problem, we assume that the load on an arc is the actual traffic on it. Moreover, the load on an arc cannot be smaller than 0.01 to prevent forming loops on the second path.

Objective and Formal Model. We search for the pair of paths X and Y between s and d such that:

- the number of common arcs is minimal (highest priority); in other words, we are interested in disjoint paths if they exist; if there is at least one solution without common arcs, we choose and return one of them; if there is not, and there is a solution with one common arc, we return it, and so forth.
- path X does not use arcs with load greater than L_X, path Y does not use arcs with load greater than L_Y; these constraints must always be satisfied; if there are no paths satisfying the constraints, then there is no solution.
- maximise a weighted sum $\xi'A + \xi''B$, where A is the reciprocal of the length of path X and B is the product of free capacities on path Y (if Y is completely free then $B = 1$; if Y consists of two arcs that are occupied in half, then $B = 0.25$); in other words, the best solution is when X is the shortest and Y is not occupied in one percent (see Assumptions above).

The problem can be formally presented using the following set of variables and constraints.

Variables

x_a binary; 1 if arc $a \in \mathcal{A}$ belongs to X
y_a binary; 1 if arc $a \in \mathcal{A}$ belongs to Y
z_a binary; 1 if arc $a \in \mathcal{A}$ is used by both X and Y
T integer; number of common arcs

Objective and constraints

$$\min \left\{ (\xi' + \xi'')T \; - \; \frac{\xi'}{\sum_{a \in \mathcal{A}} x_a} \; - \; \xi'' \prod_{a \in \mathcal{A}: y_a = 1} (1 - b(a)) \right\} \qquad (1a)$$

$$\sum_{a \in \delta^+(v)} x_a - \sum_{a \in \delta^-(v)} x_a = \begin{cases} 1 & v = d \\ 0 & v \in \mathcal{V} \setminus \{s, d\} \end{cases} \qquad (1b)$$

$$\sum_{a \in \delta^+(v)} y_a - \sum_{a \in \delta^-(v)} y_a = \begin{cases} 1 & v = d \\ 0 & v \in \mathcal{V} \setminus \{s, d\} \end{cases} \qquad (1c)$$

$$x_a = 0 \quad \forall a \in \mathcal{A} : b(a) > L_X \qquad (1d)$$

$$y_a = 0 \quad \forall a \in \mathcal{A} : b(a) > L_Y \qquad (1e)$$

$$x_a + y_a \leq 1 + z_a \quad \forall a \in \mathcal{A} \qquad (1f)$$

$$\sum_{a \in \mathcal{A}} z_a \leq T \qquad (1g)$$

Objective function (1a) consists of three elements. The first one, i.e., $(\xi' + \xi'')T$ is responsible for the priority objective, which is minimizing the number of common arcs. We will call this part of the objective function the First Criterion in Numerical Results. The second element represents the cost of path X and the third element represents the cost of path Y. They will be together called the Second Criterion in Numerical Results. Notice that the second element cannot exceed $-\xi'$ and the third element cannot reach $-\xi''$ (load cannot be smaller than 0.01), thus each change in T (First Criterion) has the absolute priority over other changes (Second Criterion).

Constraints (1b) and (1c) impose the flow conservation law on paths X and Y, respectively, while (1d) and (1e) assure that the paths do not use overloaded arcs. Finally, (1f) and (1g) set the correct number of shared arcs.

2.2   Algorithms

MIP Approach. The presented problem is not-linear and cannot be directly solved using MIP solvers. In this section, we present an algorithm that using a linear version of a simplified problem solves the considered problem invoking an MIP solver a number of times. In the simplified model, the following additional constants are used.

S length of the shortest path between s and d
P maximum number of arcs in X
T maximum number of shared arcs (variable in the base model).

The approach is presented in Algorithm 1. It is a brute force approach that checks possible values of pairs T and P. For each value T it first checks the feasibility of solutions setting P to the maximum possible value. If a solution cannot be found, the next value for T is considered. The approach uses a simplified MIP

Algorithm 1: MIP

Input: $S = shortestPath(weight(a) = 1)$
$best\_Y = -cost(shortestPath(weight(a) = -log(1 - b(a)))$
for $T=0,1,...$ **do**
    $best\_obj = 0$;
    for $P=|V| - 1,S,S+1,...,|V| - 2$ **do**
        if $best\_obj \geq \xi'/P + \xi''e^{best\_Y}$ **then**
            | return $best\_res$;
        end
        $obj, res =$ solve simplified MIP;
        if $obj=NULL$ **and** $P=|V| - 1$ **then**
            | **break**
        end
        if obj **then**
            $real\_obj = \xi'/P + \xi''e^{obj}$;
            if $real\_obj > best\_obj$ **then**
                $best\_obj = real\_obj$;
                $best\_res = res$;
            end
        end
    end
    if $best\_res$ **then**
        | return $best\_res$;
    end
end

---

model with modified objective function and an additional constraint shown in (2).

$$\max \sum_{a \in \mathcal{A}: b(a) \leq L_Y} y_a \log(1 - b(a)) \tag{2a}$$

$$\sum_{a \in \mathcal{A}} x_a \leq P \tag{2b}$$

Constraint (2b) assures that the length of path X in the obtained solution will not exceed P. Having constant T and iterating through all viable values of P, the objective function (1a) reduces to the third element, which is the product of fractions of free capacities on arcs used by path Y. The product can be made linear using a logarithm, which is visible in (2a).

Ant-Colony Approach. Algorithm 2 is based on behaviour of an ant colony. In each iteration every ant is moved by one arc from source to destination (forward) or backward, when the destination was already reached. When moving forward ant chooses an arc depending on the pheromone amount. The more pheromone lays on an arc, the more attractive it is for an ant. With some small probability an ant can choose a less attractive arc. An ant cannot use arcs with load exceeding L_X. Each ant remembers nodes visited on the path and avoids choosing arcs that could create a loop. In result, paths determined by ants fulfill requirements of path X, which should be the shortest and not overloaded. When an individual reaches the destination node, the Dijkstra algorithm finds path Y. It has the

---

Algorithm 2: Ant colony

Input: $Lx = 0.9$
Input: $Ly = 0.99$
Input: n = number of iterations
$best\_Y = -cost(shortestPath(weight(a) = -log(1 - b(a))))$
for $i=0,...n$ **do**
    for ant in ant_colony **do**
        if $ant.moving\_forward$ **then**
            if $ant.current\_node == destination$ **then**
                x_path = ant.found_path
                y_path = Dijkstra(x_path)
                $obj = \sum_{y \in y\_path} log(1 - b(y))$
                ant.evaluation = $\xi'/len(x\_path) + \xi'' e^{obj}$
            else
                arc = choose_arc_by_pheromone()
                move_ant(ant, arc)
            end
        else
            arc = pop(ant.found_path)
            move_ant(ant, arc)
            leave_pheromone(arc)
        end
    end
    if i $\%$ $evaporation\_frequency == 0$ **then**
        evaporate(pheromone_decrement, arcs)
    end
end
  return find_best_solution(ant_colony)

---

minimal number of common arcs with path X, because we explicitly set weights of arcs belonging to X found by an ant to a sufficiently great number. Arcs with load exceeding L_Y cannot be included into path Y. The selected pair of paths is evaluated by calculating weighted sum, which influences the amount of pheromone that will be left on arcs while moving backward. When specified number of iterations elapses, some of the pheromone evaporates from each arc. After the final iteration the last pairs of paths remembered by ants are compared. A pair with the highest evaluation value is returned as the solution.

Greedy Approach. Algorithm 3 presents the greedy approach based on the Dijkstra algorithm. At first it searches for path X, choosing the shortest path consisting of arcs with load less than L_X. The Dijkstra algorithm minimises a path cost, where all arcs have weight equal to 1. When choosing a neighbour to visit, arcs with the load exceeding L_X are not considered. Then, also using the Dijkstra algorithm, path Y is determined based on the chosen path X. All arcs that belong to previously found path X have much bigger weight than others that have the weight reflecting the load. Moreover, arcs with load exceeding L_Y cannot be used by path Y. The algorithm minimises the path cost, which results in finding path Y that has the least common arcs with path X and has the most free capacity on its arcs. The found pair of paths is evaluated with the weighted sum and returned as a solution.

We will refer to this approach as the DijkstraXY algorithm. Another greedy approach that we use in the research reverses the path selection order picking the optimal path Y at first and then selecting path X which is the shortest and as disjoint as possible from path Y. The latter approach will be called the DijkstraYX algorithm.

Algorithm 3: Greedy algorithm

Input: $Lx = 0.9$
Input: $Ly = 0.99$
$best\_Y = -cost(shortestPath(weight(a) = -log(1 - b(a)))$
$x\_path = Dijkstra()$
$y\_path = Dijkstra(x\_path)$
$obj = \sum_{y \in y\_path} log(1 - b(y))$
$x\_y\_evaluation = \xi'/len(x\_path) + \xi''e^{obj}$

return x_path, y_path

---

3   Experiments and Results

The common configuration parameters for AC algorithm for all simulations studies are listed in Table 1. Objective weight ξ' was set to the length of the shortest path between s and d. This way the part of the objective function that deals with the first path is scaled. The second part of the objective was also scaled by setting ξ'' to the reciprocal of e to the power of the inverse of the length of the shortest path between s and d taking $log(1 - b(a))$ as arc weights. In addition, ξ'' was multiplied by 4. In this way, we obtained a set of problems where the second path is slightly prioritized over the first path in the constant extend.

Table 1. Parameters used in AC algorithm.

Name	Short description	Value
S	$s - d$ distance	Variable
m	Size of the colony	$2|V|$
n	Number of iterations	$40\,S$
α	Pheromone dosage factor	5
β	Pheromone evaporation factor	10
γ	Pheromone evaporation frequency	$2\,S$

The calculations were carried out for all considered algorithms on a 2.1 GHz Xeon E7-4830 v.3 processor with 256 GB RAM running under Linux Debian operating system and additionally for MIP-based algorithm a linear solver engine of CPLEX 12.8.0.0 was used.

Each point in each graph presented in Figs. 2, 3, 4 (i.e. one network with a certain load level) represents the average result for 100 instances, among which there are 10 different load distributions. For each instance, different source and destination nodes are drawn uniformly at random. Each load distribution was obtained by constantly generating random demands and routing them in a network using shortest paths. Demands that could not have been satisfied were discarded. The process was repeated until the requested average load was reached. The AC algorithm, as a non-deterministic algorithm, was run for each instance 10 times and the result was averaged. On the other hand, the Dijkstra algorithm and MIP-based algorithm, as deterministic algorithms, were run one time each.

In Fig. 2, running times of the presented algorithms are displayed. We can observe the MIP-based algorithm having problems with finding optimal solutions in the time limit acceptable for dynamic environments presented in the introduction. The running times can be annoying in networks of 75 nodes and more. In networks of 150 nodes and more, the running times become unacceptable for practical implementations of the presented SDN framework. On the other hand, other presented algorithms were able to solve the problem in the assumed time limit of one second. Notice that the greedy algorithms are not depicted in the figure. They running times never exceeded 0.01 s.

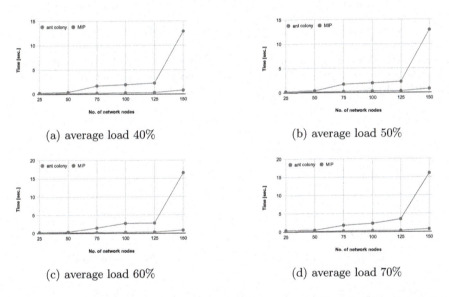

(a) average load 40% (b) average load 50%

(c) average load 60% (d) average load 70%

Fig. 2. Comparison of calculation times for different methods and different network loads: 40% - (a), 50% - (b), 60% - (c) and 70% - (d).

In Fig. 3, the results for the First Criterion are displayed. The presented greedy algorithms were not able to stand against the MIP-based approach and the Ant-Colony approach returning results with significantly more common links. The trend is clearly visible when the number of nodes reached a certain level, which in our experiments was close to 150. It is worth to notice that the Ant-Colony approach does not have serious problems with returning optimal solutions with respect to the First Criterion for any of the considered network instances.

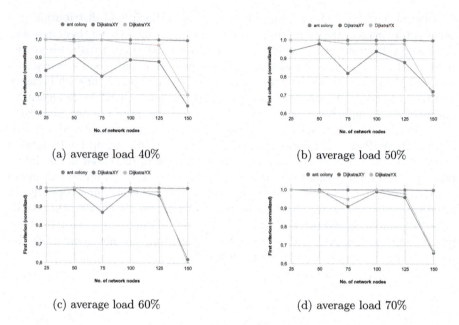

(a) average load 40%   (b) average load 50%

(c) average load 60%   (d) average load 70%

Fig. 3. Normalized objective function presenting relative number of common links (First Criterion) with respect to MIP-based method for different network loads: 40% - (a), 50% - (b), 60% - (c) and 70% - (d).

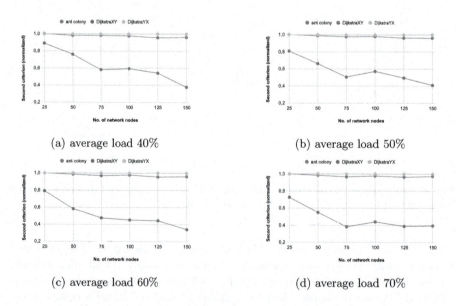

(a) average load 40%   (b) average load 50%

(c) average load 60%   (d) average load 70%

Fig. 4. Normalized objective function presenting Second Criterion with respect to MIP for different net. loads: 40% - (a), 50% - (b), 60% - (c) and 70% - (d).

Finally, in Fig. 4, the obtained results for the Second Criterion are displayed. Here it is important to explain the way the figure was prepared. Because the Second Criterion counts only when the First Criterion is optimal, we decided to include only such cases in this figure. With respect to the Second Criterion, the greedy DijkstraXY algorithm is heavily outperformed by other approaches. On the other hand, DijkstraYX algorithm always finds solutions with optimal values of the Second Criterions, thus slightly outperforming the Ant-Colony approach. However, it results from the fact that the DijkstraYX algorithm was strictly designed to optimize the second (more important, due to $4\xi' = \xi''$) part of the Second Criterion. In other words, DijkstraYX returns optimal solutions with respect to the Second Criterion, but seldom finds optimal solutions with respect to the First Criterion. On the other hand, the Ant-Colony approach repeatedly finds optimal solutions with respect to the First Criterion, which occasionally results in slightly worse performance with respect to the Second Criterion.

4   Conclusions

The paper studied the problem of traffic routing in mesh networks with a specific criteria of the objective function. The presented solution is applicable to intelligent SDN management systems with AI support, especially in programmable next generation networks (5G and beyond). Moreover, the network resource allocation in context of QoS assurance is the arising problem especially for IoT services. The designed solution based on the AC algorithm is suitable for the fast and optimal resource allocation, especially in the case of emergency and delay sensitive IoT services. In FlexNet project the solution is tested for the emergency surveillance video IoT service.

A heuristic method inspired by biology called the AC algorithm is proposed to solve the problem. The novel approach of this method is the optimal routing of a pair of paths in the network, searched simultaneously, considering three criteria described in the objective function. The considered problem has been modelled and solved as an MIP problem. Thus, there is a certainty of finding an optimal solution. We demonstrated the stability of AC through simulations. We showed that it quickly converges to the best path under situations when traffic characteristics change (among others when load on the network is increased). We have shown that with an appropriate tuning of the parameters, AC behaves better when compared to other competing approaches in mesh networks. In addition, the promising results shown in the paper underline the need for a real-life testbed evaluation on which we are currently working in FlexNet project.

Further research will be focused on more comprehensive experiments that include real scenarios and with comparisons to other competitive heuristics.

References

1. Abar, T., Letaifa, A., El Asmi, S.: Machine learning based QoE prediction in SDN networks, pp. 1395–1400 (2017). https://doi.org/10.1109/IWCMC.2017.7986488

2. Bera, S., Misra, S., Vasilakos, A.V.: Software-defined networking for internet of things: a survey. IEEE Internet Things J. **4**(6), 1994–2008 (2017). https://doi.org/10.1109/JIOT.2017.2746186

3. Bokhari, F.S., Záruba, G.V.: On the use of smart ants for efficient routing in wireless mesh networks. CoRR abs/1209.0550 (2012). http://arxiv.org/abs/1209.0550

4. Chen, B., Wan, J., Lan, Y., Imran, M., Li, D., Guizani, N.: Improving cognitive ability of edge intelligent IIOT through machine learning. IEEE Netw. **33**(5), 61–67 (2019). https://doi.org/10.1109/MNET.001.1800505

5. Choque, J., et al.: Flexnet: flexible networks for IoT based services. In: 2020 23rd International Symposium on Wireless Personal Multimedia Communications (WPMC), pp. 1–6 (2020). https://doi.org/10.1109/WPMC50192.2020.9309486

6. Dinh, K.T., Kukliński, S., Osiński, T., Wytrebowicz, J.: Heuristic traffic engineering for SDN. J. Inf. Telecommun. **4**(3), 251–266 (2020). https://doi.org/10.1080/24751839.2020.1755528

7. Dobrijevic, O., Santl, M., Matijasevic, M.: Ant colony optimization for QoE-centric flow routing in software-defined networks. In: 2015 11th International Conference on Network and Service Management (CNSM), pp. 274–278 (2015). https://doi.org/10.1109/CNSM.2015.7367371

8. Flexnet: Flexible IoT networks for value creators. (2020). https://www.celticnext.eu/project-flexnet/

9. Garey, M.R., Johnson, D.S.: Computers and Intractability; A Guide to the Theory of NP-Completeness. W. H. Freeman & Co., New York, NY, USA ©1990 (1990)

10. Guan, Y., Gao, M., Bai, Y.: Double-ant colony based UAV path planning algorithm. In: Proceedings of the 2019 11th International Conference on Machine Learning and Computing, pp. 258–262. ICMLC 2019. Association for Computing Machinery, New York, NY, USA (2019). https://doi.org/10.1145/3318299.3318376

11. Hamrioui, S., Lorenz, P.: Bio inspired routing algorithm and efficient communications within IoT. IEEE Netw. **31**(5), 74–79 (2017). https://doi.org/10.1109/MNET.2017.1600282

12. IETF: Software-defined networking: A perspective from within a service provider environment (2017). https://tools.ietf.org/html/rfc7149

13. ifstat: ifstat - linux man page. (2017). https://linux.die.net/man/1/ifstat

14. Jin, Y., Gormus, S., Kulkarni, P., Sooriyabandara, M.: Content centric routing in IoT networks and its integration in RPL. Comput. Commun. **89**(C), 87–104 (2016). https://doi.org/10.1016/j.comcom.2016.03.005

15. Kozdrowski, S., Cichosz, P., Paziewski, P., Sujecki, S.: Machine learning algorithms for prediction of the quality of transmission in optical networks. Entropy (Basel, Switzerland), **23**(1), 7 (2021). https://doi.org/10.3390/e23010007

16. Liu, X., Li, S., Wang, M.: An ant colony based routing algorithm for wireless sensor network. Int. J. Future Gen. Commun. Netw. **9**, 75–86 (2016). https://doi.org/10.14257/ijfgcn.2016.9.6.08

17. Liyanage, M., Ylianttila, M., Gurtov, A.: Securing the control channel of software-defined mobile networks. In: Proceeding of IEEE International Symposium on a World of Wireless, Mobile and Multimedia Networks, pp. 1–6 (2014). https://doi.org/10.1109/WoWMoM.2014.6918981

18. Mao, H., Alizadeh, M., Menache, I., Kandula, S.: Resource management with deep reinforcement learning. In: Proceedings of the 15th ACM Workshop on Hot Topics in Networks, pp. 50–56. HotNets 2016. Association for Computing Machinery, New York, NY, USA (2016). https://doi.org/10.1145/3005745.3005750

19. Mestres, A., et al.: Knowledge-defined networking. SIGCOMM Comput. Commun. Rev. **47**(3), 2–10 (2017). https://doi.org/10.1145/3138808.3138810
20. Mishra, P., Puthal, D., Tiwary, M., Mohanty, S.P.: Software defined IoT systems: Properties, state of the art, and future research. IEEE Wirel. Commun. **26**(6), 64–71 (2019). https://doi.org/10.1109/MWC.001.1900083
21. Municio, E., Latré, S., Marquez-Barja, J.M.: Extending network programmability to the things overlay using distributed industrial IoT protocols. IEEE Trans. Ind. Inform. **17**(1), 251–259 (2021). https://doi.org/10.1109/TII.2020.2972613
22. Municio, E., Marquez-Barja, J., Latré, S., Vissicchio, S.: Whisper: programmable and flexible control on industrial IoT networks. Sensors, **18**(11), 4048 (2018). https://doi.org/10.3390/s18114048
23. Murat Karakus, A.D.: Quality of service in software defined networking: a survey. J. Netw. Comput. Appl. **80**, 200–218 (2017). https://doi.org/10.1016/j.jnca.2016.12.019
24. de la Oliva, A., et al.: 5g-transformer: Slicing and orchestrating transport networks for industry verticals. IEEE Commun. Mag. **56**, 78–84 (2018). https://doi.org/10.1109/MCOM.2018.1700990
25. Omar, H.: Intelligent traffic information system based on integration of Internet of Things and agent technology. Int. J. Adv. Comput. Sci. Appl. **6**(2), 37–43 (2015). https://doi.org/10.14569/IJACSA.2015.060206
26. ONOS: ONOS Project. (2017). https://wiki.onosproject.org/
27. Open Networking Foundation: Software-Defined Networking: The new norm for networks. White Paper (2012)
28. Rothenberg, C.E., et al.: Revisiting routing control platforms with the eyes and muscles of software-defined networking. In: Proceedings of the First Workshop on Hot Topics in Software Defined Networks, pp. 13–18. HotSDN 2012, Association for Computing Machinery, New York, NY, USA (2012). https://doi.org/10.1145/2342441.2342445
29. sFlow: sflow-rt documentation (2017). https://sflow-rt.com/reference.php
30. Thubert, P., Palattella, M., Engel, T.: 6tisch centralized scheduling: When SDN meet IoT (2015). https://doi.org/10.1109/CSCN.2015.7390418
31. Yao, H., Mai, T., Xu, X., Zhang, P., Li, M., Liu, Y.: Networkai: an intelligent network architecture for self-learning control strategies in software defined networks. IEEE Internet Things J. **5**(6), 4319–4327 (2018). https://doi.org/10.1109/JIOT.2018.2859480
32. Zhao, Y., Le, Y., Zhang, X., Geng, G., Zhang, W., Sun, Y.: A survey of networking applications applying the software defined networking concept based on machine learning. IEEE Access, **7**95397–95417 (2019). https://doi.org/10.1109/ACCESS.2019.2928564

Scalable Computing System with Two-Level Reconfiguration of Multi-channel Inter-Node Communication

Miroslaw Hajder[1], Piotr Hajder[2]([✉]), Mateusz Liput[1],
and Janusz Kolbusz[1]

[1] University of Information Technology and Management, Rzeszow, Poland
`jkolbusz@wsiz.edu.pl`
[2] AGH University of Science and Technology, Krakow, Poland
`phajder@agh.edu.pl`

Abstract. The paper presents the architecture and organization of a reconfigurable inter-node communication system based on hierarchical embedding and logical multi-buses. The communication environment is a physical network with a bus topology or its derivatives (e.g. folded buses, mesh and toroidal bus network). In the system, multi-channel communication is forced through the use of tunable signal receivers/transmitters, with the buses or their derivatives being completely passive. In the physical environment, logical components (nodes, channels, paths) are distinguished on the basis of which logical connection networks separated from each other are created. The embedding used for this purpose is fundamentally different from the previous interpretations of this term. Improvement of communication and computational efficiency is achieved by changing the physical network architecture (e.g. the use of folded bus topologies, 2D and 3D networks), as well as the logical level by grouping system elements (processing nodes and bus channels) or their division. As a result, it is possible to ensure uniformity of communication and computational loads of system components. To enable formal design of the communication system, a method of hierarchy description and selection of its organization was proposed. In addition, methods of mathematical notation of bus topologies and the scope of their applications were analyzed. The work ends with a description of simulations and empirical research on the effectiveness of the proposed solutions. There is high flexibility of use and relatively low implementation price.

Keywords: Hierarchical embedding · Multi-channel architecture · Scalable systems

1 Introduction

Until recently, it was thought that personal computers would meet the needs for most users in the field of science and industry. There is a possibility of remote use

M. Paszynski et al. (Eds.): ICCS 2021, LNCS 12745, pp. 540–553, 2021.
https://doi.org/10.1007/978-3-030-77970-2_41

of supercomputers, however, at the time of the almost universal use of artificial intelligence and the mass of cyber attacks, such assumptions became incorrect. It is a result of the fact, that performing calculations using AI methods may turn out to be unsafe from the perspective of the possible occurrence of cyber attacks. This situation can significantly limited possibility the correctness of the results or even disturb the proper functioning of the system based on them. The security aspect is extremely important for users, so sometimes during remote access there are concerns about the integrity or confidentiality of the transmitted data. The construction of cheap, easy-to-use and scalable computer units with increased resilience to possible damage from cyber attacks has become deliberate.

There may be several solutions to the above problem. The first of them is usage of commonly available computing units based on Arduino and Raspberry. The creation of an appropriate infrastructure consisting of a dozen or so devices of this type can significantly improve the computing power needed to perform more complex tasks. The idea behind the authors of the text is to create a supercomputer structure made of the aforementioned devices together with the definition of the method of communication.

The second solution to the problem involves the use of Internet of Things devices. The authors assume that the communication structure created by them will allow for the connection of various types of devices used every day, which are elements of the Internet of Things. Each of them has its own computing powers, which, when combined with each other, make it possible to obtain additional resources needed to perform complex calculations.

The authors decided to look for solutions to the above problem in the area of application of passive optical communication to connect computational units based on Arduino or Raspberry platforms. Previous studies in the design, construction and operation of multi-machine units assumed, as in most cluster systems, the integration of computational nodes by switching electrical signal in the network or transport layer [23,24]. This solution provided scalability of computational power, however, balancing communication channels was difficult to achieve.

Another solution described in the literature was the use of optical technologies similar to those used, among others in WDM switches, however, many times cheaper [30]. A special optical system sent a light wave to the input of the appropriate receiver. Also in this case, the solution was effective in increasing the computational power of parallel computer.

The proposed solution is based on passive optical technologies. The basis of the whole system is a multi-channel optical bus network, based on which various derivative bus networks are built using hierarchical embedding. The network configuration is performed from the level of processing nodes, which greatly simplifies its operation. The use of a multi-channel optical connection network allows to efficiently connect or disconnect computational nodes from the subsystem. Depending on the formally determined need, the subsystem autonomously changes its architecture, organizing its resources to ensure maximum efficiency, reliability or minimum response time to service requests.

The result of the research is to prepare the technical basis for the functioning of the infrastructure consisting of Internet of Things devices or Arduino and Raspberry units. Regardless of the chosen method of managing computing power, proper communication should be ensured. The proposed idea assumes improvement in the following areas: **a.** availability for everyone (due to the low cost of production), **b.** scalability - the ability to connect additional devices in the event of a need for additional computing power, **c.** safety - in the event of a threat, you can turn off specific devices that have been infected without having to disconnect the entire system.

2   Multi-channel, Embedding and Hierarchy in the Construction of Connection Networks

The basis for new communication solutions is hierarchy understood in a slightly different way. So far, analyzing data transmission systems, the focus has been on two-level architectures based on the projection of logical topology on physical network resources [7,19]. Nowadays, when using hierarchy to build scaled computational environments based on IoT, this approach trivializes the problem and has several significant disadvantages. The most important of them are:

1. Focusing on logical paths as the basic elements of connection network, which simplifies the problem solved too much;
2. Ignoring the hierarchy of functional relationships of communication architecture components;
3. Limitation to two amounts of analyzed levels, preventing the synthesis and analysis of real systems;
4. Lack of parameters determining the level of physical and logical correlation of connection characteristics;
5. Application of classic graph or matrix representations to the morphological description of the network, limiting thus the possibilities of its study.

These disadvantages can be minimized by using graph theory methods [6,8,17, 29] and multi-level hierarchical systems [12,27,28]. Therefore, connection architecture should be reduced to a multi-level hierarchical structure. Then, combinatorial optimization methods implemented to solve the tasks of multi-level hierarchical systems theory can be used for its synthesis and analysis. Hierarchization will be performed using embedding implemented through multi-channeling. This approach assumes that the primary concept of further consideration is multi-channeling offered, among others by communication technologies used to build IoT network. In order to limit the area of interest, it was assumed that the channel hierarchy has a multi-echelon organization [9,10,26].

Analysis of the use of IoT technology in hierarchical networks requires an original definition of the embedding concept. In classic works [1,2,21] embedding consists in placing guest graph in the host graph nodes, for which the mapping function is used. The result of embedding is a new network whose parameters are inherited from the source graphs. Until now, Cartesian product was most often used as a mapping function, ensuring additive and multiplicative inheritance of parameters. This embedding is of theoretical nature and has been called topology embedding. Previous research focused on using the Cartesian product to build scalable topological organizations, sorting networks and fault-tolerant connection networks. Standard topologies, such as ring, tree, hypercube, toroid, etc. were used as bases. From the point of view of considered issues, topology embedding is of secondary importance.

Embedding is proposed to be expanded with new types: embedding of communication components, embedding of physical networks, and embedding of logical connection networks. The concept of communication component is defined as real or virtual transmission environment, connecting a pair of nodes (logical or physical) at the same level of the technology hierarchy. Examples of components are logical channels and paths, as well as physical components analogous to them. It is assumed that multi-channeling is available in all technologies used and is the basis for this type of embedding. As a result, there may be many subchannels of lower $(k+1)$ level in the k-th layer's channel. A set of channels of k-th level can be embedded in the logical channel of $(k-1)$ layer or directly in the physical channel of level 1 (so-called vertical embedding). The logical path of the k-th level can be a set of its logical channels, based on various channels of the $(k-1)$ layer (so-called horizontal embedding). Embedding components is described by a hierarchy that reflects the interrelationship between inter-node communication means, called the hierarchy of communication components. An example of a hierarchy of logical (virtual) components is shown in Fig. 1a. The logical channel is created by connecting two logical nodes (or more for group transmission) using a logical connection. In turn, a logical path is a sequential combination of a set of logical channels, and a logical topology is a set of paths.

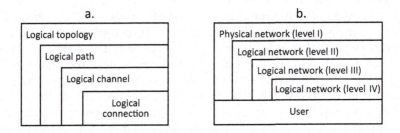

Fig. 1. Hierarchical organization of communication

A hierarchy of physical components could be presented in a similar way. Its levels are respectively: connections, channels, paths and physical topologies. A distinctive feature of the presentation from previous works is the narrowing of the set of logical topologies used only to the bus and its derivatives. Logical networks are built on their basis, the range of which is very wide.

Embedding an hierarchization can also be used to build a hierarchy of communication networks, which in general, consists of a core, a set of access networks, and end-user serving nodes. In this case, we are talking about embedding physical connection networks. This type of embedding is similar to the traditional one, described among others in [1,2,25]. The lower level connection network is embedded in a higher layer network node or in a set of them. From the graph theory point of view, access networks are represented by a forest of disconnected graphs. Because embedding may not apply to all nodes and can only include a limited set of them, the hierarchy of physical connection networks is heterogeneous. An example of this hierarchy is shown in Fig. 1b. Although in real solutions the end user is primarily connected to level IV access networks, there is nothing to prevent him from joining to other layers, including directly to the core of the network.

Another new type of embedding is connection networks embedding. Unlike the classical approach, the lower-level connection network is inscribed in all or a significant part of the upper-layer network, and not only in its communication node or channel. This is accomplished by embedding logical paths of k-th layer in the network of $(k-1)$ level, while maintaining the equivalence relationship of nodes of different hierarchy levels. The result of such embedding will not be new networks, but only implementation of lower-level networks on network resources located higher in the hierarchy of connection networks. The graph theory properties of the embedded network are determined solely by itself, and the transport characteristics passed through levels higher than the one on which the network was created. This embedding can be considered as a functional development of embedding communication components. In the literature, network embedding has a two-level character (the logical network is embedded on physical network resources) [1,2,22]. The conducted research did not limit the number of embedding levels constituting the hierarchy.

The last type of embedding described in this paper is the embedding of technology, involving the introduction in the environment of one communication technology, another, using the resources of the first. This solution is already known and described in the literature [3,11]. An example of technology embedding is widely used Ethernet network allocation on MPLS network resources functioning in the WDM physical network environment. Also in this case, on the basis of embedding, a hierarchy of technology is created, whose layers are associated with a specific level of communication, and not with a specific type.

According to the classification presented in Fig. 2, the basic element forming any hierarchy is the communication channel, understood as the environment of information transmission at the physical level. Based on this, using hierarchy of components, basic communication paths are created from the transmission point of view. These, in turn, constitute the basis of hierarchically connected interconnection networks that are used as the environment for functioning of communication technologies. Summarizing, the global hierarchical structure is built by embedding the hierarchy of components in the network hierarchy, and the result obtained in the hierarchy of communication technologies.

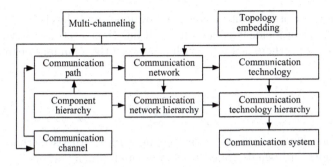

Fig. 2. Relationship between hierarchical communication components

The synthesis and analysis of systems described as multi-level hierarchical structures is one of the known directions in the study of large-scale systems [12,27]. For every hierarchy considered (components, networks, technologies), it is recommended to formally determine the number of levels, the selection of elements for every individual layer and the ways of their connection, so that the resulting structure is characterized by minimal construction and operation costs and maximum efficiency. Until now, based on a compact description, a set of acceptable structures and criteria for their assessment was defined. In addition, the task of hierarchy synthesis is performed only at the qualitative level, and quantitative models are either not considered at all, or have only a special character [4,5]. To formally solve the above task, it was proposed to use methods of construction an optimal hierarchy, previously used mainly in management, control and bioinformatics [18,20].

3   Bus Networks and Their Mathematical Representation

In the further part of the paper it is assumed that the physical network of the communication system is based solely on the bus, so let's consider its definitions. A bus network is a combination of two types of equal objects N nodes and B buses. Each node can be incidental with any number of buses. The incident means that each node can be connected with any number of buses. A network of n nodes and m buses is usually marked as $[n, m]$ and describes the incidence matrix $A = \{i, h\}$ with size $n \times m$. Element $a_{ij} \in A$ is equal to 1 if with node of number $i = 1, \ldots, n$ there is incidental bus with number of $j = 1, \ldots, m$, otherwise $a_{ij} = 0$. From the definition of the incident matrix it follows that bus networks do not allow loops for both buses and nodes. Therefore, if there are multiple connections in the designed system (i.e. the selected node will be integrated with the selected bus with several connections), the traditional way of describing the bus will not be able to be used.

If the number of incident buses with i-th node (node level) is $s_i^w = 1, \ldots, n$, and the number of incident nodes with j-th bus (bus level) as s_j^m, then for any bus network $[n, m]$ there is a relationship between the summary degree of nodes and buses:

$$\sum_{i=1}^{n} s_i^w = \sum_{j=0}^{m} s_j^m = s \qquad (1)$$

Expression (1) is the basis of the bus network synthesis method developed by the authors with single connections presented, among others in [13]. Unlike networks with direct connections, for the bus network $[n, m]$ with the incidence matrix A, there is always a network with the transposed incidence matrix A^T, where $[n, m]^T = [n, m]$.

Expression (1) also shows that bus networks and hypergraphs are structurally equivalent. To analyze the buses represented as hypergraphs, graph theory tools can be used to present the hypergraphs as bipartite graphs.

Definition 1. *Undirected graph $G = (V, E)$ is a bipartite graph, if the set of its vertices can be divided into two parts $X_0 \cup X_1 = V$ in such a way that:no vertex from the part X_0 is connected with the vertices of part X_0; no vertex from the part X_1 is connected to the vertices of part X_1. In this case, the subsets X_0, X_1 are called parts of bipartite graph G.*

If additional connections appear in the analyzed system, for example additional cables integrating buses with processing nodes, PBL neighborhood graphs can be used to describe the bus processing system. Let's assume that the number of nodes in parts X_0, X_1 of the bipartite graph is equal to n and m respectively. Its edges are local connections $l_{p,q}$, where: p, q - number of processing nodes and buses, respectively, whose task is to connect the computational node with a bus. This description is in fact a description of the PBL neighborhood graph.

Definition 2. *The PBL graph $G = (V, B, L)$ containing $|V_G| = n$ nodes, $|B_G| = m$ buses and L_G link set is the bipartite graph G_{PBL} which can be described by following pair: $(V_{G_{PBL}}, B_{G_{PBL}})$ and $V_{G_{PBL}} = VV_{G_{PBL}} \cup VB_{G_{PBL}}$, where $VV_{G_{PBL}} = V_G$ and $VB_{G_{PBL}} = B_G$. Connections in graph G between buses and nodes are represented by $B_{G_{PBL}}$. The nodes $v_{G_{PBL},i}$ and $b_{G_{PBL},j}$ are connected by the edge $l_{G_{PBL},k}$ if and only if in the source bus system, the bus b_i is connected to the node v_j with link l_k.*

The representation of the bus system according to Definition 2 is shown in Fig. 3a. In the IIoT and in traditional information systems, buses will connect two types of nodes: service provides K (measuring sensors, thin clients, etc.) and service recipients S (computational servers, data). Then, tripartite graphs should be used to represent the bus system. A graphical representation of the connection network with a single channel bus is shown in Fig. 3b.

In order to analyze the characteristics of the bus connection network, algebraic topology notation based on algebra of connected finite non-directed graphs has been proposed. Graph algebra is defined as follows:

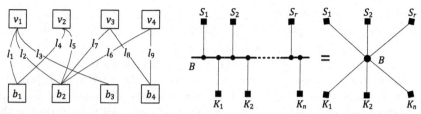

(a) Bus system presented as a neighborhood graph

(b) A connection network with single-channel bus presented as tripartite graph

Fig. 3. Possible bus system representations as different graphs

Definition 3. *The pair $A = (D, \Omega)$ will be called the universal graph algebra over the universe U, if D is a set of graph with vertices from the U set, and the Ω signature enables the zero operation Λ, binary operations for adding a vertex, adding edges, remove the vertex and remove edges.*

First, let's define the theorem, which will be then used to write the topology of selected networks.

Theorem 1. *The minimal elements of the algebra $AG = (A, \Omega$ will be trees of the form G_u^0, where: $G_u^0 = (V = \{u\}, E = \emptyset)$ - an empty tree composed of the vertex u.*

Proof. In the graph algebra, trees play a special role, because of them any connected graph is built. However, if the end vertex removal operation is applied to the tree, it is reduced to the tree of T_u^0 form. That is why empty trees can act as minimal elements in algebra. Empty trees, consisting of only one vertex, cannot be further reduced.

Using the Theorem 1, following trees will be analyzed:

1. $T_{u,v} = (V = \{u, v\}, E = \{(u, v)\})$;
2. $T_{u,v}^w = (V = \{u, v, w\}, E = \{(u, w), (w, v)\})$.

Using algebra, the above trees can be represented by the minimum elements:

1. $T_{u,v} = (V = \{u, v\}, E = \{(u, v)\}) = w_{ik}(G_u^0 \cup G_v^0, u, v) = T_{u,v} = G_u^0 * G_v^0 = G_u^0 *_f G_v^0(f(u) = v)$;
2. $T_{u,v}^w = (V = \{u, w, v\}, E = \{(u, w), (w, v)\}) = T_{u,w} \cup T_{w,v} = w_{ik}(T_u^0 \cup T_w^0, u, w) \cup w_{ik}(T_w^0 \cup T_v^0, w, v) = (T_u^0 * T_w^0) \cup (T_v^0 * T_w^0)$.

where: w_{ik} - the operation of adding an edge into connected graph; $*$ - the operation of combining two connected graphs.

In addition, the result of combining two graphs will be a connected graph, if even one or two of them does not meet connectivity condition. In particular, for a connected graph G, graphs corresponding to the expressions $(T_u^0 \cup T_v^0) * G$ and

$(T_u^0 \cup T_v^0) * (T_u^0 \cup T_v^0) * (T_w^0 \cup T_s^0)$, where: u, v, w, s in pairs different vertices, will be connected graphs. The algebraic expression describing the network in Fig. 3b in graph algebra notation has the following form: $(T_{S_1}^0 \cup \ldots \cup T_{S_r}^0 \cup T_{K_1}^0 \cup \ldots T_{K_n}^0) * T_B^0$.

Technical solutions developed on the basis of research are based solely on the multi-channel bus. Suppose the complete multi-bus (i.e. each service provider and recipient is connected to each of the buses) consists of m channels, S_r recipients and K_n service providers. Then its physical form and its notation in the form of a tripartite graph have the form presented in Fig. 4. It can be assumed from a technical point of view, that the $B_1, \ldots, B_m$ buses are logical channels functioning in the one common physical channel.

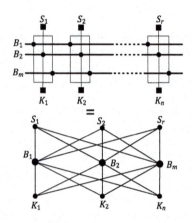

Fig. 4. A network with multi-channel bus presented as a tripartite graph

The algebraic expression describing the above network has the following form:
$$(T_{S_1}^0 \cup \ldots \cup T_{S_r}^0) * (T_{B_1}^0 \cup \ldots \cup T_{B_m}^0) \cup (T_{B_1}^0 \cup \ldots \cup T_{B_m}^0) * (T_{K_1}^0 \cup \ldots \cup T_{K_n}^0) =$$
$$(T_{S_1}^0 \cup \ldots \cup T_{S_r}^0 \cup T_{K_1}^0 \cup \ldots \cup T_{K_n}^0) * (T_{M_1}^0 \cup \ldots T_{M_m}^0)$$

Graph algebra also allows to describe solutions in which the preference relationship appears. The formalization of the above process has been extensively described in other works of the authors [14,15]. Informally, however, connections between service providers K and service recipients S can be described by the following rules:

1. The recipient prefers the selected service provider. Preferences are not permanent and can be changed without restrictions during work;
2. The connection of service recipients with service providers is performed by means of logical bus channels.

4   Modification of the Multi-bus Networks

The modifications to the bus networks described are intended to improve their functional parameters, in particular to compensate for bus communication loads and computational processing nodes. They all require a specific physical network organization.

In traditional bus networks, the communication network is single-channel and each of the system components is connected to it once. The bus usually carries out broadcasts on one common channel. In the case of traditional multi-channel buses, permanent user assignment to a specific logical channel is most common. In the offered solution, the logical communication channel is selected by the transceiver, which can be tuned. Thanks to this, the communication system can highlight the channel through which the node will communicate with the outside world. In addition, it is possible to equip computational nodes with a variable number of transceivers. It allows one or many times to join any logical bus or their set.

Because physical bus lengths are small, device prices are relatively low. Built groups (node clusters) have their own communication environment and can be isolated from external interference. For example, a separate group can be created by users using services insensitive to communication delays, another generating low, traffic, yet another having bursty character of the generated traffic. Logical buses are also grouped (clustered). In this way, not only computational power, but also the bandwidth of logical communication channels is scaled. Mixed grouping, in which computational nodes and logical communication buses are used, is the most effective. The multiplicity of node interfaces allows the use of folded buses (single, double or triple) ensuring better usage of the bus. Another method consists in the hierarchization of logical channels where channels are combined into groups that support different sets of servers. It is also possible to divide overloaded buses into smaller parts.

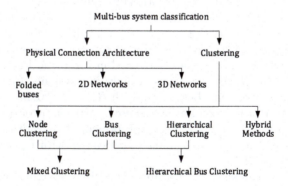

Fig. 5. Classification of the multi-bus system organization

Thanks to the variety of methods proposed, the connection architecture can be adapted to the traffic patterns present in the network (characteristic values of the communication load of the channels) and communication requirements of the clients. Because in optical systems, the change in wavelength used by the transceiver takes milliseconds, adapting the connection architecture to the current requirements of users can be dynamic and can be implemented in real time. The classification of methods for improving the efficiency of bus connection system architecture is shown in Fig. 5.

5   Simulations and Empirical Studies

In order to validate the concept of building new communication environment, simulation and empirical studies were performed. Various organizations of the communication system and the resulting computational system were examined, the generalized architecture was presented in [16]. Simulation studies concerned the optical multi-bus communication environment, and empirical were calculated for the electrical one.

A model simulation basic operational parameters has been prepared for each of the architectures. The models are based on the probability and queue theory. The organization of analyzed solutions is defined below. In complete bus systems, each service provider and recipient is connected to each of the logical buses once. In systems with single connections, they are made only to one of the buses. Usually this applies only to service providers. In a multi-channel hierarchical bus system, logical buses are divided into parts and then combined into groups with identical division into parts. Division parameters are defined by: u - number of hierarchy levels; k_u^i - the number of bus components of the i-th hierarchy level; κ_j - a way to connect service providers to j-th level buses. In hierarchical buses with a limited partition coefficient at the recipient's output, the number of their connections is limited in advance. In a homogeneous request model, ω means the like hood of making requests. Additionally, the following symbols have been introduced for the hierarchical request model: ω_p - probability of creating a request to the preferred service provider; ω_k - probability of requesting the same group of service providers; ω_O - probability of request to other groups; p - priority of the request. Obtained results are presented in Fig. 6 and Fig. 7.

In the Fig. 6 we can see a features of multi-bus systems. Figure 6a presents hardware complexity of multi-bus system with $B_m = 16$. We have a five different bus. The bus (e) is a Hierarchical bus limited partition at output $u = 4, k_u^2 = k_u^3 = k_u^4 = 2, \kappa_2 = \kappa_3 = \kappa_4 = 2$. Figure 6b presents bandwidth dependence on number of logical elements for a complete bus system with $K_s = K_k = 16, \omega_p = 0.6, \omega_k = 0.3, \omega_O = 0.1$.

In the Fig. 7 we can also see a features of multi-bus systems, but this indicate a bandwidth of transmission. Figure 7a presents dependence of bandwidth on the number of processing elements for the performance of computational system with a hierarchical and homogeneous request model, $B_m = 16, u = 3, k_u^2 = k_u 3 = 2, \kappa_1 = \kappa_2 = 2, \kappa_3 = 4, \omega_1 = 0.5, \omega_2 = \omega_3 = 0.25$. Figure 7b presents dependence

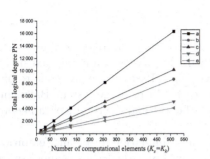

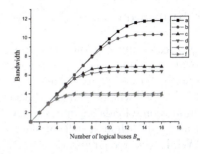

(a) a. Complete bus; b. Single bus; c.
Grouped bus g = 4; d. Hierarchical bus;
e. Hierarchical bus with limitation

(b) Request models: a. Hierarchical, $\omega = 1$;
b. Homogeneous, ω = 1; c. Hierarchical,
ω = 0.5; d. Homogeneous, ω = 0.5; e.
Hierarchical, ω = 0.25; f. Homogeneous,
$\omega = 0.25$

Fig. 6. Features of multi-bus system

of the probability of handling the request on the number of logical buses. At the
beginning, when the number of computational elements are low, the bandwidth
for (a)(c) and (b)(d) buses (pairwise) is identical. The difference can be seen
after the buses become saturated - they are loaded with traffic. The hierarchical
version offers then much better performance than homogeneous one.

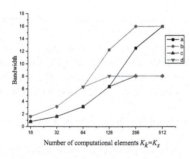

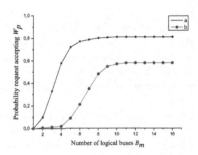

(a) Models: a. Hierarchical, ω = 0.5; b.
Hierarchical, ω = 0.1; c. Homogeneous,
$\omega = 0.5$; d. Homogeneous, $\omega = 0.1$

(b) a. $p = 15, \omega = 0.2$; b. $p = 15, \omega = 0.5$
where p - priority of requests

Fig. 7. Features of multi-bus systems

6 Summary

The research showed the desirability of construction and operating multi-bus optical communication environments both in the case of computational power scalability as well as its communication efficiency. Particularly favorable results were obtained when balancing the communication loads of the system. Balancing computational loads depends on many other factors and the results obtained are not as good, although still satisfactory.

While the reconfiguration process during simulation tests and empirical observations went smoothly, communication within a single bus channel seems unsatisfactory. This applies to the number of interfaces connected to the bus channel - the channel functioned efficiently with a smaller number of interfaces connected than it resulted from earlier calculations based on the conditional probability. The findings made indicate the need to modify the access protocol.

Further improvement of the system properties (in particular, communication efficiency) should be seen in the use of physical network with grid or toroid bus topology, as well as similar 3D topologies. This will require the development of original routing algorithms.

Acknowledgements. The work was realized as a part of fundamental research financed by the polish Ministry of Science and Higher Education.

References

1. Bagherzadeh, N., Dowd, M., Nassif, N.: Embedding an arbitrary binary tree into the star graph. IEEE Trans. Comput. **45**(4), 475–481 (1996)
2. Barak, A., Schenfeld, E.: Embedding classical communication topologies in the scalable opam architecture. IEEE Trans. Parallel Distrib. Syst. **7**(9), 979–992 (1996)
3. Bulent, Y.: Virtual embeddings on regular topology networks. In: Eighth IEEE Symposium on Parallel and Distributed Processing, pp. 562–565. New Orleans (1996)
4. Chang, B.J., Hwang, R.H.: Performance analysis for hierarchical multirate loss networks. IEEE J. Mag. **12**(1), 187–199 (2004)
5. Chang, S.Y., Wu, H.C.: Joint optimization of complexity and overhead for the routing in hierarchical networks. IEEE J. Mag. **22**(6), 1034–1041 (2011)
6. Chartrand, G., Zhang, P.: A First Course in Graph Theory. Dover Publications, Boston (2012)
7. Chen, B., Dutta, R.: On hierarchical traffic grooming in wdm networks. IEEE/ACM Trans. Netw. **16**(5), 1226–1238 (2008)
8. Diestel, R.: Graph Theory. 4, Springer, New York (2010)
9. Dutta, R., Kamal, A.E., Rouskas, G.N.: Traffic Grooming for Optical Networks: Foundations. Techniques and Frontiers. Springer, New York (2010)
10. Dutta, R., Rouskas, G.N.: On optimal traffic grooming in WDM rings. IEEE J. Sel. Areas Commun. **20**(1), 110–121 (2002)
11. Fang, H., Xuechun, W.: Research on embedded network scheme based on quantum key distribution. In: Cross Strait Quad-Regional Radio Science and Wireless Technology Conference, pp. 1689–1691. Taipei (2011)

12. Gubko, M.V.: Mathematical models of optimization of hierarchical structures. LENAND (2006)
13. Hajder, M., Bolanowski, M.: Connectivity analysis in the computational systems with distributed communications in the multichanel environment. Polish J. Environ. Stud. **17**(2A), 14–18 (2008)
14. Hajder, M., Kolbusz, J., Bartczak, T.: Effective method for optimizing hierarchical design process for interactive systems. In: 6th International Conference on Human System Interactions. Sopot, Poland (2013)
15. Hajder, M., Kolbusz, J., Bartczak, T.: Undirected graph algebra application for formalization description of information flows. In: 6th International Conference on Human System Interactions. Sopot, Poland (2013)
16. Hajder, P., Rauch, L.: Reconfiguration of the multi-channel communication system with hierarchical structure and distributed passive switching. In: Rodrigues, J.M.F., Cardoso, P.J.S., Monteiro, J., Lam, R., Krzhizhanovskaya, V.V., Lees, M.H., Dongarra, J.J., Sloot, P.M.A. (eds.) ICCS 2019. LNCS, vol. 11537, pp. 502–516. Springer, Cham (2019). https://doi.org/10.1007/978-3-030-22741-8_36
17. Harris, J., Hirst, J.L., Mossinghoff, M.: Combinatorics and Graph Theory. Springer, New York (2008). https://doi.org/10.1007/978-0-387-79711-3
18. Ilic, M.D., Liu, S.: Hierarchical Power Systems Control: Its Value in a Changing Industry. Springer, New York (1996)
19. Kamal, A.E., Ul-Mustafa, R.: Multicast traffic grooming in wdm networks. In: Proceedings of Opticomm 2003 (2003)
20. Kulish, V.: Hierarchical Methods: Undulative Electrodynamical Systems. Springer, New York (2002). https://doi.org/10.1007/0-306-48062-X
21. Lin, T.J., Hsieh, S.Y.: Embedding cycles and paths in product networks and their applications to multiprocessor systems. IEEE Trans. Parallel Distrib. Syst. **23**(6), 1081–1089 (2012)
22. Livingston, M., Stout, Q.F.: Embeddings in hypercubes. Math. Comput. Model. **11**, 222–227 (1988)
23. Morrison, C.R.: Build Supercomputers with Raspberry Pi 3. Packt, Birmingham (2017)
24. Pajankar, A.: Raspberry Pi Supercomputing and Scientific Programming. A press, Nashik (2017)
25. Rowley, R.A., Bose, B.: Fault-tolerant ring embedding in de bruijn networks. IEEE Trans. Comput. **42**(12), 1480–1486 (1993)
26. Saaty, T.L.: Models, Methods, Concepts and Applications of the Analytic Hierarchy Process. Kluwer Academic Publishers, Boston (2000)
27. Saaty, T.L.: The Analytic Hierarchy and Analytic Network Processes for the Measurement of Intangible Criteria and for Decision-Making, pp. 345–407. Multiple Criteria Decision Analysis, Springer (2005)
28. Smith, N.J., Sage, A.P.: An Introduction to Hierarchical Systems Theory. Information and Control Sciences Center, SMU Institute of Technology, Dallas, Texas (2005)
29. van Steen, M.: Graph Theory and Complex Networks: An Introduction. Maarten van Steen, Belgium (2010)
30. Stetsyura, G.G.: Extending functions of switched direct optical connections in digital system. Upravlenie Bol'shimi Sistemami **56**, 211–223 (2015)

Real-Time Object Detection for Smart Connected Worker in 3D Printing

Shijie Bian[1,3], Tiancheng Lin[1], Chen Li[1,3], Yongwei Fu[1,2],
Mengrui Jiang[1], Tongzi Wu[1], Xiyi Hang[4], and Bingbing Li[1,2(✉)]

[1] Autonomy Research Center for STEAHM, California State University,
Northridge, CA 91324, USA
bingbing.li@csun.edu
[2] Department of Manufacturing Systems Engineering and Management,
California State University, Northridge, CA 91330, USA
[3] Department of Mathematics, University of California, Los Angeles, CA 90095, USA
[4] Department of Electrical and Computer Engineering, California State University,
Northridge, CA 91330, USA

Abstract. IoT and smart systems have been introduced into the advanced manufacturing, especially 3D printing with the trend of the fourth industrial revolution. The rapid development of computer vision and IoT devices in recent years has led the fruitful direction to the development of real-time machine state monitoring. In this study, computer vision technology was adopted into the Smart Connected Worker (SCW) system with the use case of 3D printing. Specifically, artificial intelligence (AI) models were investigated instead of discrete labor-intensive methods to monitor the machine state and predict the errors and risks for the advanced manufacturing. The model achieves accurate supervision in real-time for twenty-four hours a day, which can reduce human resource costs significantly. At the same time, the experiments demonstrate the feasibility of adopting AI technology to more aspects of the advanced manufacturing.

Keywords: Object detection · Machine state monitoring · Smart connected worker · 3D printing

1 Introduction

1.1 Background and Motivation

New computer chips are constantly refreshing the computing efficiency per unit area within the framework of Moore's Law [1]. At the same time, GPUs designed for parallel computing have accelerated the development of artificial intelligence (AI). Among these achievements in AI algorithms, the development of computer vision models has been particularly remarkable. With the advent of computer vision models, numerous real-time object detection and classification algorithms

© Springer Nature Switzerland AG 2021
M. Paszynski et al. (Eds.): ICCS 2021, LNCS 12745, pp. 554–567, 2021.
https://doi.org/10.1007/978-3-030-77970-2_42

have been proposed. In particular, real-time image processing, due to its ability to extract high-level information from digital inputs with great efficiency, is widely used in realms ranging from facial recognition to self-driving cars. Since the processing of visual data is mainly performed by human labor in the current manufacturing realm, computer vision-based real-time monitoring was developed to replace the traditional discrete labor-intensive monitoring. The filtering algorithm with a trained vision model was combined into an automated system capable of monitoring a 3D printer through an internal camera in real-time to recognize objects. This paper describes in detail the comprehensive process including object analysis, data collection, algorithm implementation, model training, and result evaluation. Utilizing this system, the defined 9 working states of the 3D printer were successfully identified with high accuracy and low cost.

1.2   Related Works

Computer vision enables the extraction of features from image and video content for the purpose of identification and inference. Recent advances in this field sparked the development of various algorithms and pushed the manufacturing processes to be more efficient and intelligent [2]. An intelligent system using the computer vision method was tested to raise production efficiency under a cloud-based additive manufacturing setting [3]. Using the computer vision method combined with machine learning techniques, an autonomous system was established to perform powder classification from images containing different powder features [4].

Object detection is one of the techniques within the computer vision field, and much effort has been centered around developing algorithms and methods for its efficient applications in the manufacturing of various kinds. Recent endeavors have exhibited the potential of applying deep learning algorithms in object detection. For example, Convolutional Neural Network (CNN) was demonstrated to outperform traditional methods in extracting features from raw data with promising accuracy [5]. Deep Convolutional Neural Networks (DCNN) is a state-of-art approach for object detection, but it is suggested that further models need to be established to allow real-time scenario applications [4]. To improve the performance of object detection models, a technique of generating synthetic training data to train CNN was proposed, which significantly reduced the training time [6]. An algorithm was introduced to generate high accuracy object detection based on joint impedance control, which resulted in high flexibility of manufacturing devices [7]. In addition, remarkable progress was also made in making the object detection model robust to domain shifting and obtaining high training accuracy and efficiency [8]. A Hybrid Smart Region-Based Detection (SRBD) combined several existing object detection algorithms such as YOLO, R-CNN, to accommodate their shortcomings and obtained a high detection accuracy [9].

In recent years scholars and professionals have also made efforts to deploy object detection methods in smart systems through IoT devices. For instance, an object detection algorithm was employed in IoT-based embedded devices [10], while maintaining the minimal impact that comes from the variation of environmental conditions. Automated object detection in urban surveillance systems extracted vehicle license plates from images accurately while reducing data storage in the systems [11]. A deep learning-based object detection system [12] successfully sent keywords through Raspberry Pi to Alexa smart speakers using JSON script, thus realizing the implementation of a camera-based smart speaker system. Object detection was partnered with a robot activity support system to enable everyday activity monitoring and assessment in a smart home environment [13].

1.3   Research Contribution

This paper proposes an automated system for the real-time monitoring of 3D printers. The major contributions in this work are:

Object detection and image processing techniques were adopted from the realm of computer vision and integrated into the advanced manufacturing systems with a use case in 3D printing. Using a carefully selected dataset from diverse experimental configurations, a YOLO-based model was trained and is capable of identifying the exact positions of major components in real-time with promising efficiency and accuracy.

Based on the results acquired from the pre-trained machine learning model, a real-time filtering algorithm for both the identification and classification of the 3D printer's machine states was developed. By constantly recording the positions of 3 major components of the 3D printer, the algorithm serves as a supervisor that checks for faulty behaviors in the 3D printing processes. By recognizing the start of each printing section, and by remembering the past actions of the printer's components, the algorithm can not only identify the current machine status but also predict the future ones.

This work encapsulated the pre-trained model and the filtering algorithm into one automated system for the behavior-supervision and status-monitor of 3D printers. Since all components are working at low cost and in real-time, the proposed work would fit smoothly into the advanced manufacturing systems that are related to 3D printing and may serve as either a replacement of human labor or as an auxiliary unit.

2   Background

2.1   3D Printer Monitoring

For a traditional 3D printer, there are three essential interior components that work together during printing: the extruder that ejects the material for printing, the build plate that serves as the supporting platform, and the motor axis that controls the movement of the extruder.

Meanwhile, during the process of printing a part, a 3D printer must go through the following main states in sequential order: the **initialized state**, where the printer starts up and processes the printing command(s); the **testing state**, where all components ready their positions; the **calibration state**, where the extruder calibrates its location and finds the starting position; the **heating state**, where the nozzle and chamber heat up; the **printing state**, where the extruder ejects the model and support materials onto the build plate; the **ending state**, where all components are reset to their original positions. The goal is to predict the 3D printer's machine state in real-time through the position of three essential interior components by analyzing the interior camera image via an object detection algorithm.

2.2 Object Detection Algorithms

Region Based Convolutional Neural Networks (R-CNN). Traditional methods which utilize the CNN [14] are suited for classifying images by extracting features. However, real-world inputs that compress numerous objects with distinct characteristics will render these methods computationally expensive. Therefore, the R-CNN architecture [15] combines a regional selective search with the CNN model to solve this problem. In the R-CNN's architecture, the input image is first segmented into numerous small regions. Based on certain features, such as the similarity in color and texture, these small regions are combined together into larger pieces via a greedy algorithm. After warping these pieces into a single region, Graph Neural Networks (GNN) is applied to extract feature vectors that are later classified by a Support Vector Machine (SVM) [16]. Even though the R-CNN method serves as a promising baseline for object detection, the slow testing process greatly limits it in real-time image analysis. Therefore, other methods, such as the Single-shot Detectors (SSD) have been proposed.

Single-Shot Detectors (SSD). Instead of generating proposed regions with a dedicated algorithm, SSD [17] utilize predetermined bounding boxes for training, thus eliminating the time spent on region proposals, and achieving much higher inference speed as compared to the R-CNN model. In terms of structure, SSD generally consists of two main components, as shown in Fig. 1: a base neural network for performing general feature extractions, and auxiliary network layers with decreasingly sized filters for the final classification.

During the training process of the SSD, digital images with predetermined bounding boxes around each target object are fed into the first component of the model. By passing through the convolutional and pooling layers of a pre-trained classification neural network, such as the Visual Geometry Group from Oxford (VGG16) [18], features of the input image can be extracted into mappings. For each portion of a feature mapping, default bounding boxes (anchor boxes) with diverse shapes and scales are assigned. Using matching strategies such as the maximization of Intersection over Union (IoU), the anchor boxes most similar to the ground truth are filtered out and treated as positive samples, while the rest are identified as negatives. Finally, by passing the processed data through

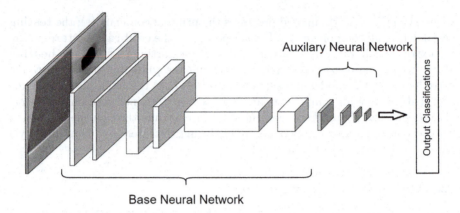

Auxilary Neural Network

Output Classifications

Base Neural Network

Fig. 1. Illustration of the SSD structure

an auxiliary set of convolution layers with decreasing sizes, the SSD model is able to produce final classification results for objects of multiple scales.

As a single-shot method, the SSD is able to achieve object detection with high accuracy and competent inference speed. Therefore, the SSD model, namely the You Only Look Once (YOLO) algorithm, was adopted for the construction of a real-time monitoring system.

3   Methodologies

3.1   An YOLO-based Object Detection Model

YOLO [19] is a state-of-the-art real-time object recognition algorithm based on the SSD method. By applying a single neural network to the full input image, YOLO is able to predict both the class and position of bounding boxes within one evaluation, hence achieving great accuracy and efficiency. Therefore, the proposed object detection model is based on the YOLO algorithm for acquiring the locations of three critical components in the 3D printer in real-time.

Data Preprocessing. Alongside the digital image that serves as the input, predetermined class labels, as well as bounding boxes, are also fed into the model as the ground truth. Specifically for each ground-truth bounding box, four descriptors were recorded: the center x-coordinate, the center y-coordinate, the width (b_w), and the height (b_h). For each input image encoded in the RGB color model, the scale was standardized by resizing it into 416×416 (unit: pixel) before plugging it into the model. During data preprocessing as shown in Fig. 2, a one-hot encoding of the class labels was concatenated after the four descriptors of the ground-truth bounding boxes, and arrive at a single vector consisting of all the information for each of the input images. In the next step, the input image was split into 19×19 grid cells for identifying the exact location of each detection item. An object of interest is only considered as belonging to a certain cell if its center is within the boundaries of that cell.

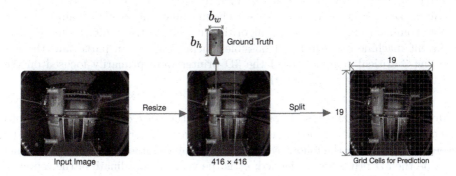

Fig. 2. Illustration of the preprocessing stage

Training. For the training process, 3 predefined anchor bounding boxes are assigned to each of the grid cells. Subsequently, the anchor boxes were passed with the predetermined input information into deep CNN architecture of YOLO version 3 [20]. The 13×13 scaled feature map was achieved by down-sampling the input image with a stride of 32. For each of the 3 anchor boxes, the probability of containing a certain class of item (i.e., the bounding box probability) using the Intersection over Union (IoU) was extracted. Similar to the previously mentioned steps of an SSD, the anchor boxes most similar to the ground truth are filtered out, while the others are omitted. The same procedure was applied to the 16×16 and 8×8 down-sampled sizes to capture information of various scales.

Data Output. After being fed into the model for inference, the original input image, resized into 416×416, is delivered as the output, followed by the predicted bounding boxes acquired from the network. For each object of interest, the following parameters were acquired: the center x-coordinate, the center y-coordinate, the width, the height, and a confidence score measured as a probability that indicates how certain the model believes that the bounding box contains the correctly predicted object. Specifically, the confidence score is calculated by taking the product of the IoU (Intersection over Union between the ground truth and the bounding box) and the probability that an object is in the bounding box (Pr(Object)). This detected information serves as the input to the filtering algorithm for the proceeding process of machine state identification.

3.2   A Filtering Algorithm for Machine State Identification

With the output gained from the YOLO-based object detection model, a filtering algorithm was developed for identifying the machine states of the 3D printer. By comparing the positions of the machine components against all possible combinations, the filtering algorithm is able to filter out the most likely current machine state from all possible machine states. In particular, the following 3 critical components of the 3D printer were primarily focused on the extruder, the build plate, and the motor axis.

Defining the Variables. In order to predict the state of the 3D printer, both the precise and relative positions of each and every major component need to be located. Therefore, the following numerical variables for locating the components inside the 3D printer were defined: e_x is the vertical-coordinate of the center of the extruder (identifying its horizontal position), e_y is the horizontal-coordinate of the bottom of the extruder (identifying its height and relative position to the camera), b_x is the horizontal coordinate of the top of the build plate (identifying its height). All of these three variables can be easily calculated from the bounding box outputs of the object detection algorithm.

Predicting the States. With the object detection's output, the predefined variables were calculated to predict the 3D printer's machine states. As introduced in the background, 6 main states of the 3D printer (listed in logical order) were mainly predicted: the initialized state, the testing state, the calibration state, the heating state, the printing state, and the ending state. In particular, the printing state, being the most complex one, has four sub-states: the printing and wiping of the support material, as well as the printing and wiping of the model material. For ease of representation, the printing state was considered as one unified state in the subsequent sections.

Since the printing process of almost any 3D model has to go through the same order of steps as introduced in the background, the detailed logic on how to predict the machine states based on the positions of the printer components obtained from the YOLO-based object detection model was provided as follows:

Initialized State: The extruder is at the left-back corner, while the build plate is at the bottom;

Testing State: The extruder remains at the same position, the build plate elevates to the top;

Calibration State: The extruder moves forward, while the build plate remains at the same position;

Heating State: The extruder returns to the left back corner, while the build plate remains at the same position;

Printing State: The extruder moves at the front positions, while the build plate gradually descends;

Ending State: The extruder returns to the left back corner, while the build plate descends to the bottom.

Even though the position of the motor axis is not directly included in the prediction of the machine state, it can be used as an auxiliary piece of information for validation.

For a real-time image acquired from the camera and processed by the object detection algorithm, whether the positions of the printer components match the prior-listed combinations aided us in identifying the machine state of the 3D printer at that exact time. This algorithm was considered as a filter that extracts the machine state from only looking once at the position results obtained from the object detection algorithm.

After predicting the current machine state, the information was stored in a record to retrieve the past states that a 3D printer went through for further processing and analysis, such as energy disaggregation. In addition, by storing the past machine states and following the logical order of the printing steps, the proposed model is able to check whether an unexpected error occurs by predicting the future machine states. Since both our object detection model, as well as our filtering algorithm, runs in real-time, a single workflow can be combined for predicting the machine state of a 3D printer in real-time.

4   Experimental Evaluation

In this section, the YOLO-based object detection model was trained and validated on a carefully selected dataset to assess its feasibility to be utilized in real-time machine state predictions of the realm of manufacturing systems. By testing on real-time data acquired during a real-world working scenario, the accuracy and efficiency of the proposed model were further evaluated.

4.1   Dataset

Camera Setting. For collecting image frames from the 3D printer, an SVPRO Fisheye Lens 180° USB heat-resistant camera with resolution 1080P and frame rate 30 frames per second (fps) was installed inside the Stratasys uPrint SE 3D Printer. To capture critical objects as needed, the camera was fixed on the interior door of the 3D printer. To increase the image quality, the LED lights inside the 3D printer were kept on with the lights on in the laboratory during image collection. Finally, a python program was developed using the OpenCV package [21] to get frames from the camera at a rate of 5 fps, and then saved as the initial training dataset. This python program also implements "start capturing" and "stop capturing" as a graphical user interface to facilitate data collection.

Image Collection. After setting up the dataset collection process, the 3D printer was set up to print a sample model within approximately 8 min to finish. When the 3D printer showed "finished", image capturing was stopped right away. As a result, there were a total of around 2400 image frames collected. However, as shown in Fig. 5, there were around 100 images among this training set that was very obscure because of the extremely fast movement of the extruder during the printing process, which made them invaluable for training. To filter a valid training dataset, these images were dismissed.

4.2 Training and Experimental Configurations

Image Labeling. For each output image from the training set, the YOLO Visual Object Tagging Tool (VoTT) v1 software was utilized to give the bounding boxes as well as labels that serve as the ground truths. As shown in Fig. 3, for each of the 2300 training images, rectangle bounding boxes were drawn for each object on observation of our eyes. The output folder from VoTT contains three parts: the first part is a folder including original images with corresponding text files storing each class's bounding box coordinates; the second part has two text files, which separates the original training dataset into training data and validation data (specifically after performing random shuffle on the entire dataset, the first 80% of the dataset was split and selected for training, while the remaining was selected for validation); the third part is a text file containing the reference of the first two parts.

Fig. 3. Training images with the ground-truth bounding boxes and labels.

Model Training Configurations. For the training process of the YOLO-based object detection algorithm, the configurations as listed in Fig. 4 were adopted. Furthermore, 4 GPUs from the Pacific Research Platform (PRP) Kubernetes Nautilus cluster cloud platform [22] were deployed to increase the computing speed during training.

Number of Classes: 3	Image Height: 416 pixels	Image Width: 416 pixels
Momentum: 0.9	Batch Size: 64	Decay Rate: 0.0005
Learning Rate: 0.0001	Steps: 100, 25000, 35000	Scales: 10, 0.1, 0.1

Fig. 4. Model configurations

4.3   Results and Discussion

Object Detection. After a training time of 3 h using the Nautilus GPU cluster cloud platform, the object detection model converged after 10000 iterations. The average class detection accuracy is 0.999 and the average IoU is 0.922. The average inference speed is approximately 0.010 s per frame with GPU, which is a sufficient efficiency for real-time object detection.

For testing and validation, a high-resolution sample video from the internal camera of the 3D printer was recorded, which consists of all the stages of the printing process. A total of 2122 image frames were exacted from the video and fed into the trained object detection model. Among the 2122 testing images, the model accurately predicted the bounding box positions of 2011 frames with an average confidence score of 0.87, thus achieving an average prediction accuracy of approximately 94.8%. This accuracy was considered to be sufficient for the subsequent machine state prediction.

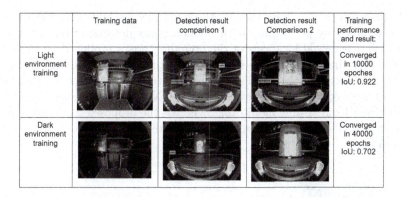

Fig. 5. Training environment comparison

Machine State Prediction. A web-based Graphical User Interface (GUI) was developed for displaying the results obtained from the machine state prediction algorithm, as shown in Fig. 6. Specifically, dark blue marks the past machine states that have been recorded, green marks the current machine state, while light blue marks the machine state that should follow the current one. In a well-lit testing environment that is similar to the training configurations of the object detection model, our algorithm achieves a perfect prediction of the machine states, provided that the object detection results are accurate.

Discussion. The YOLO-based object detection model achieved an average prediction accuracy of approximately 94.8% as well as a 100 fps frame rate during real-time evaluation. Other object detection models that may serve as candidates for the proposed approach include the traditional R-CNN methods. As demonstrated by Redmon et al. [23] with the Common Objects in Context (COCO) dataset [24], even though R-CNN models may achieve satisfactory or even better detection accuracy, their inference speed seldom exceed 20 fps and is significantly slower than YOLO, which can achieve an inference speed up to 220 fps.

In order to further evaluate the performance of the proposed model, Mask R-CNN [25], a state-of-the-art R-CNN object detection model, was adopted as the baseline. Specifically, the same 2300 training images were labeled and split into the training and validation dataset and were trained in the Mask R-CNN with default configurations. After 50 epochs, the model converged and was tested with the same 2122 image frames that were collected during the experiment. Among the total 2122 test images, 2032 frames predicted all bounding box positions correctly, achieving an average accuracy of approximately 95.8%, which is only slightly higher than the accuracy of the proposed model (94.8%). However, the average inference speed for Mask R-CNN on the test dataset was approximately 0.17 s per frame (equivalent to 6 fps), which is significantly slower than the frame rate of the proposed model (100 fps).

As shown in Fig. 7, even though Mask R-CNN was able to achieve slightly higher accuracy by performing both bounding box detection and instance segmentation, the actual locations of the bounding boxes produced by both models were very similar. Since the filtering algorithm for machine state prediction only requires accurate bounding box coordinates, the two models' accuracy is approximately equivalent. However, Mask R-CNN's slow inference speed (6 fps) is insufficient for real-time monitoring, whereas the proposed model's competent accuracy and efficiency are more suitable for integration into smart manufacturing systems.

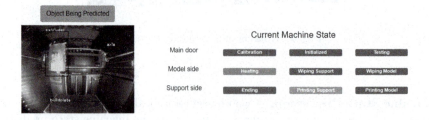

Fig. 6. Machine state predictions as shown on GUI

Fig. 7. Test results of Mask R-CNN (left) and the proposed model (right)

5 Conclusion

In this paper, the object detection model, as well as the filtering algorithm, were developed and serve as a real-time workflow for monitoring the 3D printer's machine states. By feeding real-time image frames directly into a pre-trained YOLO-based object detection model, the positions of the critical components of the 3D printer were acquired to predict and record the machine states via the filtering algorithm. The proposed model and algorithm achieved exceptional accuracy and efficiency while testing in an environment that is similar to the training set. Future efforts in this direction would include detecting the machine states of more sophisticated machinery, as well as under more diverse environments. Our work suggests the possibility of adopting state-of-the-art computer vision algorithms for more advanced manufacturing systems, such as the smart connected worker.

Acknowledgement. This research was mainly supported by the Technical Roadmap Project "Establishing Smart Connected Workers Infrastructure for Enabling Advanced Manufacturing: A Pathway to Implement Smart Manufacturing for Small to Medium Sized Enterprises (SMEs)" funded by the Clean Energy Smart Manufacturing Innovation Institute (CESMII) sponsored through the U.S. Department of Energy's Office of Energy Efficiency and Renewable Energy (EERE) under the Advanced Manufacturing Office (Award Number DOE: DE-EE0007613). This work was also supported by the project "Autonomy Research Center for STEAHM" sponsored through the U.S. NASA Minority University Research and Education Project (MUREP) Institutional Research Opportunity (MIRO) program (Award Number: 80NSSC19M0200).

References

1. Schaller, R.R.: Moore's law: past, present and future. IEEE Spectrum **34**(6), 52–59 (1997)
2. Weinstein, B.G.: A computer vision for animal ecology. J. Anim. Ecol. **87**(3), 533–545 (2018)

3. Wang, Y., Zheng, P., Xun, X., Yang, H., Zou, J.: Production planning for cloud-based additive manufacturing–a computer vision-based approach. Robot. Comput.-Integr. Manuf. **58**, 145–157 (2019)
4. Pathak, A.R., Pandey, M., Rautaray, S., Pawar, K.: Assessment of object detection using deep convolutional neural networks. In: Bhalla, S., Bhateja, V., Chandavale, A.A., Hiwale, A.S., Satapathy, S.C. (eds.) Intelligent Computing and Information and Communication. AISC, vol. 673, pp. 457–466. Springer, Singapore (2018). https://doi.org/10.1007/978-981-10-7245-1_45
5. Zhang, B., Jaiswal, P., Rai, R., Guerrier, P., Baggs, G.: Convolutional neural network-based inspection of metal additive manufacturing parts. Rapid Prototyping J. **25**(3), 530–540 (2019)
6. Li, J., Götvall, P., Provost, J., Åkesson, K.: Training convolutional neural networks with synthesized data for object recognition in industrial manufacturing. In: 2019 24th IEEE International Conference on Emerging Technologies and Factory Automation (ETFA), pp. 1544–1547 (2019)
7. Beschi, M., Villagrossi, E., Molinari Tosatti, L., Surdilovic, D.: Sensorless model-based object-detection applied on an underactuated adaptive hand enabling an impedance behavior. Robot. Comput.-Integr. Manuf. **46**, 38–47 (2017)
8. Khodabandeh, M., Vahdat, A., Ranjbar, M., Macready, W.G.: A robust learning approach to domain adaptive object detection. In: Proceedings of the IEEE/CVF International Conference on Computer Vision (ICCV), October 2019
9. Anitha, R., Jayalakshmi, S.: A systematic hybrid smart region based detection (SRBD) method for object detection. In: 2020 3rd International Conference on Intelligent Sustainable Systems (ICISS), pp. 139–145 (2020)
10. Mehmood, F., Ullah, I., Ahmad, S., Kim, D.: Object detection mechanism based on deep learning algorithm using embedded IOT devices for smart home appliances control in cot. Journal of Ambient Intelligence and Humanized Computing (2019)
11. Hu, L., Ni, Q.: Iot-driven automated object detection algorithm for urban surveillance systems in smart cities. IEEE Internet of Things J. **5**(2), 747–754 (2018)
12. Sudharsan, B., Kumar, S.P., Dhakshinamurthy, R.: Ai vision: smart speaker design and implementation with object detection custom skill and advanced voice interaction capability. In: 2019 11th International Conference on Advanced Computing (ICoAC), pp. 97–102 (2019)
13. Wilson, G., et al.: Robot-enabled support of daily activities in smart home environments. Cognitive Syst. Res. **54**, 258–272 (2019)
14. LeCun, Y., et al.: Backpropagation applied to handwritten zip code recognition. Neural Comput. **1**(4), 541–551 (1989)
15. Girshick, R., Donahue, J., Darrell, T., Malik, J.: Rich feature hierarchies for accurate object detection and semantic segmentation (2014)
16. Cortes, C., Vapnik, V.: Support-vector networks. Mach. Learn. **20**(3), 273–297 (1995)
17. Liu, W., et al.: SSD: single shot multibox detector. In: Leibe, B., Matas, J., Sebe, N., Welling, M. (eds.) ECCV 2016. LNCS, vol. 9905, pp. 21–37. Springer, Cham (2016). https://doi.org/10.1007/978-3-319-46448-0_2
18. Simonyan, K., Zisserman, A.: Very deep convolutional networks for large-scale image recognition (2015)
19. Redmon, J., Divvala, S., Girshick, R., Ali, F.: Unified, real-time object detection, You only look once (2016)

20. Chen, S.L., Lin, S.C., Huang, Y., Jen, C.W., Lin, Z.L., Su, S.F.: A vision-based dual-axis positioning system with yolov4 and improved genetic algorithms. In: 2020 Fourth IEEE International Conference on Robotic Computing (IRC), pp. 127–134 (2020)
21. Bradski, G.: The OpenCV library. Dr. Dobb's J. Softw. Tools **25**, 120–125 (2000)
22. San Diego Pacific Research Platform University of California. Nautilus
23. Redmon, J., Farhadi, A.: Yolov3: An incremental improvement. arXiv (2018)
24. Lin, T.-Y., et al.: Microsoft coco: Common objects in context (2015)
25. He, K., Gkioxari, G., Dollár, P., Girshick, R.: Mask r-cnn (2018)

Object-Oriented Internet Cloud Interoperability

Mariusz Postół$^{(\boxtimes)}$ (ID) and Piotr Szymczak (ID)

Institute of Information Technology, Lodz University of Technology, Łódź, Poland
mariusz.postol@p.lodz.pl

Abstract. Optimization of industrial processes requires further research on the integration of machine-centric systems with human-centric cloud-based services in the context of new emerging disciplines, namely the fourth industrial revolution coined as Industry 4.0 and Industrial Internet of Things. The following research aims at working out a new generic architecture and deployment scenario applicable to that integration. A reactive interoperability relationship of the communication parties is proposed to deal with the network traffic propagation asymmetry or assets' mobility. Described solution based on the OPC Unified Architecture international standard relaxes issues related to the real-time multi-vendor environment. The discussion concludes that the embedded gateway software component best suits all requirements and thus has been implemented as a composable part of the selected reactive OPC UA framework which promotes separation of concerns and reusability.

The proposals are backed by proof-of-concept reference implementations confirming the possibility of integrating selected cloud services with the OPC UA based cyber-physical system by applying the proposed architecture and deployment scenario. It is contrary to interconnecting cloud services with the selected OPC UA Server limiting the PubSub role to data export only.

Keywords: Industry 4.0 · Internet of Things · Object-Oriented Internet · Cloud computing · Industrial communication · Reactive networking · Machine to Machine communication · OPC Unified Architecture · Azure

1 Introduction

All the time, Information and Communication Technology is providing society with a vast variety of new distributed applications aimed at micro and macro optimization of the industrial processes. The design foundation of this kind of application must focus primarily on communication technologies. Based on the role humans take while using those applications they can be grouped as follows:

- **human-centric** - information origin or ultimate information destination is an operator,

© Springer Nature Switzerland AG 2021
M. Paszynski et al. (Eds.): ICCS 2021, LNCS 12745, pp. 568–581, 2021.
https://doi.org/10.1007/978-3-030-77970-2_43

– **machine-centric** - information creation, consumption, networking, and processing are achieved entirely without human interaction.

A typical **human-centric** approach is a web-service supporting, for example, a web user interface (UI) to monitor conditions and manage millions of devices and their data in a typical cloud-based IoT approach [3,9,13,29]. In this case, it is characteristic that any uncertainty and necessity to make a decision can be relaxed by human interaction. Coordination of robot behaviors in a work-cell (automation islands) is a **machine-centric** example. In this case, any human interaction must be recognized as impractical or even impossible. The interconnection scenario requires machine to machine communication (M2M) [10,12,22,25,28] demanding the integration of multi-vendor devices.

From the M2M communication concept, a broader idea of a smart factory can be derived. In this M2M deployment approach, the mentioned robots are only executive assets of an integrated supervisory control system responsible for macro optimization of an industrial process composed as one whole. Deployment of the smart factory concept requires a hybrid solution and interconnection of the above mentioned heterogeneous environments. This approach is called the fourth industrial revolution and was coined as Industry 4.0. It is worth stressing that interconnection of machines - or more general assets - is not enough, and additionally, assets interoperability must be expected for the deployment of this concept. In this case, multi-vendor integration makes communication standardization especially important, namely, it is required that the payload of the message is standardized to be factored on the data-gathering site and consumed on the ultimate destination site.

Highly-distributed solutions used to control any real-time process aggregating islands of automation (e.g. virtual power plants producing renewable energy) must, additionally, leverage public communication infrastructure, namely the Internet. The Internet is a demanding environment for highly distributed process control applications designed atop the M2M communication paradigm because it is globally shareable and can be also used by malicious users. Furthermore, it offers only non-deterministic communication making the integration of islands of automation designed against the real-time requirements a demanding task.

Today both obstacles can be overcome, and as examples, we have bank account remote control and voice over IP in daily use. The first application must be fine-tuned in the context of data security, and the second is very sensitive concerning time constraints. Similar approaches could be applied to adopt the concepts well known in process control industry, namely Human Machine Interface (HMI), Supervisory Control and Data Acquisition (SCADA), and Distributed Control Systems (DCS). It is worth stressing that, by design, all of them are designed based on interactive communication. Interactive communication is based on a data polling relationship. If that is the case, the application must follow the interactive behavioral model, because it actively polls the data source for more information by pulling data from a sequence that represents the process state in time. The application is active in the data retrieval process - it controls the pace of the retrieval by sending the requests at its convenience.

After dynamically attaching a new island of automation the control application (responsible for the data pulling) must be reconfigured for this interoperability scenario. In other words the interactive communication relationship cannot be directly applied because the control application must be informed on how to pull data from a new source. As a result, a plug and produce scenario [16] cannot be seamlessly applied. A similar drawback must be overcome if for security reasons suitable protection methods have been applied to make network traffic propagation asymmetric. It is accomplished using intermediary devices, for example, firewalls, to enforce traffic selective availability based on predetermined security rules against unauthorized access.

Going further, we shall assume that the islands of automation are mobile, e.g. autonomous cars passing a supervisory controlled service area. Here, the behavior of the interconnected assets is particularly important concerning the environment in which they must interact. This way we have entered the Internet of Things domain of Internet-based applications.

If we must bother with the network traffic propagation asymmetry or mobility of the asset network attachment-points the reactive relationship could relax the problems encountered while the interactive approach is applied [25]. In this case, the sessionless publisher-subscriber communication relationship is a typical pattern to implement the reactive interoperability paradigm. The sessionless relationship is a message distribution scenario where senders of messages, called publishers, do not send them directly to specific receivers, called subscribers, but instead, categorize the published messages into topics without knowledge about which subscribers, if any, there may be. Similarly, subscribers express interest in one or more topics and only receive messages that are of interest, without knowledge about which publishers, if any, there are. In this scenario, the publishers and subscribers are loosely coupled, i.e. they are decoupled in time, space and synchronization [6].

If the **machine-centric** Cyber-Physical System (CPS) - making up islands of automation - must be monitored and/or controlled by a supervisory system, the cloud computing concept may be recognized as a beneficial solution to replace or expand the above mentioned applications, i.e. HMI, SCADA, DCS, etc. Cloud computing is a method to provide the requested functionality as a set of services. There are many examples that cloud computing is useful to reduce costs and increase robustness. It is also valuable in case the process data must be exposed to many stakeholders. Following this idea and offering control systems as a service, there is required a mechanism created on the service concept and supporting abstraction and virtualization - two main pillars of the cloud computing paradigm. In the cloud computing concept, virtualization is recognized as the possibility of sharing the services by many users, and abstraction hides implementation details.

This article addresses further research on the integration of the multi-vendor **machine-centric** CPS designed atop of M2M communication and emerging cloud computing as a **human-centric** front-end in the context of the Industry 4.0 (I4.0) and Industrial Internet of Things (IIoT) disciplines. For this integration, a new architecture is proposed to support the reactive relationship of communicating parties. To support the multi-vendor environment the OPC Unified

Architecture [7,17,18] interoperability standard has been selected. The proposals are backed by proof of concept reference implementations - the outcome has been just published on GitHub as an open-source (MIT licensed) [24]. Prototyping addresses Microsoft Azure Cloud [4] as an example. The proposed solutions have been harmonized with the more general concept called the Object-Oriented Internet (OOI) [20,23,24].

The main goal of this article is to provide proof that:

- Reactive M2M interoperability based on the OPC UA standard can be implemented as a powerful standalone library without dependency on the Client/Server session-oriented archetype,
- Cloud interoperability can be implemented as an external part employing out-of-band communication without dependency on the OPC UA implementation,
- The proposed generic architecture allows that the gateway functionality is implemented as a composable part at run-time - no programming required.

The remainder of this paper is structured as follows. Section 2 presents the proposed open and reusable software model. It promotes a reactive interoperability pattern and a generic approach to establishing interoperability-context. A reference implementation of this archetype is described in Sect. 3. The most important findings and future work are summarized in Sect. 4.

2   Sensors to Cloud Interconnection - Architecture

To follow the Industry 4.0 concept a hybrid environment integrating reactive Machine to Machine interconnection and the interactive web-based user interface is required (Sect. 1). The main challenge of the solution in concern is to design a generic but reusable architecture that addresses interoperability of those diverse interconnection scenarios ruled by different requirements, namely:

- **machine-centric** machine to machine real-time mobile interoperability
- **human-centric** cloud-based front-end

Interconnection of the reactive **machine-centric** and interactive **human-centric** environments can be implemented by applying one of the following scenarios:

- **direct interconnection** (tightly coupled)-using a common protocol stack
- **gateway based interconnection** (loosely coupled)-using an out-of-bound protocol stack

By design, the **direct interconnection** approach requires that the cloud has to be compliant with the interoperability standard the CPS uses. As a result, it becomes a consistent communication node of the CPS. The decision to follow the **direct interconnection** scenario must be derived from an analysis of the capabilities of available services in concern. However, for the development strategy of this type of solution, the analysis can be done partially taking into account the following features that can be considered invariable. By design, the cloud-based services must be virtual - they are used to handle many solutions at the same time. Furthermore, M2M communication is usually constrained by real-time requirements. The virtualization of cloud services means that they must be very flexible to handle the attachment of new assets proactively (acting in advance) at run time. As a result, the cloud services must be responsible to register and authenticate devices by exposing endpoints in the public network to allow the device to access a provisioning cloud service. It requires that a session over the Internet has to be established by the data holding asset at a preparation step. To meet the requirements of real-time distributed control the CPS may use protocols applicable only to local computer networks (e.g. multicast IP, Ethernet, TSN[1], etc.). Because the cloud services support only protocols handling interconnection over the Internet the **direct interconnection** cannot be applied in a general case.

To support also a local network attachment point, the interaction with the cloud requires remote agents implemented by applying one of the following archetypes:

- **edge device** - a remote cloud agent acting as an intermediary for nodes of the CPS
- **field level gateway** - a dedicated custom agent acting as an intermediary for nodes of the CPS
- *Embedded Gateway* - a software part composed into a selected node of the *Cyber-physical network* (Fig. 1)

Edge device connects directly to the cloud services but acts as an intermediary for other devices called leaf devices. Additionally, it allows the selection of initial data and their processing using local resources. The **edge device** may be located close to the leaf devices and attached to the *Cyber-physical network* using protocols applicable only to local computer networks. In this scenario, it is possible to use a custom protocol stack to get connected to the **edge device** with the cloud and to help to save the bandwidth thanks to sending only the results of local processing. In this approach, the **edge device** is part of cloud vendor products and cannot be recognized as a generic solution that can be used to connect to other clouds supporting a many-to-many relationship.

The **field level gateway** is also built atop of the middleware concept [27]. The only difference as compared with the **edge device** is a necessity to use services officially supported by the cloud vendor to get connected. In this scenario, the process data may be transferred to many clouds simultaneously.

Proposition 1. *Unlike the above-described solutions, the* Embedded Gateway *is not derived from the middleware concept. A generic domain model for this interconnection is presented in the Fig. 1. Promoting separation of concerns*

---

[1] Time-Sensitive Networking (TSN) Task Group https://1.ieee802.org/tsn/.

design principle, the gateway functionality should be implemented as a self-contained software part embedded in the Networking *service of the* Cyber-physical node. *The main functionality of this component is to transfer selected data between* Cyber-physical network *using* Networking *services of an existing* Cyber-physical node *and* Cloud-based front-end *using interconnection services officially supported by the cloud vendor.*

The interconnection of assets is not enough hence their interoperability is expected. In this case, using the same communication stack must be recognized as only a necessary condition. To support interoperability, common data understanding is required. Additionally, to meet this requirement, the cloud and CPS have to establish directly the same semantic-context and security-context. The possibility of establishing a common semantic-context in the multivendor environment makes communication standardization especially important. If that is the case, it is required that the encoding of the payload exchanged over the network (Data Transfer Object) is standardized so that the appropriate messages can be factored on the data-gathering site and consumed on the ultimate destination data processing sites. Security between the data origin and ultimate data destination refers to the protection of messages (security-context) against malicious users. It is required that communicating parties are using the same cyber-security measures. To comply with the Industry 4.0 communication criterion, it is required that any product must be addressable over the network via TCP/UDP or IP and has to support the OPC UA Information Model [15,19,21]. As a result, any product being advertised as Industry 4.0 enabled must be OPC UA-capable somehow.

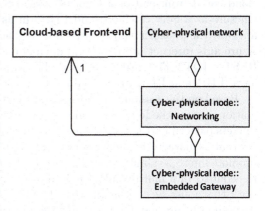

Fig. 1. Generic interconnection concept

The OPC Unified Architecture interoperability standard has been selected to support the multi-vendor environment. OPC UA supports the following two patterns to be used to transfer data between communicating parties:

- **session-oriented**: requires a session that must be established before any data can be sent between sender and receiver
- **sessionless-oriented**: the sender may start sending messages (called packets or datagrams) to the destination without any preceding handshake procedure

Using the session-oriented communication pattern it is difficult or even impossible to gather and process mobile data (Sect. 1), which is one of the Internet of Things paradigms. OPC UA Part 14 PubSub [2,26] offers the sessionless approach as an additional option to session-based client-server interoperability relationship and is a consistent part of the OPC UA specifications suit. As a result, it can be recognized as the IoT ready technology.

The proposals presented in the article are backed by proof of concept reference implementations [24]. For this study, prototyping addresses Microsoft Azure cloud products. There are many reasons for selecting Azure to accomplish the cloud-based front-end of a Cyber-Physical System (CPS). Azure offers Infrastructure as a Service (IaaS) and Platform as a Service (PaaS) capabilities. As a result, the platform can be used not only as a cloud-based front-end for CPS. Azure aids Internet protocols and open standards such as JSON, XML, SOAP, REST, MQTT [5], AMQP [8], and HTTP. Software development kits for C#, Java, PHP, and Ruby are available for custom applications.

Based on the sessionless and session-oriented communication patterns examination against the IoT requirements [25] it could be concluded that the connectionless pattern better suites issues related to assets mobility and traffic asymmetry that is characteristic for the application domains in concern. Additionally, to promote interoperability and address the demands of the M2M communication in the context of a multi-vendor environment, the prototyping should use a framework that must be compliant with the OPC UA Part 14 PubSub specification. According to proposed generic architecture (Fig. 1) to implement the *Embedded Gateway* as a composable part of the *Cyber-physical node* a library implementing *Networking* functionality in compliance with above mentioned specification is a starting point for further development. Additionally, it must be assumed that the library used to deploy *Embedded Gateway* supports dependency injection and is capable of composing an external part supporting cloud/PubSub gateway functionality. The composition process must be available without modification of the core code of an existing library. As a result, the prototyping is to be limited to implementation of the *Embedded Gateway* software part only.

To promote interoperability and address the demands of the M2M communication in the context of a multi-vendor environment the prototyping should use a framework that must be compliant with the OPC UA Part 14 PubSub and support the *Reactive Interoperability* (Sect. 1) concept. A framework compliant with these requirements has been implemented as an open-source library[2] named *UAOOI.Networking* (*Networking* for short) under an umbrella of the Object-Oriented Internet project [24]. The library is designed to be a foundation for developing application programs that are taking part in a message-centric communication pattern and interconnected using the reactive networking concept. The diagram in Fig. 2 shows the relationship between the library (*SDK*) and external parts composing any reactive networking application (*Reactive Application*). The *Reactive Application* is an aggregation of parts implementing the *Producer* and *Consumer* roles. By design, they support access to real-time process data, hence they are recognized as an extension of *DataRepository* class. To implement the *DataRepository* dedicated implementation of the *IBindingFactory* interface should be provided to create a bridge between CPS and an external

---

[2] https://github.com/mpostol/OPC-UA-OOI.

raw data represented by the *LocalResources* class. A more in-depth description of the *OOI Reactive Application* library enabling data exchange over a network using the reactive networking pattern is covered in [25].

To promote the polymorphic approach, the library has a factory class called *DataManagementSetup* that is a placeholder to gather all injection points used to compose external parts. To be injected, the part supporting data exchange with the underlying process must be compliant with the *IBindingFactory* interfaces. It is expected that the functionality implementation expressed by this interface is provided as an independent external composable part. The composition is accomplished at run time, and the effective application functionality depends essentially on reusable

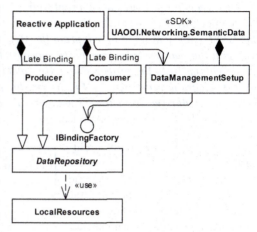

Fig. 2. Reactive interoperability architecture

loosely coupled parts composed applying the dependency injection software engineering concept.

The *DataRepository* represents data holding assets in the *Reactive Application* implementing the *IBindingFactory* interface. It captures functionality responsible for accessing the external process data from *LocalResources*. The *LocalResources* represents an external part that has a very broad usage purpose. For example, it may be any kind of the process data source/destination, i.e. raw data (e.g. PLC internal registers), OPC UA Address Space Management [25], cloud, file, database, graphical user interface, to name only a few.

Depending on the expected network role the library supports the implementation of:

- *Consumer* - entities processing data from incoming messages,
- *Producer* - entities gathering process data and populating outgoing messages.

The *Consumer* and *Producer* classes are derived from the *DataRepository* (Fig. 2). The *Consumer* uses the *IBindingFactory* to gather the data recovered from the *Message* instances pulled from a network. The received data may be processed or driven to any data destination, e.g. cloud-based front-end. The *Producer* mirrors the *Consumer* functionality and, after reading data from an associated source, populates the *Message* using the gathered data. By design, the *DataRepository* and associated entities, i.e. *Local Resources, Consumer, Producer* are embedded in external parts, and, consequently, the application scope may cover practically any concern that can be separated from the core *Reactive Application* implementation.

3   Cloud - OOI Interoperability Implementation

Proposition 2. *A generic domain model presenting interconnection architecture between the* Cloud-based Front-end *and* Cyber-physical node *attached to the* Cyber-physical network *is presented in Fig. 1. It is proposed to implement the* Cyber-physical node *by adopting the* Reactive Application *archetype compliant with the reactive interoperability concept (Fig. 2). Merging selected entities of this archetype into the proposed domain model (Fig. 1) leads to a model expressed as the diagram presented in Fig. 3. In the proposed approach the* Embedded Gateway *is derived from the* Consumer *role implemented as a composable part aggregated by the* Reactive Application.

In the final deployment architecture (Fig. 4) the *Consumer*
role has been realized by the
PartDataManagementSetup that
is derived from *DataManagementSetup* provided by the library.
Networking (SDK) was removed
from this diagram for the sake
of simplicity. Instantiating *Part-
DataManagementSetup* is the first
step for bootstrapping process of
the *Consumer* role functionality.
This class provides an entry point
to initialize all properties, which
are injection points of all parts
composing this role. It extends
the functionality of the *DataManagementSetup* based on the following associated classes: *Part-
BindingFactory* and *CommunicationContext*.

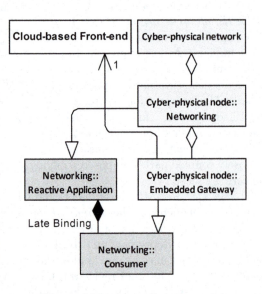

Fig. 3. Architecture domain model

The *PartBindingFactory* implements the *IBindingFactory* to gather the data recovered from the *Message* instances pulled from a network. The received data is driven to *CommunicationContext* for further encoding and finally pushing it to the cloud services using the configured out-of-band protocol.

The cloud interconnection is realized using the *CommunicationContext*. It implements a message encoder and establishes the out-of-band communication stack.

The data recovered from the *Message* is obtained from the *PartBindingFactory* using the *IDTOProvider* that defines a contract used to pull the Data Transfer Object created from a subscription by the *PartBindingFactory*. The transfer process requires data conversion from source to destination encoding, i.e. replacing bitstreams used by the CPS with equivalent ones for the cloud-based services. The Azure offers a vast variety of built-in types ready to be used in common cases, but not necessarily there are equivalent counterparts in use by the CPS. The Azure uses JSON based

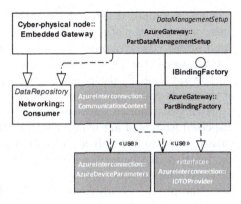

Fig. 4. Implementation architecture

Data Transfer Object encoding and schema defined based on the solution metadata. The PubSub uses JSON and binary Data Transfer Object encodings. In any case, the data recovered from the Message pulled from a subscription is stored locally using the object model based on standard .NET types. *PartBindingFactory* maps selected object graph onto the JSON message required by the cloud services.

The encoded JSON messages must be transferred to cloud over the network using the selected protocol stack. The Azure supports HTTP, AMQP, and MQTT protocol stacks, which are all standard ones. Consequently, it is possible to apply any available implementation compliant with an appropriate specification to achieve connectivity. In this case, all parameters required to establish semantic and security contexts are up to the gateway responsibility. Alternatively, the API offered by the dedicated frameworks (libraries) may be used. Using a framework, the configuration process may be reduced significantly, and the communication protocol selection has only an indirect impact on the interoperability features. In the proposed implementation, the Azure interconnection has been obtained using the above mentioned frameworks.

Azure and PubSub use different security mechanisms so in the proposed solution establishing security-context is realized independently. The *CommunicationContext* is responsible for establishing this context as an embedded negotiation phase tightly coupled with establishing interconnection.

4   Conclusion

Nowadays, the macro optimization of the industrial processes requires an integration of a vast variety of distributed applications provided by Information and Communication Technology. It requires further research on the integration of **machine - centric** Cyber-Physical Systems (CPS) with **human - centric** front-end in the context of new emerging disciplines, i.e. Industry 4.0 and the Industrial Internet of Things (Sect. 1).

CPS is composed using the multi-vendor components (data holding assets) interconnected atop of the Machine To Machine (M2M) communication. In many applications, the dynamic nature of the CPS must be considered. Dynamic nature means that interconnected assets may be added/removed from the network at any time. By design, CPS must typically fulfill the real-time and mobility of the assets requirements.

Highly-distributed solutions used to control/monitor a set of geographically dispersed islands of automation (e.g. virtual power plants producing renewable energy) must additionally leverage public communication infrastructure, namely the Internet. If islands of automation must be controlled over the Internet, the cloud computing concept may be recognized as a reasonable answer. Following this concept, the cloud-based supervisory control functionality is applied as a set of services employing abstraction and virtualization - two main pillars of the cloud computing paradigm. In the cloud computing concept, virtualization is recognized as the possibility of sharing the services by many users, and abstraction hides implementation details.

The main goal of this research is working out a new generic architecture and deployment scenario applicable for the integration of the **machine-centric** CPS and emerging cloud computing as a **human-centric** front-end.

If we must bother with the network traffic propagation asymmetry or mobility of the asset network attachment points the reactive relationship [25] could alleviate the challenges posed by the interactive approach. The real-time multivendor environment makes communication standardization especially important. To support this environment, the OPC Unified Architecture [11] interoperability standard has been selected. As it was pointed out in Sect. 2 using OPC UA PubSub [2] the aggregation of nodes by the network is loosely coupled, i.e. nodes can be added and removed from the network dynamically, and nodes may represent mobile data holding assets.

From the analysis covered by Sect. 2 it is concluded that the *Embedded Gateway* archetype best suits all requirements described above. It relaxes most of the issues related to **direct interconnection** and solutions inferred from the middleware concept, i.e. **edge device** - a remote cloud agent and **field level gateway** - a dedicated custom agent. Additionally, it could be used to connect to many independent clouds at the same time. The generic domain model for the proposed interconnection archetype is presented in Fig. 1. We have highlighted that to promote the separation of concerns design principle, the gateway functionality should be implemented as a self-contained dedicated software part embedded in the core *Networking* service of the *Cyber-physical node*. The main functionality of this component is to transfer process data between *Cyberphysical network* using *Networking* services of an existing *Cyber-physical node* and *Cloud-based front-end* using interconnection services officially supported by the cloud vendor.

To promote interoperability and address the demands of the M2M communication in the context of a multi-vendor environment the prototyping should use a framework that must be compliant with the OPC UA Part 14 PubSub

(Sect. 2) and support the *Reactive Interoperability* (Sect. 1) concept. We proposed to use an open-source library named *UAOOI.Networking* (*Networking* for short) (Fig. 2) for this purpose. It is worth stressing that based on this approach only dedicated functionality related to the communication with the cloud must be implemented.

We derived the final model presented in Fig. 4 by merging selected entities from the *Networking* library (Fig. 2) into the generic interconnection domain model (Fig. 1). In the proposed approach the *Embedded Gateway* is derived from the *Consumer* role implemented as a composable part aggregated by the *Reactive Application*.

The proposals are backed by proof of concept reference implementations. Prototyping addresses Microsoft Azure cloud as an example. The outcome has been just published on GitHub as the open-source (MIT licensed) repository. The proposed solutions have been harmonized with the more general concept called the Object-Oriented Internet.

The described results prove that the *Embedded Gateway* archetype implementation is possible based on the existing standalone framework supporting reactive interoperability atop the M2M communication compliant with the OPC UA PubSub standard. It is worth stressing that there is no dependency on the Client/Server session-oriented relationship. This relationship is in contrast to the architecture described in the OPC UA Part 1 [14] specification where the publisher role is tightly coupled with the **Address Space** [1] embedded component of the OPC UA Server. The real challenge of the future work is to prove that the proposed solution is flexible enough to be used as an archetype to inject the *Embedded Gateway* part into the OPC UA Client/Server to get connected with the cloud addressing the interactive relationship.

In the proposed approach the cloud interoperability is obtained by implementing a dedicated part employing out-of-band communication only without dependency on the OPC UA functionality at all. It is worth stressing that the gateway functionality is implemented as a part composable to the whole without programming skills. Because the part is composed at the runtime it makes it possible to modify its functionality later after releasing the library or even deploying the application program in the production environment.

Concluding, the paper describes a proof of concept that applying the proposed architecture and deployment scenario it is possible to integrate cloud services (e.g. **Azure IoT Central**) with the *Cyber-physical network* interconnected as one whole atop of the OPC UA PubSub. It is in contrast to interconnecting cloud-based front-end services with the *Address Space* instance exposed by a selected OPC UA server limiting the PubSub role to data exporter transferring the data out of the OPC UA ecosystem.

References

1. OPC unified architecture specification part 3: Address space model. Specification 10000–3, OPC Foundation (2017). https://opcfoundation.org/developer-tools/specifications-unified-architecture/part-3-address-space-model/

2. OPC unified architecture specification part 14 - pubsub. Specyfication 10000–14, OPC Foundation (2018). https://opcfoundation.org/developer-tools/ specifications-unified-architecture/part-14-pubsub/

3. Ashton, K.: That 'internet of things' thing. RFID JOURNAL 2009, pp. 1–1, 22 Jun 2009. https://www.rfidjournal.com/articles/pdf?4986

4. Bansal, N.: Designing Internet of Things Solutions with Microsoft Azure: A Survey of Secure and Smart Industrial Applications, chap. Microsoft Azure IoT Platform, pp. 33–48. Apress; 1st ed. edition (September 8, 2020) (2020)

5. Cohn, R.J., Coppen, R.J.: Mqtt version 3.1.1 plus errata 01. Technical report, OASIS, 10 Dec 2015. http://docs.oasis-open.org/mqtt/mqtt/v3.1.1/mqtt-v3.1.1.html

6. Eugster, P.T., Felber, P.A., Guerraoui, R., Kermarrec, A.M.: The many faces of publish/subscribe. ACM Comput. Surv. **35**(2), 114–131 (2003). https://doi.org/ 10.1145/857076.857078

7. González, I., Calderón, A., Figueiredo, J., Sousa, J.: A literature survey on open platform communications (OPC) applied to advanced industrial environments. Electronics **8**, 510 (2019). https://doi.org/10.3390/electronics8050510

8. Jeyaraman, R., Telfer, A.: Oasis advanced message queuing protocol (AMQP) version 1.0. Technical report, OASIS, 29 October 2012. http://docs.oasis-open.org/ amqp/core/v1.0/os/amqp-core-overview-v1.0-os.html

9. Koziolek, H., Burger, A., Platenius-Mohr, M., Rückert, J., Stomberg, G.: Openpnp: a plug-and-produce architecture for the industrial internet of things. In: 2019 IEEE/ACM 41st International Conference on Software Engineering: Software Engineering in Practice (ICSE-SEIP), pp. 131–140 (2019). https://doi.org/10.1109/ ICSE-SEIP.2019.00022

10. Lawton, G.: Machine-to-machine technology gears up for growth. Computer **37**(9), 12–15 (2004). https://doi.org/10.1109/MC.2004.137

11. Mahnke, W., Leitner, S.H., Damm, M.: OPC Unified Architecture. 1 edn. Springer, Berlin (2009). DOIurlhttps://doi.org/10.1007/978-3-540-68899-0

12. Meng, Z., Wu, Z., Muvianto, C., Gray, J.: A data-oriented m2m messaging mechanism for industrial IOT applications. IEEE Internet of Things J. **4**(1), 236–246 (2017). https://doi.org/10.1109/JIOT.2016.2646375

13. Mrozek, D., Milik, M., Małysiak-Mrozek, B., Tokarz, K., Duszenko, A., Kozielski, S.: Fuzzy intelligence in monitoring older adults with wearables. In: Krzhizhanovskaya, V.V., Závodszky, G., Lees, M.H., Dongarra, J.J., Sloot, P.M.A., Brissos, S., Teixeira, J. (eds.) ICCS 2020. LNCS, vol. 12141, pp. 288–301. Springer, Cham (2020). https:// doi.org/10.1007/978-3-030-50426-7_22

14. OPC unified architecture specification part 1 - overview and concepts. Specyfication 10000–1, OPC Foundation (2017). https://opcfoundation.org/developer-tools/specifications-unified-architecture/part-1-overview-and-concepts/

15. OPC unified architecture specification part 5: Information model. Specification 10000–5, OPC Foundation (2017). https://opcfoundation.org/developer-tools/ specifications-unified-architecture/part-5-information-model/

16. Pfrommer, J., Stogl, D., Aleksandrov, K., Navarro, S.E., Hein, B., Beyerer, J.: Plug and produce by modelling skills and service-oriented orchestration of reconfigurable manufacturing systems. At-Automatisierungstechnik **63**(10), 790–800 (2015). https://doi.org/10.1515/auto-2014-1157

17. Postół, M.: OPC From Data Access to Unified Architecture, sec. UA Specifications, pp. 94–99. VDE VERLAG GMBH, 4th revised edition edn. (2010)

18. Postół, M.: OPC From Data Access to Unified Architecture, sec. Main Technological Features, pp. 99–104. VDE VERLAG GMBH, 4th revised edition edn. (2010)

19. Postół, M.: OPC From Data Access to Unified Architecture, sec. Information model, pp. 111–130. VDE VERLAG GMBH, 4th revised edition edn. (2010)
20. Postół, M.: Object oriented internet. In: 2015 Federated Conference on Computer Science and Information Systems (FedCSIS), pp. 1069–1080 (2015). https://doi.org/10.15439/2015F160
21. Postół, M.: OPC ua information model deployment. Technical report, CAS, 18 April 2016. https://doi.org/10.5281/zenodo.2586616
22. Postół, M.: Computer Game Innovations 2018, chap. Machine to Machine Semantic-Data Based Communication: Comprehensive Survey, pp. 83–101. Lodz University of Technology Press, Łódź Poland (2018)
23. Postół, M.: Object-oriented internet. Technical report 5.1.0, GitHub (2019). https://doi.org/10.5281/zenodo.3345043
24. Postół, M.: Object oriented internet; azure gateway implementation 1.0. Technical report, GitHub (2020). https://doi.org/10.5281/zenodo.4361640, https://github.com/mpostol/OPC-UA-OOI
25. Postół, M.: Object-oriented internet reactive interoperability. In: Krzhizhanovskaya, V.V., et al. (eds.) ICCS 2020. LNCS, vol. 12141, pp. 409–422. Springer, Cham (2020). https://doi.org/10.1007/978-3-030-50426-7_31
26. Postół, M.: Object-oriented internet; ua part 14: Pubsub main technology features. Technical report, GitHub (2020). https://doi.org/10.5281/zenodo.4361640, https://commsvr.gitbook.io/ooi/reactive-communication/readme.pubsubmtf
27. Sunyaev, A.: Internet Computing: Principles of Distributed Systems and Emerging Internet-Based Technologies, chap. Middleware, pp. 125–154. Springer International Publishing, Cham (2020)
28. Verma, P.K., et al.: Machine-to-machine (m2m) communications: a survey. J. Netw. Comput. Appl. **66**, 83–105 (2016). https://doi.org/10.1016/j.jnca.2016.02.016
29. Xu, L.D., He, W., Li, S.: Internet of things in industries: a survey. IEEE Trans. Ind. Inform. **10**(4), 2233–2243 (2014). https://doi.org/10.1109/TII.2014.2300753

Static and Dynamic Comparison of Pozyx and DecaWave UWB Indoor Localization Systems with Possible Improvements

Barbara Morawska, Piotr Lipiński⬤, Krzysztof Lichy(✉) ⬤, Piotr Koch, and Marcin Leplawy

Lodz University of Technology, 116 Żeromskiego Street, 90-924 Lodz, Poland
krzysztof.lichy@p.lodz.pl

Abstract. This paper investigates static and dynamic localization accuracy of two indoor localization systems using Ultra-wideband (UWB) technology: Pozyx and DecaWave DW1000. We present the results of laboratory research, which demonstrates how those two UWB systems behave in practice. Our research involves static and dynamic tests. A static test was performed in the laboratory using the different relative positions of anchors and the tag. For a dynamic test, we used a robot that was following the EvAAL-based track located between anchors. Our research revealed that both systems perform below our expectations, and the accuracy of both systems is worse than declared by the system manufacturers. The imperfections are especially apparent in the case of dynamic measurements. Therefore, we proposed a set of filters that allow for the improvement of localization accuracy.

Keywords: UWB · Indoor localization · EvAAL

1 Introduction

Localization in GPS-denied areas, such as indoor localization, has attracted much attention due to its possible commercial applications [1]. There are several technologies which allow objects localization in such areas: Bluetooth [2], WiFi [3], Ultra-wideband (UWB) [4], inertial navigation [5], Li-Fi [6], LiDAR [7], visual monitoring [8].

This article focuses on UWB technology, which is economically justified and comprehensive as it combines reasonable costs and relatively good localization accuracy [9, 10]. Such a kind of indoor localization system is based on anchors (beacons), placed in known locations, and tags which positions are determined relative to those anchors. The position of tags is calculated using signal flight time between devices (TOF) [10]. Two nodes exchange messages and based on received and sent timestamps of those messages, devices can calculate the round trip time of the signal. The final tag position is obtained based on the trilateration algorithm that combines data acquired from the

---

The original version of this chapter was revised: the surname of the first author in reference 34 was incorrect. This has been corrected. The correction to this chapter is available at https://doi.org/10.1007/978-3-030-77970-2_51

M. Paszynski et al. (Eds.): ICCS 2021, LNCS 12745, pp. 582–594, 2021.
https://doi.org/10.1007/978-3-030-77970-2_44

anchors [12]. Several commercial products use this technology to localize objects, such as Pozyx [13], DecaWave [14], Zebra UWB Technology [15], Ubisense [16], BeSpoon [17], NXP's automotive UWB [18]. Interesting examples on this topic can also be found in [20, 21].

This paper compares the new commercially available UWB system — Pozyx and DecaWave DW1000, which is the best UWB localization system to our knowledge. The localization accuracy is compared under the same conditions. Experiments include both static and dynamic localization of the UWB tag relative to fixed anchors in a two-dimensional area. Tag node is localized in one of nineteen fixed points inside or at a short distance from the triangle marked by anchors in static experiments. A line-follower robot moves along the known EvAAL-based [34] track in close to constant linear speed motion in dynamic ones. Due to the fact, the theoretical position at any moment of the robot is known, we can evaluate the difference between the measured and actual position. The reason behind these differences are reflections, signal attenuation, clock delays, and imperfect characteristics of antennas mounted in the UWB device generating measurement noise, which influences localization accuracy [19]. Since interferences depend on both environmental and hardware factors, it is almost impossible to avoid them. Several methods can improve wireless localization accuracies, such as optimizing sensors number and position [22], antenna gain [23], machine learning [24], and filters. We have decided to apply the last option because we believe it is the most universal; it can work in real-time with a very short delay and often gives satisfactory results. We use the following filtering algorithms: median, ARMA, and Kalman filters [27], preceded by trilateration techniques.

The research, which is similar to the one presented in this article, has been performed by Simedroni et al. In [28], the authors studied DecaWave-family UWB device – MDEK 1001 [31] with 16 cm static average error results. In [29] the variance of static experiments is about 10 cm, and the accuracy of dynamic experiments is 65 cm for 100% samples (4 anchors). Both static [28, 29] and dynamic [29] measurements differ from our research in the way the antennas are arranged, the size of the scene, the devices used, and the track shape. Most importantly, the results are also difficult to reproduce, as the ground truth trajectory has a very irregular shape. Jian Wang et al. in [30] conducted the experiment in which they investigated the UWB system accuracy. Still, they focused on improving the accuracy by fusing the data from separate anchors, treating them as independent systems. Unfortunately, the method of obtaining the ground-truth path is unclear. Here we decided to perform UWB systems' comparison using the robot following the EvAAL track to ensure ground truth localization.

The organization of this paper is given as follows. In Sect. 2, Pozyx UWB and DecaWave TREK1000 are described shortly. In Sect. 3, the experimental setup is defined. The results of real-time localization and error analysis are presented in Sect. 4. In Sect. 5, we conclude and sketch the directions for further research.

2   UWB Localization Systems

2.1   DecaWave

DecaWave DW1000 [33] is a fully integrated low power, single-chip CMOS radio transceiver IC compliant with IEEE 802.15.4-2011 ultra-wideband (UWB) standard [32]. The manufacturer distributes the chip as a part of microcontroller STM32F105 ARM Cortex M3 with LCD and USB interface, which can both power and transmit data to a computer [25, 26]. The set is called DecaWave EVK1000. The DW1000 chip and microcontroller communicate via Serial Peripheral Interface (SPI). An Evaluation Kit DecaWave TREK1000 used during experiments consists of four EVK1000 boards. To perform 2D localization, at least three anchors and one tag is required. The UWB tag is localized relative to the fixed anchors' position. Every device can act as both an anchor and a tag. The configuration is set by changing dip switches located on the PCB board. These switches also allow the change operation channel to 2 or 5 (which correspond to center frequency 3993.6 MHz or 6489.6 MHz), data rate –110 kb/s, or 6.8 Mb/s, and tag/anchor ID. Of paramount importance is the fact that anchors are not the same. The anchor with an ID equal to 0 is responsible for communication between other anchors and the tag. As a result, when it is powered off, interfered with obstacles, or exposed to excessive dumping, the localization system does not work. That is why anchor 0 location should be chosen wisely.

According to the documentation, the localization accuracy of the tag is ±30 cm. The maximum measurable distance is about 250 m, but it depends on the data rate and channel [28]. Moreover, in most countries, it is not possible to use that range due to legal restrictions on maximal mean Effective Isotropical Radiated Power (EIRP) of the antenna, which is usually –41.3 dBm/Mhz. As a result, the system measurable distance is even lower.

2.2 Pozyx

Pozyx is a new localization system that also uses the DW1000 chip, same as in EVK1000, but it takes advantage of STM32F401 ARM Cortex M4, which is more advanced than STM32F105. The development kit used during experiments consists of 8 transceivers. Although every node can act as both anchor and tag, the manufacturer has drawn a clear line between them and divided the set among four anchors and five tags. The tag device is constructed to be a shield compatible with Arduino. It connects the microcontroller board using long wire-wrap headers that extend through the shield. The Arduino communicates with Pozyx via Inter-Integrated Circuit (I^2C). Moreover, the tag device does not need an Arduino to work. It can be used by connecting a computer to Micro USB as a virtual COM port, which is also a power source. The tag can determine the position and report on motion data thanks to the built-in accelerometer, gyroscope, magnetometer, and pressure sensor. Unlike the tag, the anchor node is boxed and not compatible with Arduino. The only possibility to receive data from the anchor is to use the serial port.

While TREK1000 kit configuration is based on dip switches, in Pozyx, all options are programmable. There are two dedicated libraries: C-like Arduino (Pozyx-Arduino_library) and Python (pypozyx), which allow changing the same parameters as in DecaWave, namely: channel, bitrate, and the board function to act as a tag or an anchor. There are seven independent channels that do not interfere and communicate with each other, transmitting center frequencies from 3244.8 MHz to 5948.8 MHz. Bitrate can be set to 110 kb/s, 859 kb/s or 6.8 Mb/s. Although Pozyx has many more configuration options than EVK1000, the setting up process is more complicated and longer.

The manufacturer claims that Pozyx accuracy is ±10 cm, which is better than TREK1000. The maximum measurable distance is 100 m, which is calculated, considering the legal restrictions of EIRP.

3 Experimental Setup

The experiment was divided into two parts: static and dynamic. DecaWave and Pozyx sets were evaluated in both parts with the same data rate −6.8 Mb/s and on the 5th channel. DecaWave and Pozyx systems localization was measured separately since they were operating at the same frequencies. Namely, when the robot was driving at the track, only one system was operating at a time.

In static experiments, the robot was placed at selected measuring points inside, on, or outside the track. These points are marked with yellow crosses in Fig. 1. The purpose of the experiment was to verify the static localization accuracy of both systems. To do it, we collected 200 coordinates with both Pozyx and DecaWave for each measuring point. This allows us to evaluate error distribution in different tag positions relative to both systems' anchors.

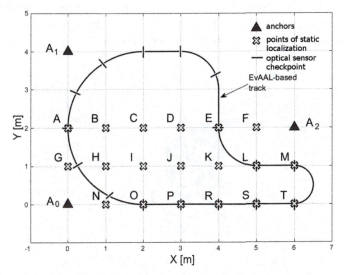

Fig. 1. Test track based on EvAAL used in experiments. Anchors A_0, A_1, A_2 are marked with blue triangles, and the track is marked with the black line. Yellow crosses mark the points of static localization. Black lines perpendicular to the track (in fact, they are white) are placed where the optical system is notified about robot localization. (Color figure online)

Dynamic localization tests are, in general, much more challenging to reproduce than static tests because they are often conducted using the testing environment, which is lab-specific. To avoid it, we used the experimental setup inspired by [34]. Our laboratory was too small to build a full test track from [34], so we decided to use only half of that track.

Please note that the same as in the static experiment, in our dynamic experiment, the anchors were placed out of the track in such a way that they formed a triangle that intersects the track. In this approach, some measurement points were inside and some outside of the triangle. To make the analysis more straightforward, we introduced a coordinate system with A_0 anchor located at $(0; 0)$, A_1 located at $(0; 4)$, and A_2 located in $(6; 2)$. The track and anchor localization is shown in Fig. 1. Anchors A_0, A_1, A_2 are marked with blue triangles, while the track is marked with a black line.

In dynamic experiment we used a robot – MakeBlock Robot mBot V1.1 from Fig. 2 – following the line on the floor using an optical sensor at constant linear speed. In each dynamic experiment, the robot passed the track twice, which allowed us to collect about 2400 (DecaWave) and 3200 (Pozyx) position samples in less than 4 min. This is equivalent to capturing 11 samples per second in DecaWave and 14 samples per second in Pozyx. We used a second optical sensor installed on the robot to find dynamic reference robot localization, which was triggered when crossing reference points. The reference points are marked with black lines perpendicular to the track in Fig. 1. Each time the robot was crossing the reference point, it recorded a timestamp. The average robot speed was calculated for each section between reference points basing on timestamp and section length. As a result, we were able to calculate the robot speed with precision, which was sufficient for this experiment. Using the average robot speed and the timestamps, we were able to find the robot localization at any moment. We used this localization as a reference for dynamic UWB localization in our experiment.

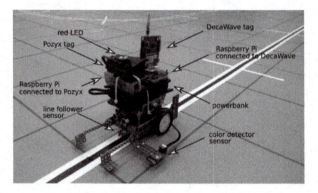

Fig. 2. Line follower robot used in experiments.

4 Experiment

This section describes the results of experiments that were performed along with the explanation of the obtained results. The experiments include a static and dynamic part. Firstly, raw data from both UWB devices are compared in static and dynamic experiments. Then we demonstrate how the localization accuracy can be improved by raw data filtration. We believe that the best way to illustrate the experiment results is to present the localization coordinates on the EvAAL track and the measurement accuracy using the empirical cumulative distribution function (ECDF).

On the top, there are the DecaWave tag and Pozyx tag connected to separate Raspberry Pi. A power bank powers both computers. The robot uses the line follower sensor to track the line and color sensor to record reference point in the robot's memory and blink the red LED when it crosses the white line to notify the optical system about robot localization. Blue arms form a counterweight, which prevents the robot from tilting backward.

4.1 Static Experiments

As described in the previous section, in static experiments, we manually place the robot in each of A-T measurement points presented in Fig. 1 and collect 200 localization samples from both localization systems. Figure 3 shows the results of a stationary measurement in which the data collected by DecaWave is marked in red, and the data collected by Pozyx is marked in green. As illustrated in Fig. 3, the measurements from both devices accumulate at designated points. One can notice that the variance of the raw data obtained from DecaWave is, in general, much smaller than in Pozyx. Simultaneously, the center of gravity of the resultant localization samples strongly depends on the measurement point localization. For example, for measurement points I and J, the center of gravity of localization results collected by Pozyx are closer to the reference point than in the case of DecaWave. In R, S, and T points, it is the opposite, the center of gravity of points collected by DecaWave is closer to the reference points than in the case of Pozyx. Such deviations are probably the result of reflections from the walls closely surrounding the track. Moreover, the variance of DecaWave localization points is similar regardless of the measuring point is inside, on, or outside of the A_0, A_1, A_2 triangle. In contrast to DecaWave, the measurement variance for Pozyx is slightly higher when the measurement point is located outside of the A_0, A_1, A_2 triangle. For example, at point T situated outside of the track, the variance reaches 0.17 on X-axis and 0.003 on Y-axis, while for the same point in DecaWave, it is only 0.04 on X-axis and 0.0003 on Y-axis. The smallest variance of the measurement samples was obtained in point B for DecaWave (values in both axes of around 0.0001) and G for Pozyx (values around 0.001). The worst result was in point T in both cases.

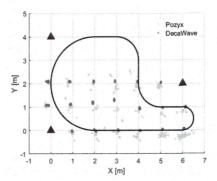

Fig. 3. Measurement points of Pozyx (marked with green dots) and DecaWave (marked with red dots) in static experiments. Anchors A_0, A_1, A_2 are marked with blue triangles, and the track is marked with the black line. (Color figure online)

4.2 Dynamic Experiments

In dynamic measurements, the robot follows the EvAAL-based track with constant velocity acquiring its position. Every correct measurement (3–5% of measurements is corrupted in the communication channel) for Pozyx and DecaWave is presented in Fig. 4. The data collected from DecaWave is marked with red dots, and the data collected by Pozyx are marked with green dots. As illustrated in Fig. 4, localization points acquired by the DecaWave system are very close to the reference track, except for the curve at the top. We suspect that this is the result of the reflection from a thick, brick wall parallel to OX, which was located next to the track. As for the Pozyx system, the measurement accuracy is significantly worse than DecaWave. Namely, most measurement points are shifted relative to the track, and the variance of the localization is much higher. On the other hand, surprisingly, the results obtained in the upper part of the track fit the reference track slightly better than in DecaWave. In general, the mean localization error for the DecaWave system is around 11 cm, and a median is approximately 7 cm. The same for Pozyx is respectively: 21 cm and 20 cm. ECDF (Fig. 5) illustrates that DecaWave mapped the track better than Pozyx, obtaining an accuracy of 32 cm when the same accuracy was achieved for Pozyx at 160 cm. However, more than 95% of the samples do not exceed a distance of 50 cm, and this value better describes the accuracy of Pozyx.

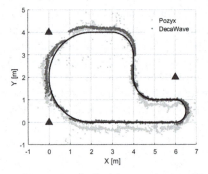

Fig. 4. Measurement points of Pozyx (marked with green dots) and DecaWave (marked with red dots) in dynamic experiments. Anchors A_0, A_1, A_2 are marked with blue triangles, and the track is marked with the black line. (Color figure online)

4.3 Accuracy Improvement

The raw measurement presented in Sect. 4.2 appears to be far from the declared measurement accuracy of a maximum of 30 cm in Pozyx and close to it in the case of DecaWave. The problem for the Pozyx system is a high variance of measurement error. Subsequent position samples do not reflect the robot movement in detail. On the other hand, the DecaWave system has many more failed acquisitions that must be removed in the localization process. The ideal solution would be to achieve Decawave accuracy with Pozyx reliability. To meet our expectations, we decided to improve measurement

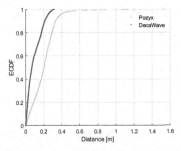

Fig. 5. The empirical cumulative distribution function of the localization error for Pozyx (green dots) and DecaWave (red dots) error accuracy in dynamic experiments. (Color figure online)

accuracy for Pozyx applying filters. At the same time, we check how the same filters affect DecaWave and whether it is possible to further improve the results.

As a data source for the experiment, we take the raw measurement results from our experiment described in Sect. 4.2 as filtration inputs. Then we process data with median filter, ARMA filter, and Kalman filter. Both series are filtered the same algorithms that, in our opinion, should improve location accuracy by reducing distribution error. To present the results, the measuring points on the track are presented, as well as the empirical cumulative distribution functions that allow checking the filtration effect.

Median Filter

The median filter is a nonlinear method for noise suppression. It operates as neighborhood averaging, which allows for the removal of out-of-range measurement samples. The number of averaging measurements is defined with window size w. For a sufficiently large window, the median filter can suppress random noise. The side effect of this is the delay, which is the result of the filter construction. In our case, the average delay is $\frac{100w}{2}$ ms, and the robot's maximum speed is around 1 m/s. In the series of experiments, we used window size $w = 7$, assuming that the delay of 350 ms is still acceptable in our case. The median filtering results are illustrated in Fig. 7 for Pozyx and Fig. 8 for DecaWave (green dots). In the Pozyx system, the filter results in the elimination of random measurement errors and thus an improvement of accuracy up to 42 cm, taking into account all samples and 32 cm at 95% of the samples. With DecaWave, the improvement is unnoticeable, and the localization accuracy improves by 2 cm for 100% samples and less than 1 cm for 95% of samples.

ARMA Filter

The autoaggressive (AR) and moving average (MA) model specifies that the output variable depends linearly on its previous values and also linearly on a set of previous inputs. In general form, ARMA filter is an infinite impulse response filter (IIR) defined as follows (1):

$$Y(z) = H(z)X(z) = \frac{\beta(1) + \beta(2)z^{-1} + \ldots + \beta(n+1)z^{-n}}{\alpha(1) + \alpha(2)z^{-1} + \ldots + \alpha(n+1)z^{-m}}X(z) \qquad (1)$$

where $\beta(i)$ (MA part) and $\alpha(i)$ (AR part) are filter coefficients and n, m define filter order. Looking for desired filter properties (low pass filter with a gain of 1), we decided to simplify it to the form (2):

$$Y(z) = \frac{\beta}{1 - \alpha(z^{-1})} X(z) \tag{2}$$

where $\beta = \alpha - 1$.

Filtering in such a way is based on fitting the appropriate α coefficient. For small α (close to zero), the filtration effect is stronger, but it is impossible to detect rapid movements of tracking objects. Unlike the median filter, the ARMA filter has zero-delay, and its computing performance makes it an efficient way to improve localization accuracy. In the experiments, we use $\alpha = 0.8$. The results of ARMA filtering are illustrated in Fig. 9 for Pozyx and Fig. 10 for DecaWave (red dots). In the case of Pozyx, the result is very similar to that obtained for median filtration. 95% of samples obtain an accuracy of 32 cm, and 100% of samples have an accuracy of 46 cm. In DecaWave, the difference is even less visible and almost impossible to note.

Kalman Filter

Kalman filter used in the experiment estimates a state vector based on a dynamic model. The model assumes that the velocity is constant in the sampling interval, and the system can measure only the target's position (have no sensors to measure velocity) [27]. According to the assumption, the Kalman filter can remove the signal noise when the covariance coefficients are properly selected. We propose a method for determining the covariance based on the function of the variance value from the distance of the anchor to the tag - $f(d)$, where d is the distance. Given the function f, it is possible to calculate the covariance matrix in a two-dimensional area. Supposing that variance is a vector, one can create three vectors from measured point T directed towards each anchor A_0, A_1, A_2. The length of each vector is the value of function f. Projecting each vector on X-axis and Y-axis and averaging vector components give two variance vectors that are the basis for the covariance matrix (3) and (4).

$$\delta_x = \frac{\delta_x^{A_o} + \delta_x^{A_1} + \delta_x^{A_2}}{3} \tag{3}$$

$$\delta_y = \frac{\delta_y^{A_o} + \delta_y^{A_1} + \delta_y^{A_2}}{3} \tag{4}$$

Figure 6 illustrates the method of determining variance in the X-axis and Y-axis.

Kalman filtering results are illustrated in Fig. 9 for Pozyx and Fig. 10 for DecaWave (purple dots). The Pozyx system is similar to median and ARMA filtering (around 32 cm for 95% of samples and 42 cm for 100% samples). The significantly worse improvement occurs in DecaWave, where the accuracy of localization decreases by around 6 cm to 38 cm for 100% samples and 34 cm for 95% samples.

While the Kalman filter reduces error in static measurement to 0.5 cm [35], the improvement works not as expected in dynamic measurements. Filter attracts trajectory

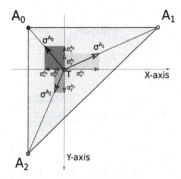

Fig. 6. Method of determining variance in X-axis and Y-axis.

to the most probable position, which results in smooth tracking considering inertia (Fig. 9 and Fig. 10), but it does not improve measurement accuracy. It contradicts the widespread belief that the Kalman filter always allows for a reduction in position measurement error. This is mainly because the measurement noise has non-Gaussian characteristics.

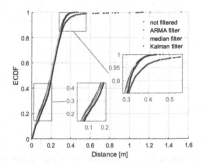

Fig. 7. Empirical cumulative distribution function of the localization error for not-filtered Pozyx data (blue dots), ARMA filtered Pozyx data (red dots), median filtered Pozyx data (green dots), and Kalman filtered Pozyx data (purple dots) in dynamic experiments. (Color figure online)

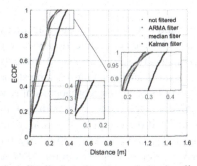

Fig. 8. Empirical cumulative distribution function of the localization error for not-filtered DecaWave data (blue dots), ARMA filtered DecaWave data (red dots), median filtered Pozyx data (green dots), and Kalman filtered DecaWave data (purple dots) in dynamic experiments. (Color figure online)

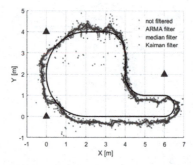

Fig. 9. Measurement points of not-filtered Pozyx data (blue dots), ARMA filtered Pozyx data (red dots), median filtered Pozyx data (green dots), and Kalman filtered Pozyx data (purple dots) in dynamic experiments. Anchors A_0, A_1, A_2 are marked with blue triangles, and the track is marked with the black line. (Color figure online)

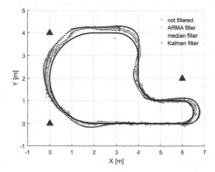

Fig. 10. Measurement points of not-filtered DecaWave data (blue dots), ARMA filtered DecaWave data (red dots), median filtered DecaWave data (green dots), and Kalman filtered DecaWave data (purple dots) in dynamic experiments. Anchors A_0, A_1, A_2 are marked with blue triangles, and the track is marked with the black line. (Color figure online)

5   Summary

The paper investigates static and dynamic localization accuracy of two indoor localization systems based on ultra-wideband communication: Pozyx and DecaWave DW1000. The final tag position is obtained in these systems based on the trilateration algorithm that combines data acquired from three anchors. Our research shows that in both: static and dynamic experiments, the Pozyx system has a broader sample variance than DecaWave. Our research has also revealed that the accuracy of both systems is worse than declared by the manufacturers (30 cm for 100% samples). Position accuracy is 38 cm for 95% of Pozyx samples and over 1.5 m for 100% samples. Similarly, the accuracy of DecaWave is 25 cm for 95% of DecaWave samples and 32 cm for 100% samples.

To improve localization accuracy, we have applied ARMA, median, and Kalman filters. The median filter improves localization accuracies of Pozyx to 32 cm for 95% of samples. In the DecaWave, improvement is almost invisible. Similar results have been obtained in ARMA filtration. Kalman filter improves Pozyx accuracy of 95% samples to

32 cm. In the case of DecaWave, the result is significantly worse than not filtered signal – the accuracy of localization decreases by around 8 cm to 34 cm for 95% samples. It should also be noted that the filter attracts trajectory to the most probable position, which results in smooth tracking, but it does not improve measurement accuracy.

In the future, we consider extending our research by applying artificial neural networks to remove the non-gaussian error and reducing the impact of signal reflections on localization results.

References

1. Mautz, R.: Indoor positioning technologies, Habilitation Thesis, Institute of Geodesy and Photogrammetry, Department of Civil, Environmental and Geomatic Engineering, ETH Zurich (2012)
2. Obreja, S.G., Vulpe, A.: Evaluation of an indoor localization solution based on bluetooth low energy beacons. In: 2020 13th International Conference on Communications (COMM), Bucharest, Romania, pp. 227–231 (2020)
3. Xue, J., Liu, J., Sheng, M., Shi, Y., Li, J.: A WiFi fingerprint based high-adaptability indoor localization via machine learning. China Commun. **17**(7), 247–259 (2020)
4. Che, F., Ahmed, A., Ahmed, S.G., Zaidi, R., Shakir, M.Z.: Machine learning based approach for indoor localization using Ultra-Wide Bandwidth (UWB) system for Industrial Internet of Things (IIoT). In: 2020 International Conference on UK-China Emerging Technologies (UCET), Glasgow, United Kingdom (2020)
5. Barbour, N.M., Stark Draper, C.: Inertial Navigation Sensors, Laboratory (P-4994), Cambridge, MA 02139, USA (2011)
6. Lam, E.W., Little, T.D.C.: Indoor 3D localization with low-cost lifi components. In: 2019 Global LIFI Congress (GLC), Paris, France, pp. 1–6 (2019)
7. Opromolla, R., Fasano, G., Rufino, G., Grassi, M., Savvaris, A.: LIDAR-inertial integration for UAV localization and mapping in complex environments. In: 2016 International Conference on Unmanned Aircraft Systems (ICUAS), pp. 649–656. Arlington, VA (2016)
8. Taira, H., et al.: InLoc: indoor visual localization with dense matching and view synthesis. In: The IEEE Conference on Computer Vision and Pattern Recognition (CVPR), pp. 7199–7209 (2018)
9. Zimmerman, T., Zimmermann, A.: Magic Quadrant for Indoor Location Services, Global Published 13 January 2020 - ID G00385050 (2020)
10. Zhang, W., Zhu, X., Zhao, Z., Liu, Y., Yang, S.: High accuracy positioning system based on multistation UWB time-of-flight measurements. In: 2020 IEEE International Conference on Computational Electromagnetics (ICCEM), Singapore (2020)
11. Decawave, APS011 Application Note, Sources of Error in DW1000 Based Two-Way Ranging (TWR) Schemes (2014)
12. Asmaa, L., Hatim, K.A., Abdelaaziz, M.: Localization algorithms research in wireless sensor network based on multilateration and trilateration techniques. In: 2014 Third IEEE International Colloquium in Information Science and Technology (CIST), Tetouan, pp. 415–419 (2014)
13. Pozyx Homepage. https://www.pozyx.io. Accessed 01 Feb 2021
14. Decawave DW1000 product homepage. https://www.decawave.com/product/dw1000-radio-ic/. Accessed 01 Feb 2021
15. Zebra Homepage. https://www.zebra.com/us/en/products/location-technologies/ultra-wideband.html. Accessed 01 Feb 2021
16. Ubisense Home Site. https://ubisense.com/dimension4/. Accessed 01 Feb 2021

17. BeeSpoon Mek 1 Product Homepage. https://bespoon.xyz/produit/mek1-ultra-wideband-module-evaluation-kit/. Accessed 01 Feb 2021
18. NXP Homepage. https://www.nxp.com/applications/enabling-technologies/connectivity/ultra-widebanduwb:UWB. Accessed 01 Feb 2021
19. Decawave, APS006 Application Note Channel effects on communications range and time stamp accuracy in DW1000 based systems. https://www.decawave.com/wpcontent/uploads/2018/10/APH001_DW1000-HW-Design-Guide_v1.1.pdf. Accessed 01 Feb 2021
20. Glonek, G., Wojciechowski, A.: Kinect and IMU sensors imprecisions compensation method for human limbs tracking. In: International Conference on Computer Vision and Graphics, ICCVG 2016. Poland (2016)
21. Daszuta, M., Szajerman, D., Napieralski, P.: New emotional model environment for navigation in a virtual reality. Open Phys. **18**(1), 864–870 (2020)
22. Zhao, Y., Li, Z., Hao, B., Wan, P., Wang, L.: How to select the best sensors for TDOA and TDOA/AOA localization? China Commun. **16**(2), 134–145 (2019)
23. Sinha, P., Yapici, Y., Guvenc, I.: Impact of 3D antenna radiation patterns on TDOA-based wireless localization of UAVs. In: IEEE INFOCOM 2019 - IEEE Conference on Computer Communications Workshops (INFOCOM WKSHPS) (2019)
24. Bibb, D.A., Yun, Z., Iskander, M.F.: Machine learning for source localization in urban environments. In: MILCOM 2016 - IEEE Military Communications (2016)
25. Decawave, APS006 Part 2 Application Note, Non Line of Sight operation and optimization to improve performance in DW1000 Based systems, version 1.5 (2014)
26. Decawave, APH001 Application Note, DW1000 hardware design guide, version 1.1 (2018)
27. Saho, K.: Kalman filter for moving object tracking: performance analysis and filter design. Kalman Filters, Theory for Advanced Applications (2017)
28. Simedroni, X.L.: Indoor positioning using decawave MDEK1001. In: 2020 International Workshop on Antenna Technology (iWAT), Bucharest, Romania (2020)
29. Delamare, Y., Boutteau, M., Savatier, R., Iriart, N.: Static and dynamic evaluation of an UWB localization system for industrial applications. Science **2**(2), 23 (2020)
30. Wang, J., Wang, M., Yang, D., Liu, F., Wen, Z.: UWB positioning algorithm and accuracy evaluation for different indoor scenes. International Journal of Image and Data Fusion (2021)
31. MDEK1001 Kit User Manual Module Development & Evaluation Kit for the DWM1001 Version 1.2
32. IEEE Standard for Local and metropolitan area networks— Part 15.4: Low-Rate Wireless Personal Area Networks (LR-WPANs)
33. DecaWave, DW1000 User Manual, version 2.11 (2017)
34. Potortì, F., Sangjoon, F., Ruiz, A.R., Barsocchi, P.: Comparing the performance of indoor localization systems through the EvAAL framework. Sensors **17**, 23–27 (2017)
35. Morawska, B.: Reduction of measurement error in spatial objects' positioning, BSc Thesis, Faculty of Technical Physics, Information Technology and Applied Mathematics of the Technical University of Lodz (2020)

Challenges Associated with Sensors and Data Fusion for AGV-Driven Smart Manufacturing

Adam Ziebinski[1]([✉]), Dariusz Mrozek[1], Rafal Cupek[1], Damian Grzechca[1], Marcin Fojcik[2], Marek Drewniak[3], Erik Kyrkjebø[2], Jerry Chun-Wei Lin[2], Knut Øvsthus[2], and Piotr Biernacki[1]

[1] Silesian University of Technology, Gliwice, Poland
{aziebinski,dmrozek,rcupek,dgrzechca,pbiernacki}@polsl.pl
[2] Western Norway University of Applied Sciences, Bergen, Norway
{marcin.fojcik,erik.kyrkjebo,knut.ovsthus}@hvl.no,
jerrylin@ieee.org
[3] AIUT Sp. z o.o. (Ltd.), Gliwice, Poland
mdrewniak@aiut.com.pl

Abstract. Data fusion methods enable the precision of measurements based on information from individual systems as well as many different subsystems to be increased. Besides, the data obtained in this way enables additional conclusions drawn from their work, e.g., detecting degradation of the work of subsystems. The article focuses on the possibilities of using data fusion to create Autonomous Guided Vehicles solutions in increasing precise positioning, navigation, and cooperation with the production environment, including docking. For this purpose, it was proposed that information from other manufacturing subsystems be used. This paper aims to review the current implementation possibilities and to identify the relationship between various research sub-areas.

Keywords: Autonomous Guided Vehicles (AGV) · Data fusion · Machine to Machine Communication (M2M) · Sensor fusion

1 Introduction

Contemporary production systems require that many stringent requirements be fulfilled, including flexibility, dynamic re-engineering processes, and production quality. Many changes enable the implementation of Industry 4.0 [1] functionalities to be made to meet these requirements. The production has to be adjusted to specific products, and the process organization must follow these changes. Avoiding non-productive time gaps reduces production losses [2]. The industrial environment consists of several Cyber-Physical Production Systems (CPPS) [3], IoT [4], and mobile subsystems. More and more efficient internal transport systems rely on solutions that use Autonomous Guided Vehicles (AGV) [5]. The logistics tasks must be performed in a distributed, dynamic, and autonomous manner, and therefore, they require the online information exchange between the AGVs and an industrial manufacturing environment. The new generation

© Springer Nature Switzerland AG 2021
M. Paszynski et al. (Eds.): ICCS 2021, LNCS 12745, pp. 595–608, 2021.
https://doi.org/10.1007/978-3-030-77970-2_45

of the manufacturing ecosystems requires the regular supply and movement of various components. To maintain the appropriate level of organization and quality of production, the precise execution of production orders is required. The currently used robotic systems enable the production of high-quality products. To work appropriately and above all efficiently [6], an AGV must be precisely docked to the assembly station (AS), and the loading and unloading operations have to be performed collaboratively. Integrating the information from an AGV and various sensors available from other production subsystems such as the IoT and CPPS [7] requires data fusion methods to be used [8] to achieve docking functionality to recalibrate an AGV to a specific AS if needed.

The aim of the article is to summarize the existing possibilities of using sensor and data fusion for the effective use of the AGVs that cooperate with the IoT subsystems and industrial manufacturing environments. The main contribution of this paper is the analysis of the challenges facing the implementation of internal logistics systems based on AGV, with a particular focus on the challenges related to data fusion:

(i)   Dynamic configuration of the data fusion methods - AGVs require a dynamic change in the way of cooperation with various ASs (Sect. 2). Different sensors on each AS and AGV necessitate the usage of suited methods of data fusions.

(ii) Wireless real-time communication – the specific docking example (Sect. 2.2) requires real-time data exchange. Otherwise, data fusion will not support the accuracy of the docking procedure,

(iii) The integration of data streams produced by many IoT devices, AGVs, ASs, and extracted from other systems in a smart factory provides a broader view of the processed data and supports efficient and accurate data mining.

The main challenge is to find a way to prepare data fusion that depends on the available sensors on the AGV and AS, which could be recognized by the M2M communication methods and improved by the data analysis methods based on real-time information from data streams.

The paper is organized as follows: the second section presents the research challenges associated with the fusion between an AGV, the IoT subsystems, and the environment. The third section describes the ontology-based approach to implementing data fusion. The fourth section presents the methods of data fusion for AGV solutions. The conclusions are presented in the fifth section.

2   Research Challenges Related to the Fusion Between an AGV, the IoT Subsystems, and the Manufacturing Environment

The internal transport systems for routing and supervising AGVs often use navigation systems. However, when an AGV reaches the specified production station, it usually has to dock there automatically. The industrial manufacturing environment uses many different types of sensors implemented in the IoT subsystems [9], CPPS [10], and AGV solutions. Although these sensors have different properties, some of them can be used to determine the position of an AGV and its distance to the objects in an industrial environment. Using data from several sensors can increase the precision of determining the position of an AGV. However, precise positioning is not required at all times, e.g., if an AGV is moving in relatively vast halls, the speed may increase, but the current position can be acquired using odometry. If the position accuracy is not high enough, then the Inertial Measurement Unit (IMU) can be switched on to support the less accurate odometry sensors in localizing it.

The possibility of powering some modules on and off leads to another big issue associated with the power management unit (PMU). It must be emphasized that a platform has a battery with a limited capacity – one of the goals is to increase the operating time for a platform. Having an up-to-date position and the platform, the IMU can be supplied by or disconnected from the battery. When connected, the dead reckoning algorithm should be used. It is assumed that there are some reference points for positioning in the local coordinate system in the working area of the platform. There is a wide range of positioning systems available for industrial use. They are starting from high-cost fast video systems and ending with ultrasound low-cost distance devices. Thus, the sensor selection is important from two aspects: one is the battery saving aspect, and the second is the positioning accuracy aspect. There are several advantages and disadvantages of both odometry and the dead reckoning algorithm [11]. The authors believe that the most crucial disadvantage is the lack of an absolute position. Therefore, other indoor positioning techniques [12] can be used for this purpose. Taking into consideration the navigation of a platform, wireless sensors based on RSSI (Radio Signal Strength Indicator) or ToF (Time of Flight) can be investigated [13]. Moreover, systems based on the methods mentioned above should also be assessed on their battery consumption. This can be achieved by selecting the most accurate sensor/system or combining the data obtained from several sensors when determining a position.

The main challenge is to support the optimal use of all available sources of information that can complement each other through sensor fusion. The goal is to integrate an AGV [14] and sensors in the manufacturing environment and to prepare a functionality to determine the position of an AGV, docking [15], and to provide support to Machine to Machine (M2M) communication [16, 17] with other subsystems.

2.1 Determining the Position of an AGV

Precisely determining the position of an AGV enables the navigation system [18] to give orders correctly and increases the accuracy of the movement of an AGV in an industrial manufacturing environment [19]. It also leads to a reduction in costs in terms of battery consumption and human interference in the path correction of an AGV. The time required for human support reduces usability and increases the overall cost of implementing a system. Hence, the next challenge is to have a precise enough position and location of a platform. Therefore, one of the first steps in precise positioning is the kinematics of an AGV [5]. Next, other systems may be used to support any other assumptions or limitations caused by the AGV platform. For the dead reckoning algorithm, information obtained from the accelerometer, gyroscope, encoders, and sometimes a magnetometer is most often used in the navigation system [20]. Encoders or hall sensors enable the speed [21] of each wheel to be measured. If an AGV has a differential drive system, information about the speed of each wheel can be used to determine the overall speed of the AGV and its relative position and heading. An accelerometer can be used to measure the speed of an AGV, which can be obtained by integrating the AGV acceleration in time. To avoid a quantization error, it is recommended that an appropriate filtration method be used [22] (the method of filtration is another big issue with the system) as well as, e.g., Simpson's rule as the method for the numerical integration rather than simply multiplying the acceleration by the elapsed time. However, the speed of the vehicle that is determined by the accelerometers has a relatively large error relative to the accuracy of the encoders. Therefore, it is better to have some other system assumptions for positioning (or introducing higher-quality sensors). Theoretically, it is possible to calculate the distance based on the accelerometer data, but it is rarely used in practice due to its inaccuracy. The gyroscope can be used to measure the angular speed of an AGV to compensate for the yaw rate errors of an AGV caused by wheel slip. The measurements from the gyroscopes and magnetometers can be fused to estimate the accurate heading of an AGV and compensate for errors caused by the high electromagnetic pollution in an industrial manufacturing environment. Dead reckoning navigation tends to accumulate the errors such as inaccuracies in the encoder readings due to quantization, wheel slip, or IMU noise. Therefore, they need to be fused to estimate the AGV state better.

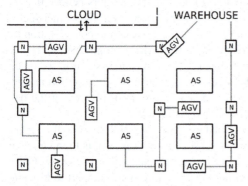

Fig. 1. Autonomous guided vehicles in manufacturing environment (N - NFC module).

Each AGV is additionally equipped with a set of distance sensors, e.g., single-beam LiDAR and 2D LiDAR. To create the map and improve the position's determination, the SLAM technique [23], based on LiDAR technology, can be used. The information from other subsystems (the IoT, CPPS) can also be used to correct the calculated position. These subsystems are often equipped with sensors to determine the position, e.g., RFID tags, NFC modules, magnetic or color markers. The trilateration method [24] enables the position of the object to be calculated in a two-dimensional plane by referring to three specified points. The most advanced subsystems (e.g., AS) enable distance measurements by LiDARs or cameras. An AGV can obtain this additional information in a manufacturing environment, thus recognizing specific markers in space and receiving information about its position [13] (Fig. 1).

2.2 Precisely Docking an AGV to an Assembly Station

Docking in the specified manner and site enables the orientation time of a robot or an automatic production loading station associated with the delivered parts to be shortened. This approach creates further challenges, the solution of which should enable increasing the precision of the orientation of the robot to be increased and any errors that require handling by production staff to be eliminated. Because of the possible presence of surface irregularities, the location of an AGV relative to the AS in both the vertical and horizontal positions must be determined (Fig. 2).

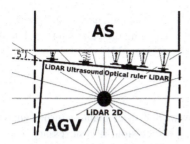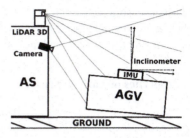

Fig. 2. Docking an AGV to an assembly station (top view on the left, side view on the right)

The distance of an AGV from an AS can be measured using, e.g., single-beam LiDAR, 2D LiDAR, an optical ruler, or ultrasound. The angle of the deviation of an AGV to an AS can be additionally measured based on these measurements. However, these sensors have different properties regarding the measurement and accuracy of distance [25]. Therefore, methods that enable the data obtained from a sensor to be selected with the highest accuracy or using data fusion methods [26] to obtain the highest degree of accuracy from several sensors. Using these methods will determine the range of motion of an AGV, ensuring a more accurate docking. Determining the horizontal position of an AGV enables an inclinometer to be used. The information about position of an AGV can be made available to the AS, which will speed up the orientation of a robot or an automatic production loading station to the delivered parts.

Integrating sensors into manufacturing systems enables multisensor data fusion to be prepared [27]. An AS can also be equipped with several sensors that enable the working status of the production, machines, and the surrounding environment, including the docking of an AGV or loading process, to be verified. An AS is usually at ground level. A constant point of reference can be obtained by precisely leveling it, e.g., 3D LiDAR or camera sensors [28]. In fact, it is possible to obtain more accurate measurements of the angle of an AGV to an AS. In addition to measuring the distance, a camera also enables the orientation to be determined and the delivered parts to be recognized. Data fusion, which is based on information from both the AGV and AS systems can be used to increase the accuracy of positioning an AGV. Both enable distance and angle measurements to be obtained and enable the vertical and horizontal position of an AGV to an AS to be determined.

A mobile AGV can also be equipped with collaborative robot (CR) manipulators to perform different types of operations at an AS, e.g., picking up objects for transport or manipulating or placing new parts into the AS. These CRs can be used as additional sensor systems to aid in the precise docking of an AGV to an AS. A CR has very accurate proprioceptive sensors for determining the position and orientation of the different links and joints on a robot. By accurately positioning the robot tool-center-point (TCP) on predefined points on an AS, the position and orientation of an AGV can be determined. This positioning does not necessarily have to be physical but can be performed using the camera systems on a robot [29].

Developing methods that enable changes in the area of production lines to be detected makes it possible to monitor the industrial process using data fusion techniques [30] and virtual sensing techniques [31]. This non-invasive method enables an operation or product quality in an industry to be optimized by measuring the parameters in dynamic systems in which stationary and mobile systems cooperate [32].

2.3 Properties of Various Sensors

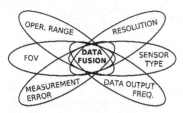

To move in an industrial manufacturing environment with high precision, it is necessary to use a location engine based on the positioning system being used. On a fundamental level, the distance between some reference points and an AGV must be determined. Although the distance measurements can be investigated on several levels, the authors focused on the commonly used sensors in this paper. Positioning and localization systems can be used to map the environment and to navigate and avoid col-

Fig. 3. Data fusion diagram – the problem of sensors' different specifications and features.

lisions. There are many different sensors for measuring distance, and each of them has various features (Fig. 3). Although that variety enables the suitable sensors for a specific task to be selected, on the other hand, there is a challenge when fusing many sensors with different specifications. Some of them (e.g., LiDARs) are very sensitive to ambient light or the color of the surface. They have different operating ranges, fields of view (FOV), measuring resolution, and accuracy, which can differ across the operating range. Ultrasound and radar sensors are not sensitive to ambient light or the color of any obstacles, but they have lower measurement and FOV resolutions.

Cameras with depth-sensing can also be used for AGV navigation and docking. They can help to map the environment and take part in understanding it using image recognition techniques. IMUs (Inertial Measurement Unit) can be used to navigate and position an AGV system. IMUs and inclinometers can also be used to determine the orientation of an AGV, which is also crucial for the accuracy of an AGV performing specific processes. An off-balance AGV, in some situations, must be leveled. Otherwise, it can lead to errors when performing a task or even cause damage to other systems near an AGV. Like distance measuring sensors, IMUs and inclinometers have different operating ranges, measurement resolution, and accuracy. To obtain an accurate estimate of an orientation and position, sensor fusion should be used. Thus, one of the challenges is to develop a distributed computer system architecture [33] to integrate AGVs [14] and the required sensors, which will enable data fusion.

3   The Ontology-Based Approach for Implementing Data Fusion

To enable the data fusion proposed in Sect. 2, it is necessary to collect and combine information from many different sensors. This information is available in various production subsystems, which means that it is represented in various formats with different communication capabilities and services. An ontological approach for information modeling can be used to exchange and merge information in heterogeneous distributed information systems, ensuring the unambiguous determination of the meaning of the available information and services and enables automatic communication between the individual system nodes. Such models can be used for flexible and dynamic communication between the system nodes according to the Machine to Machine (M2M) paradigm [34]. Ontology is how specific information such as a model of the entities and interactions in a specific area of knowledge is represented. Ontology enables the machine (independent of a human decision) interpretability of information containing the parameters and the relations between data [35]. Regarding the use case here, the ontology should describe the data and services used to navigate an AGV. The data and services must be selected according to the current position of the AGV and the tasks that are to be performed. The position of an AGV limits the list of available sources of information. The sensors must be selected based on their physical properties such as the detection method (1D, 2D, 3D), range, accuracy, scanning frequency, etc. [36]. On the other hand, the services must be adjusted to the operation to be performed by an AGV, e.g., avoiding an obstacle, preventing a collision with another AGV, or docking to a production station.

The ontology should be compatible with the contemporary models that are used in agile manufacturing, such as the Reference Architecture Model for Industry 4.0 (RAMI4.0), which defines the high-level schemas for manufacturing systems that are currently being developed [37, 38]. Moreover, the communication middleware should support a seamless connection between the entities and support the meta-information that enables the information to be interpreted correctly by considering the required presentation context. OPC UA is one of the communication solutions that has been widely accepted in the industry. It offers an object-based and service-oriented communication middleware that supports the exchange of information and organizes the information models [39]. OPC UA considers RAMI4.0 to be one of the key enabling technologies. The information and metainformation are organized in an object-oriented manner in which the relevant type definitions determine the structure and meaning of each data item. The basic OPC UA types are defined by the standard and are used to arrange the variables, objects, data, and references that show the relationships between the pieces of information. The model can easily be expanded according to the requirements of the application by using the inheritance mechanism [40].

3.1 M2M Communication

To navigate an AGV, the ontology should describe services (functions) that are offered, the available data, including any online measurements and sensor's properties, and possible communication modes. Data from individual CPPS, the IoT, and AGV subsystems are often collected and processed in supervisory systems, data centers, or cloud platforms [41]. These systems also enable the required information to be exchanged between subsystems based on the M2M communication methods [42]. Additionally, individual subsystems, e.g., an AGV or AS, can exchange information directly based on reliable and time-determined M2M methods of communication [43].

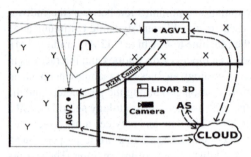

Fig. 4. M2M communication between AGVs and an industrial manufacturing environment

Using 2D LiDAR enables information about the surroundings and objects in the nearest neighborhood to be obtained. On the other hand, a single-beam LiDAR enables the continuous observation of the road, often over a broader range than 2D LiDAR, and enables any objects that suddenly appear on the path of an AGV to be detected. There will always be cases where some of the measured areas are obscured by other objects, which means that the AGV will not be able to detect any other approaching objects or other AGVs. For this reason, the route of an AGV is mapped to the superordinate navigation system. Additionally, this system may take into account information about the movement of objects from other subsystems. For example, an AS and other IoT subsystems can also share information about moving objects detected in their environment using the IoT cloud solutions [44]. As a result, the navigation system will have information about moving AGVs and other objects. This information can reconcile the route, speed, and sequence of movement corrections and warn AGVs about the possibility of a collision because of other approaching objects or AGVs. AGVs can also communicate directly with other AGVs, AS, the IoT devices and share information about warnings or even moving objects based on M2M communication (Fig. 4). As a result, a local AGV navigation system will be able to combine data from other subsystems, map additional knowledge about the nearest surroundings, and use this information to predict possible collisions.

3.2 OPC UA-Based Communication for LiDAR

The use case of sensor fusion presented in Fig. 4 requires that information be exchanged between two AGVs and a production station. The M2M communication can be performed at a low level according to the communication services available for specific sensors or be changed into high-level services defined according to the ontological approach. In the first case, the LiDAR will provide information about a cloud of points that includes the angle of the LiDAR beam, the distance to the obstacle, and the reflectance factor. This information must be processed by the recipient, which will have to convert the data that describes the cloud of points into useful information about the location of the object. To use information from an external sensor, it is necessary to know the location of the remote LiDAR and the format in which the data is presented. Moreover, the head of a LiDAR rotates several times per second, and during this time, several hundred to several thousand measurements are taken. In the case of low-level M2M communication, some high throughput, real-time transmission channels have to be built on wireless communication.

The second approach replaces the data exchange of the raw data measured by sensors with ontology-defined location services (Fig. 5). In the first step, the AGV interested in the location data sends a location request using the Locate service provided by LiDAR's communication middleware. If it is possible to locate the AGV, the Locate service creates a new variable containing the AGV's position and this is identified by the ID used in the service request. Otherwise, the LiDAR returns an error code. In this case, no new measurement is created.

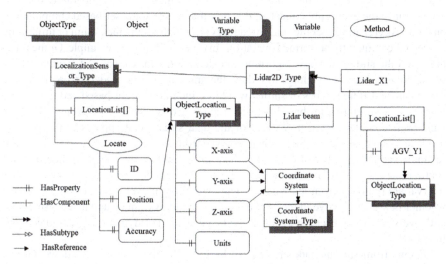

Fig. 5. Use case of OPC UA based ontology – localization sensor.

4 Data Fusion Methods

The sensors located on AGVs, ASs, and other IoT devices generate a series of events that form continuous data streams. Various sensor systems provide separate data streams. One of the challenges identified for AGVs logistics is the integration of the data streams. This integration can be performed within the data fusion process [45] to produce a complete set of data for further analysis within data mining processes. Data fusion can be achieved by joining events from the individual data streams that are produced directly by AGVs, ASs, and the IoT devices, and also those that are collected in a CPPS, a Distributed Control System (DCS), and Supervisory Control And Data Acquisition (SCADA) systems by using a stream joining operation. This operation can be implemented in various ways and in many places, i.e. (1) on the IoT device that are located on the AGV (2) another IoT device that aggregates the data streams or plays a role in the Edge/fog gateway for transmitting the data to a data center, or (3) in the data center itself.

In all of these cases, the joined data streams should include a common attribute that can be used in the join operation. Frequently, the fusion of data streams is performed in the time domain, and therefore the timestamp that accompanies all of the collected events is used as a common attribute to assess the proximity of the events. Based on the timestamp, the real-time fusion of specific events can be performed. If the events are generated in close synchrony, it is enough to pair them (in the case of two data streams) with any new events that appear.

Asynchronous data streams raise another challenge and require more sophisticated methods that rely on timestamps, but the fusion of events is performed in time windows. One of the approaches for solving this problem proposed in the scientific literature is to use various variants of the sliding window algorithm [46, 47]. The window-based algorithms group sensor events along the timeline, which simplifies operating on these events. Once the events are collected, it is possible to use some set-based computations on them. Aggregating specific sensor readings (e.g., finding the maximum or minimum) is one of the frequently performed operations in time windows. For example, Gomes et al. [48] used the sliding window algorithm to calculate the maximum and average values of the collected sensor readings to reduce the amount of data needed to be processed. There can be different types of sliding windows, including count-based and time-based windows. A sliding count-based window retains a fixed number of events. Once the window is complete, each new event of the data stream that appears in the window displaces the oldest event, which is then removed from the window. A time-based window retains a variable number of events that had arrived within the specified time interval. As time passes, the IoT events that have been in the window longer than the specified interval are removed from the window. The expiration of events happens regardless of whether new events arrive in the window or not. Windows can be updated continuously with every incoming event or cyclically with a specified or dynamically assigned cycle time [49].

Events from various data streams can be joined in these sliding windows. Gomes et al. [48] proposed the XGreedyJoin algorithm, which operates on the sensors and joins data streams. To do this, the algorithm uses a join tree with a count-based sliding window for every stream. The algorithm can run in a single data stream processing unit (e.g., in a data center) or in a distributed sensor network. However, stream fusion can be computationally demanding in distributed sensor networks, which was observed and reported by Zhuang et al. [50]. The challenge especially appears when joining more than two data streams for a sensor network. Multiple data streams may increase the pressure on the IoT devices and field gateways, which usually have limited computational resources (CPU and memory). Therefore, Zhuang et al. proposed two approaches for solving the problem, i.e., the All In One (AIO) and Step By Step (SBS) approach. The AIO approach assumes that all data streams are processed and combined in a single stream processing job. The SBS approach distributes the joining operation into many steps that combine pairs of data streams in each step. Both approaches are implemented in the Apache Samza framework.

Recent works in the area of fusing asynchronous sensor events show that using fuzzy sets could bring several benefits. For example, Malysiak et al. [51] proposed using a fuzzy umbrella join algorithm to combine the sensor data from separate IoT data lakes for Cyber-Physical Systems. Wachowicz et al. [52] proposed a fuzzy join algorithm for merging readings from various sensors while monitoring the performance of sports workouts and studying the correlations between the performance and weather conditions. The authors applied the concept of the fuzzy umbrella to join the sensor data that had been obtained from the smartwatches worn by sports amateurs with the atmospheric parameters that had been obtained from the weather services. The umbrella was spread out in overlapping time windows. However, in both solutions, the sensor events were processed after all of the collected data (i.e., not in real-time). These methods can be used, e.g., for fusing sensor data from SCADA and DCS systems in an AGV-equipped factory. An alternative solution was proposed for data in motion, i.e., for joining data streams [8], where the authors proposed a hopping umbrella to join asynchronous data streams. The hopping umbrella-based join enables the importance of specific events in the data streams to be assessed based on defined membership functions, sensor readings of higher importance to be selected in a specific case and reducing the size of the output stream. It is essential that the algorithm be used on the Edge (i.e., on an IoT device) or in a data center (the authors tested it in the cloud). Implementing the algorithm on an IoT device enables the data that is transmitted in the merged data stream to be reduced, which reduces the network traffic and the storage space consumed in a data center. These properties make the joining method suitable for the real-time fusing of data streams in the IoT devices mounted on an AGV and the events produced by the sensors located in a smart factory environment. However, the number of data streams that must be merged in such a way is challenging and requires dedicated approaches to the process implementation. Furthermore, the variety of integrated data coming from different sensors, systems, and IoT devices and the volume of data provided by these sources raise challenges of Big Data. The methods mentioned above address these challenges only partially, so there is still space for developing new techniques in the view of the Big Data problems.

5   Conclusions

Today's production ecosystems use many active subsystems that enable data about the operation of production systems and their environment to be obtained from CPPSs and IoT devices. The AGV systems used in this environment should actively participate in the exchange of data between these systems. Sharing additional information may be used by an AGV to increase the quality of its service. Such a solution can be obtained using data fusion methods based on information from multiple subsystems. However, achieving such solutions poses many challenges and related research. In this article, we presented some considerations on the issues associated with the data fusion methods for integrating an AGV and an AS with the required sensors and determining the position, docking, and communication methods of an AGV with other subsystems a smart industry environment. These challenges lead to reorganizing the methods used and adjustments of particular algorithms to be implemented in real-time industrial environments.

Acknowledgments. The research leading to these results received funding from the Norway Grants 2014–2021, which is operated by the National Centre for Research and Development under the project "Automated Guided Vehicles integrated with Collaborative Robots for Smart Industry Perspective" (Project Contract no.: NOR/POLNOR/CoBotAGV/0027/2019 -00) and partially by the Polish Ministry of Science and Higher Education Funds for Statutory Research.

References

1. Wang, S., Wan, J., Li, D., Zhang, C.: Implementing smart factory of Industrie 4.0: an outlook. Int. J. Distrib. Sens. Netw. **12**(1), 3159805 (2016)

2. European Commission. A Manufacturing Industry Vision 2025, European Commission (Joint Research Centre) Foresight study (2013)
3. Shafiq, S.I., Sanin, C., Szczerbicki, E., Toro, C.: Virtual engineering object/ virtual engineering process: a specialized form of cyber physical system for Industrie 4.0. Procedia Comput. Sci. **60**, 1146–1155 (2015)
4. Botta, A., de Donato, W., Persico, V., Pescapé, A.: Integration of cloud computing and internet of things: a survey. Future Gener. Comput. Syst. **56**, 684–700 (2016)
5. Shi, D., Mi, H., Collins, E.G., Wu, J.: An indoor low-cost and high-accuracy localization approach for AGVs. IEEE Access **8**, 50085–50090 (2020)
6. Realyvásquez-Vargas, A., et al.: Introduction and configuration of a collaborative robot in an assembly task as a means to decrease occupational risks and increase efficiency in a manufacturing company. Robot. Comput.-Integr. Manuf. **57**, 315–328 (2019)
7. Kuc, M., Sułek, W., Kania, D.: FPGA-oriented LDPC decoder for cyber-physical systems. Mathematics **8**, 723 (2020)
8. Mrozek, D., Tokarz, K., Pankowski, D., Małysiak-Mrozek, B.: A hopping umbrella for fuzzy joining data streams from IoT devices in the cloud and on the edge. IEEE Trans. Fuzzy Syst. **28**, 916–928 (2019)
9. Ji, Z., Ganchev, I., O'Droma, M., Zhao, L., Zhang, X.: A cloud-based car parking middleware for IoT-based smart cities: design and implementation. Sensors **14**, 22372–22393 (2014)
10. Opara, A., Kubica, M., Kania, D.: Methods of improving time efficiency of decomposition dedicated at FPGA structures and using BDD in the process of cyber-physical synthesis. IEEE Access **7**, 20619–20631 (2019)
11. Grzechca, D., et al.: How accurate can UWB and dead reckoning positioning systems be? comparison to SLAM using the RPLidar system. Sensors **20**, 3761 (2020)
12. Paszek, K., Grzechca, D., Tomczyk, M., Marciniak, A.: UWB positioning system with the support of MEMS sensors for indoor and outdoor environment. Journal of Communications, vol. 15 (2020)
13. Grzechca, D.E., Pelczar, P., Chruszczyk, L.: Analysis of object location accuracy for ibeacon technology based on the RSSI path loss model and fingerprint map. Int. J. Electron. Telecommun. **62**(4), 371–378 (2016). https://doi.org/10.1515/eletel-2016-0051
14. Grzechca, D., Paszek, K.: Short-term positioning accuracy based on mems sensors for smart city solutions (2019). https://doi.org/10.24425/MMS.2019.126325
15. Roth, H., Schilling, K.: Navigation and docking manoeuvres of mobile robots in industrial environments. In: IECON 1998 Proceedings of the 24th Annual Conference of the IEEE Industrial Electronics Society (Cat. No. 98CH36200), pp. 2458–2462. IEEE, Aachen, Germany (1998)
16. Thota, P., Kim, Y.: Implementation and comparison of M2M protocols for internet of things. In: 2016 International Conference ACIT-CSII-BCD, pp. 43–48. IEEE, Las Vegas, NV, USA (2016)
17. Hanzel, K., Paszek, K., Grzechca, D.: The influence of the data packet size on positioning parameters of UWB system for the purpose of tagging smart city infrastructure. Bulletin of the Polish Academy of Sciences. Technical Sciences, vol. 68 (2020)
18. Tokarz, K., Czekalski, P., Sieczkowski, W.: Integration of ultrasonic and inertial methods in indoor navigation system. Theor. Appl. Inform. **26**, 107–117 (2015)
19. Ziebinski, A., Cupek, R., Nalepa, M.: Obstacle avoidance by a mobile platform using an ultrasound sensor. In: Nguyen, N.T., Papadopoulos, G.A., Jędrzejowicz, P., Trawiński, B., Vossen, G. (eds.) ICCCI 2017. LNCS (LNAI), vol. 10449, pp. 238–248. Springer, Cham (2017). https://doi.org/10.1007/978-3-319-67077-5_23
20. Han, Y., Wei, C., Li, R., Wang, J., Yu, H.: A novel cooperative localization method based on IMU and UWB. Sensors **20**, 467 (2020)

21. Ziebinski, A., Bregulla, M., Fojcik, M., Kłak, S.: Monitoring and controlling speed for an autonomous mobile platform based on the hall sensor. In: Nguyen, N.T., Papadopoulos, G.A., Jędrzejowicz, P., Trawiński, B., Vossen, G. (eds.) ICCCI 2017. LNCS (LNAI), vol. 10449, pp. 249–259. Springer, Cham (2017). https://doi.org/10.1007/978-3-319-67077-5_24

22. Guan, H., Li, L., Jia, X.: Multisensor fusion vehicle positioning based on Kalman Filter, pp. 296–299. IEEE (2013)

23. Wen, S., Othman, K., Rad, A., Zhang, Y., Zhao, Y.: Indoor SLAM using laser and camera with closed-loop controller for NAO humanoid robot. Abstr. Appl. Anal. **2014**, 1–8 (2014)

24. Fang, B.T.: Trilateration and extension to Global Positioning System navigation. J. Guid. Control Dyn. **9**, 715–717 (1986)

25. Grzechca, D., Hanzel, K., Paszek, K.: Accuracy analysis for object positioning on a circular trajectory based on the UWB location system. In: 2018 14th International Conference on Advanced Trends in Radioelecrtronics, Telecommunications and Computer Engineering (TCSET), pp. 69–74. IEEE, Lviv, Ukraine (2018)

26. Sidek, O., Quadri, S.A.: A review of data fusion models and systems. Int. J. Image Data Fusion **3**, 3–21 (2012)

27. Liggins II, M., Hall, D., Llinas, J.: Handbook of Multisensor Data Fusion: Theory and Practice. CRC Press, Boca Raton (2017)

28. Budzan, S., Kasprzyk, J.: Fusion of 3D laser scanner and depth images for obstacle recognition in mobile applications. Opt. Lasers Eng. **77**, 230–240 (2016)

29. Bjerkeng, M., Pettersen, K.Y., Kyrkjebø, E.: Stereographic projection for industrial manipulator tasks: theory and experiments, pp. 4676–4683. IEEE (2011)

30. Błachuta, M., Czyba, R., Janusz, W., Szafrański, G.: Data fusion algorithm for the altitude and vertical speed estimation of the VTOL platform. J. Intell. Robot. Syst. **74**, 413–420 (2014)

31. Liu, L., Kuo, S.M., Zhou, M.: Virtual sensing techniques and their applications. In: 2009 International Conference on Networking, Sensing and Control, pp. 31–36. IEEE, Okayama, Japan (2009)

32. Lee, M.C., Park, M.C.: Artificial potential field based path planning for mobile robots using a virtual obstacle concept. In: Proceedings 2003 IEEE/ASME International Conference on Advanced Intelligent Mechatronics AIM 2003, pp. 735–740. IEEE, Kobe, Japan (2003)

33. Ziebinski, A., Cupek, R., Piech, A.: Distributed control architecture for the autonomous mobile platform. Thessaloniki, Greece, p. 080012 (2018)

34. Weyrich, M., Schmidt, J.-P., Ebert, C.: Machine-to-Machine communication. IEEE Softw. **31**, 19–23 (2014)

35. Cupek, R., Ziebinski, A., Fojcik, M.: An ontology model for communicating with an autonomous mobile platform. In: Kozielski, S., Mrozek, D., Kasprowski, P., Małysiak-Mrozek, B., Kostrzewa, D. (eds.) BDAS 2017. CCIS, vol. 716, pp. 480–493. Springer, Cham (2017). https://doi.org/10.1007/978-3-319-58274-0_38

36. Kohlbrecher, S., von Stryk, O., Meyer, J., Klingauf, U.: A flexible and scalable SLAM system with full 3D motion estimation. In: 2011 IEEE International Symposium on Safety, Security, and Rescue Robotics, pp. 155–160. IEEE, Kyoto, Japan (2011)

37. Hankel, M., Rexroth, B.: The reference architectural model industrie 4.0 (rami 4.0). ZVEI (2015)

38. Cupek, R., Drewniak, M., Ziebinski, A.: Information models for a new generation of manufacturing systems - a case study of automated guided vehicle. In: 2019 IEEE International Conference on Systems, Man and Cybernetics SMC, pp. 858–864. IEEE, Bari, Italy (2019)

39. Lang, J., Iwanitz, F., Burke, T.: OPC from Data Access to Unified Architecture. OPC Found. Softing (2010)

40. Cupek, R., Drewniak, M., Ziebinski, A., Fojcik, M.: Digital twins for highly customized electronic devices – case study on a rework operation. IEEE Access **7**, 164127–164143 (2019)

41. Varghese, A., Tandur, D.: Wireless requirements and challenges in Industry 4.0. In: 2014 International Conference on Contemporary Computing and Informatics (IC3I), pp. 634–638. IEEE, Mysore, India (2014)

42. Elgazzar, M.H.: Perspectives on M2M protocols. In: 2015 IEEE Seventh International Conference on Intelligent Computing and Information Systems (ICICIS), pp. 501–505. IEEE, Cairo, Abbassia, Egypt (2015)

43. Fadlullah, Z.M., Fouda, M.M., Kato, N., Takeuchi, A., Iwasaki, N., Nozaki, Y.: Toward intelligent machine-to-machine communications in smart grid. IEEE Commun. Mag. **49**, 60–65 (2011)

44. Cheng, Y., Tao, F., Xu, L., Zhao, D.: Advanced manufacturing systems: supply–demand matching of manufacturing resource based on complex networks and Internet of Things. Enterp. Inf. Syst. **12**(7), 1–18 (2016)

45. Trawiński, B., Smętek, M., Lasota, T., Trawiński, G.: Evaluation of fuzzy system ensemble approach to predict from a data stream. In: Nguyen, N.T., Attachoo, B., Trawiński, B., Somboonviwat, K. (eds.) ACIIDS 2014. LNCS (LNAI), vol. 8398, pp. 137–146. Springer, Cham (2014). https://doi.org/10.1007/978-3-319-05458-2_15

46. Golab, L., Özsu, M.T.: Processing sliding window multi-joins in continuous queries over data streams. Elsevier, pp. 500–511 (2003)

47. Hammad, M.A., Aref, W.G., Elmagarmid, A.K.: Stream window join: tracking moving objects in sensor-network databases. In: 15th International Conference on Scientific and Statistical Database Management 2003, pp. 75–84. IEEE Computer Society, Cambridge, MA, USA (2003)

48. Gomes, J., Choi, H.-A.: Adaptive optimization of join trees for multi-join queries over sensor streams. Inf. Fusion **9**, 412–424 (2008)

49. Ji, Y., Liu, S., Lu, L., Lang, X., Yao, H., Wang, R.: VC-TWJoin: a stream join algorithm based on variable update cycle time window. In: 2018 IEEE 22nd International Conference on Computer Supported Cooperative Work in Design (CSCWD), pp. 178–183. IEEE, Nanjing, China (2018)

50. Zhuang, Z., Feng, T., Pan, Y., Ramachandra, H., Sridharan, B.: Effective multi-stream joining in apache samza framework. In: 2016 IEEE International Congress on Big Data, pp. 267–274. IEEE, San Francisco, CA, USA (2016)

51. Malysiak-Mrozek, B., Lipinska, A., Mrozek, D.: Fuzzy join for flexible combining big data lakes in cyber-physical systems. IEEE Access **6**, 69545–69558 (2018)

52. Wachowicz, A., Małysiak-Mrozek, B., Mrozek, D.: Combining data from fitness trackers with meteorological sensor measurements for enhanced monitoring of sports performance. In: Rodrigues, J.M.F., Cardoso, P.J.S., Monteiro, J., Lam, R., Krzhizhanovskaya, V.V., Lees, M.H., Dongarra, J.J., Sloot, P.M.A. (eds.) ICCS 2019. LNCS, vol. 11538, pp. 692–705. Springer, Cham (2019). https://doi.org/10.1007/978-3-030-22744-9_54

Dynamic Pricing and Discounts by Means of Interactive Presentation Systems in Stationary Point of Sales

Marcin Lewicki[1]([✉]) [ID], Tomasz Kajdanowicz[2] [ID], Piotr Bródka[2] [ID],
and Janusz Sobecki[2] [ID]

[1] Poznań University of Business and Economics, Poznań, Poland
marcin.lewicki@ue.poznan.pl
[2] Wrocław University of Science and Technology, Wrocław, Poland

Abstract. The main purpose of this article was to create a model and simulate the profitability conditions of an interactive presentation system (IPS) with the recommender system (RS) used in the kiosk. 90 million simulations have been run in Python with SymPy to address the problem of discount recommendation offered to the clients according to their usage of the IPS.

Keywords: Consumer behaviour · Recommendation system · Discount · Price discrimination

1 Introduction

Nowadays, with the constantly increasing competition from the Internet stores, convincing a consumer to buy something from a stationary point of sale (PoS) is becoming a lot harder. However, with the help of new technologies: tablet recommendation algorithms, automatic customer profiling, and machine learning, even a stationary PoS can present an offer which is tailored to specific consumer needs. When these Pos form a network, then we also could profit from the application of Internet technologies to exchange online specific marketing information. Another advantage of modern technology is the possibility to build models which allow, using the knowledge about a consumer behaviour, to recommend the seller appropriate marketing actions in a real environment. Using these models, we can estimate the potential value of a new technology or an approach without conducting expensive trials, which is especially important during the earliest stages of idea development.

Recommender systems (RS) are designed to deliver customised information for very differentiated users in many different domains [21]. In [29] the following

This research was partially founded by Wroclaw University of Science and Technology and Poznan University of Economics and Business statutory funds, and BiWISS grant agreement no INNOTECHK3/IN3/56/225874/NCBR/15 of European Union Regional Development Fund Third Programme Innotech, IN-TECH.

M. Paszynski et al. (Eds.): ICCS 2021, LNCS 12745, pp. 609–622, 2021.
https://doi.org/10.1007/978-3-030-77970-2_46

areas of explainable RS have been distinguished: e-commerce, POI, social, and multimedia recommendations. It is worth to note that the first online recommender system (ORS) was Tapestry [11], which was designed to filter emails. In the recommender systems taxonomy proposed by Montaner et al. [21] we can consider the following dimensions: profile generation and maintenance, and profile exploitation. The first dimension contains the following elements: user profile representation, initial profile generation, profile learning technique, and relevance feedback. The second dimension considers information filtering method, user profile–item matching technique, user profile matching technique, and profile adaptation technique. One of these elements is of special importance, i.e., the filtering method. We can distinguish three basic filtering methods: demographic filtering (DF), content-based filtering (CBF), and collaborative filtering (CF). Nowadays also a hybrid approach, using two or more basic methods is quite often used. In the e-commerce applications of ORS that are based on customers' online history, the most common filtering methods are CBF and CF [15]. Besides research on RS development, we can also find examples of studies concerning the impact of RS on the customer's purchase decision. For example, in a recent study [13] the comparative analysis of pricing promotions that formulates the willingness to buy in fashion e-shops from two distinct markets in Russia and Sweden is given. In other study [15] application of price promotion together with product recommendation should be considered for optimal profits. It tries to find the answers to the following questions: (i) how to determine the recommended product and how to price the discounted product given the specific customers shopping attitudes, (ii) what is the impact of recommendation on the discount, (iii) how changes in the cost of the product, recommendation accuracy and complementarity influence the e-tailer's profit. In previous years, research concentrated on the application of a shopbot that finds savings for a customer on product promotion [10] as well as a recommendation system that enhances the profit of e-tailers [28].

The main research objective of this article is to use the present state of knowledge about consumer behaviour in the context of dynamic pricing and discounts (including primarily their influence on a consumer intention to buy) to build a model which will allow to understand the possible impact of the interactive presentation system (IPS) developed within the Polish National Centre for Research and Development grant entitled "Business Intelligence Tools for Virtual Network of Points of Sale" (BIWiSS) on the increase of profits in the stationary point of sales where the system is installed.

The article is structured as follows. In Sect. 2 a general concept of interacting presentation systems is illustrated. Next, the role of discounts in retail is addressed (Sect. 3) and problems of price discrimination (Sect. 4) The experimental environment is presented (Sect. 5). In Sect. 5 the key findings of the experiment are described and discussed. The paper is concluded in Sect. 7.

2   Interactive Presentation System

Presently, we are surrounded by modern digital interactive media almost everywhere, especially in public places. There are many types of interactive media applications and configurations, which are used in many application areas such as advertisement, commerce, entertainment, and education. These applications are using many different technologies such as touch screens, kinetic interfaces, interactive walls and floors, Augmented Reality, Virtual Reality, and many others [3]. One of the forms of modern interactive media is digital signage that's main application is advertisement targeted to large audiences at public venues [5]. Many authors report that they deliver an effective form of media and offer many improvements in consumers perceptions of the presented information. The Digital Signage Installations may be characterized by the following factors [3]: optimal communicativeness, utilitarianism of solutions, synergic links of physical form with software, hybrid character of the structure, localisation in architectonic spaces, the phenomenon of attracting spectators, and ambient character. The hybrid character of structure from the above-mentioned property list defines the mixing of building materials and the information technology that is placed in the real space. A sample Digital Signage installed in a shopping mall produces the revenue stream by charging advertisers improving 'atmosphere' and image by delivering interesting, informative and entertaining content at the same time [22]. Application of Digital Signage systems has many advantages over traditional presentation media: possibility to present dynamic multimedia content, scalability, flexibility, interactivity of presentation and reduction of costs in a long term usage. Taking all this into account, the multimedia shopping information system has been developed. The main purpose of this system was to deliver and monitor marketing information to the customers of the network of expert points of sale of GSM industry products that includes sale and service of mobile phones and their accessories.

The system is built out of two applications: content management and presentation. The main functionality of the first application consists of the following elements: presentation project design, presentation project management, presentation system management that consists of multiple consoles, usage, data gathering, rapport generation, and system administration and authorization. The second application is quite simple, and its functionality consists of two main elements: presentation of the marketing offer on touch screens and the tracking of the particular offer selection. The system also enables real-time monitoring of the customer usage of the presentation console by the salesmen. They can monitor which offers the customer paid attention to and how long he or she were looking at the information on the particular products. The system also enables the ex-post analysis of viewing data in the whole network of presentation consoles considering not only the localization of the console but also the time of the event. These data may also be analysed together with the sales data coming from the cash register.

3   Discounts in Retail

The role of discounts in retail has been widely discussed in the literature over the years. Undoubtedly, one of the main reasons for this interest is the simple fact that despite the possibility of using numerous promotion tools, discounts remain the most frequently used one. It is no exaggeration to say that modern consumers live in the world of permanent discounts. An obvious problem is the answer to the question about the impact of discounts on sales. Studies have shown that price reductions affect consumer's behaviour and are often used by companies to simply boost their sales [14], nevertheless, it should be emphasised that the answer to the above issue is relatively complex, becoming a major factor stimulating further discussion in the area. It is also one of the main reasons why the extent of the individual publications on the topic is highly diversified.

Gupta and Cooper [12], based on previous studies in the 80s, examined the consumer's response to the retailer's price promotions, showing that consumers actually discount price discounts (i.e. consumer's perceptions of discounts are typically less than the advertised discount) depending on the discount level, store image, and whether the advertised product is a name brand or a store brand. Key findings from their research included the following: (i) consumer's intentions to buy do not change unless the promotional discount is above a threshold level, (ii) the threshold level for a name brand is lower than that for a store brand, (iii) there is a promotion saturation point above which the effect of discounts on changes in consumer's intention to buy is minimal.

A few years later, in 1998, Chen, Monroe and Lou [6] attempted to determine the significance of framing price promotion and how does it affect the consumer's perception and intentions to buy. The authors created an original framework providing recommendations for framing price discounts, depending on the relative price level of the promoted product and the size of a price decrease. It was suggested that: **(i)** for relatively low-price products and small price decrease, emphasis should be on providing consumers the absolute amount of price reduction, **(ii)** for relatively large price reductions on low-price items, the emphasis should be on the relative savings, **(iii)** for relatively high-price items, the emphasis should be on providing absolute savings, and if the price reduction is large, the relative amount of reduction as well.

Interesting conclusions were also presented by Alford and Biswas [2]. They studies focused on a discussion about the effects of discount level, price consciousness and sale proneness on consumer's price perception and behavioural intention. They found that a higher discount level is significantly reducing search intentions and increases perceptions of the value of the offer and intention to buy (which was consistent with some of the previous findings from the authors). Moreover, the results from their research showed that consumers with higher levels of price consciousness perceive a high level of benefits of an additional search regardless of the discount level or their level of sale proneness. Whereas this finding could be expected (the more consumers focus on paying a low price, the more they should search for the best discount), surprisingly it was noted that there was no effect of the price consciousness variable on buying intention.

The authors suggested that it is possible that consumer judgement of value and buying intention could be addressed only after the most direct judgement, i.e., whether consumers can obtain the product for a lower price at another store. Finally, Alford and Biswas suggested that a dual strategy of offering a discount to affect value perceptions and buying intention and using low price guarantees to affect search intention might be an effective strategy for retail merchants.

In 2005, Drozdenko [8] addressed important questions within the subject of discount levels, i.e., what is the risk and maximum acceptable discount levels. The study of 453 consumers who were asked to choose their own optimal discount levels (from 0–80%) for eight categories across two distribution channels (physical stores and on-line merchants) revealed strong consumer perceptions about discount risks and the trade-offs consumers make between risk and financial benefits across different product categories (regardless whether it is and on-line or off-line sale) which undoubtedly could help retailers in setting optimal discount levels. It was found that setting the highest possible level of reduction does not mean that a retailer will sell more products/services. As it turned out, only 13% of respondents selected the 80% discount level for each product and each channel, despite seeing the exact price they would pay at each level. One of the reasons for such results is the risk perceived by consumers. 88% of them attributed at least one cause for the deepest discounts. Most frequently cited were concerns about quality problems, damaged goods, or stolen goods. Additionally, it was also found that consumers opted for lower discount levels from the on-line merchant than from the physical store. Finally, there was a wide divergence by product category, with consumers selecting smaller discounts on tires and cereals and the deepest discounts on shirts. Thus, it is impossible to set universal rules about optimal discount levels that will have the same effect on the consumer's intention to buy or a perceived risk regardless of the product or the market.

A brief overview of the recent publications on the topic shows some of the directions of nowadays research. Regardless of whether the article is focused on: the analysis of the impact of price promotion strategies as a motivating factor for the manufacturer and his sales performance [7], the use of fake discounts [23],the profitability of stacked discounts [9] or the use of gambled price discounts to enhance subscription-based e-commerce services [25], it is quite obvious that a room for a discussion on the subject is still present. Moreover, taking into account all possible factors that could influence the effectiveness of discounts, it is really hard to imagine a situation in which the existing cognitive gap within this topic will be ever closed.

4   Price Discrimination

The phenomenon of selling the same commodity at different prices to different consumers, which is considered the optimal method of pricing (from the economic point of view [1,18,27]) has been well described within the economic literature over the years. Some of the most significant studies in the area came from 80's (e.g. [24, 26, 27] and did serve as a foundation for further research in the area. There are three different types of price discrimination (based on Pigou classification from 1920 [20]) i.e.: **(i)** First-degree, or perfect price discrimination involves the seller charging a different price for each unit of the good in such a way that the price charged for each unit is equal to the maximum willingness to pay for that unit, **(ii)** Second-degree price discrimination, or nonlinear pricing, occurs when prices differ depending on the number of units of the good bought, but not across consumers. That is, each consumer faces the same price schedule, but the schedule involves different prices for different amounts of goods purchased, **(iii)** Third-degree price discrimination means that different purchasers are charged different prices, but each purchaser pays a constant amount for each unit of the good bought.

Pricing strategy depends on the seller's ability to take advantage of the information exchanged during the commercial relationship. A first-degree price discrimination strategy requires that the firm is able to uniquely identify each consumer. It also requires a lot of information about the consumer's tastes and the highest willingness to pay to tailor a price to an individual consumer. The second one requires general information about the dispersion of price-sensitivities among consumers to construct an efficient menu of options. The third one requires that the seller can identify at least whether the consumer has the relevant group trait that is used for discrimination [20]. Looking at the problem of price discrimination from the perspective of the past few years it is hard not to mention the role of the new information technologies. Despite the fact that shoppers gained access to new price comparison tools, the vision of fully informed and empowered consumers did not materialise. It is mostly due to the fact that new information technologies also continue to create new opportunities for tailoring prices to individual consumers – one of them being dynamic pricing [20]. Undoubtedly the extensive information collected online from consumers stimulates price discrimination, moreover, the information itself often leads to price discrimination [19]. Some recent studies further explore personalised pricing [4] or the subject of behavioural constraints on price discrimination [16] and therefore are leading to a conclusion that price discrimination is a rather fixed point in modern marketing strategies.

5   Experimental Setup

The simulation process is representing the context of a purchase process in a mall kiosk. Mall kiosk is a small retail outlet located in the aisle of a shopping mall. In essence, a customer of a kiosk can browse the on-site visible goods and approach to the counter to purchase them.

We can assume the most straightforward scenario of the possible customer's actions in the purchase process. It can be characterized by the following steps:

1. Customer enters the kiosk shopping area.
2. It may directly specify its buying needs to the salesmen, who prepares the goods and serves with the payment process, which ends up the purchase.

(a)

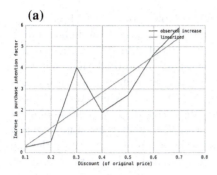

(b)

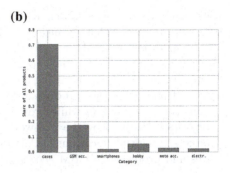

Fig. 1. Purchase intention increase as a function of discount(a) and share of particular product categories (b).

3. However, the customer can approach the interactive display prior to the counter and browse through kiosk offer.
4. While checking the available goods, the customer has the possibility to browse the list of categories of goods, listing items in the category, as well as getting to details on particular products.
5. The interactive display registers all actions, as well as the time, spend on each screen with product details. On the other side of the system, the salesmen get presented the good (list of goods) that gathered the most attention of the customer on display, measured in time units.
6. While leaving the display the customer either goes to the register or is offered a special offer by the salesmen, according to the preference revealed in the interactive display. The offer has the form of discount in original price.
7. As an effect customer can buy or not the product taking into account the offered discount.

To represent the purchase process with proper expression of customers needs and profiles it has been decided to build a model that consists of seven following parameters: (i) indicator whether a customer approaching the shopping area uses an interactive display, (ii) a category of a product that was the most observed by the customer at interactive display, (iii) price of the product that the customer watched the most at interactive display and probably is willing to buy, (iv) customer's initial purchase intention, (v) a discount given to the customer for the discovered product of interest, (vi) linearised form of discount related purchase intention increase law, based on [12], (vii) kiosk's overhead (called here margin) included in the price.

In general, the model can represent a population of customers by assuming appropriate distributions of the parameters. In the model the customer that receives a discount increases his initial purchase intention, and therefore, rises the chance to buy the product. The outcome of the parametrized model is a quantification of the overheads (sum of margins).

The idea to parametrize the model for simulation purposes is to define the probability whether a customer approaching the shopping area uses an interactive display by a binomial distribution sampled for p = (0.1, 0.7) with 0.02 step. It means that by means of distribution it is tested the behaviour of customers from 10% of them using the interactive display to 70% of the population with simulation step equal to 2%. The category of product that was most interesting for the customer using an interactive display is modeled as a random choice of one out of six product categories. It is the one which the customer spends the most time in while browsing the interactive display. The number of categories is inspired by the real-world kiosk case that sells telecommunication products in multiple locations in Poland (www.teletorium.pl). The categories are 'cases and protectors', 'GSM accessories', 'smartphones and tablets', 'hobby & sport', 'moto accessories', 'electronics'. Random choice of category is weighted by a share of a number of products in particular category to the sum of all products in all categories. It was assumed that popularity of particular category is related to the number of products in it, Fig. 1(a). The price is sampled from the previously estimated normal distribution of prices of products within each category - similarly based on mentioned real-world kiosk. It means that each category was checked regarding price distribution and Gaussian approximation of that distribution was proposed. The parameters of the distribution were the following: 'cases and protectors': $N(29, 8)$, 'GSM accessories': $N(35, 8)$, 'smartphones and tablets': $N(700, 200)$, 'hobby & sport': $N(45, 10)$, 'moto accessories': $N(80, 21)$, 'electronics': $N(50, 13)$. Customer's initial purchase intention was modeled by a parameter interpreted as the probability the product will be bought and was taken from the range (0.1, 0.7) with step 0.02. The discount parameter that is an offer of the salesmen were taken from the range (0.1, 0.7) with step 0.02. This same was simulated discounts from 10% to 70% of original prices. The linearised form of purchase intention increase thanks to discount is shown in Fig. 1(b) and expressed in **PII = 8.52 × D − 0.57**, where PII denotes purchase intention increase, D is a discount. The simulation was performed for three distinct values of margin (overhead) $\{0.3, 0.4, 0.5\}$, based on the average real-world kiosk case. The aim of the simulation was to test the proposed model for all permutations of the parameters' values. Each simulation assumes that the kiosk is visited by 1000 customers. They generate some turnover and depending on the configuration of the model (especially the discount and the margin) generate some profit or loss.

6   Results and Discussion

The simulation had to consider 89,373 permutations of parameters, for each permutation of the model parameters, it performed a behaviour simulation of 1000 customers. The simulation was performed by a programme written in Python language with the usage of SymPy module (www.sympy.org). Through careful consideration of the results, it was desired to acquire the profitability conditions of interactive display utilization in the kiosk. Therefore, the results contained roughly 90 millions of records. The presentation of them is split into the analysis of loss in margin (overhead), analysis of customers' number as well as a

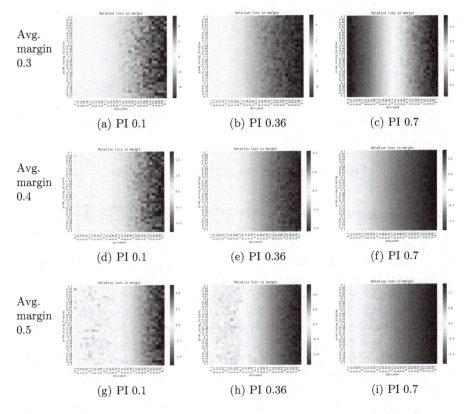

Avg. margin 0.3 — (a) PI 0.1   (b) PI 0.36   (c) PI 0.7

Avg. margin 0.4 — (d) PI 0.1   (e) PI 0.36   (f) PI 0.7

Avg. margin 0.5 — (g) PI 0.1   (h) PI 0.36   (i) PI 0.7

Fig. 2. The results of relative loss in margin for distinct permutations of parameters (PI - Purchase intention)

generalized profitability statement. End to end, the last part of this section shows when it worth to deploy the proposed interactive display.

6.1 Loss in Margin

The results presented in Fig. 2 shows the relative loss that was generated in the kiosk for distinct combinations of parameters. In the Fig. 2, we can observe the relative loss (or in some cases the profit) between two scenarios: the sum of overheads in a situation when the interactive display is not present in the kiosk and the sum of overheads when there is an interactive display available and used by some proportion of customers. By performing the second scenario, we must grant a discount to those customers who would potentially make a purchase but had too low purchase intention and while discounting products we raise it. In other words, we decrease the overhead (margin) by giving a discount to the customer that would not buy the product we have identified as interesting for that customer by means of the interactive display.

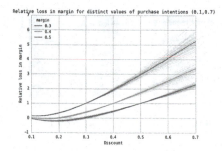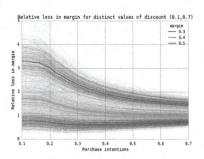

Fig. 3. Averaged results of relative loss in margin.

One straightforward observation is that regardless of what proportion of customers will be using, the interactive display loss in margin (or profit) depends only on the initial purchase intention of the customers and the granted discount. We can also observe that the higher the initial margin, the easier situation to obtain profit. Interestingly, the higher initial purchase intentions, the harder situation to obtain profit. This reveals a common-sense mechanism; you do not have to convince already convinced to purchase. We can also observe that the higher the discount, the more loss is generated. However, while discussing the findings, we must take into account that the discounting mechanism is rarely used for profit purpose [17]. It should rather gather more customers (more purchases) at a reasonable cost. Therefore, the results of the analysis of customer number should be considered in parallel.

To give an example of an interpretation of the results, we can state that in a situation when customers have very low initial purchase intention (0.1) having a high margin (0.5) giving a discount up to 30% may result in a three times higher sum of margin (overhead). On the other hand, giving the high discount (70%) with lower margin (0.3), we may loss eight times that much as it would be got without giving discounts (without interactive display).

When we average the results of relative loss in margin (in overhead) for distinct values of initial purchase intention or for distinct values of discounts, Fig. 3, we can see that only circumstances for reaching the profit regardless of initial purchase intention are achieved while having a discount lower than 20% and margin at least at 0.4 level.

6.2   Number of Customers

Similarly to the previous results, the ones presented in Fig. 4 shows the relative measure comparing two scenarios: without discounting (without interactive display) and with discounts (with interactive display). The Fig. 4 represents the relative increase of the number of purchases (number of buying customers). In general, by giving discounts, we raise the initial purchase intention and therefore increase the number of positive transactions. Generally speaking, regardless of all considered parameters, the relative increase in some customers is always

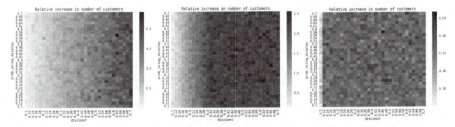

(a) Purchase intention 0.1   (b) Purchase intention 0.36   (c) Purchase intention 0.7

Fig. 4. The results of relative increase in number of buying customers for distinct permutations of parameters.

positive. Moreover, the higher discounts, the higher number of customers (up to six times more than without discounts). Similarly, the lower initial purchase intentions, the higher increase of the number of customers.

The phenomenon where giving the higher discount we achieve a higher number of purchases is observed only for some initial purchase intention threshold (around 0.4). In other words, increasing the discounts while having high initial purchase intention does not change anything, Fig. 4(c).

When we average the results of a relative increase in buying customers for distinct values of initial purchase intention or for distinct values of discounts, Fig. 5, we see that regardless of initial purchase intention increasing the discount rises the number of purchases in a logarithmic fashion and the higher the initial purchase intention the near-linearly smaller the relative increase of buying customers.

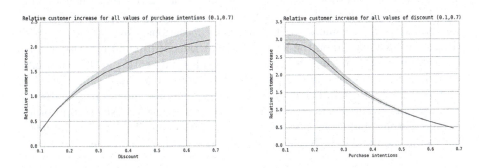

Fig. 5. Aggregated results of relative increase in number of buying customers.

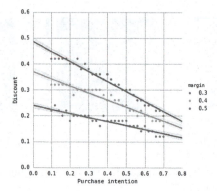

Fig. 6. Profitability results - area under the lines denotes profit while applying interactive display and discounts, above loss, in comparison to scenario without display and discounts.

6.3 Generalized Profitability

In this subsection, all results are gathered in a single Fig. 6 that can be used for assessing whether it is worth to deploy the interactive display and offer personalized discounts. As we saw in the previous subsections, the level of margin discount and initial purchase intention are important for the outcome measured in profit/loss and the number of buying customers. Figure 6 contains all these parameters and three straight lines in the figure represent the equality condition - when the cost of discounts is covered by an increase in the sum of margin caused by an increase of purchases. In other words, we do not lose comparing to a scenario with no discounts and no interactive display. The area below the line denotes the situation when we observe profit and, in contrary, the loss. One should assess a kiosk's situation according to its average margin as well as the characteristics of customers - initial purchase intentions. For instance, a kiosk with an average margin on the level of 0.3 may benefit from an increase of buying customers by deploying the interactive display only with 12–24% discounts, depending on customers' initial purchase intentions. The Fig. 6 can be used in the following way: get to know your customers' purchase intentions (e.g., count how many of the customers in your kiosk area are buying), choose an appropriate average margin, and check how much you can discount. The expected relative increase in the number of buying customers can be then read from Fig. 5(b).

7   Summary

In this paper, the application and influence of an interactive presentation system has been presented. The idea has been initiated by the the project BIWiSS, where that kind of system was designed for the Polish nation-wide network of POS with smartphones and their accessories. The system consists of two main functions: content management and presentation. The first one manages the content and usage of a network of presentation systems and the second one presents the product offer of the POS. The main problem addressed in the paper was the recommendation of discounts offered to the clients according to their usage of the presentation system. This task is observed in any class of RS, where the recommendation fact occurs after the user is firstly willing to focus his attention on a particular offer. To examine how to deliver such a recommendation, a new model for measuring profit/loss by applying discounting scenarios in the recommendation has been introduced. To verify the model, a large number of simulations (roughly 90 M) of purchases have been performed. These simulations revealed that the encountered outcome does not depend on the share of customer's using interactive display and the important parameters are customers' initial purchase intention and the level of discount. However, thanks to the application of the proposed interactive presentation system, POS can increase the number of customers by a factor of three, which is a significant outcome in building a customer base. The greater number of clients definitely builds up the potential for increasing the brand recognition and the global income of the POS network. The simulation of the proposed model has also allowed to show the profitability circumstances regarding discount and initial purchase intentions of customers to keep profitability at the same level as in the scenario without application of the system. It should be noted that performed experiments have some limitations. The behaviour of customers can be affected by a number of factors that have been not addressed within this article. Nevertheless, it is also believed that the results presented in the article are a good basis for further testing of the system in a real environment, including Social Recommender Systems.

References

1. Acquisti, A., Varian, H.R.: Conditioning prices on purchase history. Market. Sci. **24**(3), 367–381 (2005)
2. Alford, B.L., Biswas, A.: The effects of discount level, price consciousness and sale proneness on consumers' price perception and behavioral intention. J. Bus. Res. **55**(9), 775–783 (2002)
3. Anisiewicz, J., Jakubicki, B., Sobecki, J., Wantuła, Z.: Configuration of complex interactive environments. In: Zgrzywa, A., Choroś, K., Siemiński, A. (eds.) New Research in Multimedia and Internet Systems. AISC, vol. 314, pp. 239–249. Springer, Cham (2015). https://doi.org/10.1007/978-3-319-10383-9_22
4. Borgesius, F.Z., Poort, J.: Online price discrimination and EU data privacy law. J. Consum. Policy **40**(3), 347–366 (2017)
5. Chen, Q., et al.: Interacting with digital signage using hand gestures. In: Kamel, M., Campilho, A. (eds.) ICIAR 2009. LNCS, vol. 5627, pp. 347–358. Springer, Heidelberg (2009). https://doi.org/10.1007/978-3-642-02611-9_35
6. Chen, S.F.S., Monroe, K.B., Lou, Y.C.: The effects of framing price promotion messages on consumers' perceptions and purchase intentions. J. Retail. **74**(3), 353–372 (1998)
7. Cui, B., Yang, K., Chou, T.: Analyzing the impact of price promotion strategies on manufacturer sales performance. J. Serv. Sci. Manag. **9**(02), 182 (2016)
8. Drozdenko, R., Jensen, M.: Risk and maximum acceptable discount levels. J. Prod. Brand Manag. **14**(4), 264–270 (2005)

9. Ertekin, N., Shulman, J.D., Chen, H.: On the profitability of stacked discounts: identifying revenue and cost effects of discount framing. Market. Sci. **38**(2), 317–342 (2019)

10. Garfinkel, R., Gopal, R., Tripathi, A., Yin, F.: Design of a shopbot and recommender system for bundle purchases. Decis. Support Syst. **42**(3), 1974–1986 (2006)

11. Goldberg, D., Nichols, D., Oki, B.M., Terry, D.: Using collaborative filtering to weave an information tapestry. Commun. ACM **35**(12), 61–70 (1992)

12. Gupta, S., Cooper, L.G.: The discounting of discounts and promotion thresholds. J. Consum. Res. **19**(3), 401–411 (1992)

13. Ilicheva, E.: Discounts as a marketing tool for attraction and retention of customers in e-commerce through the example of a comparative analysis of the specificity of fashion e-shops in Russia and Sweden (2015)

14. Inman, J.J., McAlister, L.: A retailer promotion policy model considering promotion signal sensitivity. Market. Sci. **12**(4), 339–356 (1993)

15. Jiang, Y., Shang, J., Liu, Y., May, J.: Redesigning promotion strategy for e-commerce competitiveness through pricing and recommendation. Int. J. Prod. Econ. **167**, 257–270 (2015)

16. Leibbrandt, A.: Behavioral constraints on price discrimination: experimental evidence on pricing and customer antagonism. Eur. Econ. Rev. **121**, 103303 (2020)

17. Levy, M., Grewal, D., Kopalle, P.K., Hess, J.D.: Emerging trends in retail pricing practice: implications for research. J. Retail. **80**(3), xiii–xxi (2004)

18. McAfee, R.P.: Price discrimination. Issues in Competition Law and Policy, vol. 1, pp. 465–484 (2008)

19. Mikians, J., Gyarmati, L., Erramilli, V., Laoutaris, N.: Detecting price and search discrimination on the internet. In: Proceedings of the 11th ACM Workshop on Hot Topics in Networks, pp. 79–84. ACM (2012)

20. Miller, A.A.: What do we worry about when we worry about price discrimination? The law and ethics of using personal information for pricing. J. Technol. Law Policy **19**, 43–104 (2014)

21. Montaner, M., López, B., De La Rosa, J.L.: A taxonomy of recommender agents on the internet. Artif. Intell. Rev. **19**(4), 285–330 (2003)

22. Newman, A., Dennis, C., Wright, L.T., King, T., et al.: Shoppers' experiences of digital signage-a cross-national qualitative study. JDCTA **4**(7), 50–57 (2010)

23. Ngwe, D.: Fake discounts drive real revenues in retail. Harvard Business School (2018)

24. Phlips, L.: The Economics of Price Discrimination. Cambridge University Press, Cambridge (1983)

25. Tan, W.K., Chen, B.H.: Enhancing subscription-based ecommerce services through gambled price discounts. J. Retail. Consum. Serv. **61**, 102525 (2021)

26. Tirole, J.: The Theory of Industrial Organization. MIT Press, Cambridge (1988)

27. Varian, H.R.: Price discrimination. In: Handbook of Industrial Organization, vol. 1, pp. 597–654 (1989)

28. Wang, H.F., Wu, C.T.: A mathematical model for product selection strategies in a recommender system. Expert Syst. Appl. **36**(3), 7299–7308 (2009)

29. Zhang, Y., Chen, X.: Explainable recommendation: a survey and new perspectives. Found. Trends Inf. Retrieval **14**(1), 1–101 (2020)

Profile-Driven Synthetic Trajectories Generation to Enhance Smart System Solutions

Radosław Klimek$^{(\boxtimes)}$ ⓘ and Arkadiusz Olesek

AGH University of Science and Technology,
al. Mickiewicza 30, 30-059 Krakow, Poland
rklimek@agh.edu.pl

Abstract. The knowledge of the individual trajectories of citizens' mobility in the urban space is critical for smart cities. The data concerning trajectories from the providers of mobile phone services are still difficult to be obtained in practice and one of the considerable obstacles here are legal aspects. We have designed and implemented the tourist trajectories generator for objects located in a selected but arbitrary urban area. A generation process is based on the random selection of the predefined profiles of tourist activeness, including mobility patterns. It is possible to generate a practically unlimited number of trajectories, if needed, and they may also be directed at the certain specific types of behaviours. Thus obtained large sets of data may be used for understanding urban behaviours, calibrating urban models, recommending systems under construction, as well as anticipating the smart city further software testing.

Keywords: Smart system · Data management · Synthetic trajectories generation · Urban ecosystem calibration

1 Introduction

There is a large number of the possibilities for obtaining data on mobile phone individual trajectories, e.g. tracking applications in smartphones. It appears that the localisation data of mobile phone operators are most interesting. This data is not conditioned on the possession of any advanced smartphone or intentional enabling of a proper application. Nowadays, almost everyone has a mobile phone, even the simplest one; therefore, localisation data from the providers of mobile telephony is the most common in this regard, democratic and reliable. Localisation takes place based on triangulation and trilateration methods in BTSs (Base Transceiver Stations), which guarantee high accuracy in the urban area [2,5]. The spread of mobile devices in the recent years has contributed to the performance of many original researches concerning human mobility, and the mobile phone data collected by network operators have become the invaluable source of information about individual mobility patterns on a large scale. Whereas,

© Springer Nature Switzerland AG 2021
M. Paszynski et al. (Eds.): ICCS 2021, LNCS 12745, pp. 623–630, 2021.
https://doi.org/10.1007/978-3-030-77970-2_47

mobility and individual trajectory data are still poorly available although it is significant for a *smart city*, which provides supporting services and solving city problems. A main hindrance here are legal aspects referring to privacy, see for example [3,9,10].

Our goal and contribution is to design and implement a generator of individual mobility trajectories for mobile phone users. Trajectories mirror the behaviours of tourists staying in a given urban area. Trajectories are generated at random based on the prepared profiles of tourism behaviours, considering the various points of interests (POIs) located within the urban area. The respective elements of behaviour profiles are selected randomly for each object. The following parameters are considered: the prospective interests of tourists, their variability and possible habits connected with tourists mobility, intensity in respect of the implementation of a tourism plan, changing the means of transport and other aspects.

We believe that there are numerous prospective applications for thus obtained mass data concerning tourist mobility. These are: the analysis and understanding of the tourist traffic needs, the calibration of urban models and operational procedures, recommendation systems, increasing tourists' safety, in particular in dangerous regions or districts, solutions supporting the designing of the urban infrastructure, applications supporting the control of epidemic spread, solutions helping in emergency management and many others. Furthermore, possessing such data constitutes a perfect base for testing systems concerning the smart city which are being created or which will be created in the future.

2   Related Works

The legal regulations concerning mobile devices localisation are not too restrictive in the Asian countries, they are moderately restrictive in the USA and extremely restrictive in Europe, cf. Uhlirz et al. [12]. This last case is the main reason for the low availability of such data, despite their immense potential. A certain example here is Estonia, see Ahas et al. [1], where anonymised data are analysed, which refer to a certain group of users but at the same time they demonstrate huge advantages from possessing and analysing such data. The possibilities of the tourist trajectories analysis are also discussed in the paper [6]. Kwan et al. [7] describe an experiment of the limited data disclosure by Deutsche Telekom for a certain selected period from the past. Such data allow to analyse various hypotheses, verify them in terms of various initiatives, including startups, the assumed methodologies of data analysis, etc. Therefore, the objects mobility trajectories are crucial here. In the work by Gonzalez et al. [4], the following was analysed: the use of trajectory in searching for similarities, finding general patterns, the distribution of spatial probability, what is significant for the urban models calibration or understanding the spreading of e.g. an epidemic.

The aspect of trajectory generation has already been considered. Pelekis et al. [10] suggest the generator of objects which follow a specific main objective of such a generator. This differs considerably from our solution which involves object profiles; moreover, our work includes some sets of targets which are inter-related unambiguously with the pre-defined profiles. Pappalardo and Simini [9] generate trajectories taking into account some tendencies, also considering a tendency for breaking the routine. The work does not imply any basic source of obtaining mobile data. In the work by Giurlanda et al. [3], there is the trajectory simulator considered, concerning people's habits based on the behaviour model. This work affects our approach; yet, we consider behaviour profiles for tourists in the urban area. Zhou et al. [13] suggest the system of recommending points connected with travelling on the basis of preferences and journey time estimation. The aim is to match the trajectory with the users based on the objects mobility patterns. Networks are trained and trajectories are corrected as a result of the min-max strategy. In our paper, trajectories are generated without their corrections based on random behaviour profiles. Renso et al. [11] discuss various trajectory generators, comparing real data and synthetic data. The work discussed implies that our generator may have numerous applications. According to our best knowledge, it differs from the recently known generators, and a distinguishing factor is the random generation of trajectory based on various profiles which are weighted additionally in order to obtain better realism. Our paper is based on the unpublished work [8]. Nevertheless, a significant summary was performed and new results were generated.

3   User Profiles and POIs Planner

The main task of the generator is to produce trajectories which represent human behaviours as naturally as possible. The trajectory quality is understood here as the capability of imitating human behaviours in a natural manner. In order to ensure the conformity of the visited POI with the preferences of a given tourist, a *user profile* was designed and formed as a structure allowing for the description of features and the inclinations of a specific tourist in an adequate and implementable manner generator algorithms.

$$Profile = (I, f(i_n) \rightarrow w_n, S, d, M, T, f(t_n) \rightarrow v_n, a, p)$$

I is the set of profile interests. Function $f(z_n) \rightarrow w_n$ allocates weights to a category from set I. A weight means here a degree of being interested in each of the elements, expressed as a number from the interval $(0,1)$. The sum of all the weights for the profile amounts to 1. S – the geographical coordinates of the journey starting point for a given profile. d – a suggested maximum distance in meters between each POI matched with the profile, and a starting point S. M – a material status allocated to the profile, it is determined in a five degree scale. T is the set of the preferred means of transport, and function $f(t_n) \rightarrow v_n$ allocates a weight to each means of transport. The preferred activeness time is marked as a, it is used when determining time for trajectory. Value p is another structure containing additional tourist's preferences used in the process of allocating the points.

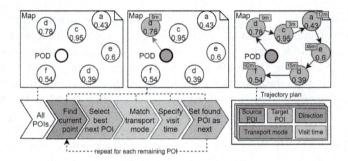

Fig. 1. The operating rule of the trajectory planner is presented as a simplified outline. (POD – Place of Departure, POI – Point of Interest). The maps depict the respective stages of trajectory generation. The first one presents an initial state, that is POIs and their assessments for the profile. A middle map is a stage repeated until all the points are planned. The last map is a ready-made trajectory plan; and there is also the outline of the ready-made trajectory plan below the last map. The outline provided below presents the steps taken at each stage; a dashed line joins given steps with the corresponding fragment of the map. Furthermore, the respective corresponding elements of the outline are marked with the same colours

The *Planner of POIs* task is to create a trajectory plan based on the pre-selected POIs. Each POI set for an urban area is obtained automatically by publicly available services, so it does not require any (manual) preparations. Firstly, the planner establishes the place of a trajectory starting point and the time of its starting. Then, searching for the best POI to be visited next takes place. It is searched in the supplied points, for which opening hours are known and for which it is possible to reach a given point timely or points for which we have no information about opening hours. In an extraordinary case, when it is not possible to create a plan consistent with opening hours, information about such opening hours may be ignored for given points. After determining another POI, the planner selects a planned transport mode matching the preferences for the profile, it estimates the travelling time to the next point and then it defines the planned visiting time in this point. Defining the visiting time takes place at random but the following factors are taken into consideration: an appropriate profile attribute and average visiting time in a given point. All the pre-calculated parameters are collected to the structure describing one fragment of the trajectory plan. At the end of each step, a selected point is marked as a current one and the process is re-implemented. While modifying the planner configuration, we can decide whether the trajectories are to be terminated in the last POI or the tourists are to return to the starting point after visiting all the points.

After implementing the steps described above for all the supplied POIs, we obtain a daily trajectory plan which consists of parts ordered depending on their planned visiting sequence. Each part contains information enabling the creation of a corresponding fragment of a trajectory. The operating rule of the planner is also presented in Fig. 1.

4   Generating Results

In order to validate our system, we have carried out several generating processes. Table 1 contains details concerning the generated sets. Figure 2 presents a fragment of the visualisation made for a generated large trajectory set (set C, 4853 trajectories, 865 unique POIs).

Table 1. The generated data sets (A, B, C)

Set	Planned number of trajectories	Number of obtained trajectories	Number of unique POIs
A	100	92	306
B	1000	976	713
C	5000	4853	865

Table 2 presents the statistics generated based on the analysed sets. The statistics refer to transport methods and parameters connected with travelling, visiting POIs and the time in which the respective trajectories were placed.

Table 3 presents the distribution of 10 most popular categories of POIs taken from 976-element set B and 4853-element set C.

We believe we have built an important and universal tool. Our system is characteristic in this regard as compared to the existing ones [3,9–11,13]. The results obtained prove the credibility and realism in reference to the trajectories

Fig. 2. The fragments of the visualisation made for Krakw City for 4853 generated trajectories (set C). On account of a large volume of data, trajectories are represented by unicolour lines. POIs are represented by dots whose colours denote different object categories

Table 2. Table presenting statistical data obtained on the basis of the analysis of the generated sets

	Set A	Set B	Set C
The earliest time stamp	2021-01-03 06:00	2021-01-03 06:00	2021-01-03 06:00
The latest time stamp	2021-01-04 07:03	2021-01-04 11:08	2021-01-04 14:24
Average duration time of a trajectory [h]	7.67	7.35	7.46
Average number of POIs visited within one hour	1.18	1.25	1.27
Average number of all the visits during one day (0–24)	434.5	4393.5	22434
Average distance covered in one trajectory [km]	12.70	12.30	12.37
Average number of visited POIs within one trajectory	8.69	8.78	8.97
Average number of visits for each POI	2.83	12.32	51.87
Number of all the POIs	306	713	865
Average number of means of transport within one trajectory	1.96	2.02	1.97
Available transport mode	driving-car, cycling-regular, cycling-road, foot-walking		

Table 3. 10 most popular categories of POIs on account of the number of visits for set B and set C

Set B			Set C		
Category	Visits	Percentage	Category	Visits	Percentage
Total	8787	100.0%	*Total*	44868	100.0%
Restaurant	2572	29.27%	Restaurant	13728	30.60%
Bar	1298	14.77%	Memorial	6642	14.80%
Memorial	1254	14.27%	Bar	6410	14.29%
Monument	682	7.76%	Monument	3542	7.89%
Museum	547	6.23%	Museum	2603	5.80%
Arts_centre	459	5.22%	Arts_centre	2268	5.05 %
Viewpoint	423	4.81%	Viewpoint	2029	4.52%
Place_of_worship	398	4.53 %	Place_of_worship	1946	4.34%
Theatre	368	4.19%	Theatre	1943	4.33%
Cinema	356	4.05%	Cinema	1744	3.89%

which thus are reliable. The trajectories generated are synthetic but much effort was made so that they could look realistic and so that they could consider both various profiles of human behaviours and a randomness factor.

5   Conclusions

By using our system, it is possible to generate practically any number of tourist trajectories, depending on demand and in the real time. Further works may comprise other groups of people instead of tourists, e.g. regular citizens.

References

1. Ahas, R., Aasa, A., Roose, A., Ülar Mark, Silm, S.: Evaluating passive mobile positioning data for tourism surveys: an Estonian case study. Tour. Manag. **29**(3), 469–486 (2008). https://doi.org/10.1016/j.tourman.2007.05.014
2. Calabrese, F., Colonna, M., Lovisolo, P., Parata, D., Ratti, C.: Real-time urban monitoring using cell phones: a case study in Rome. IEEE Trans. Intell. Transp. Syst. **12**(1), 141–151 (2011). https://doi.org/10.1109/TITS.2010.2074196
3. Giurlanda, F., Perazzo, P., Dini, G.: HUMsim: a privacy-oriented human mobility simulator. In: Kanjo, E., Trossen, D. (eds.) S-CUBE 2014. LNICST, vol. 143, pp. 61–70. Springer, Cham (2015). https://doi.org/10.1007/978-3-319-17136-4_7
4. Gonzalez, M.C., Hidalgo, C.A., Barabasi, A.L.: Understanding individual human mobility patterns. Nature **453**(7196), 779–782 (2008). https://doi.org/10.1038/nature06958
5. Klimek, R.: Exploration of human activities using message streaming brokers and automated logical reasoning for ambient-assisted services. IEEE Access **6**, 27127–27155 (2018). https://doi.org/10.1109/ACCESS.2018.2834532
6. Klimek, R.: Towards recognising individual behaviours from pervasive mobile datasets in urban spaces. Sustainability **11**(6), 1–25 (2019). https://doi.org/10.3390/su11061563
7. Kwan, M., Cartwright, W., Arrowsmith, C.: Tracking movements with mobile phone billing data: a case study with publicly-available data. In: Gartner, G., Ortag, F. (eds.) Advances in Location-Based Services, pp. 109–117. Springer, Berlin (2012). https://doi.org/10.1007/978-3-642-24198-7_7
8. Olesek, A.: Mobile phone trajectory generator using user profiles [in Polish]. Master thesis. AGH University of Science and Technology (2020). Supervisor: Radosław Klimek
9. Pappalardo, L., Simini, F.: Data-driven generation of spatio-temporal routines in human mobility. Data Min. Knowl. Disc. **32**(3), 787–829 (2017). https://doi.org/10.1007/s10618-017-0548-4
10. Pelekis, N., Ntrigkogias, C., Tampakis, P., Sideridis, S., Theodoridis, Y.: Hermoupolis: a trajectory generator for simulating generalized mobility patterns. In: Blockeel, H., Kersting, K., Nijssen, S., Železný, F. (eds.) ECML PKDD 2013. LNCS (LNAI), vol. 8190, pp. 659–662. Springer, Heidelberg (2013). https://doi.org/10.1007/978-3-642-40994-3_49
11. Renso, C., et al.: Wireless network data sources: tracking and synthesizing trajectories. In: Giannotti, F., Pedreschi, D. (eds.) Mobility, Data Mining and Privacy, pp. 73–100. Springer, Berlin (2008). https://doi.org/10.1007/978-3-540-75177-9_4

12. Uhlirz, M.: A market and user view on LBS. In: Gartner, G., Cartwright, W., Peterson, M.P. (eds.) Location Based Services and TeleCartography, pp. 47–58. Springer, Berlin (2007). https://doi.org/10.1007/978-3-540-36728-4_4
13. Zhou, F., Yin, R., Trajcevski, G., Zhang, K., Wu, J., Khokhar, A.: Improving human mobility identification with trajectory augmentation. GeoInformatica 1–31 (2019)

Augmenting Automatic Clustering with Expert Knowledge and Explanations

Szymon Bobek[1,2]([✉]) [iD] and Grzegorz J. Nalepa[1,2] [iD]

[1] Jagiellonian Human-Centered Artificial Intelligence Laboratory (JAHCAI)
and Institute of Applied Computer Science, Jagiellonian University,
31-007 Kraków, Poland
{szymon.bobek,grzegorz.j.nalepa}@uj.edu.pl
[2] AGH University of Science and Technology, Kraków, Poland

Abstract. Cluster discovery from highly-dimensional data is a challenging task, that has been studied for years in the fields of data mining and machine learning. Most of them focus on automation of the process, resulting in the clusters that once discovered have to be carefully analyzed to assign semantics for numerical labels. However, it is often the case that such an explicit, symbolic knowledge about possible clusters is available prior to clustering and can be used to enhance the learning process. More importantly, we demonstrate how a machine learning model can be used to refine the expert knowledge and extend it with an aid of explainable AI algorithms. We present our framework on an artificial, reproducible dataset.

Keywords: Data mining · Explainable AI · Clustering

1 Introduction

An effective analysis of data can often pose a major challenge in cases where data is highly-dimensional, produced in fast rate and large volumes (i.e. big data). This is where methods of Artificial Intelligence prove to be useful. Moreover, besides the use of data mining techniques in order to build machine learning models, the human expert knowledge regarding the specificity of domain of interest should be used.

Such a case most often arises in Industry 4.0 that aims at using number of information and communications technology solutions for the monitoring and optimization of industrial processes. The installations in modern factories are equipped with many of sensors gathering data about the operation of the machines involved in these processes.

In this work we focus on automated discovery of device states from machinery sensor log to enrich the expert knowledge about machinery operational states. Our main goal was to develop a workflow, which would provide a mechanism for detecting device states that can be applied to different types of industrial machinery. We confronted it with states that were discovered with knowledge-based approach to

© Springer Nature Switzerland AG 2021
M. Paszynski et al. (Eds.): ICCS 2021, LNCS 12745, pp. 631–638, 2021.
https://doi.org/10.1007/978-3-030-77970-2_48

prove its validity and expand the knowledge-base itself with an usage of eXplainable Artificial Intelligence (XAI) algorithms. For this purpose we propose two algorithms. The first one is for *splitting* clusters that possibly scramble within two or more concepts. The second one is for *merging* expert clusters possibly represent the same concept, and therefore can be considered redundant. The recommendations are justified by rules, explaining why such suggestions were made. The final decision on whether to trust recommendation or discard them is left to an expert. The whole process is iterative and can be repeated until the convergence is achieved. In this paper we limit the discussion only to artificially generated samples to provide full reproducibility of the experiments.

This work is carried out in the CHIST-ERA Pacmel project[1]. The project is oriented at the development of novel methods of knowledge modeling and intelligent data analysis in Industry 4.0.

The rest of the paper is organized as follows: In Sect. 2 we discuss selected challenges regarding clustering. Then in Sect. 3 we introduce our approach regarding knowledge augmented clustering. We summarize the paper in Sect. 4.

2   Clustering Highly-Dimensional Data

Clustering aims at unfolding hidden patterns in data to discover similar instances and group them under common cluster labels. This task is often performed to either discover unknown groups, to automate the process of discovering possibly known groups or for segmentation of data points into arbitrary number of segments. Either of the above can be done in unsupervised, semi-supervised or supervised manner.

The problem of effective analysis of clustering results, and bringing semantics into the clustering results has also been investigated. In [1] authors focus on solutions which assist users to understand long time-series data by observing its changes over time, finding repeated patterns, detecting outliers, and effectively labeling data instances. It is performed mostly via visualization layer over data that dimensionality was reduced with UMAP [6] allowing 2D/3D plotting. However, no explicit knowledge is used in this method to enhance the process of clusters analysis.

In [4] the Grouper framework was presented which is an interactive approval, refinement or decline toolkit for analysis of results of clustering. It combines the strength of algorithmic clustering with the usability of visual clustering paradigm. In [11] similar approach was presented, however it assumes more interactions with visualized clusters that alters the cluster layout. Yet, neither of these use any kind of formalized knowledge neither for clustering nor after it for refinement. Therefore, the knowledge input by en expert in a form of interactions in the system is lost for further re-use.

The human-in-the-loop paradigm was also investigated in the clustering algorithms. In [12] a similar approach was used as in our solution, where the contextual information and user feedback is used to merge clusters of photographs into

---

[1] See the project webpage at http://PACMEL.geist.re.

larger groups. However, no prior knowledge is used in clustering, nor extended at the end.

The biggest disadvantage of all of the above methods is that they do not explicitly use nor update domain knowledge. Therefore new knowledge, even if discovered by cooperation of AI and expert, is hidden into complex models and not reusable for future.

Using background knowledge in DM has been proposed in the area of semantic data mining, where the formalized expert knowledge is used for domain-specific configuration to improve the overall results of DM/ML algorithms [3]. Initial approaches reusing knowledge and experiences in DM for configuring DM tasks have been discussed in [2]. However, in those approaches, domain knowledge is only used in a very specific setting, and there is lack of feedback loop that allows for knowledge flow in opposite direction.

Therefore, in our approach we use explainable AI algorithms, that aim to inverse the process of encoding knowledge into black-box models and allow for more insight into decisions made by machine learning or data mining algorithms. This closes the feedback loop between domain experts and machine-learning algorithms by allowing knowledge exchange between an expert and ML algorithm. Although there exist a variety of XAI methods that allow explaining ML models decision [5,7–10], in our work we will focus on Anchor [10] which produces rule-based, explanations that can easily be integrated with domain knowledge encoded with the same formalism.

In the following sections we will describe our framework in more details.

3 Knowledge Augmented Clustering

Although obtaining high quality automated clustering of data is an important initial step for utilizing our framework, we skip this step in the discussion. The main goal of the work presented in this paper is to refine the initial clustering with XAI methods and expert knowledge via splits and merges of existing clusters.

In both of the cases for cluster splits and merges we assume that there exists a set of clusters obtained with an utilization of expert knowledge, denoted as:

$$E = \{E_1, E_2, \ldots, E_n\}$$

This set of clusters needs to be refined with complementary clustering performed with automated clustering algorithms. This clustering forms separate set of clusters, possibly of different size then E and is denoted as:

$$C = \{C_1, C_2, \ldots, C_m\}$$

For both sets we calculate confusion matrix M of size (n, m) where number at the intersection of i-th row and k-th column holds number of data points assigned both by cluster labeling to cluster E_j and automated clustering to cluster C_k.

Based on confusion matrix M we calculate two helper matrices for splitting and merging strategies defined respectively by Eqs. (1) and (2).

$$H_{i,j}^{split} = \frac{M_{i,j}}{H(M_i) \sum\limits_{j \in 1...m} M_{i,j}} \tag{1}$$

Where $H(M_j)$ is entropy calculated for j-th column (i.e. C_j cluster). The measure defines consistency of automated clustering with expert clustering, normalized by number of points. The H^{merge} measure is $l2$ normalized matrix M along row axis.

$$H_{i,j}^{merge} = \frac{M_{i,j}}{||M_i||_2} \tag{2}$$

These two matrices are later used for the purpose of generation of split and merge recommendations. Figure 1 depicts the two simplified datasets with two possible scenarios covered by our method. These datasets will be used to better explain mechanisms for splitting and merging recommendations discussed in following sections.

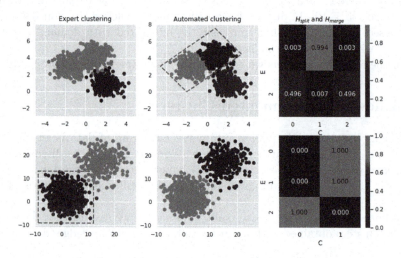

Fig. 1. Synthetic datasets with clusters to split (top row) and clusters to merge (bottom row). Columns in the figure represent clustering preformed with expert knowledge, automated clustering, and H_{split} matrix (upper) and H_{merge} matrix (lower). Dotted lines define bounding boxes for the decision stump explanation mechanism.

3.1 Recommendation Generation

Having created the H^{split} and H^{merge} matrices, we generate two types recommendations out of it: *splitting* and *merging*.

Splitting. This recommendation aim at discover clusters that were incorrectly assigned by expert knowledge. Such a case was depicted in Fig. 1 in the upper left plot. This operation can be performed using H^{split} matrix in a straightforward way. The cluster that is recommended for splitting is chosen by investigating values corresponding to it in H^{split} matrix. Values that lie on the intersection of investigated expert cluster and automated cluster, and that are greater than defined threshold ϵ_s are marked as candidates for splitting:

$$Candidates_i = \left\{ C_j : H_{i,j}^{split} > \epsilon_s \right\}$$

For the example in Fig. 1 the recommendation will look as follows:

```
SPLIT  EXPERT CLUSTER  E_2
INTO CLUSTERS  [(C_0, C_2)]  (Confidence 0.98)
```

The confidence of split is an average of $H_{i,j}^{split}$ values associated with candidates for splitting normalized by the maximum entropy. The maximum entropy depends on the number of expert clusters to merge, and equals 0.5 in this recommendation case.

Merging. This recommendation goal is to detect concepts that were incorrectly labelled by an expert knowledge as two clusters. Such a case is depicted in Fig. 1 in lower left plot. Candidates for merging are chosen using H^{merge} matrix. Because the matrix is $l2$ normalized along rows, calculating dot product of selected rows produces cosine similarity between them. This cosine similarity reflect the similarity in the distribution of data points spread over the automated discovered clusters. If two expert clusters have similar distribution of points over automatically discovered clusters this *might* be a premise that they share the same concept and should be merged. Such a case was depicted in Fig. 1 in lower right plot. Similarly as in the case of splitting a threshold ϵ_m is defined arbitrarily that denotes the lower bound on the cosine similarity between clusters to be considered as merge candidates.

$$Candidates_m = \left\{ E_j, E_k : sim(H_j^{merge}, H_j^{merge}) \geq \epsilon_m \right\}$$

The merge recommendation is generated as follows. The confidence value is calculated as a cosine similarity between rows associated to candidates E_0, E_1 in H^{merge} matrix.

```
MERGE  EXPERT CLUSTER E_0  WITH  EXPERT CLUSTER E_1
INTO  CLUSTER C_1 # (Confidence 1.0)
```

In the next section, the justification of the split and merge recommendation are discussed.

3.2   Recommendation Explanation

Once the recommendation is generated it is augmented with an explanation. Depending on the recommendation type, the explanation is created differently.

Splitting Recommendation. In case of this type of recommendation we transform the original task from clustering to classification, taking automatically discovered cluster labels as target values for the classifier.

Then, we explain the decision of a classifier to present an expert *why* and *how* two (or more) clusters $C_i, C_j, \ldots, C_n$ that were formed by splitting original one are different from each other. An explanation is formulated in a form of a rule that uses original features as conditional attributes, to help expert better understand the difference between splitting candidates. We use the Anchor algorithm for that [10].

The explanation for the splitting of cluster E_2 presented in Fig. 1 looks as follows:

```
C_0: x1 > 0.88 AND x2 > 3.61 (Precision: 1.00, Coverage: 0.21)
C_2: x1 <= -1.51 (Precision: 1.00, Coverage: 0.25)
```

If the difference is important, the clusters can be split and the rules generated above can be added to knowledge base. The final decision on weather splitting E_2 into C_0 and C_2 is needed, is left to the expert.

Alternatively a decision stump can be build on the dataset narrowed to points that form clusters candidates C_0 and C_1 and its visual form can also be presented to an expert. It is worth noting that the decision stump can be different than Anchor rule, as the former is build on just a fraction of data, while Anchor takes into consideration whole dataset.

The decision stump for the case discussed in this section is given in Fig. 2a. The cyan bounding-box in Fig. 1 in the upper middle plot roughly defines the dataset used for building the decision stump.

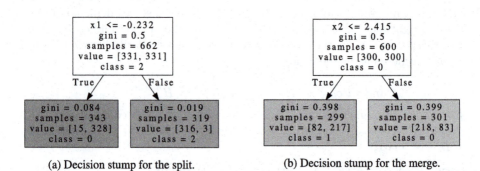

(a) Decision stump for the split.   (b) Decision stump for the merge.

Fig. 2. Decision stumps for explanations of recommendations.

Merging Recommendation. In explanation for merging recommendation we use the same approach as previously described. The difference is that the classification models are now trained with expert labels as target.

After that the explanation that answers the questions how two expert clusters E_i and E_j are different from each other while looking at them not through the definition in knowledge base, but in data.

The answer to this question is given in a form of Anchor rules, and the final decision left to an expert. The explanation for the case presented in Fig. 1 is given below:

```
E_0: x2 <= 5.16 AND x1 > 0.23 (Precision: 0.74, Coverage: 0.32)
E_1: x1 <= 3.73 (Precision: 0.58, Coverage: 0.50)
```

Similarly to splitting explanation a decision stump can be created as presented in Fig. 2b. The results are again slightly different than in case of the Anchor explanation, as the decision stump is build only on a fraction of the data and can omit other variable dependencies. The cyan bounding-box in Fig. 1 in the lower left plot roughly defines the dataset used for building decision stump.

4 Summary

In this paper we presented a framework for expert knowledge extension with a usage of clustering algorithms for multidimensional time series. We described how automated mechanism for labeling device operational states can be used to refine expert-based labeling and demonstrated its functionality on a synthetic, reproducible scenario. These refinements was defined by us as *splits* and *merges* of expert labeling and were augmented with detailed explanations. The explanations were formulated as rules and therefore can be easily interpreted incorporated with expert knowledge.

For the future works, we plan to extend the framework with additional methods supporting splits and merges. In particular we would like to exploit different linkage methods known form hierarchical clustering for merges and clustering metrics, such as silhouette score for splits.

Acknowledgements. The paper is funded from the PACMEL project funded by the National Science Centre, Poland under CHIST-ERA programme (NCN 2018/27/Z/ST6/03392). The authors are grateful to ACK Cyfronet, Krakow for granting access to the computing infrastructure built in the projects No. POIG.02.03.00-00-028/08 "PLATON - Science Services Platform" and No. POIG.02.03.00-00-110/13 "Deploying high-availability, critical services in Metropolitan Area Networks (MAN-HA)".

References

1. Ali, M., Jones, M.W., Xie, X., Williams, M.: TimeCluster: dimension reduction applied to temporal data for visual analytics. Vis. Comput. **35**(6–8), 1013–1026 (2019)
2. Atzmueller, M.: Experience management with task-configurations and task-patterns for descriptive data mining. In: Proceedings of KESE 2007, 30th German Conference on Artificial Intelligence (KI-2007) (2007)

3. Atzmueller, M., Seipel, D.: Declarative specification of ontological domain knowledge for descriptive data mining (extended version). In: Proceedings of 18th International Conference on Applications of Declarative Programming and Knowledge Management (2008)
4. Coden, A., Danilevsky, M., Gruhl, D., Kato, L., Nagarajan, M.: A method to accelerate human in the loop clustering, pp. 237–245
5. Lundberg, S.M., et al.: Explainable AI for trees: from local explanations to global understanding. CoRR. arXiv:1905.04610 (2019)
6. McInnes, L., Healy, J., Melville, J.: UMAP: uniform manifold approximation and projection for dimension reduction (2020)
7. Mujkanovic, F., Doskoč, V., Schirneck, M., Schäfer, P., Friedrich, T.: timeXplain - a framework for explaining the predictions of time series classifiers (2020)
8. Pope, P.E., Kolouri, S., Rostami, M., Martin, C.E., Hoffmann, H.: Explainability methods for graph convolutional neural networks. In: 2019 IEEE/CVF Conference on Computer Vision and Pattern Recognition (CVPR), pp. 10764–10773 (2019)
9. Ribeiro, M.T., Singh, S., Guestrin, C.: "Why should i trust you?": explaining the predictions of any classifier. In: Proceedings of the 22nd ACM SIGKDD International Conference on Knowledge Discovery and Data Mining, KDD 2016, pp. 1135–1144. Association for Computing Machinery, New York (2016)
10. Ribeiro, M.T., Singh, S., Guestrin, C.: Anchors: high-precision model-agnostic explanations. In: AAAI (2018)
11. Wenskovitch, J., North, C.: Observation-level interaction with clustering and dimension reduction algorithms. In: HILDA 2017. Association for Computing Machinery, New York (2017)
12. Zhang, L., Kalashnikov, D.V., Mehrotra, S.: Context-assisted face clustering framework with human-in-the-loop. Int. J. Multimedia Inf. Retrieval **3**(2), 69–88 (2014). https://doi.org/10.1007/s13735-014-0052-1

Renewable Energy-Aware Heuristic Algorithms for Edge Server Selection for Stream Data Processing

Tomasz Szydlo[1]([⊠]) and Chris Gniady[2]

[1] Institute of Computer Science, AGH University of Science and Technology,
Krakow, Poland
tomasz.szydlo@agh.edu.pl
[2] Department of Computer Science, University of Arizona, Tucson, AZ 85719, USA

Abstract. Internet of Things and Edge computing are evolving, bringing data processing closer to the source and a result closer to the network's edge. This distributed processing can increase energy consumption and carbon footprint. One solution to overcome the environment's impact is using renewable energy sources such as photovoltaic panels to power both cloud and edge servers. Since solar energy is not available at night or can vary with cloudiness, the centers still rely on conventional energy sources. Any solar energy that exceeds the demand power of the computing infrastructure is put back to the grid. Fluctuations in energy output due to moving clouds can have a negative impact on conventional energy suppliers as they have to maintain a constant energy supply. This paper presents heuristic algorithms for selecting edge servers for data stream processing to manage renewable energy utilization and smooth out energy fluctuations.

Keywords: IoT · Edge computing · Renewable energy

1 Introduction

Internet of Things consists of smart homes, smart cities, and the fourth industrial revolution based on cyber-physical systems is becoming ubiquitous. The IoT devices surround us and generate a vast amount of data that requires analysis to reap the benefits of smart infrastructures. Processing data directly on the end device provides a short response time but may have limited scope due to a global view. This scope vs. response challenge is faced by Industry 4.0 that requires response times in the millisecond range. Pushing processing into the network increases response time but broadens the decision-making process's scope by enabling aggregation of data from multiple sensors in the factory. Finally, centralized cloud processing enables the development of truly global systems at the cost of larger data transfer delays from the cloud. The challenge is exacerbated by large bandwidth requirements from hundreds or even thousands of

© Springer Nature Switzerland AG 2021
M. Paszynski et al. (Eds.): ICCS 2021, LNCS 12745, pp. 639–646, 2021.
https://doi.org/10.1007/978-3-030-77970-2_49

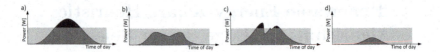

Fig. 1. PV energy generation: a) summer, b)summer during cloudy weather, c) spring with dynamic cloud movements, d) winter. (Color figure online)

concurrent data streams to the cloud. The approach to solve the challenge relies on intelligent processing distribution between the edge and the cloud to minimize delays and maximize the global decision-making process.

Such distributed computing can have a negative impact on energy consumption and carbon footprint since the computing hardware has to be distributed between multiple sites. Energy-efficient software and hardware help curb carbon footprint, and once combined with renewable energy sources, the carbon footprint can be eliminated. While solar energy is most promising, it is only available during the day and can have large variability due to cloud movements. Subsequently, utilization of solar energy in computing is challenging as it required adaptation from computing infrastructure to the changing energy supply. Real-time stream processing in Industry 4.0 or other smart environments make it even more challenging as unexpected delays can result in severe consequences. In this paper, we focus on data stream processing, also called dataflow processing, and how we can maximize solar energy utilization based on its availability.

2   Related Work

Computing infrastructure is expected to generate up to 14% of global gas emissions by 2040, with datacenters and networking infrastructure accounting for 33% by 2025 and becoming more dominant in the future. The use of renewable energy can contribute to the sustainable development of computing infrastructure. However, renewable energy sources are characterized by high dynamicity of change and uncertainty in their supply, making their use in computing infrastructure a challenge. The challenges are tackled by datacenter design to accommodate renewable energy [3,5] and to manage workflow scheduling based on energy availability [6]. Alternatively, the workload can be migrated between datacenters based on renewable energy availability [8].

In this paper, we focus on stream data processing systems based on a programming paradigm that emphasizes data flows [1,4] generated by IoT devices. This class of systems' characteristic property is that they define applications as directed graphs where vertices correspond to processes while edges represent data streams. Flow-based programming (FBP) [7] and proposes architectures consistent with this paradigm such as the staged event-driven architecture (SEDA) [10], or reactive programming [2] try to tackle challenges associated with distributed stream processing to provide scalable and efficient systems. Subsequently, asynchronous communication between processes coupled with the SEDA

Table 1. Notations used in the system model.

Symbol	Meaning
G	Edge processing network
k	Number of edge processing locations
n	Number of stream data sources
M	Set of edge datacenters
S	Set of stream data sources
e_j^m	Power necessary for processing on m_j
e_j^{re}	Renewable energy available at m_j
e_i^s	Power necessary to process stream data from s_i
$x_{i,j}$	binary variable indicating assignment of s_i to m_j
$d_{i,j}$	network latency between s_i and m_j

architecture ensures scalability and reduces response time. Flow-based processing makes energy optimizations challenging as it is not delay-tolerant and can't be postponed until renewable energy becomes available.

Solar energy is one of the most cost-efficient and readily deployable renewable energy sources. Figure 1 shows solar energy production and consumption as related to seasons and cloudiness. The grey color on a chart represents the grid's consumed energy; green shows the solar energy produced and consumed by the system; while blue is the energy sent to the grid and as it is overproduced. As we observe, winter and cloudiness result in low energy production that cannot satisfy the demand at any time. Overproduction of energy by the solar system can be stored in the grid and drawn from the grid as needed. While such a solution seems practical as overproduction can be delivered somewhere else, solar energy production variability makes it challenging for utility companies to provide high quality of energy, requiring them to either overproduce or keep expensive natural gas generators on standby in case the solar output drops. Furthermore, grid usage cost is usually structured into the customer's energy cost and solar energy production, and resulting grid usage needs to be accounted for, leading to complex purchase/sale agreements. Subsequently, maximization of solar energy usage is critical to mitigating costs associated with energy trade.

3   Problem and Algorithms Description

In the context of edge processing, the problem of selecting the appropriate edge location of stream data processing can be considered as an undirected graph $G = (V, E)$ consisting of many stream data sources, and locations of edge datacenters thus $V = M \cup S$, where M is a set of edge datacenters and S is a set of stream data sources (Table 1). E represents the links enabling network communication between M and S. Computing centers use backbone computer networks with high capacity, usually based on fiber optic technology. Assume that there are k

edge server locations, and each of them has a similar performance for processing data streams. We use the notation e_j^m to denote the power used by the datacenter and e_j^{re} to denote the amount of energy from photovoltaic panels. To process data stream $s_i \in S$ in the edge datacenter it is necessary to consume e_i^s of power. Every edge datacenter can process streams from several sources as long as it does not exceed the computing capacity. Each data stream can be processed only by one edge datacenter at a particular time. Nevertheless, the processing location's choice impacts communication delays $d_{i,j}$ due to the geographical distance and the computer network's extent.

Local consumption of renewable energy generated by solar panels used to power edge datacenter m_j is calculated as:

$$local\ consumption = \begin{cases} 1 & e_j^{re} < e_j^m \\ \frac{e_j^m}{e_j^{re}} & e_j^{re} \geqslant e_j^m \end{cases} \tag{1}$$

When local consumption is less than 100%, the surplus of energy is sent to the grid. To increase the local energy consumption, it is necessary to process more data to increase the e_j^m. It can be achieved by moving stream processing from other edge datacenters. In general, choosing an edge datacenter is an NP optimization problem[9] solvable only for small cases.

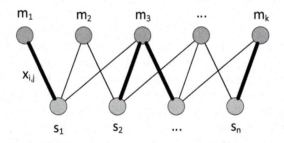

Fig. 2. Adaptability mechanisms for stream processing.

The problem of edge selection can be represented graphically, as in Fig. 2. We introduce the binary decision variable $x_{i,j} \in \{0,1\}$ to indicate whether the data stream from source s_i is processed by m_j edge datacenter. The decision to choose an edge datacenter can impact the quality of data processing and the processing delays. In the paper, we propose three heuristic algorithm, shown below, for edge server selection: 1) BEAS selects the processing center that has the lowest latency; 2) GEAS aims to distribute the load on edge servers evenly, potentially suffering from increased latency; 3) k-REAS takes into account information about the availability of renewable energy and selects the edge server with the highest surplus of renewable energy (to control latency k specifies that no more than the k-th server can be selected in terms of distance).

Algorithm 1. Best Effort Assignment Strategy (BEAS).

1: **function** BEAS(G)
2:     $x \leftarrow \varnothing$
3:     **for** each $s_i \in S$ **do**
4:         sort $m_j \in M$ in order of increasing distance $d_{i,j}$
5:         $x[i, j] \leftarrow 1$ for m_j with the smallest distance $d_{i,j}$
6:     **return** x

Algorithm 2. Greedy equal assignment strategy (GEAS).

1: **function** GEAS(G)
2:     $x \leftarrow \varnothing$
3:     $max \leftarrow \lceil n/k \rceil$
4:     $counter \leftarrow \varnothing$
5:     **for** each $s_i \in S$ **do**
6:         sort $m_j \in M$ in order of increasing distance $d_{i,j}$
7:         **for** each $m_j \in M$ with increasing distance $d_{i,j}$ **do**
8:             **if** $counter[i, j] < max$ **then**
9:                 $x[i, j] \leftarrow 1$
10:                 $counter[i, j] = counter[i, j] + 1$
    return x

Algorithm 3. k-Renewable Energy-aware Assignment Strategy (k-REAS).

1: **function** K-REAS(G, k)
2:     $x \leftarrow \varnothing$
3:     $energy \leftarrow \varnothing$
4:     **for** each $s_i \in S$ **do**
5:         sort $m_j \in M$ in order of increasing distance $d_{i,j}$
6:         $temp \leftarrow \varnothing$
7:         **for** k stations $m_j \in M$ with lowest distance $d_{i,j}$ **do**
8:             $temp[j] \leftarrow e_j^{re} - (e_j^m + energy[j] + e_i^s)$
9:         select m_j with highest value of $temp[j]$
10:         $energy[j] = energy[j] + e_i^s$
11:         $x[i, j] \leftarrow 1$
    return x

---

4   Evaluation

To evaluate the solution, we set up a synthetic application representing a typical
IoT system's operation that processes the sensors' data streams and requires
real-time interactions between sensors and actuators. We consider a system with
data stream sources in 50 US states. The location of edge data processing centers was adopted from Amazon Web Services - CloudFront services computing
centers. We assume that edge servers do not process other tasks than those currently analyzed. We also assume that the photovoltaic installation provides the
annual energy budget for processing 3 data streams at each edge datacenter,

with the energy necessary to process a single data stream is equal to 100 W. This leads to variable sizes of solar panels at each location. We used PVGIS, an European Commissions' Project, that gives information about solar radiation and photovoltaic system performance based on satellite image analysis, with hourly resolution. Network delays are calculated based on the distance between stream data sources and the datacenters, assuming that the connection is based on the fiber technology.

Table 2. Detailed results of the experiments

Algorithm	Latency				Local consumption			
	Min	Max	Avg	Stdev	Min	Max	Avg	Stdev
GEAS	0.48	66.29	10.03	11.50	0.0	41.0	29.57	13.06
BEAS	0.47	66.29	7.40	10.80	0.0	73.0	28.0	20.91
1-REAS	0.47	66.29	7.40	10.69	0.0	73.0	28.0	20.91
2-REAS	0.47	66.56	8.67	10.70	6.0	57.0	31.28	15.90
3-REAS	0.47	66.61	10.35	11.06	5.0	52.0	32.76	14.58
4-REAS	0.47	70.18	11.45	11.23	12.0	48.0	33.47	12.85
5-REAS	0.47	72.41	12.53	11.50	13.0	46.0	34.14	10.61

We performed simulation of the IoT system based on the presented assumptions. For the k-$REAS$ algorithm, we analyzed the operation for $k = \{1..5\}$. Detailed results are presented in Table 2. In the case of the $BEAS$ algorithm, the lowest communication delay of 7.40 ms was obtained. Local consumption of solar energy was in the range of $[0\%; 73\%]$. This means that some processing centers were underutilized, delivering energy to the grid, and some were overloaded. We note that the decision to choose the edge datacenter was made once - at the beginning of the simulation. The $GEAS$ algorithm distributed the load evenly among the edge servers, resulting in a maximum local consumption of 41%. As a result, this led to an increase in the average latency of 10.03 ms. As before, the selection of the edge datacenters was performed once at the beginning of the simulation.

The third algorithm, k-$REAS$, is based on the current and time-varying energy balance available in each edge data center. It aimed to increase local consumption of solar energy at the expense of increased processing latency. The algorithm's aggressiveness depends on the parameter k and increases the local consumption of solar energy. As a result, for $k = 5$, the lowest achieved level of local consumption was 13%. Contrary to the previous algorithms, k-$REAS$ dynamically changes the edge server's selection based on information about the instantaneous load on the servers and the amount of available energy. Figure 3 visualizes energy informations from Table 2. We were able to increase the average local consumption by 6.14% (BEAS/5-REAS), but resulting delays increased by 5.13 ms. Notably, the choice between processing delays and the local energy

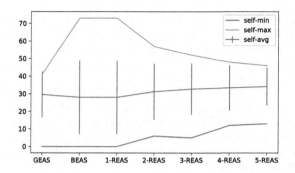

Fig. 3. Average local consumption of renewable-energy at the edge datacenters.

consumption is a trade-off and should be selected according to the system application. At the same time, the increase in the delay by a few milliseconds, as was the case with the *k-REAS* algorithms, is in many cases insignificant but should be consciously chosen.

5 Summary and Future Work

In this paper, we have presented the concept of heuristic management algorithms. The results show that the dynamic choice of where to process the streaming data impacts the use of energy obtained from renewable sources. The use of this class of algorithms as presented in the paper is a step towards designing and constructing sustainable computer systems to minimize the impact on the natural environment. We have focused on the problem of selecting the place for processing data streams. An interesting direction for further work would be the analysis of transferring processing between devices, edge computing, and cloud computing. This would provide more heterogeneous systems with different performance and energy trade-offs, potentially improving the overall system efficiency.

Acknowledgment. The research presented in this paper was supported by the National Centre for Research and Development (NCBiR) under Grant No. Gospostrateg1/385085/21/NCBR/2019.

References

1. Al-Fuqaha, A., Guizani, M., Mohammadi, M., Aledhari, M., Ayyash, M.: Internet of things: a survey on enabling technologies, protocols, and applications. IEEE Commun. Surv. Tutor. **17**(4), 2347–2376 (2015)
2. Bainomugisha, E., Carreton, A.L., Cutsem, T.V., Mostinckx, S., Meuter, W.D.: A survey on reactive programming. ACM Comput. Surv. **45**(4), 1–34 (2013)
3. Gu, C., Li, Z., Liu, C., Huang, H.: Planning for green cloud data centers using sustainable energy. In: 2016 IEEE Symposium on Computers and Communication (ISCC), pp. 804–809 (2016)

4. Giang, N.K., Blackstock, M., Lea, R., Leung, V.C.M.: Developing IoT applications in the fog: a distributed dataflow approach. In: 2015 5th International Conference on the Internet of Things (IOT), pp. 155–162, October 2015

5. Gill, S.S., Buyya, R.: A taxonomy and future directions for sustainable cloud computing: 360 degree view. ACM Comput. Surv. **51**(5), 1–33 (2018)

6. Grange, L., Costa, G.D., Stolf, P.: Green IT scheduling for data center powered with renewable energy. Future Gener. Comput. Syst. **86**, 99–120 (2018)

7. Morrison, J.P.: Flow-Based Programming, 2nd Edition: A New Approach to Application Development. CreateSpace, Paramount (2010)

8. Shuja, J., Gani, A., Shamshirband, S., Ahmad, R.W., Bilal, K.: Sustainable cloud data centers: a survey of enabling techniques and technologies. Renew. Sustain. Energy Rev. **62**, 195–214 (2016)

9. Wang, S., Zhao, Y., Xu, J., Yuan, J., Hsu, C.: Edge server placement in mobile edge computing. J. Parallel Distributed Comput. **127**, 160–168 (2019)

10. Welsh, M., Culler, D., Brewer, E.: SEDA: an architecture for well-conditioned, scalable internet services. In: Proceedings of the Eighteenth ACM Symposium on Operating Systems Principles (SOSP 2001), pp. 230–243. ACM, New York (2001)

Dataset for Anomalies Detection in 3D Printing

Tomasz Szydlo$^{(\boxtimes)}$, Joanna Sendorek, Mateusz Windak,
and Robert Brzoza-Woch

Institute of Computer Science, AGH University of Science and Technology,
Krakow, Poland
tomasz.szydlo@agh.edu.pl

Abstract. Nowadays, the Internet of Things plays a significant role in many domains. Especially, Industry 4.0 is making significant usage of concepts like smart sensors and big data analysis. IoT devices are commonly used to monitor industry machines and detect anomalies in their work. This paper presents and describes a set of data streams coming from a working 3D printer. Among others, it contains accelerometer data of printer head, intrusion power and temperatures of the printer elements. In order to gain data, we lead to several printing malfunctions applied to the 3D model. The resulting dataset can therefore be used for anomalies detection research.

Keywords: Internet of Things · Industry 4.0 · Anomaly detection

1 Introduction

The use of 3D printers is becoming more and more desirable. They are used not only in professional production plants, but also among home users. As a result, methods of automatic fault detection during their operation are gaining importance. This paper presents the data that we have gathered from the 3D printer during the printing process. Among all, data samples include a temperature of working elements of the printer, intrusion force and the acceleration of printing head. The data has been gathered using two types of sources - custom-made measurement devices and the printer's internal software. In order to enable the dataset to serve as an example of anomalies detection for intelligent Industry 4.0 systems, we provoked several types of failures during the printing process. All of the files are placed in the repository[1] and can be used under the `Creative Commons Attribution 4.0 International` license.

The rest of the paper is organised as follows. Section 2 presents related work. In Sect. 3 we describe the characteristics of the printer machine used for gathering data samples. In Sect. 4 we characterize each type of data source while printing failures that we created are presented in Sect. 5. Section 6 contains sample data analysis. The last section sums up the paper.

---

[1] https://github.com/joanna-/3D-Printing-Data.

© Springer Nature Switzerland AG 2021
M. Paszynski et al. (Eds.): ICCS 2021, LNCS 12745, pp. 647–653, 2021.
https://doi.org/10.1007/978-3-030-77970-2_50

2   Related Work

As interest in 3D printing increases in various applications, anomaly detection systems are gaining in importance [1]. Moreover, data sets which facilitate development of anomaly detection systems are made available, for instance as described in [2]. We can distinguish several types of systems used to monitor the operation of machines for manufacturing processes. The industrial device monitoring systems may use data from various types of sensors [3], e.g. kinematic, visual, inertial, as well as auditory [4]. Systems, which are specialized for monitoring 3D printers' operation, are also designed, developed, and described, e.g. systems based on image analysis [5–7] or on sensor data analysis [8,9].

Proper preparation of the 3D printer and retrofitting it with sensors requires some time and equipment expenditure. In the article, we present a set of test data that were obtained using devices built using the FogDevices platform [10]. The presented set of test data can be used to develop new algorithms for detecting anomalies in the work of 3D printers and, what is important, to compare them.

3   3D Printer Characteristics

The 3D printer utilized for collecting its operation data was Monkeyfab Spire manufactured by Monkeyfab[2] - its basic properties are listed in Table 1. It is a *delta printer* in which the printing head is mounted on magnetic ball joints. The Monkeyfab Spire uses the *RepRapFirmware* and is controlled over the network via *Duet Web Control*[3] interface.

Table 1. Basic parameters of the utilized 3D printer according to manufacturer's specifications.

Maximum printed object dimensions	150 mm diameter
	165 mm height
Default nozzle diameter	0.4 mm
Minimum layer height	0.05 mm
Filament diameter	1.75 mm
Maximum hotend temperature	262 °C
Maximum platform temperature	120 °C

4   Data Sources Characteristics

The sensor data comes from two sources - (i) internal electronics that control the operation of the printer and from (ii) additionally mounted sensors. They are described in more details in the next subsections.

---

[2] http://www.monkeyfab.com.
[3] https://duet3d.dozuki.com.

4.1   Duet Web Control

The 3D printer is controlled by the Duet3D electronics which expose user interface (UI) called *Duet Web Control Interface* that is accessible via a web browser allowing to monitor and change printer state. Among others, it includes such features as emergency stop, monitoring temperatures of printer parts, changing filament and selecting 3D models to print. The aforementioned information is also exposed via API in *json* format which can be accessed remotely via a network.

4.2   Data Acquisition Hardware

The printer has been equipped with additional custom sensors developed as part of the FogDevices[4] research project. Data from the sensors were collected using a device assembled using modular hardware components.

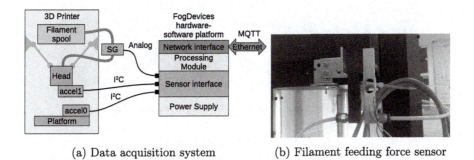

(a) Data acquisition system   (b) Filament feeding force sensor

Fig. 1. Sensors attached to the printer

The printer has been equipped with two inertial measurement unit (IMU) sensors LSM9DS1 that can measure acceleration, angular rate and magnetic field in 3 axes but only linear acceleration was used. First of the sensors, called *accel0* is attached to the printing platform and *accel1* is on the print head. Both sensors use the I^2C digital interface and are connected to the FogDevices hardware platform.

The method of measuring the filament feeding force is based on indirect measurement of the force acting on the Bowden tube during printer operation as presented in Fig. 1b. This was possible since the extruder is located on the printer's body, not at the print head. Therefore, a force sensor *SG* based on a strain gauge was developed. Its operation is based on the Wheatstone bridge and it produces small voltage output. The voltage is amplified in FogDevices sensor interface module with INA128 instrumentation amplifiers and then measured using an analog-to-digital converter (ADC) with 12-bit resolution.

---

[4] http://fogdevices.agh.edu.pl.

Block diagram of the hardware is presented in Fig. 1a. The FogDevices hardware platform has been utilized to collect data from three sensors: *SG*, *accel0*, and *accel1*. Data collected by the device was being sent through the MQTT protocol over the Ethernet interface. The data were then saved by a data logger running on a PC. The acquisition system collects and processes 200 samples per second[5].

5 Test Prints

We have used two variants of the same five towers print to collect data. In the variant (a), presented in Fig. 2a, towers have printed base that is an integral part of the print and in variant (b) presented in Fig. 2b, towers do not have a base - they are placed only on the raft.

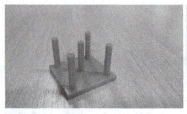

(a) towers with the base - photo   (b) towers without base - photo

Fig. 2. Models used in the experiments

For both variants, we have collected data from the undisturbed, properly made print. Apart from that, we provoked six printing anomalies presented in Fig. 3:

1. **Variant (a):**
   - Figure 3b - printer ran out of plastic before the print was finished;
   - Figure 3c - part of the print unstuck from the printing base, but the rest of print remained undisturbed;
   - Figure 3d - the speed of the retraction has been set too low (to 0.5);
   - Figure 3e - during the printing, the Bowden tube fell out from its place;
   - Figure 3f - during the printing, the arm of printer head has been detached from magnets holding it in the place;
2. **Variant (b):**
   - Figure 3a - during the printing, part of the print has been removed.

---

[5] Additional sensors and devices are provided by the FogDevices platform. The video showing printing process is available online https://youtu.be/SFBInVsVDgk.

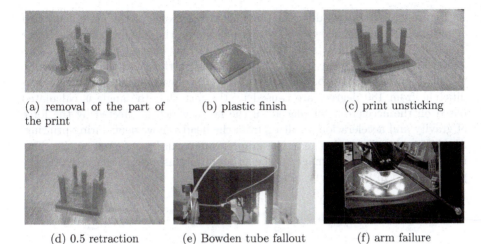

(a) removal of the part of the print   (b) plastic finish   (c) print unsticking

(d) 0.5 retraction   (e) Bowden tube fallout   (f) arm failure

Fig. 3. Various malfunctions of the print

6 Sample Data Analysis

Provoked failures cause different symptoms that can be detected with the data analysis. Different failures may have similar symptoms depending on their type and therefore inferring the initial cause can require more complex analysis. In this section, we present very basics of analysis and show three types of symptoms related to five types of failures. The summary of failure-symptom correlation is presented in the Table 2.

Table 2. Symptoms characteristic of the printing failures.

Failure type	Symptoms	Brief explanation
Finish of plastic	Decrease of intrusion power	There is no more plastic to intrude
Bowden tube fallout		There is no friction with the print - plastic doesn't reach printed model
Wrong retraction (0.5)	Printing base jolting	Too much plastic hooks on the next layers
Unsticking of the model		Printing head hooks on the rolled print
Arm failure	Printing head angle change	Detachment of arm causes head to tilt

Figure 4 presents two different plots that show some of the aforementioned symptoms. Figure 4a shows the situation where the filament feeding force dropped rapidly at time 11:40. That symptom may suggest that the filament is over or there was severe mechanical problem - in this case, the Bowden tube fallout. Figure 4b shows the tilt angle of the print head during printing. Values different from 180 degrees are caused by the fact that the angle is calculated based on the accelerometer placed on the head, which is affected by the force of gravity and acceleration resulting from the head's movement during printing. At 11:00 a significant change in the graph's value can be observed on the chart indicating mechanical damage to the printer. In this case, the arm fixing the printing head in the delta system is damaged.

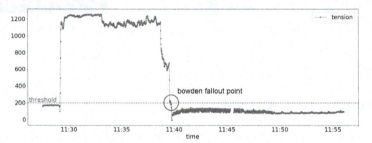

(a) Tension values for the print with Bowden tube fallout.

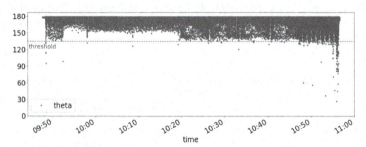

(b) Tilt angle values for the print with head arm detachment.

Fig. 4. Plots presenting symptoms of printing anomalies

7   Summary

The article presents data collected during the operation of the 3D printer, including typical errors observed during the printing process. The collected data can be used to develop advanced algorithms for detection and prediction of failures. The automatic detection of 3D printing machines failure can be useful for owners of printers farms allowing them to decrease the maintenance expenditures.

The paper presents also the possibilities offered by the use of IoT devices in Industry 4.0. Retrofitting machines with additional sensors and devices analyzing their work in real-time can provide valuable information about their work. IoT devices such as those offered by *FogDevices Platform* allow simplifying the process of adding sensors and analyzing data on the edge, near the sensors without sending them to the computational clouds.

Data Usage

The dataset is under Creative Commons Attribution 4.0 International license. Please cite this paper if you use it.

Acknowledgment. The research presented in this paper was supported by the National Centre for Research and Development (NCBiR) under Grant No. LIDER/15/0144/L-7/15/NCBR/2016.

References

1. Graß, A., Beecks, C., Soto, J.A.C.: Unsupervised anomaly detection in production lines. Machine Learning for Cyber Physical Systems. TA, vol. 9, pp. 18–25. Springer, Heidelberg (2019). https://doi.org/10.1007/978-3-662-58485-9_3
2. Purohit, H., et al.: Mimii dataset: sound dataset for malfunctioning industrial machine investigation and inspection. arXiv preprint arXiv:1909.09347 (2019)
3. Su, Y., Zhao, Y., Niu, C., Liu, R., Sun, W., Pei, D.: Robust anomaly detection for multivariate time series through stochastic recurrent neural network. In: Proceedings of the 25th ACM SIGKDD International Conference on Knowledge Discovery & Data Mining, pp. 2828–2837 (2019)
4. Koizumi, Y., Saito, S., Uematsu, H., Kawachi, Y., Harada, N.: Unsupervised detection of anomalous sound based on deep learning and the Neyman-Pearson lemma. IEEE/ACM Trans. Audio Speech Lang. Process. 27(1), 212–224 (2018)
5. Wu, G., Shen, Z., Shang, X., Wu, H., Xiong, G., Yang, J.: 3D printer optical detection system based on DLP projection technology. In: Chinese Automation Congress (CAC), vol. 2019, pp. 1040–1045 (2019)
6. Baumann, F., Roller, D.: Vision based error detection for 3D printing processes. In: MATEC Web of Conferences, vol. 59, p. 06003 (2016)
7. Tonnaer, L., Li, J., Osin, V., Holenderski, M., Menkovski, V.: Anomaly detection for visual quality control of 3D-printed products. In: 2019 International Joint Conference on Neural Networks (IJCNN), pp. 1–8. IEEE (2019)
8. Kim, C., Espalin, D., Cuaron, A., Perez, M.A., MacDonald, E., Wicker, R.B.: A study to detect a material deposition status in fused deposition modeling technology. In: IEEE International Conference on Advanced Intelligent Mechatronics (AIM), vol. 2015, pp. 779–783 (2015)
9. Windau, J., Itti, L.: Inertial machine monitoring system for automated failure detection. In: IEEE International Conference on Robotics and Automation (ICRA), vol. 2018, pp. 93–98 (2018)
10. Brzoza-Woch, R., Szydło, T., Windak, M., Sendorek, J.: The FogDevices platform-a comprehensive hardware solution for IoT applications. IFAC-PapersOnLine 52(27), 44–49 (2019)

Correction to: Computational Science – ICCS 2021

Maciej Paszynski⊙, Dieter Kranzlmüller⊙,
Valeria V. Krzhizhanovskaya⊙, Jack J. Dongarra⊙,
and Peter M. A. Sloot⊙

Correction to:
M. Paszynski et al. (Eds.): *Computational Science – ICCS 2021*,
LNCS 12745, https://doi.org/10.1007/978-3-030-77970-2

Chapter 18, "Modelling and Forecasting Based on Recurrent Pseudoinverse Matrices" was previously published non-open access. This have now been changed to open access under a CC BY 4.0 license and the copyright holders updated to 'The Author(s)' and the acknowledgement section added. The book has also been updated with this change.

In chapter 44, in reference 34, the surname of the first author was incorrect. The surname has been corrected from "Porti" to "Potortì."

The updated version of these chapters can be found at
https://doi.org/10.1007/978-3-030-77970-2_18
https://doi.org/10.1007/978-3-030-77970-2_44

Author Index